STUDY GUIDE

to accompany

ORGANIC CHEMISTRY

Fourth Edition

T. W. GRAHAM SOLOMONS
JACK E. FERNANDEZ

University of South Florida

JOHN WILEY & SONS

NEW YORK CHICHESTER BRISBANE TORONTO SINGAPORE

Library of Congress Cataloging in Publication Data:

Solomons, T. W. Graham
Study guide to accompany organic chemistry.

1. Chemistry, Organic. I. Fernandez, Jack E., 1930- II. Title. III. Title: organic chemistry.

ISBN 0-471-83661-3

Printed in the United States of America

10 9 8 7 6 5

TO THE STUDENT

This study guide contains several items to aid you in your study of organic chemistry; these include the following:

Answers to the Problems. Solutions are given for all the problems in the text, including the end-of-chapter problems. In many instances we have given not only the answer, but also an explanation of the reasoning that leads to the solution. Although many problems have more than one solution, we have generally given only one. Thus, you should not necessarily assume that your answer is incorrect if it differs from the one given here.

The heart of organic chemistry lies in problem solving. This is as true for the practicing organic chemist as for the beginning student. But, a problem in organic chemistry is like a riddle; once you have seen the answer, it is impossible for you to go through the process of solving it. The essential value of a problem lies in the mental exercise of the problem-solving process, and you cannot get this exercise if you already know the answer. The best way to use this manual, therefore, is to check the problems you have worked, or to find explanations for unsolved problems *only* after you have made serious attempts at working them.

See Tables 13.1, 13.2, and 13.3 inside the front and back covers; they are reproduced to help you solve problems.

Flow Diagrams of Reactions Covered. These flow diagrams should serve to give you a one-page overview of the reactions in the chapter as well as a way to view the interrelations of the reactions. These diagrams should be helpful to you in the early Additional Problems on reactions as well as in the problems that involve multistep syntheses.

Section References to Additional Problems. If you have trouble working the Additional Problems, you may wish to refer to the section in the chapter that deals with the concept being examined. Generally, you should seek help in the chapter before going to the solution in this study guide.

Self-Tests. A Self-Test is given for each chapter of the text. After you think you have mastered the material in each chapter you should then take the Self-Test. When you have finished you can check your answers against those given in Appendix D.

Supplementary Problems. These problems are designed to help you study for tests after you have worked the Additional Problems and Self-Tests.

Answers to Review Problems. Two sets of review problems occur in the Study Guide after Chapters 10 and 20. These problems require you to use material covered in the preceding chapters. These problems should help you review for examinations. The answers to these review problems are also given in this Study Guide.

A Section on the Calculation of Empirical and Molecular Formulas. This topic, usually included in the general chemistry course that precedes the study of organic chemistry, has been included as Appendix A. If you missed it in general chemistry or need a review, you should study it early in the course.

Molecular Model Set Exercises. Appendix B is a set of exercises with solutions. These exercises are designed to help you gain facility with molecular models and to help you understand the relationship between formulas on the page and the three-dimensional molecules that these formulas represent. The chapter corresponding to each exercise is given in Appendix B.

Glossary of Important Terms. Important terms and concepts are collected in Appendix C. These terms and concepts are defined and a reference to the text is given.

ACKNOWLEDGMENTS

We thank George R. Jurch for reviewing and proofreading this edition of the study guide.

We wish to acknowledge Ronald Starkey of the University of Wisconsin for providing the Molecular Model Set Exercises and Solutions of Appendix B.

We also wish to thank the following persons who graciously read an earlier version of this study guide and who made many helpful suggestions: George R. Wenzinger, University of South Florida; Prof. Darrell Berlin, Oklahoma State University; Prof. John Mangravite, West Chester State College; Prof. J. G. Traynham, Louisiana State University; and Prof. Desmond M. S. Wheeler, University of Nebraska. We are also much indebted to Jeannette Stiefel for editing and proofreading the entire study guide.

T. W. Graham Solomons

Jack E. Fernandez

CONTENTS

1 CARBON COMPOUNDS AND CHEMICAL BONDS

SOLUTIONS TO PROBLEMS

Another Approach to Writing Lewis Structures

When we write Lewis structures using this method we assemble the molecule or ion from the constituent atoms showing only the valence electrons (i.e., the electrons of the outermost shell). By having the atoms share electrons, we try to give each atom the electronic structure of a noble gas. For example, we give hydrogen atoms two electrons because this gives them the structure of helium. We give carbon, nitrogen, oxygen, and fluorine atoms eight electrons because this gives them the electronic structure of neon. The number of valence electrons of an atom can be obtained from the periodic table because it is equal to the group number of the atom. Carbon, for example, is in group IVA and has four valence electrons; fluorine, in group VIIA has seven; hydrogen in group IA, has one. As an illustration let us write the Lewis structure for CH_3F. In the example below we will at first show a hydrogen's electron as an x, carbon's electrons as o's and fluorine's electrons as dots.

Example A

```
      o        ..
3 Hx, oCo, and •F: are assembled as
      o        ..

 H             H
 xo ..         ..  ..
HxCoF:   or  H:C:F:
 xo ..         ..  ..
 F             H
```

If the structure is an ion we add or subtract electrons to give it the proper charge. As an example consider the chlorate ion, ClO_3^-.

Example B

```
 ..          oo
:Cl•, and   oOo  and an extra electron x are assembled as
 ..          oo

[      oo       ]-       [     ..      ]-
[     oOo       ]        [    :O:      ]
[  oo  ..  oo   ]   or   [ ..  ..  ..  ]
[ oO:Cl xOo     ]        [:O:Cl:O:     ]
[  oo  ..  oo   ]        [ ..  ..  ..  ]
```

1.1

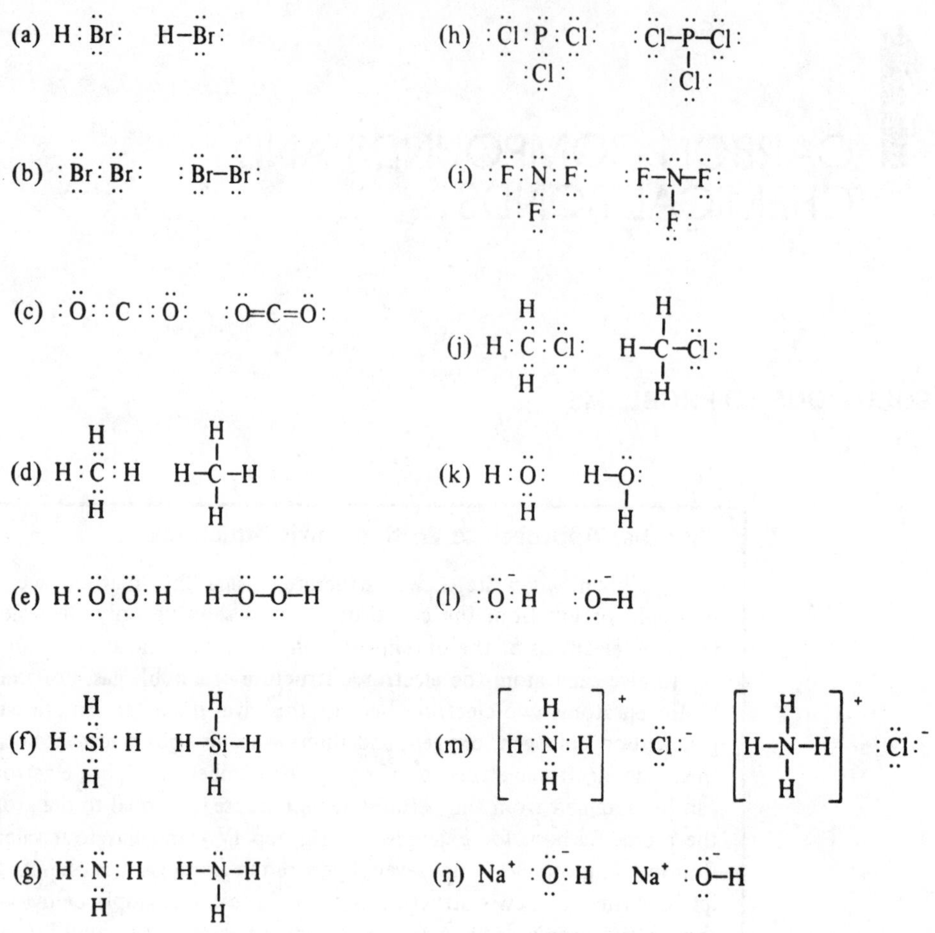

1.2 Formal charge = group number − [½ (number of shared electrons) + (number of unshared electrons)]

Charge on ion = sum of all formal charges;

		Formal Charge	Total Charge
(a) 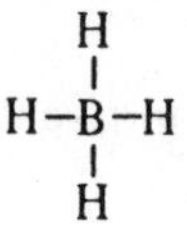	H	1-(1 + 0) = 0	-1
	B	3-(4 + 0) = -1	
(b) 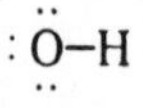	H	1-(1 + 0) = 0	-1
	O	6-(1 + 6) = -1	
(c) 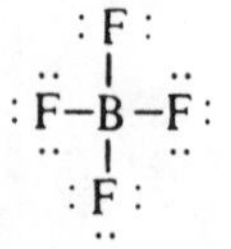	F	7-(1 + 6) = 0	-1
	B	3-(4 + 0) = -1	

Structure	Atom	Formal charge	Charge
(d) $H-\ddot{O}-H$ with H on O	H	$1-(1+0)=0$	+1
	O	$6-(3+2)=+1$	
(e) $:\ddot{O}=C$ with two $:\ddot{O}:$ below	top O	$6-(2+4)=0$	−2
	C	$4-(4+0)=0$	
	bottom O's	$6-(1+6)=-1$	
(f) $H-\ddot{C}-H$ with H on C	H	$1-(1+0)=0$	−1
	C	$4-(3+2)=-1$	
(g) $H-C-H$ with H on C	H	$1-(1+0)=0$	+1
	C	$4-(3+0)=+1$	
(h) $H-\dot{C}-H$ with H on C	H	$1-(1+0)=0$	0
	C	$4-(3+1)=0$	
(i) $H-\ddot{C}-H$	H	$1-(1+0)=0$	0
	C	$4-(2+2)=0$	
(j) $:\ddot{N}-H$ with H on N	H	$1-(1+0)=0$	−1
	N	$5-(2+4)=-1$	

1.3 Zero formal charges are not shown.

(a) No formal charges (d) No formal charges (g) CH_3-C with $=\ddot{O}:$ and $-\ddot{O}:^-$

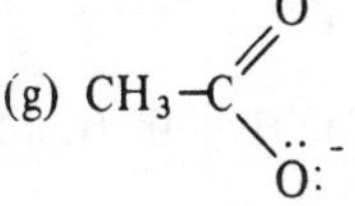

(b) No formal charges (e) No formal charges (h) $CH_3CH_2-\ddot{O}^+-H$ with H on O

(c) $CH_3-N^+(CH_3)-CH_3$ with $:\ddot{O}:^-$ on N   (f) $CH_3-\overset{+}{N}=\ddot{O}:$ with $:\ddot{O}:^-$ on N   (i) $CH_3CH-CHCH_3$ bridged by $:\overset{+}{Br}:$

1.4 (a) $^-:\ddot{O}-\overset{+}{\ddot{S}}=\ddot{O}: \longleftrightarrow :\ddot{O}=\overset{+}{\ddot{S}}-\ddot{O}:^-$

(b) Yes. Both O–S bonds are hybrids of a single and a double bond. Therefore the two O–S bonds are equivalent and of equal length.

1.5 (a) H–Br (dipole arrow pointing toward Br) (c) H–H (dipole moment = 0)

(b) I–Cl (dipole arrow pointing toward Cl) (d) Cl–Cl (dipole moment = 0)

1.6 Each electron experiences less repulsion from other electrons if it is in an orbital by itself because the electrons can be further apart. Consider the three $2p$ orbitals as an example (see Fig. 1.6). With one electron in each $2p$ orbital each electron occupies a different region of space. This would not be true if two electrons were in the same $2p$ orbital.

1.7 (a) In its ground state the valence electrons of carbon might be disposed as shown in the following figure.

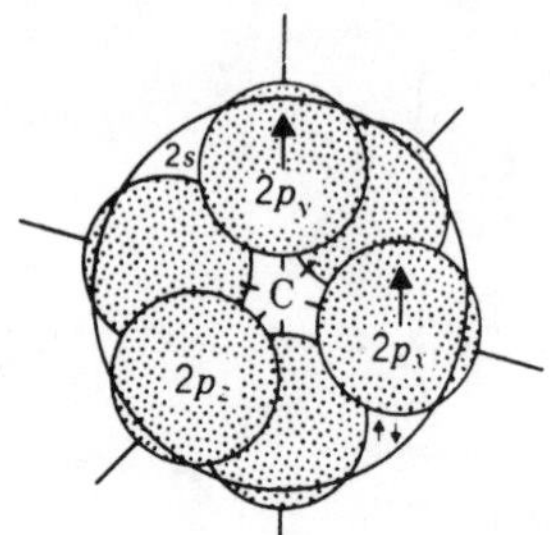

The electronic configuration of a ground state carbon atom. The p orbitals are designated $2p_x$, $2p_y$, and $2p_z$ to indicate their respective orientations along the x, y, and z axes. The assignment of the unpaired electrons to the $2p_y$ and $2p_x$ orbitals is arbitrary. They could also have been placed in the $2p_x$ and $2p_z$ or $2p_y$ and $2p_z$ orbitals. (To have placed them both in the same orbital would not have been correct, however, for this would have violated Hund's rule.) (Section 1.11)

The formation of the covalent bonds of methane *from individual atoms* requires that the carbon atom overlap its orbitals containing *single electrons* with $1s$ orbitals of hydrogen atoms (which also contain a single electron). If a ground-state carbon atom were to combine with hydrogen atoms in this way, the result would be that depicted below. *Only two carbon-hydrogen bonds would be formed, and these would be at right angles to each other.*

The Hypothetical formation of CH_2 from a carbon atom in its ground state.

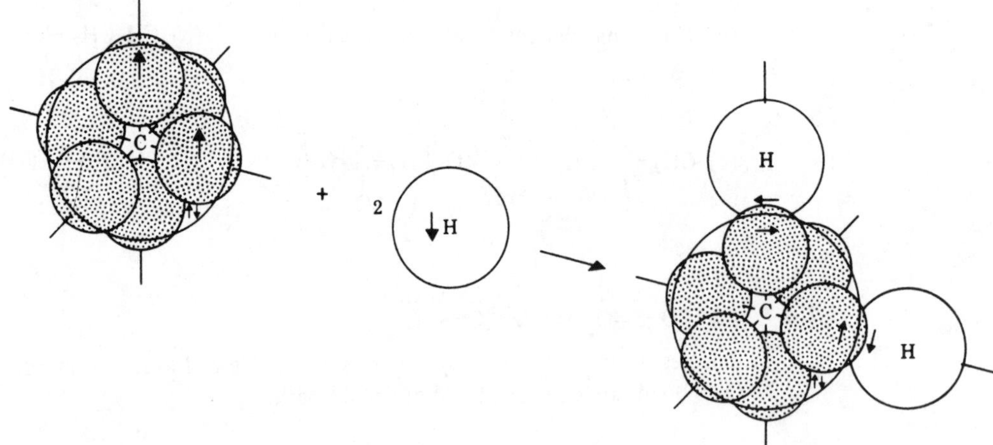

(b) An excited-state carbon atom might combine with four hydrogen atoms as shown in the following figure.

The promotion of an electron from the $2s$ orbital to the $2p_z$ orbital requires energy. The amount of energy required has been determined and is equal to 96 kcal/mole.

This expenditure of energy can be rationalized by arguing that the energy released when two additional covalent bonds form would more than compensate for that required to excite the electron. No doubt this is true, but it solves only one problem. The problems that cannot be solved by using an excited-state carbon as a basis for a model of methane are the problems of the carbon-hydrogen bond angles and the apparent equivalence of all four carbon-hydrogen bonds. Three of the hydrogens—those overlapping their 1*s* orbitals with the three *p* orbitals—would, in this model, be at angles of 90° with respect to each other; the fourth hydrogen, the one overlapping its 1*s* orbital with the 2*s* orbital of carbon, would be at some other angle, probably as far from the other bonds as the confines of the molecule would allow. Basing our model of methane on this excited state of carbon gives us a carbon that is tetravalent *but one that is not tetrahedral,* and it predicts a structure for methane in which one carbon-hydrogen bond differs from the other three.

The hypothetical formation of CH_4 from an excited-state carbon atom.

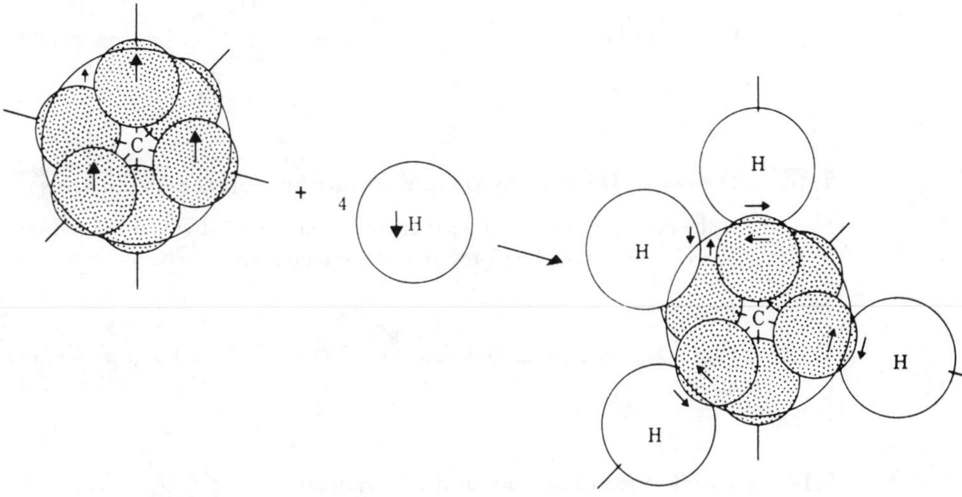

1.8 (a) Monovalent because only one orbital (2*p*) contains a single electron; the 2*s* orbital is filled. (b) The two *p* orbitals lie at 90° to one another; the resulting bonds would also lie at 90° to each other. Thus BF_3, based on an excited state of boron, would have the following structure. The angles of 135° result by dividing (360–90) by 2.

90° F 135° F–B F 135°

1.9 (a) Yes. (b) No. Two square planar structures are possible:

X H C X H and X H C H X

Therefore, if CH_2X_2 had a square planar structure we should observe two compounds (isomers) with the formula CH_2X_2

1.10 (a) Tetrahedral (c) Trigonal planar (e) Tetrahedral (g) Tetrahedral

(b) Tetrahedral (d) Tetrahedral (f) Linear (h) Linear

(i) Trigonal planar

1.11 (a) $H_2C{=}CH_2$ (120°, 120°), trigonal planar at each carbon atom.

(b) H–C≡C–H (180°) linear

(c) H–C≡N: (180°) linear

(d) $\ddot{S}^{+}$(=O)–O^{-} angular

(e) O=S^{+2}(O^{-})(O^{-}) trigonal planar, 120° bond angles

1.12 The two C=O bond moments are opposed and cancel each other: O = C = O

If the bond angle were other than 180°, then the individual bond moments would not cancel. There would be a resultant dipole moment.

1.13 That SO_2 is an angular molecule O=$\ddot{S}$=O. Its S–O bond moments do not cancel each other.

1.14 The direction of polarity of the N–H bond is opposite to that of the N–F bond.

In NH_3, the resultant N–H bond polarities and the polarity of the unshared electron pair are in the same direction.

In NF_3 the resultant N–F bond polarities partially cancel the polarity of the unshared electron pair.

1.15 BF_3 is trigonal planar. Its B–F bonds are all necessarily equal in polarity, and the F–B–F bond angles are all equal (120°).

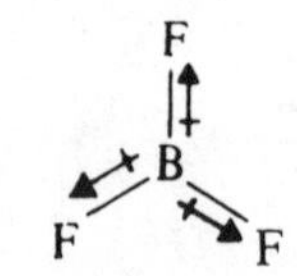

1.16 $CH_3CH_2CH_2OH$ $CH_3CH(OH)CH_3$ $CH_3OCH_2CH_3$

1.17

(a)

```
      H   :Cl:  H    H
      |    |    |    |
  H — C —  C  — C —  C — H
      |    |    |    |
      H   :Cl:  H    H
```

Cl Cl

(e)

```
      H   :OH   H    H
      |    |    |    |
  H — C —  C  — C —  C — H
      |    |    |    |
      H    H    H    H
```

OH

(b)

```
      H    H    H    H
      |    |    |    |
  H — C —  C  — C —  C — H
      |    |    |    |
      H    |    H    H
         H-C-H
           |
          :Cl:
```

Cl

(f)

```
      H    H    H    H
      |    |    |    |
  H — C —  C  — C —  C — ÖH
      |    |    |    |
      H    H    H    H
```

OH

(c)

```
           H
         H-C-H
      H    |    H    H
      |    |    |    |
  H — C —  C  — C —  C — H
      |    |    |    |
      H    |    H    H
         H-C-H
           H
```

(g)

```
                          H
                        H-C-H
      H   :O:   H         |       H
      |    ||   |         |       |
  H — C —  C  — C    —    C   —   C — H
      |         |         |       |
      H         H         H       H
```

O

(d)

```
      H   :Cl: :Cl:  H
      |    |    |    |
  H — C —  C  — C —  C — H
      |    |    |    |
      H    H    H    H
```

Cl

Cl

(h)

```
                          H
                        H-C-H
      H    H   :OH        |       H
      |    |    |         |       |
  H — C —  C  — C    —    C   —   C — H
      |    |    |         |       |
      H    H    H         H       H
```

OH

1.18 (a) and (d) are constitutional isomers. (e) and (f) are constitutional isomers.

1.19 (a) 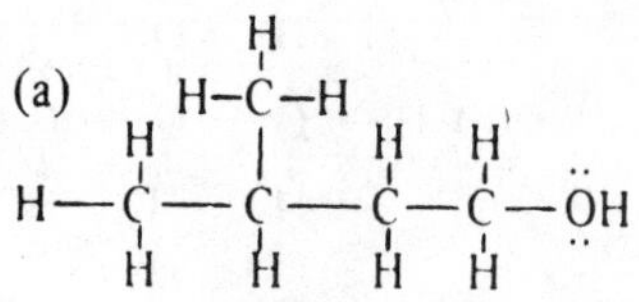 (b) 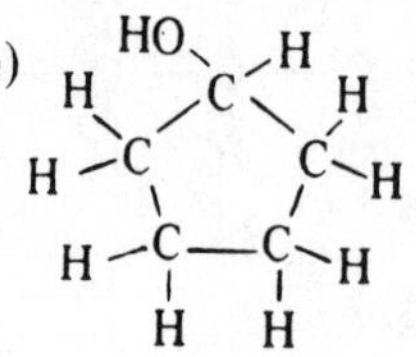 (c) 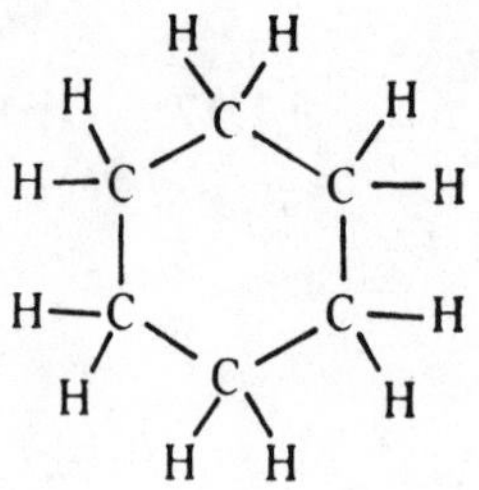 (d)

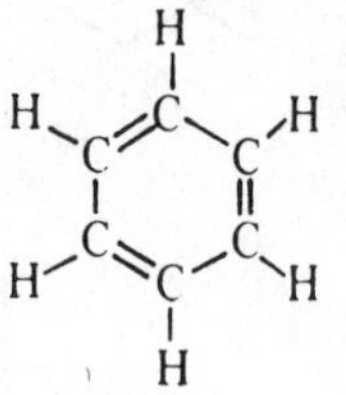

1.20 (a) 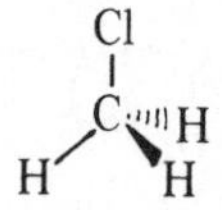 (Note that the Cl atom and the three H atoms may be written at any of the four positions.)

(b) 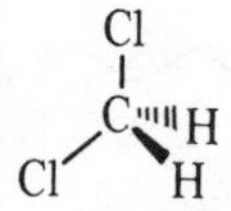 or 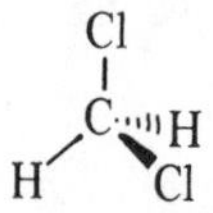 and so on

(c) 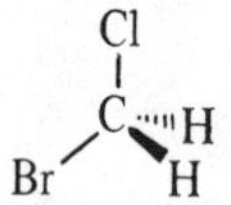 and others

(d) 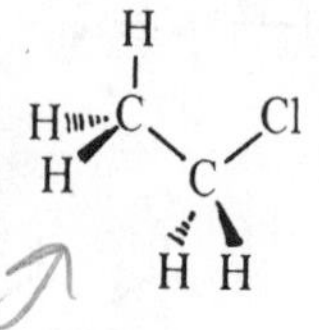 and others

1.21 (a) 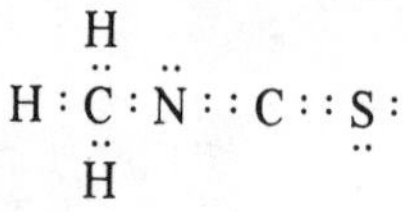 H : C : N : : C : : S :

(d) 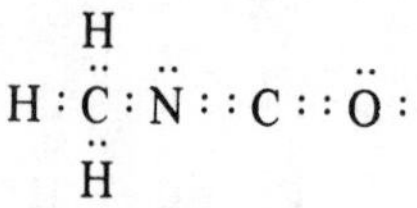 H : C : N : : C : : O :

(g) 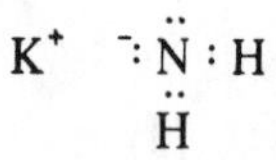 K^+ $^-$: N : H

(b) 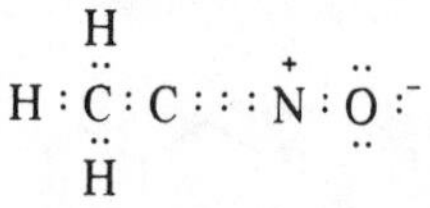 H : C : C : : : $\overset{+}{N}$: O :$^-$

(e) H : C : : C : : O :

(h) Na^+ $^-$: N : : $\overset{+}{N}$: : N :$^-$

(i) H, H C : : O :

(c) H : C : O : $\underset{+}{N}$: O :$^-$

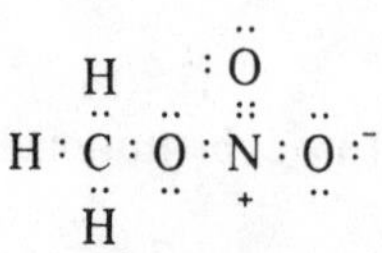

(f) 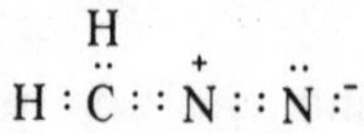 H : C : : $\overset{+}{N}$: : N :$^-$

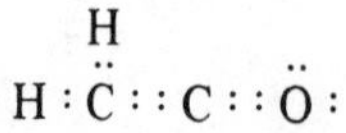

(j) 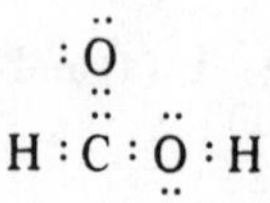 H : C : O : H

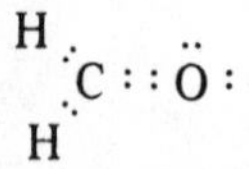

1.22 (a) Electron Configuration

(1) Be $1s^2 2s^2$

(2) B $1s^2 2s^2 2p_x^1$

(3) C $1s^2 2s^2 2p_x^1 2p_y^1$

(4) N $1s^2 2s^2 2p_x^1 2p_y^1 2p_z^1$

(5) O $1s^2 2s^2 2p_x^2 2p_y^1 2p_z^1$

(b) Orbital Arrangement

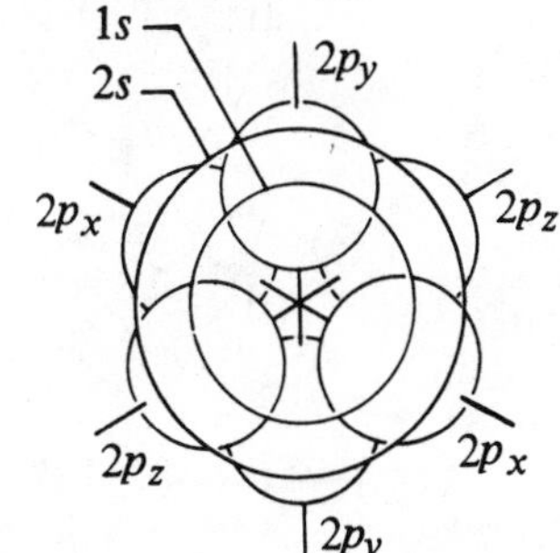

1.23

(a) $CH_3\text{—}\ddot{\underset{..}{O}}\text{—}\underset{\|}{\overset{\|}{S}}\text{—}\ddot{\underset{..}{O}}:^-$ (with $=\ddot{O}:$ above and below S)

(b) $CH_3\text{—}\overset{+}{\underset{..}{S}}\text{—}CH_3$ (with $^-:\ddot{\underset{..}{O}}:$ single-bonded above S)

(c) $^-:\ddot{\underset{..}{O}}\text{—}\underset{\|}{\overset{\|}{S}}\text{—}\ddot{\underset{..}{O}}:^-$ (with $=\ddot{O}:$ above and below S)

(d) $CH_3\text{—}\underset{\|}{\overset{\|}{S}}\text{—}\ddot{\underset{..}{O}}:^-$ (with $=\ddot{O}:$ above and below S)

1.24 (a) $(CH_3)_2CHCH_2OH$

(b) $(CH_3)_2CH\overset{O}{\overset{\|}{C}}CH(CH_3)_2$

(c)
```
HC—CH2
‖   |
HC—CH2
```

(d) $(CH_3)_2CHCH_2CH_2OH$

1.25 (a) $C_4H_{10}O$ (c) C_4H_6

(b) $C_7H_{14}O$ (d) $C_5H_{12}O$

1.26

(a) Different compounds
(b) Constitutional isomers
(c) Same compound
(d) Same compound
(e) Same compound
(f) Constitutional isomers
(g) Different compounds
(h) Same compound
(i) Different compounds
(j) Same compound
(k) Constitutional isomers
(l) Different compounds
(m) Same compound
(n) Same compound
(o) Same compound
(p) Constitutional isomers

1.27

(a) 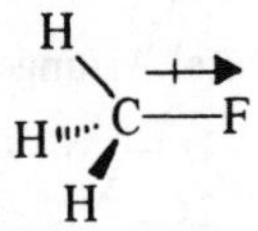 CH_3F (tetrahedral; dipole toward F)

(b) 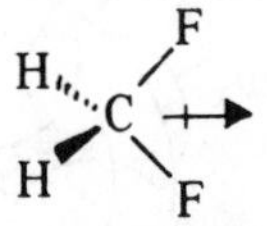 CH_2F_2 (dipole shown)

(c) 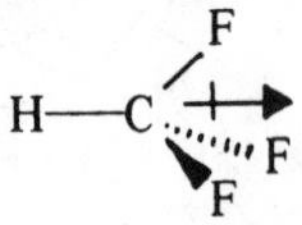 CHF_3 (dipole shown)

(d) 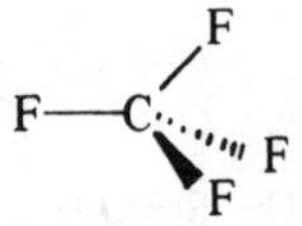 CF_4

No dipole moment

(e) 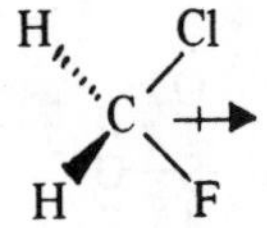 CH_2ClF (dipole shown)

(f) 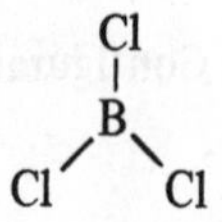 BCl_3 (trigonal planar)

No dipole moment

(g) F–Be–F

No dipole moment

(h) 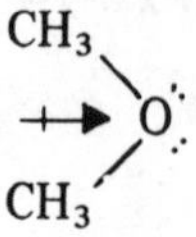 $CH_3–\ddot{O}–CH_3$ (dipole toward O)

(i) 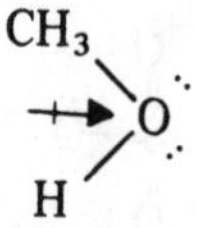 $CH_3–\ddot{O}–H$ (dipole toward O)

(j) 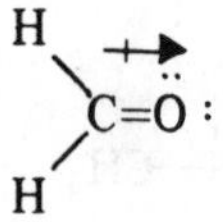 $H_2C=\ddot{O}:$ (dipole toward O)

1.28

(a) 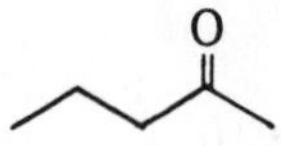 (bond-line structure of a ketone, C=O)

(b) (bond-line structure of a branched alkane)

(c) (bond-line structure of an alcohol, –OH)

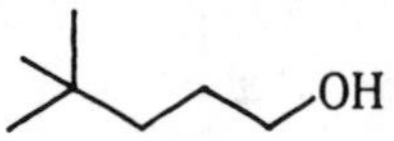

(d) 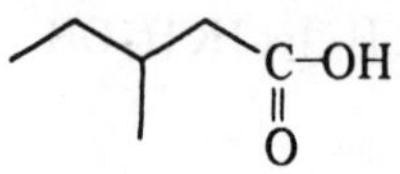 (bond-line structure of a carboxylic acid, C(=O)–OH)

(e)  (bond-line structure of a diene)

(f) 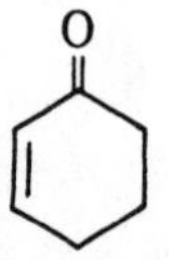 (cyclohexenone ring, C=O)

1.29

(a) 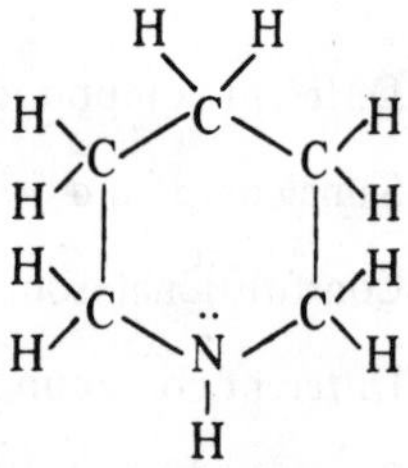 (piperidine ring: five CH_2 groups and N–H, with a lone pair on N)

(b) 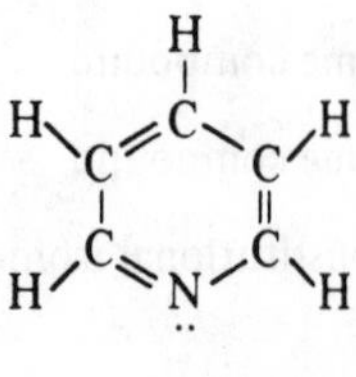 (pyridine ring: five C–H and N, with a lone pair on N)

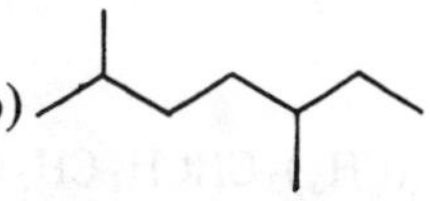

(c) $H_3C–\ddot{N}(CH_3)–CH_2–CH_3$

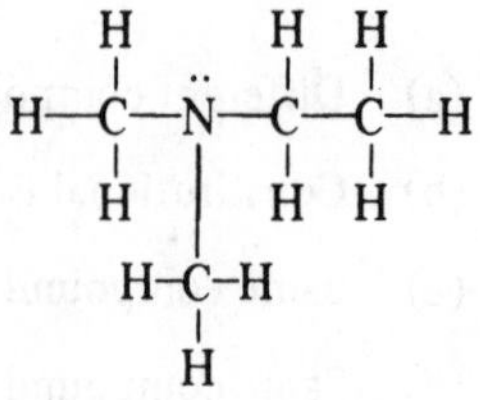

(d) 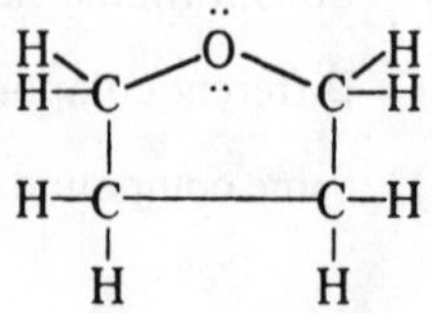 (ring of one O, with lone pairs, and four C: two CH_2 next to O and two C–H)

1.30 $CH_2{=}CHCH_2CH_3$ $CH_3CH{=}CHCH_3$ $CH_2{=}C(CH_3)CH_3$

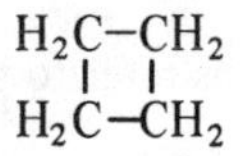

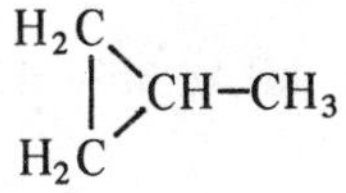

$H_2C{-}CH_2$ / $H_2C{-}CH_2$ (cyclobutane) H_2C / H_2C $CH{-}CH_3$ (methylcyclopropane)

1.31 $^{-}:\ddot{O}{-}\ddot{N}{=}\ddot{O}:$ $:\ddot{O}{=}\ddot{N}{-}\ddot{O}:^{-}$ Yes because the two O–N bonds are equivalent hybrids of a single and a double bond.

1.32 (a) An sp^3 orbital. (b) sp^3 Orbitals. In ammonia and in water the bond angles are close to the tetrahedral angle of 109½°; therefore the N and O atoms must be sp^3 hybridized.

1.33 A carbon-chlorine bond is longer than a carbon-fluorine bond because chlorine is a larger atom than fluorine. Thus in $\overset{\delta+}{C}H_3{-}\overset{\delta-}{Cl}$ the distance, d, that separates the charges is greater than in $\overset{\delta+}{C}H_3{-}\overset{\delta-}{F}$. The greater value of d for CH_3Cl more than compensates for the smaller value of e and thus the dipole moment ($e \times d$) is larger.

1.34 (a) While the structures differ in the position of their electrons they also differ in the positions of their nuclei and thus *they are not resonance structures.* (In cyanic acid the hydrogen nucleus is bonded to oxygen; in isocyanic acid it is bonded to nitrogen.)

(b) The anion obtained from either acid is a resonance hybrid of the following structures: $^{-}:\ddot{O}{-}C{\equiv}N: \longleftrightarrow :\ddot{O}{=}C{=}\ddot{N}:^{-}$

1.35 (a) BF_3 has an empty orbital which can accommodate the electron pair of $:NH_3$; also, formation of the new B–N bond gives B an octet. (b) −1, (c) +1, (d) sp^3, (e) sp^3.

1.36 Acid strength increases with increasing (positive) formal charge on the central atom.

Acid Strength: $H_3O^+ > H_2O > OH^-$; $NH_4^+ > NH_3$; and $H_2S > HS^-$.

1.37 (a) 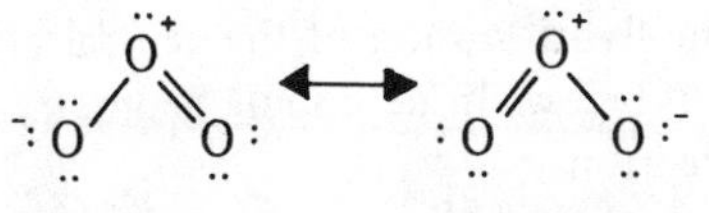 $^{-}:\ddot{O}{-}\ddot{O}^{+}{=}\ddot{O}: \longleftrightarrow :\ddot{O}{=}\ddot{O}^{+}{-}\ddot{O}:^{-}$

(b) Yes. (c) The ozone molecule is angular, thus the two O–O dipoles do not cancel. (d) Yes. The unshared electron pair on the central atom occupies space and repels the electrons of the oxygen-oxygen bonds.

1.38 The carbon atom in CH_3^+ utilizes only three of its four valence orbitals; therefore it is sp^2 hybridized. The vacant orbital is a p orbital.

SECTION REFERENCES FOR ADDITIONAL PROBLEMS

If you have trouble solving the Additional Problems refer to the sections in the text next to the problem numbers.

Problem	Sections	Problem	Sections
1.21	1.6C, 1.7	**1.30**	1.3
1.22	1.11, 1.14	**1.31**	1.7, 1.8
1.23	1.7	**1.32**	1.14, 1.18
1.24	1.20	**1.33**	1.9
1.25	1.2B, 1.20	**1.34**	1.8
1.26	1.3, 1.20	**1.35**	1.14, 1.7
1.27	1.19, 1.20	**1.36**	1.7
1.28	1.20D	**1.37**	1.8, 1.18, 1.19
1.29	1.20D	**1.38**	1.15

SELF-TEST

1.1 Using the atomic arrangements given below, draw two *different* and valid Lewis dot structures for the nitrous oxide molecule (N_2O). Show all formal charges that are not zero.

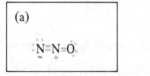

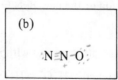

1.2 In the spaces provided give (a) the hybridization of the central atom and (b) the overall geometry of each molecule or ion below. In describing geometry take into account the atomic nuclei and the unshared electrons.

Molecule	*Hybridization*	*Geometry*
(a) BCl_3	sp^2	trig planar

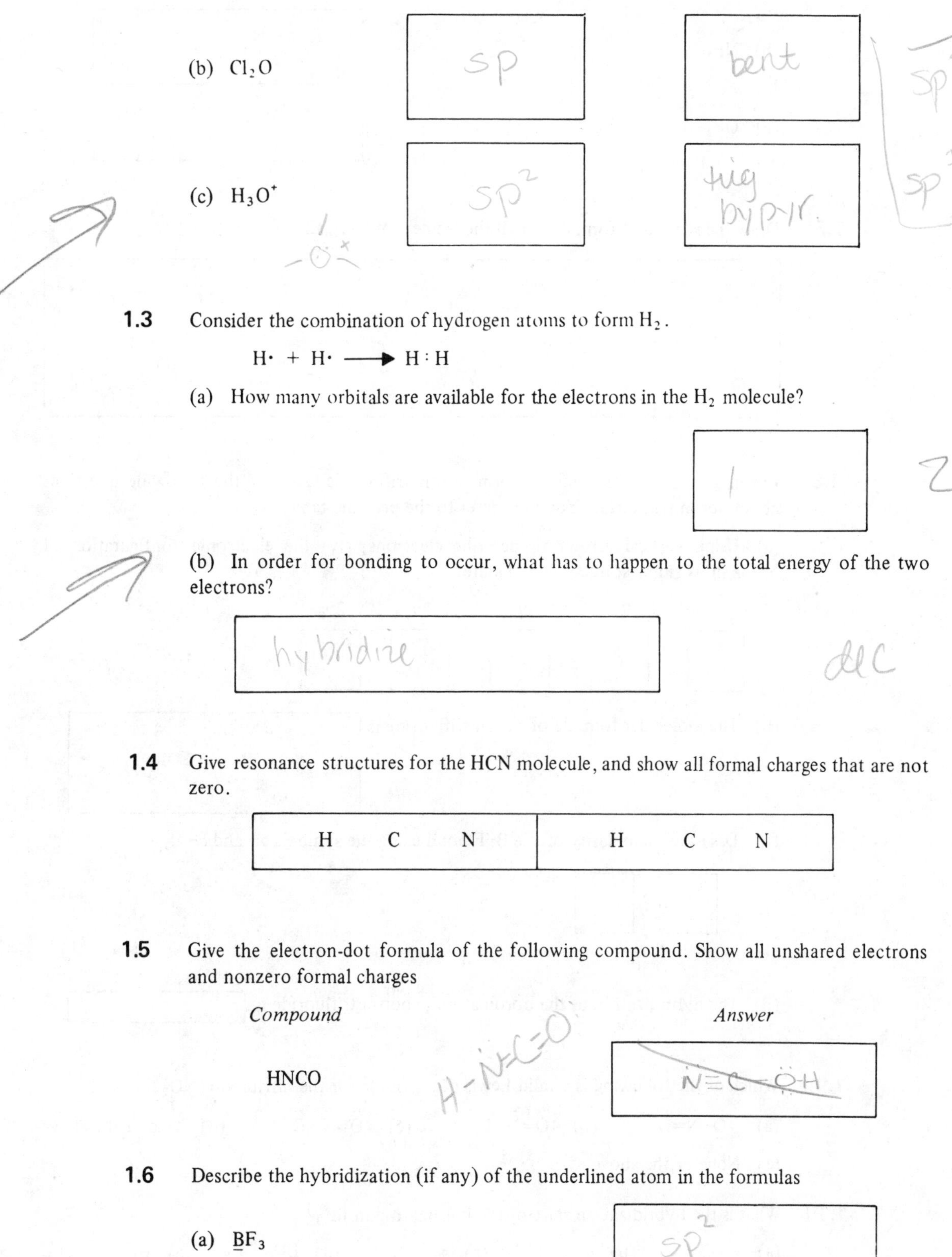

(b) Cl_2O

(c) H_3O^+

1.3 Consider the combination of hydrogen atoms to form H_2.

$$H\cdot + H\cdot \longrightarrow H:H$$

(a) How many orbitals are available for the electrons in the H_2 molecule?

(b) In order for bonding to occur, what has to happen to the total energy of the two electrons?

1.4 Give resonance structures for the HCN molecule, and show all formal charges that are not zero.

H C N	H C N

1.5 Give the electron-dot formula of the following compound. Show all unshared electrons and nonzero formal charges

Compound	*Answer*
HNCO	

1.6 Describe the hybridization (if any) of the underlined atom in the formulas

(a) $\underline{B}F_3$

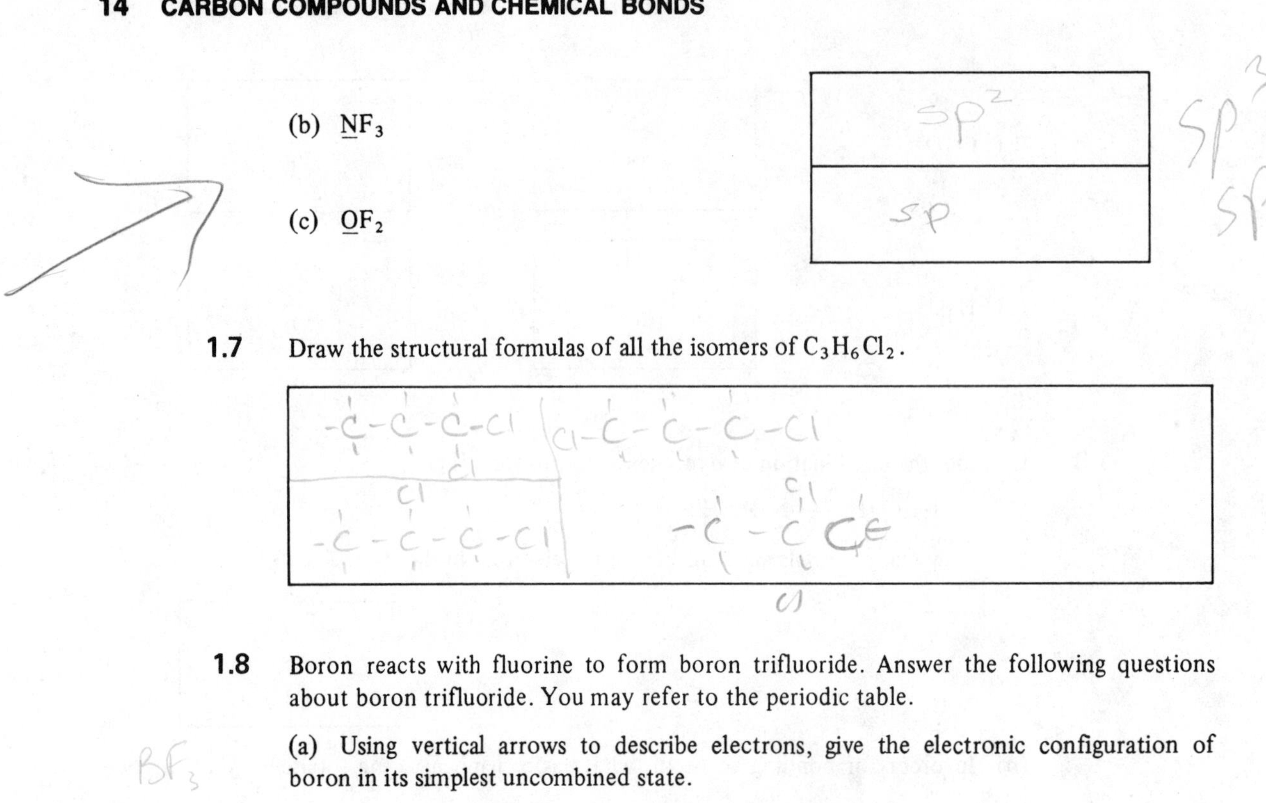

(b) $\underline{N}F_3$

(c) $\underline{O}F_2$

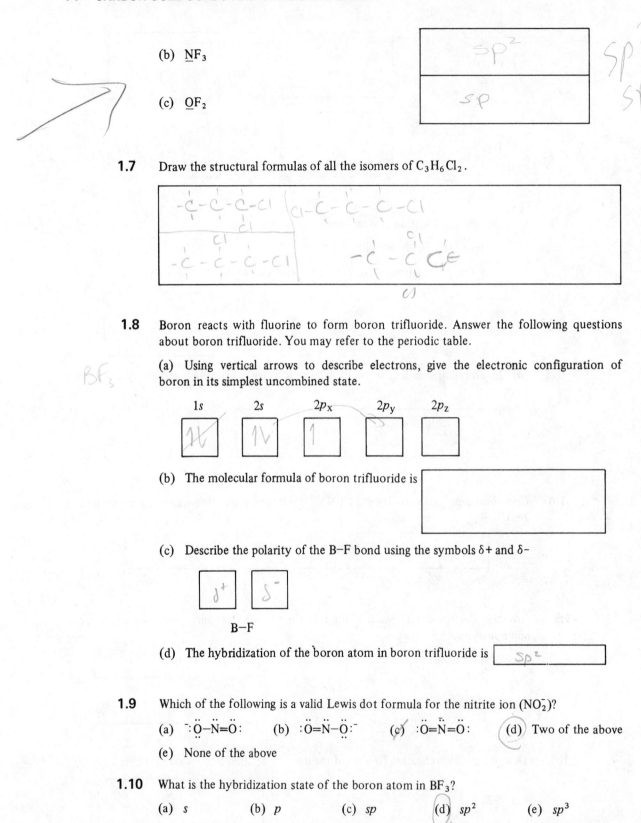

1.7 Draw the structural formulas of all the isomers of $C_3H_6Cl_2$.

1.8 Boron reacts with fluorine to form boron trifluoride. Answer the following questions about boron trifluoride. You may refer to the periodic table.

(a) Using vertical arrows to describe electrons, give the electronic configuration of boron in its simplest uncombined state.

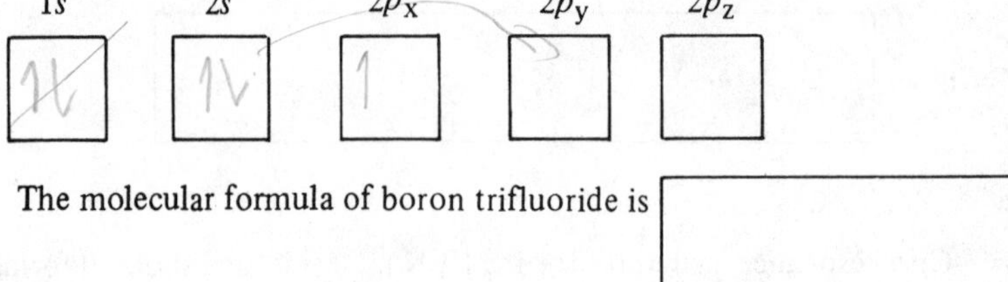

(b) The molecular formula of boron trifluoride is

(c) Describe the polarity of the B–F bond using the symbols δ+ and δ−

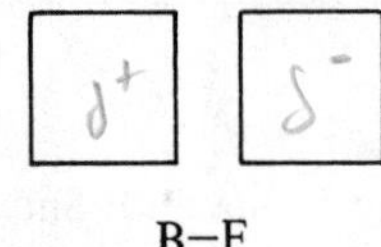

B–F

(d) The hybridization of the boron atom in boron trifluoride is

1.9 Which of the following is a valid Lewis dot formula for the nitrite ion (NO_2^-)?

(a) $^-\!:\ddot{\underset{..}{O}}-\ddot{N}=\ddot{O}:$ (b) $:\ddot{O}=\ddot{N}-\ddot{\underset{..}{O}}:^-$ (c) $:\ddot{O}=\ddot{N}=\ddot{O}:$ (d) Two of the above

(e) None of the above

1.10 What is the hybridization state of the boron atom in BF_3?

(a) s (b) p (c) sp (d) sp^2 (e) sp^3

1.11 BF_3 reacts with NH_3 to produce a compound

```
   F  H
   |  |
F—B—N—H
   |  |
   F  H
```

The hybridization on B is

(a) s (b) p (c) sp (d) sp^2 (e) sp^3

1.12 The formal charge on N in the compound given in Problem 1.11 is

(a) −2 (b) −1 (c) 0 (d) +1 (e) +2

1.13 The correct bond-line formula of the compound whose condensed formula is $CH_3CHClCH_2CH(CH_3)CH(CH_3)_2$ is

(a) Cl (b) Cl (c) Cl (d) Cl

(e) Cl

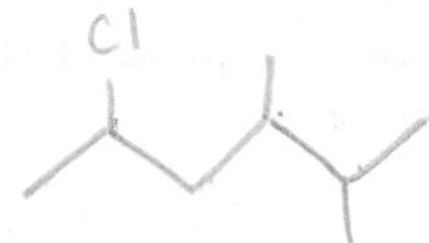

SUPPLEMENTARY PROBLEMS

S1.1 Write the dash formulas for all the isomers of $C_2H_3Cl_3$.

S1.2 Given the following electronegativities, explain the difference between the dipole moments of the molecules BF_3 and NF_3.

Electronegativity		μ, (D)
B = 2.0	BF_3	0
N = 3.0	NF_3	0.23
F = 4.0		

S1.3 Which of the following statements apply *both* to orbital hybridization and to molecular orbital formation?

(1) The process involves the combination of atomic orbitals.

(2) The combination of n atomic orbitals produces n new orbitals.

(3) The process involves the combination of atomic orbitals on the same atom.

SOLUTIONS TO SUPPLEMENTARY PROBLEMS

S1.1 The essential information here is the number of bonds that each atom must have:

```
  |
 -C-     -H     -C̈l:
  |              ··
```

The problem is then to assemble molecules that contain 2 C atoms, 3 H atoms, and 3 Cl atoms. Only two isomers are possible:

```
   H Cl               H H
   | |                | |
 H-C-C-Cl    and   Cl-C-C-Cl
   | |                | |
   H Cl               H Cl
```

Note that the direction of the bonds in these formulas is unimportant. That is, the following structures are equivalent:

```
   H H          H Cl         H Cl
   | |          | |          | |
Cl-C-C-Cl = Cl-C-C-H  =  H-C-C-Cl
   | |          | |          | |
   H Cl         H Cl         Cl H
```

These are equivalent structures because each carbon atom is tetrahedral and not square planar as drawn.

S1.2 The dipole moment is given by the sum of the individual bond dipole moments. The BF_3 molecule is trigonal planar, and the three bond dipole moments cancel each other. The NF_3 molecule is pyramidal, so the three bond dipole moments do not cancel each other.

S1.3 (1) and (2) are true for both orbital hybridization and molecular orbital formation. (3) is true only for hybridization because molecular orbital formation involves the combination of atomic orbitals on *different* atoms.

2 REPRESENTATIVE CARBON COMPOUNDS

SOLUTIONS TO PROBLEMS

2.1 (a) Cis-trans isomers are not possible

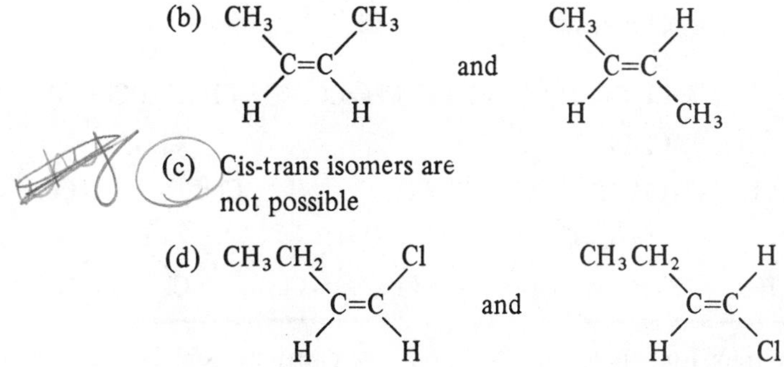

2.2 (a) The C–X bond moments in the trans isomers point in opposite directions and therefore cancel:

H X
C=C
X H

trans

X X
C=C
H H

cis

X = Cl or Br

In the cis isomers the bond moments are additive.

(b) The C–Cl bond moment is larger than the C–Br bond moment because Cl is more electronegative than Br, and this effect is not compensated for by greater bond distance. (See Problem 1.33.)

2.3 The structure of propene is

CH_3 H
C=C
H H

Note that in some cases cis–trans isomers are formed when hydrogen atoms are successively replaced by chlorine atoms.

(a) $(CH_3)(H)C{=}C(Cl)(H)$ $\quad$ $(CH_3)(H)C{=}C(H)(Cl)$ $\quad$ $(CH_3)(Cl)C{=}CH_2$ $\quad$ $Cl{-}CH_2CH{=}CH_2$

cis–trans Isomers (first two structures)

(b) $(CH_3)(Cl)C{=}C(H)(Cl)$ $\quad$ $(CH_3)(Cl)C{=}C(Cl)(H)$ — cis–trans Isomers

$(Cl{-}CH_2)(H)C{=}C(Cl)(H)$ $\quad$ $(Cl{-}CH_2)(H)C{=}C(H)(Cl)$ — cis–trans Isomers

(c) $CH_3CH{=}CCl_2$ $\quad$ $Cl{-}CH(Cl)CH{=}CH_2$ $\quad$ $Cl{-}CH_2C(Cl){=}CH_2$

$Cl_3CCH{=}CH_2$ $\quad$ $Cl{-}CH(Cl)C(Cl){=}CH_2$ $\quad$ $Cl{-}CH_2CH{=}CCl_2$ $\quad$ $CH_3C(Cl){=}CCl_2$

$(Cl_2CH)(H)C{=}C(Cl)(H)$ $\quad$ $(Cl_2CH)(H)C{=}C(H)(Cl)$ — cis–trans Isomers

$(Cl{-}CH_2)(Cl)C{=}C(H)(Cl)$ $\quad$ $(Cl{-}CH_2)(Cl)C{=}C(Cl)(H)$ — cis–trans Isomers

(d) $Cl_3CC(Cl){=}CH_2$ $\quad$ $(Cl_3C)(H)C{=}C(Cl)(H)$ $\quad$ $(Cl_3C)(H)C{=}C(H)(Cl)$ $\quad$ $Cl_2CHCH{=}CCl_2$

cis–trans Isomers (the two middle structures)

$(Cl_2CH)(Cl)C{=}C(H)(Cl)$ $\quad$ $(Cl_2CH)(Cl)C{=}C(Cl)(H)$ $\quad$ $ClCH_2C(Cl){=}CCl_2$

cis–trans Isomers (first two structures)

(e) $(Cl_3C)(Cl)C{=}C(H)(Cl)$ $\quad$ $(Cl_3C)(Cl)C{=}C(Cl)(H)$ $\quad$ $Cl_3CCH{=}CCl_2$ $\quad$ $Cl_2CHC(Cl){=}CCl_2$

cis–trans Isomers (first two structures)

(f) See (a–e) above.

2.4

(a) RCH_2X (b) $\underset{\displaystyle R}{RCHX}$ (c) $\overset{\displaystyle R}{\underset{\displaystyle R}{RCX}}$ (d) RX

2.5 (a) $CH_3CH_2CH_2Cl$ (b) $CH_3\underset{\displaystyle Br}{C}HCH_3$ (c) Ethyl fluoride

(d) Isopropyl iodide (e) Methyl iodide

2.6

(a) RCH_2OH (b) $\underset{\displaystyle R}{RCHOH}$ (c) $\overset{\displaystyle R}{\underset{\displaystyle R}{RCOH}}$

2.7 (a) $CH_3CH_2CH_2OH$ (b) $CH_3\underset{\displaystyle OH}{C}HCH_3$

2.8

(a) $CH_3{-}O{-}CH_2CH_3$ (b) $CH_3CH_2CH_2{-}O{-}CH_2CH_2CH_3$ (c) $CH_3{-}O{-}\overset{\displaystyle CH_3}{C}HCH_3$

(d) Ethyl propyl ether (e) Isopropyl propyl ether

2.9 (a) $CH_3{-}\underset{\displaystyle CH_3}{N}{-}H$ (b) $CH_3CH_2{-}\underset{\displaystyle CH_2CH_3}{N}{-}CH_2CH_3$ (c) $CH_3CH_2{-}\underset{\displaystyle CH_3}{N}{-}CH_2CH_2CH_3$

(d) Isopropylmethylamine (e) Methyldipropylamine (f) Isopropylamine

2.10 (a) (f) only (b) (a, d) (c) (b, c, e)

2.11 (a) In each case the oxygen atom has to accommodate eight electrons, so four orbitals are required. To obtain four hybrid orbitals we must mix one 2*s* and three 2*p* atomic orbitals. The result is that four sp^3 orbitals are used.

(b) sp^3 orbitals.

2.12

(a) $$K_a = \frac{[H_3O^+]\ [CF_3CO_2^-]}{[CF_3CO_2H]} = 1$$

let $[H_3O^+] = [CF_3CO_2^-] = X$

then, $[CF_3CO_2H] = 0.1 - X$

$$\therefore \quad \frac{(X)(X)}{0.1 - X} = 1 \quad \text{or} \quad X^2 = 0.1 - X$$

$$X^2 + X - 0.1 = 0$$

Using the quadratic formula, $X = \dfrac{-b \pm \sqrt{b^2 - 4ac}}{2a}$,

$$X = \frac{-1 \pm \sqrt{1 + 0.4}}{2} = \frac{-1 \pm \sqrt{1.4}}{2} = \frac{-1 \pm 1.183}{2} = \frac{+0.183}{2}$$

$X = 0.0915$ (We can exclude negative values of X.)

$[H_3O^+] = [CF_3CO_2^-] = 0.0915\ M$

(b) Percentage ionized $= \dfrac{[H_3O^+]}{0.1} \times 100 = \dfrac{(0.0915)(100)}{0.1}$

Percentage ionized = 91.5%

2.13

(a) $HC{\equiv}CH + NaH \xrightarrow{\text{hexane}} HC{\equiv}CNa + H_2$

(b) $HC{\equiv}CNa + D_2O \xrightarrow{\text{hexane}} HC{\equiv}CD + NaOD$

(c) $CH_3CH_2Li + D_2O \xrightarrow{\text{hexane}} CH_3CH_2D + LiOD$

(d) $CH_3CH_2OH + NaH \xrightarrow{\text{hexane}} CH_3CH_2ONa + H_2$

(e) $CH_3CH_2ONa + T_2O \xrightarrow{\text{hexane}} CH_3CH_2OT + NaOT$

(f) $CH_3CH_2CH_2Li + D_2O \xrightarrow{\text{hexane}} CH_3CH_2CH_2D + LiOD$

2.14

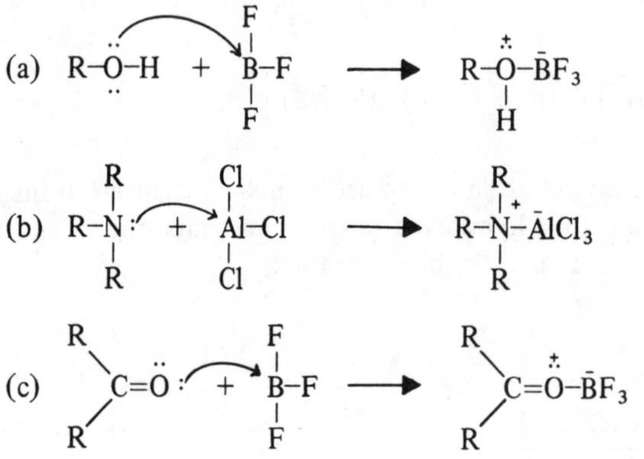

(a) $R{-}\ddot{\underset{..}{O}}{-}H + BF_3 \longrightarrow R{-}\overset{+}{O}(H){-}\overset{-}{B}F_3$

(b) $R_3N{:} + AlCl_3 \longrightarrow R_3\overset{+}{N}{-}\overset{-}{Al}Cl_3$

(c) $R_2C{=}\ddot{O}{:} + BF_3 \longrightarrow R_2C{=}\overset{+}{O}{-}\overset{-}{B}F_3$

2.15

(a) $CH_3{-}\ddot{\underset{..}{Cl}}{:} + AlCl_3 \longrightarrow CH_3{-}\overset{+}{Cl}{-}\overset{-}{Al}Cl_3$

Lewis base   Lewis acid

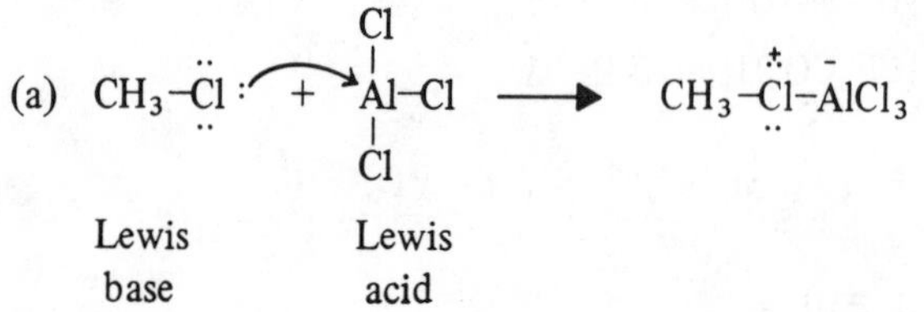

(b) $$R\text{-}\ddot{\underset{..}{O}}\text{-}H + H^+ \longrightarrow R\text{-}\overset{\overset{H}{|+}}{\underset{..}{O}}\text{-}H$$

Lewis base Lewis acid

(c) $$:\ddot{\underset{..}{Cl}}:^- + {}^+C(CH_3)_3 \longrightarrow :\ddot{\underset{..}{Cl}}\text{-}C(CH_3)_3$$

Lewis base Lewis acid

(d) $$H\ddot{\underset{..}{O}}:^- + CH_3\overset{\overset{:\ddot{O}}{\|}}{C}\text{-}OCH_2CH_3 \longrightarrow CH_3\overset{\overset{:\ddot{O}:^-}{|}}{\underset{\underset{H\ddot{\underset{..}{O}}:}{|}}{C}}\text{-}OCH_2CH_3$$

Lewis base Lewis acid

(e) $$CH_2{=}CH_2 + H^+ \longrightarrow \overset{+}{C}H_2\text{-}CH_3$$

Lewis base Lewis acid

(f) $$CH_3CH_2:^- + CH_3\text{-}\overset{\overset{:\ddot{O}}{\|}}{C}\text{-}H \longrightarrow CH_3\overset{\overset{:\ddot{O}:^-}{|}}{\underset{\underset{CH_2CH_3}{|}}{C}}\text{-}H$$

Lewis base Lewis acid

2.16 Molecules of propylamine can form hydrogen bonds to each other

$$CH_3CH_2CH_2N(H)\text{-}\overset{\delta+}{H}\cdots\overset{\delta-}{:N}(H)(H)CH_2CH_2CH_3$$

whereas molecules of trimethylamine, because they have no hydrogen atoms attached to a nitrogen atom, cannot form hydrogen bonds to each other.

2.17 Cyclopropane, because its cyclic structure makes it more rigid and symmetrical, permitting stronger crystal lattice forces.

2.18 (a) Alkyne (b) Carboxylic acid (c) Alcohol

(d) Aldehyde (c) Alkane (f) Ketone

2.19 (a) Carbon–carbon double bonds, primary alcohol group

(b) Ketone group, secondary alcohol group, carbon–carbon double bond

(c) Carbon–carbon double bond, ester group

(d) Amide groups

(e) Aldehyde group, primary and secondary alcohol groups

(f) Carbon–carbon double bond, ether linkage

(g) Carbon–carbon double bond, primary alcohol group

(h) Carbon–carbon double bond, aldehyde group

(i) Carbon–carbon double bond, esther groups

2.20 $CH_3CH_2CH_2CH_2Br$
1° Alkyl halide

$CH_3CH_2\underset{\displaystyle Br}{\underset{|}{C}}HCH_3$
2° Alkyl halide

$CH_3\underset{\displaystyle CH_3}{\underset{|}{C}}HCH_2Br$
1° Alkyl halide

$CH_3\text{–}\overset{\displaystyle CH_3}{\overset{|}{\underset{\displaystyle Br}{\underset{|}{C}}}}\text{–}CH_3$
3° Alkyl halide

2.21 $CH_3CH_2CH_2CH_2OH$
1° Alcohol

$CH_3CH_2\underset{\displaystyle OH}{\underset{|}{C}}HCH_3$
2° Alcohol

$CH_3\underset{\displaystyle CH_3}{\underset{|}{C}}HCH_2OH$
1° Alcohol

$CH_3\text{–}\overset{\displaystyle CH_3}{\overset{|}{\underset{\displaystyle OH}{\underset{|}{C}}}}\text{–}CH_3$
3° Alcohol

$CH_3OCH_2CH_2CH_3$
Ether

$CH_3O\underset{\displaystyle CH_3}{\underset{|}{C}}HCH_3$
Ether

$CH_3CH_2OCH_2CH_3$
Ether

2.22 Any four of the following:

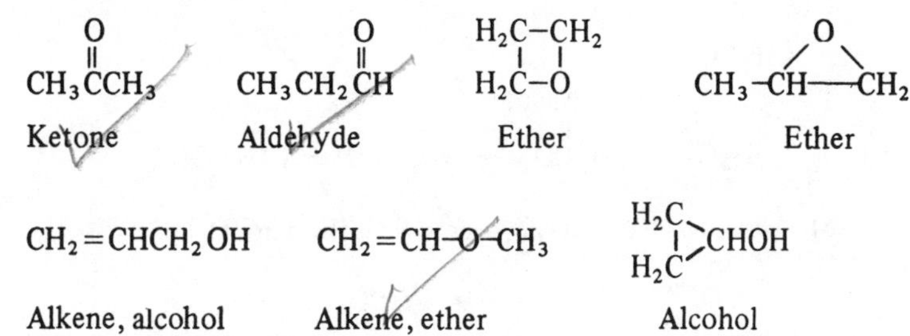

2.23 (a) Primary (b) Secondary (c) Tertiary (d) Secondary
(e) Secondary (f) Tertiary

2.24 (a) Secondary (b) Primary (c) Tertiary (d) Primary
(e) Secondary

2.25

(a) $CH_3OCH_2CH_3$ (b) $CH_3CH_2CH_2OH$ (c) $CH_3CH(OH)CH_3$

(d) $CH_3C(=O)OCH_2CH_3$ $CH_3CH_2C(=O)OCH_3$ (e) $CH_3CH_2CH_2CH_2X$

(f) $CH_3CH_2CHXCH_3$ (g) $CH_3C(CH_3)(X)CH_3$ (h) $CH_3CH(CH_3)CH{=}O$ or $CH_3CH_2CH_2CH{=}O$

(i) $CH_3C(=O)CH_2CH_3$ (j) $CH_3CH_2CH(CH_3)NH_2$ (k) $CH_3CH_2CH_2NHCH_3$

(l) $CH_3CH_2N(CH_3)_2$ (m) $CH_3CH_2CH_2C(=O)NH_2$ (n) $CH_3C(=O)NHCH_2CH_3$

(o) 1-methylcyclopropanol (cyclopropane ring bearing CH_3 and OH on the same carbon)

2.26 To write *net ionic* equations, our first task is to determine which ions are actually present in the solution.

(a) Both HCl and Na_2CO_3 are completely dissociated in aqueous solution, so the ions are $H_3O^+ + Cl^-$ and $Na^+ + CO_3^{-2}$. The Cl^- and Na^+ ions are spectator ions because they occur in the same form on both sides of the equation. The net ionic equation is thus

$$2H_3O^+ + CO_3^{-2} \longrightarrow [H_2CO_3] + 2H_2O \longrightarrow 3H_2O + CO_2$$

(b) Here again the ions that interact, excluding the spectator ions Br^- and Na^+, are H_3O^+ and $CH_3\overset{O}{\overset{\|}{C}}O^-$

$$H_3O^+ + CH_3\overset{O}{\overset{\|}{C}}O^- \longrightarrow H_2O + CH_3\overset{O}{\overset{\|}{C}}OH$$

(c) Here, the base, CO_3^{-2}, reacts with the acid, H_2O

$$CO_3^{-2} + H_2O \longrightarrow HCO_3^- + OH^-$$

(d) NaH is a very strong base because it releases the very basic $H:^-$

$$:H^- + H_2O \longrightarrow H_2 + OH^-$$

(e) Here the base is $:CH_3^-$

$$:CH_3^- + H_2O \longrightarrow CH_4 + OH^-$$

(f) In this reaction, the acid is $HC{\equiv}CH$

$$:CH_3^- + HC{\equiv}CH \longrightarrow HC{\equiv}C:^- + CH_4$$

(g) The basic species in aqueous NH_3 is $:NH_3$

$$H_3O^+ + :NH_3 \longrightarrow NH_4^+ + H_2O$$

(h) The acid is NH_4^+; the base is NH_2^-. As before, Cl^- and Na^+ are spectator ions.

$$NH_4^+ + NH_2^- \longrightarrow 2NH_3$$

(i) $CH_3CH_2O^- + H_2O \longrightarrow CH_3CH_2OH + OH^-$

2.27 Oxygen-containing compounds contain either $=\ddot{O}:$ or $-\ddot{\underset{..}{O}}-$. Both of these are Brønsted–Lowry bases in the presence of the strong proton donor, sulfuric acid. The equation for the reaction using an ether as an example is

$$R{-}\ddot{\underset{..}{O}}{-}R + H_2SO_4 \rightleftharpoons \underbrace{R{-}\overset{H}{\underset{..}{\overset{|}{O}}}{}^{+}{-}R + HSO_4^-}_{\text{Salt}}$$

The salt is soluble in the highly polar H_2SO_4.

2.28 (a) Ethyl alcohol because its molecules can form hydrogen bonds to each other. Methyl ether molecules have no hydrogen atoms attached to oxygen atoms.

(b) Ethylene glycol because its molecules have more OH groups and will therefore participate in more extensive hydrogen bonding.

(c) Heptane because it has a higher molecular weight. (Neither compound can form hydrogen bonds.)

(d) Propyl alcohol because its molecules can form hydrogen bonds to each other. Acetone molecules have no hydrogen atoms attached to oxygen atoms.

(e) *cis*-1,2-Dichloroethene because its molecules have a higher dipole moment.

(f) Propionic acid because its molecules can form hydrogen bonds to each other.

2.29 (a) 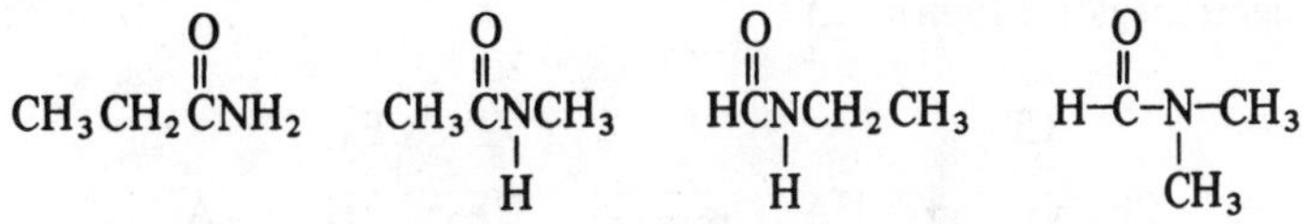 $CH_3CH_2\overset{O}{\overset{\|}{C}}NH_2$ $\quad CH_3\overset{O}{\overset{\|}{C}}\underset{H}{\underset{|}{N}}CH_3$ $\quad H\overset{O}{\overset{\|}{C}}\underset{H}{\underset{|}{N}}CH_2CH_3$ $\quad H{-}\overset{O}{\overset{\|}{C}}{-}\underset{CH_3}{\underset{|}{N}}{-}CH_3$

(b) The last one given above [i.e., $H\overset{O}{\overset{\|}{C}}N(CH_3)_2$] because it does not have a hydrogen that is covalently bonded to nitrogen, and, therefore, its molecules cannot form hydrogen bonds to each other. The other molecules all have a hydrogen covalently bonded to nitrogen, and, therefore, hydrogen bond formation is possible. With the first molecule, for example, hydrogen bonds could form in the following way:

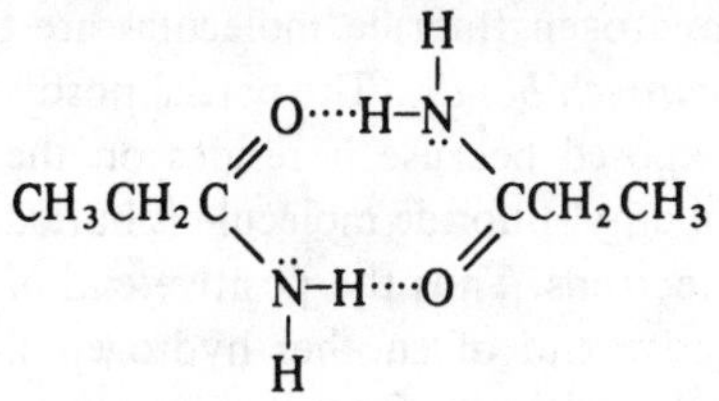

```
                  H
                  |
           O····H-N
          //       \
CH3CH2C             CCH2CH3
          \        //
           N-H····O
           |
           H
```

2.30 An acid–base reaction as follows:

$$CH_3CH_2OH + HC{\equiv}C\!:^- Na^+ \xrightarrow[NH_3]{liq.} CH_3CH_2O^-Na^+ + HC{\equiv}CH$$

2.31 (a) $pK_a = -\log K_a$

$pK_a = -\log 1.8 - \log 10^{-5}$

$pK_a = -0.25 + 5 = 4.75$

(b) $K_a = 10^{-13}$

2.32 (a) For acid HA, $K_a = 10^{-20}$; for acid HB, $K_a = 10^{-10}$. Because 10^{-10} is larger than 10^{-20}, acid HB is the stronger acid.

(b) Yes because the equilibrium lying to the right (below) yields the weaker acid and weaker base.

$$\underset{\text{Stronger acid}}{HB} + \underset{\text{Stronger base}}{A\!:^-} \rightleftharpoons \underset{\text{Weaker acid}}{HA} + \underset{\text{Weaker base}}{:B^-}$$

2.33 Basic strength depends on ability to accept a proton. In $(CF_3)_3N:$, the high electronegativity of fluorine reduces the availability of the lone electron pair on nitrogen to accept a proton. In $(CH_3)_3N:$, the lone pair is more available to bond with a proton. (An alternative view is that the conjugate acid is rendered less stable by the presence of the electronegative fluorines.)

$$(CF_3)_3N: + HA \rightleftharpoons (CF_3)_3N^{+}{-}H + A^-$$

2.34 An ester group, $-C(=O)-O-$

2.35 The attractive forces between hydrogen fluoride molecules are the very strong dipole–dipole attractions that we call *hydrogen bonds.* (The partial positive charge of a hydrogen fluoride molecule is relatively exposed because it resides on the hydrogen nucleus. By contrast, the positive charge of an ethyl fluoride molecule is buried in the ethyl group and is shielded by the surrounding electrons. Thus the positive end of one hydrogen fluoride molecule can approach the negative end of another hydrogen fluoride molecule much more closely with the result that the attractive force between them is much stronger.)

SECTION REFERENCES FOR ADDITIONAL PROBLEMS

2.18 2.3, 2.5–2.6, 2.8–2.14
2.19 2.3, 2.5–2.6, 2.8–2.14
2.20 2.9
2.21 2.10, 2.11
2.22 2.10, 2.11, 2.13
2.23 2.10
2.24 2.12
2.25 2.4–2.14
2.26 2.15
2.27 2.15A
2.28 2.17
2.29 2.14B, 2.17C
2.30 2.15C
2.31 2.15B
2.32 2.15B
2.33 1.9, 2.15B
2.34 2.14
2.35 2.17B, 2.17C

SELF-TEST

2.1 Supply the appropriate formula for each of the following:

(a) The isomer of $C_2H_2Br_2$ that does *not* exhibit cis–trans isomerism.

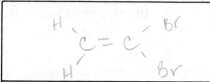

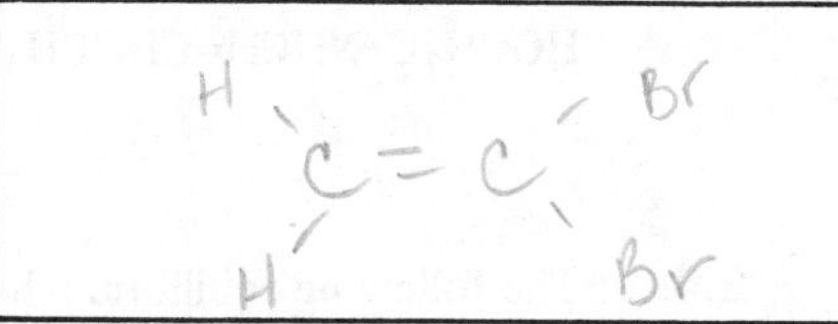

(b) A hydroxyl group containing compound that is *not* an alcohol.

(c) The bond-line formula of a secondary alcohol that has four carbon atoms.

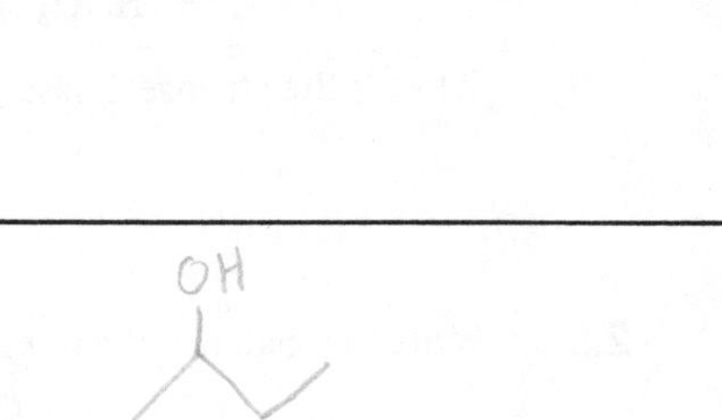

(d) The bond-line formula of a tertiary amine that has four carbon atoms.

(e) An ester that has three carbon atoms.

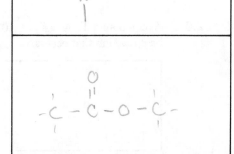

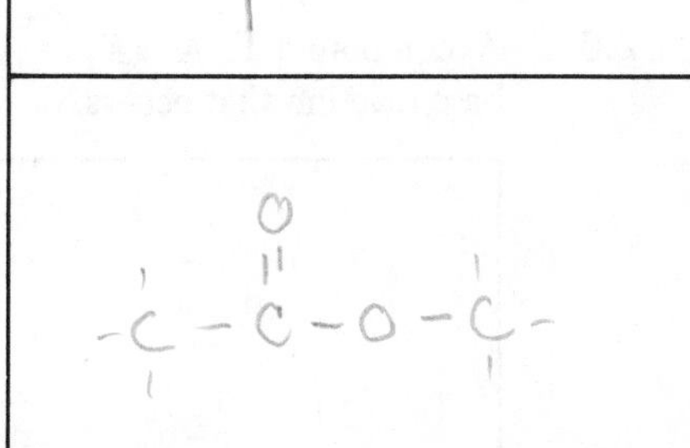

2.2 Classify the following alcohols and amines as primary (1°), secondary (2°), or tertiary (3°).

(a) $CH_3CH_2-C(CH_3)_2-OH$ [3°]

(b) CH_3–(cyclohexene ring)–OH [2°]

(c) $CH_3CH_2-C(CH_3)_2-CH_2OH$ [1°]

(d) (piperidine ring) N–H [2°]

(e) $CH_3CH_2N(CH_3)_2$ [3°]

(f) $CH_3-C(CH_3)_2-NH_2$ [1°]

2.3 Name the functional groups in this structure. Give their names in the order in which they occur (left to right).

$$HOCH_2\overset{\overset{\displaystyle O}{\|}}{C}{-}NHCH{=}CHOCH_3$$

2.4 The following equilibrium has $K_{eq} >> 10$:

$$HClO_4 + HNO_3 \rightleftharpoons H_2NO_3^+ + ClO_4^-.$$

What is the strongest base present in the mixture?

2.5 Write and balance the *net ionic equation* for the acid–base reaction in the space provided.

$$HBr_{(aq)} + CH_3\overset{\overset{\displaystyle O}{\|}}{C}{-}ONa_{(aq)} \longrightarrow CH_3\overset{\overset{\displaystyle O}{\|}}{C}{-}OH + NaBr_{(aq)}$$

2.6 A compound $H{:}\ddot{\underset{\cdot\cdot}{A}}{:}$ ($K_a = 10^4$) is dissolved in water. Write the equation for the acid–base reaction that occurs.

2.7 Circle the compound in the following pair that has the higher boiling point.

(a) $CH_2{=}CHCH_2OH$ (b) $CH_3\overset{\overset{\displaystyle O}{\|}}{C}CH_3$

2.8 Which of the following pairs of compounds is *not* a pair of constitutional isomers?

(a) $CH_3{-}O{-}CH{=}CH_2$ and $CH_3CH_2\overset{\overset{\displaystyle O}{\|}}{C}H$

(b) (cyclopentane) and $CH_3CH{=}CHCH_2CH_3$

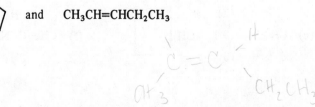

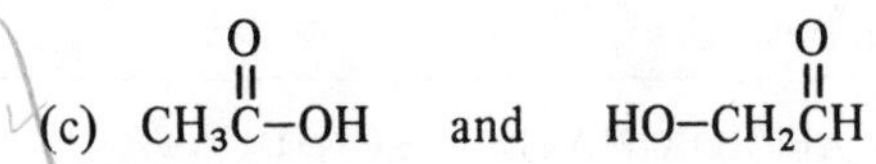

(c) $CH_3\overset{O}{\overset{\|}{C}}-OH$ and $HO-CH_2\overset{O}{\overset{\|}{C}}H$

(d) $CH_3CH_2C{\equiv}CH$ and $CH_3CH{=}C{=}CH_2$

(e) $CH_3\underset{CH_3}{\underset{|}{C}}HCH(CH_3)_2$ and $(CH_3)_2CHCH(CH_3)_2$

2.9 Which of the answers to Problem 2.8 contains an ether?

2.10 Which of the following pairs of structures represents a pair of isomers?

(a) $(CH_3)(CH_3)C{=}C(CH_3)(H)$ and $(CH_3)(CH_3)C{=}C(H)(CH_3)$

(b) $CH_3C{\equiv}CCH_3$ and $CH_3CH_2C{\equiv}CH$

(c) $H,Cl,F\text{–}C\text{–}C\text{–}F,Cl,H$ and $H,Cl,F\text{–}C\text{–}C\text{–}Cl,H,F$

(d) $CH_3CH_2\underset{}{\overset{CH_3}{\overset{|}{C}}}HCH_2CH_3$ and $CH_3CH_2\underset{CH_2CH_3}{\underset{|}{C}}HCH_3$

(e) More than one of these pairs are isomers.

2.11 Using the data of Table 2.2, predict what will happen when $NaNH_2$ is dissolved in CH_3CH_2OH.

(a) The solution will consist primarily of solvated Na^+ and NH_2^- ions.

(b) The solution will consist primarily of metallic Na and $CH_3CH_2O^-$ ions.

(c) The solution will consist primarily of solvated Na^+ and $CH_3CH_2O^-$ ions and NH_3 molecules.

(d) The solution will consist of approximately equal amounts of NH_2^- and $CH_3CH_2O^-$ ions and corresponding Na^+ ions.

(e) None of the above.

TABLE 2.2 Scale of Acidities and Basicities

INCREASING STRENGTH OF ACID (↓)	ACID	APPROXIMATE K_a	pK_a	CONJUGATE BASE	INCREASING STRENGTH OF CONJUGATE BASE (↑)
	CH_3CH_3	10^{-50}	50	$CH_3CH_2^-$	
	$CH_2{=}CH_2$	10^{-44}	44	$CH_2{=}CH^-$	
	H_2	10^{-35}	35	H^-	
	NH_3	10^{-33}	33	NH_2^-	
	$HC{\equiv}CH$	10^{-25}	25	$HC{\equiv}C^-$	
	CH_3CH_2OH	10^{-18}	18	$CH_3CH_2O^-$	
	H_2O	10^{-16}	16	HO^-	
	CH_3CO_2H	10^{-5}	5	$CH_3CO_2^-$	
	CF_3CO_2H	1	0	$CF_3CO_2^-$	
	HNO_3	20	−1.3	NO_3^-	
	H_3O^+	50	−1.7	H_2O	
	HCl	10^7	−7	Cl^-	
	H_2SO_4	10^9	−9	HSO_4^-	
	HI	10^{10}	−10	I^-	
	$HClO_4$	10^{10}	−10	ClO_4^-	
	$SbF_5{\cdot}FSO_3H$	$>10^{12}$	−12	$SbF_5{\cdot}FSO_3^-$	

SUPPLEMENTARY PROBLEMS

S2.1 Noting the direction in which the following acid–base equilibria are displaced, circle the strongest acid and base in each.

(a) $HA + HB \rightleftharpoons H_2A^+ + B^-$ (equilibrium displaced to the right)

(b) $HD + HE \rightleftharpoons D^- + H_2E^+$ (equilibrium displaced to the left)

S2.2 Classify the alcohol and amine groups as primary (1°), secondary (2°), or tertiary (3°) in the following compounds.

(a) (cyclobutyl)$NCH(CH_3)CH(CH_3)CH_2OH$

(b) $CH_3C(CH_3)_2NHCH(CH_3)CH(CH_3)OH$ — i.e., $CH_3\overset{CH_3}{\underset{CH_3}{C}}NH\overset{CH_3}{C}H\underset{CH_3}{C}HOH$

S2.3 Name the functional groups in each of the following molecules.

(a) $CH_3\overset{O}{\overset{\|}{C}}CH_2OH$ (b) $CH_3CH_2\overset{O}{\overset{\|}{C}}OH$ (c) $H\overset{O}{\overset{\|}{C}}CH_2CH_2NH\overset{O}{\overset{\|}{C}}H$

SOLUTIONS TO SUPPLEMENTARY PROBLEMS

S2.1 In any acid–base reaction, the weaker acid and weaker base will predominate at equilibrium.

(a) Since equilibrium is displaced to the right, we know that H_2A^+ is a weaker acid than HB and that HA acts as a base. B^- is a weaker base than HA.

(b) HD is a weaker acid than H_2E^+, and HE is a weaker base than D^-.

S2.2 (a) 3° Amine, 1° alcohol (b) 2° Amine, 2° alcohol

S2.3 (a) Ketone, 1° alcohol (b) Carboxyl (c) Aldehyde, amide

3 ALKANES AND CYCLOALKANES. CONFORMATIONAL ANALYSIS

SOLUTION TO PROBLEMS

3.1 The condensed structural formulas are shown below each line-and-circle formula.

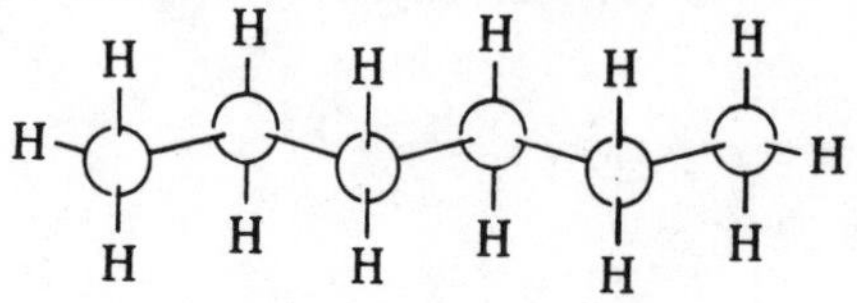

(1) $CH_3CH_2CH_2CH_2CH_2CH_3$

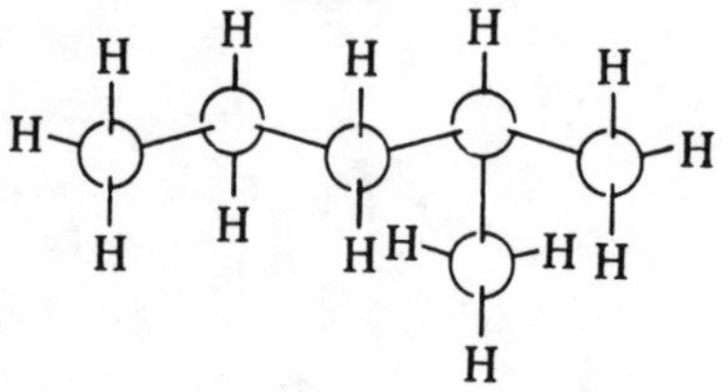

(2) $CH_3CH_2CH_2CH(CH_3)CH_3$

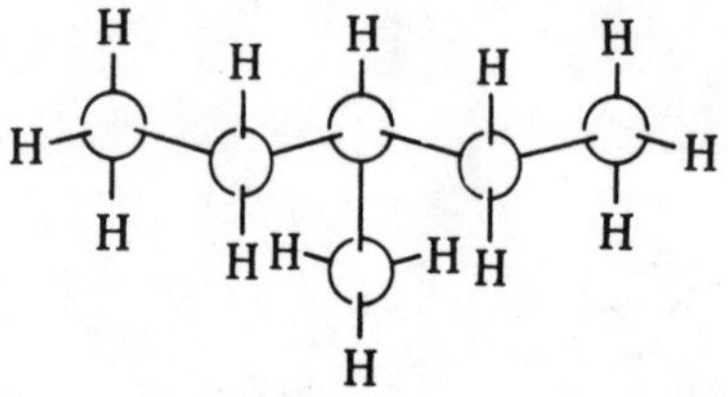

(3) $CH_3CH_2CH(CH_3)CH_2CH_3$

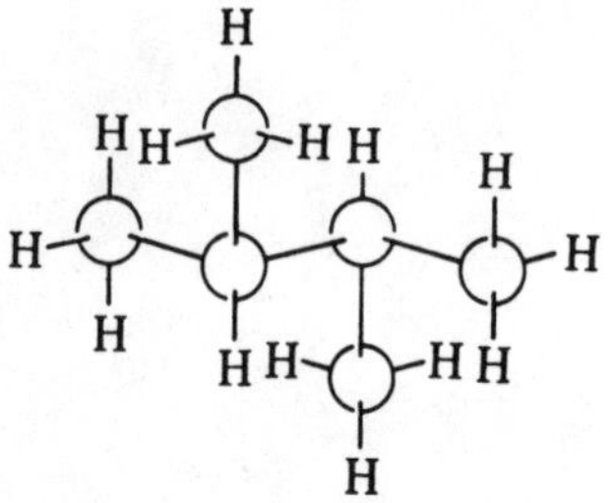

(4) $CH_3CH(CH_3)CH(CH_3)CH_3$

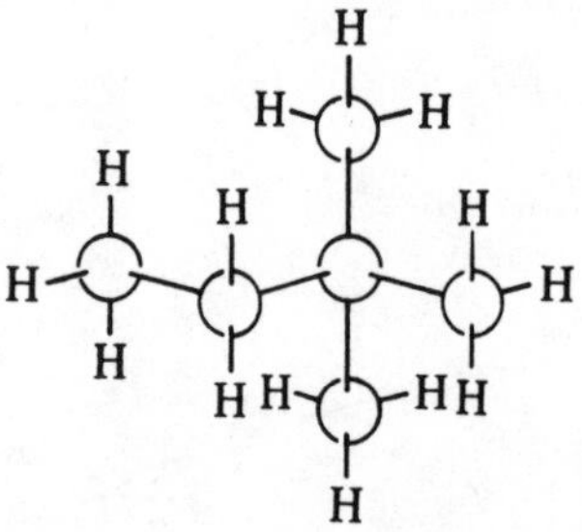

(5) $CH_3CH_2C(CH_3)_2CH_3$

3.2 (a) Refer to Problem 3.1:

(1) Hexane, (2) 2-Methylpentane, (3) 3-Methypentane, (4) 2,3-Dimethylbutane, (5) 2,2-Dimethylbutane.

(b) $CH_3CH_2CH_2CH_2CH_2CH_2CH_3$ — Heptane

$CH_3CH_2CH_2CH_2CHCH_3$ (with CH_3 on the CH) — 2-Methylhexane

$CH_3CH_2CH_2CHCH_2CH_3$ (with CH_3 on the CH) — 3-Methylhexane

$CH_3CH_2CH_2CCH_3$ (with two CH_3 on the C) — 2,2-Dimethylpentane

$CH_3CH_2CHCHCH_3$ (with CH_3 on each CH) — 2,3-Dimethylpentane

$CH_3CHCH_2CHCH_3$ (with CH_3 on each CH) — 2,4-Dimethylpentane

$CH_3CH_2CCH_2CH_3$ (with two CH_3 on the C) — 3,3-Dimethylpentane

$CH_3CH{-}CCH_3$ (with CH_3 on the CH and two CH_3 on the C) — 2,2,3-Trimethylbutane

$CH_3CH_2CHCH_2CH_3$ (with CH_2CH_3 on the CH) — 3-Ethylpentane

3.3 (a) $CH_3CH_2CH_2CH_2Cl$
1-Chlorobutane

CH_3CHCH_2Cl (with CH_3 on the CH)
1-Chloro-2-methylpropane

$CH_3CH_2CHCH_3$ (with Cl on the CH)
2-Chlorobutane

$CH_3{-}C{-}CH_3$ (with CH_3 and Cl on the C)
2-Chloro-2-methylpropane

(b) $CH_3CH_2CH_2CH_2CH_2Br$
1-Bromopentane

$CH_3CHCH_2CH_2Br$
|
CH_3
1-Bromo-3-methylbutane

$CH_3CH_2CH_2CHCH_3$
|
Br
2-Bromopentane

$CH_3CH_2CHCH_2Br$
|
CH_3
1-Bromo-2-methylbutane

$CH_3CH_2CHCH_2CH_3$
|
Br
3-Bromopentane

CH_3
|
$CH_3CHCHCH_3$
|
Br
2-Bromo-3-methylbutane

CH_3
|
CH_3CCH_2Br
|
CH_3
1-Bromo-2,2-dimethylpropane

CH_3
|
$CH_3CH_2CCH_3$
|
Br
2-Bromo-2-methylbutane

3.4 (a) $CH_3CH_2CH_2CH_2OH$
1-Butanol

CH_3CHCH_2OH
|
CH_3
2-Methyl-1-propanol

$CH_3CH_2CHCH_3$
|
OH
2-Butanol

CH_3
|
CH_3COH
|
CH_3
2-Methyl-2-propanol

(b) $CH_3CH_2CH_2CH_2CH_2OH$
1-Pentanol

$CH_3CHCH_2CH_2OH$
|
CH_3
3-Methyl-1-butanol

$CH_3CH_2CH_2CHCH_3$
|
OH
2-Pentanol

$CH_3CH_2CHCH_2OH$
|
CH_3
2-Methyl-1-butanol

$CH_3CH_2CHCH_2CH_3$
|
OH
3-Pentanol

CH_3
|
$CH_3CHCHCH_3$
|
OH
3-Methyl-2-butanol

$CH_3\overset{\displaystyle CH_3}{\underset{\displaystyle CH_3}{C}}CH_2OH$

2, 2-Dimethyl-1-propanol

$CH_3CH_2\overset{\displaystyle CH_3}{\underset{\displaystyle OH}{C}}CH_3$

2-Methyl-2-butanol

3.5 (a) 1-*tert*-Butyl-3-methylcyclohexane

(b) 1, 3-Dimethylcyclobutane

(c) 1-Butylcyclohexane

(d) 1-Chloro-2, 4-dimethylcyclohexane

(e) 2-Chlorocyclopentanol

(f) 3-*tert*-Butylcyclohexanol

3.6 (a) Bicyclo[2.2.0] hexane

(b) Bicyclo[4.4.0] decane

(c) Bicyclo[2.2.2] octane

(d) 3-Methylbicyclo[3.2.0] heptane

(e) 7-Methylbicyclo[4.2.1] nonane

(f) Bicyclo[3.1.1] heptane; or

Bicyclo[4.1.0] heptane

3.7

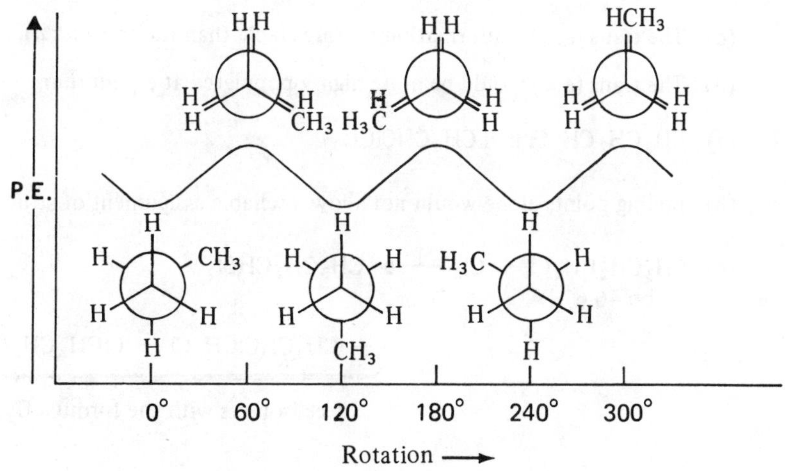

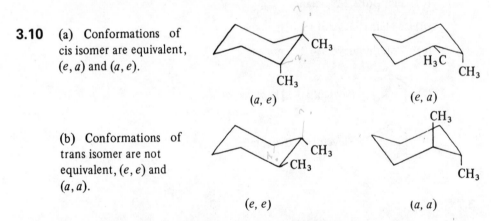

3.8

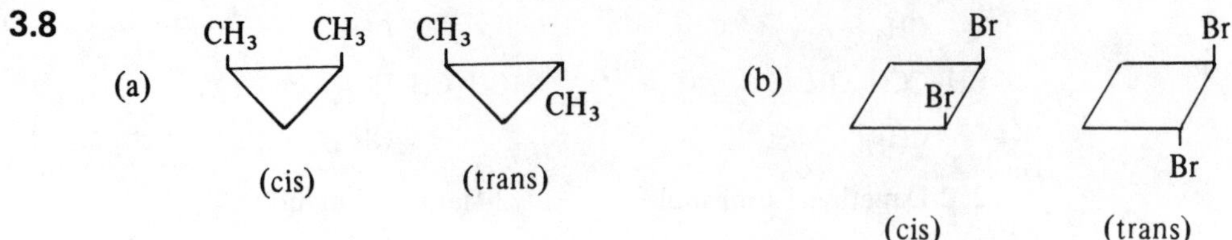

3.9 (a)

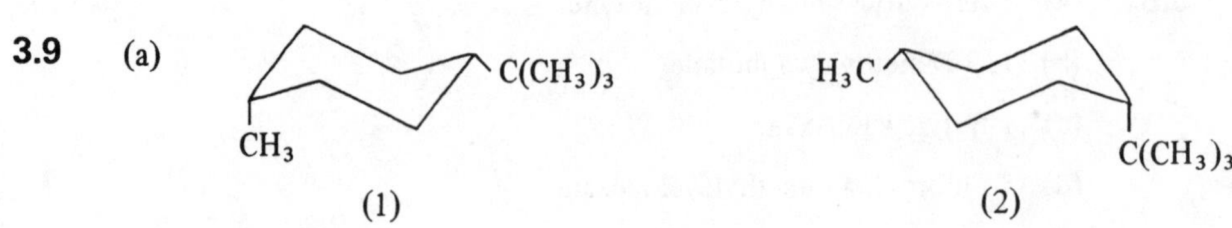

(b) No. In (1), the methyl group is axial and the *tert*-butyl group is equatorial; in (2) the situation is reversed.

(c) The *tert*-butyl group is larger than the methyl; conformation (1) is more stable because the *tert*-butyl group is equatorial.

(d) The preferred conformation at equilibrium is (1).

3.10 (a) Conformations of cis isomer are equivalent, (e, a) and (a, e).

CH_3 CH_3 (a, e) H_3C CH_3 (e, a)

(b) Conformations of trans isomer are not equivalent, (e, e) and (a, a).

CH_3 CH_3 (e, e) CH_3 CH_3 (a, a)

(c) The trans (e, e) conformation is more stable than the trans (a, a).

(d) The trans (e, e) would be more highly populated at equilibrium.

3.11 (a) $CH_3CH_2CH_2Cl$ and $CH_3CHClCH_3$

(b) Boiling points alone would not allow a reliable assignment of structures.

(c) $CH_3CH_2CH_2Cl + Cl_2 \longrightarrow CH_3CH_2CHCl_2$
(bp 46.6°C)

$+ CH_3CHClCH_2Cl + ClCH_2CH_2CH_2Cl$

Three isomers with the formula $C_3H_6Cl_2$

$$CH_3CHClCH_3 + Cl_2 \longrightarrow CH_3CHClCH_2Cl + CH_3CCl_2CH_3$$

(bp 36.5°C) — Two isomers with the formula $C_3H_6Cl_2$

The number of isomers produced in each reaction allows us to assign the structures without ambiguity.

(d) See (c) above

3.12

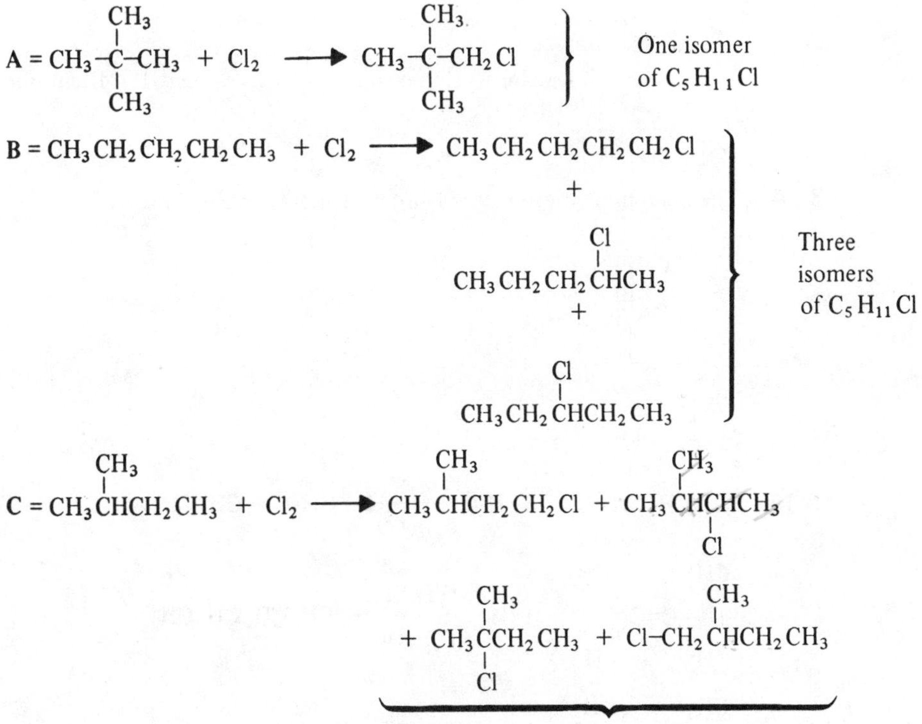

$$A = CH_3-C(CH_3)_2-CH_3 + Cl_2 \longrightarrow CH_3-C(CH_3)_2-CH_2Cl$$

One isomer of $C_5H_{11}Cl$

$$B = CH_3CH_2CH_2CH_2CH_3 + Cl_2 \longrightarrow CH_3CH_2CH_2CH_2CH_2Cl + CH_3CH_2CH_2CHClCH_3 + CH_3CH_2CHClCH_2CH_3$$

Three isomers of $C_5H_{11}Cl$

$$C = CH_3CH(CH_3)CH_2CH_3 + Cl_2 \longrightarrow CH_3CH(CH_3)CH_2CH_2Cl + CH_3C(CH_3)HCHClCH_3 + CH_3CCl(CH_3)CH_2CH_3 + Cl-CH_2CH(CH_3)CH_2CH_3$$

Four isomers of $C_5H_{11}Cl$

3.13

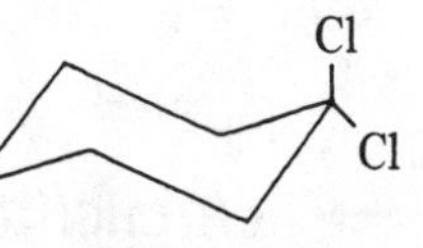

1, 1-Dichlorocyclohexane

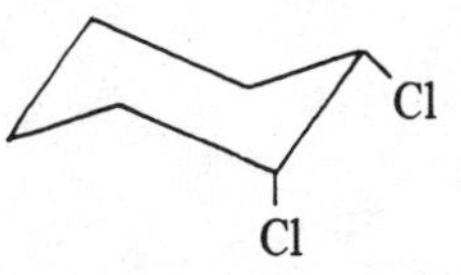

cis-1, 2-Dichlorocyclohexane

trans-1, 2-Dichlorocyclohexane

cis-1, 3-Dichlorocyclohexane

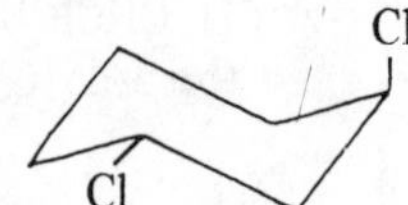

trans-1, 3-Dichlorocyclohexane

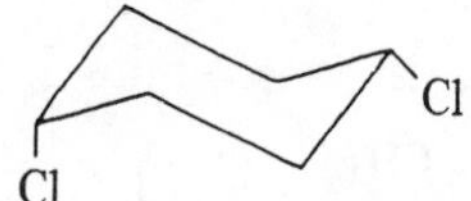

cis-1, 4-Dichlorocyclohexane

Cl Cl

trans-1, 4-Dichlorocyclohexane

3.14 The predominant product is 1-bromo-1-methylcyclopentane.

CH_3 Br

3.15 $CH_3CH_2CH{=}CH_2 + H_2 \xrightarrow[C_2H_5OH]{\text{Pt or Ni}} CH_3CH_2CH_2CH_3$

$$(CH_3)(H)C{=}C(CH_3)(H) \text{ (cis)} + H_2 \xrightarrow[C_2H_5OH]{\text{Pt or Ni}} CH_3CH_2CH_2CH_3$$

$$(CH_3)(H)C{=}C(H)(CH_3) \text{ (trans)} + H_2 \xrightarrow[C_2H_5OH]{\text{Pt or Ni}} CH_3CH_2CH_2CH_3$$

3.16 $CH_3CH(CH_3)CH(Br)CH_3 \xrightarrow[Zn]{H^+} CH_3CH(CH_3)CH_2CH_3$

$$CH_3C(Br)(CH_3)CH_2CH_3 \xrightarrow[Zn]{H^+} CH_3CH(CH_3)CH_2CH_3$$

$$BrCH_2CH(CH_3)CH_2CH_3 \xrightarrow[Zn]{H^+} CH_3CH(CH_3)CH_2CH_3$$

3.17

(a) $CH_3CHCHCH_2CH_3$
$\quad Cl\ Cl$

(b) CH_3
CH_3CCH_3
I

(c) $CH_3CH_2CHCH_2CH_3$
CH_2
CH_3

(d) $CH_3CH{-}CH{-}CHCH_2CH_2CH_2CH_2CH_2CH_3$
$CH_3\ CH_3\ CH_3$

(e) $CH_3CH_2CH_2CHCH_2CH_2CH_2CH_2CH_3$
CH
$CH_3 \quad CH_3$

(f) CH_3, CH_3

(g) 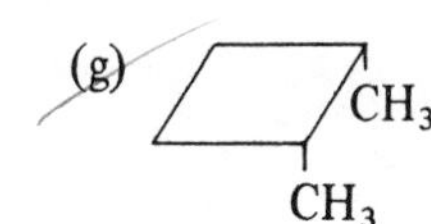

(h) H_3C, CH_3

(i) H, $CH(CH_3)_2$

(j) CH_3, H, H, $CH(CH_3)_2$

(k) $CH_3CHCH_2CH_2CH_2Cl$
CH_3

(l) $CH_3\ \ CH_3$
$CH_3CCH_2CCH_2CH_2CH_2CH_3$
$CH_3\ \ CH_3$

(m) CH_3
CH_3CCH_2Cl
CH_3

(n) CH_3
$CH_3CHCH_2CH_3$

3.18

(a) 3, 4-Dimethylhexane
(b) 2-Methylbutane
(c) 2, 4-Dimethylpentane
(d) 3-Methylpentane
(e) Ethylcyclohexane
(f) Cyclopentylcyclopentane
(g) 6-Isobutyl-2-methyldecane

3.19

(a) CH_3
CH_3CCH_3 2, 2-Dimethylpropane (neopentane)
CH_3

(b) CH_3
$CH_3CHCH_2CH_3$ 2-Methylbutane (isopentane)

(c) $CH_3CH_2CH_2CH_2CH_3$ Pentane

(d) Cyclopentane (ring of five CH_2 groups: $H_2C—CH_2$, H_2C, CH_2, $C H_2$)

(e) $CH_3CH(CH_3)—CH(CH_3)CH_3$ 2, 3-Dimethylbutane

3.20 Each of the desired alkenes must have the same carbon skeleton as 2-methylbutane,

$$C–C(–C)–C–C$$

they are therefore

$$\left.\begin{array}{l} CH_2{=}C(CH_3)CH_2CH_3 \\ CH_3C(CH_3){=}CHCH_3 \\ CH_3CH(CH_3)CH{=}CH_2 \end{array}\right\} + H_2 \xrightarrow[C_2H_5OH]{Ni} CH_3CH(CH_3)CH_2CH_3$$

3.21 Only one isomer of C_6H_{14} can be produced from five isomeric hexyl chlorides ($C_6H_{13}Cl$).

The alkane is 2-methylpentane, $CH_3CH(CH_3)CH_2CH_2CH_3$. The five alkyl chlorides are

$ClCH_2CH(CH_3)CH_2CH_2CH_3$ $CH_3CCl(CH_3)CH_2CH_2CH_3$ $CH_3CH(CH_3)CHClCH_2CH_3$

$CH_3CH(CH_3)CH_2CHClCH_3$ and $CH_3CH(CH_3)CH_2CH_2CH_2Cl$

3.22 $CH_3CH(CH_3)—CH(CH_3)CH_3$ 2, 3-Dimethylbutane

From two alkyl chlorides

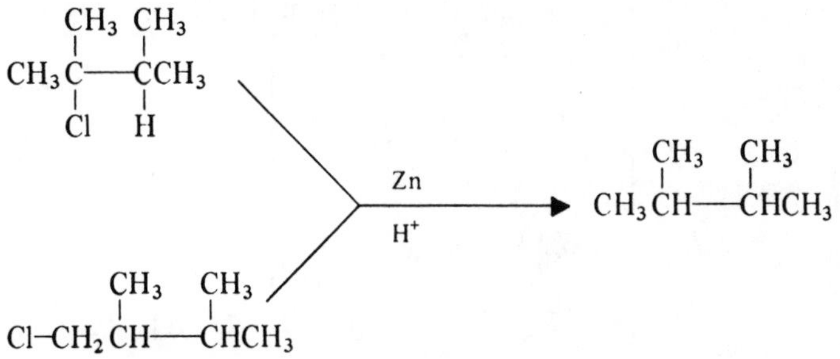

$$CH_3C(CH_3)(Cl){-}C(CH_3)(H)CH_3 + Cl{-}CH_2CH(CH_3){-}CH(CH_3)CH_3 \xrightarrow[H^+]{Zn} CH_3CH(CH_3){-}CH(CH_3)CH_3$$

From two alkenes

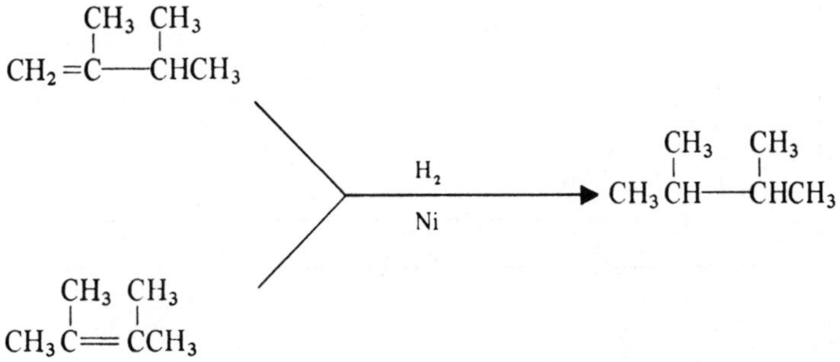

$$CH_2{=}C(CH_3){-}CH(CH_3)CH_3 + CH_3C(CH_3){=}C(CH_3)CH_3 \xrightarrow[Ni]{H_2} CH_3CH(CH_3){-}CH(CH_3)CH_3$$

3.23 Methylenecyclopentane (CH_2) $\xrightarrow[Ni]{H_2}$ methylcyclopentane (CH_3)

1-Methylcyclopentene (CH_3) $\xrightarrow[Ni]{H_2}$ methylcyclopentane (CH_3)

3-Methylcyclopentene (CH_3) $\xrightarrow[Ni]{H_2}$ methylcyclopentane (CH_3)

4-Methylcyclopentene (CH_3) $\xrightarrow[Ni]{H_2}$ methylcyclopentane (CH_3)

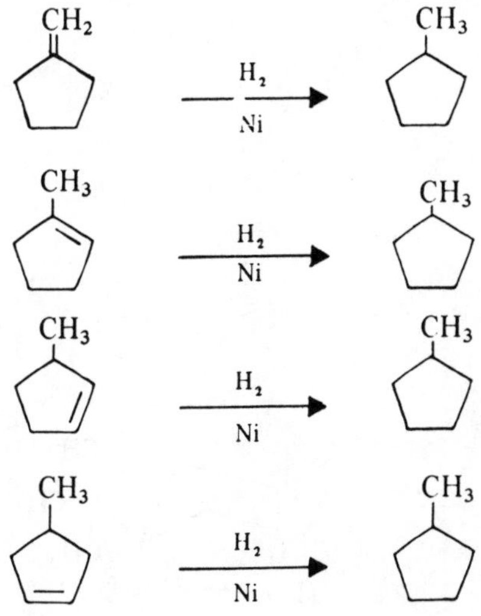

3.24 $(CH_3)_3CCH_3$ is the most stable isomer (i.e., it is the isomer with the lowest potential energy) because it evolves the least amount of heat on a molar basis when subjected to complete combustion.

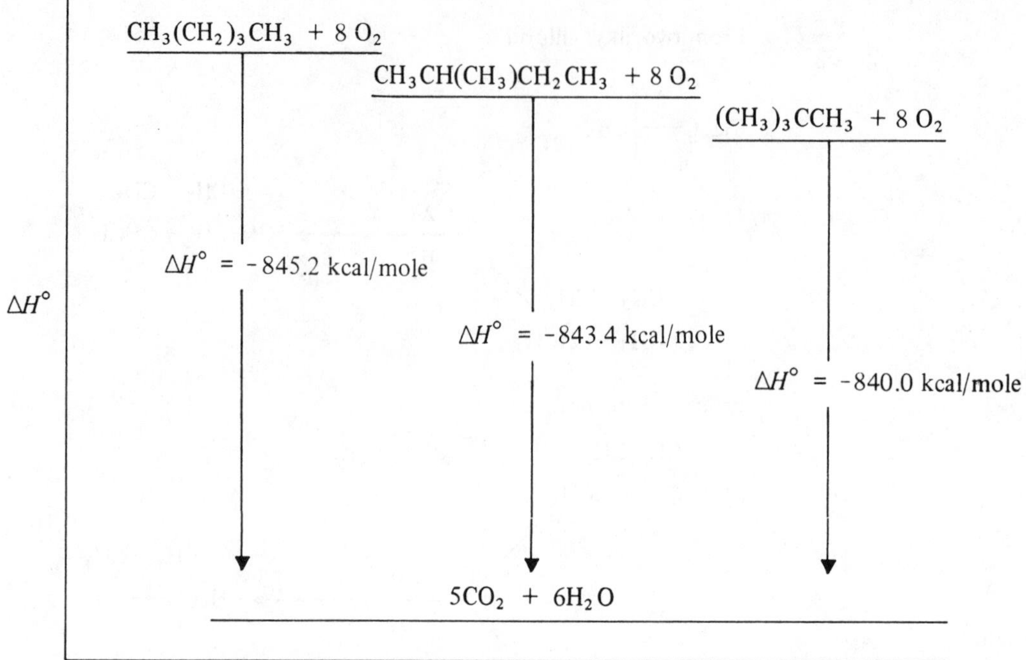

3.25 A homologous series is one in which each member of the series differs from the one preceding it by a constant amount, usually a CH_2 group. A homologous series of alkyl halides would be the following:

CH_3X
CH_3CH_2X
$CH_3(CH_2)_2X$
$CH_3(CH_2)_3X$
$CH_3(CH_2)_4X$
etc.

3.26

This conformation is *less stable* because 1, 3-diaxial interactions with the large *tert*-butyl group cause considerable repulsion

This conformation is *more stable* because 1, 3-diaxial interactions with the smaller methyl group are less repulsive

3.27

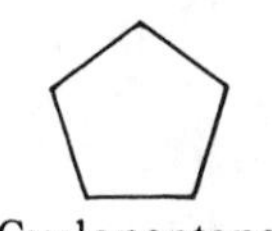

Cyclopentane

CH$_3$

Methylcyclobutane

CH$_3$ CH$_3$

cis-1, 2-Dimethylcyclopropane

CH$_3$ CH$_3$

trans-1, 2-Dimethylcyclopropane

CH$_3$ CH$_3$

1, 1-Dimethylcyclopropane

CH$_2$CH$_3$

Ethylcyclopropane

3.28 (a)

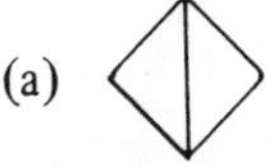

(b)

(c)

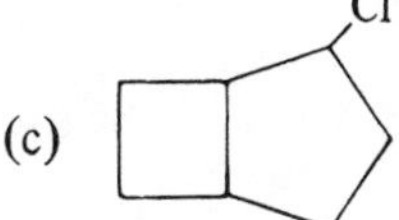

(d)

3.29

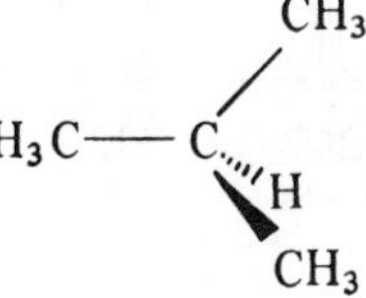

The methyl groups are larger than the hydrogen atom. The resulting mutual repulsions among the methyl groups cause a larger than tetrahedral bond angle

3.30 (a)

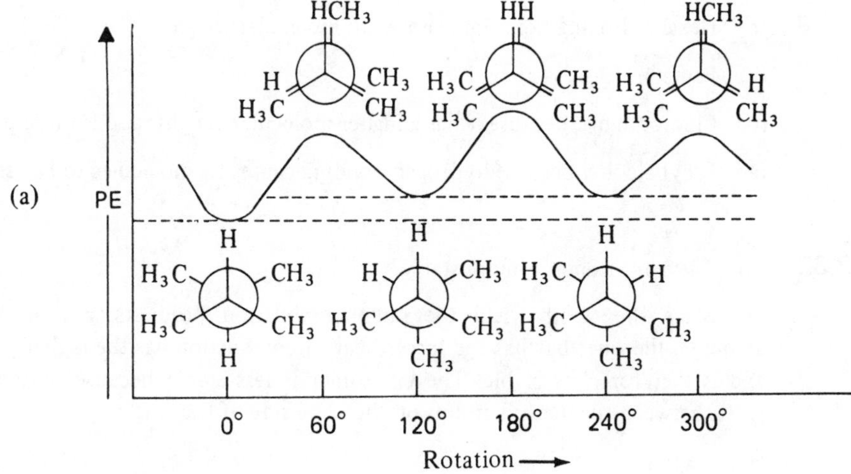

(b)

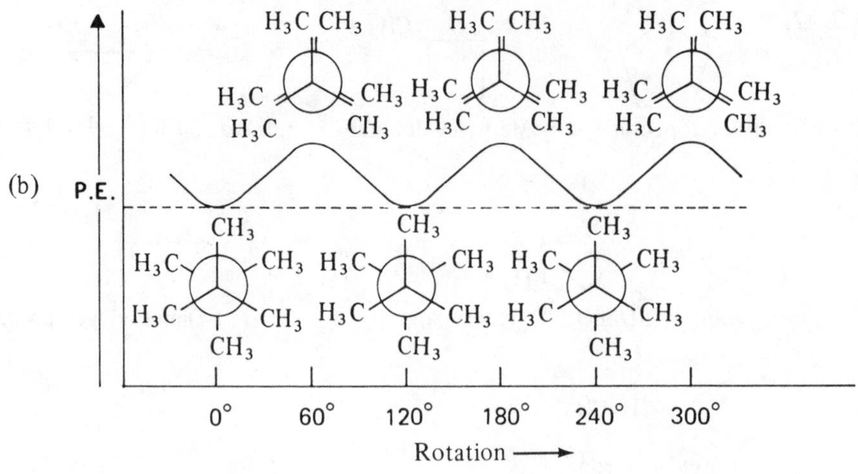

(c)

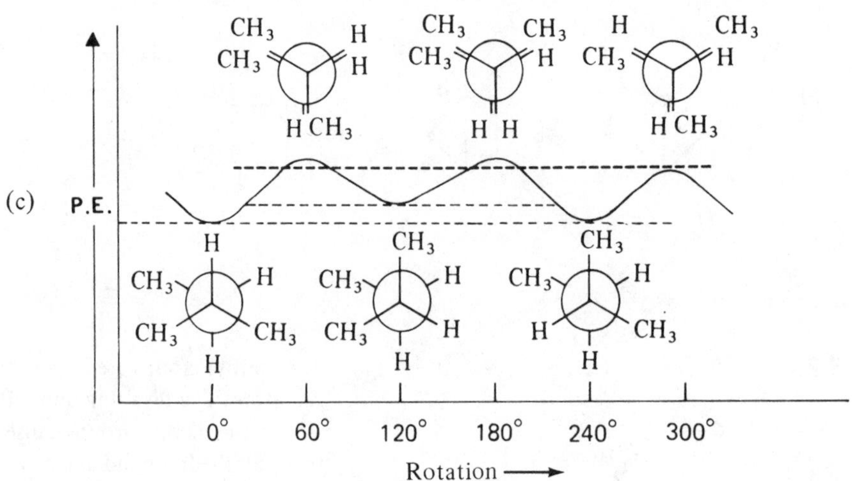

3.31 (a) Hexane. Branched chain hydrocarbons have lower boiling points than their unbranched isomers.

(b) Hexane. Boiling point increases with molecular weight.

(c) Pentane. [See (a) above].

(d) Chloroethane, because it has a higher molecular weight, and is more polar.

(e) Ethyl alcohol because hydrogen bonding causes its molecules to be associated.

3.32 (a) The trans isomer is more stable.

(b) Since they both yield the same combustion products and in the same molar amounts, the one that has the larger heat of combustion has the higher potential energy, and is therefore less stable. The cis isomer is less stable because of the crowding that exists between the methyl groups on the same side of the ring.

3.33 (a)

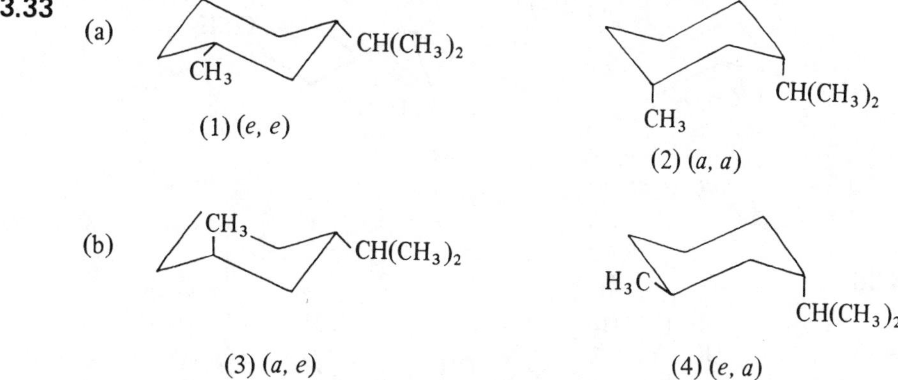

(c) (1) is more stable than (2) because in (1), both substituents are equatorial. (3) is more stable than (4) because in (3), the larger group [$CH(CH_3)_2$] is equatorial.

3.34 (a) The trans isomer is more stable because both methyl groups can be equatorial in one conformation (below). In both conformations of *cis*-1, 2-dimethylcyclohexane, one methyl must be axial.

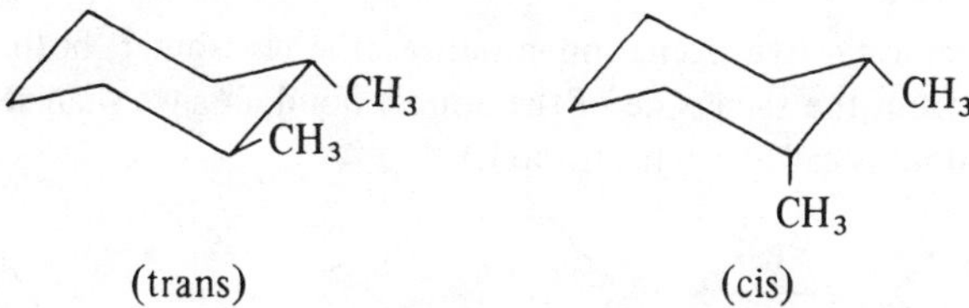

(b) The cis isomer is more stable because both methyl groups are equatorial in one conformation . In the trans isomer, one methyl must be axial in either conformation.

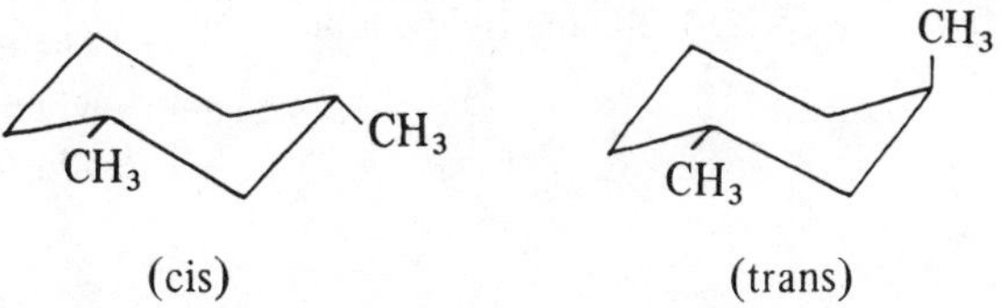

(c) The trans isomer is more stable for the same reason as in (a).

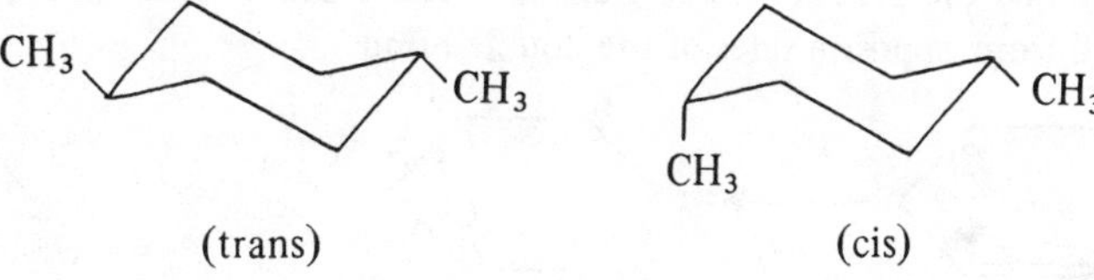

3.35 In *cis*-1,3-di-*tert*-butylcyclohexane, the two substituents are both equatorial [see Problem 3.34 (b)], whereas in the trans isomer, one of the *tert*-butyl groups must be axial. The instability of a chair conformation with such a large group in an axial position forces the molecule into a less strained twist conformation:

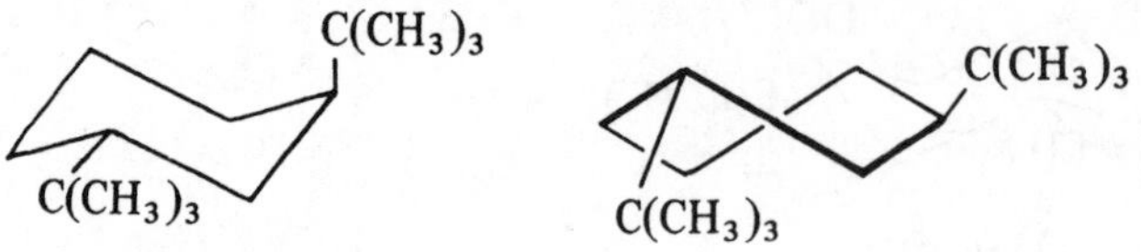

trans *(chair conformation)*

3.36

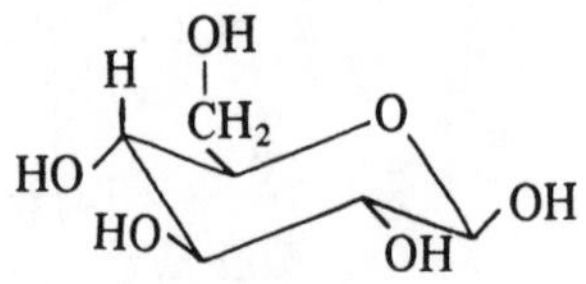

β-Glucose

3.37 (a)

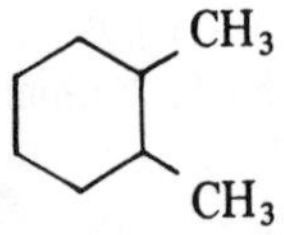

(b) From Table 3.7 we find that this is *cis*-1,2-dimethylcyclohexane.

(c) Since catalytic hydrogenation produces the cis isomer, both hydrogen atoms must have added from the same side of the double bond. (As we shall see in Section 6.6A, this type of addition is called a syn addition.)

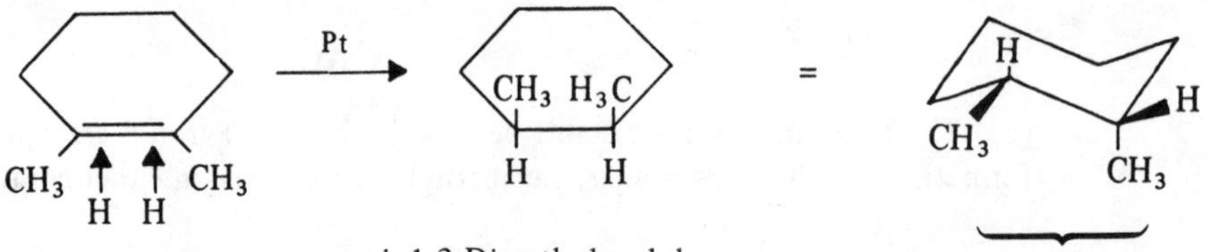

cis-1,2-Dimethylcyclohexane

The cis isomer is produced when both hydrogen atoms add from the same side

3.38 (a) From Table 3.7 we find that this is *trans*-1,2-dichlorocyclohexane.

(b) Since the product is the trans isomer we can conclude that the chlorine atoms have added from opposite sides of the double bond.

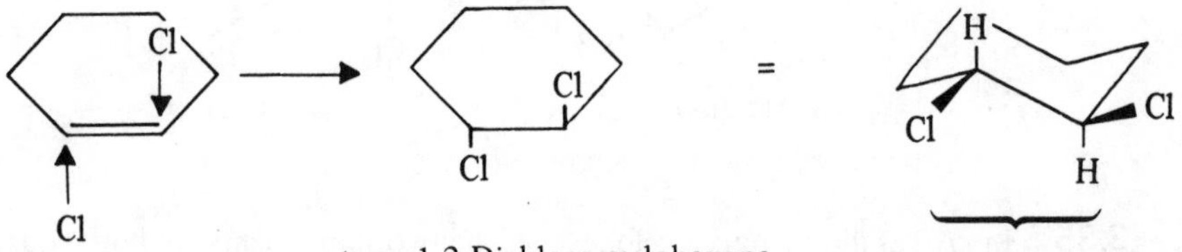

trans-1,2-Dichlorocyclohexane

The trans isomer is produced when the chlorine atoms add from opposite sides of the double bond

SECTION REFERENCES FOR ADDITIONAL PROBLEMS

If you have trouble solving the Additional Problems refer to the sections in the text next to the problem numbers:

3.17	3.3, 3.4	**3.28**	3.4B
3.18	3.3, 3.4	**3.29**	3.7
3.19	3.3, 3.4, 2.10	**3.30**	3.7
3.20	3.18	**3.31**	3.5
3.21	3.18	**3.32**	3.8, 3.9
3.22	3.18	**3.33**	3.10, 3.12, 3.13
3.23	3.18	**3.34**	3.13
3.24	3.8A	**3.35**	3.13
3.25	3.5	**3.36**	3.12
3.26	3.12, 3.13	**3.37**	3.18, 3.13
3.27	3.3, 3.4	**3.38**	3.18, 3.13

SELF-TEST

3.1 Give the IUPAC name of the following compound.

```
          CH3
           |
CH3CHCHCH2CHCH3
    |          |
    CH2       CH3
    |
    CH3
```

3.2 Draw the Newman projection formula of the molecule below using the partial structure given here.

H, CH3, H, CH3

3.3 (a) Give the other chair conformation of molecule I.

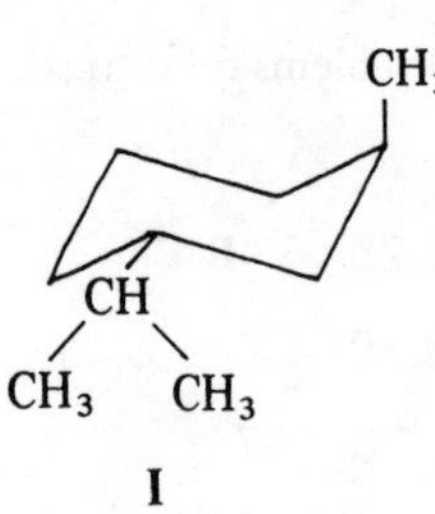

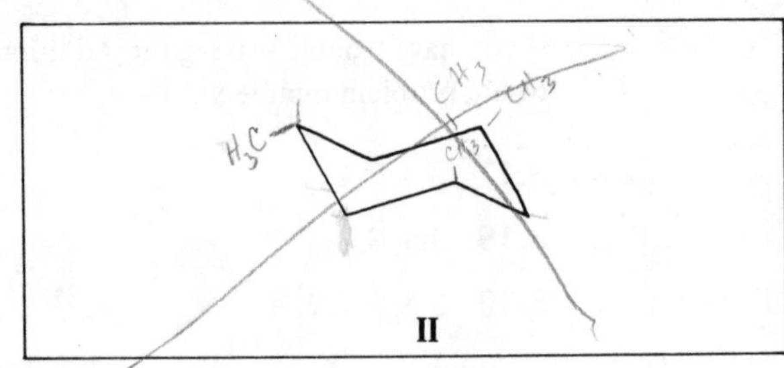

(b) Which conformation (I or II) is present in greater concentration in the equilibrium mixture?

3.4 Write the Newman projection formula for each of the following:

(a) The anti conformation of $ClCH_2CH_2Cl$

(b) A staggered conformation of $ClCH_2CH_2Cl$

(c) The most stable conformation of $CH_3CH(CH_3)CH(CH_3)CH_3$

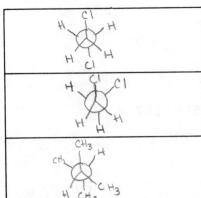

3.5 The following names may be incorrect. Write the correct IUPAC name in the space provided. If the name is correct as given, write OK.

(a) 2-Ethylpentane

(b) 3-Dimethylheptane

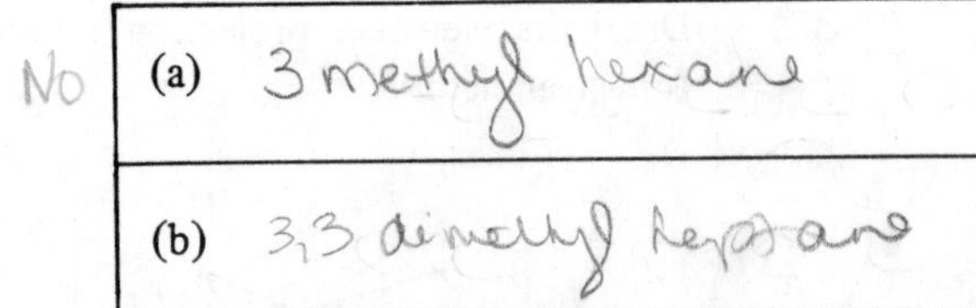

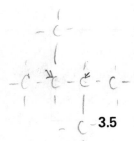

3.6 Complete the line formula given for the most stable conformation of *cis*-1,3-dimethylcyclohexane.

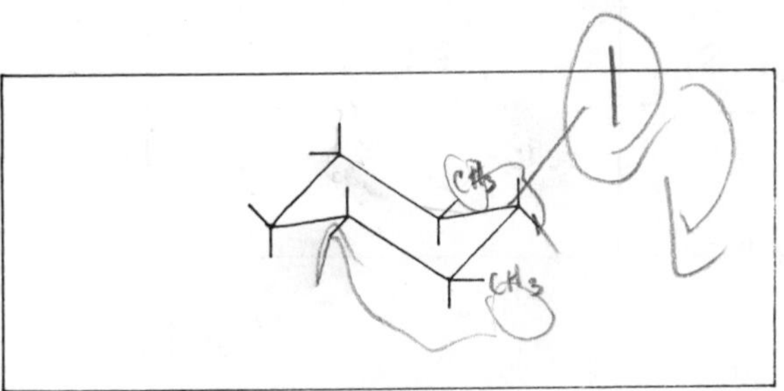

3.7 Consider the formula shown on the right.

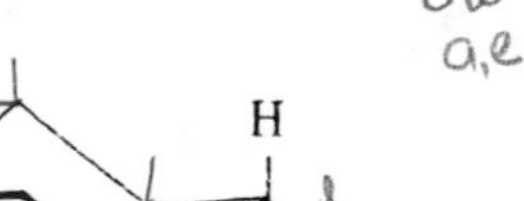

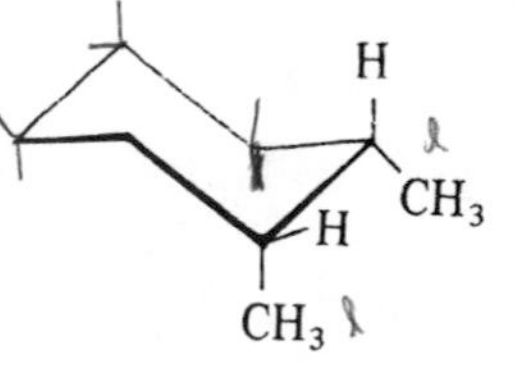

(a) Is the conformation given a cis or trans isomer?

cis

(b) How many different chair conformations exist for the other (cis or trans) isomer? 1 2

(c) Draw the Newman projection formula of the structure shown above.

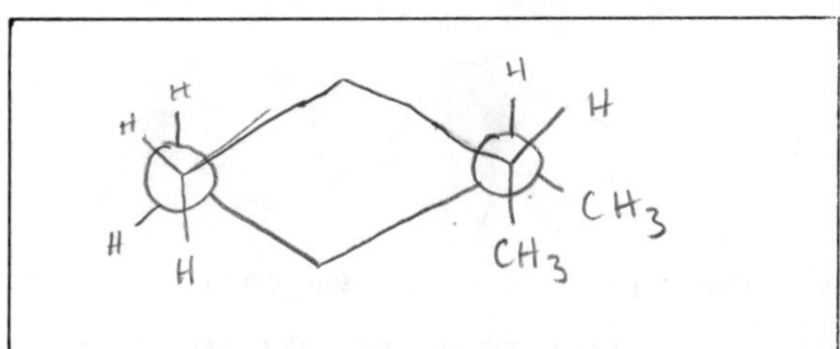

3.8 Give the systematic name of the compound,

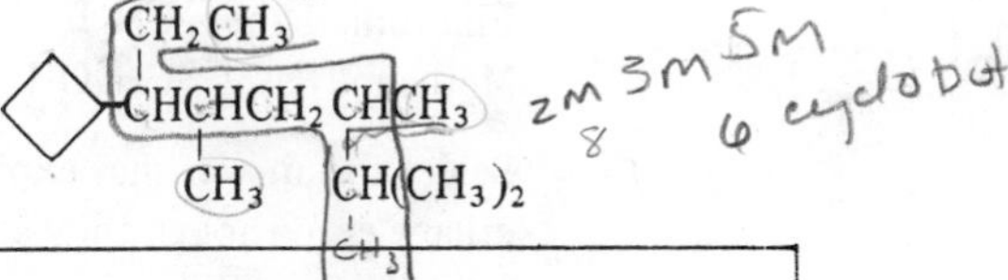

6-cyclobutyl-2,3,5-trimethyl octane

3.9 Give the missing organic product(s) or reactant(s) in each of the following reactions. Use the type of formula that shows the appropriate stereochemical features.

(a) $CH_3CH_2Cl + Cl_2$ (1 mole) $\xrightarrow{h\nu}$

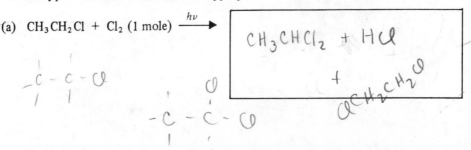

$CH_3CHCl_2 + HCl$ + $ClCH_2CH_2Cl$

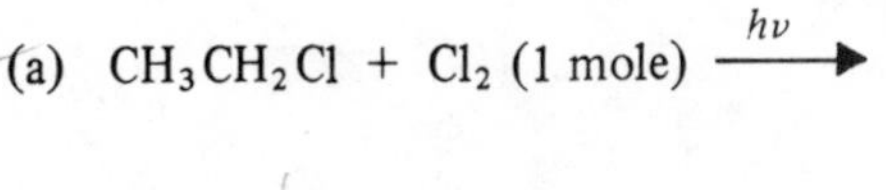

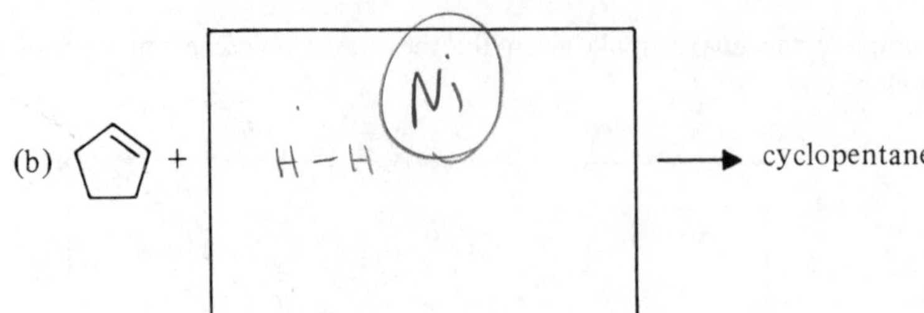

(c) $CH_3CH_2CH(Br)CH_3 + Zn + H_3O^+ \longrightarrow$

(d) $+ H_2 \xrightarrow{Pt}$ cyclohexane

3.10 Consider the properties of the following compounds:

NAME	FORMULA	BOILING POINT (°C)	MOLECULAR WEIGHT
Ethane	CH_3CH_3	-88.2	30
Fluoromethane	CH_3F	-78.6	34
Methanol	CH_3OH	+64.7	32

Select the answer that explains why methanol boils so much higher than ethane or fluoroethane even though they all have nearly equal molecular weights.

(a) Ion–ion forces between molecules.

(b) Weak dipole–dipole forces between molecules.

(c) Hydrogen bonding between molecules.

(d) van der Waals forces between molecules.

(e) Covalent bonding between molecules.

3.11 Select the correct name of the compound whose structure is

$$\begin{array}{c} \quad CH_3 \qquad\qquad\qquad\qquad\qquad \\ \quad | \qquad\qquad\qquad\qquad\qquad \\ CH_3CH_2CHCHCH_2CH_2CHCH_3 \\ \qquad\qquad | \qquad\qquad\quad | \\ \qquad\qquad CH_2CH_3 \quad CH_2CH_3 \end{array}$$

(a) 2,5-Diethyl-6-methyloctane

(b) 4,7-Diethyl-3-methyloctane

(c) 4-Ethyl-3,7-dimethylnonane

(d) 6-Ethyl-3,7-dimethylnonane

(e) More than one of the above

3.12 Select the best name for the compound whose structure is $CH_3\overset{\displaystyle CH_3}{\overset{|}{C}}HCH_2Cl$

(a) Butyl chloride

(b) Isobutyl chloride

(c) *sec*-Butyl chloride

(d) *tert*-Butyl chloride

(e) More than one of the above

3.13 The structure shown in Problem **3.11** has:

(a) 1°, 2°, and 3° carbon atoms

(b) 1° and 2° carbon atoms only

(c) 1° and 3° carbon atoms only

(d) 2° and 3° carbon atoms only

(e) None of the above

3.14 How many isomers are possible for C_3H_7Br?

(a) 1 (b) 2 (c) 3 (d) 4 (e) 5

3.15 Which isomer of 1,3-dimethylcyclohexane is more stable?

(a) cis (b) trans (c) Both are equally stable

(d) Impossible to tell

3.16 Which is the lowest energy conformation of *trans*-1,4-dimethylcyclohexane?

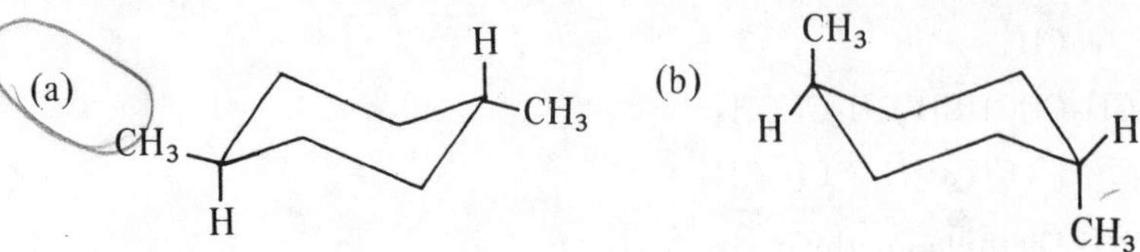

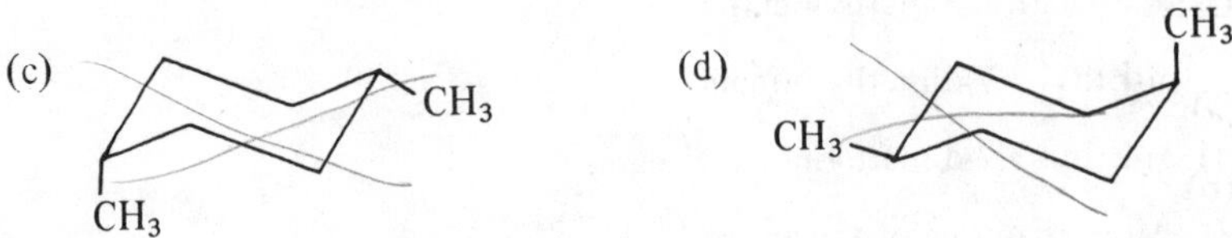

(e) More than one of the above

SUPPLEMENTARY PROBLEMS

S3.1 Give the IUPAC name for the following compounds.

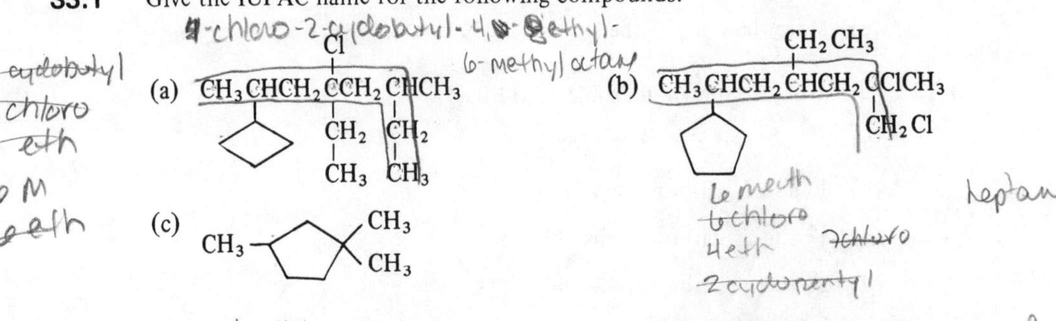

S3.2 Give the Newman projection formula of the (a) highest energy, and (b) lowest energy, staggered conformations of 1,2-dichloroethane.

S3.3 Give the formula of the most stable conformation of (a) *cis*-1,2-dichlorocyclohexane, and (b) *cis*-1,3-dichlorocyclohexane.

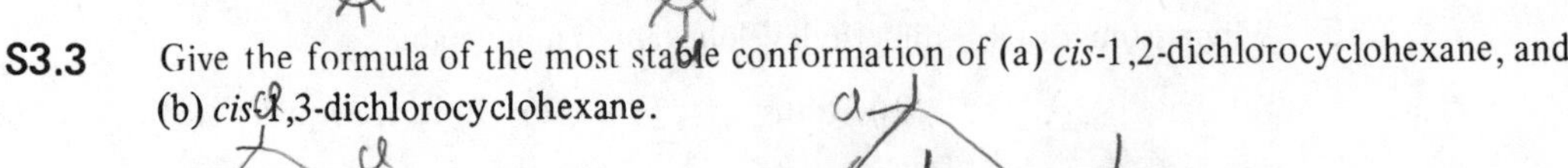

S3.4 Complete the reaction sequences by giving possible formulas for the missing compounds (A, B).

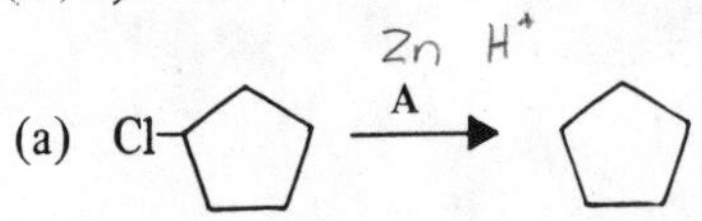

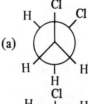

(b) B

SOLUTIONS TO SUPPLEMENTARY PROBLEMS

S3.1 (a) (The longest chain is octane.) 4-Chloro-2-cyclobutyl-4-ethyl-6-methyloctane

(b) 1,2-Dichloro-6-cyclopentyl-4-ethyl-2-methylheptane

(c) 1,1,3-Trimethylcyclopentane

S3.2

(a) The gauche form is higher in energy than the anti form (b).

(b)

S3.3

(a)

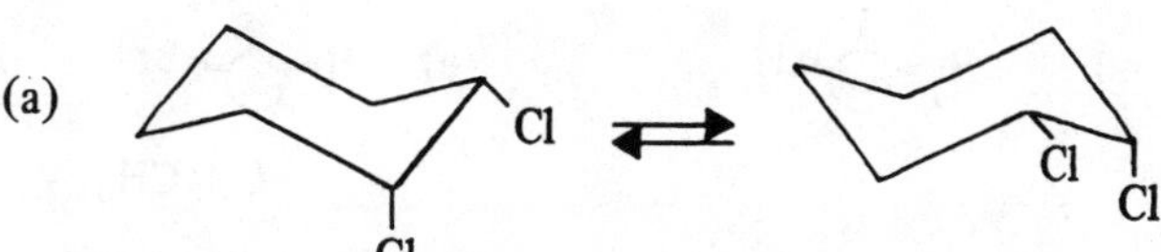

The two cis forms are equivalent because they are both *a, e*.

(b) In this case the *e, e* form shown has lower energy than the *a, a* form.

S3.4 (a) A = Zn/H_3O^+

(b) B = H_2/Ni

4 STEREOCHEMISTRY. CHIRAL MOLECULES

SOLUTIONS TO PROBLEMS

4.1 Chiral (a) Screw, (e) Foot, (f) Ear, (g) Shoe, (h) Spiral staircase
Achiral (b) Plain spoon, (c) Fork, (d) Cup

4.2 (b) Yes (c) No (d) No

4.3 (a) Yes (b) Yes (c) No (d) No

4.4 (a) 1-Chloropropane, (c) 1-Chloro-2-methylpropane, (d) 2-Chloro-2-methylpropane (f) 1-Chloropentane, and (h) 3-Chloropentane are all achiral

(b) Cl; H, Br; I — I Cl; Br, H; I — II

(e) CH_3; H, Br; CH_2CH_3 — I CH_3; Br, H; CH_2CH_3 — II

(g) CH_3; H, Cl; $CH_2CH_2CH_3$ — I CH_3; Cl, H; $CH_2CH_2CH_3$ — II

4.5

(a) 1. X; H, H; H 2. H; X, X; H or H; X, Y; H 3. H; Z, X; Y H; X, Z; Y

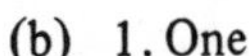

(b) 1. One

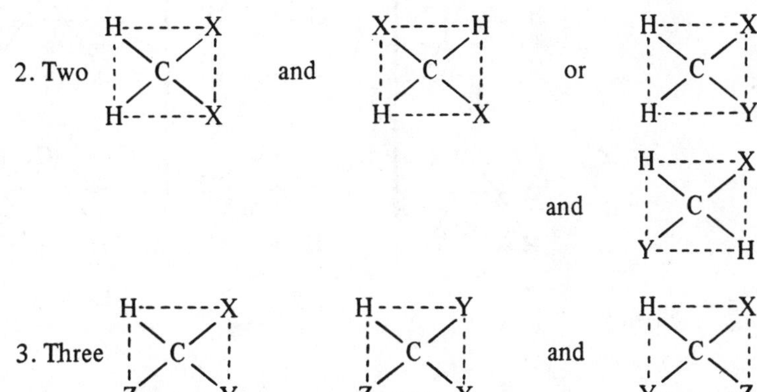

(c) 1. One

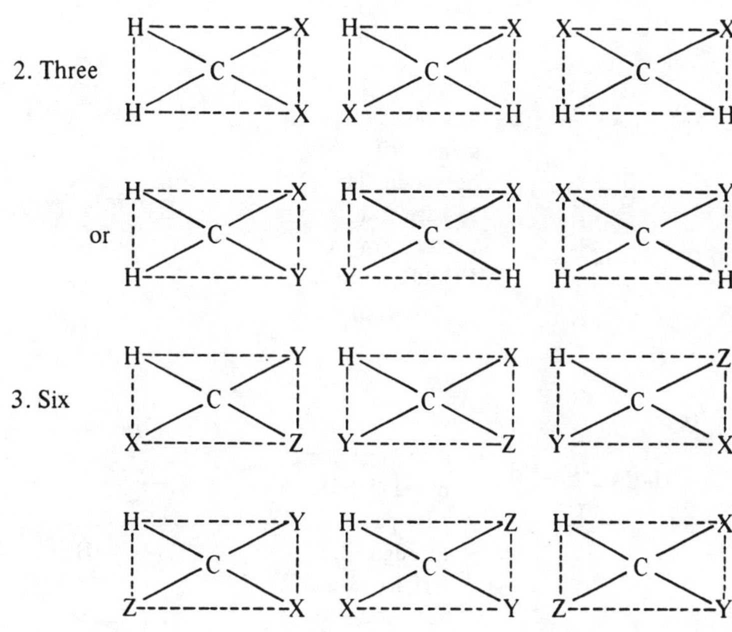

(d) 1. One

2. Two or three

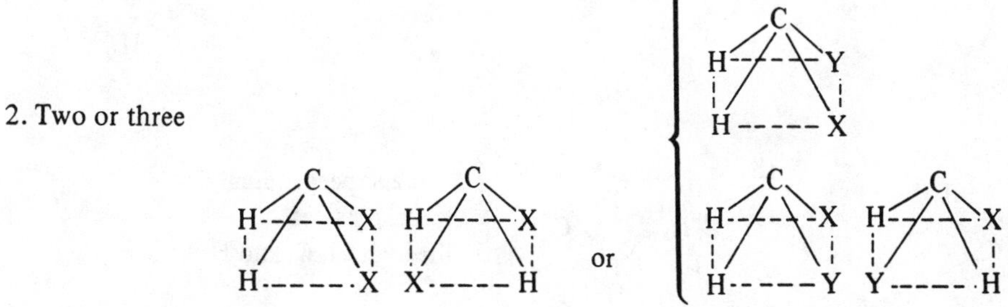

3. Six

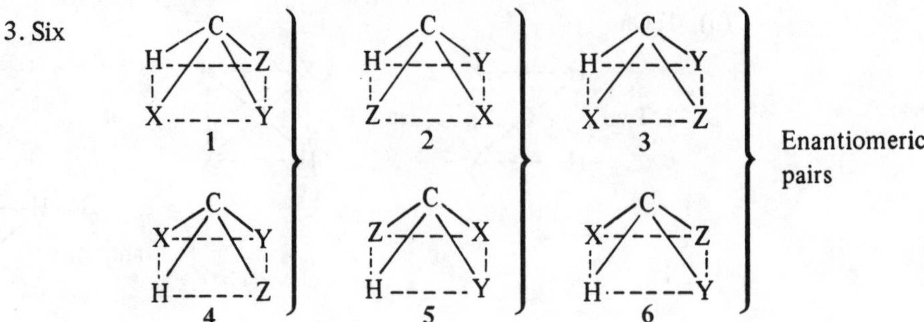

4.6 (b) Plain spoon (c) Fork (d) Cup all possess a plane of symmetry

4.7

(a) The plane of symmetry is perpendicular to page and passes through Cl and 3C atoms

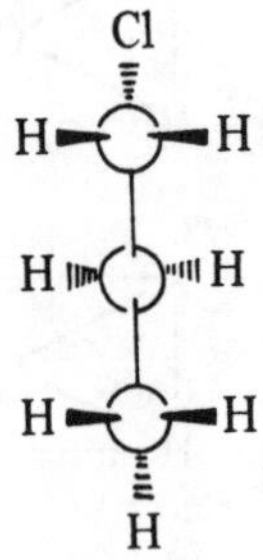

(c) The plane of symmetry is perpendicular to page and passes through Cl and 2C atoms

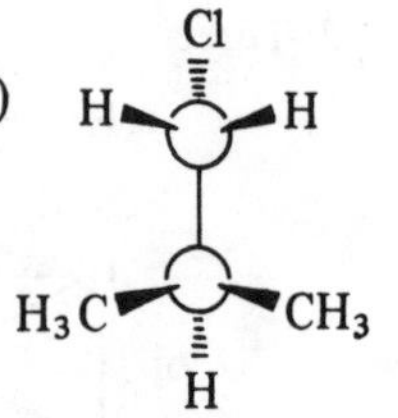

(d) A vertical plane perpendicular to page passes through Cl, tertiary C, and CH_3 at bottom

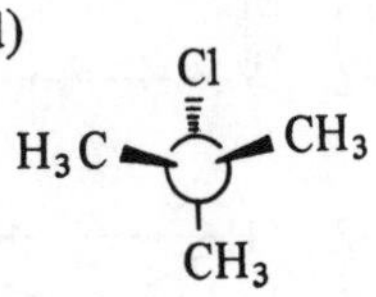

(f) A plane perpendicular to page passes through Cl and 5C atoms

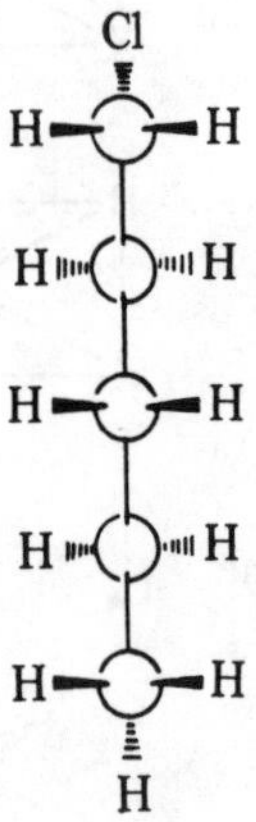

(h) A plane perpendicular to page passes through Cl, C, and H

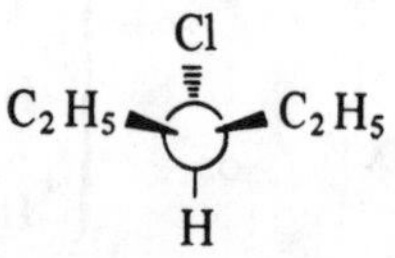

4.8

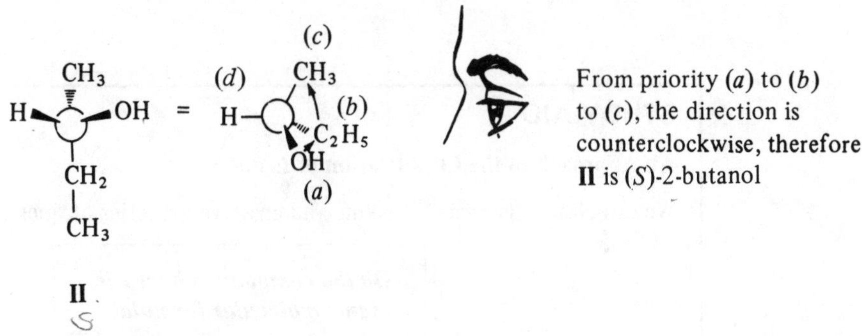

4.9

(b)

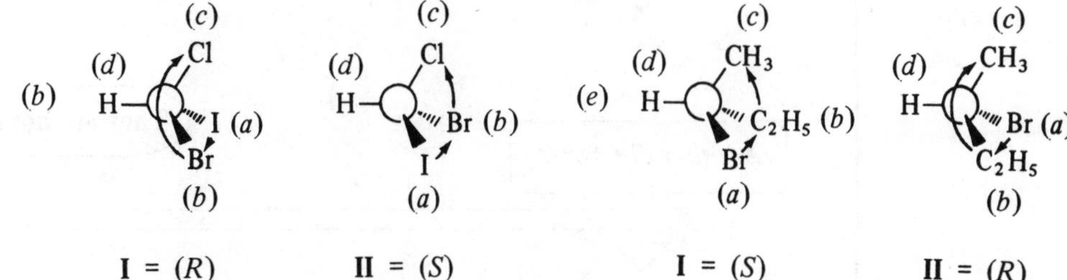

(g)

(c) (d) CH_3 H— C_3H_7(b) Cl (a) — I = (S)

(c) (d) CH_3 H— Cl (a) C_3H_7 (b) — II = (R)

4.10

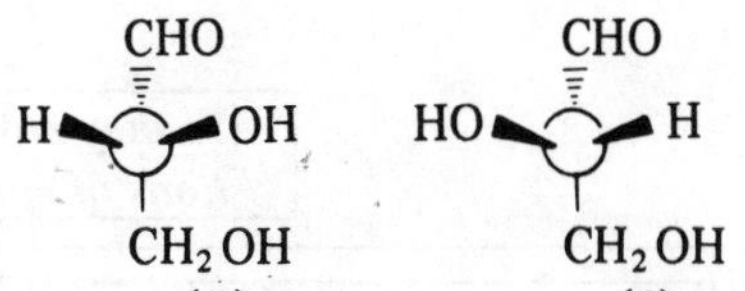

CHO, H, OH, CH_2OH — (R)

CHO, HO, H, CH_2OH — (S)

4.11 (a) (R) (b) (R) (c) (R)

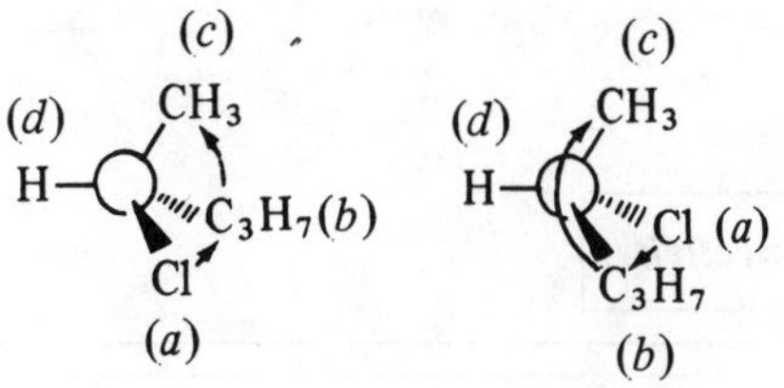

4.12 (a) Enantiomers

(b) Two molecules of the same compound

(c) Enantiomers

STUDY AID

An Approach to the Classification of Isomers

We can classify isomers by asking and answering a series of questions:

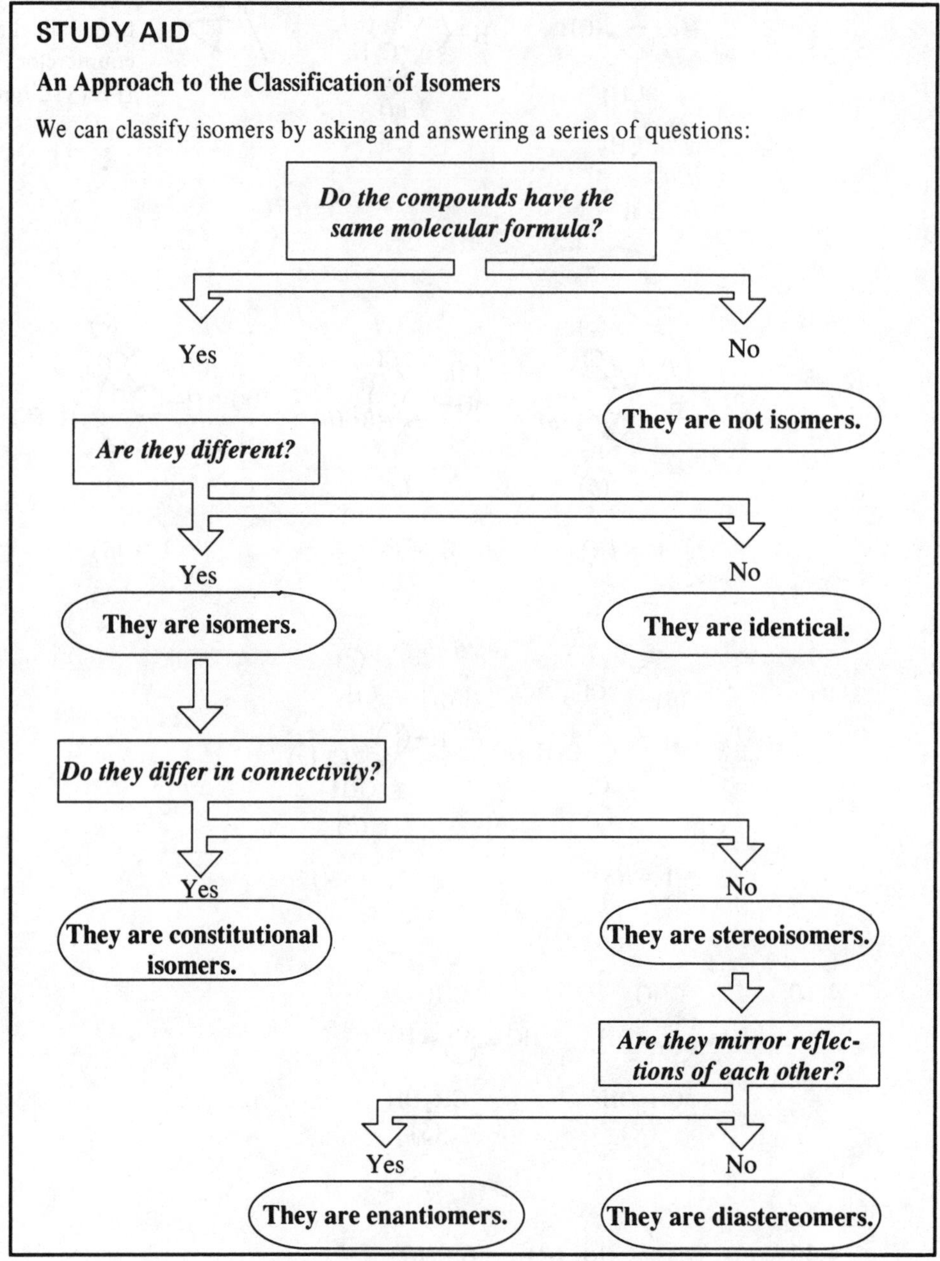

4.13 The optical purity is 50% (see previous paragraph in text). That means that the sample contains 50% of the (*S*) enantiomer and 50% of the racemic form. The racemic form is 50% (*S*) and 50% (*R*). Therefore the total percentage of (*S*) enantiomer in the sample is 75%, the percentage of (*R*) enantiomer is 25%.

4.14 $CH_3CH_2CH_2\underset{\displaystyle CH_3}{\underset{|}{C}}HCH_2CH_3$ (as a 50:50 mixture of enantiomers because reactants are achiral)

4.15 (a) Diastereomers (b) Diastereomers (c) Diastereomers

(d)

	1	2	3	4
1		Enantiomers	Diastereomers	Diastereomers
2	Enantiomers		Diastereomers	Diastereomers
3	Diastereomers	Diastereomers		Enantiomers
4	Diastereomers	Diastereomers	Enantiomers	

(e) Yes (f) No

4.16 (a) **A** alone would be optically active.

(b) **B** alone would be optically active.

(c) **C** would be optically inactive because it is a meso compound.

(d) An equimolar mixture of **A** and **B** would be optically inactive because it is a racemic form.

4.17 (1) **C** (2) **A** (3) **B**

4.18

(a)

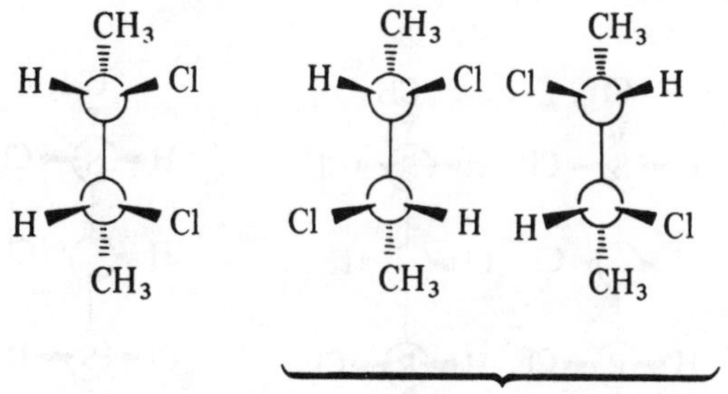

CH_3 H Cl H Cl CH_3 CH_3 H Cl Cl H CH_3 CH_3 Cl H H Cl CH_3

(meso) Enantiomers

(b)

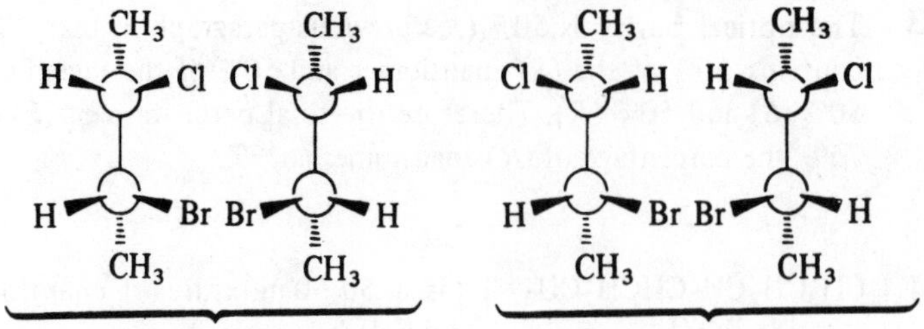

I	II	III	IV
CH_3; H, Cl; H, Br; CH_3	CH_3; Cl, H; Br, H; CH_3	CH_3; Cl, H; H, Br; CH_3	CH_3; H, Cl; Br, H; CH_3

I and II: Enantiomers; III and IV: Enantiomers

(c)

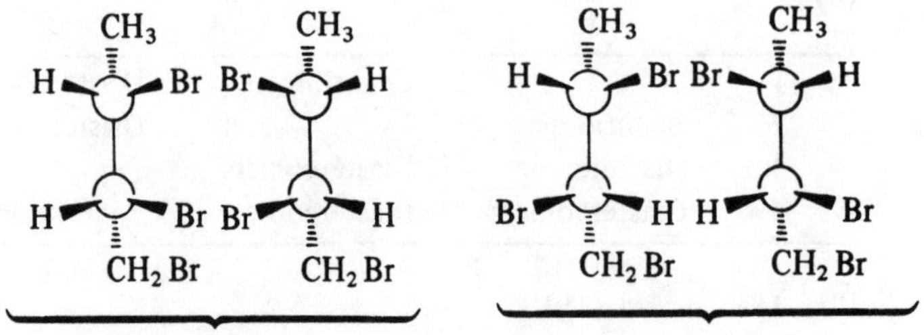

CH_3; H, Br; H, Br; CH_2Br	CH_3; Br, H; Br, H; CH_2Br	CH_3; H, Br; Br, H; CH_2Br	CH_3; Br, H; H, Br; CH_2Br

Enantiomers; Enantiomers

(d)

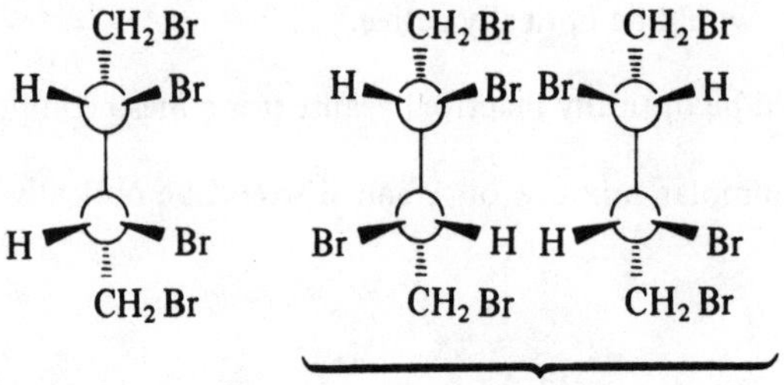

CH_2Br; H, Br; H, Br; CH_2Br	CH_2Br; H, Br; Br, H; CH_2Br	CH_2Br; Br, H; H, Br; CH_2Br
(meso)	Enantiomers	Enantiomers

(e)

CH_3; H–(S)–Cl; H, Cl; H–(R)–Cl; CH_3	CH_3; H–(S)–Cl; Cl, H; H–(R)–Cl; CH_3	CH_3; H–(S)–Cl; H, Cl; Cl–(S)–H; CH_3	CH_3; Cl–(R)–H; Cl, H; H–(R)–Cl; CH_3
(meso)	(meso)	Enantiomers	Enantiomers

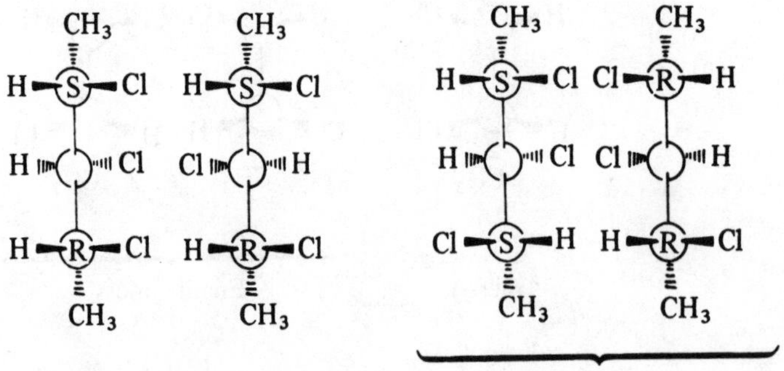

4.19 **B** (2*S*,3*S*)-2,3-Dibromobutane, **C** (2*R*,3*S*)-2,3-dibromobutane

4.20 (a) The meso compound is (2*R*,3*S*)-2,3-dichlorobutane, the two enantiomers are (2*S*,3*S*)-2,3-dichlorobutane and (2*R*,3*R*)-2,3-dichlorobutane.

(b) **I** (2*R*,3*S*)-2-bromo-3-chlorobutane
II (2*S*,3*R*)-2-bromo-3-chlorobutane
III (2*R*,3*R*)-2-bromo-3-chlorobutane
IV (2*S*,3*S*)-2-bromo-3-chlorobutane

4.21 (a) No (b) Yes (c) No (d) No (e) Diastereomers (f) Diastereomers

4.22 Me Me H H (meso) Me H H Me H Me Me H (Enantiomers)

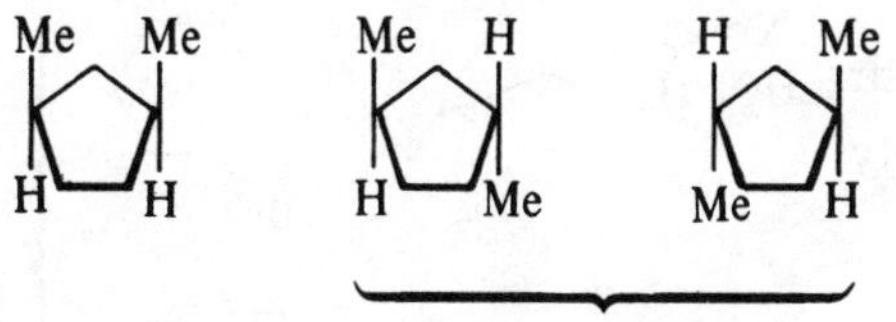

4.23 (a) (1*R*,2*R*) (1*S*,2*S*) Br Cl Br Cl Enantiomers (both trans)

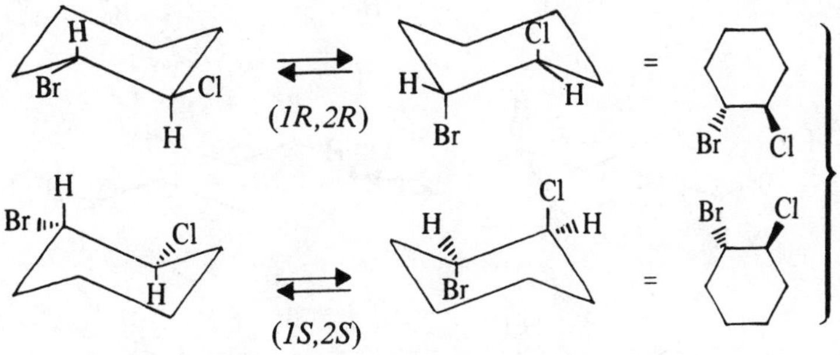

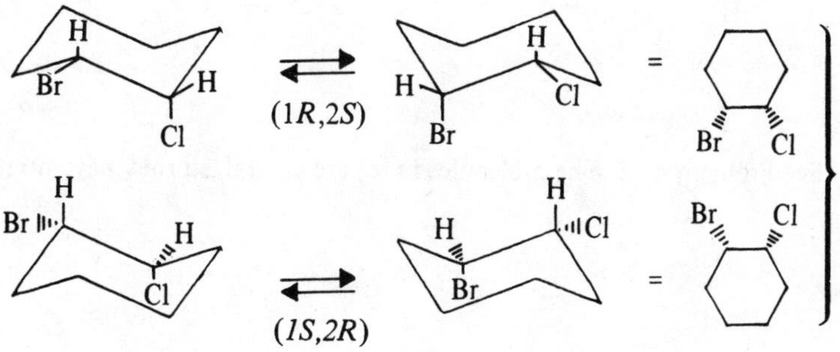

(1*R*,2*S*) (1*S*,2*R*) Br Cl Br Cl Enantiomers (both cis)

(b)

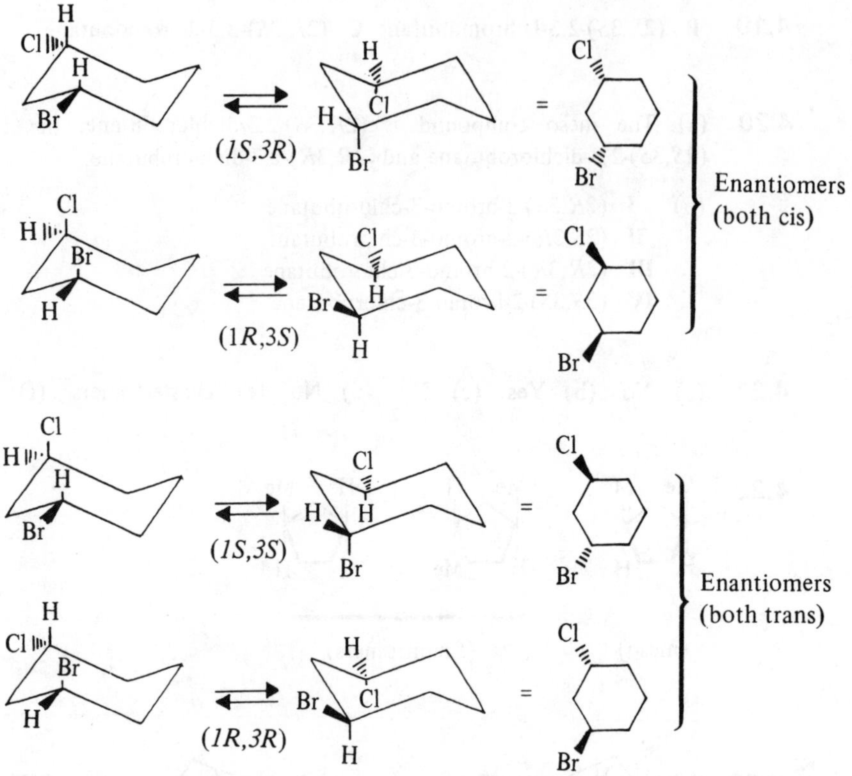

(1S,3R)

(1R,3S)

Enantiomers (both cis)

(1S,3S)

(1R,3R)

Enantiomers (both trans)

(c)

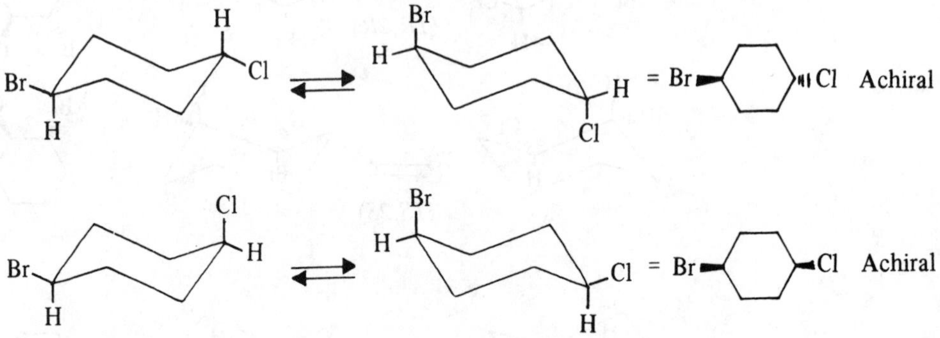

Achiral

Achiral

4.24 See Problem 4.23. The molecules in (c) are achiral, so they have no (R–S) designation.

4.25 (a) (b) O, OH, C, H, OH, CH_2OH — (*R*)-(−)-Glyceric acid

(a) (d) O, OH, C, H, OH, CH_2Br — (*S*)-(−)-3-Bromo-2-hydroxypropanoic acid

(c) O, OH, C, H, OH, CH_2NH_2 — (*R*)-(+)-Isoserine

4.26 (a) Isomers are different compounds that have the same molecular formula. C_2H_6O: CH_3CH_2OH and CH_3OCH_3

(b) Constitutional isomers are isomers that differ because their atoms are joined in a different order. C_4H_{10}: $CH_3CH_2CH_2CH_3$ and CH_3CHCH_3 (with CH_3 on the middle C)

(c) Stereoisomers are isomers that differ only in the arrangement of their atoms in space: *cis*- and *trans* -2-butene.

(d) Diastereomers are stereoisomers that are not mirror reflections of each other: *cis*- and *trans* -2-butene, or (2*S*, 3*S*)- and (2*S*, 3*R*)-2, 3-dibromobutane.

(e) Enantiomers are stereoisomers that are nonsuperposable mirror reflections of each other: (2*S*,3*S*)-and (2*R*,3*R*)- 2, 3-dibromobutane.

(f) A meso compound is made up of achiral molecules that contain atoms with four different attached groups: (2*S*,3*R*)- 2, 3-dibromobutane.

(g) A racemic form is an equimolar mixture of enantiomers.

(h) A plane of symmetry is an imaginary plane that bisects a molecule in such a way that the two halves of the molecule are mirror reflections of each other. (See Fig. 4.7.)

(i) A tetrahedral atom that has four different groups attached to it is a stereocenter. Interchanging two groups at a stereocenter produces a stereoisomer.

(j) A chiral molecule is one that is not superposable on its mirror reflection.

(k) An achiral molecule is superposable on its mirror reflection.

(l) Optical activity is the rotation of the plane of polarization of plane polarized light by a substance placed in the light path.

(m) A dextrorotatory substance is one that rotates the plane of polarization of plane polarized light in a clockwise direction.

(n) A reaction occurs with retention of configuration when all the groups around the chiral atom retain the same relative configuration after the reaction that they had before the reaction.

4.27 (a) Enantiomers (b) Same (c) Enantiomers (d) Diastereomers (e) Same (f) Constitutional isomers (g) Same (h) Diastereomers (i) Same (j) Enantiomers (k) Same (l) Enantiomers (m) Same (n) Constitutional isomers (o) Same (p) Diastereomers (q) Enantiomers

4.28 (a)

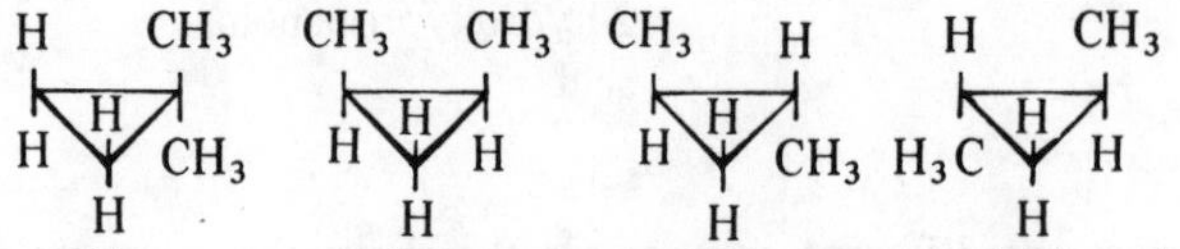

(I) (II) (III) (IV)

Enantiomers

(b) **III** and **IV** (c) Three: **I**, **II**, and a mixture of **III** and **IV**. Enantiomers cannot be separated from one another by distillation. (d) None, since the only chiral molecules are **III** and **IV**, and they would be obtained in the same amounts as a racemic form.

4.29 (a)

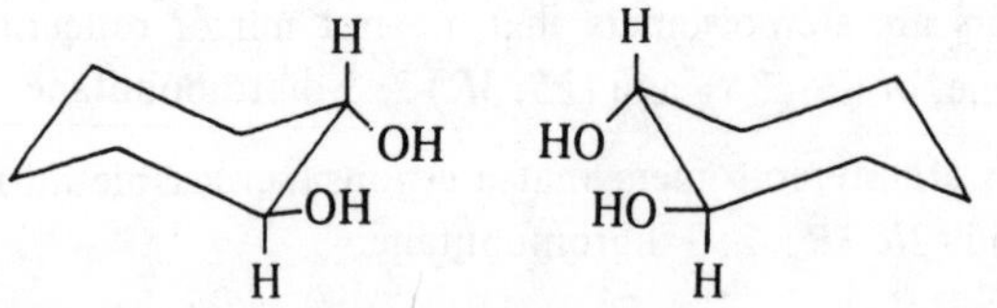

(b) No, they are not superposable.

(c) No, and they are, therefore, enantiomers of each other.

(d)

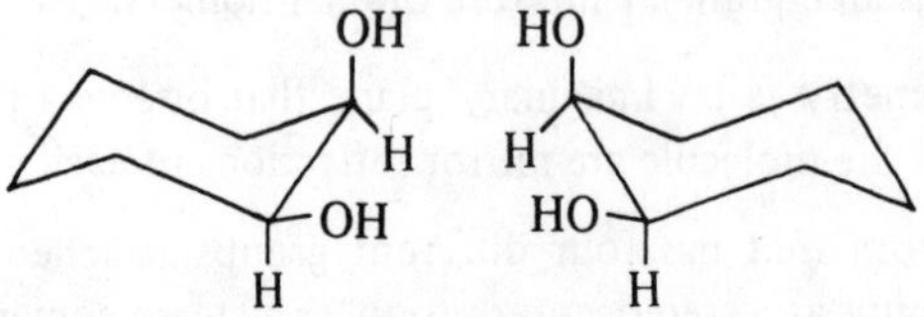

(e) No, they are not superposable.

(f) Yes, and they are, therefore, just different conformations of the same molecule.

4.30 (a)

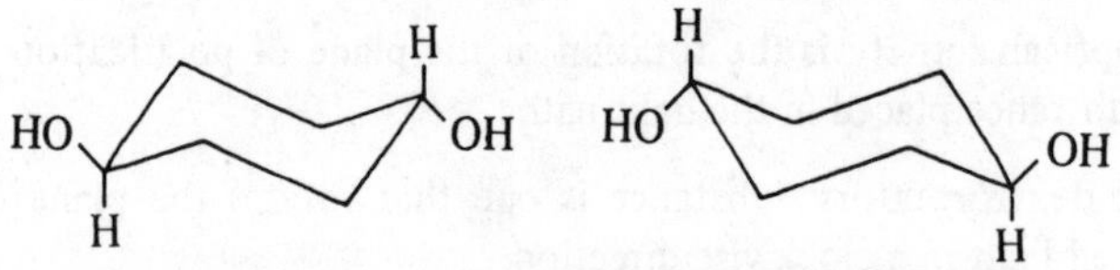

(b) Yes, and therefore *trans*-1,4-cyclohexanediol is achiral.

(c) No, they are different orientations of the same molecule.

(d) Yes, *cis*-1,4-cyclohexanediol is a stereoisomer (a diastereomer) of *trans*-1,4-cyclohexanediol.

cis-1, 4-Cyclohexanediol

(e) No, it, too, is superposable on its mirror image. (Notice, too, that the plane of the page constitutes a plane of symmetry for both *cis*-1,2-cyclohexanediol and for *trans*-1,2-cyclohexanediol as we have drawn them.)

4.31 *trans*-1,3-Cyclohexanediol can exist in the following enantiomeric forms.

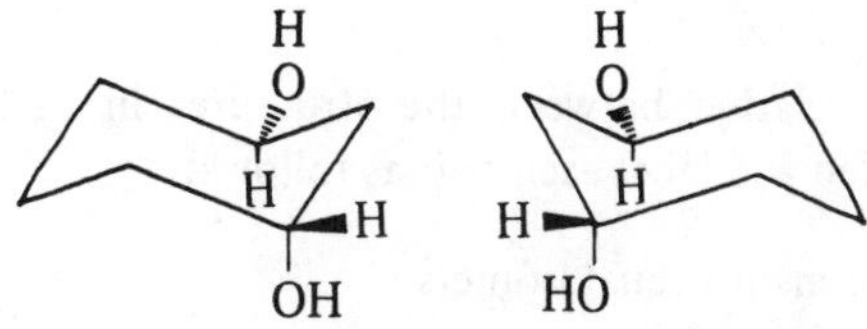

trans-1,3-Cyclohexanediol enantiomers

cis-1,3-Cyclohexanediol consists of achiral molecules because they have a plane of symmetry. [The plane of the page (below) is a plane of symmetry.]

cis-1, 3-Cyclohexanediol

4.32 (a) Since it is optically inactive and not resolvable, it must be the meso form:

(meso)

(b)

(c) No (d) A racemic modification

SECTION REFERENCES FOR ADDITIONAL PROBLEMS

4.26 4.2, 4.3, 4.4

4.27 4.2, 4.3, 4.4

4.28 4.12

4.29 4.12

4.30 4.12

4.31 4.12

4.32 4.9

SELF-TEST

4.1 Identify the relation between the structures in each of the following pairs. Use the spaces provided and label each pair as follows:

S if they are constitutional isomers.
E if they are a pair of enantiomers.
D if they are diastereomers.
I if they are two molecules of the same compound (not isomers).
X if they are different compounds that are not isomers.

(a) and =CH_2 ☐

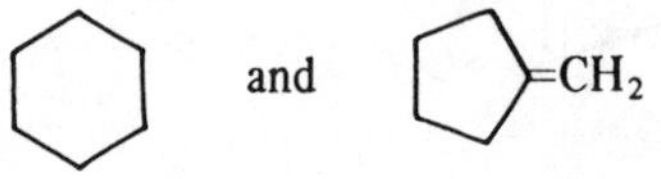

(b) H, CH_3, CH_2CH_3, OH and CH_2CH_3, HO, H, CH_3 ☐

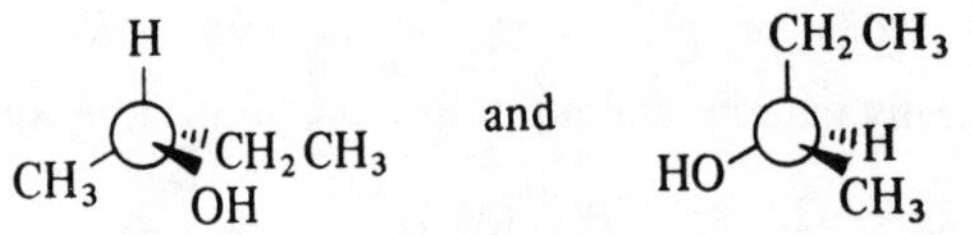

(c) CH_3, H, OH, H, OH, CH_2CH_3 and CH_3, HO, H, HO, H, CH_2CH_3 ☐

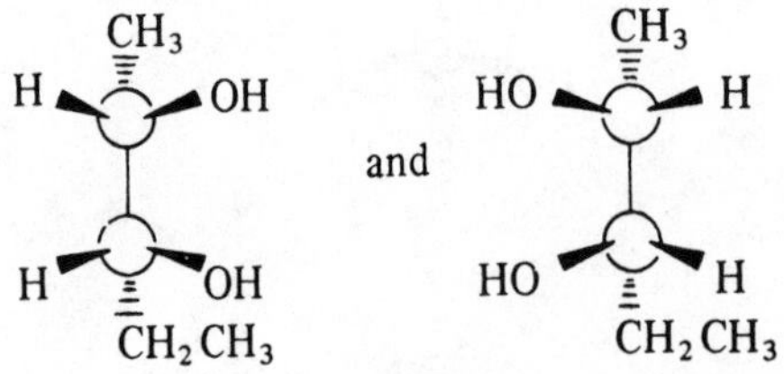

(d) 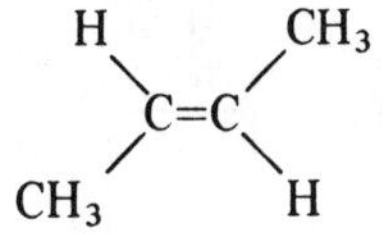 H, CH_3 / C=C / CH_3, H and H, H / C=C / CH_3, CH_3 □

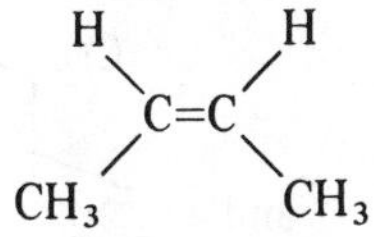

(e) 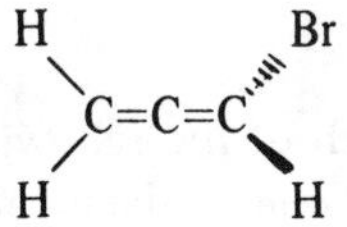 H, Br / C=C=C / H, H and Br, H / C=C=C / H, H □

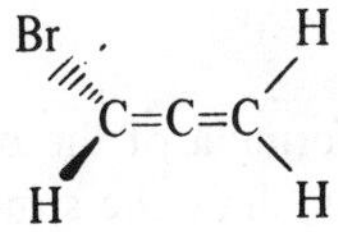

(f) $CH_2{=}C{=}CHCH_3$ and $CH_2{=}CHCH{=}CH_2$ □

(g) 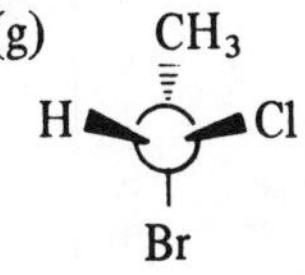 CH_3, H, Cl, Br and CH_3, Br, H, Cl □

(h) Cl, H, H, Cl and Cl, H, H, Cl □

(i) H, H, Cl, CH_3 and H, Cl, H, CH_3 □

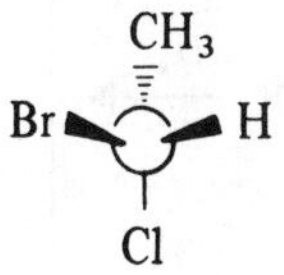

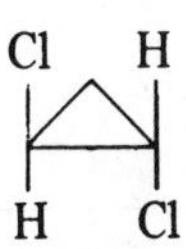

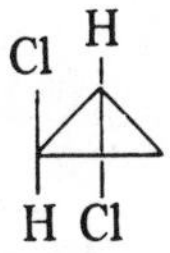

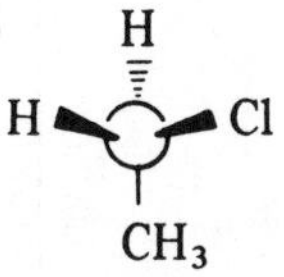

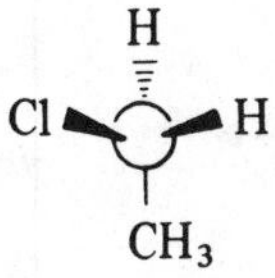

(j) 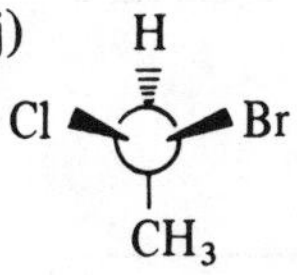 H, Cl, Br, CH_3 and Cl, H, H, CH_2Br □

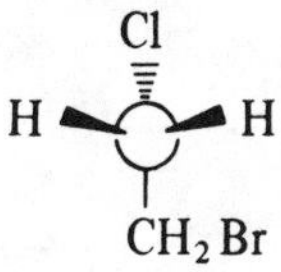

(k) 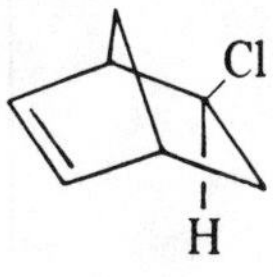 Cl, H and Cl, H □

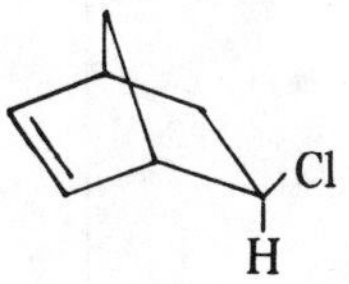

(l) 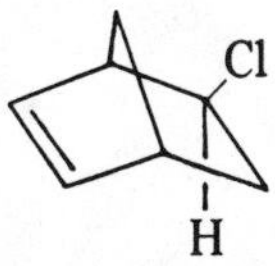 Cl, H and H, Cl □

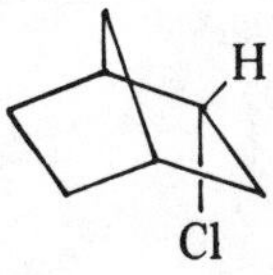

(m) H Cl and Cl H

4.2 Give the structural formula of the product of each of the following reactions. If a pair of enantiomers results, draw the structure of only *one* of the enantiomers. Show stereochemistry where appropriate.

(a) 1,2-Dimethylcyclopentene $\xrightarrow[Pt]{H_2}$

Would the product be optically active?

(b) Cyclohexene $\xrightarrow[25°C,\ neutral]{KMnO_4}$

Would the product be optically active?

(c) Cyclobutene $\xrightarrow[CCl_4]{Br_2}$

Would the product be optically active?

4.3 Tell whether each of the following statements is true (+) or false (-).

(a) If all of the molecules in a sample are chiral, the sample is optically active.

(b) The terms chiral and optically active mean the same thing.

(c) A chiral molecule is any molecule that has one or more stereocenters.

(d) There are four stereoisomers of 1,1-dibromo-2-methylcyclopentane.

(e) A pure sample of [structure: cyclohexane with H and Cl on adjacent carbons; labels H, Cl, Cl, H] is optically active.

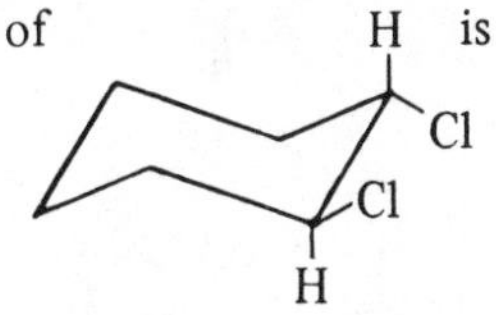

(f) There are three stereoisomers 2,3-diphenylbutane.

(g) There are two stereoisomers of 1,1-dibromo-1,2-propadiene.

(h) The formula $CH_2{=}CH{-}C(CH_3)(NH_2)H$ has the (*S*) configuration.

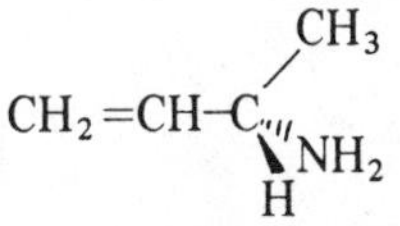

(i) The formula [structure: phenyl–C with CH_3, OH, C_2H_5] has the (*R*) configuration.

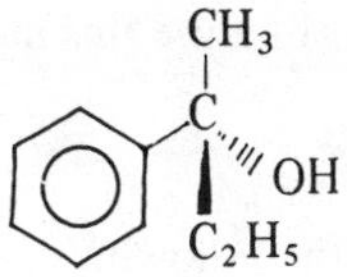

(j) 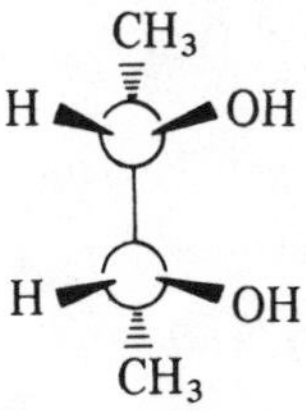 [Newman-type structures: CH_3; H, OH; H, OH; CH_3] and [CH_3; HO, H; HO, H; CH_3] are related as an object and its mirror image, however, they are not enantiomers.

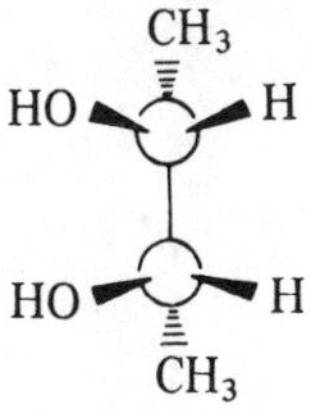

(k) 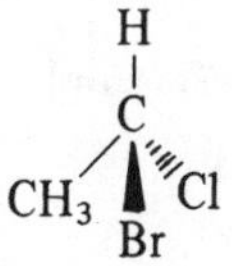 [structure: C with H, CH_3, Cl, Br] and [structure: C with Cl, Br, H, CH_3] are enantiomers.

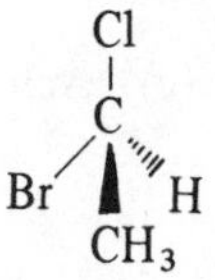

4.4 Describe the relationship between the two structures shown.

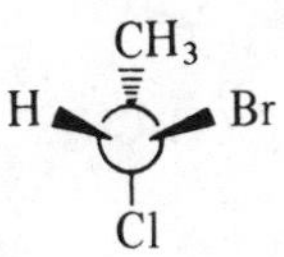

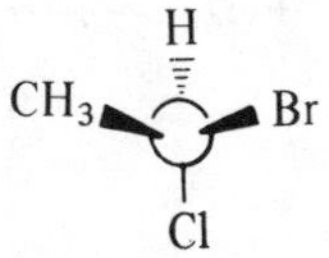

[structures: CH_3; H, Br; Cl] [structures: H; CH_3, Br; Cl]

(a) Enantiomers (b) Diastereomers (c) Constitutional isomers

(d) Conformations (e) Two molecules of the same compound

4.5 Which of the following molecule(s) possess(es) a plane of symmetry?

(a) 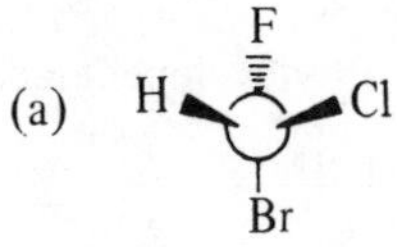

(b) 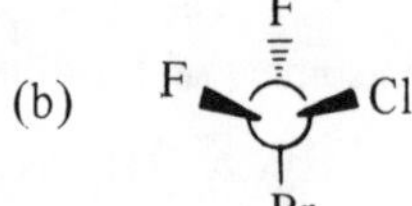

(c) 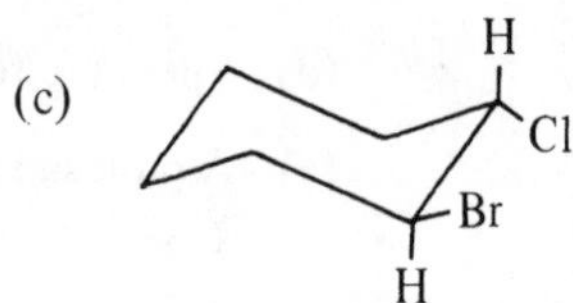

(d) More than one of these (e) None of these

4.6 Give the (R)-(S) designation of the structure shown.

O
‖
HO–C CH$_3$ H

Cl

(a) (R) (b) (S) (c) Neither because this molecule has no stereocenter

(d) Impossible to tell

4.7 Select the words that best describe the following structure:

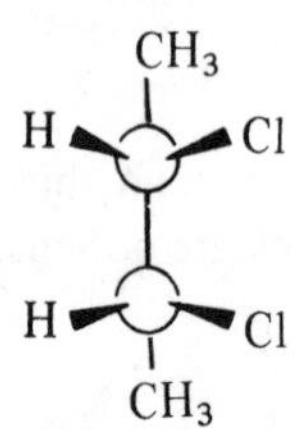

(a) Chiral (b) Meso form (c) Achiral (d) Has a plane of symmetry

(e) More than one of these

4.8 Select the words that best describe what happens to the optical rotation of the alkene shown when it is hydrogenated to the alkane according to the following equation:

H H
(R)–CH_3CH_2 CH_3 + H_2 $\xrightarrow{\text{Ni}}$ CH_3CH_2 CH_3
CH=CH$_2$ CH$_2$CH$_3$

(a) Increases (b) Drops to zero (c) Changes sign

(d) Stays the same (e) Impossible to predict

SUPPLEMENTARY PROBLEMS

S4.1 Which word best describes the following pairs of compounds: enantiomers, diastereomers, constitutional isomers, not isomers, identical.

(a) 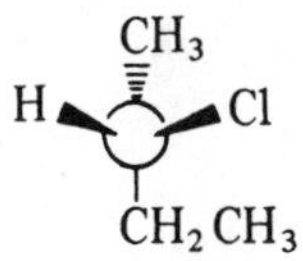 and

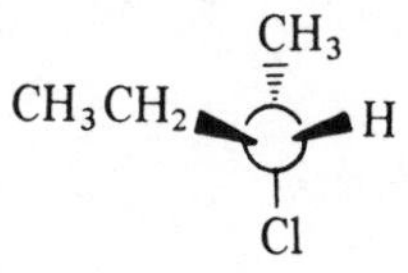

(b) 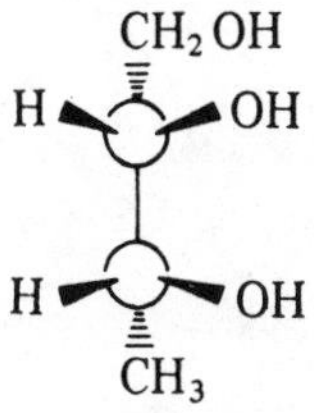 and

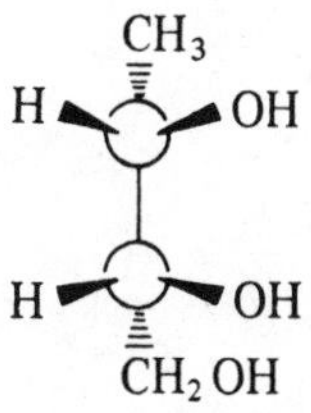

(c) 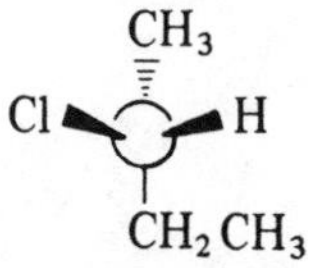 and

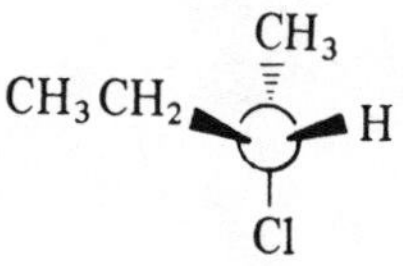

(d) 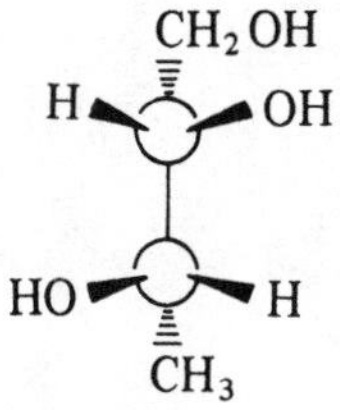 and

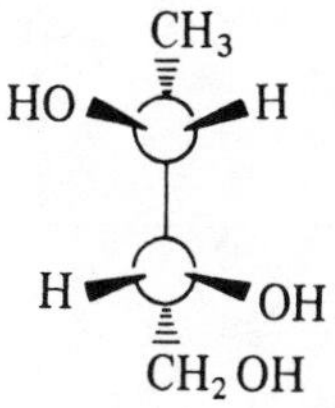

S4.2 Give the (*R*)–(*S*) designation of all the stereocenters in the following morphine molecule.

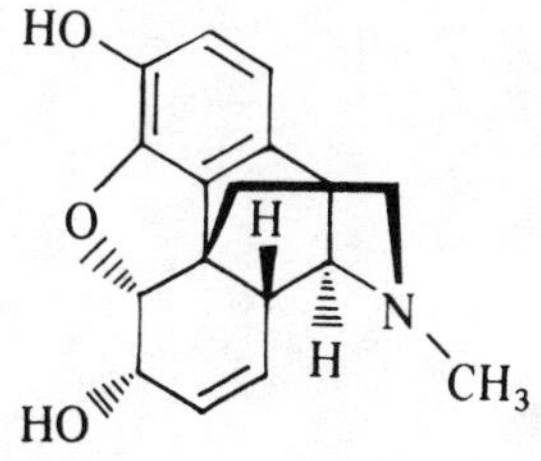

SOLUTIONS TO SUPPLEMENTARY PROBLEMS

S4.1 (a) Identical (b) Enantiomers (c) Enantiomers (d) Enantiomers.

S4.2

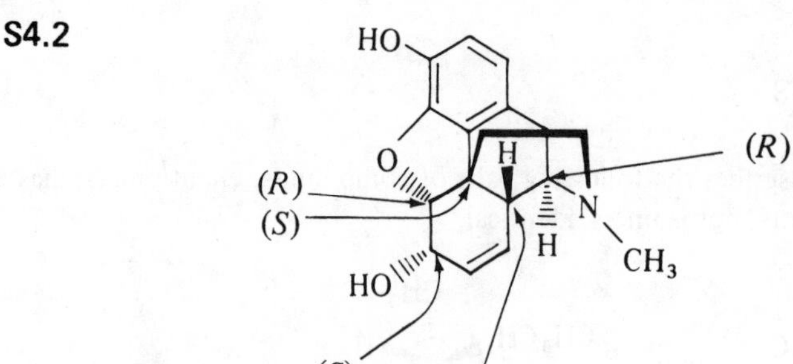

5 IONIC REACTIONS— NUCLEOPHILIC SUBSTITUTION AND ELIMINATION REACTIONS OF ALKYL HALIDES

SUMMARY OF MECHANISMS

Mechanism of S_N2 Reaction

$$Nu:^- + C{-}L \longrightarrow \overset{\delta-}{Nu} \cdots C \cdots \overset{\delta-}{L} \longrightarrow Nu{-}C + :L^-$$

Transition state

Mechanism of E2 Reaction

$$B:^- + H{-}C{-}C{-}L \longrightarrow B{-}H + C{=}C + :L^-$$

Mechanism of S_N1/E1 Reaction

$$-\overset{H}{C}-\overset{R}{\underset{R}{C}}-L \xrightarrow{-L^-} -\overset{H}{C}-\overset{R}{\underset{R}{C^+}}$$

$$\xrightarrow[S_N1]{R'OH} -\overset{H}{C}-\overset{R}{\underset{R}{C}}-\underset{H}{O}-R'^+$$

$$\xrightarrow[E1]{-H^+} C{=}C\begin{matrix} R \\ R \end{matrix}$$

SUMMARY OF IMPORTANT REACTION PATHWAYS ACCORDING TO THE TYPE OF SUBSTRATE

CH_3X Methyl	RCH_2X 1°	R–CHX (RCHX with R) 2°	R_3C–X 3°
← Bimolecular reactions only →			← S_N1/E1 or E2 →
Gives S_N2 reactions.	Gives mainly S_N2 except with a hindered strong base [e.g., $(CH_3)_3CO^-$] and then gives mainly E2.	Gives mainly S_N2 with weak bases (e.g., I^-, CN^-, RCO_2^-) and mainly E2 with strong bases (e.g., RO^-)	No S_N2 reaction. In solvolysis gives S_N1/E1, and at lower temperatures S_N1 is favored. When a strong base (e.g., RO^-) is used, E2 predominates.

SOME SYNTHETICALLY USEFUL S_N2 REACTIONS

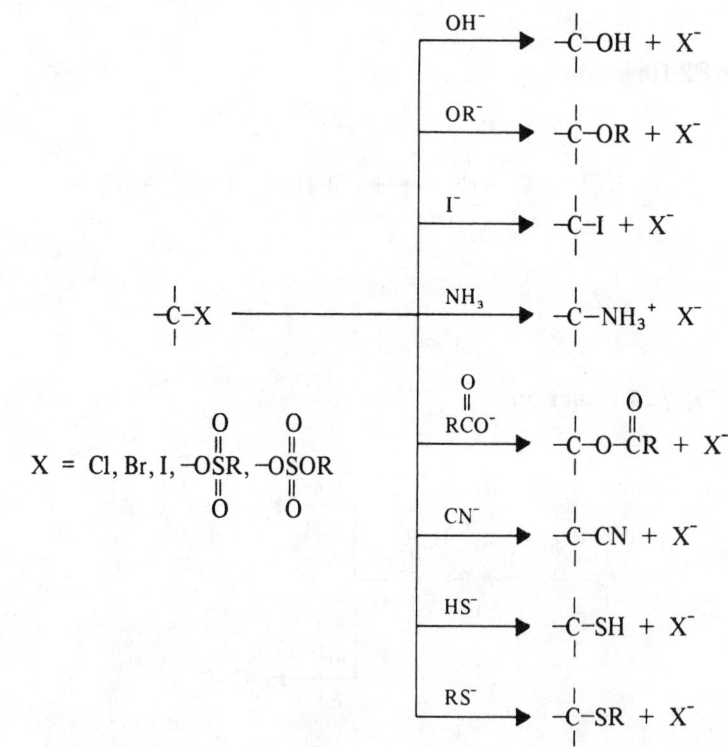

SOLUTIONS TO PROBLEMS

5.1 (a) $CH_3CH_2\text{–}\ddot{\underset{..}{O}}\text{–}H$ (b) $CH_3CH_2\text{–}\ddot{\underset{..}{O}}:^-$

(c) $H\text{–}\ddot{N}\text{–}H$ with H below N (d) $CH_3\text{–}\ddot{N}\text{–}H$ with H below N

(e) $^-:CN:$ (f) $CH_3C(=\ddot{O}:)\text{–}\ddot{\underset{..}{O}}\text{–}H$

(g) $CH_3\text{–}C(=\ddot{O}:)\text{–}\ddot{\underset{..}{O}}:^-$ (h) $H\text{–}C(=\ddot{O}:)\text{–}\ddot{\underset{..}{O}}\text{–}H$

(i) $H\text{–}C(=\ddot{O}:)\text{–}\ddot{\underset{..}{O}}:^-$ (j) $CH_3CH_2\text{–}\ddot{\underset{..}{S}}\text{–}H$

(k) $CH_3CH_2\text{–}\ddot{\underset{..}{S}}:^-$ (l) $^-:\ddot{N}=\overset{+}{N}=\ddot{N}:^-$

5.2

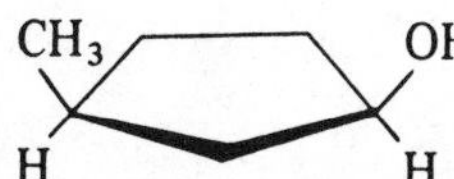

cis-3-Methylcyclopentanol

5.3 (a) We know that when a secondary alkyl halide reacts with hydroxide ion by substitution, the reaction occurs with *inversion of configuration* because the reaction is S_N2. If we know that the configuration of (–)-2-butanol (from Section 4.6C) is that shown here, then we can conclude that (+)-2-chlorobutane has the opposite configuration.

(*R*)-(–)-2-Butanol $\xleftarrow[S_N2]{OH^-}$ (*S*)-(+)-2-Chlorobutane

(*R*)-(–)-2-Butanol
$[\alpha]_D^{25°} = -13.52°$

(*S*)-(+)-2-Chlorobutane
$[\alpha]_D^{25°} = +36.00°$

(b) Again the reaction is S_N2. Because we now know the configuration of (+)-2-chlorobutane to be (*S*) [cf., part (a)], we can conclude that the configuration of (–)-2-iodobutane is (*R*).

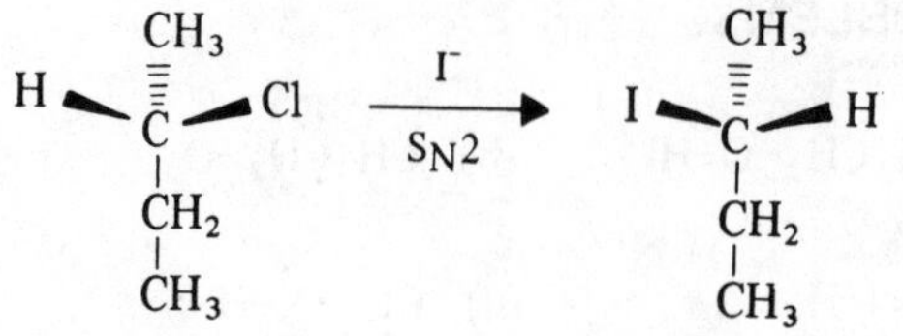

$$\underset{(S)\text{-}(+)\text{-2-Chlorobutane}}{\text{H}\!\blacktriangleleft\!\overset{\text{CH}_3}{\underset{\underset{\text{CH}_3}{|}}{\underset{\text{CH}_2}{|}}}\!\!\text{C}\!\blacktriangleright\!\text{Cl}} \xrightarrow[\text{S}_\text{N}2]{\text{I}^-} \underset{(R)\text{-}(-)\text{-2-Iodobutane}}{\text{I}\!\blacktriangleleft\!\overset{\text{CH}_3}{\underset{\underset{\text{CH}_3}{|}}{\underset{\text{CH}_2}{|}}}\!\!\text{C}\!\blacktriangleright\!\text{H}}$$

(+)-2-Iodobutane has the (*S*) configuration.

5.4

(a) [cyclopentyl carbocation: ring carbon bearing CH_3 and H; cationic carbon C^+ bearing CH_3]

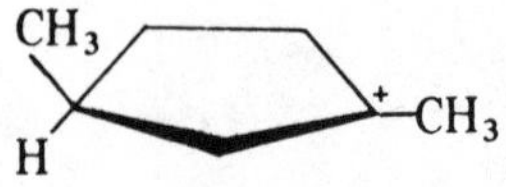

(b) Because the carbocation is planar, the nucleophile, H_2O, may approach from above or below the plane:

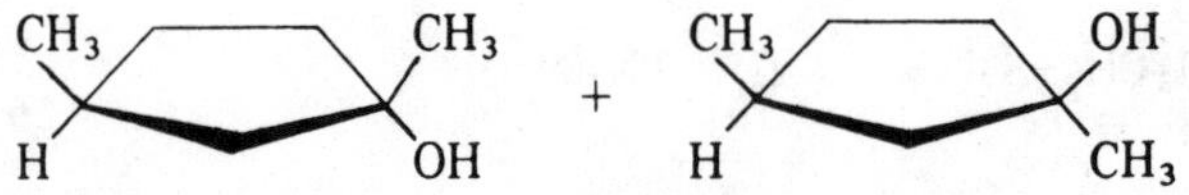

[cyclopentane: CH_3/H on one carbon, CH_3 (up)/OH (down) on the other] + [cyclopentane: CH_3/H on one carbon, OH (up)/CH_3 (down) on the other]

5.5

(a) $CH_3\overset{CH_3}{\underset{CH_3}{\overset{|}{\underset{|}{C}}}}\text{–}OCH_2CH_3$

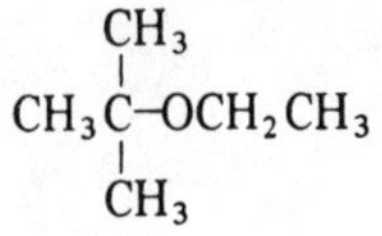

(b) $CH_3\overset{CH_3}{\underset{CH_3}{\overset{|}{\underset{|}{C}}}}\text{–}Cl \overset{\text{slow}}{\rightleftharpoons} CH_3\overset{CH_3}{\underset{CH_3}{\overset{|}{\underset{|}{C^+}}}} + Cl^-$

$$CH_3\overset{CH_3}{\underset{CH_3}{\overset{|}{\underset{|}{C^+}}}} + CH_3CH_2\ddot{O}H \overset{\text{fast}}{\rightleftharpoons} CH_3\overset{CH_3}{\underset{CH_3}{\overset{|}{\underset{|}{C}}}}\text{—}\overset{H}{\overset{|}{\underset{+}{\ddot{O}}}}CH_2CH_3$$

$$CH_3\overset{CH_3}{\underset{CH_3}{\overset{|}{\underset{|}{C}}}}\text{—}\overset{H}{\overset{|}{\underset{+}{\ddot{O}}}}CH_2CH_3 + CH_3CH_2OH \overset{\text{fast}}{\rightleftharpoons} CH_3\overset{CH_3}{\underset{CH_3}{\overset{|}{\underset{|}{C}}}}\text{–}OCH_2CH_3 + CH_3CH_2\overset{+}{O}H_2$$

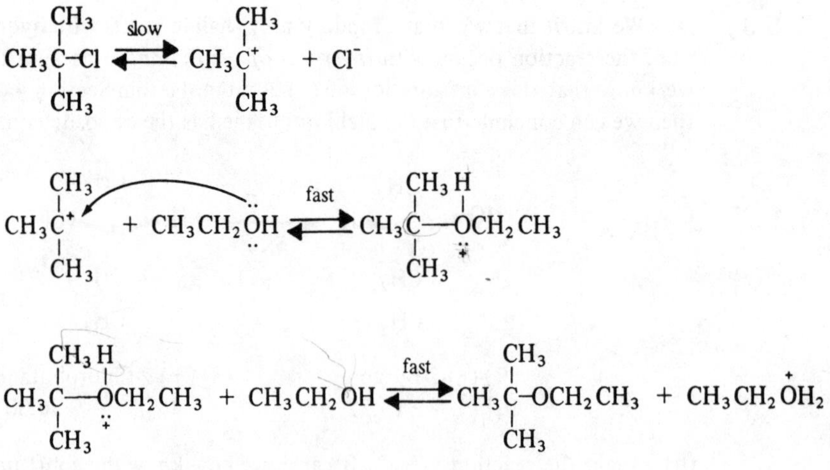

5.6 **Protic solvents** are those that have an H bonded to an oxygen or nitrogen (or to another strongly electronegative atom). Therefore, the protic solvents are formic acid, $HC(=O)OH$; formamide, $HC(=O)NH_2$; ammonia, NH_3, and ethylene glycol, $HOCH_2CH_2OH$.

Aprotic solvents lack an H bonded to a strongly electronegative element. Aprotic solvents in this list are Acetone, $CH_3C(=O)CH_3$; acetonitrile $CH_3C{\equiv}N$; sulfur dioxide, SO_2; and trimethylamine, $N(CH_3)_3$.

5.7 The reaction is an S_N2 reaction. In the polar aprotic solvent (DMF), the nucleophile (CN^-) will be relatively unencumbered by solvent molecules, and, therefore, it will be more reactive than in ethanol. As a result, the reaction will occur faster in *N,N*-dimethylformamide.

5.8 In (a) and (b) the base is a better nucleophile than its conjugate acid. In (c) the determining factor is the size of the atoms in the same group in the periodic table: P > N.

(a) NH_2^- (b) RS^- (c) PH_3

5.9 (a) Increasing the percentage of water in the mixture increases the polarity of the solvent. (Water is more polar than methanol.) Increasing the polarity of the solvent increases the rate of the solvolysis because separated charges develop in the transition state. The more polar the solvent, the more the transition state is stabilized (Section 5.13D, page 222).

(b) In an S_N2 reaction of this type, the charge becomes dispersed in the transition state:

$$I^- + CH_3CH_2{-}Cl \longrightarrow \overset{\delta-}{I}\cdots\overset{CH_3}{\underset{}{CH_2}}\cdots\overset{\delta-}{Cl} \longrightarrow ICH_2CH_3 + Cl^-$$

Reactants — *Charge is concentrated*

Transition state — *Charge is dispersed*

Increasing the polarity of the solvent increases the stabilization of the reactant I^- more than the stabilization of the transition state, and thereby increases the energy of activation, thus decreasing the rate of reaction.

5.10 In the forward reaction, Cl^- is the leaving group; in the reverse reaction, OH^- would have

$$HO^- + CH_3{-}Cl \rightleftarrows Cl^- + CH_3OH$$

to be the leaving group. OH^- is very basic and therefore is such a poor leaving group that, for all practical purposes, the reverse reaction does not occur.

5.11 (a) $CH_3CH_2O^- + (S)\text{-}CH_3(CH_3CH_2)CH{-}Br \xrightarrow[S_N2]{\text{inversion}} (R)\text{-}CH_3CH_2O{-}CH(CH_3)CH_2CH_3 + Na^+Br^-$

(S) (R)

(b) $CH_3\overset{O}{\overset{\|}{C}}O^-\ Na^+ + (S)\text{-}CH_3(CH_3CH_2)CH{-}Br \xrightarrow[S_N2]{\text{inversion}} (R)\text{-}CH_3\overset{O}{\overset{\|}{C}}O{-}CH(CH_3)CH_2CH_3 + Na^+Br^-$

(S) (R)

(c) $HS^-\ Na^+ + (S)\text{-}CH_3(CH_3CH_2)CH{-}Br \xrightarrow[S_N2]{\text{inversion}} (R)\text{-}HS{-}CH(CH_3)CH_2CH_3 + Na^+Br^-$

(S) (R)

(d) $CH_3S^-\ Na^+ + (S)\text{-}CH_3(CH_3CH_2)CH{-}Br \xrightarrow[S_N2]{\text{inversion}} (R)\text{-}CH_3S{-}CH(CH_3)CH_2CH_3 + Na^+Br^-$

(S) (R)

5.12 (a) $CH_3CH_2CH_2Br + NaOH \longrightarrow CH_3CH_2CH_2OH + NaBr$

(b) $CH_3CH_2CH_2Br + NaI \longrightarrow CH_3CH_2CH_2I + NaBr$

(c) $CH_3CH_2CH_2Br + CH_3CH_2ONa \longrightarrow CH_3CH_2CH_2{-}O{-}CH_2CH_3 + NaBr$

(d) $CH_3CH_2CH_2Br + CH_3SNa \longrightarrow CH_3CH_2CH_2{-}S{-}CH_3 + NaBr$

(e) $CH_3CH_2CH_2Br + CH_3\overset{O}{\overset{\|}{C}}{-}ONa \longrightarrow CH_3CH_2CH_2{-}O{-}\overset{O}{\overset{\|}{C}}CH_3$

(f) $CH_3CH_2CH_2Br + NaN_3 \longrightarrow CH_3CH_2CH_2N_3 + NaBr$

(g) $CH_3CH_2CH_2Br + :N(CH_3)_3 \longrightarrow CH_3CH_2CH_2{-}\overset{CH_3}{\overset{|}{\underset{CH_3}{\underset{|}{N^+}}}}{-}CH_3\ \ Br^-$

(h) $CH_3CH_2CH_2Br + NaCN \longrightarrow CH_3CH_2CH_2CN + NaBr$

(i) $CH_3CH_2CH_2Br + NaSH \longrightarrow CH_3CH_2CH_2SH + NaBr$

5.13 (a) $CH_3CH_2CH_2CH_2Br$ because 1° halides are less hindered than 2° halides.

(b) $CH_3CH_2CH(Br)CH_3$ because 2° halides are less hindered than 3° halides.

(c) $CH_3CH_2CH_2Br$ because a bromide ion is a better leaving group than a chloride ion.

(d) $CH_3CH(CH_3)CH_2CH_2Br$ because it is less hindered than $CH_3CH_2CH(CH_3)CH_2Br$

(e) CH_3CH_2Cl because vinyl halides ($CH_2{=}CHCl$) are very unreactive.

5.14 (a) The second because CH_3O^- is a better nucleophile than CH_3OH.

(b) The second because SH^- is a better nucleophile than OH^- (in protic solvents).

(c) The second because CH_3SH is a better nucleophile than CH_3OH (in protic solvents).

(d) The second, CH_3S^- (2.0 *M*), because the rate is proportional to $[CH_3S^-]$ as well as to $[CH_3CH_2I]$.

5.15 (a) The first because I^- is a better leaving group than Cl^-.

(b) The first because H_2O is a more polar solvent than CH_3OH.

(c) Both the same because $[CH_3O^-]$ does not affect the rate of an S_N1 reaction.

(d) The first because vinylic halides are unreactive.

5.16 Possible methods are given here.

(a) $\underset{\text{(excess)}}{CH_4} \xrightarrow[h\nu,\ \text{heat}]{Cl_2} CH_3Cl \xrightarrow[\substack{CH_3OH \\ (S_N2)}]{I^-} CH_3I$

(b) $\underset{\text{(excess)}}{CH_3CH_3} \xrightarrow[h\nu,\ \text{heat}]{Cl_2} CH_3CH_2Cl \xrightarrow[\substack{CH_3OH \\ (S_N2)}]{I^-} CH_3CH_2I$

(c) $CH_3Cl \xrightarrow[\substack{CH_3OH/H_2O \\ (S_N2)}]{OH^-} CH_3OH$

(d) $CH_3CH_2Cl \xrightarrow[\substack{CH_3OH/H_2O \\ (S_N2)}]{OH^-} CH_3CH_2OH$

(e) $CH_3Cl \xrightarrow[\substack{CH_3OH \\ (S_N2)}]{SH^-} CH_3SH$

(f) $CH_3CH_2Cl \xrightarrow[\substack{CH_3OH \\ (S_N2)}]{SH^-} CH_3CH_2SH$

(g) $CH_3Cl \xrightarrow[\text{DMF}]{CN^-} CH_3CN$

(h) $CH_3CH_2Cl \xrightarrow[\text{DMF}]{CN^-} CH_3CH_2CN$

(i) $CH_3OH \xrightarrow[(-H_2)]{\text{NaH}} CH_3ONa \xrightarrow[CH_3OH]{CH_3I} CH_3OCH_3$

(j) $CH_3CH_2OH \xrightarrow[(-H_2)]{\text{NaH}} CH_3CH_2ONa \xrightarrow{CH_3I} CH_3CH_2OCH_3$

(k) cyclopentane (excess) $\xrightarrow[h\nu,\ \text{heat}]{Cl_2}$ chlorocyclopentane $\xrightarrow[CH_3CH_2OH]{CH_3CH_2ONa}$ cyclopentene

5.17 (a) $H:^-$ is a very strong base and therefore is an extremely poor leaving group.

(b) $:CH_3^-$ is a very strong base and therefore is an extremely poor leaving group.

(c) $-\ddot{C}H_2^-$ is a very strong base and an extremely poor leaving group.

(d) With a relatively strong base like CN^-, elimination would predominate to yield $CH_2{=}C(CH_3)_2 + HCN + Br^-$. S_N2 attack cannot take place at the 3° carbon.

(e) Vinylic halides are unreactive in S_N1 and S_N2 reactions.

(f) CH_3O^- is a strong base and therefore a poor leaving group.

(g) $CH_3CH_2\overset{+}{O}H_2$ is a strong acid and would react with NH_3 to convert it to NH_4^+ which is not a nucleophile.

(h) $CH_3:^-$ would react with the acidic proton in CH_3CH_2OH to form $CH_4 + CH_3CH_2O^-$.

5.18 $CH_3CHBrCH_3$ because a 2° halide is less likely to give an S_N2 reaction than a 1° halide, and therefore an E2 reaction (dehydrohalogenation) would be more likely to predominate.

5.19 Method (2) would be better because the substrate for the S_N2 reaction is a methyl halide. In method (1), because the substrate is a 2° halide, considerable (predominant) elimination (E2) would accompany the S_N2 reaction.

5.20 (a) $CH_3CH_2CH_2CH_2OCH_3$ (major) by S_N2; $CH_3CH_2CH{=}CH_2$ (minor) by E2.

(b) $CH_3CH_2CH_2CH_2{-}OC(CH_3)_3$ (minor) by S_N2; $CH_3CH_2CH{=}CH_2$ (major) by E2.

(c) $CH_3{-}O{-}C(CH_3)_3$ (only product) by S_N2.

(d) $CH_2{=}C(CH_3)_2$ (only product) by E2.

(e) 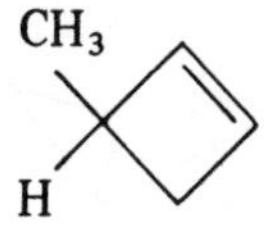 CH_3, H, cyclobutene (major) by E2 CH_3, H, H, OCH_3 (minor) by S_N2

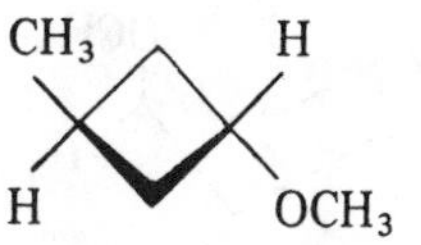

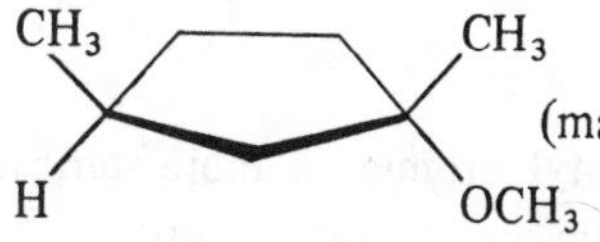

(f) 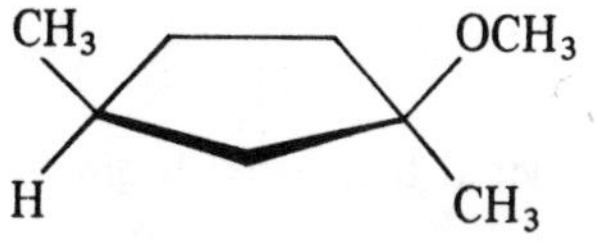 CH_3, H, OCH_3, CH_3 + CH_3, H, CH_3, OCH_3 (major products) by S_N1

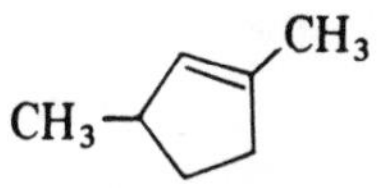

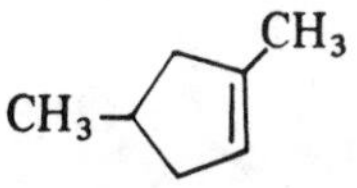

CH_3, CH_3, CH_3, CH_3 and CH_3, CH_2 (minor products) by E1 reaction

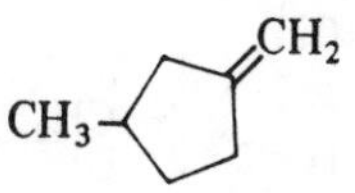

(g) $CH_3CH{=}CHCH_2CH_3$ (major) by E2 $CH_3CH_2CHCH_2CH_3$ with OC_2H_5 (minor) by S_N2

(h) $CH_2{=}CHCH_3$ (major) by E2 CH_3CHCH_3 with $OC(CH_3)_3$ (minor) by S_N2

(i) $CH_3CH{=}CHCH_3$ (major) and $CH_2{=}CHCH_2CH_3$ (minor) by E2
(S)-$CH_3CH(OH)CH_2CH_3$ (minor) by S_N2

(j) ($\pm$)$CH_3CH_2C(CH_3)CH_2CH_2CH_3$ with OCH_3 (major) by S_N1 $CH_3CH_2C(CH_3){=}CHCH_2CH_3$

$CH_3CH{=}C(CH_3)CH_2CH_2CH_3$ and $CH_3CH_2C({=}CH_2)CH_2CH_2CH_3$ (all minor) by E1

(k) (R)-$CH_3CHIC_6H_{13}$ (only product) by S_N2

5.21 (a), (b), and (c) are all S_N2 reactions and, therefore, proceed with inversion of configuration: The products are

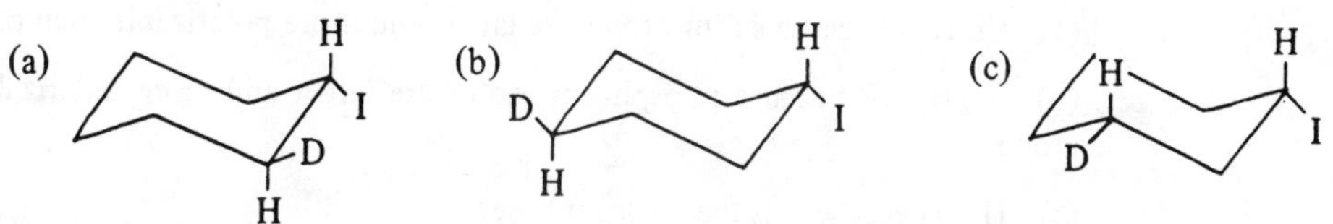

(d) is an S_N1 reaction. The carbocation that forms can react with either nucleophile (H_2O or CH_3OH) from either the top or bottom side of the molecule. Four substitution products (below) would be obtained. (Considerable elimination by an E1 path would also occur.)

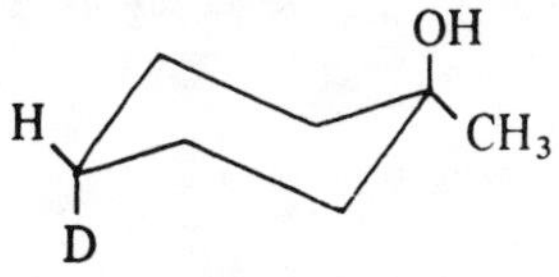

H, D, OH, CH_3 and H, D, CH_3, OH

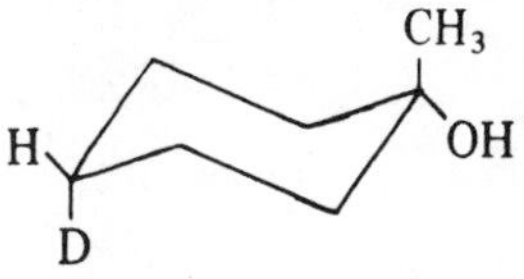

OCH$_3$ / H, D / CH$_3$ and CH$_3$ / H, D / OCH$_3$

5.22 Isobutyl bromide is more sterically hindered than ethyl bromide because of the methyl groups on the β carbon atom.

CH_3, H, CH_3 on β C—α CH_2–Br — Isobutyl bromide

H, H, H on C—CH_2–Br — Ethyl bromide

This steric hindrance causes isobutyl bromide to react more slowly in S_N2 reactions and to give relatively more elimination (by an E2 path) when a strong base is used.

5.23 (a) S_N2 because the substrate is a 1° halide.

(b) Rate $= k\,[CH_3CH_2Cl]\,[I^-]$

$= 5 \times 10^{-5}$ L mole^{-1}s^{-1} $\times$ 0.1 mole/L $\times$ 0.1 mole/L

Rate $= 5 \times 10^{-7}$ mole/L s

(c) 1×10^{-6} mole/L s

(d) 1×10^{-6} mole/L s

(e) 2×10^{-6} mole/L s

5.24 (a) CH_3NH^- because it is the stronger base.

(b) CH_3O^- because it is the stronger base.

(c) CH_3SH because sulfur atoms are larger and more polarizable than oxygen atoms.

(d) $(C_6H_5)_3P$ because phosphorus atoms are larger and more polarizable than nitrogen atoms.

(e) H_2O because it is the stronger base.

(f) NH_3 because it is the stronger base.

(g) HS^- because it is the stronger base.

(h) OH^- because it is the stronger base.

5.25

(a) $HOCH_2CH_2Br + OH^- \rightleftarrows H_2C(O^-)\text{—}CH_2Br \longrightarrow H_2C\text{—}CH_2(\text{—O—}) + Br^-$

(b) $H_2C\text{—}CH_2$, H_2C, $CH_2\text{—}Br$, $\ddot{N}$, H H $\longrightarrow$ $H_2C\text{—}CH_2$, H_2C, CH_2, N^+, H H + Br^- $\xrightarrow{OH^-}$ $H_2C\text{—}CH_2$, H_2C, CH_2, N, H $+ H_2O + Br^-$

5.26 Iodide ion is a good nucleophile and a good leaving group; it can rapidly convert an alkyl chloride or alkyl bromide into an alkyl iodide, and the alkyl iodide can then react rapidly with another nucleophile. With methyl bromide in water, for example, the following reaction can take place:

$CH_3Br \xrightarrow[\text{(slower)}]{H_2O\text{ alone}} CH_3OH_2^+ + Br^-$

$CH_3Br \xrightarrow[\text{(faster)}]{H_2O\text{ containing }I^-} CH_3I \xrightarrow[\text{(faster)}]{H_2O} CH_3\overset{+}{O}H_2 + I^-$

5.27 The rate of formation of *tert*-butyl alcohol does not increase with increasing $[OH^-]$ because the reaction is S_N1 and is therefore independent of $[OH^-]$. Increasing $[OH^-]$, however, increases the rate of the competing E2 reaction which consumes OH^- through the conversion of *tert*-butyl chloride into $CH_2{=}C(CH_3)_2$.

5.28 (a) Use a strong, hindered base such as $(CH_3)_3COK$ in a solvent of low polarity in order to bring about an E2 reaction.

(b) Here we want an S_N1 reaction. We use ethanol as the solvent *and as the nucleophile,* and we carry out the reaction at a low temperature so that elimination will be minimized.

5.29 (a) Backside attack by the nucleophile is prevented by the cyclic structure. (Notice, too, that the carbon atom bearing the leaving group is tertiary.)

(b) The bridged cyclic structure prevents the carbon atom bearing the leaving group from assuming the planar trigonal conformation required of a carbocation.

5.30 The products are

$H{:}\underset{\ddot{H}}{\overset{H}{\ddot{C}}}{:}C{:}{:}{:}N{:}$ and $H{:}\underset{\ddot{H}}{\overset{H}{\ddot{C}}}{:}\overset{+}{N}{:}{:}{:}\overset{-}{C}{:}$

The nucleophile can be described by resonance structures that place a pair of electrons and a formal negative charge on either atom $^{-}:C:::N: \longleftrightarrow :C::\ddot{N}:^{-}$. Thus both atoms are nucleophilic.

5.31 (a) Since the halides are all primary, these are almost certainly S_N2 reactions with ethanol acting as the nucleophile.

$$C_2H_5\ddot{O}H + \text{R(H)(H)}C-Br \xrightarrow[-HBr]{} C_2H_5O-C\text{(R)(H)(H)}$$

(b) Increasing the size of the R group increases steric hindrance to the approaching ethanol molecule and decreases the rate of reaction.

5.32 (a) This is another example of the relation between reactivity and selectivity that we first encountered in Chapter 4: generally speaking, highly reactive species are relatively unselective while less reactive species are more selective. In an S_N1 reaction the species that reacts with the nucleophile is a *carbocation*–a species that is electron deficient and thus is *highly reactive.* A carbocation, therefore, shows little tendency to discriminate between weak and strong nucleophiles–most often it simply reacts with the first nucleophile that it encounters. In S_N2 reactions, on the other hand, the species that reacts with the nucleophile is an alkyl halide or an alkyl tosylate. Such compounds are far less reactive toward nucleophiles than carbocations and they show much greater nucleophilic selectivities. An alkyl halide molecule, for example, might collide with a weak nucleophile thousands of times before a reaction takes place because few of the collisions will have sufficient energy to allow the weak nucleophile to displace the leaving group. On the other hand, an alkyl halide molecule might collide with a strong nucleophile only a few times before a collison leads to a reaction. This will be true because the strong nucleophile is better able to displace the leaving group and therefore a larger fraction of collisions will have sufficient energy to be fruitful.

(b) The reaction of $CH_3CH_2CH_2CH_2Cl$ is an S_N2 reaction and thus $CH_3CH_2CH_2CH_2Cl$ discriminates very effectively between the strongly nucleophilic CN^- ions and the weakly nucleophilic solvent molecules. By contrast, the reaction of $(CH_3)_3CCl$ is an S_N1 reaction and the carbocation that is formed shows little tendency to discriminate between solvent molecules and CN^- ions. Since solvent molecules are present in a much higher concentration the major product is $(CH_3)_3C{-}OCH_2CH_3$.

5.33 The rate-determining step in the S_N1 reaction of *tert*-butyl bromide is the following:

$$(CH_3)_3C{-}Br \xrightarrow{\text{slow}} (CH_3)_3C^+ + Br^-$$

$$(CH_3)_3C^+ \xrightarrow{H_2O} (CH_3)_3COH_2^+$$

$(CH_3)_3C^+$ is so unstable that it reacts almost immediately with one of the surrounding water molecules and, for all practical purposes, no reverse reaction with Br^- takes place. Adding a common ion (Br^- from NaBr) therefore, has no effect on the rate.

Because the $(C_6H_5)_2CH^+$ cation is more stable, a reversible first step occurs and

$$(C_6H_5)_2CHBr \rightleftharpoons (C_6H_5)_2CH^+ + Br^-$$

$$(C_6H_5)_2CH^+ \xrightarrow{H_2O} (C_6H_5)_2CHOH_2^+$$

adding a common ion (Br^-) slows the overall reaction by increasing the rate at which $(C_6H_5)_2CH^+$ is converted back to $(C_6H_5)_2CHBr$.

5.34 Two different mechanisms are involved. $(CH_3)_3CBr$ reacts by an S_N1 mechanism and apparently this reaction takes place fastest. The other three alkyl halides react by an S_N2 mechanism and their reactions are slower because the nucleophile (H_2O) is weak. The reaction rates of CH_3Br, CH_3CH_2Br, and $(CH_3)_2CHBr$ are affected by the steric hindrance and thus their order of reactivity is $CH_3Br > CH_3CH_2Br > (CH_3)_2CHBr$.

SECTION REFERENCES FOR ADDITIONAL PROBLEMS

5.12 5.2
5.13 5.13A, 5.14A
5.14 5.13B
5.15 5.13B, 5.14A
5.16 3.17, 3.18, 5.2
5.17 5.13, 5.14A
5.18 5.18
5.19 5.18
5.20 5.8A, 5.10A, 5.18
5.21 5.8, 5.10
5.22 5.13
5.23 5.2, 5.13
5.24 5.13B
5.25 5.2
5.26 5.13E
5.27 5.18B
5.28 5.18B
5.29 5.13A
5.30 1.6, 5.2
5.31 5.13A

SELF-TEST

5.1 Using CF_4 as reactant write equations using Lewis dot formulas to show each of the following:

(a) Homolytic cleavage

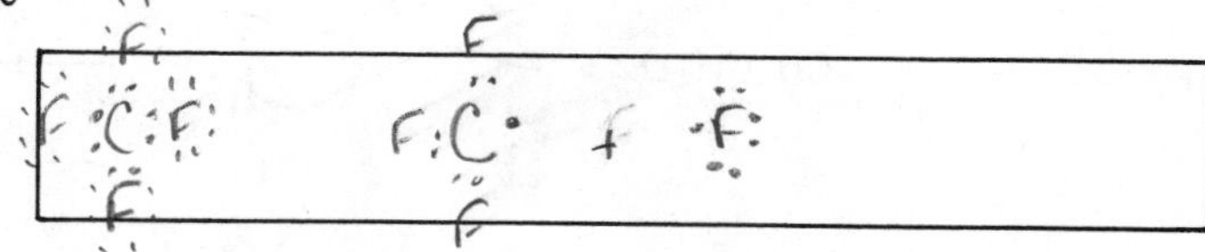

(b) The most reasonable type of heterolytic cleavage

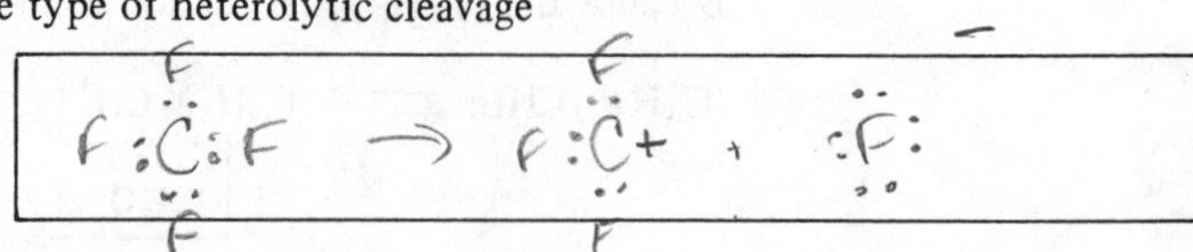

(c) The least reasonable type of heterolytic cleavage

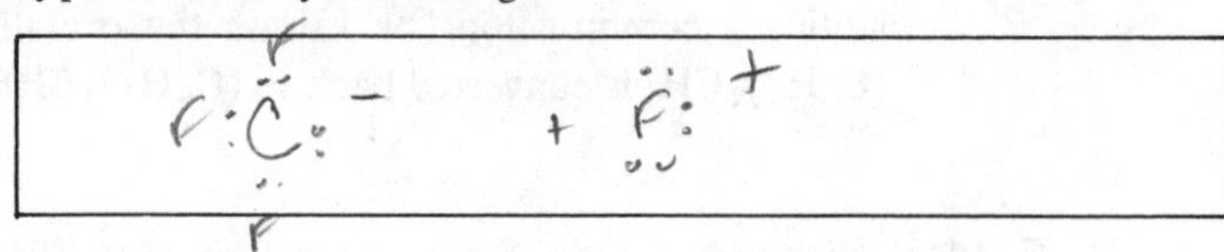

5.2 Mark the following statements true or false for each of the four mechanisms shown. (Use + for true, – for false, and 0 for impossible to tell.)

	S_N1	S_N2	E1 (carbocation)	E2
(a) The reaction shows first-order kinetics	+ ✓	–	+ ✓	–
(b) The rate of reaction depends markedly on the nucleophilicity or basicity of the attacking nucleophile	–	+	–	+

	S_N1	S_N2	E1 (carbocation)	E2
(c) The mechanism involves one step	–	+	–	+
(d) Carbocations are intermediates	+	–	+	–
(e) The rate of reaction is proportional to the concentration of the attacking nucleophile or base	–	+	–	+
(f) The rate of reaction depends on the nature of the leaving group	+	+	+	+

5.3 (a) Give the structural formula of the product of the following S_N2 reaction:

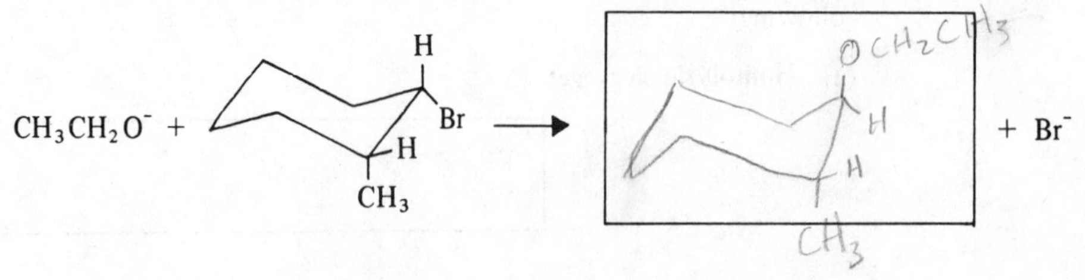

$CH_3CH_2O^-$ + (H, Br, H, CH_3) → (OCH_2CH_3, H, H, CH_3) + Br^-

(b) Complete the table below for the preceding reaction.

Experimental Number	$[CH_3CH_2O^-]$	[RBr]	Initial Rate of Formation of S_N2 Product
1	0.1 M	0.1 M	0.01 mole $L^{-1}s^{-1}$
2	0.2 M	0.2 M	
3	0.1 M	0.2 M	

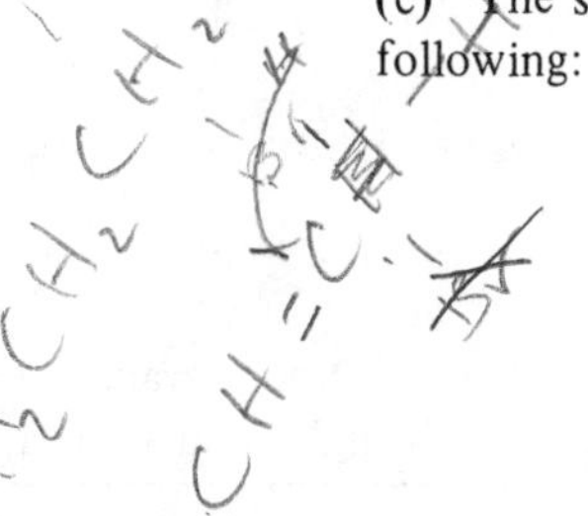

(c) The structural formula of the other organic product in the preceding reaction is the following:

(d) The favored reaction is

5.4 Mark the following statements true (+) or false (-).

(a) If we want to convert *tert*-butyl chloride into *tert*-butyl alcohol with the least amount of by-products, we should use a polar solvent like water and a very weak base, preferably water itself.

(b) If we heat 1-bromobutane with NaOH in ethanol as solvent, the major product will be 1-butene.

(c) The conditions described in (a) should encourage the S_N2 mechanism rather than the S_N1 mechanism.

(d) The conditions described in (b) should favor bimolecular mechanisms rather than unimolecular mechanisms.

5.5 Which set of conditions would you use to obtain the best yield in the reaction shown?

$$CH_3C(CH_3)_2{-}Br \xrightarrow{?} CH_2{=}C(CH_3)_2$$

(a) H_2O, heat (b) CH_3CH_2ONa/CH_3CH_2OH, heat (c) Heat alone

(d) H_2SO_4 (e) None of the above

5.6 Which of the following reactions would give the best yield?

(a) $CH_3ONa + (CH_3)_2CHBr \longrightarrow CH_3OCH(CH_3)_2$

(b) $(CH_3)_2CHONa + CH_3Br \longrightarrow CH_3OCH(CH_3)_2$

(c) $CH_3OH + (CH_3)_2CHBr \xrightarrow{\text{heat}} CH_3OCH(CH_3)_2$

5.7 A kinetic study yielded the following reaction rate data:

Experimental Number	Initial Concentrations $[OH^-]$	$[R{-}Br]$	Initial Rate of Disappearance of R—Br and Formation of R—OH
1	0.50	0.50	1.00
2	0.50	0.25	0.50
3	0.25	0.25	0.25

Which of the following statements best describes this reaction?

(a) The rate is second order.

(b) The rate is first order.

(c) The reaction is S_N1.

(d) Increasing the concentration of OH^- has no effect on the rate.

(e) More than one of the above.

SUPPLEMENTARY PROBLEMS

S5.1 Which of the following alkyl halides would you expect to react more rapidly in (a) an S_N2 mechanism, (b) an S_N1 mechanism?

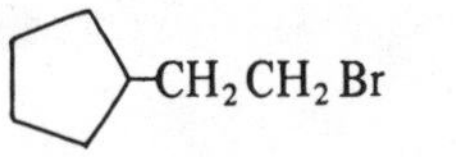

I: cyclopentyl–CH_2CH_2Br

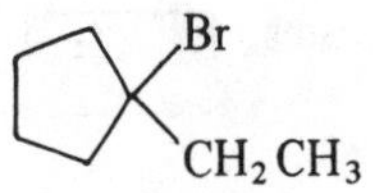

II: 1-bromo-1-ethylcyclopentane (Br, CH_2CH_3)

S5.2 Which mechanism (S_N1, S_N2, E1, or E2) would you expect to predominate in the following reactions? Predict the major product in each.

(a) CH_3 / I (cyclopentane) $\xrightarrow[CH_3CH_2OH]{CH_3CH_2ONa}$

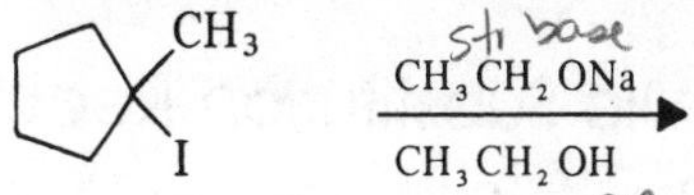

(b) CH_3 / I (cyclopentane) $\xrightarrow[CH_3CH_2OH]{H_2O}$

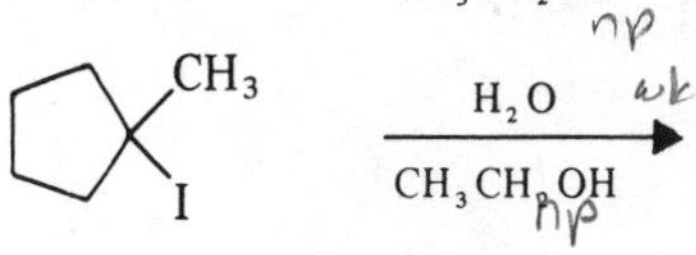

S5.3 Which of the following reactions would you expect to occur at a faster rate? Explain.

(a) $CH_3CH_2CH_2Cl \xrightarrow[C_2H_5OH,\ 25°C]{C_2H_5ONa} CH_3CH_2CH_2OC_2H_5 + NaCl$

(b) $CH_3CH_2CH_2I \xrightarrow[C_2H_5OH,\ 25°C]{C_2H_5ONa} CH_3CH_2CH_2OC_2H_5 + NaI$

SOLUTIONS TO SUPPLEMENTARY PROBLEMS

S5.1 The order of reactivity of alkyl halides is different in the two mechanisms:

S_N2 $CH_3X > 1°RX > 2°RX$ (3°RX do not react)

S_N1 3°RX only. (2°RX and 1°RX do not react appreciably)

The answers are therefore (a) I and (b) II.

S5.2 (a) The conditions of a strong base (nucleophile) and relatively nonpolar solvent favor the bimolecular mechanism. Since S_N2 does not occur with 3° RX's, the expected reaction is E2 to yield CH_3 (1-methylcyclopentene) as the major product.

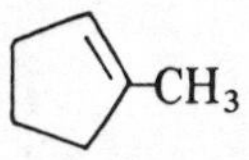

(b) The conditions of a weak base (nucleophile) and a highly polar solvent (H_2O) favor the monomolecular mechanism. S_N1 usually predominates over E1, so the major products are CH_3 / OH (cyclopentane) and CH_3 / OCH_2CH_3 (cyclopentane)

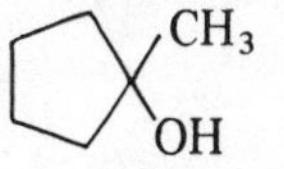

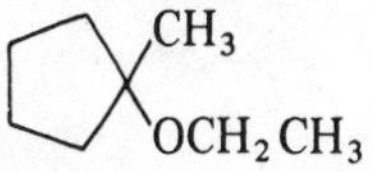

S5.3 Reaction (b) is faster because I^- is a better leaving group than Cl^-. All other factors are the same in the two reactions.

SPECIAL TOPIC

A Biological Nucleophilic Substitution Reaction: Biological Methylation

A.1 (a) $^{-}OOCCHCH_2CH_2-\ddot{S}-CH_2$ (with NH_3^+ on the CH; CH_2 attached to a ribose ring: ring O, H, H, H, H, OH, OH, Adenine)

(b) $^{-}OOCCHCH_2CH_2-\ddot{S}:^{-}$ (with NH_3^+ on the CH)

(c) The leaving group (a) is a weaker base than (b), therefore (a) is the better leaving group. The reaction with methionine would be much slower than the reaction with *S*-adenosylmethionine.

B SPECIAL TOPIC
Elementary Thermodynamics: $\Delta H°$, $\Delta S°$, and $\Delta G°$

SOLUTIONS TO PROBLEMS

B.1

(a) $X \longrightarrow Y \quad K_{eq} = \dfrac{[Y]}{[X]} = 10$

Initial $[X] = 1.0$
Equilibrium $[Y] = a$
Equilibrium $[X] = 1.0 - a$

$$\text{then } K_{eq} = 10 = \frac{a}{1.0 - a}$$

$$10 - 10a = a$$
$$-11a = -10$$
$$a = \frac{10}{11} = 0.91 \text{ mole/L}$$

At equilibrium, $[Y] = 0.91$ mole/L
$[X] = 0.09$ mole/L
and 91% of X is converted to product, Y.

(b) If $K_{eq} = 1$, $1 = \dfrac{a}{1.0 - a}$

$$1 - a = a$$
$$-2a = -1$$
$$a = 0.5$$

$\therefore$At equilibrium, $[Y] = 0.5$ mole/L
$[X] = 0.5$ mole/L
and 50% of X is converted to product, Y.

(c) If $K_{eq} = 10^{-3}$, $10^{-3} = \dfrac{a}{1 - a}$

$$10^{-3} - 10^{-3}a = a$$
$$-1.001a = -10^{-3}$$
$$a = \frac{10^{-3}}{1.001} \cong 10^{-3}$$

At equilibrium, $[Y] = 10^{-3}$ mole/L
$[X] = 0.999$ mole/L
and 0.1% of X is converted to product, Y.

B.2 (a) The majority of the molecules (~99.99%) are in the chair form at equilibrium because the equilibrium, chair $\rightleftharpoons$ boat, has a $\Delta G^\circ = -5$ to -6 kcal/mole. See Table B.1, last column under -5.5 kcal/mole.

(b) For $\Delta G^\circ \sim -1.8$, Table B.1 tells us that 95% of ethylcyclohexane is in the equatorial form.

B-3 (a)
$$\Delta G^\circ = \Delta H^\circ - T\Delta S^\circ$$
$$\Delta G^\circ = -41{,}700 \text{ cal/mole} - 300 \text{ deg } (-\ 26.6 \text{ cal/ deg mole})$$
$$\Delta G^\circ = -41{,}700 + 7980 = -33{,}720 \text{ cal/mole}$$
or
$$\Delta G^\circ = -33.72 \text{ kcal/mole}$$

(b) Yes, because a negative value of ΔG° tells us that the products are favored at equilibrium.

(c) No, a negative entropy tells us that the products are more ordered, and therefore less favored than the reactants.

(d) There are fewer degrees of freedom in the product molecule, ethene, than in the separate and independent molecules, ethyne and hydrogen.

6 ALKENES AND ALKYNES I. PROPERTIES AND SYNTHESIS

SUMMARY OF SYNTHESES OF ALKENES

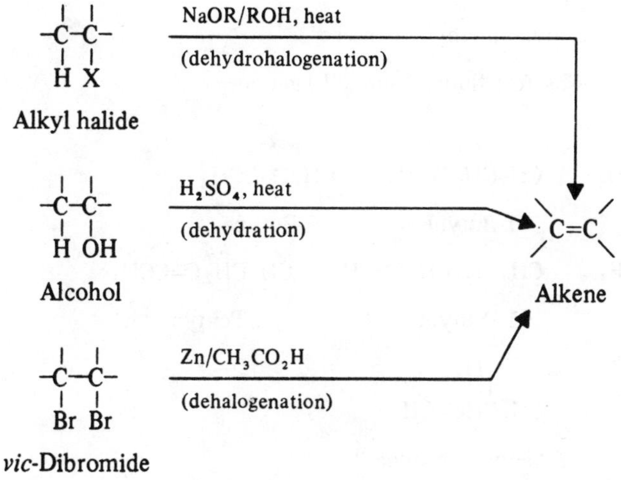

Alkyl halide (—C(H)—C(X)—) $\xrightarrow[\text{(dehydrohalogenation)}]{\text{NaOR/ROH, heat}}$ Alkene ($>C=C<$)

Alcohol (—C(H)—C(OH)—) $\xrightarrow[\text{(dehydration)}]{H_2SO_4\text{, heat}}$ Alkene ($>C=C<$)

vic-Dibromide (—C(Br)—C(Br)—) $\xrightarrow[\text{(dehalogenation)}]{Zn/CH_3CO_2H}$ Alkene ($>C=C<$)

SOLUTIONS TO PROBLEMS

6.1 (a) 2-Methyl-2-butene

(b) *cis*-4-Octene

(c) 1-Bromo-2-methylpropene

(d) 4-Methylcyclohexene

6.2 (a) $CH_3CH_2(H)C=C(H)CH_2CH_3$ (cis: CH_3CH_2 and CH_2CH_3 on same side, H and H on same side)

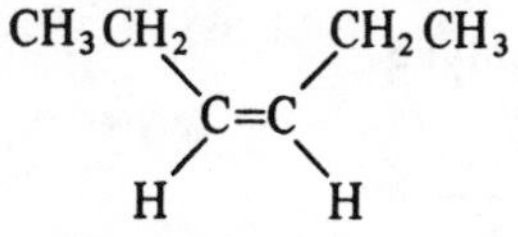

(b) $CH_3(H)C=C(H)CH_2CH_3$ (trans: CH_3 and CH_2CH_3 on opposite sides)

(c) 3-ethylcyclohexene (cyclohexene ring with CH_2CH_3)

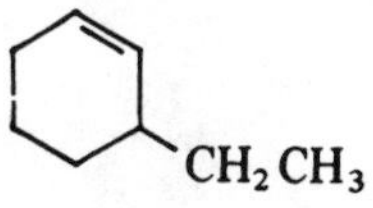

(d) cyclohexyl–$CH=CH_2$

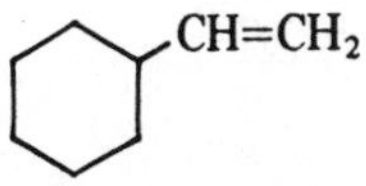

(e) $CH_2=CHCH_2C(CH_3)_2CH_2CH_3$

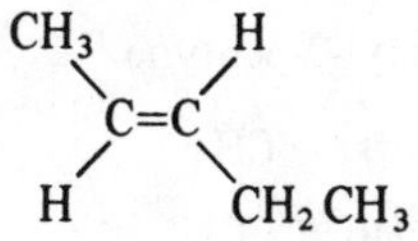

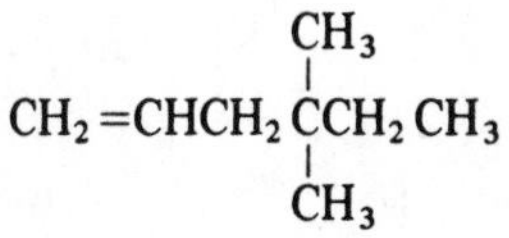

(f) 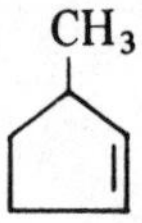 3-methylcyclopentene structure (ring with CH_3)

(g) $CH_3(CH_2)_4CHCH{=}CH_2$ with Cl on CH

(h) 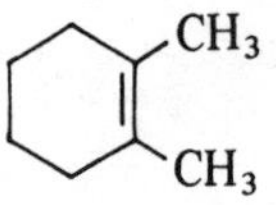 ring with two CH_3 groups on the double bond

(i) 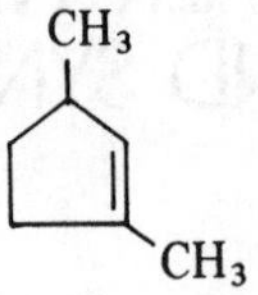 ring with CH_3 at the top and CH_3 on the double bond

(j) 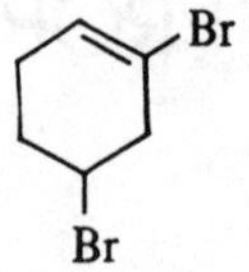 ring with Br on the double bond and a second Br

6.3 (a) (*Z*) - 1-Bromo-1-chloro-1-butene

(b) (*Z*) - 2-Bromo-1-chloro-1-iodopropene

(c) (*E*) - 3-Ethyl-4-methyl-2-pentene

(d) (*E*) - 1-Chloro-1-fluoro-2-methyl-1-butene

6.4 (a) C_4H_6 $CH_3CH_2C{\equiv}CH$ (1-Butyne); $CH_3C{\equiv}CCH_3$ (2-Butyne)

(b) C_5H_8 $CH_3CH_2CH_2C{\equiv}CH$ (1-Pentyne); $CH_3CH_2C{\equiv}CCH_3$ (2-Pentyne)

$CH_3CH(CH_3)C{\equiv}CH$ (3-Methyl-1-butyne)

(c) $CH_3CH_2CH_2CH_2C{\equiv}CH$ (1-Hexyne); $CH_3CH_2CH_2C{\equiv}CCH_3$ (2-Hexyne)

$CH_3CH_2C{\equiv}CCH_2CH_3$ (3-Hexyne); $CH_3CH(CH_3)CH_2C{\equiv}CH$ (4-Methyl-1-pentyne)

$CH_3C{\equiv}CCH(CH_3)CH_3$ (4-Methyl-2-pentyne); $HC{\equiv}CCH(CH_3)CH_2CH_3$ (3-Methyl-1-pentyne)

$HC{\equiv}CC(CH_3)_2CH_3$ (3, 3-Dimethyl-1-butyne)

6.5 (a) C_6H_{14} = formula of alkane
C_6H_{12} = formula of 2-hexene

H_2 = difference = 1 pair of hydrogen atoms
Index of hydrogen deficiency = 1

(b) C_6H_{14} = formula of alkane
C_6H_{12} = formula of methylcyclopentane

H_2 = difference = 1 pair of hydrogen atoms
Index of hydrogen deficiency = 1

(c) No, all isomers of C_6H_{12}, for example, have the same index of hydrogen deficiency.

(d) No

(e) C_6H_{14} = formula of alkane
C_6H_{10} = formula of 2-hexyne

H_4 = difference = 2 pairs of hydrogen atoms
Index of hydrogen deficiency = 2

(f) $C_{10}H_{22}$ (alkane)
$C_{10}H_{16}$ (compound)

H_6 = difference = 3 pairs of hydrogen atoms
Index of hydrogen deficiency = 3

The structural possibilities are thus

3 double bonds
1 double bond and one triple bond
2 double bonds and 1 ring
1 double bond and 2 rings
3 rings
1 triple bond and one ring

6.6 (a) $C_{15}H_{32}$ = formula of alkane
$C_{15}H_{24}$ = formula of zingiberene

H_8 = difference = 4 pairs of hydrogen atoms
Index of hydrogen deficiency = 4

(b) Since 1 mole of zingiberene absorbs 3 moles of hydrogen, one molecule of zingiberene must contain three double bonds. (We are told that molecules of zingiberene do not contain any triple bonds.)

(c) If a molecule of zingiberene has three double bonds and an index of hydrogen deficiency equal to 4, it must have one ring. (The structural formula for zingiberene can be found in Problem 22.2.)

Study Aid

More on Calculating the Index of Hydrogen Deficiency (IHD)

Calculating the index of hydrogen deficiency (IHD) for compounds other than hydrocarbons is relatively easy.

For compounds containing halogen atoms we simply count the halogen atoms as though they were hydrogen atoms. Consider a compound with the formula $C_4H_6Cl_2$. To calculate the IHD, we change the two chlorine atoms to hydrogen atoms, considering the formula as though it were C_4H_8. This formula has two hydrogen atoms fewer than the formula for a saturated alkane (C_4H_{10}), and this tells us that the compound has an IHD = 1. It could, therefore, have either one ring or one double bond. [We could tell which it has from a hydrogenation experiment: If the compound adds one molar equivalent of hydrogen (H_2) on catalytic hydrogenation at room temperature, then it must have a double bond; if it does not add hydrogen, then it must have a ring.]

For compounds containing oxygen we simply ignore the oxygen atoms and calculate the IHD from the remainder of the formula. Consider as an example a compound with the formula C_4H_8O. For the purposes of our calculation we consider the compound to be simply C_4H_8 and we calculate an IHD = 1. Again, this means that the compound contains either a ring or a double bond. Some structural possibilities for this compound are shown next. Notice that the double bond may be present as a carbon-oxygen double bond.

$CH_2{=}CHCH_2CH_2OH$ $\quad CH_3CH{=}CHCH_2OH \quad CH_3CH_2\overset{O}{\overset{\|}{C}}CH_3$

$CH_3CH_2CH_2\overset{O}{\overset{\|}{C}}H$ (tetrahydrofuran ring) (2-methyloxetane ring, CH_3) and so on

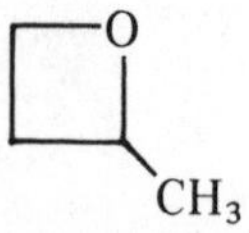

For compounds containing nitrogen atoms we ignore the nitrogen atoms and subtract one hydrogen for each nitrogen. For example, we treat a compound with the formula C_4H_9N as though it were C_4H_8, and again we get an IHD = 1. Some structural possibilities are the following:

$CH_2{=}CHCH_2CH_2NH_2 \quad CH_3CH{=}CHCH_2NH_2 \quad CH_3CH_2\overset{NH}{\overset{\|}{C}}CH_3$

$CH_3CH_2CH_2CH{=}NH$ (pyrrolidine ring, N–H) (2-methylazetidine ring, N–H, CH_3) and so on

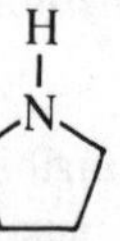

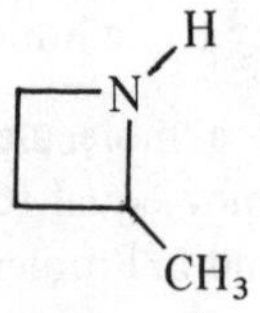

6.7 (a), (b)

$$\underset{\text{2-Methyl-1-butene (disubstituted)}}{CH_2{=}\overset{\overset{CH_3}{|}}{C}CH_2CH_3} \xrightarrow[Pt]{H_2} CH_3\overset{\overset{CH_3}{|}}{C}HCH_2CH_3 \qquad \Delta H^\circ = -28.5 \text{ kcal/mole}$$

$$\underset{\text{3-Methyl-1-butene (monosubstituted)}}{CH_3\overset{\overset{CH_3}{|}}{C}HCH{=}CH_2} \xrightarrow[Pt]{H_2} CH_3\overset{\overset{CH_3}{|}}{C}HCH_2CH_3 \qquad \Delta H^\circ = -30.3 \text{ kcal/mole}$$

$$\underset{\text{2-Methyl-2-butene (trisubstituted)}}{CH_3\overset{\overset{CH_3}{|}}{C}{=}CHCH_3} \xrightarrow[Pt]{H_2} CH_3\overset{\overset{CH_3}{|}}{C}HCH_2CH_3 \qquad \Delta H^\circ = -26.9 \text{ kcal/mole}$$

(c) Yes, because hydrogenation converts each alkene to the same product.

(d) $$\underset{\text{(trisubstituted)}}{CH_3\overset{\overset{CH_3}{|}}{C}{=}CHCH_3} > \underset{\text{(disubstituted)}}{CH_2{=}\overset{\overset{CH_3}{|}}{C}CH_2CH_3} > \underset{\text{(monosubstituted)}}{CH_3\overset{\overset{CH_3}{|}}{C}HCH{=}CH_2}$$

Notice that this predicted order of stability is confirmed by the heats of hydrogenation. 2-Methyl-2-butene evolves the least heat, therefore, it is the most stable; 3-methyl-1-butene evolves the most heat, therefore, it is the least stable.

(e) $CH_2{=}CHCH_2CH_2CH_3$ (1-Pentene); $(CH_3)(H)C{=}C(CH_2CH_3)(H)$ with CH_3 and CH_2CH_3 on the same side (*cis*-2-Pentene); $(CH_3)(H)C{=}C(H)(CH_2CH_3)$ with CH_3 and CH_2CH_3 on opposite sides (*trans*-2-Pentene)

1-Pentene   *cis*-2-Pentene   *trans*-2-Pentene

(f) Heats of combustion, because complete combustion would convert all of the alkenes to the same products. (All of these alkenes have the formula C_5H_{10}.)

$$C_5H_{10} + 7\tfrac{1}{2}\,O_2 \longrightarrow 5CO_2 + 5H_2O$$

6.8 (a) $(CH_3)(H)C{=}C(CH_2(CH_2)_2CH_3)(H)$ (cis) $>$ $CH_2{=}CH(CH_2)_4CH_3$

cis-2-Heptene (disubstituted) *More stable*   1-Heptene (monosubstituted) *Less stable*

(b)  $\underset{H}{\overset{CH_3}{}}\!\!>\!C{=}C\!<\!\!\underset{CH_2(CH_2)_2CH_3}{\overset{H}{}}$ > $\underset{H}{\overset{CH_3}{}}\!\!>\!C{=}C\!<\!\!\underset{H}{\overset{CH_2(CH_2)_2CH_3}{}}$

trans-2-Heptene
More stable

cis-2-Heptene
Less stable

(c) 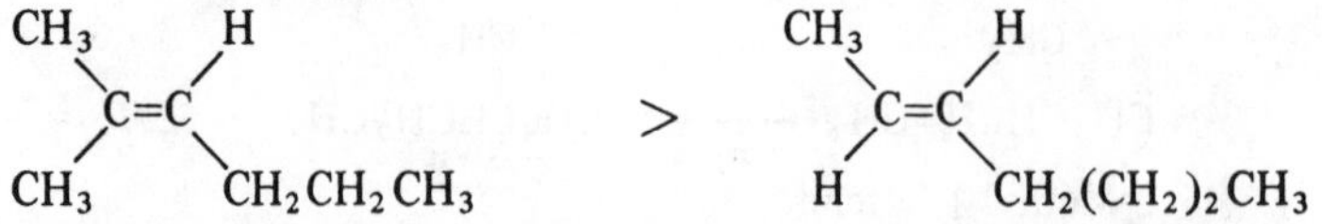 $\underset{CH_3}{\overset{CH_3}{}}\!\!>\!C{=}C\!<\!\!\underset{CH_2CH_2CH_3}{\overset{H}{}}$ > $\underset{H}{\overset{CH_3}{}}\!\!>\!C{=}C\!<\!\!\underset{CH_2(CH_2)_2CH_3}{\overset{H}{}}$

2-Methyl-2-hexene
(trisubstituted)
More stable

trans-2-Heptene
(disubstituted)
Less stable

(d) 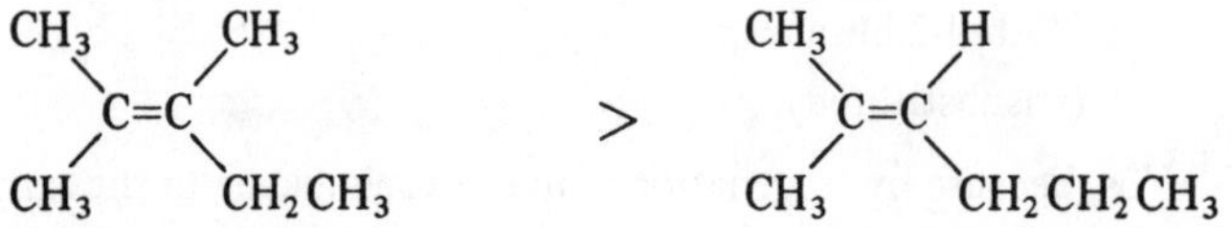 $\underset{CH_3}{\overset{CH_3}{}}\!\!>\!C{=}C\!<\!\!\underset{CH_2CH_3}{\overset{CH_3}{}}$ > $\underset{CH_3}{\overset{CH_3}{}}\!\!>\!C{=}C\!<\!\!\underset{CH_2CH_2CH_3}{\overset{H}{}}$

2,3-Dimethyl-2-pentene
(tetrasubstituted)
More stable

2-Methyl-2-hexene
(trisubstituted)
Less stable

6.9 You could use heats of hydrogenation to determine the relative stabilities of pairs (a) and (b). You would be required to use heats of combustion for pairs (c) and (d) because the members in pairs (c) and (d) give different alkanes on hydrogenation.

6.10 (a) 2-Butene, the more highly substituted alkene. (b) *trans*-2-Butene.

6.11 An anti periplanar transition state allows the molecule to assume the more stable staggered conformation;

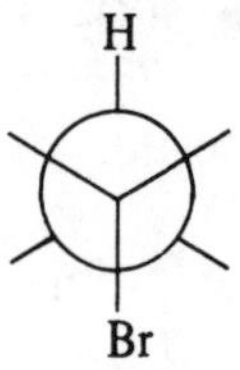

H

Br

whereas, a syn periplanar transition state requires the molecule to assume the less stable eclipsed conformation:

Br H

6.12 *cis*-1-Bromo-4-*tert*-butylcyclohexane can assume an anti periplanar transition state in which the bulky *tert*-butyl group is equatorial:

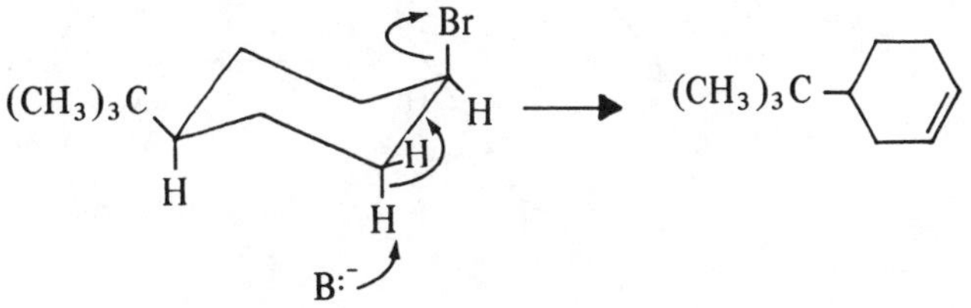

$$(CH_3)_3C\text{-cyclohexane (Br axial, anti H)} \xrightarrow{B:^-} (CH_3)_3C\text{-cyclohexene}$$

The conformation (above), because it is relatively stable, is assumed by most of the molecules present, and, therefore, the reaction is rapid.

On the other hand, for *trans*-1-bromo-4-*tert*-butylcyclohexane to assume an anti periplanar transition state, the molecule must assume a conformation in which the large *tert*-butyl group is axial:

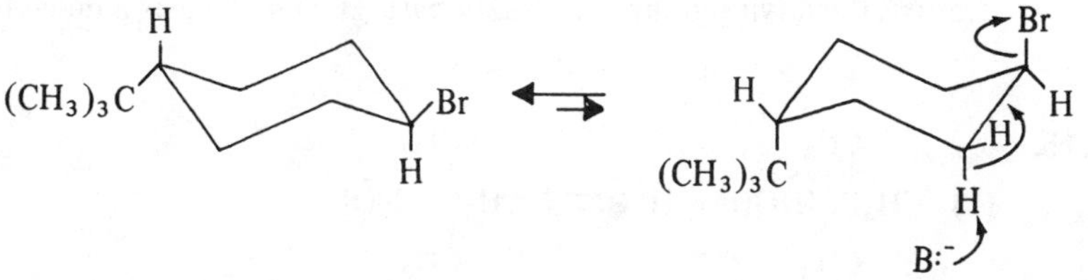

$$(CH_3)_3C\text{ (equatorial), Br (equatorial)} \rightleftharpoons (CH_3)_3C\text{ (axial), Br (axial)} \xrightarrow{B:^-}$$

Such a conformation is of high energy; therefore very few molecules assume this conformation. The reaction, consequently, is very slow.

6.13 (a) Anti periplanar elimination can occur in two ways with the cis isomer.

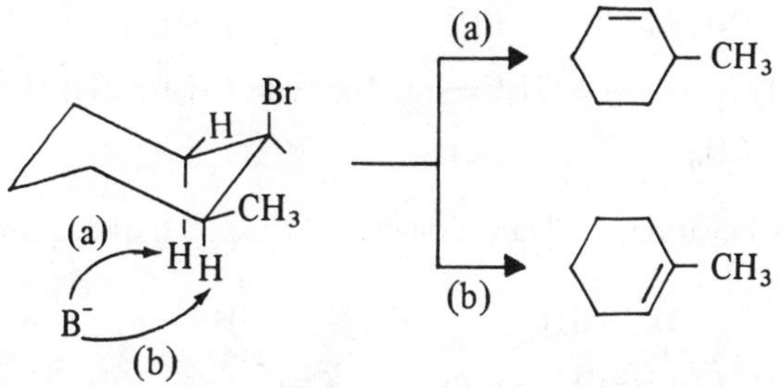

cis-1-Bromo-2-methylcyclohexane $\xrightarrow{(a)}$ 3-methylcyclohexene-type product (CH_3); $\xrightarrow{(b)}$ 1-methylcyclohexene (CH_3)

(major products)

(b) Anti periplanar elimination can occur in only one way with the trans isomer.

$$\text{(chair, Br axial; H and CH}_3\text{ on adjacent carbon; B}^- \text{ removes H)} \longrightarrow \text{1-methylcyclohexene (ring-CH}_3)$$

trans-Bromo-2-methylcyclohexane

6.14 (a) OH^-, a strong base and an extremely poor leaving group. (b) The acid catalyst reacts with the alcohol to form the protonated alcohol $R\overset{+}{O}H_2$. When this ion undergoes dehydration, the leaving group is a weakly basic H_2O molecule—a much better leaving group.

6.15

$$\text{(1)}\quad CH_3\overset{CH_3}{\underset{CH_3}{C}}CH_2OH + H^+ \rightleftharpoons CH_3\overset{CH_3}{\underset{CH_3}{C}}CH_2\overset{+}{O}H_2$$

$$\text{(2)}\quad CH_3\overset{CH_3}{\underset{CH_3}{C}}CH_2\overset{+}{O}H_2 \rightleftharpoons CH_3\overset{CH_3}{\underset{CH_3}{C}}\overset{+}{C}H_2 + H_2O$$

1° Carbocation

$$\text{(3)}\quad CH_3\overset{CH_3}{\underset{CH_3}{C}}\overset{+}{C}H_2 \longrightarrow H_3\overset{CH_3}{C}\text{—}CH_2\ (\text{bridging } \overset{\delta+}{CH_3}) \longrightarrow CH_3\overset{CH_3}{\underset{+}{C}}\text{—}CH_2\text{—}CH_3$$

1° Carbocation Transition state 3° Carbocation

$$\text{(4)}\quad CH_3\text{—}\overset{+}{\underset{CH_3}{C}}\text{—}\overset{H\ :OH_2}{CH}\text{—}CH_3 \longrightarrow \overset{CH_3}{\underset{CH_3}{\;}}C{=}C\overset{H}{\underset{CH_3}{\;}} + H_3\overset{+}{O}$$

2-Methyl-2-butene

6.16

$$CH_3\overset{CH_3}{\underset{CH_3}{C}}CH_2\text{—}I \xrightarrow{(-I^-)} CH_3\overset{CH_3}{\underset{CH_3}{C}}CH_2^+ \longrightarrow CH_3\overset{CH_3}{C}\text{—}\underset{H}{\overset{+}{C}}HCH_3 \xrightarrow{(-H^+)} CH_3\overset{CH_3}{C}{=}CHCH_3$$

6.17

Isoborneol $\xrightarrow[(-H_2O)]{H^+}$ carbocation $\longrightarrow$ rearranged carbocation $\xrightarrow{-H^+}$ Camphene

6.18 (a) $HC{\equiv}CH + :\ddot{N}H_2^- \rightleftharpoons HC{\equiv}C:^- + :NH_3$

Stronger acid	Stronger base	Weaker base	Weaker acid

(No appreciable amount of reactants are present at equilibrium.)

(b) $CH_2{=}CH_2 + :\ddot{N}H_2^- \rightleftharpoons CH_2{=}\ddot{C}H^- + :NH_3$

Weaker acid	Weaker base	Stronger base	Stronger acid

(No appreciable amount of products are present at equilibrium.)

(c) $CH_3CH_3 + :\ddot{N}H_2^- \rightleftharpoons CH_3\ddot{C}H_2^- + :NH_3$

Weaker acid	Weaker base	Stronger base	Stronger acid

(No appreciable amount of products are present at equilibrium.)

(d) $HC{\equiv}C:^- + CH_3CH_2\ddot{O}H \rightleftharpoons HC{\equiv}CH + CH_3CH_2\ddot{O}:^-$

Stronger base	**Stronger acid**	Weaker acid	Weaker base

(No appreciable amount of reactants are present at equilibrium.)

(e) $HC{\equiv}C:^- + H{-}\ddot{O}{:}(H) \rightleftharpoons HC{\equiv}CH + {}^-{:}\ddot{O}H$

Stronger base	Stronger acid	Weaker acid	Weaker base

(No appreciable amount of reactants are present at equilibrium.)

6.19

$$CH_3{-}C(CH_3)_2{-}C{\equiv}CH + NaNH_2 \longrightarrow CH_3{-}C(CH_3)_2{-}C{\equiv}C:^-Na^+ + NH_3$$

$$CH_3{-}C(CH_3)_2{-}C{\equiv}C:^-Na^+ \xrightarrow{CH_3CH_2Br} CH_3{-}C(CH_3)_2{-}C{\equiv}C{-}CH_2CH_3$$

A reaction between $CH_3CH_2C{\equiv}C:^- \overset{+}{Na}$ and $CH_3-\underset{CH_3}{\overset{CH_3}{C}}-Br$ would result in elimination to produce $CH_2{=}\underset{CH_3}{C}-CH_3$ + $CH_3CH_2C{\equiv}CH$.

6.20 (a) $CH_3\overset{O}{\overset{\|}{C}}CH_3 \xrightarrow[0°C]{PCl_5} CH_3CCl_2CH_3 + POCl_3 \xrightarrow[\text{mineral oil, heat; (2) } H^+]{(1)\ 3NaNH_2,} CH_3C{\equiv}CH$

(b) $CH_3CH_2CHBr_2 \xrightarrow[\text{mineral oil, heat; (2) } H^+]{(1)\ 2NaNH_2,} CH_3C{\equiv}CH$

(c) $CH_3CHBrCH_2Br \xrightarrow{\text{[same as (b)]}} CH_3C{\equiv}CH$

(d) $CH_3CH{=}CH_2 \xrightarrow[CCl_4]{Br_2} CH_3\underset{Br}{C}HCH_2Br \xrightarrow{\text{[same as (b)]}} CH_3C{\equiv}CH$

6.21 (a) One must use the lower number to designate the location of the double bond.

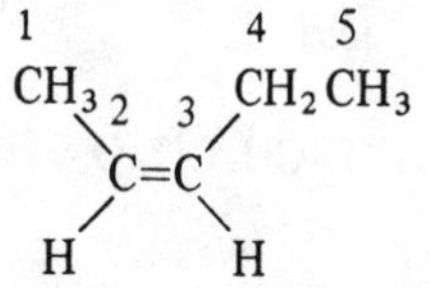

$\overset{1}{CH_3}$(2)(3)$\overset{4\ \ 5}{CH_2CH_3}$ / H, H on C=C — *cis*-2-Pentene (*not cis*-3 pentene)

(b) One must select the longest chain as the base name.

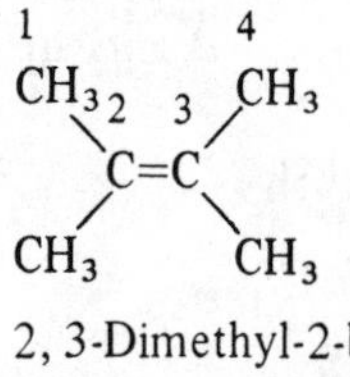

$\overset{1}{CH_3}$(2)(3)$\overset{4}{CH_3}$ / CH_3, CH_3 on C=C

2, 3-Dimethyl-2-butene
(*not* 1,1,2,2-tetramethylethene)

(c) One must number the ring so as to give the carbon atoms of the double bond numbers 1 and 2 *and to give the substituent the lower number.*

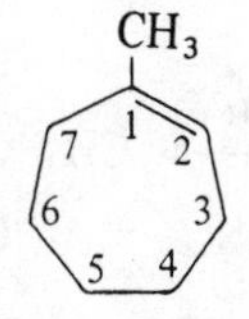

CH_3 on C1 of seven-membered ring (1–7), double bond C1=C2

1-Methylcycloheptene
(*not* 2-methylcycloheptene)

(d) One must select the longest chain.

$$\overset{1}{CH_3}\overset{2}{CH}{=}\overset{3}{CH}\overset{4}{CH_2}\overset{5}{CH_2}\overset{6}{CH_2}\overset{7}{CH_2}\overset{8}{CH_3}$$

2-Octene
(*not* 1-methylheptene)

(e) One must number the chain from the other end. This choice gives the double bond the same number but it gives the methyl group a *lower* number.

$$\underset{4}{CH_3}\underset{3}{CH}{=}\underset{2}{\overset{CH_3}{\overset{|}{C}}}\underset{1}{CH_3}$$

2-Methyl-2-butene
(*not* 3-methyl-2-butene)

(f) One must number the ring the other way. This choice gives the substituents lower numbers while retaining positions 1 and 2 for the double bond.

Cl, Cl on ring positions 3 and 4; double bond C1=C2; 5

3,4-Dichlorocyclopentene
(*not* 4,5-dichlorocyclopentene)

6.22 (a) 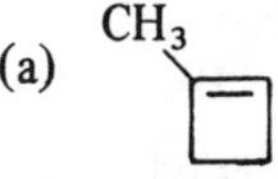

(b) (cyclopentene ring) $-CH_3$

(c) $CH_3\underset{CH_3}{\underset{|}{C}}{=}\underset{CH_3}{\underset{|}{C}}CH_2CH_3$

(d) $(CH_3)(H)C{=}C(H)(CH_2CH_2CH_3)$ — CH_3 and H on top; H and $CH_2CH_2CH_3$ on bottom

(e) $(CH_3CH_2)(H)C{=}C(CH_2CH_2CH_3)(H)$ — CH_3CH_2 and $CH_2CH_2CH_3$ on top; H and H on bottom

(f) $CH_2{=}CHCCl_3$ (g) $CH_2{=}\underset{CH_3}{\underset{|}{C}}CH_3$ (h) $CH_3CH{=}CH_2$

(i) $CH_2{=}CHCH_2\underset{\text{cyclopentyl}}{\underset{|}{CH}}CH_3$ (j) $CH_2{=}CH-$(cyclopropyl)

6.23

(a) $CH_2{=}CHCH_2CH_2CH_3$ — 1-Pentene

$(CH_3)(H)C{=}C(CH_2CH_3)(H)$ — CH_3 and CH_2CH_3 on top; H and H on bottom — *cis*-2-Pentene

$(CH_3)(H)C{=}C(H)(CH_2CH_3)$ — CH_3 and H on top; H and CH_2CH_3 on bottom — *trans*-2-Pentene

$(CH_3)_2C{=}CHCH_3$ — 2-Methyl-2-butene

$CH_2{=}C(CH_3)CH_2CH_3$ — 2-Methyl-1-butene

$CH_2{=}CH{-}CH(CH_3)CH_3$ — 3-Methyl-1-butene

(b) $CH_2{=}CHCH_2CH_2CH_2CH_3$ — 1-Hexene

$CH_3CH{=}CHCH_2CH_2CH_3$ (CH₃ and CH₂CH₂CH₃ cis) — *cis*-2-Hexene

$CH_3CH{=}CHCH_2CH_2CH_3$ (CH₃ and CH₂CH₂CH₃ trans) — *trans*-2-Hexene

$CH_3CH_2CH{=}CHCH_2CH_3$ (cis) — *cis*-3-Hexene

$CH_3CH_2CH{=}CHCH_2CH_3$ (trans) — *trans*-3-Hexene

$CH_2{=}C(CH_3)CH_2CH_2CH_3$ — 2-Methyl-1-pentene

$CH_2{=}CHCH(CH_3)CH_2CH_3$ — 3-Methyl-1-pentene

$CH_2{=}CHCH_2CH(CH_3)CH_3$ — 4-Methyl-1-pentene

$CH_3C(CH_3){=}CHCH_2CH_3$ — 2-Methyl-2-pentene

$CH_3(H)C{=}C(CH_2CH_3)(CH_3)$ — *trans*-3-Methyl-2-pentene

$CH_3(H)C{=}C(CH_3)(CH_2CH_3)$ — *cis*-3-Methyl-2-pentene

$CH_3CH{=}CHCH(CH_3)CH_3$ (cis) — *cis*-4-Methyl-2-pentene

$CH_2{=}C(CH_2CH_3)_2$ — 2-Ethyl-1-butene

$CH_3CH{=}CHCH(CH_3)CH_3$ (trans) — *trans*-4-Methyl-2-pentene

$CH_2{=}C(CH_3)CH(CH_3)CH_3$ — 2,3-Dimethyl-1-butene

$CH_2{=}CHC(CH_3)_2CH_3$ — 3, 3-Dimethyl-1-butene

$(CH_3)_2C{=}C(CH_3)_2$ — 2, 3-Dimethyl-2-butene

(c) C_5H_{10}: cyclopentane; methylcyclobutane (CH_3); 1,1-dimethylcyclopropane (H_3C, CH_3); *cis*-1,2-dimethylcyclopropane (H_3C, CH_3, H, H); *trans*-1,2-dimethylcyclopropane (H_3C, H, H, CH_3); ethylcyclopropane (CH_2CH_3)

C_6H_{12}

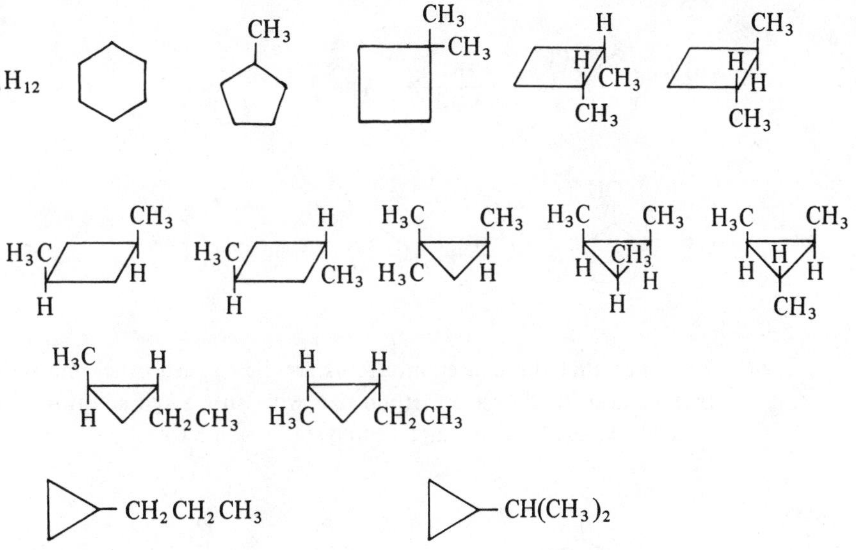

CH₃ CH₃ CH₃ H CH₃ H CH₃ H H CH₃ CH₃

CH₃ H H₃C CH₃ H₃C CH₃ H₃C CH₃ H₃C H₃C H CH₃ H₃C H CH₃ H CH₃ H H H H H CH₃

H₃C H H H H CH₂CH₃ H₃C CH₂CH₃

—$CH_2CH_2CH_3$ —$CH(CH_3)_2$

6.24 (a) 1,3-Dimethylcyclohexene (b) 2-Ethyl-1-pentene (c) 2-Ethyl-1-pentene
(d) 1-Ethyl-2-methylcyclopentene

6.25

(a) $CH_3CH_2CH_2Cl \xrightarrow[(CH_3)_3COH]{(CH_3)_3CONa} CH_3CH{=}CH_2$

(b) $CH_3\underset{\displaystyle Cl}{\underset{|}{C}}HCH_3 \xrightarrow[CH_3CH_2OH]{CH_3CH_2ONa} CH_3CH{=}CH_2$

(c) $CH_3CH_2CH_2OH \xrightarrow{H^+,\ heat} CH_3CH{=}CH_2$

(d) $CH_3\underset{\displaystyle OH}{\underset{|}{C}}HCH_3 \xrightarrow{H^+,\ heat} CH_3CH{=}CH_2$

(e) $CH_3\underset{\displaystyle Br}{\underset{|}{C}}HCH_2Br \xrightarrow[\text{or NaI acetone}]{Zn,\ CH_3CO_2H} CH_3CH{=}CH_2$

6.26

(a) 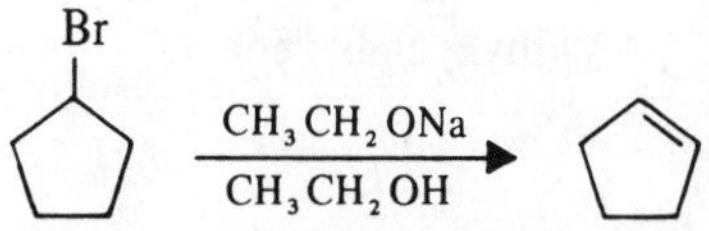 Br $\xrightarrow[CH_3CH_2OH]{CH_3CH_2ONa}$

(b) 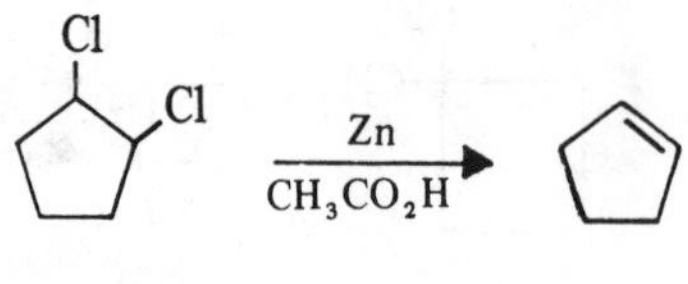 Cl, Cl (dichlorocyclopentane) $\xrightarrow[CH_3CO_2H]{Zn}$ cyclopentene

(c) 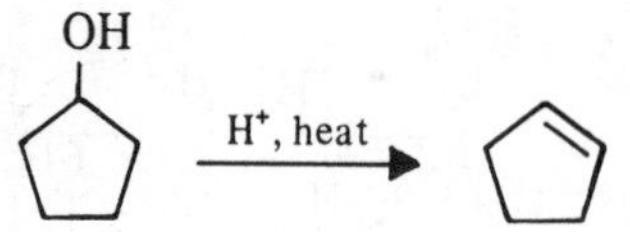 OH (cyclopentanol) $\xrightarrow{H^+, \text{heat}}$ cyclopentene

6.27 We notice that the deuterium atoms are cis to each other and we conclude, therefore, that we need to choose a method that will cause a syn addition of deuterium. One way would be to use D_2 and a metal catalyst (Section 6.6)

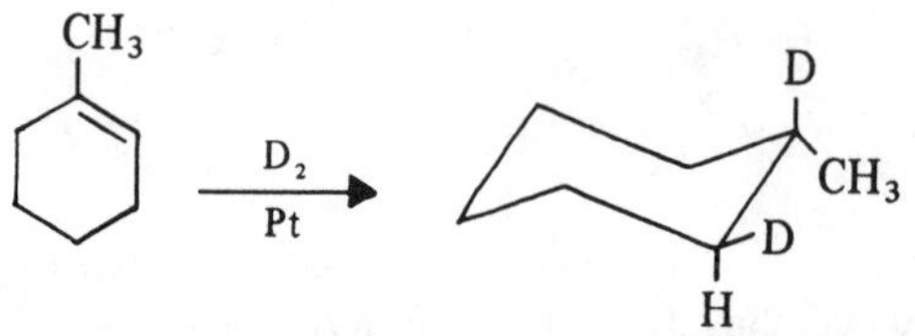

CH_3 $\xrightarrow[Pt]{D_2}$ D, CH_3, D, H

6.28 Dehydration of *trans*-2-methylcyclohexanol proceeds through the formation of a carbocation (through an El reaction of the protonated alcohol) and leads preferentially to the more stable alkene. 1-Methylcyclohexene (below) is more stable than 3-methylcyclohexene (the minor product of the dehydration) because its double bond is more highly substituted.

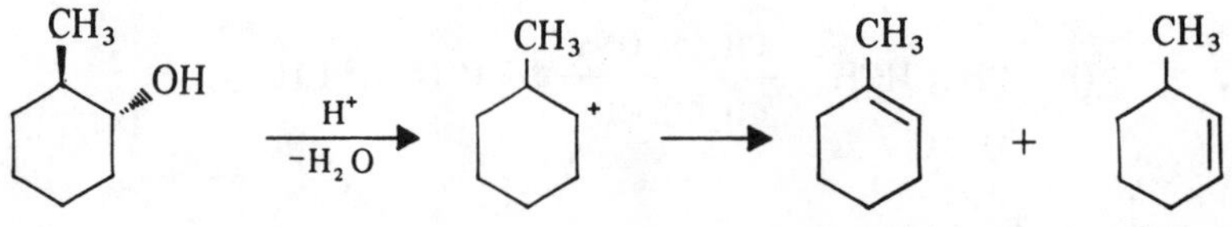

CH_3, OH $\xrightarrow[-H_2O]{H^+}$ CH_3, + $\longrightarrow$ CH_3 + CH_3

(major) Trisubstituted double bond

(minor) Disubstituted double bond

Dehydrohalogenation of *trans*-1-bromo-2-methylcyclohexane is an E2 reaction and must proceed through an anti periplanar transition state. Such a transition state is possible only for the elimination leading to 3-methylcyclohexene (cf. Problem 6.13).

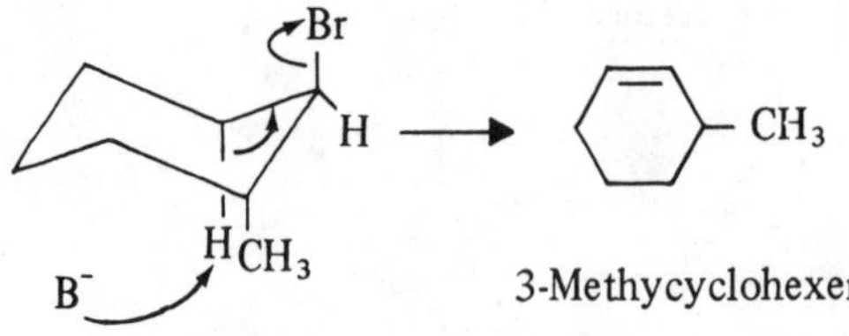

Br, H, H, CH_3, B^- $\longrightarrow$ CH_3

3-Methycyclohexene

6.29 (a) 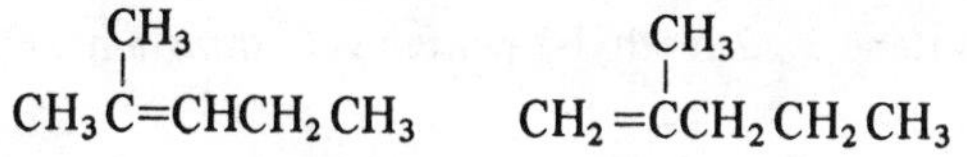 $CH_3\overset{\overset{CH_3}{|}}{C}{=}CHCH_2CH_3$ (major) $CH_2{=}\overset{\overset{CH_3}{|}}{C}CH_2CH_2CH_3$ (minor)

(b) 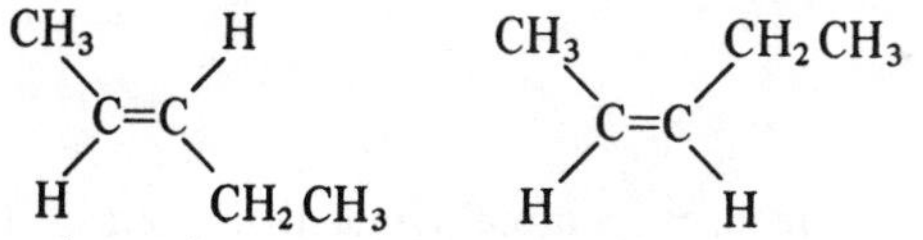 $\underset{H}{\overset{CH_3}{}}C{=}C\underset{CH_2CH_3}{\overset{H}{}}$ (major) $\underset{H}{\overset{CH_3}{}}C{=}C\underset{H}{\overset{CH_2CH_3}{}}$ (minor) $CH_2{=}CHCH_2CH_2CH_3$ (minor)

(c) $CH_3\overset{\overset{CH_3}{|}}{C}{=}CHCH_2CH_3$ (major) $CH_2{=}\overset{\overset{CH_3}{|}}{C}CH_2CH_2CH_3$ (minor)

(d) 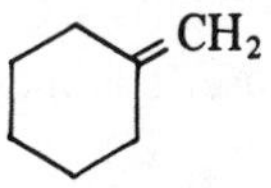 cyclohexane ring $={}CH_2$ (e) 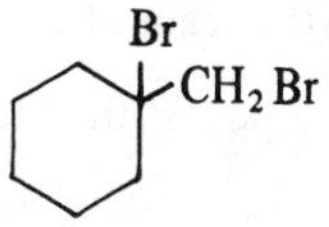 cyclohexane ring with Br and CH_2Br on the same carbon

(f) 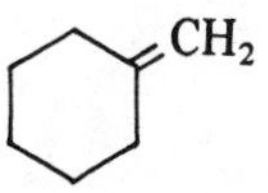 cyclohexane ring $={}CH_2$ (g) 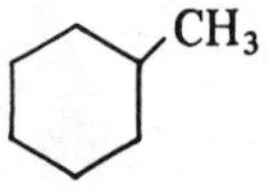 cyclohexane ring $-CH_3$

6.30 $CH_2{=}CHCH_2CH_2CH_2CH_3 + H_2 \xrightarrow{Pt} CH_3CH_2CH_2CH_2CH_2CH_3$ (colorless)

Cyclohexane does not react with H_2 and a catalyst. The subject of simple chemical tests will be treated fully in Chapter 7 (Section 7.19).

6.31 (a) No (b) No

(c) Yes $\underset{H}{\overset{CH_3}{}}C{=}C\underset{H}{\overset{CH_2CH_2CH_2CH_3}{}}$ $\underset{H}{\overset{CH_3}{}}C{=}C\underset{CH_2CH_2CH_2CH_3}{\overset{H}{}}$ (d) No

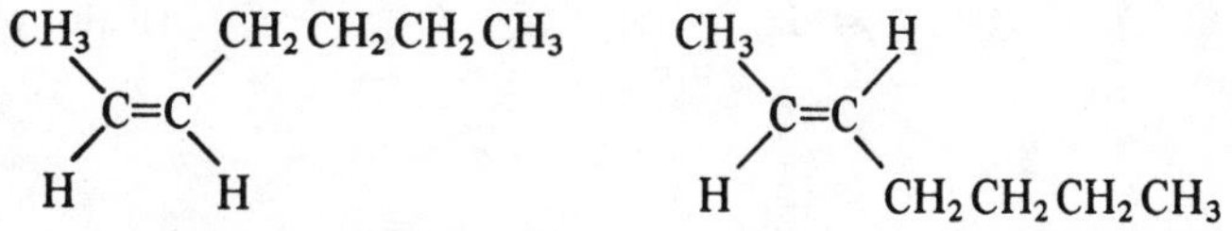

(e) Yes $\underset{Cl}{\overset{H}{}}C{=}C\underset{CH_2CH_3}{\overset{H}{}}$ $\underset{Cl}{\overset{H}{}}C{=}C\underset{H}{\overset{CH_2CH_3}{}}$ (f) No

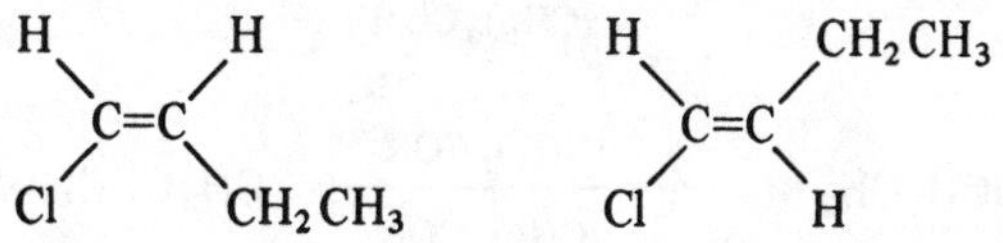

(g) No (h) No (i) Yes H, CH_3 on C $=$ cyclopropane ring with CH_3; H_3C, H on C $=$ cyclopropane ring with CH_3

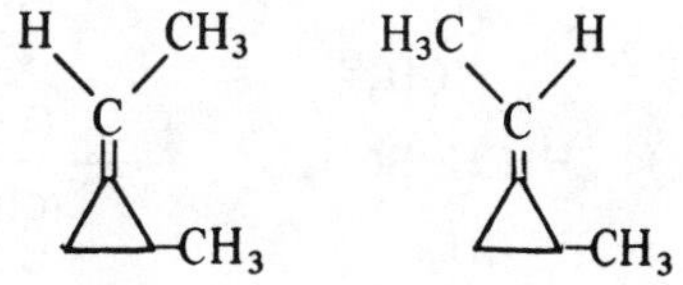

6.32 (a) 2,3-Dimethyl-2-butene > 2-methyl-2-pentene > *trans*-3-hexene > *cis*-2-hexene > 1-hexene.

(b) The only alkenes whose relative stabilities could be measured by comparative heats of hydrogenation are those that yield the same hydrogenation produce; that is, *trans*-3-hexene, 1-hexene, *cis*-2-hexene all yield hexane on hydrogenation.

6.33 Although trans molecules are usually more stable than their cis isomers, in the case of cyclooctene, the trans isomer is probably more strained than the cis isomer because the ring is too small to allow a strain-free trans configuration. Therefore we would expect the trans isomer to have the higher heat of hydrogenation.

6.34 (a) Cis-trans isomerization caused by rupture of the π bond.

(b) Equilibrium should favor the trans isomer because it is more stable than the cis isomer.

6.35

(a) $CH_3CH{=}C(CH_3)CH_3$ (major) + $CH_2{=}CHCH(CH_3)CH_3$

(b) $CH_3CH_2C(CH_3){=}CH_2$ (c) $CH_3CH{=}CHCH_2CH_3$ (trans predominates)

(d) 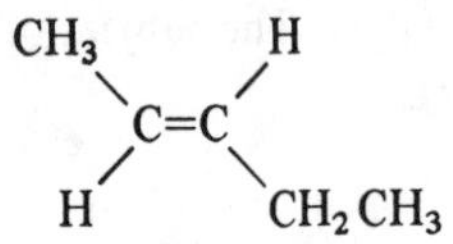 (*trans*) $CH_3(H)C{=}C(H)CH_2CH_3$ (major) + (*cis*) $CH_3(H)C{=}C(H)CH_2CH_3$ + $CH_2{=}CHCH_2CH_2CH_3$

(e) 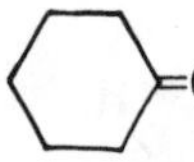 methylenecyclohexane (cyclohexane$=CH_2$) (f) 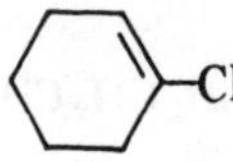 1-methylcyclohexene (cyclohexene–CH_3) (major product) + methylenecyclohexane (cyclohexane$=CH_2$)

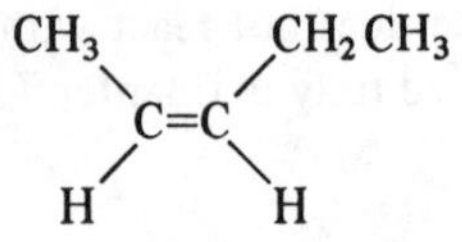

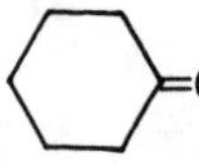

6.36 (a) $$CH_3CH_2CH_2CH_2CH_2Br \xrightarrow[(CH_3)_3COH]{(CH_3)_3COK} CH_3CH_2CH_2CH{=}CH_2$$

(b) $$CH_3CH(CH_3)CH_2CH_2Br \xrightarrow[(CH_3)_3COH]{(CH_3)_3COK} CH_3CH(CH_3)CH{=}CH_2$$

(c) $$CH_3CH(CH_3)CH(CH_3)CH_2Br \xrightarrow[(CH_3)_3COH]{(CH_3)_3COK} CH_3CH(CH_3)C(CH_3){=}CH_2$$

(d) 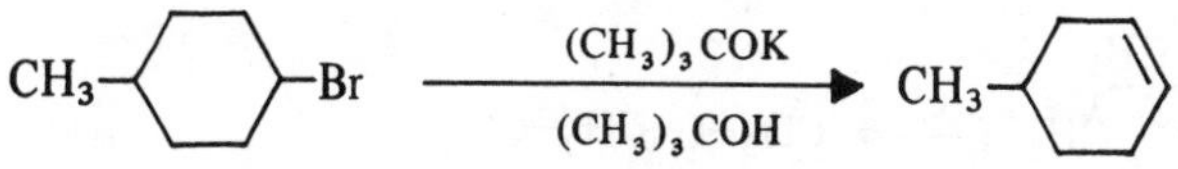 CH_3–(cyclohexyl)–Br $\xrightarrow[(CH_3)_3COH]{(CH_3)_3COK}$ CH_3–(cyclohexenyl)

(e) 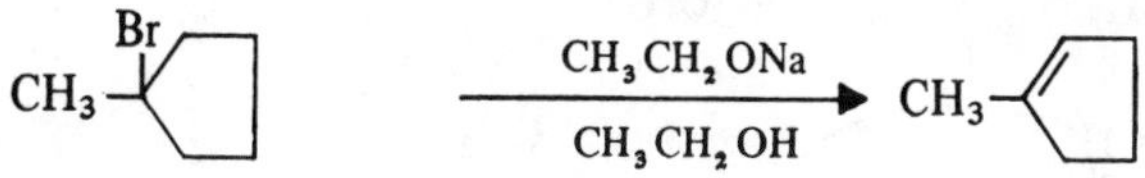 CH_3–C(Br)(cyclopentyl) $\xrightarrow[CH_3CH_2OH]{CH_3CH_2ONa}$ CH_3–(cyclopentenyl)

6.37

(a) 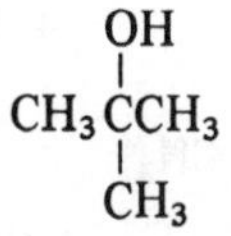 $CH_3C(OH)(CH_3)CH_3$ or $CH_3CH(CH_3)CH_2OH$

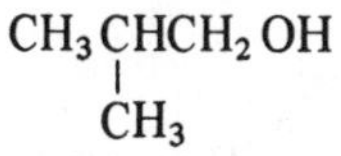

(c) 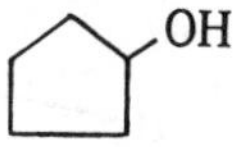 cyclopentyl–OH

(b) 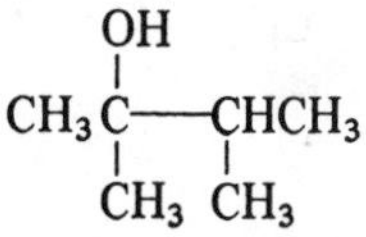 $CH_3C(OH)(CH_3)$—$CH(CH_3)CH_3$ or $CH_3C(CH_3)(CH_3)$—$CH(OH)$–CH_3

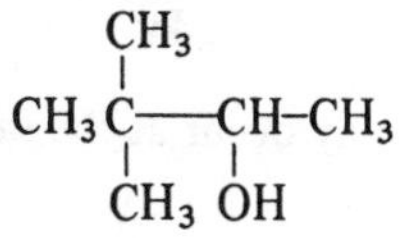

(d) 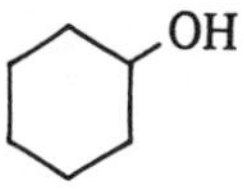 cyclohexyl–OH

(e) 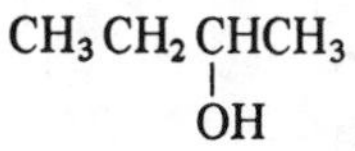 $CH_3CH_2CH(OH)CH_3$

(f) 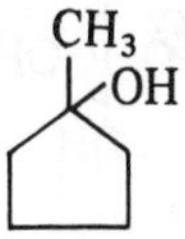 1-methylcyclopentanol (CH_3, OH on the same ring carbon)

6.38

$CH_3C(OH)(CH_3)CH_2CH_3$ > $CH_3CH(OH)CH(CH_3)CH_3$ > $CH_3CH(CH_3)CH_2CH_2OH$

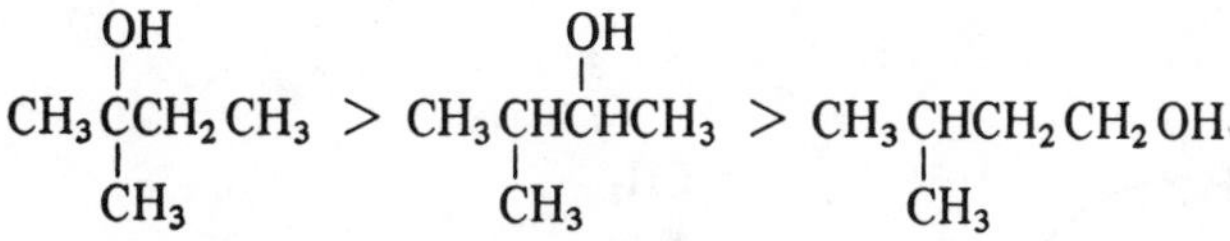

The order of reactivity is dictated by the order of stability of the intermediate carbocations: tertiary > secondary > primary.

6.39 (a) *cis*-1, 2-Dimethylcyclopentane

(b) *cis*-1, 2-Dimethylcyclohexane

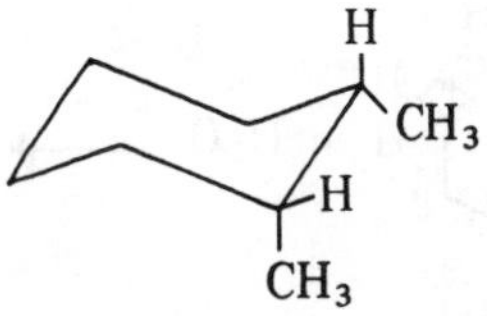

(chair: H, CH_3; H, CH_3)

6.40

(a) (1) 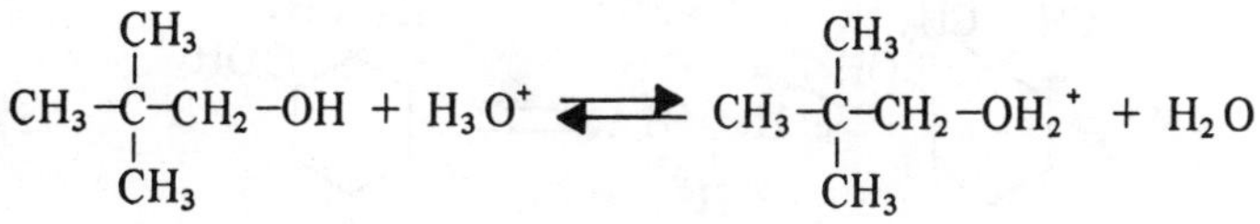 CH_3–$C(CH_3)_2$–CH_2–OH + H_3O^+ $\rightleftharpoons$ CH_3–$C(CH_3)_2$–CH_2–OH_2^+ + H_2O

(2) CH_3–$C(CH_3)_2$–CH_2–OH_2^+ ⟶ CH_3–$C(CH_3)_2$–CH_2^+ + H_2O

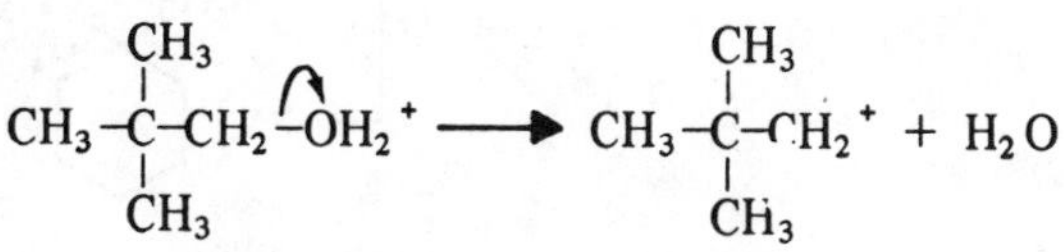

(3) $CH_3-\overset{CH_3}{\underset{CH_3}{C}}-CH_2^+ \longrightarrow CH_3-\overset{+}{\underset{CH_3}{C}}-CH_2-CH_3$

(4) $CH_3-\overset{+}{\underset{CH_3}{C}}-\overset{H}{CH}-CH_3 + :\ddot{O}H_2 \longrightarrow (CH_3)_2C{=}CHCH_3$ (more substituted alkene) $+ H_3O^+$

(4a) $\overset{H}{CH_2}-\overset{+}{\underset{CH_3}{C}}-CH_2-CH_3 + :\ddot{O}H_2 \longrightarrow CH_2{=}C(CH_2CH_3)(CH_3)$ (less substituted alkene) $+ H_3O^+$

Steps 2 and 3 may occur at the same time

(b) 1,4-dimethylcyclohexanol (CH₃, $\ddot{O}H$) $+ H-O-P(=O)(OH)-OH \rightleftarrows$ protonated alcohol (CH₃, $\overset{+}{O}H_2$) $+ H_2PO_4^- \longrightarrow$ 1,4-dimethylcyclohexyl cation $+ H_2O$

(cation, $H-C(H)(H)$ adjacent to C⁺) $+ H_2\ddot{O}: \longrightarrow$ methylenecyclohexane (CH_2, CH_3) (less substituted alkene) $+ H_3O^+$

(cation, ring H adjacent to C⁺) $+ H_2\ddot{O}: \longrightarrow$ 1,4-dimethylcyclohexene (CH_3, CH_3) (more substituted alkene) $+ H_3O^+$

(c) 2,2-dimethylcyclohexanol (CH_3 CH_3, $\ddot{O}H$) $+ H-\overset{+}{\ddot{O}}-H$ (H) $\rightleftarrows$ protonated alcohol (CH_3 CH_3, OH_2^+) $+ H_2O$

$-H_2O \longrightarrow$ 2,2-dimethylcyclohexyl cation (CH_3 CH_3, +) $\longrightarrow$ 1,2-dimethylcyclohexyl cation (CH_3, +, CH_3)

CH_3 CH_3 + :Ö–H (H) ⟶ CH_3 CH_3 + H_3O^+ (most substituted alkene)

H–C–H, CH_3 + :Ö–H (H) ⟶ CH_2 CH_3 + H_3O^+

CH_3 CH_3 H H + :Ö–H (H) ⟶ CH_3 CH_3 + H_3O^+

(less substituted alkenes)

6.41 Cholesterol $\xrightarrow[CHCl_3]{Br_2}$ (crude) dibromide (Br, Br) $\xrightarrow{\text{crystallization}}$ (pure) dibromide (Br, Br) $\xrightarrow[C_2H_5OH]{Zn}$ Cholesterol + $ZnBr_2$

6.42 (a) Caryophyllene has the same molecular formula as zingiberene (Problem 6.6), thus it, too, has an index of hydrogen deficiency equal to 4. That 1 mole of caryophyllene absorbs 2 moles of hydrogen on catalytic hydrogentation indicates the presence of two double bonds per molecule.

(b) Caryophyllene molecules must also have two rings. (See Problem 22.2 for the structure of caryophyllene.)

6.43 (a) $C_{30}H_{62}$ = formula of alkane
$C_{30}H_{50}$ = formula of squalene

H_{12} = difference = 6 pairs of hydrogen atoms

Index of hydrogen deficiency = 6

(b) Molecules of squalene contain six double bonds.

(c) Squalene molecules contain no rings. (See Problem 22.2 for the structural formula of squalene.)

6.44 (a) We are given (Section 6.9A) the following heats of hydrogenation:

cis-2-Butene + $H_2 \xrightarrow{Pt}$ butane $\Delta H° = -28.6$ kcal/mole

trans-2-Butene + $H_2 \xrightarrow{Pt}$ butane $\Delta H° = -27.6$ kcal/mole

thus for

cis-2-Butene ⟶ *trans*-2-butene $\Delta H° = -1.0$ kcal/mole

(b) Converting *cis*-2-butene into *trans*-2-butene involves breaking the π bond. Therefore we would expect the energy of activation to be at least as large as the π-bond strength, that is, at least 63 kcal/mole.

(c)

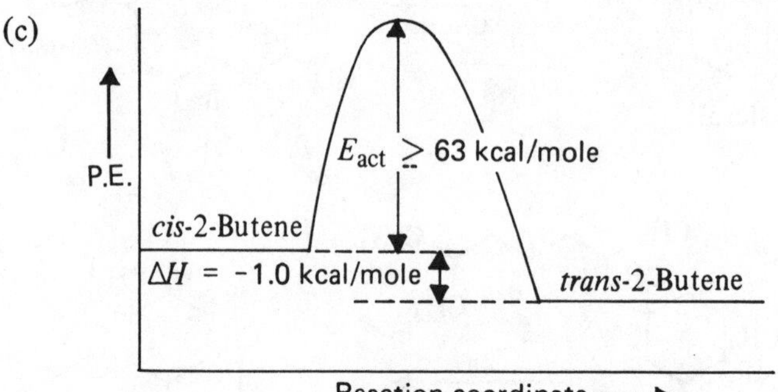

6.45

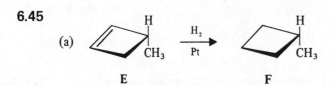

(a) H, CH_3 $\xrightarrow[Pt]{H_2}$ H, CH_3

E F

Optically active (the enantiomeric form is an equally valid answer)

Optically inactive and nonresolvable

(b) $CH_3CH_2(H)C{=}C{=}C(CH_3)(H)$ $\xrightarrow[Pt]{H_2}$ $CH_3CH_2CH_2CH_2CH_2CH_3$

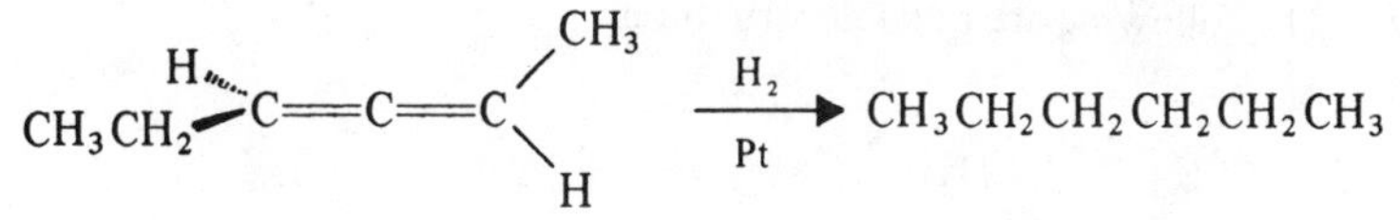

G

Optically active
(the enantiomeric form is
an equally valid answer)

H

Optically inactive and
nonresolvable

6.46 That **I** and **J** rotate plane-polarized light in the same direction tells us that **I** and **J** are not enantiomers of each other. Thus, the following are possible structures for **I**, **J**, and **K**. (The enantiomers of **I**, **J**, and **K** would form another set of structures, and other answers are possible as well.)

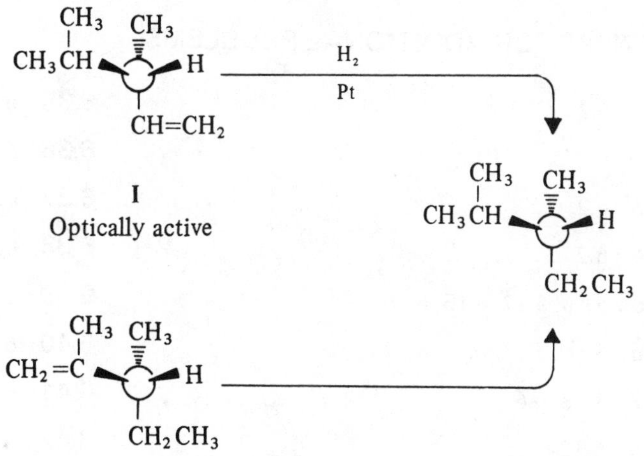

CH_3, CH_3, CH_3CH, H, $CH{=}CH_2$ $\xrightarrow[Pt]{H_2}$ K

I
Optically active

CH_3, CH_3, CH_3CH, H, CH_2CH_3

K
Optically
active

CH_3, CH_3, $CH_2{=}C$, H, CH_2CH_3 → K

J
Optically active

6.47 The following are possible structures:

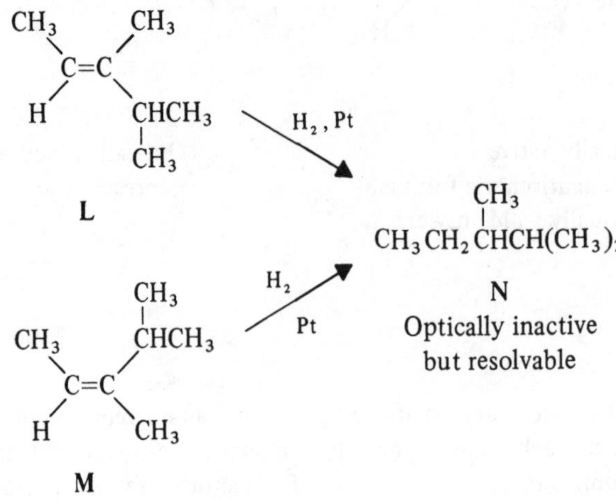

L: $CH_3(H)C{=}C(CH_3)CH(CH_3)_2$ (CH₃ and H on one carbon; CH₃ and CHCH₃ with CH₃ on the other)

M: $CH_3(H)C{=}C(CH_3)CH(CH_3)_2$ (CH₃ and H on one carbon; CHCH₃ with CH₃ and CH₃ on the other)

L —H_2, Pt→ N

M —H_2, Pt→ N

N: $CH_3CH_2CH(CH_3)CH(CH_3)_2$

Optically inactive but resolvable

(other answers are possible as well)

SECTION REFERENCES FOR ADDITIONAL PROBLEMS

6.21 6.2
6.22 6.2
6.23 6.2, 6.5
6.24 6.2
6.25 5.16, 5.17, 6.13, 6.16
6.26 5.16, 5.17, 6.13, 6.17
6.27 3.18, 6.6
6.28 6.12-6.15
6.29 6.6, 6.12-6.16
6.30 6.21B, 6.6
6.31 6.2, 6.9
6.32 6.9
6.33 6.9
6.34 6.9
6.35 6.12
6.36 6.12
6.37 6.13, 6.14, 6.15
6.38 6.13
6.39 6.6
6.40 6.14, 6.15
6.41 6.16
6.42 6.8
6.43 6.8
6.44 6.9A
6.45 4.9, 6.5, 6.6
6.46 4.9, 6.5, 6.6
6.47 4.9, 6.5, 6.6

SELF-TEST

6.1 Give an acceptable name for each of the following compounds.

(a) 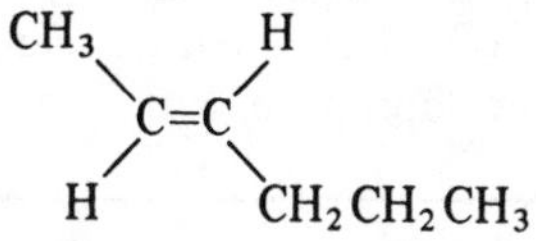 $CH_3(H)C{=}C(H)CH_2CH_2CH_3$ (CH$_3$ and H on one carbon; H and CH$_2$CH$_2$CH$_3$ on the other, H and CH$_3$ trans-positioned as drawn)

(b) 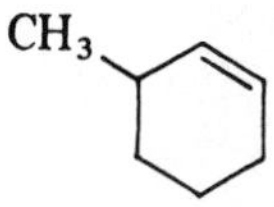 3-methylcyclohexene structure: CH_3 on a cyclohexene ring carbon adjacent to the double bond

6.2 Supply the formula of the missing reactant(s) or *major* organic product.

(a) 2,3-Dibromopentane + [ ] $\longrightarrow$ 2-pentene

(b) 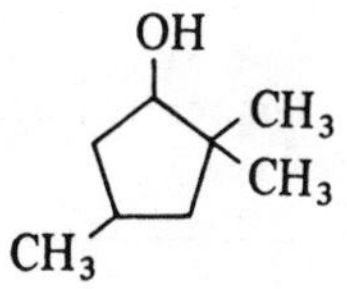 Cyclopentanol bearing OH on C-1, two CH_3 groups on C-2, and a CH_3 on C-4 $\xrightarrow{H_2SO_4,\ heat}$ [ ]

(c) [ ] $\xrightarrow[heat]{C_2H_5ONa/C_2H_5OH}$ $CH_3CH(CH_3)CH{=}CH_2$

(d) $CH_3C(CH_3)(OH)CH_2CH_2CH_3$ $\xrightarrow[heat]{H_2SO_4}$ [ ]

6.3 Which of the following compounds is capable of exhibiting cis-trans isomerism? **(Give the letters only.)**

(a) $CH_2{=}CHCH_2CH_3$ (b) $FCH{=}CHF$ (c) $F_2C{=}CH_2$ (d) 1-Chloro-2-methylpropene (e) 2-Pentene (f) 1,3-Dimethylcyclobutane (g) 1,2-Dimethylcyclobutane (h) 1,1-Dimethylcyclopropane

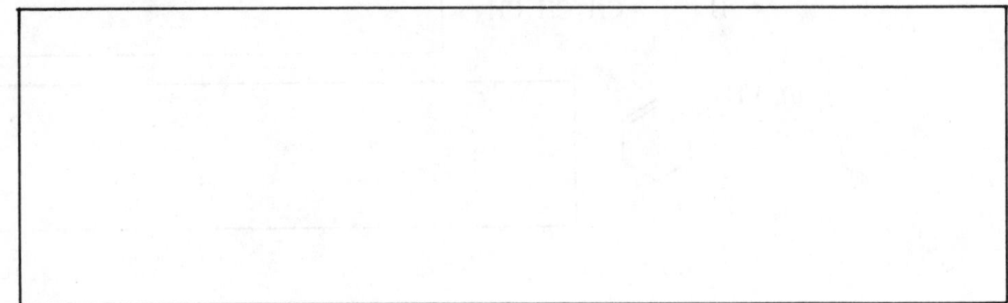

6.4 Arrange the following alkenes in order of increasing stability. Label the *most* stable 3, the least 1, and so on.

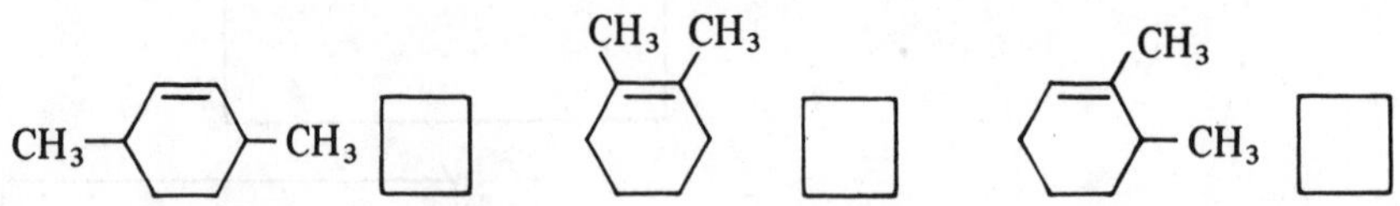

6.5 Compound A whose molecular formula is C_5H_8 undergoes hydrogenation to give C_5H_{10}. A possible structure for compound A is

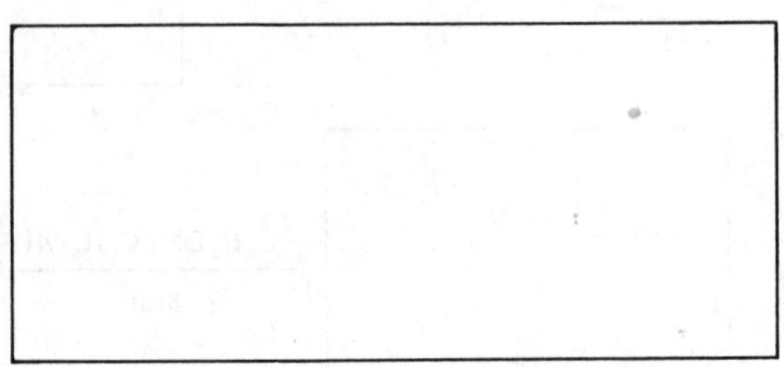

6.6 Which conditions/reagents would you employ to obtain the best yields in the following reaction?

$$CH_3CH_2\underset{\displaystyle Br}{\underset{|}{C}}HCH_3 \xrightarrow{?} CH_3CH_2CH{=}CH_2$$

(a) H_2O/heat (b) CH_3CH_2ONa/CH_3CH_2OH, heat

(c) $(CH_3)_3COK/(CH_3)_3COH$, heat (d) Reaction cannot occur as shown

6.7 Which of the following names is incorrect?

(a) 1-Butene (b) *trans*-2-Butene (c) (Z)-2-Chloro-2-pentene

(d) 1,1-Dimethylcyclopentene (e) Cyclohexene

6.8 Select the major product of the reaction

$$CH_3CH_2\overset{\overset{\displaystyle CH_3}{|}}{\underset{\underset{\displaystyle Br}{|}}{C}}-CH(CH_3)_2 \xrightarrow[C_2H_5OH]{C_2H_5ONa} ?$$

(a) $CH_3CH_2\overset{\overset{\displaystyle CH_3}{|}}{C}=C(CH_3)_2$ (b) $CH_3CH_2\overset{\overset{\displaystyle CH_2}{\|}}{C}-CH(CH_3)_2$

(c) $CH_3CH=\overset{\overset{\displaystyle CH_2}{|}}{C}-CH(CH_3)_2$ (d) $CH_2=CH-\overset{\overset{\displaystyle CH_3}{|}}{C}-CH(CH_3)_2$

(e) $CH_3CH_2\overset{\overset{\displaystyle CH_3}{|}}{\underset{\underset{\displaystyle OC_2H_5}{|}}{C}}-CH(CH_3)_2$

SUPPLEMENTARY PROBLEMS

S6.1 Name the following compounds

(a) $CH_3CH_2\overset{\overset{\displaystyle CH_2CH=CH_2}{|}}{C}HCH_2CH_2CH_3$ (b) 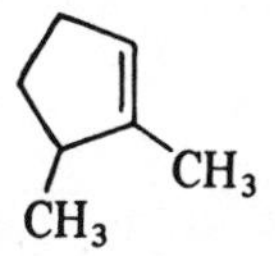 [cyclopentene ring bearing CH_3 on the double-bond carbon and CH_3 on the adjacent ring carbon]

S6.2 Compound A has the molecular formula C_7H_{12}. Hydrogenation over a Ni catalyst produces a saturated compound, B, whose molecular formula is C_7H_{14}. How many double bonds and/or rings do A and B possess?

S6.3 Predict the possible products of the following reaction.

(a) 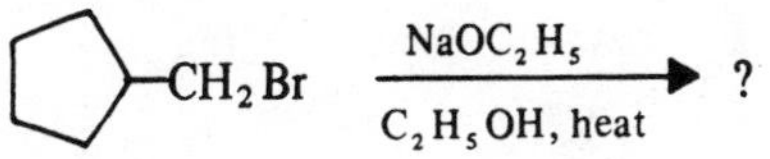 [cyclopentyl]$-CH_2Br \xrightarrow[C_2H_5OH,\ heat]{NaOC_2H_5} ?$

S6.4 1-Butyne can be prepared by dehydrobromination of what compound?

SOLUTIONS TO SUPPLEMENTARY PROBLEMS

S6.1 (a) 4-Ethyl-1-heptene (b) 2,3-Dimethylcyclopentene

S6.2 Hydrogenation of **A** produces compound **B**, which has as index of hydrogen deficiency of 1. Therefore **B** and **A** each has a ring. Compound **A** has an index of hydrogen deficiency of 2, so it must also have a double bond. Thus **A** has a double bond and a ring.

S6.3

(a) Cyclopentyl–CH_2Br $\xrightarrow{S_N2}$ Cyclopentyl–$CH_2OC_2H_5$

Cyclopentyl–CH_2Br $\xrightarrow{E2}$ Cyclopentylidene$=CH_2$

S6.4 $CH_3CH_2CH_2CHBr_2$

7 ALKENES AND ALKYNES II. ADDITION REACTIONS OF CARBON-CARBON MULTIPLE BONDS

SUMMARY OF REACTIONS OF ALKENES

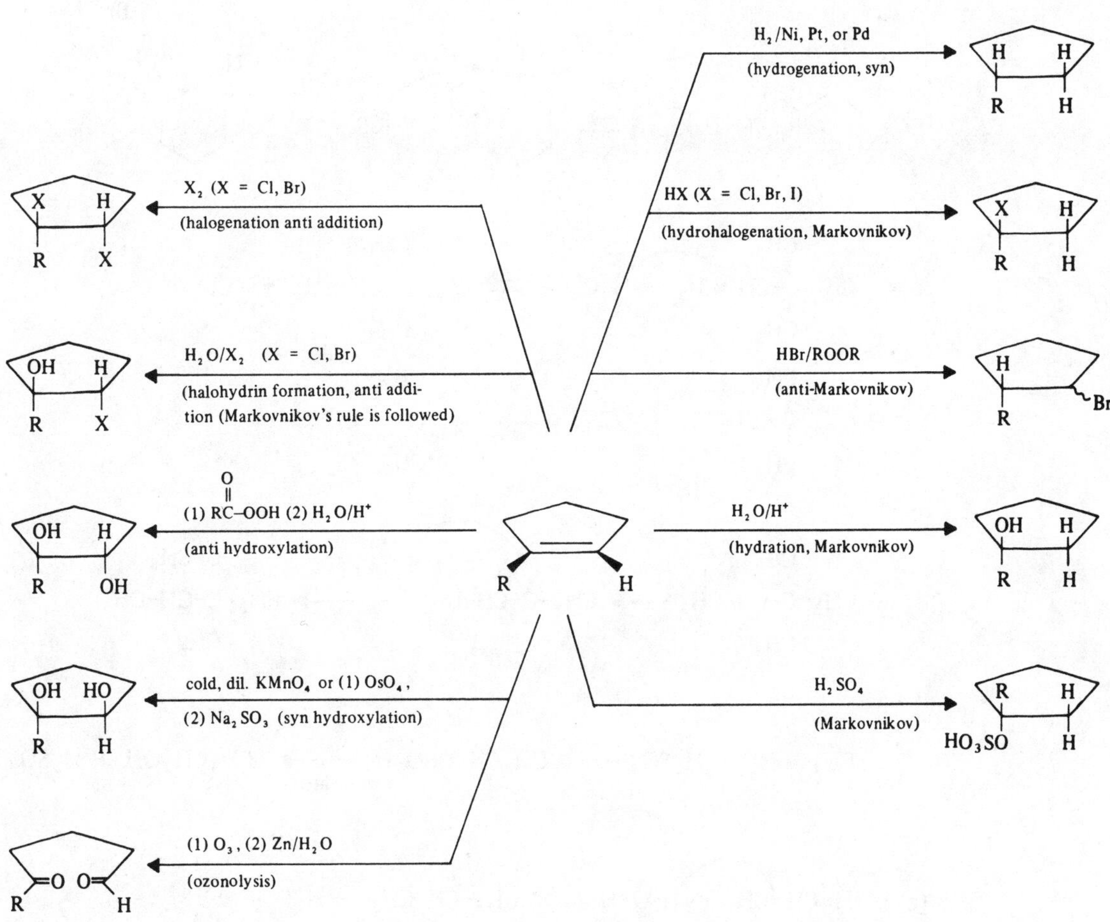

SOLUTIONS TO PROBLEMS

7.1 ICl adds as though it consisted of the ions I^+ and Cl^-:

$$\underset{\text{I}}{CH_2}-\underset{\text{Cl}}{CH}-CH_3$$

2-Chloro-1-iodopropane

7.2 (a) $CH_3CH_2CH{=}CH_2 + H{-}\ddot{I}: \rightleftharpoons CH_3CH_2\overset{+}{C}HCH_3 + :\ddot{I}:^- \longrightarrow CH_3CH_2\underset{\text{I}}{C}HCH_3$

(b) $(CH_3)_2C{=}C(CH_3)H + :\overset{\delta+}{\ddot{I}}{-}\overset{\delta-}{\ddot{Br}}: \rightleftharpoons (CH_3)_2\overset{+}{C}{-}C(CH_3)(H)(I) + :\ddot{Br}:^- \longrightarrow CH_3\underset{\text{Br}}{\overset{CH_3}{C}}{-}\underset{\text{I}}{C}HCH_3$

(c) 1-methylcyclohexene $+ H{-}\ddot{Cl}: \rightleftharpoons$ 1-methylcyclohexyl cation $+ :\ddot{Cl}:^- \longrightarrow$ 1-chloro-1-methylcyclohexane

7.3

$$CH_3{-}\overset{CH_3}{\underset{CH_3}{C}}{-}CH{=}CH_2 + H{-}\ddot{Cl}: \rightleftharpoons CH_3{-}\overset{CH_3}{\underset{CH_3}{C}}{-}\overset{+}{C}H{-}CH_3 + :\ddot{Cl}:^-$$

$$CH_3{-}\overset{CH_3}{\underset{CH_3}{C}}{-}\overset{+}{C}H{-}CH_3 + :\ddot{Cl}:^- \longrightarrow CH_3{-}\overset{CH_3}{\underset{CH_3}{C}}{-}\underset{Cl}{CH}{-}CH_3$$

or

$$CH_3{-}\overset{CH_3}{\underset{CH_3}{C}}{-}\overset{+}{C}H{-}CH_3 \longrightarrow CH_3{-}\overset{+}{\underset{CH_3}{C}}{-}\overset{CH_3}{CH}{-}CH_3 \xrightarrow{:\ddot{Cl}:^-} CH_3{-}\overset{Cl}{\underset{CH_3}{C}}{-}\overset{CH_3}{CH}{-}CH_3$$

7.4

$$CH_2{=}CH_2 + H_2SO_4 \longrightarrow CH_3CH_2OSO_3H \xrightarrow[\text{heat}]{H_2O} CH_3CH_2OH + H_2SO_4$$

7.5 (a) $CH_3{-}CH{=}CH_2 + H{-}\overset{..+}{\underset{H}{O}}{-}H \rightleftharpoons CH_3{-}\overset{+}{C}H{-}CH_3 + H_2O$

$$CH_3-\overset{+}{C}H-CH_3 + :\ddot{O}(H)-H \rightleftharpoons CH_3-CH(\overset{+}{O}H_2)-CH_3$$

$$CH_3-CH(\overset{+}{O}H_2)-CH_3 + :\ddot{O}(H)-H \rightleftharpoons CH_3-CH(OH)-CH_3 + H_3O^+$$

(b) The product is isopropyl alcohol because the more stable isopropyl carbocation is produced in the first step. The formation of propyl alcohol would require the production of the less stable propyl carbocation.

7.6

$$CH_3-C(CH_3)_2-CH=CH_2 \xrightarrow{H_3O^+} CH_3-C(CH_3)_2-\overset{+}{C}H-CH_3 \xrightarrow[\text{migration}]{\text{methanide}} CH_3-\overset{+}{C}(CH_3)-CH(CH_3)-CH_3 \xrightarrow{H_2O}$$

$$CH_3-C(CH_3)(H_2O^+)-CH(CH_3)-CH_3 \xrightarrow{-H^+} CH_3-C(CH_3)(HO)-CH(CH_3)-CH_3$$

7.7 The order reflects the relative ease with which these alkenes accept a proton and form a carbocation. $(CH_3)_2C{=}CH_2$ reacts fastest because it leads to a tertiary cation,

$$(CH_3)_2CH{=}CH_2 \xrightarrow{H^+} CH_3-\overset{+}{C}(CH_3)-CH_3 \quad 3° \text{ Carbocation}$$

$CH_3CH{=}CH_2$ leads to a secondary cation,

$$CH_3CH{=}CH_2 \xrightarrow{H^+} CH_3\overset{+}{C}HCH_3 \quad 2° \text{ Carbocation}$$

and, $CH_2{=}CH_2$ reacts most slowly because it leads to a primary carbocation.

$$CH_2{=}CH_2 \xrightarrow{H^+} CH_3CH_2^+ \quad 1° \text{ Carbocation}$$

Recall that formation of the cation is the rate-determining step in acid-catalyzed hydration and that the order of stabilities of carbocations is the following:

$$3° > 2° > 1° > \overset{+}{C}H_3$$

7.8

$$CH_3-C(CH_3){=}CH_2 \xrightarrow{H^+} CH_3-\overset{+}{C}(CH_3)-CH_3 \xrightarrow{CH_3-\ddot{O}-H} CH_3-C(CH_3)(H-\overset{+}{O}-CH_3)-CH_3 \xrightarrow{(-H^+)} CH_3-C(CH_3)(:\ddot{O}-CH_3)-CH_3$$

7.9

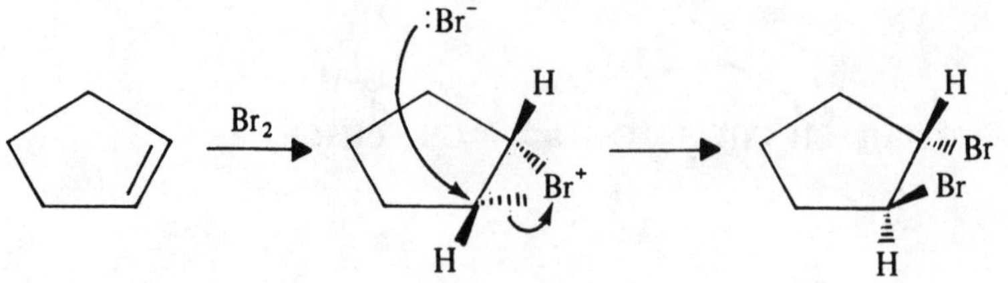

$:Br^-$; Br_2 ; H ; Br^+ ; H ; H ; Br ; Br ; H

7.10

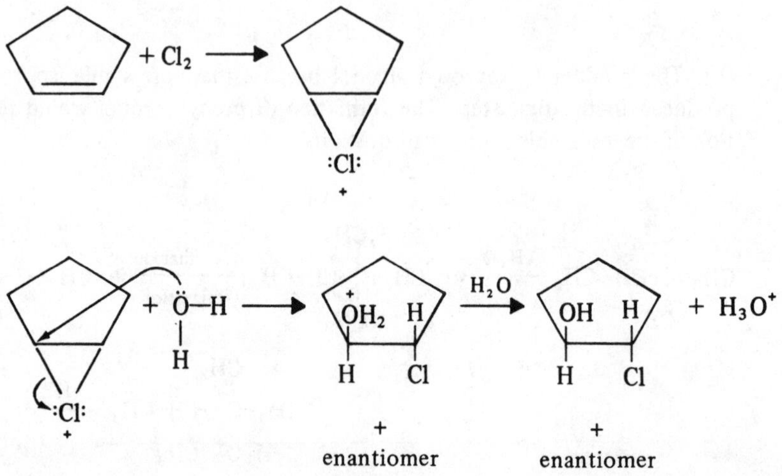

$+ Cl_2 \longrightarrow$:Cl: +

$+ \ddot{O}-H \longrightarrow$; H ; :Cl: + ; $\overset{+}{O}H_2$ H ; H Cl ; H_2O ; OH H ; H Cl ; $+ H_3O^+$

\+ enantiomer

\+ enantiomer

7.11 $CH_2{=}CH_2 + Br-Br \longrightarrow H_2C{-}CH_2 (Br^+) + Br^-$

$H_2C{-}CH_2 (Br^+)$ $\xrightarrow{H_2O}$ $Br{-}CH_2CH_2{-}\overset{+}{O}H_2 \xrightarrow{-H^+} BrCH_2CH_2OH$

$\xrightarrow{Br^-}$ $Br{-}CH_2CH_2{-}Br$

$\xrightarrow{Cl^-}$ $Br{-}CH_2CH_2{-}Cl$

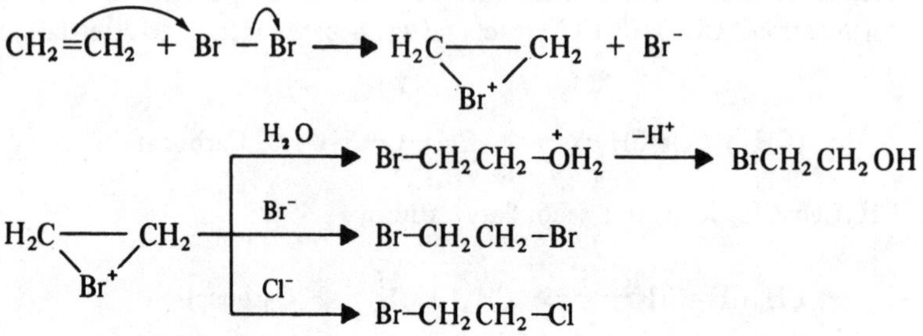

7.12

H ; H ; $+ H_3O^+ \rightleftarrows$; $H_2O:$; H ; OH^+ ; H ; H ; OH ; H ; $H_2\overset{+}{O}$

$\xrightarrow{-H^+}$ H ; OH ; H ; OH

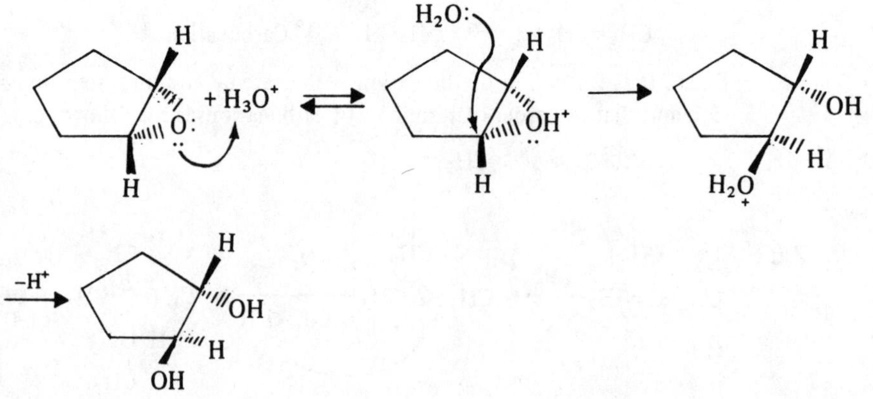

7.13 (a) *meso*-2,3-Butanediol

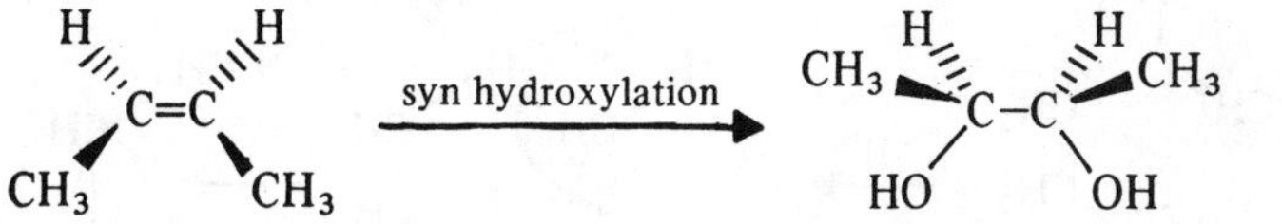

cis-2-Butene (H, H / CH_3, CH_3) $\xrightarrow{\text{syn hydroxylation}}$ CH_3(H)C–C(H)CH_3 with HO, OH

cis-2-Butene

(2*R*, 3*S*)-2,3-Butanediol (meso)

(Notice that hydroxylation at the top face yields the same product.)

(b) The enantiomeric 2,3-butanediols as a racemic form. (Here hydroxylation at the bottom face yields one enantiomer and hydroxylation at the top yields the other:)

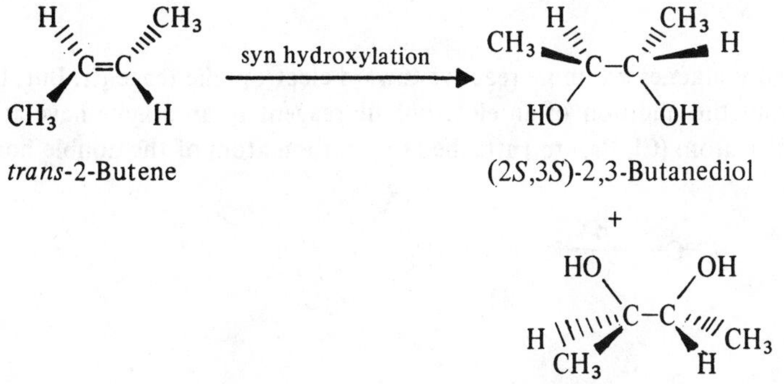

trans-2-Butene (H, CH_3 / CH_3, H) $\xrightarrow{\text{syn hydroxylation}}$ CH_3(H)C–C(CH_3)H with HO, OH

trans-2-Butene

(2*S*,3*S*)-2,3-Butanediol

+

HO, OH / H, CH_3 C–C CH_3, H

(2*R*,3*R*)-2,3-Butanediol

(c) Yes. cis-2-Butene (a given stereoisomeric form of the reactant) yields the meso compound (a specific stereoisomeric form of the product), and *trans*-2-butene yields the enantiomers.

7.14 (a) $CH_3CH_2CH{=}CHCH_3$

(b) $CH_3\overset{CH_3}{\overset{|}{C}}{=}\overset{CH_3}{\overset{|}{C}}CH_3$

(c) $CH_3CH_2\underset{CH_3}{\underset{|}{C}}HCH{=}CH_2$

(d) 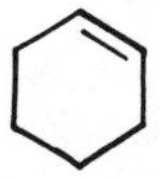 cyclohexene (ring structure)

7.15 Rewriting the starting compound, we can better see the required reaction:

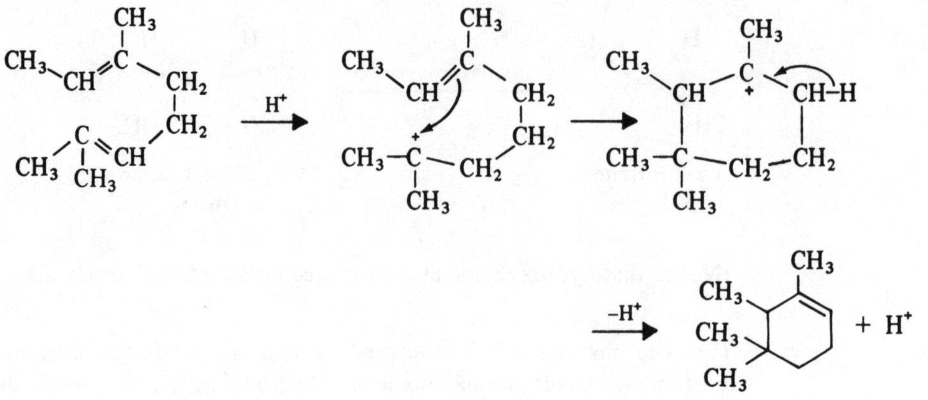

$$CH_3CH{=}C(CH_3)CH_2CH_2CH{=}C(CH_3)_2 \xrightarrow{H^+} CH_3CH{=}C(CH_3)CH_2CH_2CH_2\overset{+}{C}(CH_3)_2 \longrightarrow \text{(cyclized cation)} \xrightarrow{-H^+} \text{1,5,5,6-tetramethylcyclohexene} + H^+$$

7.16 Ordinary alkenes *are* more reactive toward electrophilic reagents. But, the alkenes obtained from the addition of an electrophilic reagent to an alkyne have at least one electronegative atom (Cl, Br, etc.) attached to a carbon atom of the double bond.

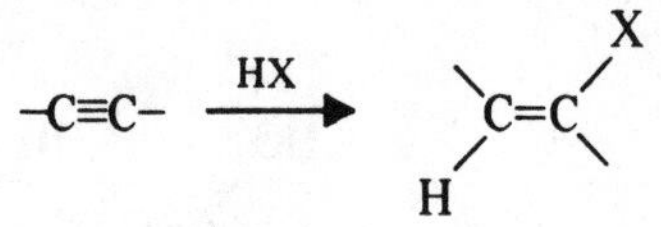

$$-C{\equiv}C- \xrightarrow{HX} \;\; \text{(H)C=C(X)}$$

or

$$-C{\equiv}C- \xrightarrow{X_2} \;\; \text{(X)C=C(X)}$$

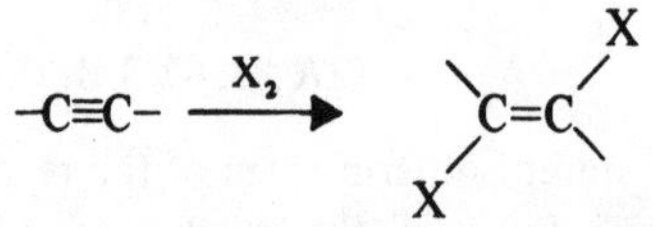

These alkenes are less reactive than alkynes toward electrophilic addition because the electronegative group makes the double bond "electron poor."

7.17 (a) $CH_3CH(CH_3)C{\equiv}CCH_2CH_3$ 2-Methyl-3-hexyne

(b) 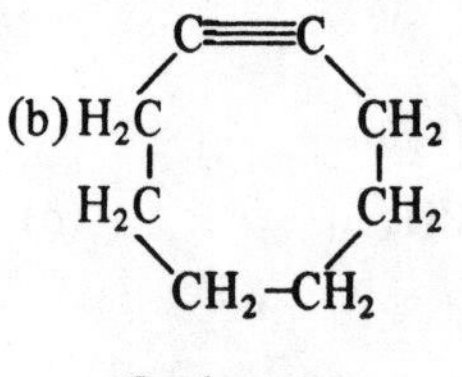 Cyclooctyne (ring: C≡C, CH_2, CH_2, CH_2–CH_2, H_2C, H_2C)

(c) $HC{\equiv}CCH_2CH_2CH_2CH_2CH_3$ 1-Heptyne

7.18 By converting the 3-hexyne to *cis*-3-hexene using H_2/Ni_2B (P-2).

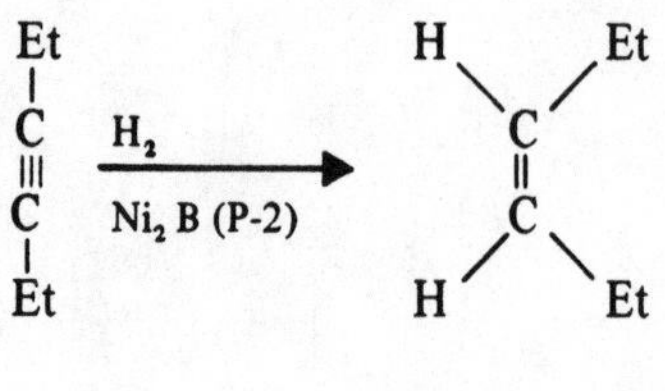

$$Et{-}C{\equiv}C{-}Et \xrightarrow[Ni_2B\ (P\text{-}2)]{H_2} \;\; \text{(H)(Et)C=C(H)(Et)}$$

Then, addition of bromine to *cis*-3-hexene will yield (3*R*,4*R*)-and (3*S*,4*S*)-3,4-dibromohexane as a racemic form.

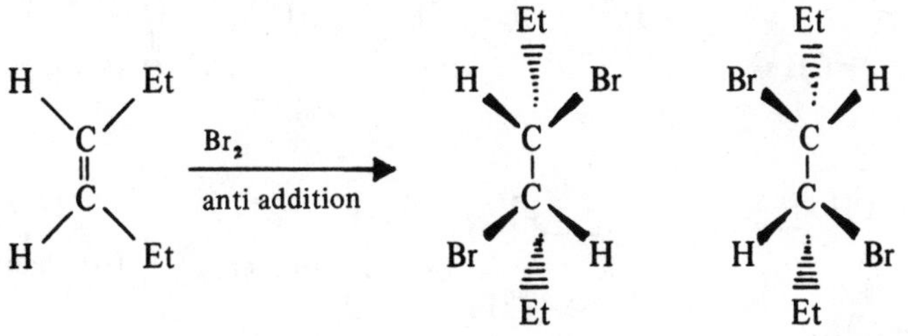

Br_2, anti addition

(3*S*, 4*S*) (3*R*, 4*R*)

Racemic 3, 4-dibromohexane

7.19 (a) $CH_3CH_2CH_2CHClCH_3$ (b) $CH_3CH_2CH_2CHBrCH_2Br$

(c) $CH_3CH_2CH_2CHOHCH_3$ (d) $CH_3CH_2CH_2CH(OSO_2OH)CH_3$

(e) Same as (c) (f) $CH_3CH_2CH_2CHBrCH_3$

(g) $CH_3CH_2CH_2CHICH_3$ (h) $CH_3CH_2CH_2CH_2CH_3$

(i) $CH_3CH_2CH_2CH{=}CH_2$ (j) $CH_3CH_2CH_2CH(OH)CH_2OH$

(k) Same as (j) (l) $CH_3CH_2CH_2\overset{O}{\overset{\|}{C}}{-}OH + CO_2$

(m) $CH_3CH_2CH_2\overset{O}{\overset{\|}{C}}H + H\overset{O}{\overset{\|}{C}}H$ (n) $CH_3CH_2CH_2CHBrCH_2Br$

(o) $CH_3CH_2CH_2CH_2CH_2Br$

7.20 (a) Cl

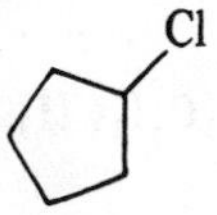

(b)

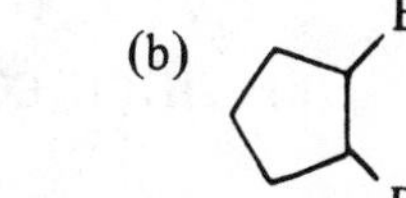

(c) OH

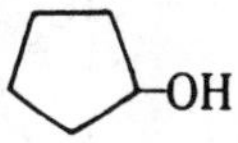

(d) OSO_2OH (e) Same as (c) (f) Br

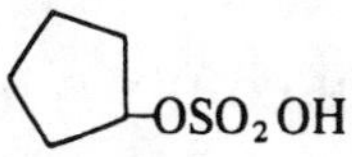

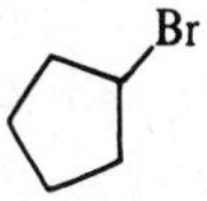

(g) I (h)  (i)

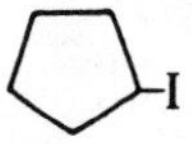

(j) cis-1,2-cyclopentanediol (OH, OH both wedged)

(k) Same as (j)

(l) $HOOC-CH_2CH_2CH_2-COOH$

(m) $OHC-CH_2CH_2CH_2-CHO$

(n) trans-1,2-dibromocyclopentane (Br wedged, Br hashed) + enantiomer

(o) Same as (f)

7.21 (a) $(CH_3CH_2CH_2)(Br)C{=}C(Br)(H)$ — $CH_3CH_2CH_2$ and Br on one carbon, Br and H on the other; the two Br atoms trans

(b) $CH_3CH_2CH_2(Cl)C{=}CH_2$

(c) $CH_3CH_2CH_2CCl_2CH_3$

(d) $CH_3CH_2CH_2CH{=}CHBr$

(e) $CH_3CH_2CH_2\overset{O}{\overset{\|}{C}}CH_3$

(f) $CH_3CH_2CH_2CH{=}CH_2$

(g) $CH_3CH_2CH_2C{\equiv}C{:}^-\ Na^+$

(h) $CH_3CH_2CH_2C{\equiv}CCH_3$

(i) $CH_3CH_2CH_2C{\equiv}CAg$

(j) $CH_3CH_2CH_2C{\equiv}CCu$

7.22 (a) $(CH_3CH_2)(Cl)C{=}C(H)(CH_2CH_3)$ — CH_3CH_2 trans to H; Cl trans to CH_2CH_3

(b) $CH_3CH_2CCl_2CH_2CH_2CH_3$

(c) $(CH_3CH_2)(Br)C{=}C(Br)(CH_2CH_3)$ — CH_3CH_2 trans to CH_2CH_3; the two Br atoms trans

(d) $CH_3CH_2CBr_2{-}CBr_2CH_2CH_3$

(e) $(CH_3CH_2)(H)C{=}C(H)(CH_2CH_3)$ — CH_3CH_2 and CH_2CH_3 cis

(f) $(CH_3CH_2)(H)C{=}C(Br)(CH_2CH_3)$ — CH_3CH_2 trans to CH_2CH_3

(g) $(CH_3CH_2)(H)C{=}C(H)(CH_2CH_3)$ — CH_3CH_2 and CH_2CH_3 trans

(h) $CH_3CH_2\overset{O}{\overset{\|}{C}}CH_2CH_2CH_3$

(i) No reaction

(j) $CH_3CH_2CH_2CH_2CH_2CH_3$

(k) $2CH_3CH_2CO_2H$

(l) $2CH_3CH_2CO_2H$

(m) No reaction

7.23 (a) $CH_3CH_2CH_2CH{=}CH_2 + Br_2 \longrightarrow CH_3CH_2CH_2\underset{\displaystyle Br}{\underset{|}{C}}HCH_2Br \xrightarrow[(2)\ H^+]{(1)\ 3NaNH_2} CH_3CH_2CH_2C{\equiv}CH$

(b) $CH_3CH_2CH_2CH_2CH_2Cl \xrightarrow[(CH_3)_3COH]{(CH_3)_3COK} CH_3CH_2CH_2CH{=}CH_2$

then proceed as in (a) above.

(c) $CH_3CH_2CH_2CH{=}CHCl \xrightarrow[(2)\ H^+]{(1)\ 2NaNH_2} CH_3CH_2CH_2C{\equiv}CH$

(d) $CH_3CH_2CH_2CH_2CHCl_2 \xrightarrow[(2)\ H^+]{(1)\ 3NaNH_2} CH_3CH_2CH_2C{\equiv}CH$

(e) $HC{\equiv}CH \xrightarrow[liq.\ NH_3]{NaNH_2} HC{\equiv}C{:}^-\ Na^+ \xrightarrow{CH_3CH_2CH_2Br} HC{\equiv}CCH_2CH_2CH_3$

7.24 (a) $CH_3\overset{\displaystyle CH_3}{\overset{|}{C}}{=}CH_2 \xrightarrow{H_3O^+,\ H_2O} CH_3\underset{\displaystyle OH}{\underset{|}{\overset{\displaystyle CH_3}{\overset{|}{C}}}}CH_3$

(b) $CH_3\overset{\displaystyle CH_3}{\overset{|}{C}}{=}CH_2 \xrightarrow{HCl} CH_3\underset{\displaystyle Cl}{\underset{|}{\overset{\displaystyle CH_3}{\overset{|}{C}}}}{-}CH_3$

(c) $CH_3\overset{\displaystyle CH_3}{\overset{|}{C}}{=}CH_2 \xrightarrow[(no\ peroxides)]{HBr} CH_3\underset{\displaystyle Br}{\underset{|}{\overset{\displaystyle CH_3}{\overset{|}{C}}}}CH_3$

(d) $CH_3\overset{\displaystyle CH_3}{\overset{|}{C}}{=}CH_2 \xrightarrow[ROOR]{HBr} CH_3\overset{\displaystyle CH_3}{\overset{|}{C}}HCH_2Br$

(e) $CH_3\overset{\displaystyle CH_3}{\overset{|}{C}}HCH_2Br \xrightarrow[S_N2]{I^-} CH_3\overset{\displaystyle CH_3}{\overset{|}{C}}HCH_2I$

(as in part d)

(f) $CH_3\overset{\displaystyle CH_3}{\overset{|}{C}}HCH_2Br \xrightarrow[S_N2]{CN^-} CH_3\overset{\displaystyle CH_3}{\overset{|}{C}}HCH_2CN$

[as in part (d)]

(g) $CH_3\overset{\displaystyle CH_3}{\overset{|}{C}}{=}CH_2 \xrightarrow{HF} CH_3\underset{\underset{\displaystyle F}{|}}{\overset{\overset{\displaystyle CH_3}{|}}{C}}CH_3$

(h) $CH_3\overset{\displaystyle CH_3}{\overset{|}{C}}{=}CH_2 \xrightarrow{Cl_2,\ H_2O} CH_3\underset{\underset{\displaystyle OH}{|}}{\overset{\overset{\displaystyle CH_3}{|}}{C}}CH_2Cl$

7.25 (a) $C_{10}H_{22}$ (saturated alkane)
$C_{10}H_{16}$ (formula of myrcene)

H_6 = 3 pairs of hydrogen atoms

Index of hydrogen deficiency (IHD) = 3

(b) Myrcene contains no rings because complete hydrogenation gives $C_{10}H_{22}$, which corresponds to an alkane.

(c) That myrcene absorbs three molar equivalents of H_2 on hydrogenation indicates that it contains three double bonds.

(d) Three structures are possible; however, only one gives 2, 6-dimethyloctane on complete hydrogenation. Myrcene is therefore

$$CH_3\overset{\displaystyle CH_3}{\overset{|}{C}}{=}CHCH_2CH_2\overset{\displaystyle CH_2}{\overset{\|}{C}}CH{=}CH_2$$

(e) $O{=}CHCH_2CH_2\overset{\displaystyle O}{\overset{\|}{C}}CH{=}O$

7.26 $CH_3CH{=}CHCH_3 + HCl \rightleftharpoons CH_3CH_2\overset{+}{C}HCH_3 + :\ddot{Cl}:^-$

$$CH_3CH_2\overset{+}{C}HCH_3 + :\ddot{O}(H){-}CH_2CH_3 \longrightarrow CH_3CH_2\underset{\displaystyle {}^{+}\!:\overset{\overset{\displaystyle H}{|}}{O}{-}CH_2CH_3}{\overset{|}{C}}HCH_3 \xrightarrow{-H^+} CH_3CH_2\overset{\displaystyle OCH_2CH_3}{\overset{|}{C}}HCH_3$$

7.27 The order of reactivity parallels the order of stability of the carbocations produced by the attack of H^+ on each alkene.

$$\underset{3^\circ}{R{-}\overset{\overset{\displaystyle R}{|}}{\overset{+}{C}}{-}CH_3} > \underset{2^\circ}{R{-}\overset{+}{C}H{-}CH_3} > \underset{1^\circ}{\overset{+}{C}H_2{-}CH_3}$$

7.28 $2\left(CH_3\overset{\displaystyle O}{\overset{\|}{C}}CH_3\right)$ $4\left(O{=}CHCH_2CH_2\overset{\displaystyle CH_3}{\overset{|}{C}}{=}O\right)$ $O{=}CHCH_2CH{=}O$

7.29 $CH_3C(CH_3){=}CH_2 > CH_3CH{=}CH_2 > CH_2{=}CH_2$

The order is the same as the order of stability of the carbocations formed by protonation of the alkenes.

$$\underset{3^\circ}{CH_3\overset{+}{C}(CH_3){-}CH_3} > \underset{2^\circ}{CH_3\overset{+}{C}H{-}CH_3} > \underset{1^\circ}{\overset{+}{C}H_2{-}CH_3}$$

7.30 (a) $CH_3CH{=}CHCH_3 + H^+ \rightleftharpoons CH_3\overset{+}{C}HCH_2CH_3$

(cis or trans)

$$CH_3{-}\overset{+}{C}H{-}CH(H){-}CH_3 \longrightarrow \begin{cases} (CH_3)(H)C{=}C(H)(CH_3)\ (trans) + H^+ \\ (CH_3)(H)C{=}C(CH_3)(H)\ (cis) + H^+ \end{cases}$$

$$CH_3{-}CH_2{-}\overset{+}{C}H{-}CH_2(H) \longrightarrow CH_3CH_2CH{=}CH_2 + H^+$$

The most stable (most substituted) alkene is formed in greatest amount; that is, 2-butenes > 1-butene, and *trans*-2-butene > *cis*-2-butene.

(b) 1-Butene, on protonation, gives the same intermediate carbocation $CH_3CH_2\overset{+}{C}HCH_3$.

(c) The carbocation ($CH_3CH_2\overset{+}{C}HCH_3$) cannot easily rearrange to the branched-chain compound because to do so would require the formation of an intermediate primary carbocation, $^+CH_2CH(CH_3)CH_3$.

7.31

$$CH_3{-}CH(OH){-}C(CH_3)_2{-}CH_3 + H{-}Cl \rightleftharpoons CH_3{-}CH(\overset{+}{O}H_2){-}C(CH_3)_2{-}CH_3 + :\ddot{Cl}:^-$$

$$CH_3{-}CH(\overset{+}{O}H_2){-}C(CH_3)_2{-}CH_3 \longrightarrow CH_3{-}\overset{+}{C}H{-}C(CH_3)_2{-}CH_3 + H_2O$$

$$CH_3{-}\overset{+}{C}H{-}C(CH_3)_2{-}CH_3 \longrightarrow CH_3{-}CH(CH_3){-}\overset{+}{C}(CH_3){-}CH_3 \xrightarrow{:\ddot{Cl}:^-} CH_3{-}CH(CH_3){-}C(Cl)(CH_3){-}CH_3$$

7.32 Reassembling the oxidation product as follows we can see where the original double bond was.

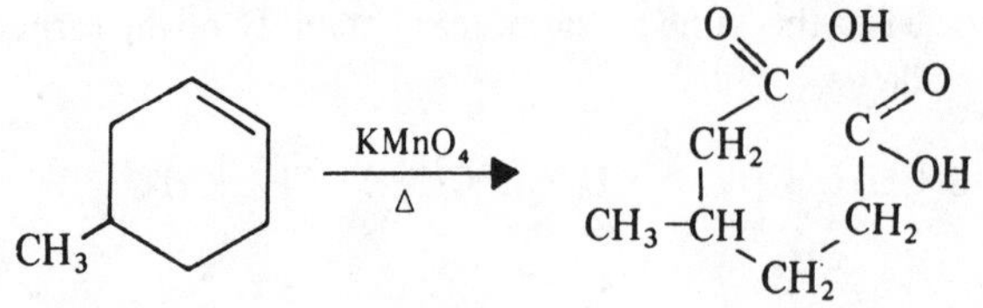

About Synthesis

Problems involving syntheses are always intriguing to chemists and students alike. These problems are probably unlike any that you have seen in other courses. In them you are asked to put together a series of reactions that will convert one compound into another. To do this, it is not enough to start your reasoning with the starting materials because you must also keep in mind the desired product. In some instances you can work from both ends simultaneously, and this is best done by trying to find an intermediate that will link the starting materials and products in a sequence of reactions. In many cases, however, a synthesis problem is best solved by reasoning backward, by starting your thinking with the product and by trying to discover a series of reactions that will lead back to the starting compounds (Section 7.18). Sherlock Holmes in *A Study in Scarlet* said:

Most people if you describe a train of events to them, will tell you what the result would be. They can put these events together in their minds, and argue from them that something will come to pass. There are a few people, however, who, if you told them a result, would be able to evolve from their own inner consciousness what the steps were which led up to that result. This power is what I mean when I talk of reasoning backward, or analytically.

This power is what the organic chemistry student must develop.

Let us illustrate this process with the synthesis of the chlorohydrin, 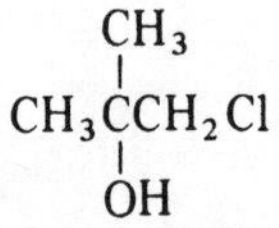

from 2-methylpropane. We start by recognizing that we can synthesize the final product from 2-methylpropene:

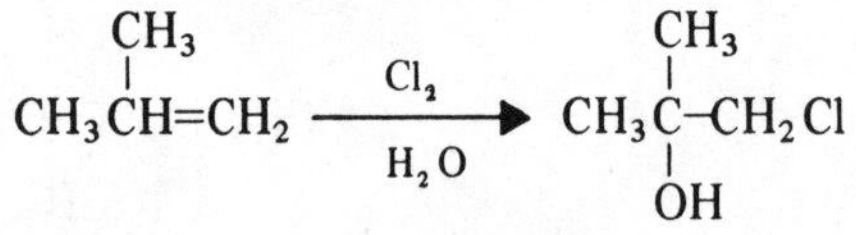

Then our task is to synthesize 2-methylpropene.

This can be done by dehydrohalogenatng either of the compounds shown here.

$$\underset{\text{tert-Butyl bromide}}{CH_3\overset{CH_3}{\underset{Br}{\overset{|}{\underset{|}{C}}}}-CH_3} \xrightarrow[(CH_3)_3COH]{(CH_3)_3COK} CH_3\overset{CH_3}{\overset{|}{C}}=CH_2$$

or

$$\underset{\text{Isobutyl bromide}}{CH_3\overset{CH_3}{\overset{|}{C}}HCH_2Br} \xrightarrow[(CH_3)_3COH]{(CH_3)_3COK} CH_3\overset{CH_3}{\overset{|}{C}}=CH_2$$

Both of these compounds can be prepared by brominating isobutane.

$$CH_3\overset{CH_3}{\overset{|}{C}}HCH_3 + Br_2 \xrightarrow[\text{light}]{\text{heat}} CH_3\overset{CH_3}{\underset{Br}{\overset{|}{\underset{|}{C}}}}CH_3 + CH_3\overset{CH_3}{\overset{|}{C}}HCH_2Br$$

That this method yields a mixture of compounds is not a problem in this synthesis because either compound yields 2-methylpropene when subjected to dehydrohalogenation. We need not even separate the components of the mixture. (As we saw in Section 3.17A, the mixture formed in this instance is primarily composed of *tert*-butyl bromide.)

7.33 (a)
$$\underset{\text{(excess)}}{CH_3CH_2CH_3} + Br_2 \xrightarrow[\text{light}]{CCl_4} \left.\begin{matrix} CH_3CH_2CH_2Br \\ + \\ CH_3\underset{Br}{\underset{|}{C}}HCH_3 \end{matrix}\right\} \xrightarrow[(CH_3)_3COH]{(CH_3)_3COK} CH_3CH=CH_2$$

(b)
$$CH_3CH=CH_2 \text{ (above)} + HBr \xrightarrow{\text{no peroxides}} CH_3\underset{Br}{\underset{|}{C}}HCH_3$$

(c)
$$CH_3CH=CH_2 \text{ [from (a)]} + HBr \xrightarrow{\text{peroxides}} CH_3CH_2CH_2Br$$

(d)
$$\underset{\text{(excess)}}{CH_3\overset{CH_3}{\overset{|}{C}}HCH_3} + Br_2 \xrightarrow[\text{light}]{\text{heat}} CH_3\overset{CH_3}{\underset{Br}{\overset{|}{\underset{|}{C}}}}CH_3 \xrightarrow[CH_3CH_2OH]{CH_3CH_2ONa} CH_3\overset{CH_3}{\overset{|}{C}}=CH_2$$

(e)
$$CH_3\overset{CH_3}{\overset{|}{C}}=CH_2 \text{ [from (d)]} + H_2O \xrightarrow{H_3O^+} CH_3\overset{CH_3}{\underset{OH}{\overset{|}{\underset{|}{C}}}}CH_3$$

(f)
$$CH_3CH_2CH_2CH_2Cl + \xrightarrow[(CH_3)_3COH]{(CH_3)_3COK} CH_3CH_2CH=CH_2 \xrightarrow[CCl_4]{Cl_2} CH_3CH_2\underset{Cl}{\underset{|}{C}}HCH_2Cl$$

(g) $CH_3CH_2Br \xrightarrow[(CH_3)_3COH]{(CH_3)_3COK} CH_2{=}CH_2 \xrightarrow{Br_2 + H_2O} \underset{OH}{CH_2}\underset{Br}{CH_2}$

(h) cyclopentane + Br_2 $\xrightarrow[\text{light}]{CCl_4}$ bromocyclopentane $\xrightarrow[CH_3CH_2OH]{CH_3CH_2ONa}$ cyclopentene

(excess)

cyclopentene + Cl_2 + H_2O $\longrightarrow$ trans-2-chlorocyclopentanol (Cl, OH) + enantiomer

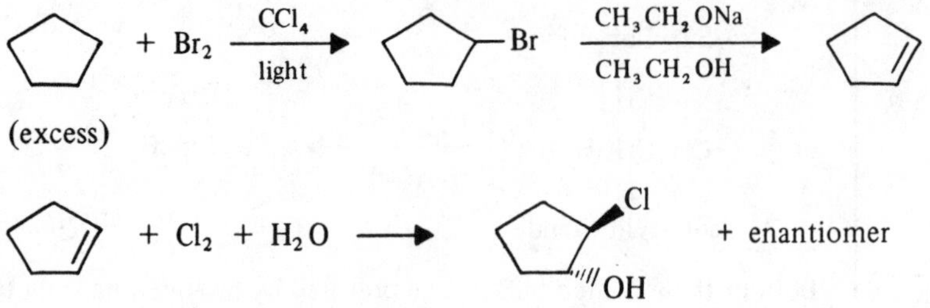

(i) $CH_3CH_2CH_2CH_2Br \xrightarrow[(CH_3)_3COH]{(CH_3)_3COK} CH_3CH_2CH{=}CH_2 \xrightarrow[\text{no peroxides}]{HBr}$

$CH_3CH_2\underset{Br}{C}HCH_3$

7.34 (a) cyclopentene + $HC(=O){-}OOH$ $\longrightarrow$ cyclopentene oxide (O) $\xrightarrow{HCl}$ trans-2-chlorocyclopentanol (Cl, OH)

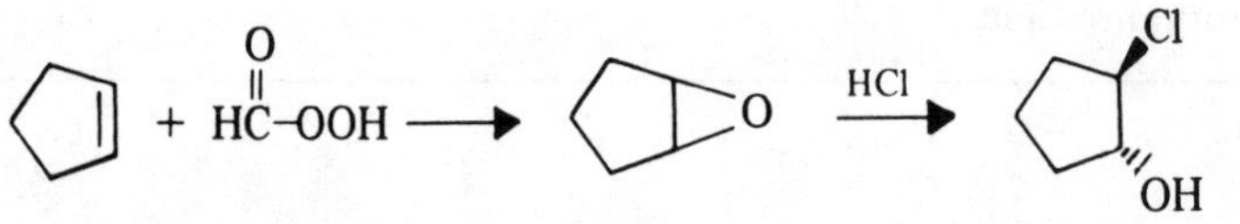

(b) The trans product because the Cl^- attacks anti to the epoxide and an inversion of configuration occurs.

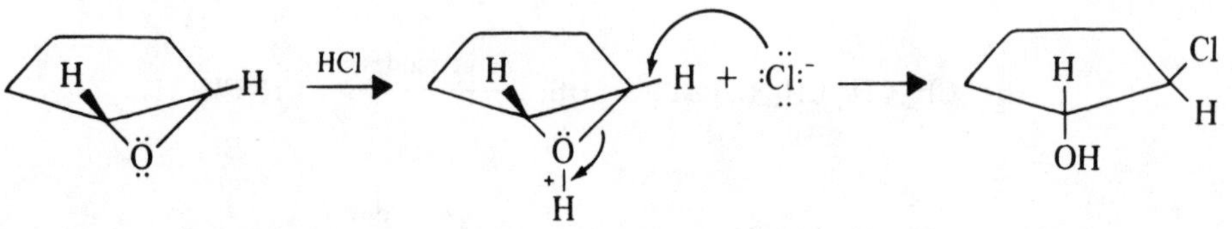

epoxide (H, H, $\ddot{O}$) $\xrightarrow{HCl}$ protonated epoxide (H, H, $\ddot{O}^+$–H) + $:\ddot{Cl}:^-$ $\longrightarrow$ trans product (H, Cl, H, OH)

7.35

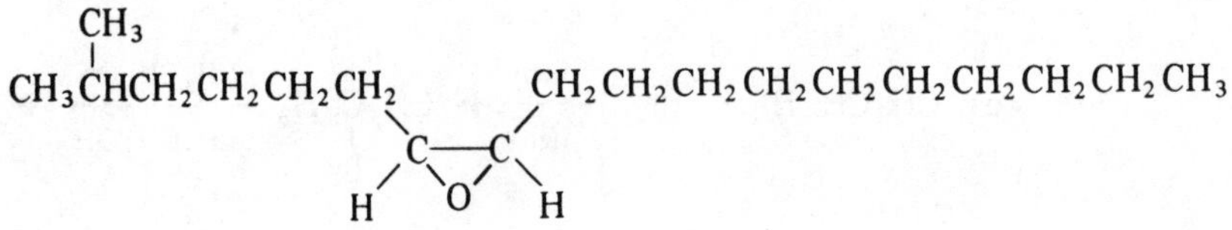

$$\underset{}{CH_3\overset{CH_3}{C}HCH_2CH_2CH_2CH_2}\,\underset{H}{C}\!\!-\!\!\overset{O}{}\!\!-\!\!\underset{H}{C}\,CH_2CH_2CH_2CH_2CH_2CH_2CH_2CH_2CH_2CH_3$$

(cis epoxide: $CH_3CH(CH_3)CH_2CH_2CH_2CH_2$ and $CH_2CH_2CH_2CH_2CH_2CH_2CH_2CH_2CH_2CH_3$ on the two epoxide carbons, each bearing H)

7.36

$$CH_3\overset{CH_3}{C}HCH_2CH_2CH_2\overset{CH_3}{C}HCH_2CH_2CH_2\overset{CH_3}{C}HCH_2CH_3$$

1 2 3 4 5 6 7 8 9 10 11 12

When subjected to ozonolysis followed by treatment with zinc and water 1 mole of the alarm pheromone produces the following compounds. Identifying the chain atoms as just shown allows us to assign the fragments.

Two moles of formaldehyde, $H\overset{O}{\overset{\|}{C}}H$, must come from $-\underset{6}{\overset{CH_2}{\overset{\|}{C}}}-$, $-\underset{10}{\overset{CH_2}{\overset{\|}{C}}}-$, or $-\underset{11}{\overset{|}{C}}{=}\underset{12}{CH_2}$

One mole of acetone, $CH_3\overset{O}{\overset{\|}{C}}CH_3$, must come from $\underset{1}{CH_3}\underset{2}{\overset{CH_3}{\overset{|}{C}}}=$

One mole of $\underset{6}{CH_3}\underset{5}{\overset{O}{\overset{\|}{C}}}\underset{4}{CH_2}CH_2\underset{3}{\overset{O}{\overset{\|}{C}}H}$ (location in the chain is shown)

One mole of $H\underset{7}{\overset{O}{\overset{\|}{C}}}\underset{8}{CH_2}\underset{9}{CH_2}\underset{10}{\overset{O}{\overset{\|}{C}}}-\underset{11}{\overset{O}{\overset{\|}{C}}H}$ (location in the chain is shown)

The assignment for the alarm pheromone is therefore

$$CH_3\overset{CH_3}{\overset{|}{C}}{=}CHCH_2CH_2\overset{CH_3}{\overset{|}{C}}{=}CHCH_2CH_2\overset{CH_2}{\overset{\|}{C}}CH{=}CH_2$$

7.37

(a) $CH_3\overset{CH_3}{\overset{|}{C}}{=}CH_2$

(b) By dehydrobromination as shown in (a), and then by adding HBr to the resulting alkene, isobutyl bromide can be converted into *tert*-butyl bromide:

$$CH_3\overset{CH_3}{\overset{|}{C}}{=}CH_2 + HBr \xrightarrow{CCl_4} CH_3\underset{Br}{\underset{|}{\overset{CH_3}{\overset{|}{C}}}}CH_3 \quad \text{(Markovnikov's rule)}$$

7.38 The intermediate **I** is competitively attacked by Cl^-, Br^-, and H_2O.

Cyclopentene + Br_2 ⇄ [cyclopentene·:Br–Br:] ⟶

Br^- + I (bromonium ion, :Br:$^+$) + :Cl:$^-$ (from NaCl) ⟶ trans-1-bromo-2-chlorocyclopentane (H, Cl, H, Br) + enantiomer

I

7.39 One dimer (the major product) gives the following products on ozonolysis.

$$CH_3-\underset{CH_3}{\overset{CH_3}{\underset{|}{\overset{|}{C}}}}-CH_2-\underset{CH_3}{\underset{|}{C}}=CH_2 \xrightarrow[(2)\ Zn,\ H_2O]{(1)\ O_3} CH_3-\underset{CH_3}{\overset{CH_3}{\underset{|}{\overset{|}{C}}}}-CH_2-\overset{O}{\overset{\|}{C}}-CH_3 + H-\overset{O}{\overset{\|}{C}}-H$$

The other dimer gives different products.

$$CH_3-\underset{CH_3}{\overset{CH_3}{\underset{|}{\overset{|}{C}}}}-CH=\underset{CH_3}{\underset{|}{C}}-CH_3 \xrightarrow[(2)\ Zn,\ H_2O]{(1)\ O_3} CH_3-\underset{CH_3}{\overset{CH_3}{\underset{|}{\overset{|}{C}}}}-C\begin{smallmatrix}{\diagup\!\!\!\!=O}\\ {\diagdown H}\end{smallmatrix} + CH_3-\overset{O}{\overset{\|}{C}}-CH_3$$

By isolating and identifying the products of each reaction, Whitmore and his students were able to deduce the structures of the diisobutylenes.

7.40 The isomers are propene tetramers formed by an acid-catalyzed reaction:

$$CH_3CH=CH_2 + H_3PO_4 \rightleftarrows CH_3\overset{+}{C}HCH_3 + H_2PO_4^-$$

$$\underset{CH_3}{CH_3CH^+} + CH_2=CHCH_3 \longrightarrow \underset{CH_3}{CH}\underset{CH_3}{CHCH_2CH^+}$$

$$CH_3\underset{CH_3}{CH}CH_2\underset{CH_3}{CH^+} + CH_2=CHCH_3 \longrightarrow CH_3\underset{CH_3}{CH}CH_2\underset{CH_3}{CH}CH_2\underset{CH_3}{CH^+}$$

$$CH_3\underset{CH_3}{CH}CH_2\underset{CH_3}{CH}CH_2\underset{CH_3}{CH^+} + CH_2=CHCH_3 \longrightarrow CH_3\underset{CH_3}{CH}CH_2\underset{CH_3}{CH}CH_2\underset{CH_3}{CH}CH_2\overset{+}{C}HCH_3$$

$$CH_3\underset{CH_3}{CH}CH_2\underset{CH_3}{CH}CH_2\underset{CH_3}{CH}CH_2\overset{+}{C}HCH_3 \xrightarrow{-H^+} CH_3\underset{CH_3}{CH}CH_2\underset{CH_3}{CH}CH_2\underset{CH_3}{CH}CH=CHCH_3$$

$$+$$

$$CH_3\underset{CH_3}{CH}CH_2\underset{CH_3}{CH}CH_2\underset{CH_3}{CH}CH_2CH=CH_2$$

7.41

(a) Newman-type structure: front carbon with CH_3, H, OH; rear carbon with H, OH, CH_2CH_3 — and enantiomer through syn addition

(b) Newman-type structure: front carbon with CH_3, H, OH; rear carbon with HO, H, CH_2CH_3 — and enantiomer through anti addition

(c) CH$_3$, H, OH / HO, H, CH$_2$CH$_3$ and enantiomer through syn addition

(d) CH$_3$, H, OH / H, OH, CH$_2$CH$_3$ and enantiomer through anti addition

(e) CH$_3$, H, Br / H, Br, CH$_2$CH$_3$ and enantiomer through anti addition

(f) CH$_3$, H, Br / Br, H, CH$_2$CH$_3$ and enantiomer through anti addition

7.42 (a) (2*S*,3*R*)- [enantiomer is (2*R*,3*S*)]; (b) (2 *S*,3 *S*) - [the enantiomer is (2 *R*,3 *R*)-]
(c) Same as (b); (d) Same as (a); (e) (2 *S*,3 *R*) - [the enantiomer is (2 *R*,3 *R*)-]
(f) (2 *S*,3 *S*) - [the enantiomer is (2 *R*,3 *R*)-]

7.43 (a) Propyne decolorizes Br_2/CCl_4; propane does not.

(b) $Ag(NH_3)_2^+ OH^-$ gives a precipitate with propyne, not with propene.

(c) Dilute $KMnO_4$ oxidizes 1-bromopropene and not 2-bromopropane.

(d) $Ag(NH_3)_2^+ OH^-$ gives a precipitate with 1-butyne, not with 2-bromo-2-butene.

(e) $Ag(NH_3)_2^+OH^-$ gives a precipitate with 1-butyne not with 2-butyne.

(f) Br_2/CCl_4 is decolorized by 2-butyne, not by butyl alcohol.

(g) $AgNO_3/C_2H_5OH$ gives a AgBr precipitate with 2-bromobutane, not with 2-butyne.

(h) Br_2/CCl_4 is decolorized by $CH_3C{\equiv}CCH_2OH$, not by $CH_3CH_2CH_2CH_2OH$.

(i) Br_2/CCl_4 is decolorized by $CH_3CH{=}CHCH_2OH$, not by $CH_3CH_2CH_2CH_2OH$.

(In many cases other tests are possible.)

7.44 The following are some tentative conclusions:

(1) **A** and **B** have the skeleton C–C–C–C–C (they both yield pentane on hydrogenation.)

(2) **A** has the HC≡C– group (gives ppt with ammoniacal $AgNO_3$).

(3) **C** has a ring because hydrogenation yields C_5H_{10} and not C_5H_{12}.

(4) All three compounds have carbon-carbon multiple bonds because they all react with Br_2/CCl_4 and Baeyer's reagent, and are soluble in cold, conc. H_2SO_4.

Assignments follow:

(a) **A** = $CH_3CH_2CH_2C{\equiv}CH$, **B** = $CH_3CH_2C{\equiv}CCH_3$, **C** = cyclopentene

(b) Yes, **B** may also be $CH_3CH{=}CH{-}CH{=}CH_2$ or $CH_2{=}CH{-}CH_2{-}CH{=}CH_2$.

C may also be 1-methylcyclobutene ($-CH_3$) or 3-methylcyclobutene ($-CH_3$) or methylenecyclobutane ($=CH_2$) or 3-ethylcyclopropene (CH_2CH_3) etc.

(c) **B** = $CH_3CH_2C{\equiv}CCH_3$

(d) cyclopentene

7.45 (a) $CH_3CH(CH_3)C{\equiv}CH$ + HCl(1 molar equivalent) $\longrightarrow$ $CH_3CH(CH_3)C(Cl){=}CH_2$

(b) $CH_3CH(CH_3)C{\equiv}CH \xrightarrow{H_2/Ni_2B\ (P\text{-}2)} CH_3CH(CH_3)CH{=}CH_2 \xrightarrow[\text{peroxides}]{HBr} CH_3CH(CH_3)CH_2CH_2Br$

(c) Product of (a) $\xrightarrow{H_2/Pt}$ $CH_3CH(CH_3)CH(Cl)CH_3$

(d) Product of (a) $\xrightarrow[\text{dark}]{Cl_2/CCl_4}$ $CH_3CH(CH_3){-}C(Cl)_2{-}CH_2Cl$

(e) Product of (a) $\xrightarrow[\text{no peroxides}]{HBr}$ $CH_3CH(CH_3){-}C(Cl)(Br){-}CH_3$

(f) $CH_3CH(CH_3)C{\equiv}CH \xrightarrow[\text{then } H^+]{KMnO_4/OH^-,}$ $CH_3CH(CH_3)CO_2H + CO_2$

7.46 (a) Four

(b) $CH_3(CH_2)_5{-}C(OH)(H){-}CH_2{-}CH{=}CH{-}(CH_2)_7CO_2H$ (cis; H and H on the same side of C=C) + Enantiomer

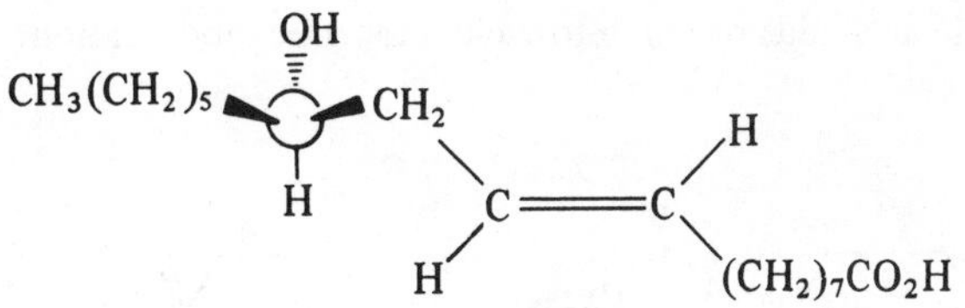

OH

$CH_3(CH_2)_5$ CH$_2$

H C=C H

H (CH$_2$)$_7$CO$_2$H

+ enantiomer

7.47 Hydroxylations by $KMnO_4$ are syn hydroxylations (cf. Section 7.10A). Thus, maleic acid must be the *cis*-dicarboxylic acid:

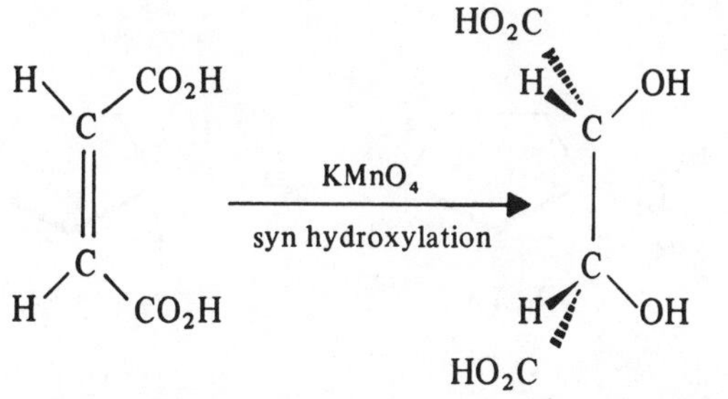

$KMnO_4$ → syn hydroxylation

Maleic acid meso-Tartaric acid

Fumaric acid must be the *trans*-dicarboxylic acid:

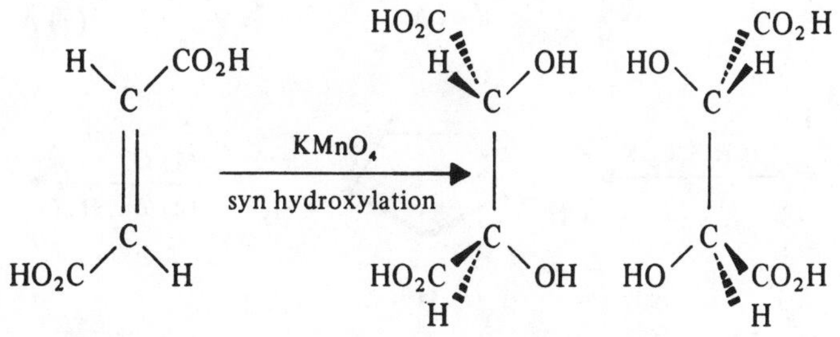

$KMnO_4$ → syn hydroxylation

Fumaric acid (±)-Tartaric acid

7.48 (a) The addition of bromine is an anti addition. Thus fumaric acid yields a meso compound.

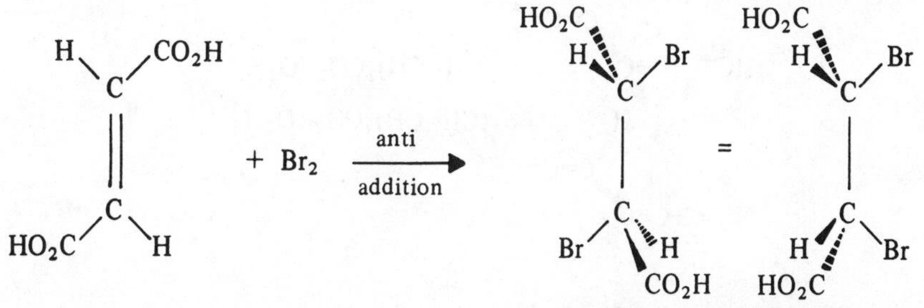

+ Br_2 anti addition →

A meso compound

(b) Maleic acid adds bromine to yield a racemic modification.

7.49

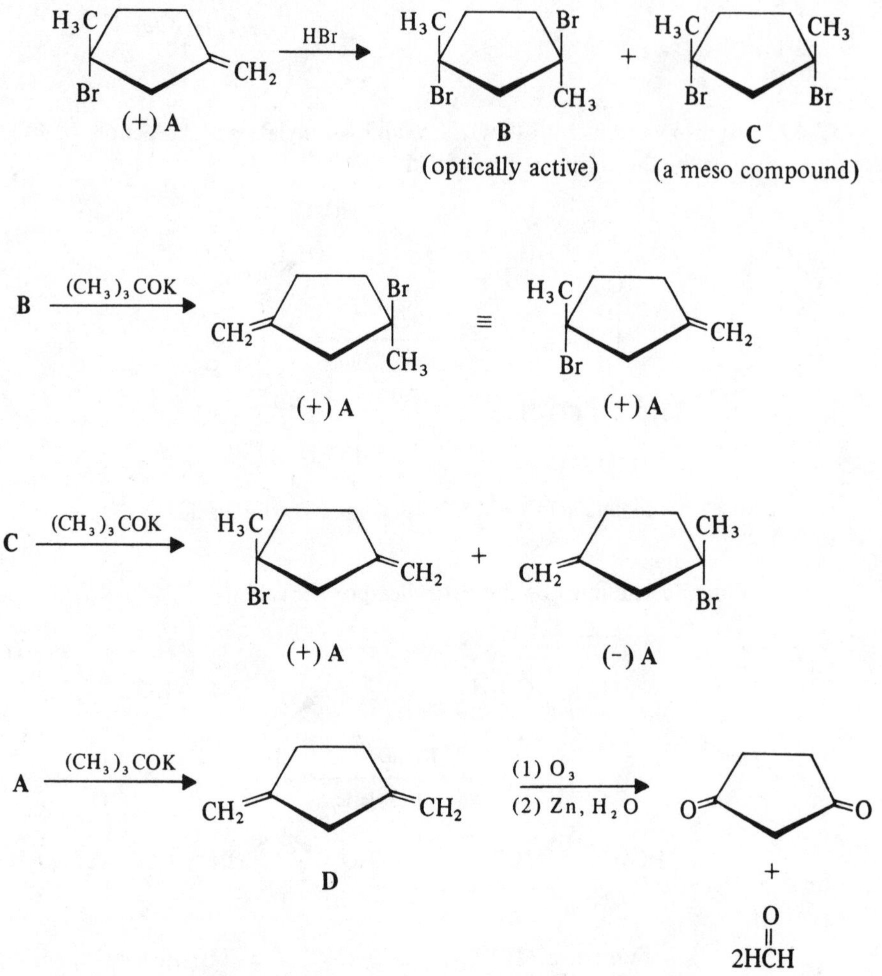

H_3C, Br, CH_2 — (+) **A** $\xrightarrow{HBr}$ H_3C, Br, Br, CH_3 — **B** (optically active) + H_3C, CH_3, Br, Br — **C** (a meso compound)

B $\xrightarrow{(CH_3)_3COK}$ CH_2, Br, CH_3 — (+) **A** ≡ H_3C, Br, CH_2 — (+) **A**

C $\xrightarrow{(CH_3)_3COK}$ H_3C, Br, CH_2 — (+) **A** + CH_2, CH_3, Br — (-) **A**

A $\xrightarrow{(CH_3)_3COK}$ CH_2, CH_2 — **D** $\xrightarrow{(1)\ O_3;\ (2)\ Zn, H_2O}$ O, O + 2HCH (with C=O)

7.50

$HC{\equiv}C{-}C{\equiv}C(H)C{=}C{=}C(H)(CH{=}CH)_2CH_2CO_2H$

$HC{\equiv}C{-}C{\equiv}C(H)C{=}C{=}C(H)(CH{=}CH)_2CH_2CO_2H$

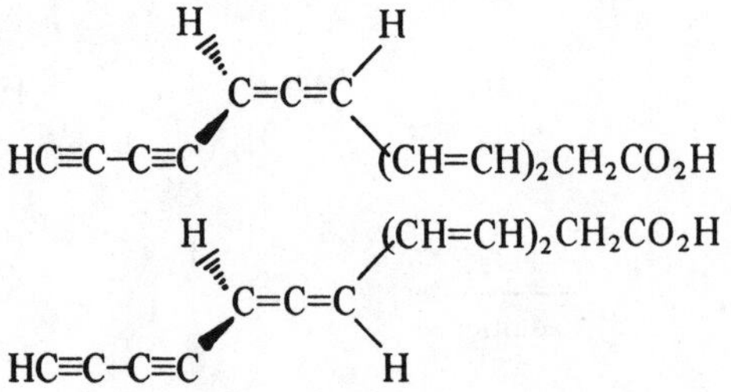

7.51

$$HC{\equiv}C\text{-}C(CH_3)(H)(CH_2CH_3) \xrightarrow[Pt]{H_2} CH_3CH_2CH(CH_3)CH_2CH_3$$

D

Optically active (the other enantiomer is an equally valid answer)

Optically inactive nonresolvable

SECTION REFERENCES FOR ADDITIONAL PROBLEMS

7.25 6.8, 7.10

7.26 7.2

7.27 7.2

7.28 7.10C

7.29 7.5

7.30 6.9, 6.15, 7.2, 7.4

7.31 6.15, 7.2

7.32 7.10B

7.33 6.13, 7.12

7.34 7.9A

7.35 7.9

7.36 7.10C

7.37 6.12, 7.2

7.38 7.8

7.39 7.10C, 7.11

7.40 7.11

7.41 7.7A, 7.9B, 7.10A

7.42 4.5

7.43 7.19

7.44 7.17-7.19

7.45 7.17

7.46 4.2, 4.3

7.47 7.10A, 7.7A

7.48 7.7

7.49 7.2, 6.12, 7.10

7.50 4.16

7.51 6.7, 6.20

SELF-TEST

7.1 Supply the missing compounds in the following equations. Show stereochemistry where appropriate. If more than one organic product results, give only the major product. If two steps are required, show them as (1) step 1, (2) step 2, and so on.

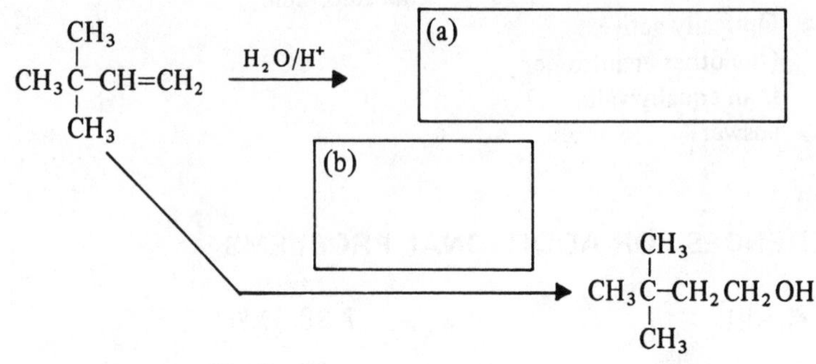

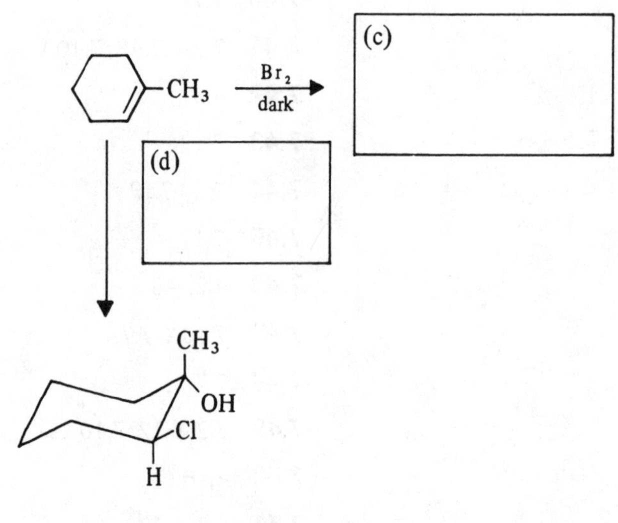

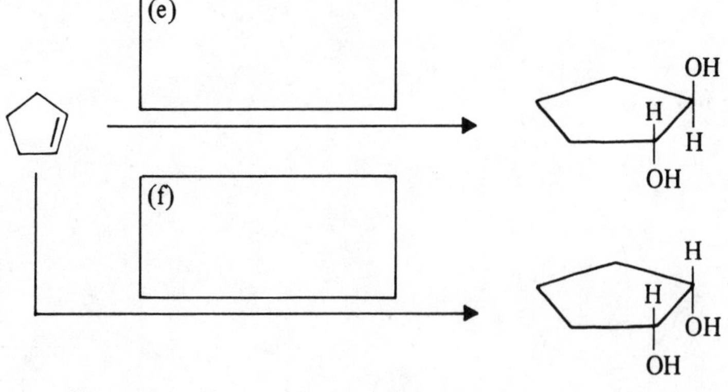

7.2 Give the structural formulas of the missing compounds. Show stereochemistry where appropriate.
Note: **A** and **B** are different

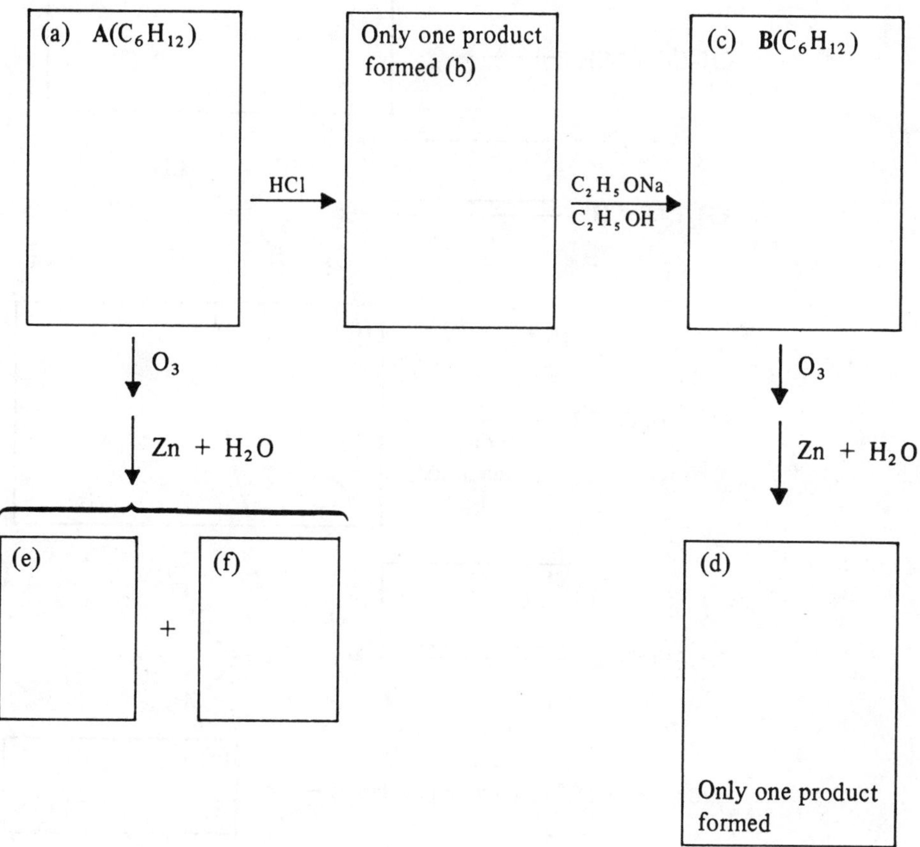

7.3 An unknown hydrocarbon, **A**, has the molecular formula C_7H_{10}. Complete hydrogenation gives **B** whose molecular formula is C_7H_{14}. Ozonolysis of A, followed by Zn + H_2O reduction of the ozonide, gives the following compounds in equal molar amounts.

$$\overset{O}{\overset{\|}{HC}}CH_2\overset{O}{\overset{\|}{C}}CH_3 \quad \text{and} \quad \overset{O}{\overset{\|}{HC}}CH_2\overset{O}{\overset{\|}{C}}H$$

A = (a) and **B** = (b)

7.4 Supply the missing reactant or major organic product in each of the following equations. Show stereochemistry where appropriate. If more than one step is needed, show them as (1), (2), and so on. If no reaction occurs. Write N.R.

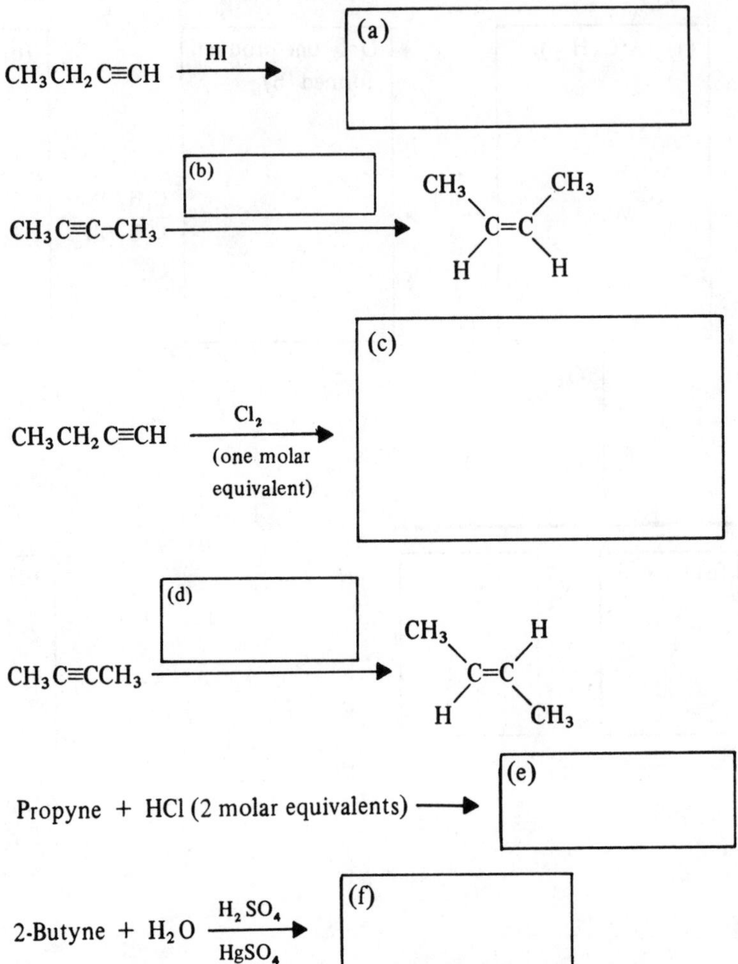

7.5 Tell whether each of the following statements is true (+) or false (−).

(a) Carbon 2 of 2-butyne is sp^2 hybridized.

(b) Carbon 1 of 1-butyne is sp hybridized.

7.6 A hydrocarbon whose molecular formula is C_7H_{12}, on catalytic hydrogenation (excess H_2/Pt), yields C_7H_{16}. The original hydrocarbon adds bromine and also reacts with $Ag(NH_3)_2^+OH^-$ to give a precipitate. Which of the following is a plausible choice of structure for the original hydrocarbon?

(a) 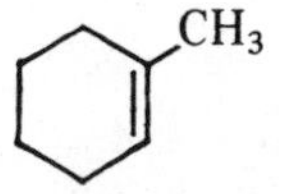 CH_3 (b) 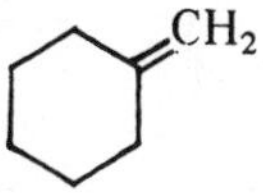 CH_2 (c) $CH_3CH{=}CHCH{=}CHCH_2CH_3$

(d) $CH_3CH_2CH_2C{\equiv}CCH_2CH_3$ (e) $CH_3CH_2CH_2CH_2CH_2C{\equiv}CH$

7.7 Select the major product of the dehydration of the alcohol,

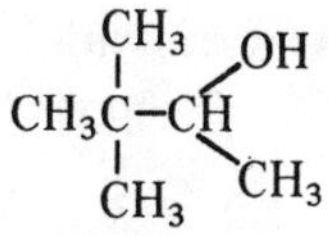

```
   CH3  OH
   |   /
CH3C—CH
   |   \
   CH3  CH3
```

```
(a)    CH3               (b)    CH3              (c)       CH3
       |                        |                          |
    CH3C—CH=CH2             CH3C=CHCH3              CH3C=C—CH3
       |                        |                      |
       CH3                      CH3                    CH3

(d)        CH3           (e)       CH3
           |                       |
    CH3CH—C=CH2             CH2=C—CHCH3
       |                         |
       CH3                       CH3
```

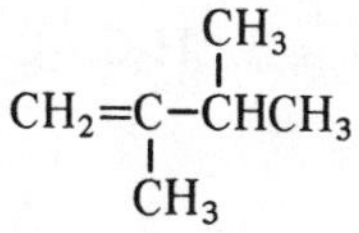

7.8 Give the major product of the reaction of *cis*-2-pentene with bromine.

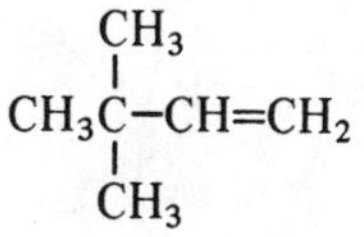

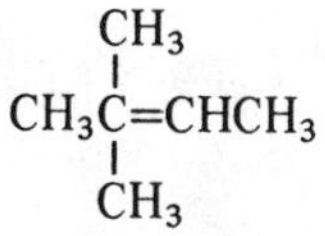

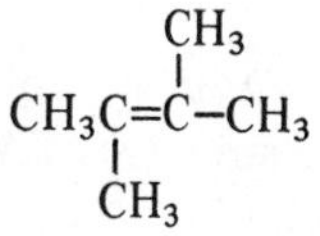

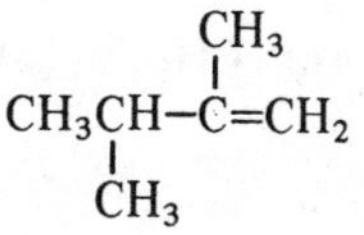

```
(a)   CH3        (b)   CH3        (c)   CH3        (d)   CH3
    H—+—Br          Br—+—H          H—+—Br          Br—+—H
    H—+—Br          Br—+—H          Br—+—H          H—+—Br
     CH2CH3          CH2CH3          CH2CH3          CH2CH3
```

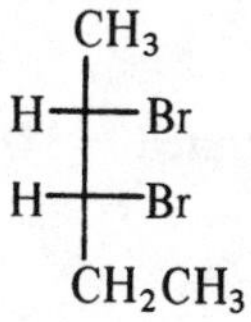

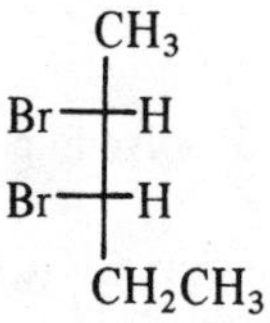

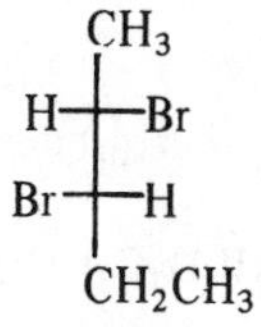

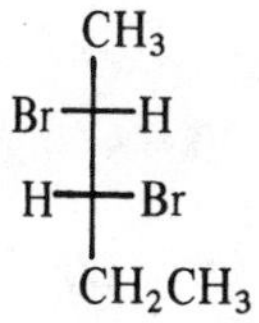

(e) A racemic mixture of (c) and (d)

7.9 The compound shown here is best prepared by which sequence of reactions?

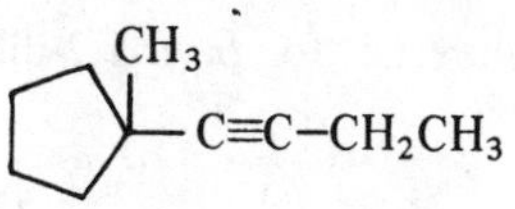

CH_3

$-C{\equiv}C{-}CH_2CH_3$

(a) 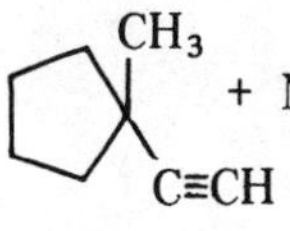 CH_3 / $C{\equiv}CH$ + $NaNH_2$ ⟶ then CH_3CH_2Br ⟶ product

(b) $CH_3CH_2C{\equiv}CH + NaNH_2 \longrightarrow$ then 1-bromo-1-methylcyclopentane (CH_3, Br) $\longrightarrow$ product

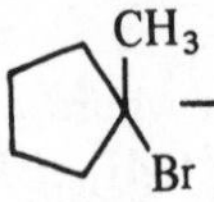

(c) 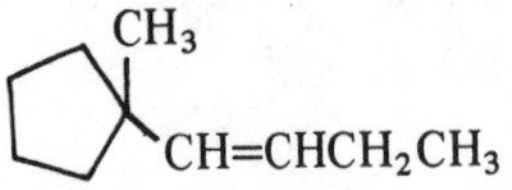 1-methylcyclopentyl–$CH{=}CHCH_2CH_3 + H_2 \xrightarrow{Pt}$ product

(d) 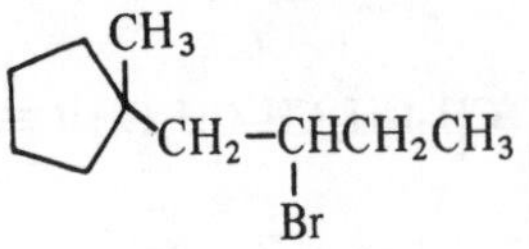 1-methylcyclopentyl–$CH_2{-}CH(Br)CH_2CH_3 \xrightarrow[C_2H_5OH]{NaOC_2H_5}$ product

7.10 A compound whose formula is C_6H_{10} (Compound **A**) reacts with H_2/Pt in excess to give a product C_6H_{12}, which does not decolorize Br_2/CCl_4. Compound **A** does not give any visible reaction with $Ag(NH_3)_2^+OH^-$.

Ozonolysis of 1 mole of **A** gives 1 mole of HCH (with $C{=}O$) and 1 mole of cyclopentanone ($=O$). Give the structure of **A**.

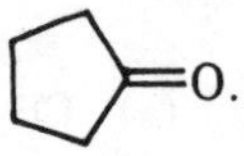

(a)  cyclohexene (b) $CH_3CH_2CH_2C{\equiv}CCH_3$ (c) $CH_3CH_2CH_2CH_2C{\equiv}CH$

(d) 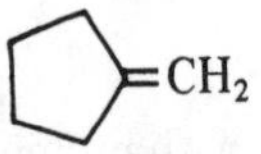 cyclopentane$=CH_2$ (e) $CH_2{=}CHCH_2CH_2CH{=}CH_2$

7.11 Compound **B** (C_5H_{10}) does not dissolve in cold, concentrated H_2SO_4. What is **B**?

(a) $CH_2{=}CHCH_2CH_2CH_3$ (b) $CH_3CH{=}CHCH_2CH_3$

(c)  cyclopentane (d)  cyclopentene

7.12 Which reaction sequence converts cyclohexene to *trans*-1,2-dihydroxycyclohexane? That is,

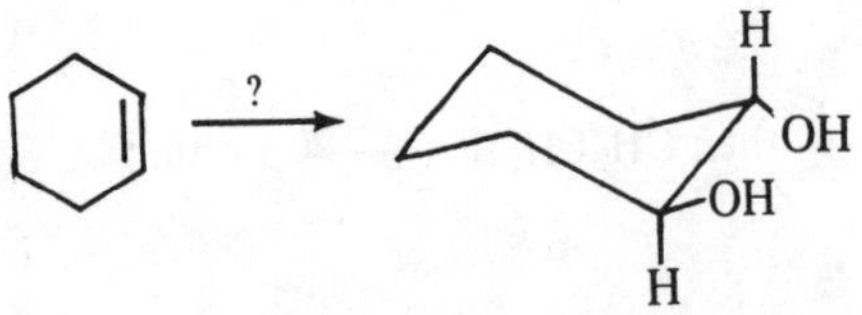

cyclohexene $\xrightarrow{?}$ chair cyclohexane with H, OH (upper carbon) and OH, H (lower carbon)

(a) Cold, Dilute, aqueous $KMnO_4$, OH^- (b) (1) O_3 (2) Zn/H_2O

(c) (1) OsO_4 (2) $NaHSO_3$ (d) (1) $R\overset{O}{\overset{\|}{C}}-OOH$ (2) H_3O^+/H_2O

(e) More than one of these

7.13 Which of the following sequences leads to the best synthesis of the compound $CH_3CH_2C{\equiv}CH$? (Assume that the quantities of reagents are sufficient to carry out the desired reaction.)

(a) $CH_3CH_2CH{=}CH_2 \xrightarrow{Br_2} \xrightarrow[H_2O]{NaOH}$

(b) $CH_3CH_2CH{=}CH_2 \xrightarrow{Br_2} \xrightarrow{NaNH_2}$

(c) $CH_3CH_2CH_2CHBr_2 \xrightarrow{H_2SO_4}$

(d) $CH_3CH_2CH_2CH_3 \xrightarrow[light]{Br_2} \xrightarrow{NaNH_2}$

(e) $CH_3CH_2CH{=}CH_2 \xrightarrow{O_3} \xrightarrow{Zn, H_2O}$

SUPPLEMENTARY PROBLEMS

S7.1 Supply the missing compound in each of the following reactions:

(a) [cyclopentene] $\xrightarrow[\text{dark } 25°C]{Br_2}$?

(b) $C_6H_{12}(?) \xrightarrow{O_3} \xrightarrow[H_2O]{Zn} CH_3\overset{O}{\overset{\|}{C}}CH_3$

S7.2 Show all the steps necessary to convert

[methylenecyclopentane, CH_2] into (a) [1-methylcyclopentene, CH_3] (b) [cyclopentylmethanol, CH_2OH] (c) [1-methylcyclopentanol, CH_3, OH]

S7.3 A certain compound has the molecular formula C_5H_8. Ozonolysis yields

$$H\overset{O}{\overset{\|}{C}}-CH_2CH_2\overset{O}{\overset{\|}{C}}CH_3$$

Give the structural formula of the original compound.

SOLUTIONS TO SUPPLEMENTARY PROBLEMS

S7.1 (a) 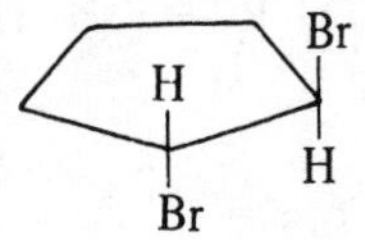 H, Br, Br, H (cyclopentane with trans Br substituents) Bromination involves anti addition

(b) 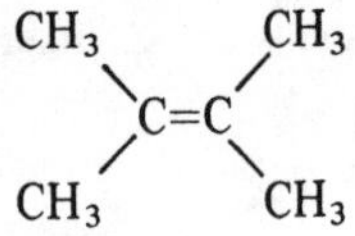 $(CH_3)_2C{=}C(CH_3)_2$

S7.2

(a) 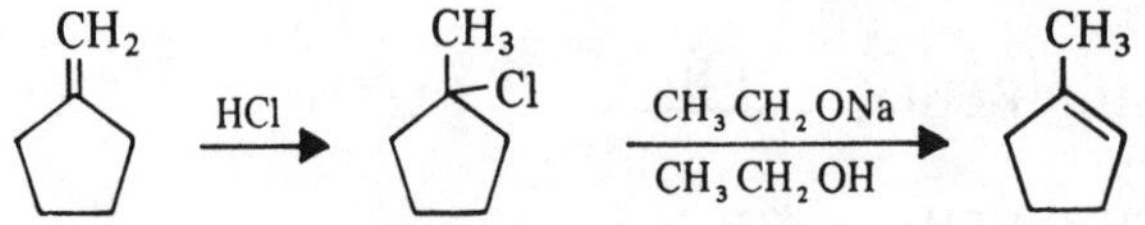 methylenecyclopentane ($=CH_2$) $\xrightarrow{HCl}$ 1-chloro-1-methylcyclopentane (CH_3, Cl) $\xrightarrow[CH_3CH_2OH]{CH_3CH_2ONa}$ 1-methylcyclopentene (CH_3)

(b) 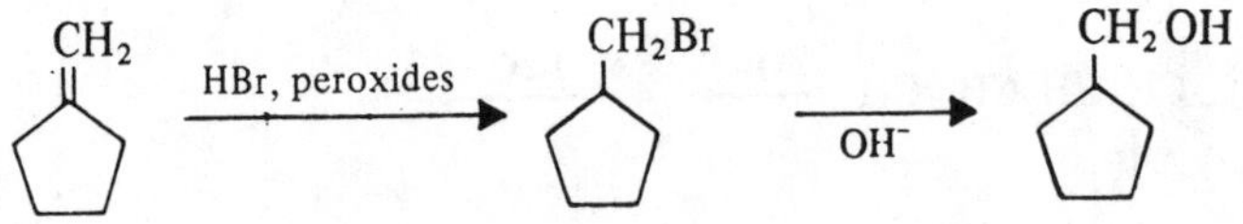 methylenecyclopentane ($=CH_2$) $\xrightarrow{\text{HBr, peroxides}}$ CH_2Br-cyclopentane $\xrightarrow[OH^-]{}$ CH_2OH-cyclopentane

(c) 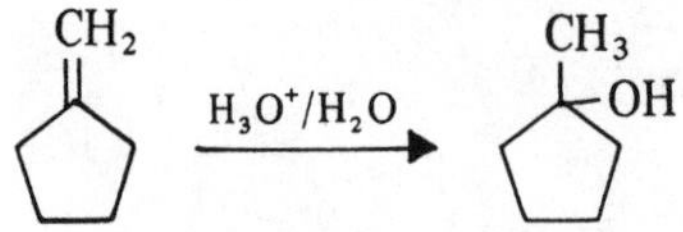 methylenecyclopentane ($=CH_2$) $\xrightarrow{H_3O^+/H_2O}$ 1-methylcyclopentanol (CH_3, OH)

S7.3 Rewriting the ozonolysis product with the carbon atoms of the two carbonyl groups joined together as a double bond reveals the structure of the original C_5H_8

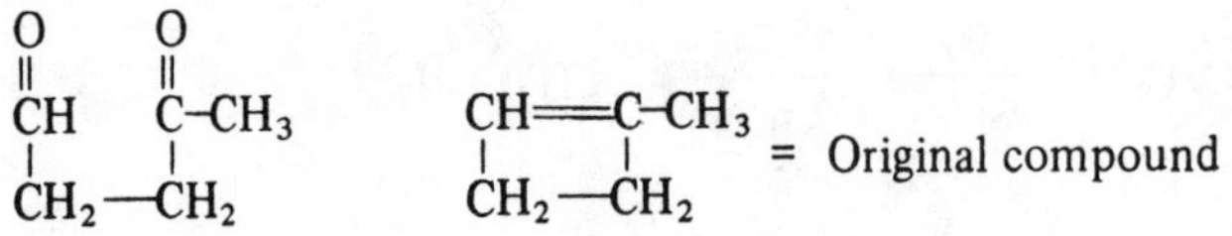

$$\underset{CH_2-CH_2}{\overset{O\quad\ \ O}{CH\quad C-CH_3}} \qquad \underset{CH_2-CH_2}{CH{=}C-CH_3} = \text{Original compound}$$

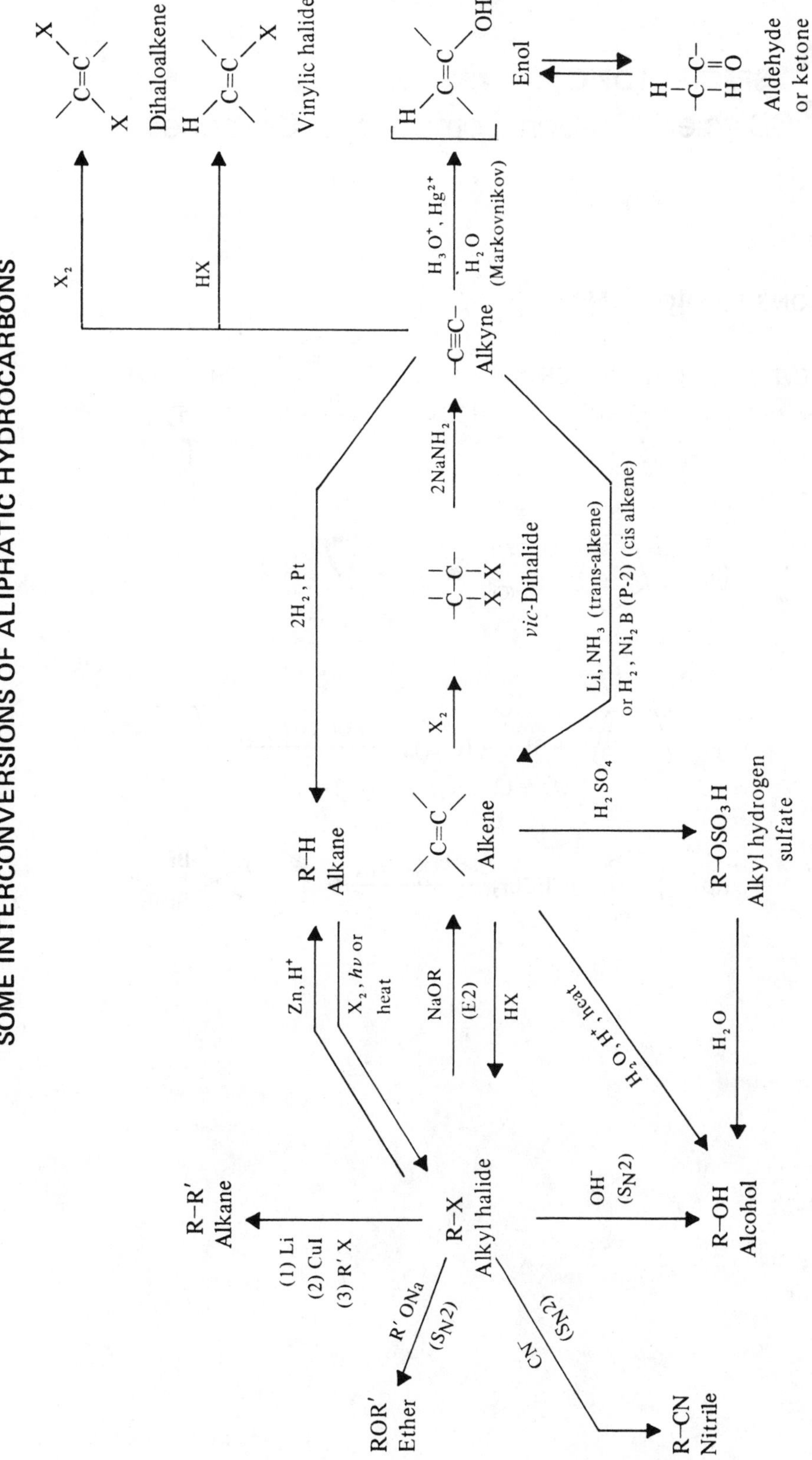
SOME INTERCONVERSIONS OF ALIPHATIC HYDROCARBONS
R–R′
Alkane
(1) Li
(2) CuI
(3) R′X
R–X
Alkyl halide
R′ONa
(S_N2)
ROR′
Ether
CN^-
(S_N2)
R–CN
Nitrile
OH^-
(S_N2)
R–OH
Alcohol
Zn, H^+
X_2, $h\nu$ or heat
R–H
Alkane
NaOR
(E2)
HX
H_2O, H^+, heat
H_2O
Alkene
H_2SO_4
$R–OSO_3H$
Alkyl hydrogen sulfate
$2H_2$, Pt
X_2
vic-Dihalide
$2NaNH_2$
Li, NH_3 (trans-alkene)
or H_2, Ni_2B (P-2) (cis alkene)
–C≡C–
Alkyne
X_2
Dihaloalkene
HX
Vinylic halide
H_3O^+, Hg^{2+}
H_2O
(Markovnikov)
Enol
Aldehyde
or ketone

C SPECIAL TOPIC
Divalent Carbon Compounds: Carbenes

SOLUTIONS TO PROBLEMS

C.1 (a) $CH_3CH{=}CHCH_3$ (cis; CH_3, CH_3 on same side, H, H) $+ HCCl_3 \xrightarrow{KOC(CH_3)_3}$ cis-1,2-dimethyl-3,3-dichlorocyclopropane (CH_3, CH_3; H, H; Cl, Cl)

(b) cyclopentene $\xrightarrow[Zn(Cu)]{CH_2I_2}$ bicyclo[3.1.0]hexane (H, CH_2, H)

(c) cyclopentene $+ H{-}C(Cl)_2{-}C(CH_3)_3 \xrightarrow{KOC(CH_3)_3}$ bicyclic product with $C(C(CH_3)_3)(Cl)$

(d) cyclohexene $+ HCBr_3 \xrightarrow{KOC(CH_3)_3}$ 7,7-dibromobicyclo[4.1.0]heptane (Br, Br)

8 ALCOHOLS AND ETHERS

SUMMARY OF ETHERS AND EPOXIDES

ETHERS

$$R\text{-}OH \xrightarrow[(-H_2O)]{H^+} R\text{-}O\text{-}R$$

$$R\text{-}OH \xrightarrow{Na} R\text{-}ONa \xrightarrow[(X=Cl,\ Br,\ I)]{R'X} R\text{-}O\text{-}R' \xrightarrow{HX} RX + R'X$$

$$R\text{-}OH \xrightarrow[H_2SO_4]{CH_2=C(CH_3)\text{-}CH_3} R\text{-}O\text{-}C(CH_3)_2\text{-}CH_3$$

EPOXIDES

$$>C=C< \xrightarrow{R\text{-}C(=O)\text{-}O\text{-}OH} >C\overset{O}{\text{—}}C< \text{ (epoxide)}$$

$$\text{epoxide} \xrightarrow[H^+]{H_2O} -C(OH)\text{-}C(OH)-$$

$$\text{epoxide} \xrightarrow{NH_3} -C(OH)\text{-}C(NH_2)-$$

$$\text{epoxide} \xrightarrow[ROH]{RONa} -C(OR)\text{-}C(OH)-$$

SUMMARY OF ALCOHOLS

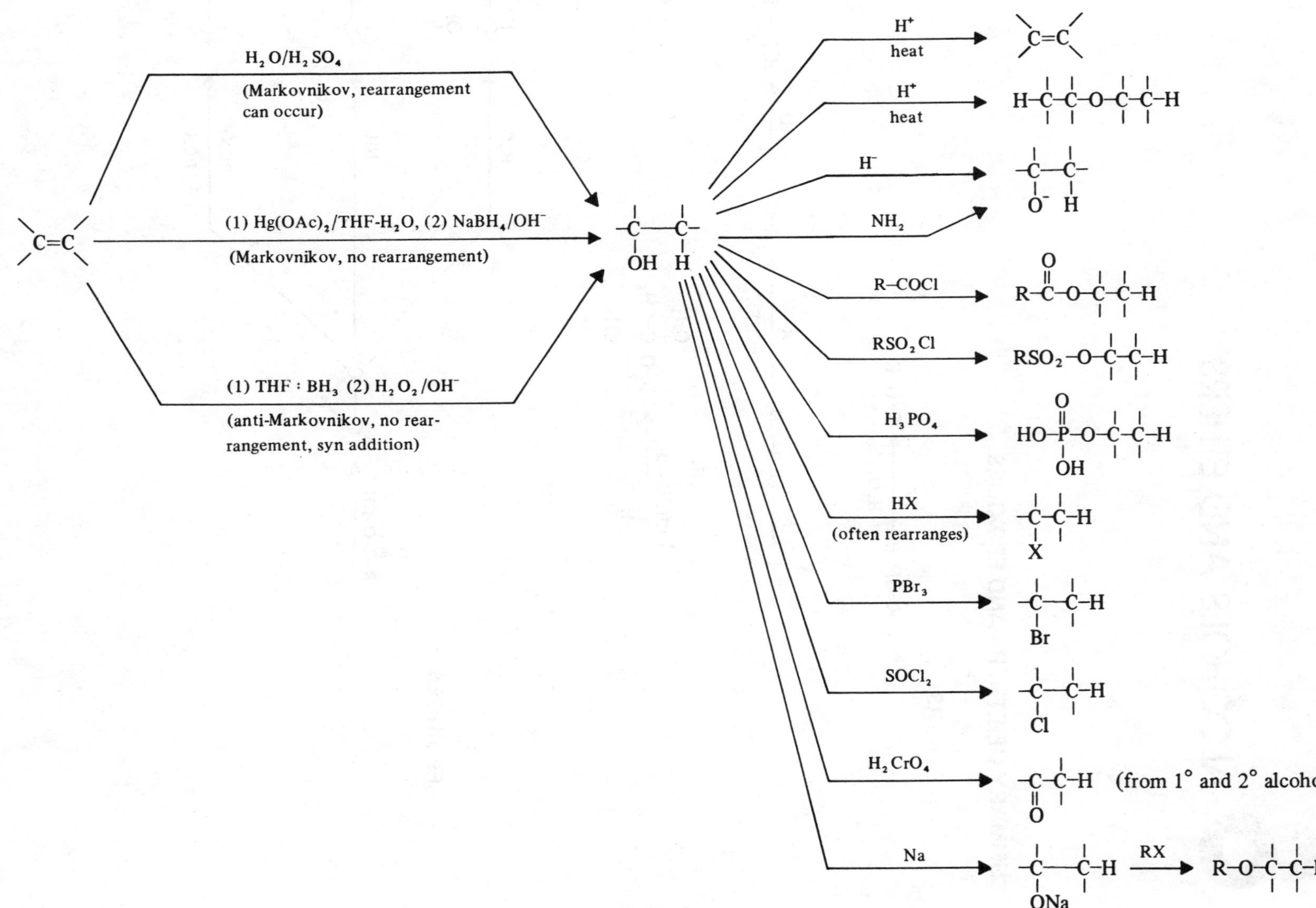

SOLUTIONS TO PROBLEMS

8.1 (a) Alcohols $CH_2{=}CHCH_2OH$
2-Propen-1-ol
(allyl alcohol)

H_2C—CH–OH (ring closed by CH_2)

Cyclopropanol

Ethers $CH_2{=}CH{-}O{-}CH_3$
Methoxyethene
(methyl vinyl ether)

(b) Alcohols $CH_2{=}CHCH_2CH_2OH$
3-Buten-1-ol

$CH_2{=}CHCH(OH)CH_3$

3-Buten-2-ol

$CH_3CH{=}CHCH_2OH$
2-Buten-1-ol (cis and trans)

$CH_2{=}C(CH_3){-}CH_2OH$
2-Methyl-2-propen-1-ol

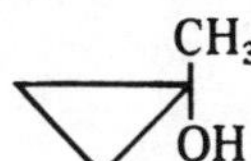

Cyclobutanol

1-Methylcyclopropanol (ring carbon bearing CH_3 and OH)

Cyclopropylmethanol (ring bearing CH_2OH)

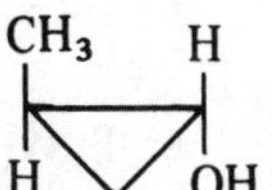

trans-2-Methylcyclopropanol

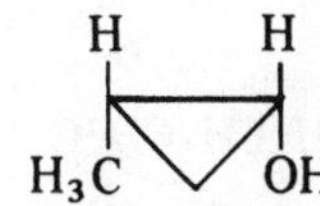

cis-2-Methylcyclopropanol

Ethers $CH_3CH{=}CH{-}OCH_3$
1-Methoxypropene

$CH_2{=}CHCH_2OCH_3$
3-Methoxypropene

$CH_2{=}C(CH_3){-}OCH_3$
2-Methoxypropene

$CH_2{=}CHOCH_2CH_3$
Ethoxyethene

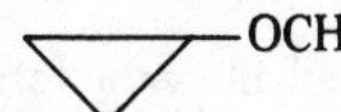

Methoxycyclopropane

(c) Alcohols $CH_3CH_2CH_2CH_2CH_2OH$ 1-Pentanol

$CH_3CH_2CH_2CH(OH)CH_3$ 2-Pentanol

	$CH_3CH_2CHCH_2CH_3$ | OH	3-Pentanol
	CH_3 | $CH_3CH_2CHCH_2OH$	2-Methyl-1-butanol
	CH_3 | $CH_3CHCH_2CH_2OH$	3-Methyl-1-butanol
	CH_3 | $CH_3CH_2CCH_3$ | OH	2-Methyl-2-butanol
	CH_3 | $CH_3CHCHCH_3$ | OH	3-Methyl-2-butanol
	CH_3 | $CH_3{-}C{-}CH_2OH$ | CH_3	2,2-Dimethyl-1-propanol (neopentyl alcohol)
Ethers	$CH_3CH_2CH_2CH_2{-}O{-}CH_3$	Butyl methyl ether (1-methoxybutane)
	CH_3 | $CH_3CHCH_2{-}O{-}CH_3$	Isobutyl methyl ether (2-methyl-1-methoxypropane)
	CH_3 | $CH_3{-}C{-}O{-}CH_3$ | CH_3	*tert*-Butyl methyl ether (2-methyl-2-methoxypropane)
	CH_3 | $CH_3CH_2CH{-}O{-}CH_3$	*sec*-Butyl methyl ether (2-methoxybutane)
	$CH_3CH_2CH_2{-}O{-}CH_2CH_3$	Ethyl propyl ether (1-ethoxypropane)
	CH_3 | $CH_3CH{-}O{-}CH_2CH_3$	Ethyl isopropyl ether (2-ethoxypropane)

8.2 The two hydroxyl groups in ethylene glycol allow the formation of more hydrogen bonds than in the monohydroxy alcohols. Thus a single diol molecule can be associated with many neighboring diol molecules.

8.3

(a) $CH_3\overset{\displaystyle CH_3}{\overset{|}{C}}{=}CH_2 + H_2O \xrightarrow{H^+} CH_3\underset{\displaystyle OH}{\underset{|}{\overset{\displaystyle CH_3}{\overset{|}{C}}}}CH_3$

(b) $CH_3CH_2CH_2CH_2CH{=}CH_2 + H_2O \xrightarrow{H^+} CH_3CH_2CH_2CH_2\overset{\displaystyle OH}{\overset{|}{C}}HCH_3$

(c) cyclopentene $+ H_2O \xrightarrow{H^+}$ cyclopentanol (OH)

(d) 1-methylcyclohexene (CH_3) $+ H_2O \xrightarrow{H^+}$ 1-methylcyclohexanol (CH_3, OH)

8.4 Rearrangement of the secondary carbocation to the more stable tertiary carbocation,

$$CH_3\underset{\displaystyle CH_3}{\underset{|}{\overset{\displaystyle CH_3}{\overset{|}{C}}}}{-}CH{=}CH_2 \underset{}{\overset{H^+}{\rightleftharpoons}} CH_3{-}\underset{\displaystyle CH_3}{\underset{|}{\overset{\displaystyle CH_3}{\overset{|}{C}}}}{-}\overset{+}{C}H{-}CH_3 \longrightarrow CH_3{-}\underset{\displaystyle CH_3}{\underset{|}{\overset{\displaystyle CH_3}{\overset{|}{\overset{+}{C}}}}}{-}CH{-}CH_3$$

followed by reaction of the resulting carbocation with water:

$$CH_3{-}\underset{\displaystyle CH_3}{\underset{|}{\overset{\displaystyle CH_3}{\overset{|}{\overset{+}{C}}}}}{-}CHCH_3 + H_2O \rightleftharpoons CH_3{-}\underset{\displaystyle CH_3}{\underset{|}{\overset{\displaystyle {}^{+}OH_2}{\overset{|}{C}}}}{-}\underset{\displaystyle CH_3}{\underset{|}{C}}HCH_3 \rightleftharpoons CH_3{-}\underset{\displaystyle CH_3}{\underset{|}{\overset{\displaystyle OH}{\overset{|}{C}}}}{-}\underset{\displaystyle CH_3}{\underset{|}{C}}HCH_3 + H^+$$

8.5

(a) $CH_3\overset{\displaystyle CH_3}{\overset{|}{C}}{=}CH_2 \xrightarrow[\text{THF–}H_2O]{Hg(OAc)_2} CH_3{-}\underset{\displaystyle HO}{\underset{|}{\overset{\displaystyle CH_3}{\overset{|}{C}}}}{-}CH_2HgOAc \xrightarrow[OH^-]{NaBH_4} CH_3{-}\underset{\displaystyle OH}{\underset{|}{\overset{\displaystyle CH_3}{\overset{|}{C}}}}{-}CH_3$

(b) $CH_3CH{=}CH_2 \xrightarrow[\text{THF–}H_2O]{Hg(OAc)_2} CH_3\underset{\displaystyle OH}{\underset{|}{C}}HCH_2HgOAc \xrightarrow[OH^-]{NaBH_4} CH_3\underset{\displaystyle OH}{\underset{|}{C}}HCH_3$

(c) $CH_3\underset{\displaystyle CH_3}{\underset{|}{C}}{=}CHCH_3 \xrightarrow[\text{THF–}H_2O]{Hg(OAc)_2} CH_3\underset{\displaystyle CH_3}{\underset{|}{\overset{\displaystyle HO}{\overset{|}{C}}}}{-}\overset{\displaystyle HgOAc}{\overset{|}{C}}HCH_3 \xrightarrow[OH^-]{NaBH_4} CH_3\underset{\displaystyle CH_3}{\underset{|}{\overset{\displaystyle OH}{\overset{|}{C}}}}CH_2CH_3$

8.6

(a) $>C=C< + {}^{+}HgOCCF_3$ (O above) $\longrightarrow$ $-\overset{|}{C}^{\delta+}-\overset{|}{C}-$ with bridging $HgOCCF_3^{\delta+}$ (O above) $\xrightarrow{R-\ddot{O}-H}$ $ROH^{+}-\overset{|}{C}-\overset{|}{C}-HgOCCF_3$ (O above)

$\xrightarrow{-H^{+}}$ $RO-\overset{|}{C}-\overset{|}{C}-HgOCCF_3$ (O above)

(b) $CH_3-\overset{CH_3}{\overset{|}{C}}=CH_2 \xrightarrow[\text{solvomercuration}]{Hg(O\overset{O}{\overset{\|}{C}}CF_3)_2/THF\text{-}CH_3OH} CH_3-\overset{CH_3}{\overset{|}{\underset{CH_3O}{\underset{|}{C}}}}-CH_2Hg\overset{O}{\overset{\|}{O}}CCF_3$

$\xrightarrow[\text{demercuration}]{NaBH_4/OH^-} CH_3\overset{CH_3}{\overset{|}{\underset{OCH_3}{\underset{|}{C}}}}CH_3 + Hg + CF_3CO_2^-$

8.7

(a) $CH_3CH_2CH{=}CH_2 \xrightarrow{THF:BH_3} (CH_3CH_2CH_2CH_2)_3B$

(b) $CH_3\overset{CH_3}{\overset{|}{C}}{=}CH_2 \xrightarrow{THF:BH_3} (CH_3\overset{CH_3}{\overset{|}{C}}HCH_2)_3B$

(c) $CH_3CH{=}CHCH_3 \xrightarrow{THF:BH_3} (CH_3CH_2\overset{CH_3}{\overset{|}{C}}H)_3B$

(d) 1-methylcyclohexene (CH_3) $\xrightarrow[\text{syn addition, anti-Markovnikov}]{THF:BH_3}$ cyclohexane chair with H, CH_3, H, $B-$

8.8

$CH_3\overset{CH_3}{\overset{|}{C}}{=}CHCH_3 \xrightarrow{THF:BH_3} \left(CH_3\overset{CH_3}{\overset{|}{C}}H-\underset{CH_3}{\underset{|}{C}}H-\right)_2BH$

Disiamylborane

8.9

$$(CH_3)_3CCH{=}CHCH_3 \xrightarrow{THF:BH_3} (CH_3)_3CCH_2\underset{\underset{|}{B-}}{\underset{|}{C}}HCH_3 +$$

$$(CH_3)_3C\underset{\underset{|}{-B}}{\underset{|}{C}}HCH_2CH_3 \xrightarrow{160°C} (CH_3)_3CCH_2CH_2CH_2{-}B\langle$$

$$\xrightarrow[160°C]{CH_3(CH_2)_7CH{=}CH_2} (CH_3)_3CH_2CH{=}CH_2 + CH_3(CH_2)_8CH_2B\langle$$

Distills out at 160°C — Boils above 160°C

8.10 (a)

$$3CH_3CH_2CH{=}CH_2 + THF:BH_3 \longrightarrow (CH_3CH_2CH_2CH_2)_3B \xrightarrow[OH^-]{H_2O_2} 3CH_3CH_2CH_2CH_2OH + H_3BO_3$$

(b)

$$2CH_3{-}\overset{\overset{CH_3}{|}}{C}{=}CH{-}CH_3 + THF:BH_3 \longrightarrow (CH_3{-}\overset{\overset{CH_3}{|}}{C}H{-}\overset{\overset{CH_3}{|}}{C}H)_2BH$$

$$\xrightarrow[OH^-]{H_2O_2} 2CH_3{-}\overset{\overset{CH_3}{|}}{C}H{-}\overset{\overset{OH}{|}}{C}H{-}CH_3 + H_3BO_3$$

(c) 3 (1-methylcyclohexene) $-CH_3$ + THF:BH_3 ⟶ (cyclohexane ring with H, CH_3, H, B)$_3$ $\xrightarrow[OH^-]{H_2O_2}$

3 (cyclohexane ring with H, OH, CH_3, H) + H_3BO_3

8.11 (a)

$$3CH_3{-}\overset{\overset{CH_3}{|}}{C}{=}CH_2 + THF:BH_3 \longrightarrow (CH_3{-}\overset{\overset{CH_3}{|}}{C}H{-}CH_2)_3B \xrightarrow[heat]{CH_3CO_2D}$$

$$3CH_3{-}\overset{\overset{CH_3}{|}}{C}H{-}CH_2D + (CH_3CO_2)_3B$$

(b) $3\ \text{(cyclohexylidene)}{=}CH_2 + THF:BH_3 \longrightarrow \left(\text{(cyclohexyl-H)}{-}CH_2\right)_3B \xrightarrow[\text{heat}]{CH_3CO_2D}$

$3\ \text{(cyclohexyl)}{-}CH_2D + (CH_3CO_2)_3B$

(c) $3\ \text{(1-methylcyclopentene, } CH_3) + THF{:}BH_3 \longrightarrow$ (cyclopentyl with CH_3, H, H, B–) $\xrightarrow[\text{heat}]{CH_3CO_2D}$

3 (cyclopentane with CH_3, H, H, D) $+ (CH_3CO_2)_3B$

(d) $3\ \text{(1-methylcyclopentene, } CH_3) + THF{:}BD_3 \longrightarrow$ (cyclopentyl with CH_3, D, H, B–) $\xrightarrow[\text{heat}]{CH_3CO_2T}$

$(BD_3)_2 \xrightarrow{THF} THF{:}BD_3$

3 (cyclopentane with CH_3, D, H, T) $+ (CH_3CO_2)_3B$

8.12 Dehydrohalogenation would lead mainly to the following alkene:

$$CH_3\underset{\underset{Br}{|}}{\overset{\overset{CH_3}{|}}{C}}CH_2CH_2CH_3 \xrightarrow[\substack{CH_3CH_2OH \\ (E2)}]{CH_3CH_2ONa} CH_3\overset{\overset{CH_3}{|}}{C}{=}CHCH_2CH_3$$

Then hydroboration, isomerization, and oxidation would produce the required alcohol

$$CH_3\overset{\overset{CH_3}{|}}{C}{=}CHCH_2CH_3 \xrightarrow{THF:BH_3} CH_3\overset{\overset{CH_3}{|}}{C}H\underset{\underset{B-}{|}}{C}HCH_2CH_3 \xrightarrow{160°C}$$

$$CH_3\overset{\overset{CH_3}{|}}{C}HCH_2CH_2CH_2{-}B\langle \xrightarrow{H_2O_2,\ OH^-} CH_3\overset{\overset{CH_3}{|}}{C}HCH_2CH_2CH_2OH$$

8.13 (a) $CH_3C{\equiv}C:^- + CH_3CH_2OH \rightleftharpoons CH_3C{\equiv}CH + CH_3CH_2O^-$

Stronger base | Stronger acid | Weaker acid | Weaker base

(b) $\overset{\delta-}{CH_3CH_2CH_2CH_2}:\overset{\delta+}{Li} + CH_3CH_2OH \rightleftharpoons CH_3CH_2CH_2CH_3 + CH_3CH_2\bar{O}Li^+$

Stronger base | Stronger acid | Weaker acid | Weaker base

(c) $NaH + CH_3CH_2OH \rightleftharpoons H_2 + CH_3CH_2\bar{O}Na^+$

Stronger base | Stronger acid | Weaker acid | Weaker base

8.14

$$CH_3CH_2\overset{*}{O}H + Cl{-}\underset{O}{\overset{O}{\overset{\|}{\underset{\|}{S}}}}{-}CH_3 + OH^- \longrightarrow CH_3CH_2\overset{*}{O}{-}\underset{O}{\overset{O}{\overset{\|}{\underset{\|}{S}}}}{-}CH_3 + Cl^- + H_2O$$

If C–O bond cleavage does not occur, then all of the isotopically labeled oxygen (O* = ^{18}O) will be found in the sulfonate ester. Otherwise all or part of the ^{18}O will be found in the water formed in the reaction.

8.15

(a) p-$CH_3C_6H_4SO_3H$ $\xrightarrow{PCl_5}$ p-$CH_3C_6H_4SO_2Cl$ $\xrightarrow{CH_3OH}$ p-$CH_3C_6H_4SO_2{-}OCH_3$

(b) p-$CH_3C_6H_4SO_2Cl$ $\xrightarrow[OH^-]{CH_3CH(OH)CH_3}$ p-$CH_3C_6H_4S(=O)_2{-}OCH(CH_3)CH_3$

(c) $CH_3SO_3H \xrightarrow{PCl_5} CH_3SO_2Cl \xrightarrow[(CH_3)_3C{-}OH]{OH^-} CH_3SO_2{-}O{-}C(CH_3)_3$

8.16

(a) $H\cdots C(CH_3)(C_2H_5)-OH + TsCl \xrightarrow[(-HCl)]{\text{retention}} H\cdots C(CH_3)(C_2H_5)-OTs$

(*R*)-2-Butanol

(b) $HO^- \; H\cdots C(CH_3)(C_2H_5)-OTs \xrightarrow[S_N2]{\text{inversion}} HO-C\cdots H(CH_3)(C_2H_5) + OTs^-$

(c) H_3C–cyclohexane–OH $\xrightarrow[\text{retention}]{TsCl}$ H_3C–cyclohexane–OTs $\xrightarrow[\text{inversion}]{Cl^-}$ H_3C–cyclohexane–Cl

cis-4-Methylcyclohexanol

trans-1-Chloro-4-methylcyclohexane

8.17 *trans*-2-Pentene because it is more stable.

8.18

(a) $CH_3-C(CH_3)_2-CH(OH)CH_3 + H^+ \rightleftharpoons CH_3-C(CH_3)_2-CH(\overset{+}{O}H_2)CH_3 \overset{-H_2O}{\rightleftharpoons} CH_3-C(CH_3)_2-\overset{+}{C}H-CH_3$

$CH_3-C(CH_3)_2-\overset{+}{C}H-CH_2-H \overset{H_2O}{\rightleftharpoons} CH_3-C(CH_3)_2-CH{=}CH_2 + H_3O^+$

I

But this secondary carbocation can also rearrange to a tertiary carbocation before losing a proton:

$CH_3-C(CH_3)_2-\overset{+}{C}H-CH_3 \longrightarrow CH_2(H_b)-\overset{+}{C}(CH_3)-C(H_a)(CH_3)-CH_3$

(b) $\xrightarrow{H_2O}$ $CH_2{=}C(CH_3)-CH(CH_3)-CH_3 + H_3O^+$

II

(a) $\xrightarrow{H_2O}$ $CH_3-C(CH_3){=}C(CH_3)-CH_3 + H_3O^+$

III

(b) 2,3-Dimethyl-2-butene (**III**) is the most substituted alkene, therefore it is most stable.

8.19 (a) Tertiary alcohols react faster than secondary alcohols because they form more stable carbocations; that is, 3° rather than 2°:

$$CH_3-\underset{CH_3}{\overset{CH_3}{C}}-\underset{H}{\overset{+}{O}}-H \rightleftharpoons CH_3-\overset{+}{C}(CH_3)_2 + HO-H$$

$$CH_3-\overset{+}{C}(CH_3)_2 \xrightarrow{Cl^-} CH_3-\underset{CH_3}{\overset{CH_3}{C}}-Cl$$

(b) CH_3OH reacts faster than 1° alcohols because it offers less hindrance to S_N2 attack. (Recall that CH_3OH and 1° alcohols must react through an S_N2 mechanism.)

8.20

$$CH_3\overset{CH_3}{C}H\underset{OH}{C}HCH_3 + HBr \rightleftharpoons CH_3\overset{CH_3}{C}H\underset{^+OH_2}{C}HCH_3 + Br^-$$

$$\xrightleftharpoons{-H_2O}$$

$$CH_3\overset{CH_3}{C}H-\overset{+}{C}HCH_3 \longrightarrow CH_3\overset{CH_3}{\underset{+}{C}}-CH_2CH_3 \xrightarrow{Br^-} CH_3\underset{Br}{\overset{CH_3}{C}}-CH_2CH_3$$

8.21

(a) $$CH_3-\underset{CH_3}{\overset{CH_3}{C}}-OH \xrightarrow{H^+} CH_3-\underset{CH_3}{\overset{CH_3}{C}}-OH_2^+ \xrightarrow{-H_2O}$$

$$CH_3-\underset{CH_3}{\overset{CH_3}{C^+}} \xrightarrow[\text{(1° only)}]{R-OH} CH_3-\underset{CH_3}{\overset{CH_3}{C}}-\underset{H}{\overset{+}{O}}-R \xrightarrow{-H^+} CH_3-\underset{CH_3}{\overset{CH_3}{C}}-O-R$$

This reaction succeeds because a 3° carbocation is much more stable than a 1° carbocation. Mixing the 1° alcohol and H_2SO_4, consequently, does not lead to formation of appreciable amounts of a 1° carbocation. However, when the 3° alcohol is added, it is rapidly converted to a 3° carbocation, which then reacts with the 1° alcohol that is present in the mixture.

8.22

(a) (1) $CH_3CH(CH_3)O^-Na^+ + CH_3{-}L \longrightarrow CH_3CH(CH_3)O{-}CH_3 + L^- + Na^+$

(L = X, OSO_2R, or OSO_2OR)

(2) $CH_3O^- + CH_3{-}CH(CH_3){-}L \longrightarrow CH_3O{-}CH(CH_3)CH_3 + L^-$

(L = X, OSO_2R, or OSO_2OR)

(b) Both methods involve S_N2 reactions. Therefore, method (1) is better because substitution takes place at an unhindered methyl carbon atom. In method (2) where substitution must take place at a relatively hindered secondary carbon atom the reaction would be accompanied by considerable elimination.

8.23 Reaction of the alcohol with K and then of the resulting salt with C_2H_5Br does not break bonds to the stereocenter, and these reactions therefore occur with retention of configuration at the stereocenter.

Reaction of the tosylate, $C_6H_5CH_2CH(OTs)CH_3$, with C_2H_5OH in K_2CO_3 solution, however, is an S_N2 reaction that takes place at the stereocenter and thus it occurs with inversion at the stereocenter.

8.24

$Cl{-}CH_2CH_2CH_2CH_2{-}OH \underset{}{\overset{OH^-}{\rightleftharpoons}} Cl{-}CH_2CH_2CH_2CH_2{-}\ddot{O}:^- + H_2O \longrightarrow$ (tetrahydrofuran ring: $CH_2{-}CH_2{-}CH_2{-}CH_2{-}O$) $+ Cl^-$

8.25

(a) $HO^- + HOCH_2{-}CH_2{-}Cl \rightleftharpoons H_2O + {}^-O{-}CH_2{-}CH_2{-}Cl \longrightarrow H_2C{-}CH_2$ (epoxide, bridged by O) $+ Cl^-$

(b) The $-\ddot{O}:^-$ group must displace the Cl^- from the backside,

trans-2-Chlorocyclohexanol (OH, H, H, Cl) $\overset{OH^-}{\rightleftharpoons}$ (O^-, H, H, Cl) $\overset{SN2}{\rightleftharpoons}$ epoxide

trans-2-Chlorocyclohexanol

Backside attack is not possible with the cis isomer (below) therefore it does not form an epoxide.

H, OH, H, Cl

cis-2-Chlorocyclohexanol

8.26

(a) $CH_2{=}C(CH_3)_2 \xrightarrow{H^+} (CH_3)_3C^+ \xrightarrow{R\text{–}OH} (CH_3)_3C\text{–}\overset{+}{O}(H)R \xrightarrow{-H^+} (CH_3)_3C\text{–}OR$

(b) The *tert*-butyl group is easily removed because, in acid, it is easily converted to a relatively stable, tertiary carbocation.

(c) $(CH_3)_3C\text{–}O\text{–}R \xrightarrow{H^+} (CH_3)_3C\text{–}\overset{+}{O}(H)\text{–}R \longrightarrow (CH_3)_3C^+ + HOR$

$(CH_3)_3C^+ \xrightarrow{H_2O} (CH_3)_3C\text{–}OH_2^+ \xrightarrow{-H^+} (CH_3)_3C\text{–}OH$

$(CH_3)_3C^+ \xrightarrow{-H^+} CH_2{=}C(CH_3)\text{–}CH_3$

8.27

(a) $CH_3O\text{–}CH(CH_3)(CH_2CH_3) + HI \longrightarrow I^- + CH_3\text{–}\overset{+}{O}(H)\text{–}CH(CH_3)(CH_2CH_3) \longrightarrow$

$ICH_3 + HO\text{–}CH(CH_3)(CH_2CH_3)$

S_N2 attack of I^- occurs at the methyl carbon atom because it is less hindered, therefore, the bond between the *sec*-butyl group and the oxygen is not broken.

(b) $CH_3{-}O{-}C(CH_3)_3 + HI \longrightarrow CH_3{-}\overset{+}{O}(H){-}C(CH_3)_3 + I^-$

$\updownarrow$

$CH_3OH + CH_3{-}C^+(CH_3)_2 \xrightarrow{I^-} CH_3{-}C(CH_3)_2{-}I$

In this reaction the much more stable *tert*-butyl cation is produced. It then combines with I^- to form *tert*-butyl iodide.

8.28

(a) $H_2C{-}CH_2$ (epoxide, O) $\xrightarrow{H^+}$ $CH_2{-}CH_2$ (protonated epoxide, $O^+{-}H$) $\xrightarrow{CH_3\ddot{O}H}$ $HO{-}CH_2{-}CH_2{-}\overset{+}{O}(H)CH_3$ $\longrightarrow$ $HOCH_2CH_2OCH_3$ Methyl Cellosolve

(b) An analogous reaction yields Ethyl Cellosolve, $HOCH_2CH_2OCH_2CH_3$

(c) $H_2C{-}CH_2$ (epoxide, O) $\xrightarrow{I^-}$ $^-O{-}CH_2CH_2{-}I$ $\xrightarrow{H_2O}$ $HOCH_2CH_2I + OH$

(d) $H_2C{-}CH_2$ (epoxide, O) $\xrightarrow{:NH_3}$ $^-O{-}CH_2CH_2{-}NH_3^+$ $\longrightarrow$ $HOCH_2CH_2NH_2$

(e) $H_2C{-}CH_2$ (epoxide, O) $\xrightarrow{CH_3O^-}$ $^-O{-}CH_2CH_2{-}OCH_3$ $\xrightarrow{CH_3OH}$ $HOCH_2CH_2OCH_3 + CH_3O^-$

8.29 The reaction is an S_N2 reaction and thus nucleophilic attack takes place much more rapidly at the primary carbon atom than at the more hindered secondary carbon atom.

$CH_3C(CH_3){-}CH_2$ (epoxide, O) $+ CH_3O^- \xrightarrow[C_2H_5OH]{fast} CH_3C(CH_3)(OH)CH_2OCH_3$ Major product

$CH_3C(CH_3){-}CH_2$ (epoxide, O) $+ CH_3O^- \xrightarrow[C_2H_5OH]{slow} CH_3C(CH_3)(OCH_3)CH_2OH$ Minor product

8.30 Ethoxide ion attacks the epoxide ring at the primary carbon because it is less hindered and the following reactions take place.

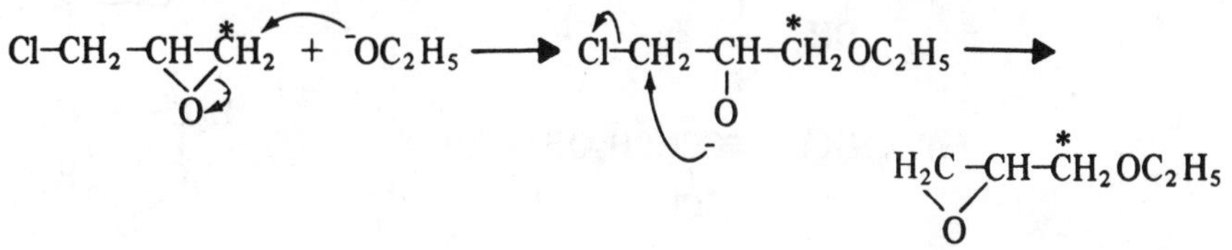

$$Cl{-}CH_2{-}\underset{\diagdown O \diagup}{CH{-}\overset{*}{C}H_2} + {}^{-}OC_2H_5 \longrightarrow Cl{-}CH_2{-}\underset{O^-}{\underset{|}{CH}}{-}\overset{*}{C}H_2OC_2H_5 \longrightarrow \underset{\diagdown O \diagup}{H_2C{-}CH}{-}\overset{*}{C}H_2OC_2H_5$$

8.31

$Na^+ CN^-$ + $R_4N^+ Cl^-$	⇌	$Na^+ X^-$ + $R_4N^+ CN^-$	Aqueous phase
↑↓		↓↑	
$R_4N^+ Cl^-$ + $CH_3(CH_2)_7CN$	⟵	$R_4N^+ CN^-$ + $CH_3(CH_2)_7Cl$	Organic phase (decane)

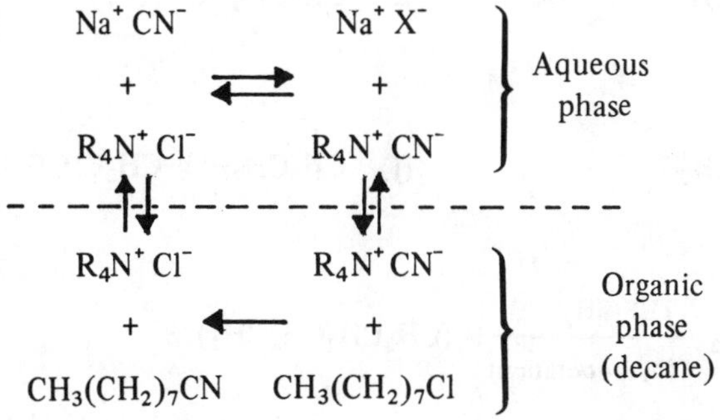

(X = Br or Cl)

8.32 (a) 15-Crown-5

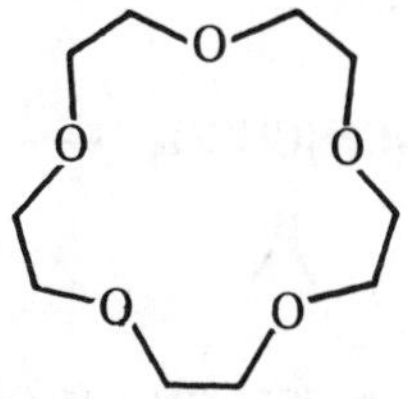

(b) Size of cation and ring. 12-Crown-4

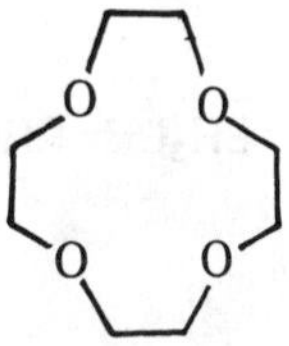

8.33 (a) 3,3-Dimethyl-1-butanol

(b) 4-Penten-2-ol

(c) 2-Methyl-1,4-butanediol

(d) 2-Phenylethanol

(e) 1-Methyl-2-cyclopenten-1-ol

(f) *cis*-3-Methylcyclohexanol

8.34 (a) $CH_3(H)C{=}C(H)CH_2OH$ (cis: CH_3 and CH_2OH on the same side)

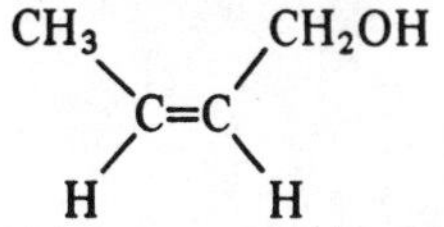

(b) $HOCH_2CH_2{-}C(OH)(H){-}CH_2OH$

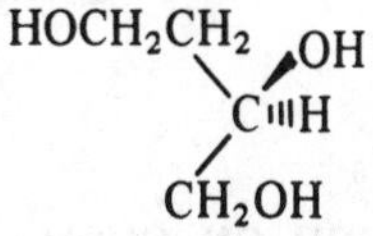

(c) 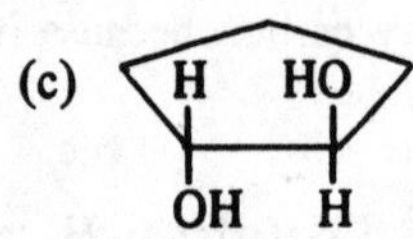

(d) 1-ethylcyclobutanol: cyclobutane ring bearing OH and CH_2CH_3 on the same carbon

(e) $CH_3CH_2C{\equiv}C\underset{\displaystyle Cl}{\underset{|}{C}}HCH_2OH$

(f) $H_2C—CH_2$ / $H_2C\quad CH_2$ / O (tetrahydrofuran ring)

(g) $CH_3\underset{\displaystyle OCH_2CH_3}{\underset{|}{C}}HCH_2CH_2CH_3$

(h) $CH_3CH_2–O–C_6H_5$

(i) $CH_3\overset{\displaystyle CH_3}{\overset{|}{C}}H–O–\overset{\displaystyle CH_3}{\overset{|}{C}}HCH_3$

(j) $CH_3CH_2–O–CH_2CH_2OH$

8.35 (a) $CH_3CH_2CH{=}CH_2 \xrightarrow[\text{(hydroboration)}]{\text{THF:BH}_3} (CH_3CH_2CH_2CH_2)_3B$

$\xrightarrow[\text{(oxidation)}]{H_2O_2/OH^-} CH_3CH_2CH_2CH_2OH$

(b) $CH_3CH{=}CHCH_3 \xrightarrow{\text{THF:BH}_3} CH_3CH_2\underset{\displaystyle B\diagup\diagdown}{\underset{|}{C}}HCH_3 \xrightarrow{160°C}$

$CH_3CH_2CH_2CH_2–B\lessgtr \xrightarrow{H_2O_2/OH^-} CH_3CH_2CH_2CH_2OH$

(c) $CH_3CH_2CH_2CH_2Cl \xrightarrow{OH^-} CH_3CH_2CH_2CH_2OH$

(d) $CH_3CH_2\underset{\displaystyle Cl}{\underset{|}{C}}HCH_3 \xrightarrow{EtO^-/EtOH} CH_3CH{=}CHCH_3$

$\xrightarrow{\text{[as in (b)]}} CH_3CH_2CH_2CH_2OH$

(e) $CH_3CH_2C{\equiv}CH \xrightarrow[Ni_2B\ (P\text{-}2)]{H_2} CH_3CH_2CH{=}CH_2$

$\xrightarrow{\text{[as in (a)]}} CH_3CH_2CH_2CH_2OH$

8.36 (a) $3CH_3CH_2\underset{\displaystyle OH}{\underset{|}{C}}HCH_3 + PBr_3 \longrightarrow 3CH_3CH_2\underset{\displaystyle Br}{\underset{|}{C}}HCH_3 + H_3PO_3$

(b) $CH_3CH_2CH_2CH_2OH \xrightarrow{PBr_3} CH_3CH_2CH_2CH_2Br \xrightarrow{(CH_3)_3COK}$

$CH_3CH_2CH{=}CH_2 \xrightarrow[\text{(no peroxides)}]{HBr} CH_3CH_2\underset{\displaystyle Br}{\underset{|}{C}}HCH_3$

(c) See (b) above.

(d) $CH_3CH_2C{\equiv}CH \xrightarrow{H_2, Ni_2B\ (P\text{-}2)} CH_3CH_2CH{=}CH_2 \xrightarrow[\text{(no peroxides)}]{HBr}$

$CH_3CH_2\underset{\displaystyle Br}{\underset{|}{C}}HCH_3$

8.37 (a) $CH_3CH_2\underset{\displaystyle OH}{\underset{|}{C}}HCH_3 \xrightarrow[\text{heat } (-H_2O)]{H^+} CH_3CH{=}CHCH_3 + CH_3CH_2CH{=}CH_2 \xrightarrow{THF:BH_3} CH_3CH_2\underset{\displaystyle B-}{\underset{|}{C}}HCH_3 + CH_3CH_2CH_2CH_2B-$

$\xrightarrow{160°C} CH_3CH_2CH_2CH_2B- \xrightarrow[OH^-]{H_2O_2} CH_3CH_2CH_2CH_2OH \xrightarrow{PBr_3} CH_3CH_2CH_2CH_2Br$

(b) $CH_3CH_2CH_2CH_2OH \xrightarrow{PBr_3} CH_3CH_2CH_2CH_2Br + H_3PO_3$

(c) $CH_3CH_2CH{=}CH_2 \xrightarrow[\text{(peroxides)}]{HBr} CH_3CH_2CH_2CH_2Br$

(d) $CH_3CH_2C{\equiv}CH \xrightarrow[\text{(peroxides)}]{HBr} CH_3CH_2CH{=}CHBr \xrightarrow[H_2]{Pt} CH_3CH_2CH_2CH_2Br$

8.38 (a) cyclohexanol (OH) $+ SOCl_2 \longrightarrow$ chlorocyclohexane (Cl) $+ SO_2 + HCl$

(b) cyclohexene $+ HCl \longrightarrow$ chlorocyclohexane (Cl)

(c) 1-methylcyclohexene (CH_3) $+ HBr \xrightarrow[\text{(peroxides)}]{\text{no}}$ 1-bromo-1-methylcyclohexane (Br, CH_3)

(d) 1-Methylcyclohexene $\xrightarrow[(2)\ H_2O_2/OH^-]{(1)\ THF{:}BH_3}$ trans-2-methylcyclohexanol (H, CH₃, H, OH shown) + enantiomer

(e) 1-Bromo-1-methylcyclohexane (Br, CH_3) $\xrightarrow[t\text{-BuOH}]{t\text{-BuOK,}}$ methylenecyclohexane (=CH_2) $\xrightarrow[(2)\ H_2O_2/OH^-]{(1)\ THF{:}BH_3}$ cyclohexylmethanol (CH_2OH)

8.39 (a) $CH_3CH_2CH_2O^-Na^+$ Sodium propoxide

(b) $CH_3CH_2CH_2{-}O{-}CH_2CH_2CH_2CH_3$ Butyl propyl ether

(c) $CH_3{-}\overset{O}{\underset{O}{\overset{\|}{\underset{\|}{S}}}}{-}OCH_2CH_2CH_3$ Propyl methanesulfonate

(d) $CH_3{-}C_6H_4{-}SO_2{-}O{-}CH_2CH_2CH_3$ Propyl *p*-toluenesulfonate (or propyl tosylate)

(e) $CH_3{-}O{-}CH_2CH_2CH_3$ 1-Methoxypropane

(f) $CH_3CH_2CH_2I$ 1-Iodopropane

(g) $CH_3CH_2CH_2Cl$ 1-Chloropropane

(h) $CH_3CH_2CH_2Cl$ 1-Chloropropane

(i) $CH_3CH_2CH_2{-}O{-}CH_2CH_2CH_3$ Dipropyl ether

(j) $CH_3CH_2CH_2Br$ 1-Bromopropane

8.40

(a) $CH_3\overset{CH_3}{\overset{|}{C}}HO^-Na^+$ Sodium isopropoxide

(b) $CH_3\overset{CH_3}{\overset{|}{C}}H{-}O{-}CH_2CH_2CH_2CH_3$ Butyl isopropyl ether

(c) $CH_3SO_2{-}O{-}\overset{CH_3}{\overset{|}{C}}HCH_3$ Isopropyl methanesulfonate

(d) $CH_3{-}C_6H_4{-}SO_2{-}O{-}\overset{CH_3}{\overset{|}{C}}HCH_3$ Isopropyl *p*-toluenesulfonate

(e) $CH_3O\overset{CH_3}{\overset{|}{C}}HCH_3$ 2-Methoxypropane

(f) 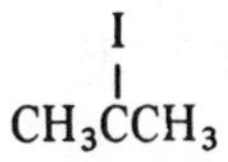 $CH_3\overset{I}{\overset{|}{C}}HCH_3$ 2-Iodopropane

(g) 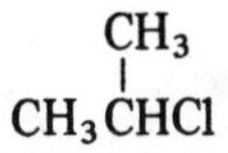 $CH_3\overset{CH_3}{\overset{|}{C}}HCl$ 2-Chloropropane

(h) Same as (g)

(i) 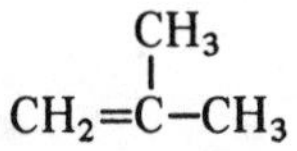 $CH_2{=}\overset{CH_3}{\overset{|}{C}}{-}CH_3$ 2-Methylpropene

(j) 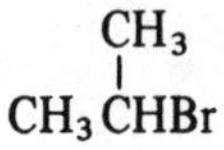 $CH_3\overset{CH_3}{\overset{|}{C}}HBr$ 2-Bromopropane

8.41 (a) $CH_3Br + CH_3CH_2Br$

(b) 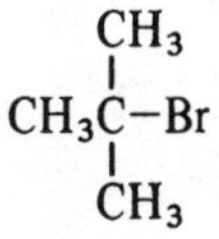 $CH_3\underset{CH_3}{\underset{|}{\overset{CH_3}{\overset{|}{C}}}}{-}Br + CH_3CH_2Br$

(c) $Br{-}CH_2CH_2CH_2CH_2{-}Br$

(d) $Br{-}CH_2CH_2{-}Br$ (2 molar equivalents)

8.42 H_3C, CH_3, OH, H (cyclohexane) $\xrightarrow{H^+}$ H_3C, CH_3, $\overset{+}{O}H_2$, H $\rightleftarrows$ H_3C, CH_3, +, H $+ H_2O$ $\longrightarrow$ CH_3, +, CH_3, H

3° Carbocation is more stable

CH_3, +, H, CH_3 $\xrightarrow{-H^+}$ CH_3, CH_3 (cyclohexene)

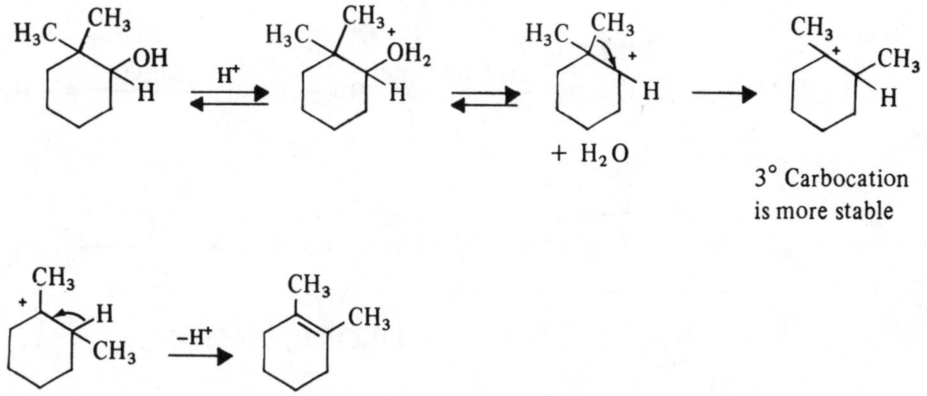

8.43

(a) 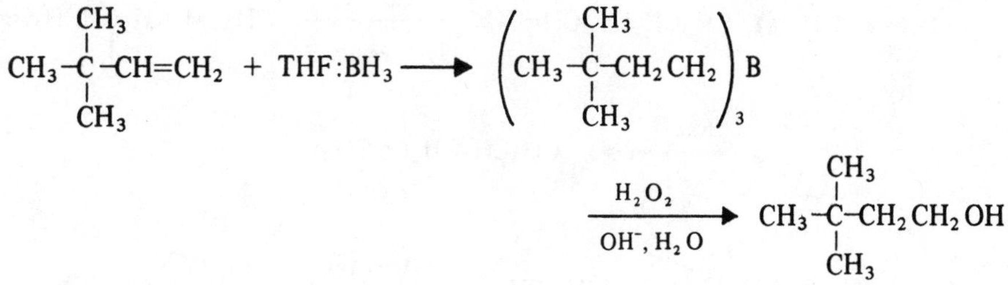 $CH_3{-}\underset{CH_3}{\underset{|}{\overset{CH_3}{\overset{|}{C}}}}{-}CH{=}CH_2 + THF{:}BH_3 \longrightarrow \left(CH_3{-}\underset{CH_3}{\underset{|}{\overset{CH_3}{\overset{|}{C}}}}{-}CH_2CH_2\right)_3B$

$\xrightarrow[OH^-,\ H_2O]{H_2O_2} CH_3{-}\underset{CH_3}{\underset{|}{\overset{CH_3}{\overset{|}{C}}}}{-}CH_2CH_2OH$

(b) $CH_3CH_2CH_2CH_2CH{=}CH_2 \xrightarrow[(2)\ H_2O_2,\ OH^-,\ H_2O]{(1)\ THF{:}BH_3} CH_3CH_2CH_2CH_2CH_2CH_2OH$

(c) 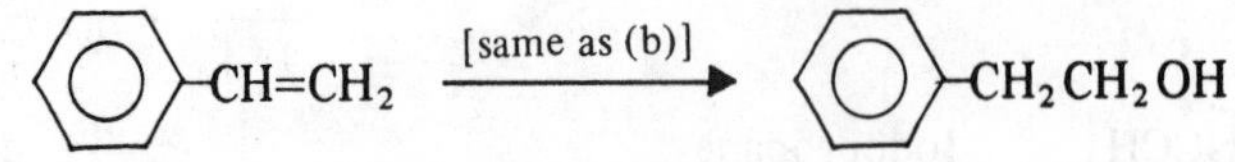 $C_6H_5-CH=CH_2 \xrightarrow{\text{[same as (b)]}} C_6H_5-CH_2CH_2OH$

(d) 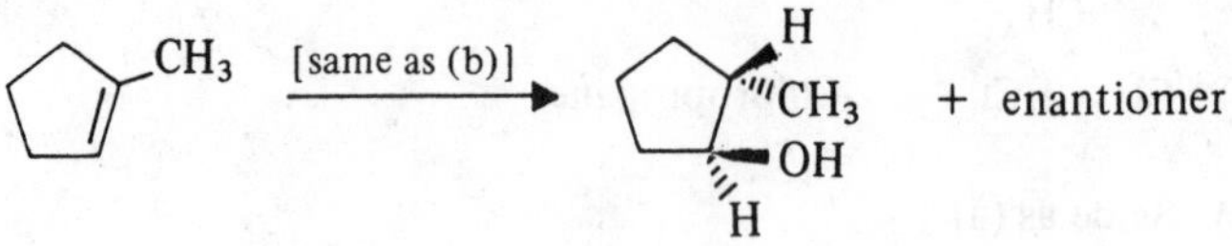 1-methylcyclopentene (CH_3) $\xrightarrow{\text{[same as (b)]}}$ trans-2-methylcyclopentanol (H, CH_3, OH, H) + enantiomer

8.44 (a) 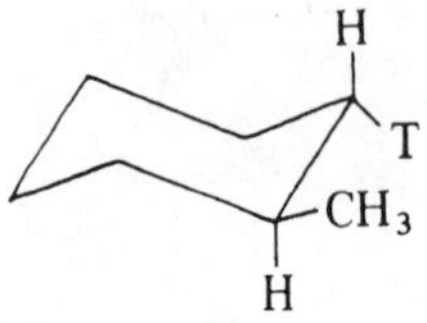 cyclohexane chair with H, T, CH_3, H

(b) 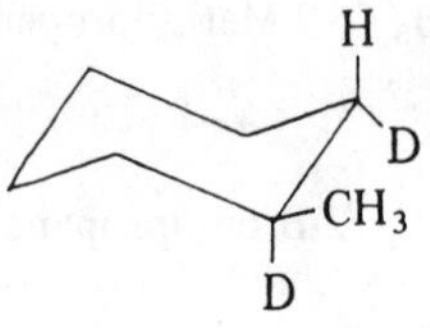 cyclohexane chair with H, D, CH_3, D

(c) 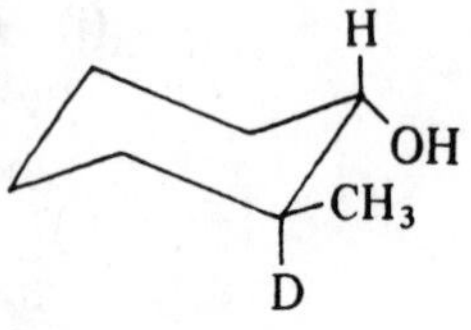 cyclohexane chair with H, OH, CH_3, D

8.45 (a) $CH_3C(CH_3)=CH_2 \xrightarrow[\text{(2) } H_2O_2/OH^-]{\text{(1) THF:BH}_3} CH_3CH(CH_3)CH_2OH$

(b) $CH_3C(CH_3)=CH_2 \xrightarrow[\text{(2) } CH_3CO_2T]{\text{(1) THF:BH}_3} CH_3CH(CH_3)CH_2T$

(c) $CH_3C(CH_3)=CH_2 \xrightarrow{\text{THF:BD}_3} CH_3C(CH_3)(D)-CH_2-B\langle \xrightarrow{CH_3CO_2T} CH_3C(CH_3)(D)CH_2T$

(d) $CH_3CH(CH_3)CH_2OH \xrightarrow{\text{Na}} CH_3CH(CH_3)CH_2ONa \xrightarrow{CH_3CH_2Br}$

$CH_3CH(CH_3)CH_2-O-CH_2CH_3$

8.46 (a) $CH_3CH_2CH_2CH=CH_2 \xrightarrow{\text{Hg(OAc)}_2} CH_3CH_2CH_2CH(OH)CH_2HgOAc$

$\xrightarrow[\text{OH}^-]{\text{NaBH}_4} CH_3CH_2CH_2CH(OH)CH_3$

(b) 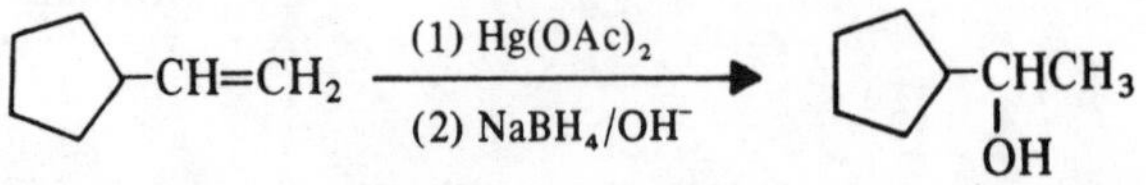 cyclopentyl$-CH=CH_2 \xrightarrow[\text{(2) NaBH}_4/\text{OH}^-]{\text{(1) Hg(OAc)}_2}$ cyclopentyl$-CH(OH)CH_3$

(c) 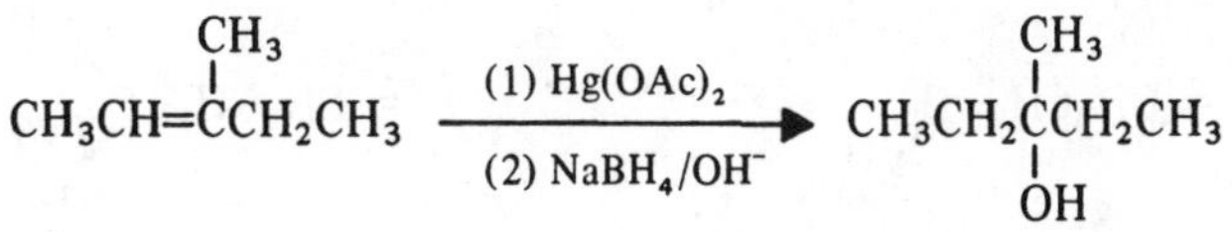 $CH_3CH{=}C(CH_3)CH_2CH_3 \xrightarrow[(2)\ NaBH_4/OH^-]{(1)\ Hg(OAc)_2} CH_3CH_2C(CH_3)(OH)CH_2CH_3$

(d) 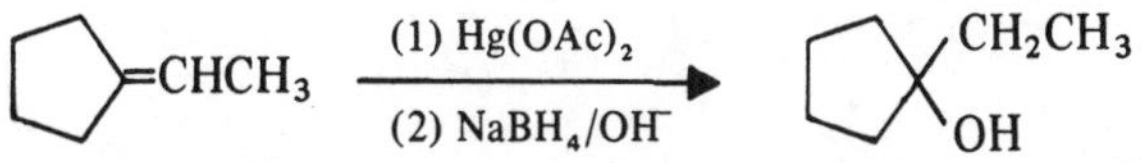 cyclopentylidene$=CHCH_3 \xrightarrow[(2)\ NaBH_4/OH^-]{(1)\ Hg(OAc)_2}$ 1-ethylcyclopentanol (CH_2CH_3, OH)

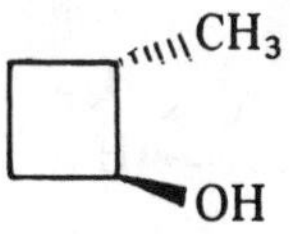

8.47 (a) **A** = (cyclobutane with CH_3 hashed, OH wedged) + enantiomer **B** = (cyclobutane with CH_3 hashed, OTs wedged) + enantiomer **C** = (cyclobutane with CH_3 hashed, OH hashed) + enantiomer

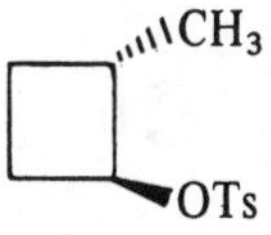

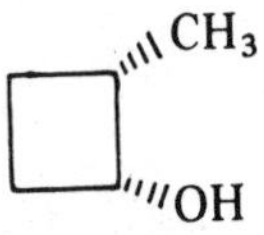

(b) Diastereomers

(c) **D** = (cyclobutane with CH_3 hashed, I hashed) + enantiomer

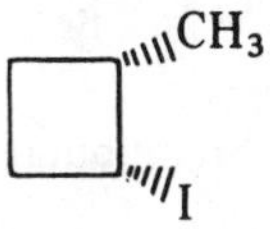

(d) **E** = (chair cyclohexane: CH_3 equatorial, H axial; H axial, OMs equatorial)

F = (chair: CH_3 equatorial, H axial; C≡CH axial, H equatorial) ⇄ (chair: H equatorial, CH_3 axial; C≡CH equatorial, H axial)

G = (chair: CH_3 axial, H equatorial; $\overset{O}{\overset{\|}{C}}CH_3$ equatorial, H axial)

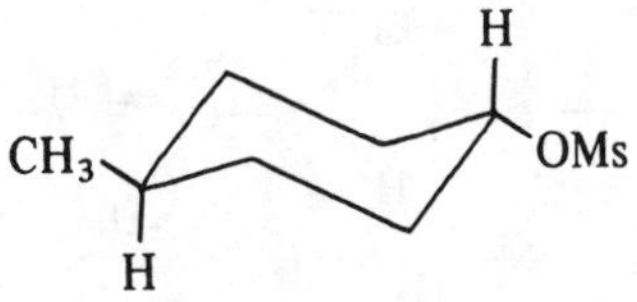

(e) **H** = $CH_3CH_2{-}C(CH_3)(H)(ONa)$ **J** = $CH_3CH_2{-}C(CH_3)(H)(OCH_3)$

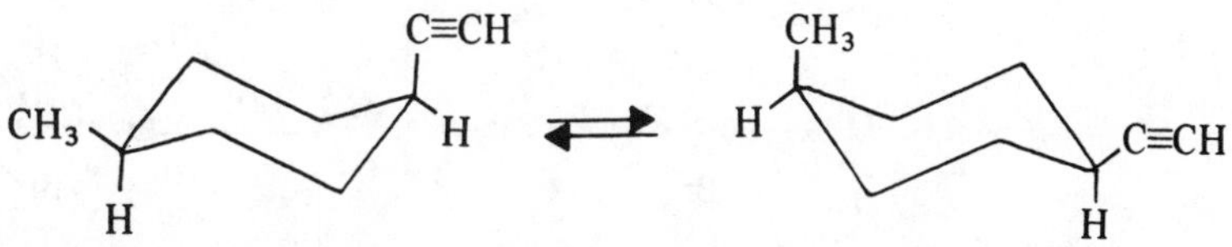

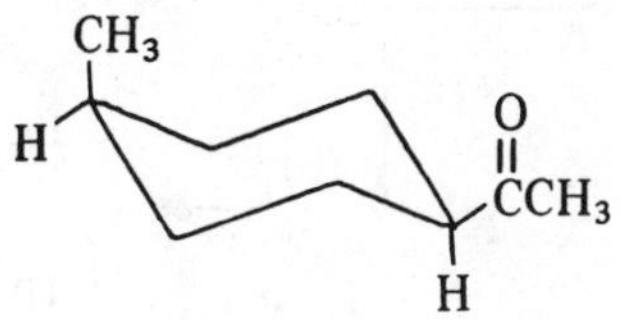

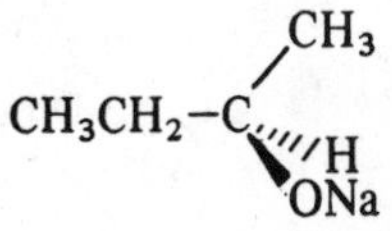

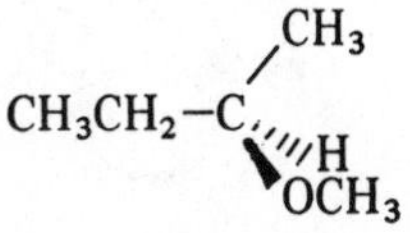

(f) K = CH_3CH_2–C(CH$_3$)(H)(OMs) L = CH_3CH_2–C(CH$_3$)(OCH$_3$)(H)

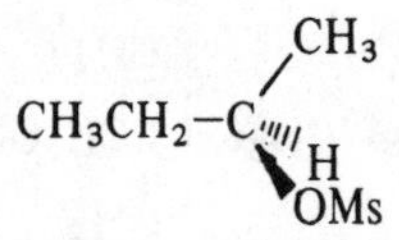

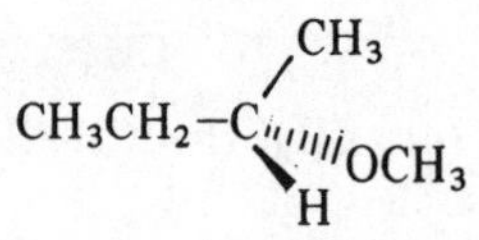

(g) Enantiomers

8.48 The reactions proceed through the formation of bromonium ions identical to those formed in the bromination of *trans*- and *cis*-2-butene (see Section 7.6A).

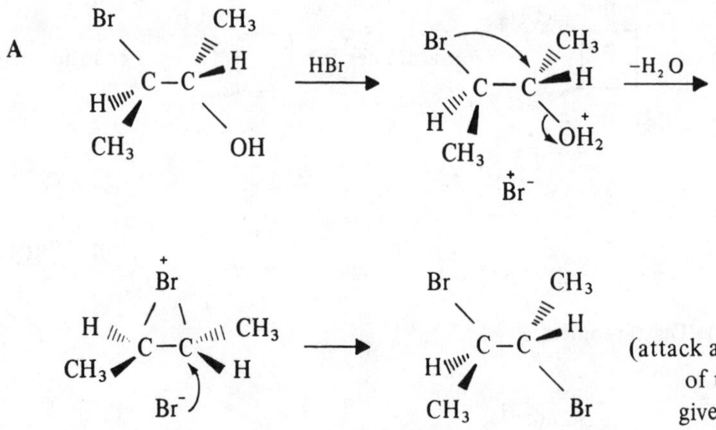

A

Br, H, CH_3 / C–C / CH_3, H, OH $\xrightarrow{HBr}$ Br, H, CH_3 / C–C / CH_3, H, $\overset{+}{O}H_2$ $\xrightarrow{-H_2O}$

Br^-

$\overset{+}{Br}$ bromonium ion: H, CH_3 / C–C / CH_3, H; Br^- $\longrightarrow$ Br, H, CH_3 / C–C / CH_3, H, Br (attack at the other carbon atom of the bromonium ion gives the same product)

meso-2,3-Dibromobutane

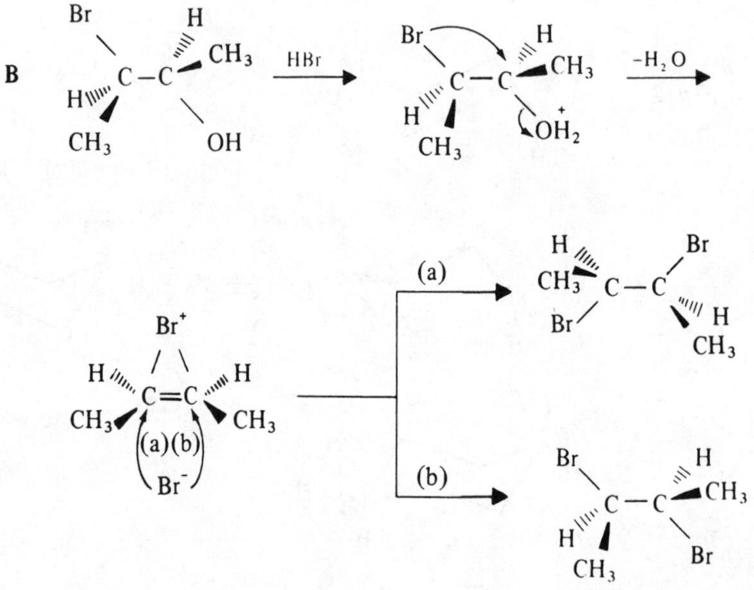

B

Br, H, CH_3 / C–C / H, CH_3, OH $\xrightarrow{HBr}$ Br, H, CH_3 / C–C / H, CH_3, $\overset{+}{O}H_2$ $\xrightarrow{-H_2O}$

Br^+ bromonium ion: H, CH_3 / C=C / H, CH_3; (a)(b) Br^-

(a) $\longrightarrow$ H, CH_3, Br / C–C / Br, H, CH_3

(b) $\longrightarrow$ Br, H, CH_3 / C–C / H, CH_3, Br

(±)-2,3-Dibromobutane

SECTION REFERENCES FOR ADDITIONAL PROBLEMS

8.33	8.1	**8.41**	8.18
8.34	8.1	**8.42**	8.14
8.35	8.4, 8.6, 8.7	**8.43**	8.7
8.36	8.7, 8.13, 8.15	**8.44**	8.7, 8.7A, 8.7B
8.37	8.8, 8.13, 8.15	**8.45**	8.7
8.38	8.4, 8.5, 8.6, 8.7, 8.8	**8.46**	8.5
8.39	8.9, 8.10, 8.11, 8.12, 8.1	**8.47**	8.7, 8.14
8.40	8.9, 8.10, 8.11, 8.12, 8.1	**8.48**	8.14

SELF-TEST

8.1 Give the structural formula of the missing reactants or major organic product in each of the following reactions. Write N.R. if no reaction occurs. If two steps are needed, label them (1), (2), and so on.

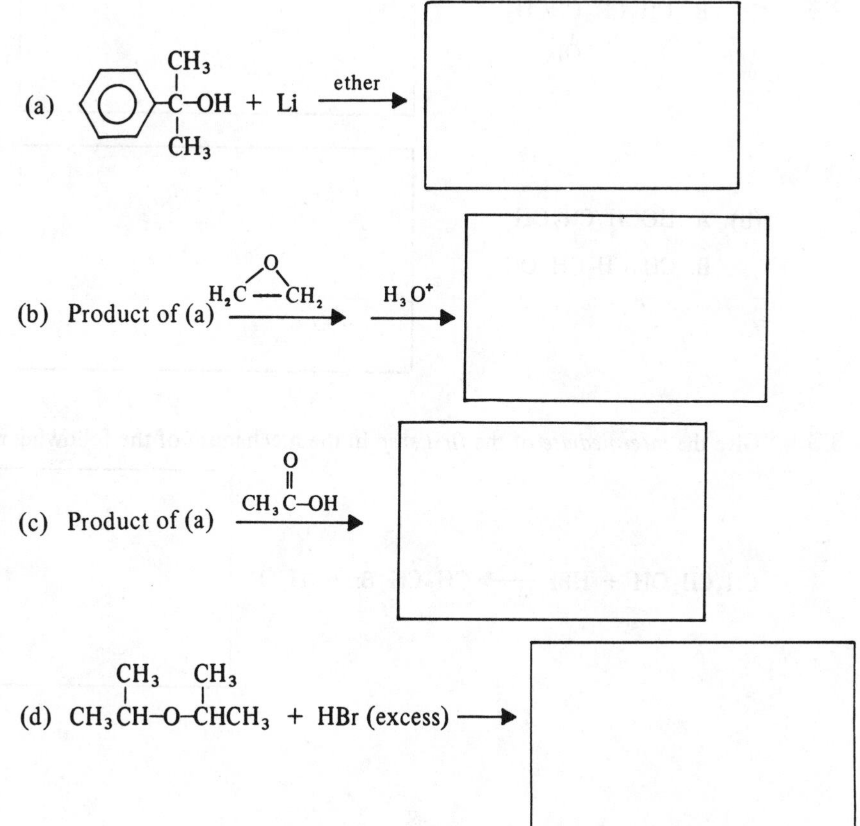

(e) OH + NaOH $\xrightarrow[H_2O]{25°C}$

(f) $CH_3CH_2CH_2OH$ + $\longrightarrow$ $CH_3CH_2CH_2{-}O{-}\overset{O}{\underset{O}{\overset{\|}{\underset{\|}{S}}}}{-}C_6H_5$

(g) $CH_3CH_2CH_2OH + CH_3Li \longrightarrow$

8.2 For each of the following pairs of compounds, tell which has the higher boiling point (write its letter, A or B).

Higher bp

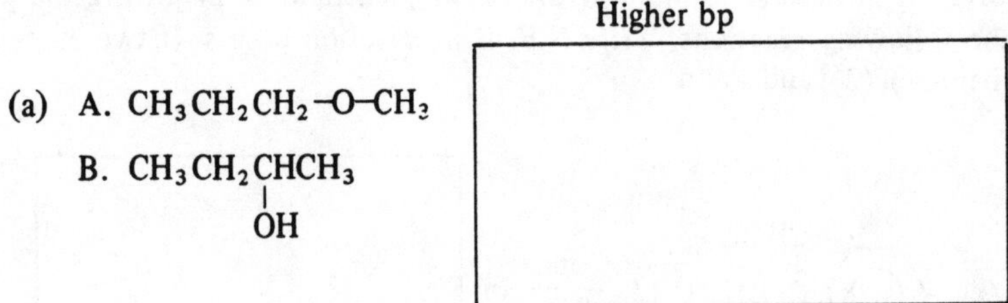

(a) A. $CH_3CH_2CH_2{-}O{-}CH_3$

B. $CH_3CH_2\underset{OH}{CH}CH_3$

(b) A. $HOCH_2CH_2OH$

B. $CH_3CH_2CH_2OH$

8.3 Give the *intermediate* of the *first step* in the mechanism of the following reaction:

$CH_3CH_2OH + HBr \longrightarrow CH_3CH_2Br + H_2O$

8.4 Which set of reagents would effect the conversion,

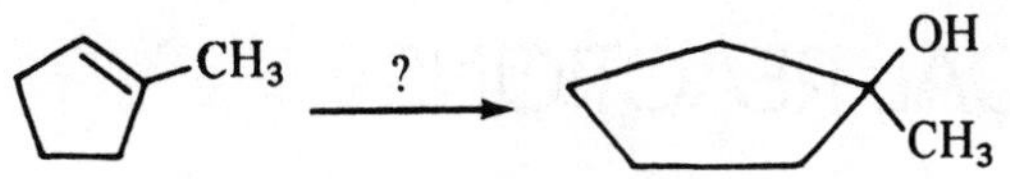

(a) BH_3: THF, then H_2O_2/OH^- (b) $H_2O/HgSO_4$, H^+, then $NaBH_4/OH^-$

(c) H_3O^+, H_2O, heat (d) More than one of these (e) None of these

8.5 Which of the reagents in item **8.4** would effect the conversion,

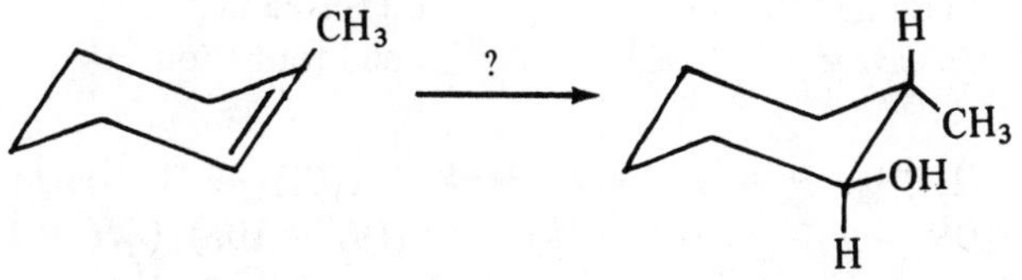

9 FREE RADICAL REACTIONS

SOLUTIONS TO PROBLEMS

9.1 (a) $H{-}H \ + \ Br{-}Br \longrightarrow 2\,H{-}Br$

$(DH^\circ = 104)$ $(DH^\circ = 46)$ $2(DH^\circ = 87.5)$

+ 150 kcal/mole is required for bond cleavage

− 175 kcal/mole is evolved in bond formation

$\Delta H^\circ = +150 - 175 = -25$ kcal/mole (exothermic)

(b) $CH_3CH_2{-}H \ + \ F{-}F \longrightarrow CH_3CH_2{-}F \ + \ H{-}F$

$(DH^\circ = 98)$ $(DH^\circ = 38)$ $(DH^\circ = 106)$ $(DH^\circ = 136)$

+ 136 kcal/mole − 242 kcal/mole $\Delta H^\circ = -106$ kcal/mole (exothermic)

(c) $CH_3CH_2{-}H \ + \ I{-}I \longrightarrow CH_3CH_2{-}I \ + \ H{-}I$

$(DH^\circ = 98)$ $(DH^\circ = 36)$ $(DH^\circ = 53.5)$ $(DH^\circ = 71)$

+ 134 kcal/mole − 124.5 kcal/mole $\Delta H^\circ = +9.5$ kcal/mole (endothermic)

(d) $CH_3{-}H \ + \ Cl{-}Cl \longrightarrow CH_3{-}Cl \ + \ HCl$

$(DH^\circ = 104)$ $(DH^\circ = 58)$ $(DH^\circ = 83.5)$ $(DH^\circ = 103)$

+ 162 kcal/mole − 186.5 kcal/mole $\Delta H^\circ = -24.5$ kcal/mole (exothermic)

(e) $(CH_3)_3C{-}H \ + \ Cl{-}Cl \longrightarrow (CH_3)_3C{-}Cl \ + \ H{-}Cl$

$(DH^\circ = 91)$ $(DH^\circ = 58)$ $(DH^\circ = 78.5)$ $(DH^\circ = 103)$

+ 149 kcal/mole − 181.5 kcal/mole $\Delta H^\circ = -32.5$ kcal/mole (exothermic)

(f) $(CH_3)_3C{-}H \ + \ Br{-}Br \longrightarrow (CH_3)_3C{-}Br \ + \ H{-}Br$

$(DH^\circ = 91)$ $(DH^\circ = 46)$ $(DH^\circ = 63)$ $(DH^\circ = 87.5)$

+ 137 kcal/mole − 150.5 kcal/mole $\Delta H^\circ = -13.5$ kcal/mole (exothermic)

(g) $CH_3CH_2{-}CH_3 \longrightarrow CH_3CH_2\cdot \ + \ CH_3\cdot$

$(DH^\circ = 85)$

+ 85 kcal/mole $\Delta H^\circ = +85$ kcal/mole (endothermic)

(h) $2CH_3CH_2\cdot \longrightarrow CH_3CH_2{-}CH_2CH_3$

$(DH^\circ = 82)$

− 82 kcal/mole $\Delta H^\circ = -82$ kcal/mole (exothermic)

9.2 $\Delta H_2° > \Delta H_1°$; therefore isopropyl is more stable than ethyl.

(a)

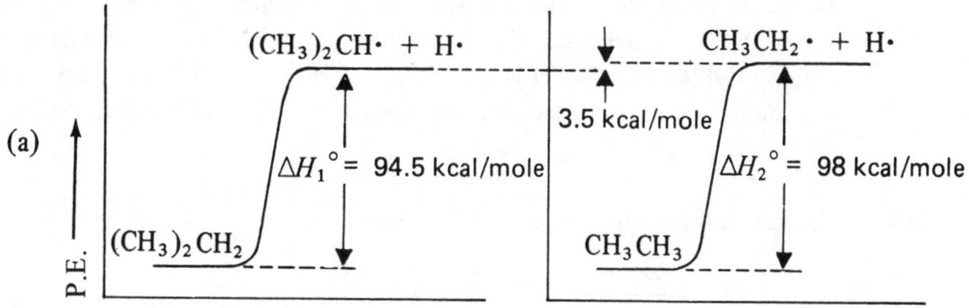

$\Delta H_3° > \Delta H_2°$; therefore ethyl is more stable than methyl.

(b)

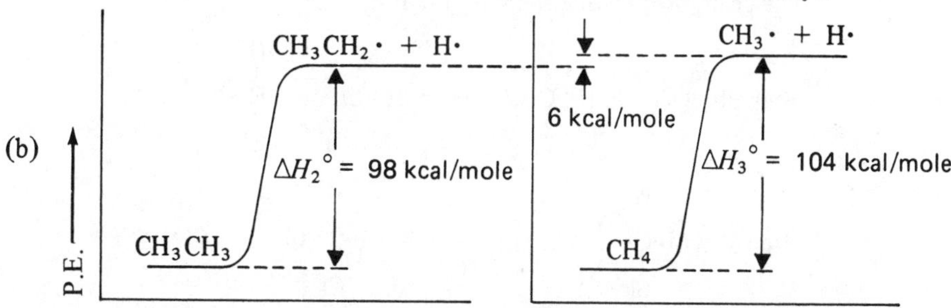

$\Delta H_2° \simeq \Delta H_4°$; therefore the two radicals have nearly equal stabilities.

(c)

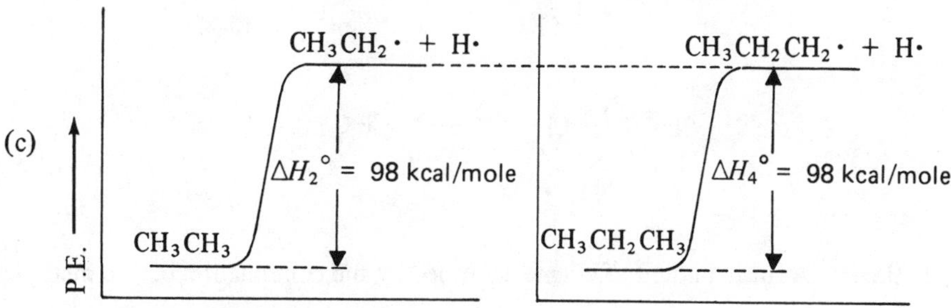

(d) The radicals produced are both primary radicals, and they are otherwise structurally similar, therefore they are of essentially equal stability.

9.3 Homolytic bond dissociation energies of the following C–Cl bonds are

$$CH_3\text{–}Cl \longrightarrow CH_3\cdot + Cl\cdot \qquad \Delta H° = 83.5 \text{ kcal/mole}$$

$$CH_3CH_2\text{–}Cl \longrightarrow CH_3CH_2\cdot + Cl\cdot \qquad \Delta H° = 81.5 \text{ kcal/mole}$$

$$(CH_3)_2CH\text{–}Cl \longrightarrow (CH_3)_2CH\cdot + Cl\cdot \qquad \Delta H° = 81 \text{ kcal/mole}$$

$$(CH_3)_3C\text{–}Cl \longrightarrow (CH_3)_3C\cdot + Cl\cdot \qquad \Delta H° = 78.5 \text{ kcal/mole}$$

Since in each case the same kind of compound (an alkyl chloride) is decomposed into the same kinds of products (an alkyl free radical and a chlorine atom), it follows that the energy required (ΔH°) is a measure of the instability of the radical relative to the alkyl halide. In other words, the less stable the free radical, the more energy will be required to break the bond between it and the chlorine atom. Bond dissociation energies for these alkyl chlorides are, respectively, 83.5, 81.5, 81, and 78.5. They are in the same order as the stabilities of the free radicals produced: $CH_3\cdot < CH_3CH_2\cdot < (CH_3)_2CH\cdot < (CH_3)_3C\cdot$

9.4 The chain-initiating step is

$$Cl_2 \xrightarrow[\text{light}]{\text{heat or}} 2\,:\ddot{\underset{..}{Cl}}\cdot$$

The chain-propagating steps are the following:

$$\text{Step 2b}\quad :\ddot{\underset{..}{Cl}}\cdot + H:\overset{H}{\underset{Cl}{C}}-Cl \longrightarrow H:\ddot{\underset{..}{Cl}}: + \cdot\overset{H}{\underset{Cl}{C}}-Cl$$

$$\text{Step 3b}\quad Cl-\overset{H}{\underset{Cl}{C}}\cdot + :\ddot{\underset{..}{Cl}}:\ddot{\underset{..}{Cl}}: \longrightarrow Cl-\overset{H}{\underset{Cl}{C}}:\ddot{\underset{..}{Cl}}: + \cdot\ddot{\underset{..}{Cl}}:$$

$$\text{Step 2c}\quad :\ddot{\underset{..}{Cl}}\cdot + H:\overset{Cl}{\underset{Cl}{C}}-Cl \longrightarrow H:\ddot{\underset{..}{Cl}}: + \cdot\overset{Cl}{\underset{Cl}{C}}-Cl$$

$$\text{Step 3c}\quad Cl-\overset{Cl}{\underset{Cl}{C}}\cdot + :\ddot{\underset{..}{Cl}}:\ddot{\underset{..}{Cl}}: \longrightarrow Cl-\overset{Cl}{\underset{Cl}{C}}:\ddot{\underset{..}{Cl}}: + \cdot\ddot{\underset{..}{Cl}}:$$

9.5 A small amount of ethane is formed by the combination of two methyl radicals:

$$2CH_3\cdot \longrightarrow CH_3:CH_3$$

This ethane then reacts with chlorine in a substitution reaction (see Section 9.5) to form chloroethane.

The significance of this observation is that it is evidence for the proposal that the combination of methyl radicals is one of the chain-terminating steps in the chlorination of methane.

9.6 The use of a large excess of chlorine allows all of the chlorinated methanes (CH_3Cl, CH_2Cl_2, and $CHCl_3$) to react with chlorine.

9.7 Chain-initiating step

$$\underset{(DH^\circ = 46)}{Br-Br} \longrightarrow 2\,Br\cdot \qquad \Delta H^\circ = +46\ \text{kcal/mole}$$

Chain-propagating steps	$Br\cdot + \underset{(DH^\circ = 104)}{CH_3{-}H} \longrightarrow \underset{}{CH_3\cdot} + \underset{(DH^\circ = 87.5)}{HBr}$	$\Delta H^\circ = +16.5$ kcal/mole
	$CH_3\cdot + \underset{(DH^\circ = 46)}{Br{-}Br} \longrightarrow \underset{(DH^\circ = 70)}{CH_3{-}Br} + Br\cdot$	$\Delta H^\circ = -24$ kcal/mole
Chain-terminating steps	$CH_3\cdot + Br\cdot \longrightarrow \underset{(DH^\circ = 70)}{CH_3{-}Br}$	$\Delta H^\circ = -70$ kcal/mole
	$CH_3\cdot + CH_3\cdot \longrightarrow \underset{(DH^\circ = 88)}{CH_3{-}CH_3}$	$\Delta H^\circ = -88$ kcal/mole
	$Br\cdot + Br\cdot \longrightarrow \underset{(DH^\circ = 46)}{Br{-}Br}$	$\Delta H^\circ = -46$ kcal/mole

9.8 It would be incorrect to include chain-initiation and chain-termination steps in the calculation of the overall value of ΔH° because addition of all these steps would not yield the overall equation for the chlorination of methane.

9.9 (a) E_{act} would equal zero for reactions (3) and (5) because radicals (in the gas phase) are combining to form molecules.

(b) E_{act} would be greater than zero for reactions (1), (2), and (4) because all of these involve bond breaking.

(c) E_{act} equals ΔH° for reaction (1) because this is a gas-phase reaction in which a bond is broken homolytically but no bonds are formed.

9.10 (1) $CH_3\cdot + \underset{(DH^\circ = 103)}{H{-}Cl} \longrightarrow \underset{(DH^\circ = 104)}{CH_3{-}H} + Cl\cdot$ $\quad \Delta H^\circ = -1$ kcal/mole

$E_{act} = +2.8$ kcal/mole

(See text, p. 408; E_{act} for the reverse reaction is 3.8 kcal/mole.)

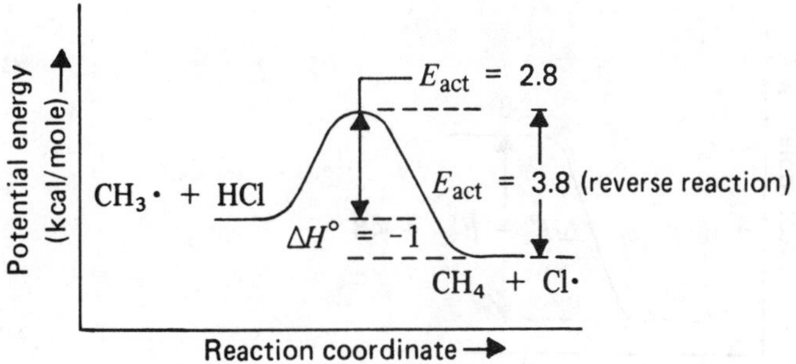

(2) $CH_3\cdot$ + H–Br ⟶ CH_3–H + Br· ΔH° = – 16.5 kcal/mole
$(DH^\circ = 87.5)$ $(DH^\circ = 104)$ E_{act} = + 2.1 kcal/mole

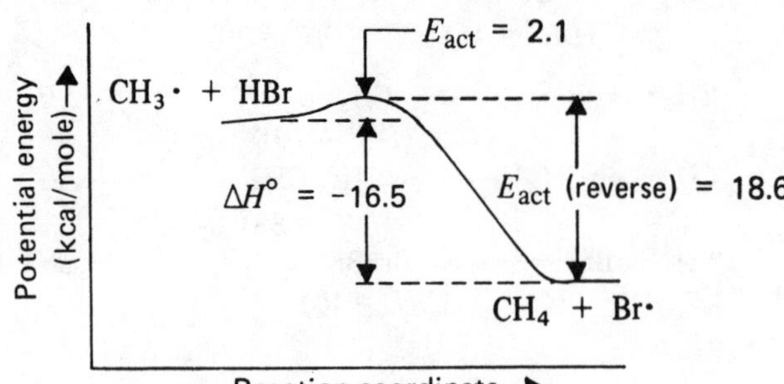

(3) CH_3–CH_3 ⟶ 2 $CH_3\cdot$ ΔH° = + 88 kcal/mole
$(DH^\circ = 88)$ E_{act} = + 88 kcal/mole

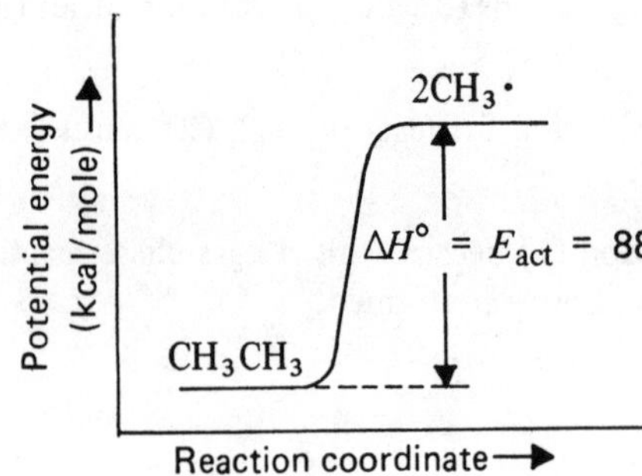

$\Delta H^\circ = E_{act}$ for any reaction in which bonds are broken but no bonds are formed

(4) Br–Br ⟶ 2 Br·
$(DH^\circ = 46)$

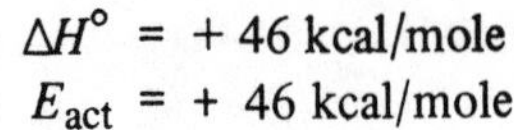

ΔH° = + 46 kcal/mole
E_{act} = + 46 kcal/mole

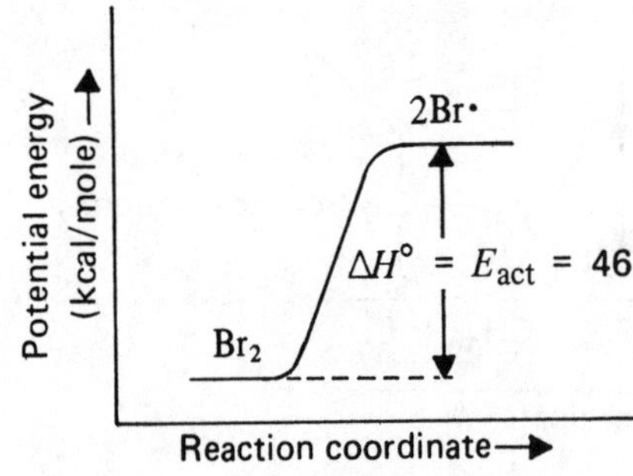

(5) $2\,Cl\cdot \longrightarrow Cl{-}Cl$ $(DH^\circ = 58)$ $\Delta H^\circ = -58$ kcal/mole, $E_{act} = 0$ kcal/mole

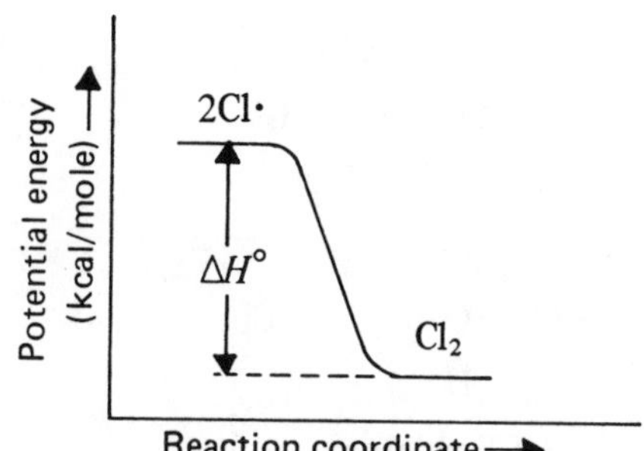

9.11 (a) $CH_3CH_2{-}H + Cl\cdot \longrightarrow CH_3CH_2\cdot + H{-}Cl$
$(DH^\circ = 98)$ $(DH^\circ = 103)$
$\Delta H^\circ = -103 + 98 = -5$ kcal/mole

(b)

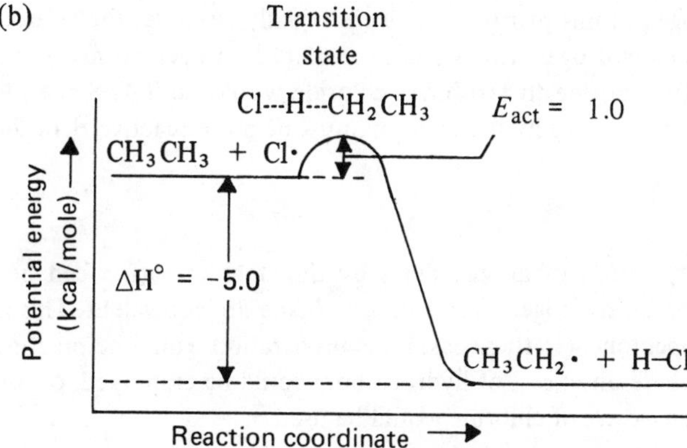

(c) The hydrogen abstraction step, for ethane,

$$CH_3CH_2{-}H + Cl\cdot \longrightarrow CH_3CH_2\cdot + HCl \qquad (E_{act} = 1.0 \text{ kcal/mole})$$

has a much lower energy of activation than the corresponding step for methane:

$$CH_3{-}H + Cl\cdot \longrightarrow CH_3\cdot + HCl \qquad (E_{act} = 3.8 \text{ kcal/mole})$$

Therefore, ethyl radicals form much more rapidly in the mixture than methyl radicals, and this, in turn, leads to the more rapid formation of ethyl chloride.

9.12

$$Cl_2 \xrightarrow[\text{or heat}]{h\nu} 2\,Cl\cdot$$

Step 2a $$Cl\cdot + H:\overset{\displaystyle Cl}{\underset{\displaystyle H}{C}}-CH_3 \longrightarrow H:Cl + \cdot\overset{\displaystyle Cl}{\underset{\displaystyle H}{C}}-CH_3$$

Step 3a $$CH_3\overset{\displaystyle Cl}{\underset{\displaystyle H}{C}}\cdot + Cl:Cl \longrightarrow CH_3-\overset{\displaystyle Cl}{\underset{\displaystyle H}{C}}-Cl + Cl\cdot$$

1,1-Dichloro-
ethane

Step 2b $Cl\cdot + H-CH_2CH_2Cl \longrightarrow H:Cl + \cdot CH_2CH_2Cl$

Step 3b $ClCH_2CH_2\cdot + Cl:Cl \longrightarrow ClCH_2CH_2Cl + Cl\cdot$

1,2-Dichloro-
ethane

9.13 If all 10 hydrogen atoms of isobutane were equally reactive, the relative amounts of reaction at primary hydrogen atoms and at tertiary hydrogen atoms would be 9/1, and the ratio of isobutyl chloride to tert-butyl chloride would be 9:1. Since the ratio is instead 63/37 ($\sim$6:4), the tertiary hydrogen atom must be more reactive than the primary hydrogen atoms.

9.14 Laboratory preparation of alkyl halides by direct chlorination can be accomplished in good yield when all hydrogen atoms in the alkane are equivalent. This is true of neopentane, and cyclopentane. (In these cases, the preparation would be practical only for monochlorination, where an excess of hydrocarbon would be employed, or for complete chlorination where an excess of chlorine would be used.

9.15 (a) $Cl\cdot + CH_3CH_2-H \longrightarrow CH_3CH_2\cdot + H-Cl$
($DH^\circ = 98$) ($DH^\circ = 103$)

$\Delta H^\circ = 98 - 103 = -5$ kcal/mole (exothermic)

(b) $Cl\cdot + (CH_3)_2CH-H \longrightarrow (CH_3)_2CH\cdot + H-Cl$
($DH^\circ = 94.5$) ($DH^\circ = 103$)

$\Delta H^\circ = 94.5 - 103 = -8.5$ kcal/mole (exothermic)

(c) $Cl\cdot + CH_3CH_2CH_2-H \longrightarrow CH_3CH_2CH_2\cdot + H-Cl$
($DH^\circ = 98$) ($DH^\circ = 103$)

$\Delta H^\circ = 98 - 103 = -5$ kcal/mole (exothermic)

9.16 The hydrogen abstraction steps in alkane fluorinations are always highly exothemic. Thus the transition states are even more reactantlike in structure and in energy than they are in alkane chlorinations. The type of C—H bond being broken (1°, 2°, or 3°) has practically no effect on the relative rates of the reactions.

9.17 (a)

$CH_3CH_2CH_2C(Cl)(H)CH_3 \xrightarrow[\text{light}]{Cl_2}$

(*S*)-2-Chloropentane

$CH_3C(Cl)(H)CH_2C(Cl)(H)CH_3$ + $CH_3C(H)(Cl)CH_2C(Cl)(H)CH_3$

(*R*,*S*)-2,4-Dichloropentane (*S*,*S*)-2,4-Dichloropentane

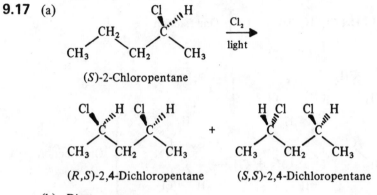

(b) Diastereomers

(c) No, the (*R*,*S*) isomer is a meso form.

(d) No, the (*R*,*S*) isomer is optically inactive.

(e) Yes, because diastereomers have different physical properties.

(f)(g)

$ClCH_2CH_2CH_2C(Cl)(H)CH_3$ (active)

$CH_3CH_2C(Cl)(H)C(Cl)(H)CH_3$ (active)

$CH_3CH_2C(H)(Cl)C(Cl)(H)CH_3$ (active)

$CH_3CH_2CH_2C(Cl)(Cl)CH_3$ (inactive)

$CH_3CH_2CH_2C(Cl)(H)CH_2Cl$ (active)

9.18 (a) $CH_3CH_2CH_2CH_2CHCl_2$ } One fraction

$CH_3CH_2CH_2C(Cl)(H)CH_2Cl$ + enantiomer } One fraction

$CH_3CH_2C(Cl)(H)CH_2CH_2Cl$ + enantiomer } One fraction

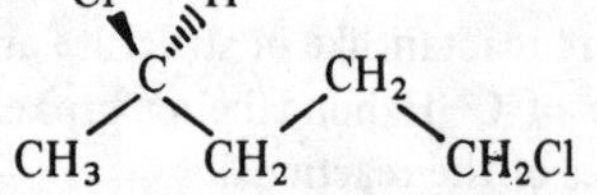

Cl H
C CH$_2$ + enantiomer } One fraction
CH$_3$ CH$_2$ CH$_2$Cl

$ClCH_2CH_2CH_2CH_2CH_2Cl$ } One fraction

$CH_3CH_2CCl_2CH_2CH_3$ } One fraction

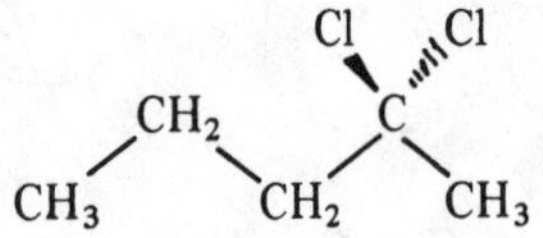

Cl Cl
CH$_2$ C } One fraction
CH$_3$ CH$_2$ CH$_3$

Cl H H Cl
CH$_2$ C CH$_2$ C
CH$_3$ C CH$_3$ + CH$_3$ C CH$_3$ } One fraction
Cl H H Cl

Cl H H Cl
CH$_2$ C CH$_2$ C
CH$_3$ C CH$_3$ + CH$_3$ C CH$_3$ } One fraction
H Cl Cl H

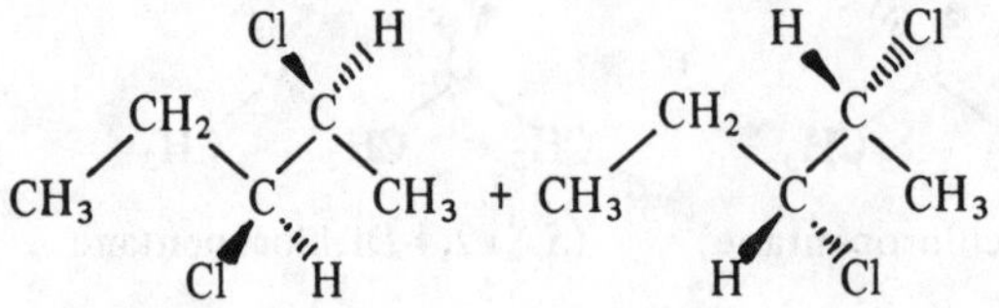

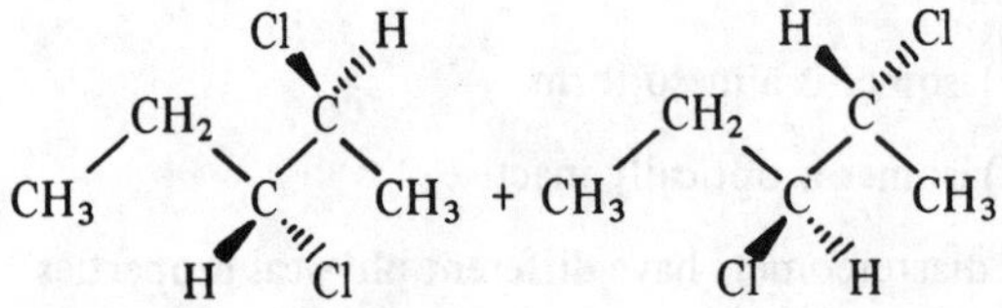

Cl H Cl H
C C (meso form) } One fraction
CH$_3$ CH$_2$ CH$_3$

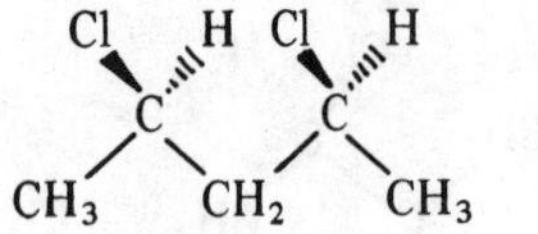

H Cl Cl H Cl H H Cl
C C + C C } One fraction
CH$_3$ CH$_2$ CH$_3$ CH$_3$ CH$_2$ CH$_3$

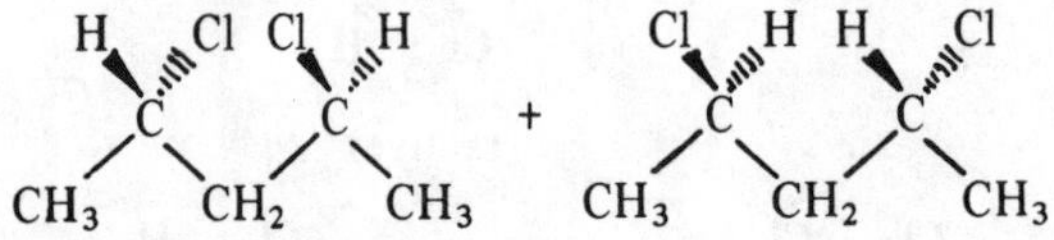

(b) All fractions are optically inactive.

9.19 (a)

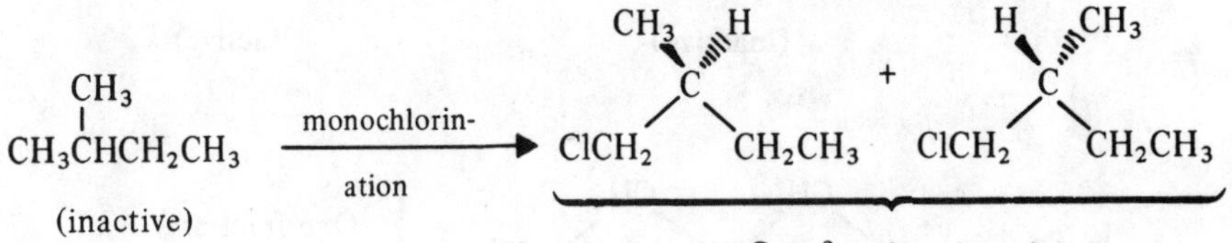

CH$_3$
CH$_3$CHCH$_2$CH$_3$ (inactive) —monochlorination→

CH$_3$ H H CH$_3$
C + C
ClCH$_2$ CH$_2$CH$_3$ ClCH$_2$ CH$_2$CH$_3$

One fraction, inactive (racemic form)

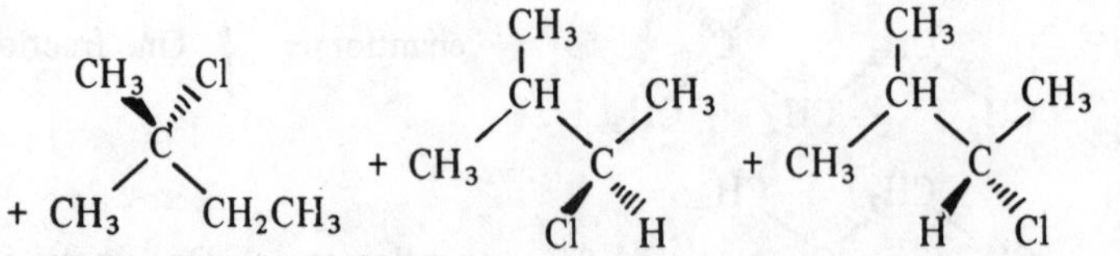

CH$_3$ Cl
C
+ CH$_3$ CH$_2$CH$_3$

One fraction, inactive

CH$_3$ CH$_3$
CH CH$_3$ CH CH$_3$
+ CH$_3$ C + CH$_3$ C
Cl H H Cl

One fraction, inactive (racemic form)

$$+\ \mathrm{CH_3\underset{}{\overset{CH_3}{\overset{|}{C}}}HCH_2CH_2Cl}$$

None of the fractions would show optical activity.

(b) The two fractions that contain racemic forms would be resolvable.

9.20 (a) **Chain Initiation**

Step 1 $\mathrm{R{-}\ddot{\underset{..}{O}}{-}\ddot{\underset{..}{O}}{-}R \xrightarrow{\text{heat}} 2\,R{-}\ddot{\underset{..}{O}}\cdot}$

Step 2 $\mathrm{R{-}\ddot{\underset{..}{O}}\cdot + H{-}CCl_3 \longrightarrow R{-}\ddot{\underset{..}{O}}H + \cdot CCl_3}$

Chain Propagation

Step 3 $\mathrm{CH_3CH_2CH_2CH{=}CH_2 + \cdot CCl_3 \longrightarrow CH_3CH_2CH_2\dot{C}H{-}CH_2CCl_3}$

Step 4 $\mathrm{CH_3CH_2CH_2\dot{C}HCH_2CCl_3 + H{-}CCl_3 \longrightarrow CH_3CH_2CH_2CH_2CH_2CCl_3 + \cdot CCl_3}$

then steps 3, 4, 3, 4, and so on.

(b) **Chain Initiation**

Step 1 $\mathrm{R{-}\ddot{\underset{..}{O}}{-}\ddot{\underset{..}{O}}{-}R \xrightarrow{\text{heat}} 2\,R{-}\ddot{\underset{..}{O}}\cdot}$

Step 2 $\mathrm{R{-}\ddot{\underset{..}{O}}\cdot + CH_3CH_2{-}\ddot{\underset{..}{S}}{-}H \longrightarrow R{-}\ddot{\underset{..}{O}}H + CH_3CH_2{-}\ddot{\underset{..}{S}}\cdot}$

Chain Propagation

Step 3 $\mathrm{CH_3\overset{CH_3}{\overset{|}{C}}{=}CH_2 + \cdot\ddot{\underset{..}{S}}CH_2CH_3 \longrightarrow CH_3\overset{CH_3}{\overset{|}{\underset{\bullet}{C}}}{-}CH_2{-}\ddot{\underset{..}{S}}{-}CH_2CH_3}$

Step 4 $\mathrm{CH_3\overset{CH_3}{\overset{|}{\underset{\bullet}{C}}}CH_2SCH_2CH_3 + HSCH_2CH_3 \longrightarrow CH_3\overset{CH_3}{\overset{|}{C}}HCH_2SCH_2CH_3 + \cdot\ddot{\underset{..}{S}}CH_2CH_3}$

then steps 3, 4, 3, 4, and so on.

(c) **Chain Initiation**

Step 1 $\mathrm{R{-}\ddot{\underset{..}{O}}{-}\ddot{\underset{..}{O}}{-}R \xrightarrow{\text{heat}} 2\,R{-}\ddot{\underset{..}{O}}\cdot}$

Step 2 $\mathrm{R{-}\ddot{\underset{..}{O}}\cdot + Cl{-}CCl_3 \longrightarrow R{-}\ddot{\underset{..}{O}}{-}Cl + \cdot CCl_3}$

Chain Propagation

Step 3 $\mathrm{CH_3CH_2\overset{CH_3}{\overset{|}{C}}{=}CH_2 + \cdot CCl_3 \longrightarrow CH_3CH_2\overset{CH_3}{\overset{|}{\underset{\bullet}{C}}}{-}CH_2CCl_3}$

Step 4 $\mathrm{CH_3CH_2\overset{CH_3}{\overset{|}{\underset{\bullet}{C}}}CH_2CCl_3 + CCl_4 \longrightarrow CH_3CH_2\overset{CH_3}{\overset{|}{\underset{\underset{Cl}{|}}{C}}}CH_2CCl_3 + \cdot CCl_3}$

then steps 3, 4, 3, 4, and so on

9.21 **Chain-Initiating Step**

$$Cl_2 \xrightarrow[\text{light}]{\text{heat, } h\nu} 2Cl\cdot$$

Chain-Propagating Steps

$$CH_3CH_2CH_3 \xrightarrow{Cl\cdot} CH_3CH_2CH_2\cdot + HCl \xrightarrow{Cl_2} CH_3CH_2CH_2Cl + Cl\cdot$$

$$CH_3CH_2CH_3 \xrightarrow{Cl\cdot} CH_3\underset{\substack{|\\CH_3}}{CH}\cdot + HCl \xrightarrow{Cl_2} CH_3\underset{\substack{|\\CH_3}}{CH}Cl + Cl\cdot$$

9.22 (a) $$CH_3\overset{\substack{CH_3\\|}}{CH}CH_3 \xrightarrow[\text{heat, } h\nu]{Br_2} CH_3\overset{\substack{CH_3\\|}}{\underset{\substack{|\\Br}}{C}}CH_3$$

(b) $$CH_3\overset{\substack{CH_3\\|}}{\underset{\substack{|\\Br}}{C}}CH_3 \xrightarrow[CH_3CH_2OH]{CH_3CH_2ONa} CH_3\overset{\substack{CH_3\\|}}{C}{=}CH_2$$

(c) $$CH_3\overset{\substack{CH_3\\|}}{C}{=}CH_2 \xrightarrow[\text{peroxides}]{HBr} CH_3\overset{\substack{CH_3\\|}}{CH}CH_2Br$$

(d) $$CH_3\overset{\substack{CH_3\\|}}{CH}CH_2Br \xrightarrow[\text{acetone}]{KI} CH_3\overset{\substack{CH_3\\|}}{CH}CH_2I$$

(e) $$CH_3\overset{\substack{CH_3\\|}}{CH}CH_2Br \xrightarrow[H_2O]{OH^-} CH_3\overset{\substack{CH_3\\|}}{CH}CH_2OH$$

or

$$CH_3\overset{\substack{CH_3\\|}}{C}{=}CH_2 \xrightarrow[(2)\ H_2O_2,\ OH^-]{(1)\ (BH_3)_2} CH_3\overset{\substack{CH_3\\|}}{CH}CH_2OH$$

(f) $$CH_3\overset{\substack{CH_3\\|}}{C}{=}CH_2 \xrightarrow[\text{heat}]{H_3O^+,\ H_2O} CH_3\overset{\substack{CH_3\\|}}{\underset{\substack{|\\OH}}{C}}CH_3$$

(g) $$CH_3\overset{\substack{CH_3\\|}}{CH}CH_2Br \xrightarrow[CH_3OH]{CH_3ONa} CH_3\overset{\substack{CH_3\\|}}{CH}CH_2OCH_3$$

(h) $$CH_3\overset{\substack{CH_3\\|}}{CH}CH_2Br \xrightarrow[CH_3\overset{\substack{O\\\|}}{C}OH]{CH_3\overset{\substack{O\\\|}}{C}ONa} CH_3\overset{\substack{CH_3\\|}}{CH}CH_2O\overset{\substack{O\\\|}}{C}CH_3$$

(i) $CH_3CH(CH_3)CH_2Br \xrightarrow{NaCN} CH_3CH(CH_3)CH_2CN$

(j) $CH_3CH(CH_3)CH_2Br \xrightarrow{CH_3SNa} CH_3CH(CH_3)CH_2SCH_3$

or

$CH_3C(CH_3){=}CH_2 \xrightarrow[\text{peroxides}]{CH_3SH} CH_3CH(CH_3)CH_2SCH_3$

(k) $CH_3C(CH_3){=}CH_2 \xrightarrow[\text{peroxides}]{CBr_4} CH_3CBr(CH_3)CH_2CBr_3$

(l) $CH_3C(CH_3){=}CH_2 \xrightarrow{\text{peroxides}} \left(CH_2C(CH_3)_2 \right)_n$

9.23 (a) Step 1 $RO{-}OR \longrightarrow 2RO\cdot$

Step 2 $RO\cdot + Br{-}CCl_3 \longrightarrow ROBr + \cdot CCl_3$

Step 3 $CH_3(CH_2)_5CH{=}CH_2 + \cdot CCl_3 \longrightarrow CH_3(CH_2)_5\dot{C}HCH_2CCl_3$

Step 4 $CH_3(CH_2)_5\dot{C}HCH_2CCl_3 + Br{-}CCl_3 \longrightarrow CH_3(CH_2)_5CHBrCH_2CCl_3 + \cdot CCl_3$

then steps 3, 4, 3, 4, and so on

(b) Step 1 $RO{-}OR \longrightarrow 2RO\cdot$

Step 2 $RO\cdot + H{-}CCl_3 \longrightarrow ROH + \cdot CCl_3$

Step 3 $CH_3(CH_2)_5CH{=}CH_2 + \cdot CCl_3 \longrightarrow CH_3(CH_2)_5\dot{C}HCH_2CCl_3$

Step 4 $CH_3(CH_2)_5\dot{C}HCH_2CCl_3 + H{-}CCl_3 \longrightarrow CH_3(CH_2)_5CH(H)CH_2CCl_3 + \cdot CCl_3$

then steps 3, 4, 3, 4, and so on

(c) Step 1 $RO{-}OR \longrightarrow 2RO\cdot$

Step 2 $RO\cdot + Cl{-}CCl_4 \longrightarrow ROCl + \cdot CCl_3$

Step 3 $CH_3(CH_2)_5CH{=}CH_2 + \cdot CCl_3 \longrightarrow CH_3(CH_2)_5\dot{C}HCH_2CCl_3$

Step 4 $CH_3(CH_2)_5\dot{C}HCH_2CCl_3 + Cl{-}CCl_3 \longrightarrow CH_3(CH_2)_5CHClCH_2CCl_3 + \cdot CCl_3$

then steps 3, 4, 3, 4, and so on

9.24 Because of the selectivity of bromine, it replaces the tertiary hydrogen almost exclusively:

$$\begin{array}{c}CH_3\\ |\\ CH_3CCH_2CH_3\\ |\\ H\end{array} + Br_2 \xrightarrow[\text{heat}]{h\nu} \begin{array}{c}CH_3\\ |\\ CH_3CCH_2CH_3\\ |\\ Br\end{array} + HBr$$

(a) $$\begin{array}{c}CH_3\\ |\\ CH_3CCH_2CH_3\\ |\\ Br\end{array} \xrightarrow[(CH_3)_3COH]{(CH_3)_3CO^-} \begin{array}{c}CH_3\\ |\\ CH_2{=}CCH_2CH_3\\ \text{2-Methyl-1-butene}\end{array}$$

(b) $$\begin{array}{c}CH_3\\ |\\ CH_3CCH_2CH_3\\ |\\ Br\end{array} \xrightarrow[CH_3CH_2OH]{CH_3CH_2O^-} \begin{array}{c}CH_3\\ |\\ CH_3C{=}CHCH_3\\ \text{2-Methyl-2-butene}\end{array}$$

(c) $$\begin{array}{c}CH_3\\ |\\ CH_2{=}CCH_2CH_3\\ \text{[from (a)]}\end{array} \xrightarrow[ROOR]{HBr} \begin{array}{c}CH_3\\ |\\ BrCH_2CHCH_2CH_3\\ \text{1-Bromo-2-methylbutane}\end{array}$$

(d) $$\begin{array}{c}CH_3\\ |\\ CH_3C{=}CHCH_3\\ \text{[from (b)]}\end{array} \xrightarrow[ROOR]{HBr} \begin{array}{c}CH_3\\ |\\ CH_3CHCHCH_3\\ |\\ Br\\ \text{2-Bromo-3-methylbutane}\end{array}$$

(e) $$\begin{array}{c}CH_3\\ |\\ CH_3CHCHCH_3\\ |\\ Br\\ \text{[from (d)]}\end{array} \xrightarrow[(CH_3)_3COH]{(CH_3)_3CO^-} \begin{array}{c}CH_3\\ |\\ CH_3CHCH{=}CH_2\\ \text{3-Methyl-1-butene}\end{array}$$

(f) $$\begin{array}{c}CH_3\\ |\\ CH_3CHCH{=}CH_2\\ \text{[from (e)]}\end{array} \xrightarrow[ROOR]{HBr} \begin{array}{c}CH_3\\ |\\ CH_3CHCH_2CH_2Br\\ \text{1-Bromo-3-methylbutane}\end{array}$$

(g) $$\begin{array}{c}CH_3\\ |\\ BrCH_2CHCH_2CH_3\\ \text{[from (c)]}\end{array} \xrightarrow[\text{acetone}]{I^-} \begin{array}{c}CH_3\\ |\\ ICH_2CHCH_2CH_3\\ \text{1-Iodo-2-methylbutane}\end{array}$$

(h) $$\begin{array}{c}CH_3\\ |\\ CH_2{=}CCH_2CH_3\\ \text{[from (a)]}\end{array} \xrightarrow{HI} \begin{array}{c}CH_3\\ |\\ CH_3CCH_2CH_3\\ |\\ I\\ \text{2-Iodo-2-methylbutane}\end{array}$$

(i) $$\begin{array}{c}CH_3\\ |\\ CH_3CHCH_2CH_2Br\\ \text{[from (f)]}\end{array} \xrightarrow[\text{acetone}]{I^-} \begin{array}{c}CH_3\\ |\\ CH_3CHCH_2CH_2I\\ \text{1-Iodo-3-methylbutane}\end{array}$$

(j) $$\begin{array}{c}CH_3\\ |\\ CH_2{=}CCH_2CH_3\\ \text{[from (a)]}\end{array} \xrightarrow{ICl} \begin{array}{c}CH_3\\ |\\ CH_2CCH_2CH_3\\ |\quad|\\ I\quad Cl\\ \text{2-Chloro-1-iodo-2-methylbutane}\end{array}$$

(k) $$\underset{\text{[from (a)]}}{CH_2{=}\overset{\overset{\displaystyle CH_3}{|}}{C}CH_2CH_3} \xrightarrow[(2)\ H_2O_2,\ OH^-]{(1)\ (BH_3)_2} HOCH_2\overset{\overset{\displaystyle CH_3}{|}}{C}HCH_2CH_3$$

(l) $$\underset{\text{[from (e)]}}{CH_3\overset{\overset{\displaystyle CH_3}{|}}{C}HCH{=}CH_2} \xrightarrow[(2)\ H_2O_2,\ OH^-]{(1)\ (BH_3)_2} CH_3\overset{\overset{\displaystyle CH_3}{|}}{C}HCH_2CH_2OH$$

(m) $$\underset{\text{[from (e)]}}{CH_3\overset{\overset{\displaystyle CH_3}{|}}{C}HCH{=}CH_2} \xrightarrow[H_2O,\ heat]{H_3O^+} CH_3\overset{\overset{\displaystyle CH_3}{|}}{C}H\underset{\underset{\displaystyle OH}{|}}{C}HCH_3$$

or

$$\underset{\text{[from (b)]}}{CH_3\overset{\overset{\displaystyle CH_3}{|}}{C}{=}CHCH_3} \xrightarrow[(2)\ H_2O_2,\ OH^-]{(1)\ (BH_3)_2} CH_3\overset{\overset{\displaystyle CH_3}{|}}{C}H\underset{\underset{\displaystyle OH}{|}}{C}HCH_3$$

(n) $$\underset{\text{[from (e)]}}{CH_3\overset{\overset{\displaystyle CH_3}{|}}{C}HCH{=}CH_2} \xrightarrow[ROOR]{CBr_4} CH_3\overset{\overset{\displaystyle CH_3}{|}}{C}H\underset{\underset{\displaystyle Br}{|}}{C}HCH_2CBr_3$$

(o) $$\underset{\text{[from (a)]}}{CH_2{=}\overset{\overset{\displaystyle CH_3}{|}}{C}CH_2CH_3} \xrightarrow[(2)\ Zn,\ H_2O]{(1)\ O_2} CH_3\overset{\overset{\displaystyle O}{\|}}{C}CH_2CH_3\quad (+\ H\overset{\overset{\displaystyle O}{\|}}{C}H)$$

(p) $$\underset{\text{[from (e)]}}{CH_3\overset{\overset{\displaystyle CH_3}{|}}{C}HCH{=}CH_2} \xrightarrow[(2)\ Zn,\ H_2O]{(1)\ O_3} CH_3\overset{\overset{\displaystyle CH_3}{|}}{C}H\underset{\underset{\displaystyle H}{|}}{C}{=}O\quad (+\ H\overset{\overset{\displaystyle O}{\|}}{C}H)$$

(q) $$\underset{\text{[from (a)]}}{CH_2{=}\overset{\overset{\displaystyle CH_3}{|}}{C}CH_2CH_3} \xrightarrow[\text{cold, dilute}]{KMnO_4} HOCH_2\overset{\overset{\displaystyle CH_3}{|}}{\underset{\underset{\displaystyle OH}{|}}{C}}CH_2CH_3$$

(r) $$\underset{\text{[from (b)]}}{CH_3\overset{\overset{\displaystyle CH_3}{|}}{C}{=}CHCH_3} \xrightarrow[(2)\ H_3O^+,\ H_2O]{(1)\ R\overset{\overset{\displaystyle O}{\|}}{C}OOH} CH_3\overset{\overset{\displaystyle CH_3}{|}}{C}\underset{O}{\diagdown\!\!\diagup}CHCH_3$$

9.25 (a) Six

$$CH_3\overset{\overset{\displaystyle CH_3}{|}}{C}HCH_2CH_3 \xrightarrow{F_2} \underbrace{\text{I} + \text{II}}_{\text{Racemic form}} + \underset{\text{III}}{CH_3\overset{\overset{\displaystyle CH_3}{|}}{\underset{\underset{\displaystyle F}{|}}{C}}CH_2CH_3} +$$

I: Newman projection with CH_2F (top), H (left), CH_3 (right); rear carbon bearing CH_2CH_3

II: Newman projection with CH_2F (top), CH_3 (left), H (right); rear carbon bearing CH_2CH_3

Racemic form

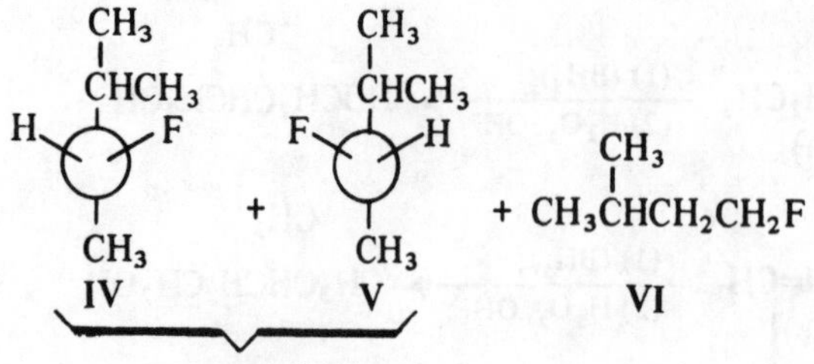

$$\text{IV} + \text{V (Newman projections: } CH_3,\ CHCH_3 \text{ front; H, F; } CH_3 \text{ back)} + CH_3CH(CH_3)CH_2CH_2F \text{ (VI)}$$

Racemic form

(b) Only four: a fraction containing the racemic form, **I** and **II**; a fraction containing **III**; a fraction containing the racemic form **IV** and **V**; and a fraction containing **VI**.

(c) All of them.

(d) The fraction containing the racemic form **I** and **II**, and the fraction containing the racemic form **IV** and **V**.

9.26 (a) Five

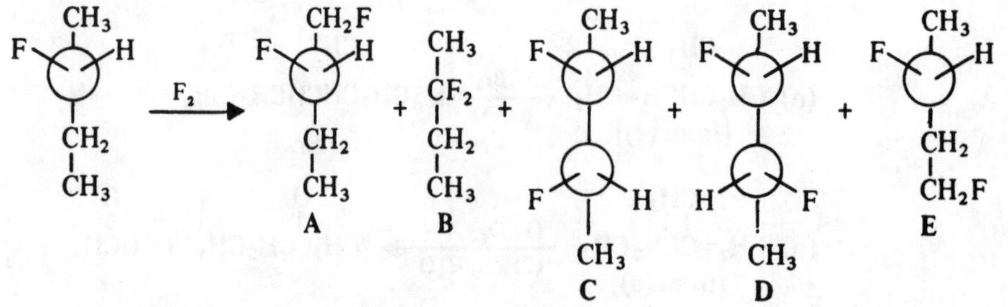

$$\text{(Newman: } CH_3, F, H / CH_2CH_3) \xrightarrow{F_2} \text{A (}CH_2F, F, H / CH_2CH_3) + \underset{\text{B}}{CH_3CF_2CH_2CH_3} + \text{C} + \text{D} + \text{E (}CH_3, F, H / CH_2CH_2F)$$

(b) Five. None of the fractions would be a racemic form.

(c) The fractions containing **A**, **D**, and **E**. The fraction containing **B** and **C** would be optically inactive. (**B** contains no stereocenter and **C** is a meso compound.)

9.27 (a) $HC{\equiv}CH \xrightarrow{NaNH_2} HC{\equiv}CNa \xrightarrow{CH_3I} CH_3C{\equiv}CH \xrightarrow[\text{Lindlar's catalyst}]{H_2}$

$CH_3CH{=}CH_2 \xrightarrow[ROOR]{HBr} CH_3CH_2CH_2Br$

(b) As in (a) except use CH_3CH_2Br in the second step.

(c) $\underset{\text{[from (a)]}}{HC{\equiv}CNa} \xrightarrow[\text{[from (b)]}]{CH_3CH_2CH_2CH_2Br} CH_3CH_2CH_2CH_2C{\equiv}CH$

(d) $\underset{\text{[from (a)]}}{CH_3C{\equiv}CH} \xrightarrow{NaNH_2} CH_3C{\equiv}CNa \xrightarrow[\text{[from (b)]}]{CH_3CH_2CH_2CH_2Br}$

$CH_3C{\equiv}CCH_2CH_2CH_2CH_3$

(e) $\underset{\text{[from (b)]}}{CH_3CH_2C{\equiv}CH} \xrightarrow{NaNH_2} CH_3CH_2C{\equiv}CNa \xrightarrow[\text{[from (a)]}]{CH_3CH_2CH_2Br}$

$CH_3CH_2C{\equiv}CCH_2CH_2CH_3$

(f) $HC\equiv CH \xrightarrow{NaNH_2} HC\equiv CNa \xrightarrow[\text{[from (a)]}]{CH_3CH_2CH_2Br} CH_3CH_2CH_2C\equiv CH \xrightarrow{NaNH_2}$

$$CH_3CH_2CH_2C\equiv CNa \xrightarrow{CH_3CH_2CH_2Br} CH_3CH_2CH_2C\equiv CCH_2CH_2CH_3$$

9.28 (a) Oxygen-oxygen single bonds are especially weak, that is,

$$HO{-}OH \qquad DH^\circ = 51 \text{ kcal/mole}$$
$$CH_3CH_2O{-}OCH_3 \qquad DH^\circ = 44 \text{ kcal/mole}$$

This means that a peroxide will dissociate into free radicals at a relatively low temperature.

$$RO{-}OR \xrightarrow{100\text{-}200^\circ C} 2RO\cdot$$

Oxygen-hydrogen single bonds, on the other hand, are very strong. (For HO–H, DH° = 119 kcal/mole.) This means that reactions like the following will be highly exothermic.

$$RO\cdot + R{-}H \longrightarrow RO{-}H + R\cdot$$

(b) Step 1 $(CH_3)_3CO{-}OC(CH_3)_3 \xrightarrow{heat} 2(CH_3)_3CO\cdot$

Step 2 $(CH_3)_3CO\cdot + R{-}H \longrightarrow (CH_3)_3COH + R\cdot$

Chain Initiation

Step 3 $R\cdot + Cl{-}Cl \longrightarrow R{-}Cl + Cl\cdot$

Step 4 $Cl\cdot + R{-}H \longrightarrow H{-}Cl + R\cdot$

Chain Propagation

9.29 (a) $H{-}SH \xrightarrow{h\nu} H\cdot + \cdot SH$ **Chain-initiating step**

$R{-}CH{=}CH_2 + \cdot SH \longrightarrow R\dot{C}H{-}CH_2SH$

$R{-}\dot{C}HCH_2SH + H{-}SH \longrightarrow RCH_2CH_2SH$

Chain-propagating steps

(b) $R\dot{C}HCH_2SH + RCH_2CH_2SH \longrightarrow RCH_2CH_2SH + RCH_2CH_2S\cdot$

$RCH{=}CH_2 + \cdot SCH_2CH_2R \longrightarrow R\dot{C}HCH_2SCH_2CH_2R$

$R\dot{C}HCH_2SCH_2CH_2R + RCH_2CH_2SH \longrightarrow (RCH_2CH_2)_2S + RCH_2CH_2S\cdot$

9.30 (a) $\underset{(DH^\circ = 104)}{CH_3{-}H} + \underset{(DH^\circ = 38)}{F{-}F} \longrightarrow CH_3\cdot + \underset{(DH^\circ = 136)}{H{-}F} + F\cdot$ $\quad \Delta H^\circ = +6$ kcal/mole, $E_{act} > +6$ kcal/mole

$CH_3\cdot + F\cdot \longrightarrow \underset{(DH^\circ = 108)}{CH_3{-}F}$ $\quad \Delta H^\circ = -108$ kcal/mole, $E_{act} = 0$

If E_{act} for the first step is not much greater than 6 kcal/mole, this mechanism is likely.

(b) $CH_3{-}H$ + $Cl{-}Cl$ $\longrightarrow$ $CH_3\cdot$ + $H{-}Cl$ + $Cl\cdot$ $\quad \Delta H° = +59$ kcal/mole
($DH° = 104$) ($DH° = 58$) ($DH° = 103$) $\quad E_{act} \geq +59$ kcal/mole

$CH_3\cdot + Cl\cdot \longrightarrow CH_3{-}Cl$ $\quad \Delta H° = -83.5$ kcal/mole
($DH° = 83.5$) $\quad E_{act} = 0$

This mechanism is highly unlikely because the E_{act} for the first step must be ≥ 59 kcal/mole.

9.31 (a) $CH_3{-}H$ $DH° = 104$; $CH_3CH_2{-}H$ $DH° = 98$ kcal/mole. (Recall that here, $E_{act} = DH°$.)

$CH_3CH_2{-}H$ bond rupture requires less energy, therefore spontaneous homolysis (cracking) occurs at a lower temperature.

(b) $CH_3{-}CH_3$ $DH° = 88$ kcal/mole $= E_{act}$

C–C bond rupture requires less energy than C–H bond rupture, therefore C–C bond rupture occurs more readily than $CH_3CH_2{-}H$ bond rupture.

(c) $CH_3CH_2{-}CH_2CH_3$ $DH° = 82$ kcal/mole $= E_{act}$

$CH_3CH_2CH_2{-}CH_3$ $DH° = 85$ kcal/mole $= E_{act}$

Here again the bond with the lower bond dissociation energy will undergo spontaneous homolysis (cracking) more readily.

9.32

Step 1 $CH_3CH_2CH_3 \longrightarrow CH_3CH_2\cdot + CH_3\cdot$ $\quad DH° = 85$ kcal/mole
Step 2 $CH_3CH_2CH_3 \longrightarrow CH_3CH_2CH_2\cdot + H\cdot$ $\quad DH° = 98$ kcal/mole
Step 3 $CH_3CH_2CH_3 \longrightarrow CH_3\dot{C}HCH_3 + H\cdot$ $\quad DH° = 94.5$ kcal/mole

(a) Since E_{act} is equal to $DH°$, we can assume that (1) is the most likely chain-initiating step.

(b) $CH_3\cdot + CH_3CH_2CH_3 \longrightarrow CH_3{-}H + \cdot CH_2CH_2CH_3$ $\quad \Delta H° = -6$ kcal/mole
($DH° = 98$) ($DH° = 104$)

Since $\Delta H°$ is negative, E_{act} need not be large.

(c) $CH_3\cdot + CH_3CH_2CH_3 \longrightarrow CH_4 + CH_3\dot{C}HCH_3$ $\quad \Delta H° = -9.5$ kcal/mole
($DH° = 94.5$) ($DH° = 104$)

On the basis of energy requirements, this is a likely alternative to step 1. On the basis of the probability factor, it is less likely because there are only two secondary hydrogen atoms compared with six primary hydrogen atoms.

9.33

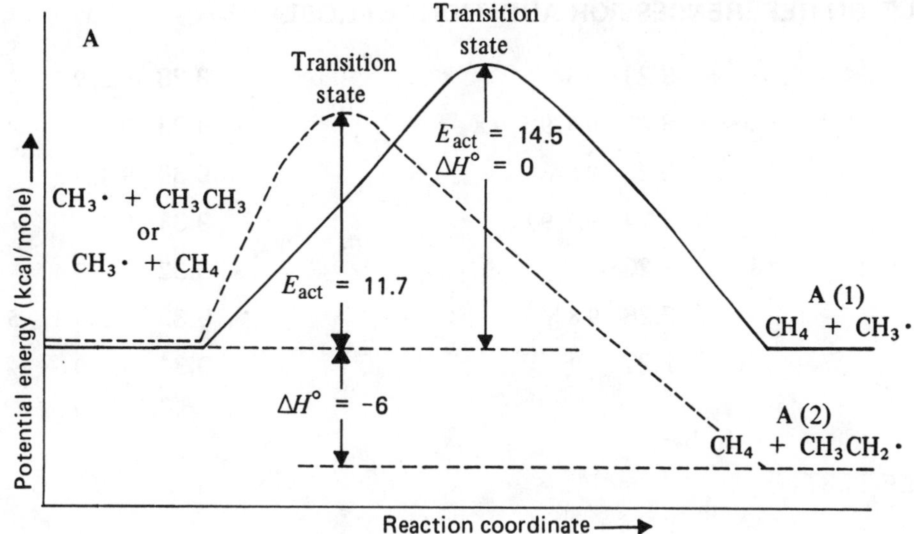

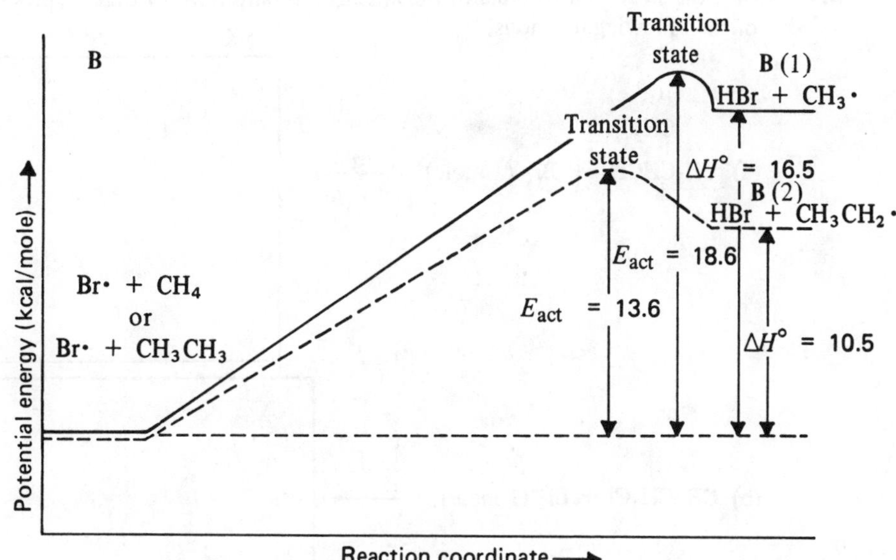

(b) Reaction **A** (2) since it is most exothermic.

(c) Reaction **B** (1) since it is most endothermic.

(d) Since $\Delta H^\circ = 0$, bond breaking should be approximately 50% complete.

(e) The reactions of set **B**.

(f) The difference in ΔH° simply reflects the difference in the C–H bond strengths of methane and ethane.

(g) Because the reactions of set **B** are highly endothermic the transition states show a strong resemblance to products in structure and *in energy*, and the products differ in energy by 6 kcal/mole. (In this instance, since the difference in E_{act} is five sixths of the difference in ΔH°, we can estimate that bond breaking is about five sixths complete when the transition states are reached.)

SECTION REFERENCES FOR ADDITIONAL PROBLEMS

9.21	9.3	**9.28**	9.2, 9.3
9.22	9.4, 9.5, 9.8	**9.29**	9.9
9.23	9.8A	**9.30**	9.4, 9.5
9.24	9.5, 9.8	**9.31**	9.2
9.25	9.8	**9.32**	9.2, 9.4
9.26	9.8	**9.33**	9.2, 9.4, 9.5
9.27	9.9	**9.34**	9.2, 9.6

SELF-TEST

9.1 Give the structural formula of the missing reactant or *major* organic product in each of the following reactions.

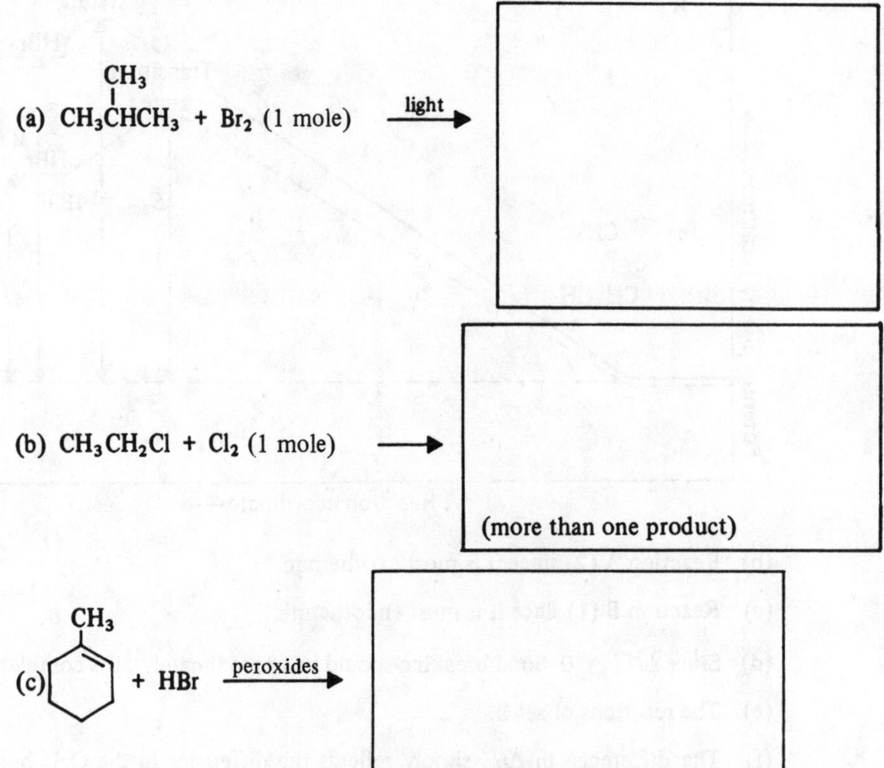

9.2 Calculate the ΔH of the following reaction.

(a) $CH_3CH_2CH_3 + Cl_2 \xrightarrow{\text{light}} CH_3CHClCH_3 + HCl$ ☐

(b) $CH_3CH_3 + \cdot Br \longrightarrow CH_3\dot{C}H_2 + HBr$ ☐

(c) $CH_3\dot{C}H_2 + Br\cdot \longrightarrow CH_3CH_2Br$ ☐

Use the single-bond dissociation energies of Table 9.1:

TABLE 9.1 Single-bond homolytic dissociation energies DH° at 25°C

$A{:}B \longrightarrow A\cdot + B\cdot$

Compound	kcal/mole	kJ/mole	Compound	kcal/mole	kJ/mole
H–H	104	435	$(CH_3)_2CH$–H	94.5	395
D–D	106	444	$(CH_3)_2CH$–F	105	439
F–F	38	159	$(CH_3)_2CH$–Cl	81	339
Cl–Cl	58	243	$(CH_3)_2CH$–Br	68	285
Br–Br	46	192	$(CH_3)_2CH$–I	53	222
I–I	36	151	$(CH_3)_2CH$–OH	92	385
H–F	136	569	$(CH_3)_2CH$–OCH_3	80.5	337
H–Cl	103	431	$(CH_3)_2CHCH_2$–H	98	410
H–Br	87.5	366	$(CH_3)_3C$–H	91	381
H–I	71	297	$(CH_3)_3C$–Cl	78.5	328
CH_3–H	104	435	$(CH_3)_3C$–Br	63	264
CH_3–F	108	452	$(CH_3)_3C$–I	49.5	207
CH_3–Cl	83.5	349	$(CH_3)_3C$–OH	90.5	379
CH_3–Br	70	293	$(CH_3)_3C$–OCH_3	78	326
CH_3–I	56	234	$C_6H_5CH_2$–H	85	356
CH_3–OH	91.5	383	CH_2=$CHCH_2$–H	85	356
CH_3–OCH_3	80	335	CH_2=CH–H	108	452
CH_3CH_2–H	98	410	C_6H_5–H	110	460
CH_3CH_2–F	106	444	HC≡C–H	125	523
CH_3CH_2–Cl	81.5	341	CH_3–CH_3	88	368
CH_3CH_2–Br	69	289	CH_3CH_2–CH_3	85	356
CH_3CH_2–I	53.5	224	$CH_3CH_2CH_2$–CH_3	85	356
CH_3CH_2–OH	91.5	383	CH_3CH_2–CH_2CH_3	82	343
CH_3CH_2–OCH_3	80	335	$(CH_3)_2CH$–CH_3	84	351
			$(CH_3)_3C$–CH_3	80	335
$CH_3CH_2CH_2$–H	98	410	HO–H	119	498
$CH_3CH_2CH_2$–F	106	444	HOO–H	90	377
$CH_3CH_2CH_2$–Cl	81.5	341	HO–OH	51	213
$CH_3CH_2CH_2$–Br	69	289	CH_3CH_2O–OCH_3	44	184
$CH_3CH_2CH_2$–I	53.5	224	CH_3CH_2O–H	103	431
$CH_3CH_2CH_2$–OH	91.5	383			
$CH_3CH_2CH_2$–OCH_3	80	335	$CH_3\overset{O}{\overset{\Vert}{C}}$–H	87	364

9.3 Draw structures of all the monobromination products of butane (reaction conditions = $h\nu$, 25°C).

Use the bond dissociation energies in Table 9.1 to answer the following questions.

9.4 (a) The most likely products of the thermal homolytic cleavage of propane are

```
   H  H  H
   |  |  |
H—C—C—C—H
   |  |  |
   H  H  H
```

[ ] and [ ]

(b) The ΔH of this reaction is [ ] kcal/mole.

(c) The most likely monobromination product of propane is [ ]

(d) The ΔH of reaction (c) is [ ] kcal/mole.

9.5 When ethane is heated to high temperatures it undergoes thermal cracking. One of the reactions it undergoes is

$$CH_3CH_3 \xrightarrow{\Delta} CH_3\dot{C}H_2 + H\cdot$$

Recombination of the ethyl radicals yields butane.

(a) Assuming that the hydrogen atoms also recombine to produce H_2, calculate the ΔH° for the overall reaction, $2CH_3CH_3 \longrightarrow CH_3CH_2CH_2CH_3 + H_2$.

ΔH° =

(b) Is this a chain reaction?

9.6 On the basis of Table 9.1, what is the order of decreasing stability of the free radicals, $HC{\equiv}C\cdot$ $CH_2{=}CH\cdot$ $CH_2{=}CHCH_2\cdot$?

(a) $HC{\equiv}C\cdot > CH_2{=}CH\cdot > CH_2{=}CHCH_2\cdot$

(b) $CH_2{=}CH\cdot > HC{\equiv}C\cdot > CH_2{=}CHCH_2\cdot$

(c) $CH_2{=}CHCH_2\cdot > HC{\equiv}C\cdot > CH_2{=}CH\cdot$

(d) $CH_2{=}CHCH_2\cdot > CH_2{=}CH\cdot > HC{\equiv}C\cdot$

(e) $CH_2{=}CH\cdot > CH_2{=}CHCH_2\cdot > HC{\equiv}C\cdot$

9.7 In the free radical chlorination of methane, one propagation step is shown as $Cl\cdot + CH_4 \longrightarrow HCl + \cdot CH_3$. Why do we eliminate the posibility that this step goes as shown?

$$Cl\cdot + CH_4 \longrightarrow CH_3Cl + H\cdot$$

(a) Because in the next propagation step $H\cdot$ would have to react with Cl_2 to form $Cl\cdot$ and HCl; this reaction is not feasible.

(b) Because this alternative step has a more endothermic ΔH° than the first.

(c) Because free hydrogen atoms cannot exist.

(d) Because this alternative step is not consistent with the high photochemical efficiency of this reaction.

9.8 Pure (*S*)-$CH_3CH_2CHBrCH_3$ is subjected to monobormination to form several isomers of $C_4H_8Br_2$. Which of the following is not produced?

(a) Fischer projection: CH_3 (top); H—C—Br; H—C—Br; CH_3 (bottom)

(b) Fischer projection: CH_3 (top); H—C—Br; Br—C—H; CH_3 (bottom)

(c) Fischer projection: CH_3 (top); Br—C—H; H—C—Br; CH_3 (bottom)

(d) $CH_3CH_2CBr_2CH_3$

(e) (*R*)-$CH_3CH_2CHBrCH_2Br$

9.9 Using the data of Table 9.1, calculate the heat of reaction, ΔH°, of the reaction,

$$CH_3CH_3 + Br_2 \longrightarrow CH_3CH_2Br + HBr$$

(a) 12.5 kcal/mol (b) -12.5 kcal/mol (c) 300.5 kcal/mol

(d) -300.5 kcal/mol (e) -58.5 kcal/mol

D SPECIAL TOPIC
Polymerization of Alkenes: Addition Polymers

SOLUTIONS TO PROBLEMS

D.1 The reaction will proceed through the most stable free radical that can be produced. Head-to-head polymerization, as shown, will lead to a primary radical,

$$R{-}CH_2{-}\underset{\displaystyle CH_3}{\underset{|}{CH}}\cdot + \underset{\displaystyle CH_3}{\underset{|}{CH}}{=}CH_2 \longrightarrow R{-}CH_2{-}\underset{\displaystyle CH_3}{\underset{|}{CH}}{-}\underset{\displaystyle CH_3}{\underset{|}{CH}}{-}CH_2\cdot$$

which is less stable than the secondary radical that is produced by head-to-tail polymerization:

$$R{-}CH_2{-}\underset{\displaystyle CH_3}{\underset{|}{CH}}\cdot + CH_2{=}\underset{\displaystyle CH_3}{\underset{|}{CH}} \longrightarrow R{-}CH_2{-}\underset{\displaystyle CH_3}{\underset{|}{CH}}{-}CH_2{-}\underset{\displaystyle CH_3}{\underset{|}{CH}}\cdot$$

D.2

(a) $$nCH_2{=}\underset{\displaystyle F}{\underset{|}{CH}} \xrightarrow[\text{peroxide}]{\text{organic}} \left(CH_2{-}\underset{\displaystyle F}{\underset{|}{CH}}\right)_n$$

(b) $$nCF_2{=}\underset{\displaystyle Cl}{\underset{|}{CF}} \xrightarrow[\text{peroxide}]{\text{organic}} \left(CF_2{-}\underset{\displaystyle Cl}{\underset{|}{CF}}\right)_n$$

(c) $$nCF_2{=}\underset{\displaystyle CF_3}{\underset{|}{CF}} + mCH_2{=}CF_2 \xrightarrow[\text{peroxide}]{\text{organic}} \left(CF_2{-}\underset{\displaystyle CF_3}{\underset{|}{CF}}\right)_n\left(CH_2{-}CF_2\right)_m$$

Note that the units are randomly ordered, and not necessarily joined to their own kind as shown.

D.3 Polymerization will occur to produce the most stable carbocation possible. The scheme shown in this problem involves formation of the primary carbocations,

$$\overset{\displaystyle CH_3}{\overset{|}{\underset{\displaystyle CH_3}{\underset{|}{CH}}}}{-}CH_2^{+} \qquad \overset{\displaystyle CH_3}{\overset{|}{\underset{\displaystyle CH_3}{\underset{|}{CH}}}}{-}CH_2{-}\overset{\displaystyle CH_3}{\overset{|}{\underset{\displaystyle CH_3}{\underset{|}{C}}}}{-}CH_2^{+} \quad \text{etc.}$$

instead of the tertiary carbocations,

$$CH_3-\underset{CH_3}{\overset{CH_3}{\underset{|}{\overset{|}{C}}}}-CH_2-\underset{CH_3}{\overset{CH_3}{\underset{|}{\overset{|}{C}}}}{+}$$

D.4 (a) By proton transfer from water to the strongly basic carbanion,

$$R-CH_2-\underset{CN}{\underset{|}{CH}}:^- + H-\underset{H}{\underset{|}{\ddot{O}}}: \longrightarrow R-CH_2-\underset{CN}{\underset{|}{CH_2}} + :\ddot{O}H^-$$

(b)

$$\left(CH_2\underset{R}{\underset{|}{CH}}\right)_n CH_2\underset{R}{\underset{|}{CH}}:^- + (m+1)\,H_2C\underset{O}{\diagdown\!\diagup}CH_2 \longrightarrow$$

$$\left(CH_2\underset{R}{\underset{|}{CH}}\right)_{n+1}\left(CH_2-CH_2-O\right)_m CH_2-CH_2-\ddot{O}:^-$$

$$\xrightarrow{H_2O} \left(CH_2\underset{R}{\underset{|}{CH}}\right)_{n+1}\left(CH_2-CH_2-O\right)_m CH_2-CH_2-OH$$

In this polymer, each chain consists of a long uninterrupted segment of the first repeating unit, $\left(CH_2\underset{R}{\underset{|}{CH}}\right)_{n+1}$ followed by a long uninterrupted segment of the second repeating unit, $\left(CH_2-CH_2-O\right)_m$.

D.5

(a) Chain substituents (above / below), left to right: H/H, CH_3/CO_2CH_3 (O=C–OCH_3), H/H, CH_3/CO_2CH_3, H/H, CO_2CH_3/CH_3, H/H, CH_3/CO_2CH_3, H/H, CO_2CH_3/CH_3, H/H, CO_2CH_3/CH_3, H/H, CO_2CH_3/CH_3, H/H, CH_3/CO_2CH_3

(b) Chain substituents (above / below), left to right: H/H, CH_3/CO_2CH_3, H/H, CO_2CH_3/CH_3, H/H, CH_3/CO_2CH_3, H/H, CO_2CH_3/CH_3, H/H, CH_3/CO_2CH_3, H/H, CO_2CH_3/CH_3, H/H, CH_3/CO_2CH_3

(c) Chain substituents (above / below), left to right: H/H, CH_3/CO_2CH_3, H/H, CH_3/CO_2CH_3, H/H, CH_3/CO_2CH_3, H/H, CH_3/CO_2CH_3, H/H, CH_3/CO_2CH_3, H/H, CH_3/CO_2CH_3, H/H, CH_3/CO_2CH_3

10 CONJUGATED UNSATURATED SYSTEMS

REACTIONS OF DIENES

1. Allylic substitution

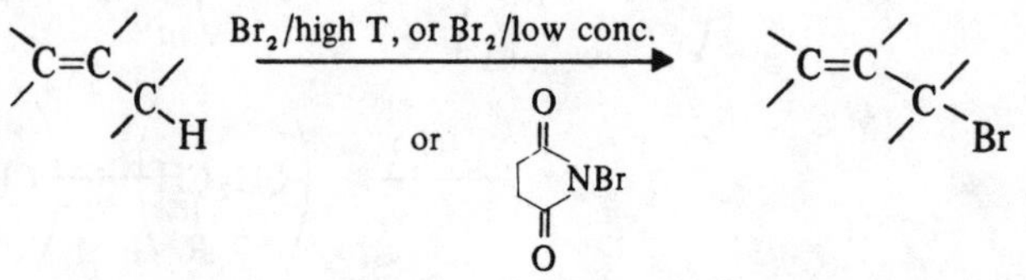

$$\mathrm{>C{=}C{<}\ \text{–}C{-}H} \xrightarrow[\text{or NBS}]{Br_2/\text{high T, or } Br_2/\text{low conc.}} \mathrm{>C{=}C{<}\ \text{–}C{-}Br}$$

(NBS shown as N-bromosuccinimide structure: O, NBr, O)

2. 1,2 and, 1,4 Addition

1, 2 Addition | 1, 4 Addition

$$\mathrm{>C{=}C{-}C{=}C{<}} \xrightarrow{HCl} \mathrm{H{-}C{-}C(Cl){-}C{=}C{<}} + \mathrm{H{-}C{-}C{=}C{-}C{-}Cl}$$

$$\xrightarrow{HBr} \mathrm{H{-}C{-}C(Br){-}C{=}C{<}} + \mathrm{H{-}C{-}C{=}C{-}C{-}Br}$$

$$\xrightarrow{Br_2} \mathrm{Br{-}C{-}C(Br){-}C{=}C{<}} + \mathrm{Br{-}C{-}C{=}C{-}C{-}Br}$$

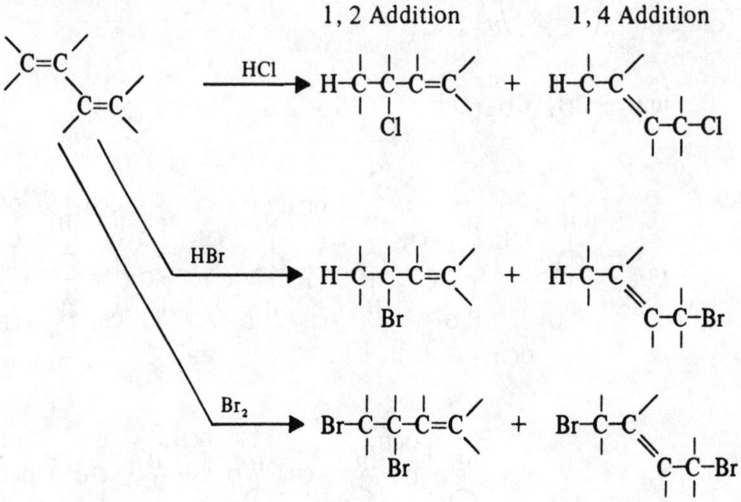

3. Diels-Alder reaction

$$\text{diene} + \mathrm{CH(X){=}CH(Y)} \longrightarrow \text{cyclohexene with X and Y}$$

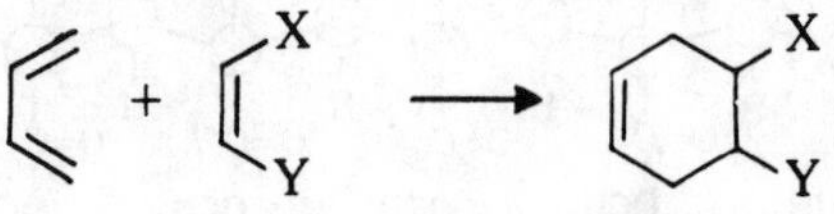

SOLUTIONS TO PROBLEMS

10.1 (a) $^{14}CH_2{=}CHCH_2X$ and $CH_2{=}CH{-}^{14}CH_2X$

(b) The reaction proceeds through the resonance-stabilized free radical.

$$^{14}CH_2{=}CH{-}\dot{C}H_2 \longleftrightarrow {}^{14}\dot{C}H_2{-}CH{=}CH_2 \quad \text{or} \quad {}^{14}\overset{\delta\cdot}{C}H_2{\overset{...}{-}}CH{\overset{...}{-}}\overset{\delta\cdot}{C}H_2$$

Thus attack on X_2 can occur by the carbon atom at either end of the chain since these atoms are equivalent.

(c) 50:50 because attack at the two ends of the chain are equally probable.

10.2

(a) $\overset{4}{C}H_3{-}\overset{3}{C}H^{+}{-}\overset{2}{C}H{=}\overset{1}{C}H_2 \longleftrightarrow \overset{4}{C}H_3{-}\overset{3}{C}H{=}\overset{2}{C}H{-}\overset{1}{C}H_2^{+}$ or $\overset{4}{C}H_3{-}\overset{3}{C}H^{\delta+}{\overset{...}{-}}\overset{2}{C}H{\overset{...}{-}}\overset{1}{C}H_2^{\delta+}$

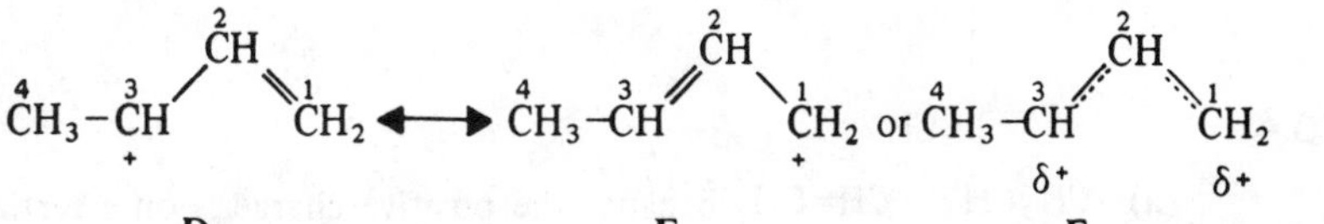

D E F

(b) We know that the allylic cation atom is almost as stable as a tertiary carbocation. Here we find not only the resonance stabilization of an allylic cation but also the additional stabilization that arises from contributor **D** in which the plus charge is on a secondary carbon atom.

(c) $CH_3{-}CH(Cl){-}CH{=}CH_2$ and $CH_3{-}CH{=}CH{-}CH_2{-}Cl$, because the Cl^- will attack the chain at the two positive centers shown in structure **F**.

10.3

(a) $CH_2{=}C(CH_3){-}\dot{C}H_2 \longleftrightarrow \cdot CH_2{-}C(CH_3){=}CH_2$

(b) $CH_2{=}CH{-}\overset{+}{C}H{-}CH{=}CH_2 \longleftrightarrow \overset{+}{C}H_2{-}CH{=}CH{-}CH{=}CH_2 \longleftrightarrow$

$CH_2{=}CH{-}CH{=}CH{-}\overset{+}{C}H_2$

(c) [cyclohexadienyl radical resonance structures] ⟷ ⟷

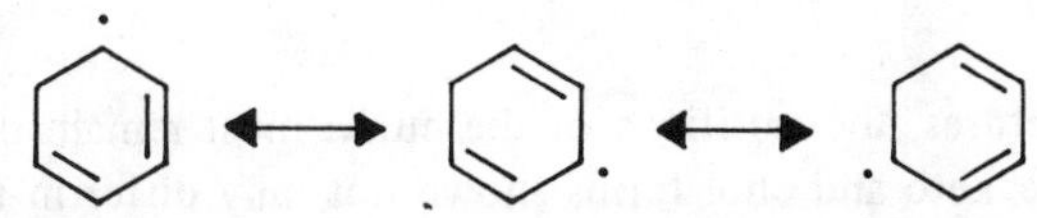

(d) [cyclohexadienyl cation resonance structures] ⟷ ⟷

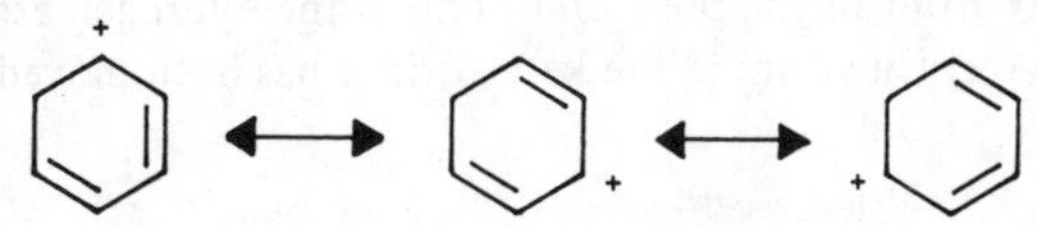

(e) $CH_3CH{=}CH{-}CH{=}\overset{+}{\ddot{O}}H \longleftrightarrow CH_3CH{=}CH{-}\overset{+}{C}H{-}\ddot{O}H \longleftrightarrow CH_3\overset{+}{C}H{-}CH{=}CH{-}\ddot{O}H$

(f) $CH_2{=}CH{-}\ddot{Cl}: \longleftrightarrow {}^{-}:CH_2{-}CH{=}\ddot{Cl}:^{+}$

(g) [benzyl cation resonance structures: $C_6H_5CH_2^+$ ⟷ $C_6H_5CH_2^+$ ⟷ ortho-cation $=CH_2$ ⟷ para-cation $=CH_2$ ⟷ ortho-cation $=CH_2$]

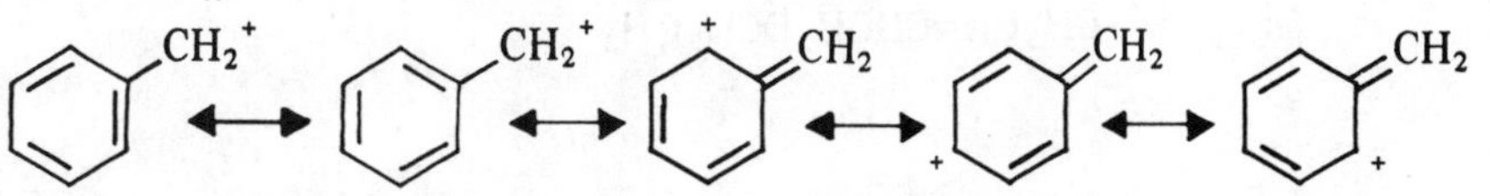

(h) $^{-}\!:\!CH_2-\overset{\overset{\ddot{O}:}{\|}}{C}-CH_3 \longleftrightarrow CH_2=\overset{\overset{:\ddot{O}:^-}{|}}{C}-CH_3$

(i) $CH_3-\ddot{\underset{\cdot\cdot}{S}}-\overset{+}{C}H_2 \longleftrightarrow CH_3-\overset{+}{\underset{\cdot\cdot}{S}}=CH_2$

(j) 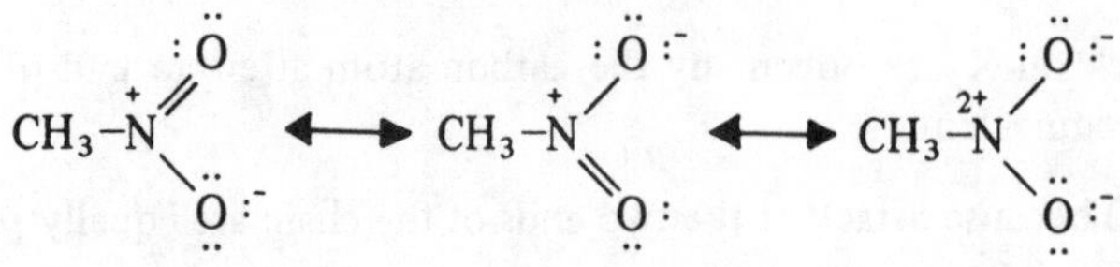 $CH_3-\overset{+}{N}(=\ddot{O}:)(-\ddot{O}:^-) \longleftrightarrow CH_3-\overset{+}{N}(-\ddot{O}:^-)(=\ddot{O}:) \longleftrightarrow CH_3-\overset{2+}{N}(-\ddot{O}:^-)(-\ddot{O}:^-)$

(minor)

10.4 (a) $CH_3CH_2\overset{CH_3}{\underset{+}{C}}-CH=CH_2$ because the positive charge is on a tertiary carbon atom rather than a primary one (rule 8).

(b) 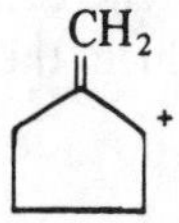 (cyclopentane ring bearing $=CH_2$, with + on the adjacent ring carbon) because the positive charge is on a secondary carbon atom rather than a primary one (rule 8).

(c) $CH_2=\overset{+}{N}(CH_3)_2$ because all atoms have a complete octet (rule 8b), and there are more covalent bonds (rule 8a).

(d) $CH_3-\overset{\overset{O}{\|}}{C}-OH$ because it has no charge separation (rule 8c).

(e) $CH_2=CH\dot{C}HCH=CH_2$ because the radical is on a secondary carbon atom rather than a primary one (rule 8).

(f) $:NH_2-C\equiv N:$ because it has no charge separation (rule 8c).

10.5 In resonance structures, the positions of the nuclei must remain the same for all structures (rule 2). The keto and enol forms shown not only differ in the positions of their electrons, they also differ in the position of one of the hydrogen atoms. In the enol form it is attached to an oxygen atom; in the keto form it has been moved so that it is attached to a carbon atom.

10.6 (a) In concentrated base and ethyl alcohol (a relatively nonpolar solvent) the S_N2 reaction is favored. Thus the rate depends on the concentration of both the alkyl halide and $NaOC_2H_5$. Since no carbocation is formed, the only product is

$$CH_3CH=CHCH_2OCH_2CH_3$$

(b) When the concentration of $C_2H_5O^-$ ion is small or zero, the reaction occurs through the S_N1 mechanism. The carbocation that is produced in the first step of the S_N1 mechanism is a resonance hybrid.

$$CH_3CH{=}CHCH_2Cl \rightleftharpoons \left[CH_3CH{=}CH\overset{+}{C}H_2 \longleftrightarrow CH_3\overset{+}{C}H{-}CH{=}CH_2 \right] + Cl^-$$

This ion reacts with the nucleophile ($C_2H_5O^-$ or C_2H_5OH) to produce two isomeric ethers

$$CH_3CH{=}CHCH_2{-}OCH_2CH_3 \quad \text{and} \quad CH_3\underset{}{C}H(OCH_2CH_3)CH{=}CH_2$$

(c) In the presence of water, the first step of the S_N1 reaction occurs. The reverse of this reaction produces two compounds because the positive charge on the carbocation is distributed over carbon atoms one and three

$$\left[CH_3CH{=}CH\overset{+}{C}H_2 \longleftrightarrow CH_3\overset{+}{C}H{-}CH{=}CH_2 \right] + Cl^- \longrightarrow CH_3CH{=}CHCH_2Cl + CH_3CH(Cl){-}CH{=}CH_2$$

10.7 (a) The carbocation that is produced in the S_N1 reaction is exceptionally stable because one resonance contributor is not only allylic but also tertiary.

$$CH_3\overset{CH_3}{\overset{|}{C}}{=}CHCH_2Cl \overset{S_N1}{\rightleftharpoons} \left[CH_3\overset{CH_3}{\overset{|}{C}}{=}CH\overset{+}{C}H_2 \longleftrightarrow CH_3{-}\overset{CH_3}{\overset{|}{\underset{+}{C}}}{-}CH{=}CH_2 \right]$$

A 3° allylic carbocation

(b) $CH_3\overset{CH_3}{\overset{|}{C}}{=}CHCH_2OH + CH_3\overset{CH_3}{\overset{|}{\underset{OH}{\underset{|}{C}}}}CH{=}CH_2$

10.8 Compounds that undergo reactions by an S_N1 path must be capable of forming relatively stable carbocations. Primary halides of the type, $ROCH_2X$ form carbocations that are stabilized by resonance:

$$R{-}\ddot{\underset{..}{O}}{-}\overset{+}{C}H_2 \longleftrightarrow R{-}\overset{+}{\underset{..}{O}}{=}CH_2$$

10.9 The relative rates are in the order of the relative stabilities of the carbocations:

$$C_6H_5\overset{+}{C}H_2 < C_6H_5\overset{+}{C}HCH_3 < (C_6H_5)_2\overset{+}{C}H < (C_6H_5)_3\overset{+}{C}$$

The solvolysis reaction involves a carbocation intermediate.

10.10 (a) *cis*-1,3-Pentadiene, *trans, trans*-2,4-hexadiene, *cis*-2-*trans*-4-hexadiene, and 1,3-cyclohexadiene are conjugated dienes.

(b) 1,4-Cyclohexadiene is an isolated diene.

(c) 1-Penten-4-yne is an isolated enyne.

10.11 (a) Recall that 1,2 and 1,4 addition refer to the conjugated system itself and not the entire carbon chain. $CH_3CH_2\underset{\displaystyle Cl}{\underset{|}{C}}HCH{=}CHCH_3$ and $CH_3CH_2CH{=}CH\underset{\displaystyle Cl}{\underset{|}{C}}HCH_3$

(b) The most stable cation is a hybrid of equivalent forms:

$$CH_3\underset{+}{C}HCH{=}CHCH_3 \longleftrightarrow CH_3CH{=}CH\underset{+}{C}HCH_3.$$

Thus 1,4 and 1,2 addition yield the same product.

$$CH_3\underset{\displaystyle Cl}{\underset{|}{C}}HCH{=}CHCH_3$$

10.12 (a) Addition of the proton gives the resonance hybrid

$$\underset{\textbf{I}}{CH_3{-}\overset{+}{C}H{-}CH{=}CH_2} \longleftrightarrow \underset{\textbf{II}}{CH_3{-}CH{=}CH{-}\overset{+}{C}H_2}$$

The inductive effect of the methyl group in **I** stabilizes the positive charge on the adjacent carbon. Such stabilization of the positive charge does not occur in **II**. Because **I** contributes more heavily to the resonance hybrid than does **II**, C-2 bears a greater positive charge and reacts faster with the bromide ion.

(b) In the 1,4-addition product, the double bond is more highly substituted than in the 1,2-addition product, hence it is the more stable alkene.

10.13 (a) (c)

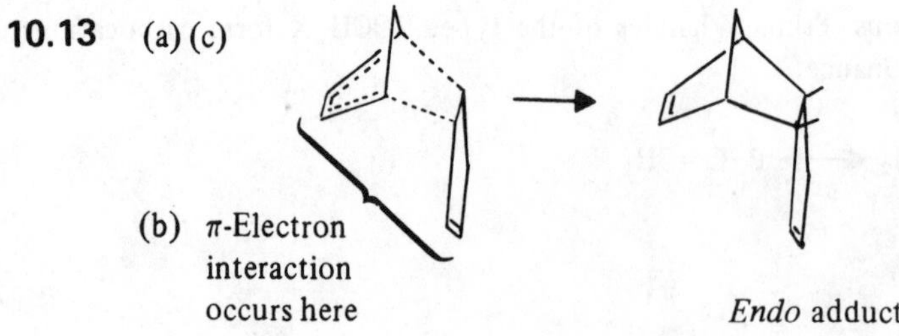

(b) π-Electron interaction occurs here

Endo adduct

10.14

(a) 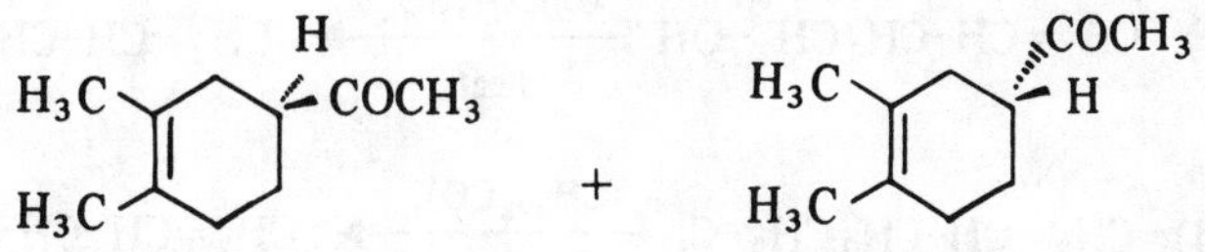 [structure: 3,4-dimethylcyclohexene ring with H and $COCH_3$ at C-1] + [enantiomer with $COCH_3$ and H]

(b) 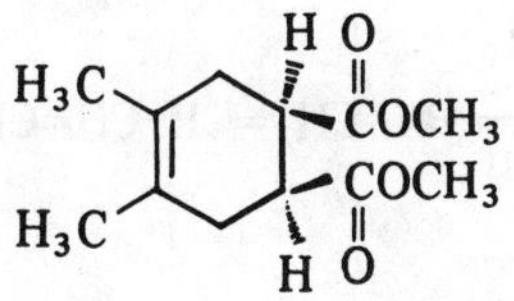 [structure: 4,5-dimethylcyclohex-4-ene-1,2-dicarboxylate with H, $\overset{O}{\|}COCH_3$ and $COCH_3$ groups]

(c) 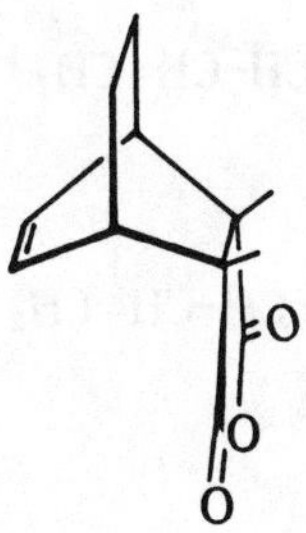 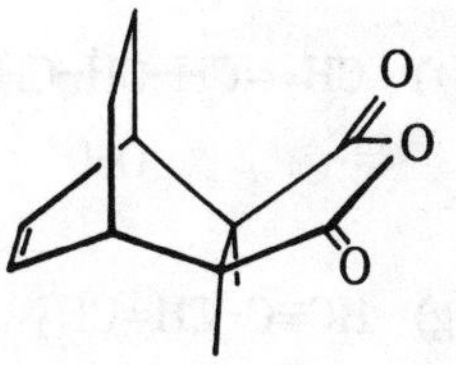 (major product) + (minor product)

10.15 Use the trans diester because the stereochemistry is retained in the adduct.

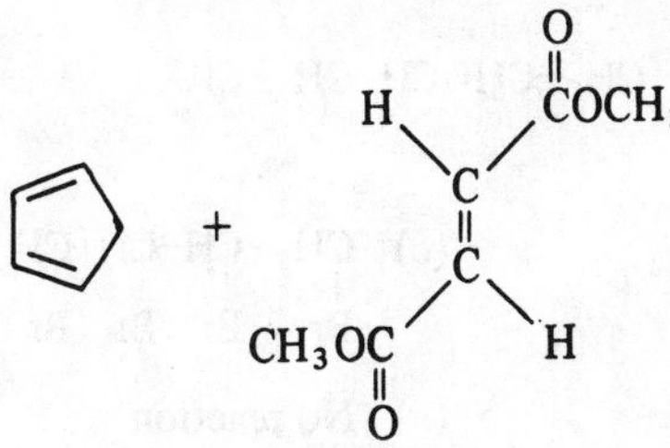

[cyclopentadiene] + trans-$CH_3OOC{-}CH{=}CH{-}COOCH_3$ (H, $\overset{O}{\|}COCH_3$; $CH_3OC(=O)$, H)

10.16

[cyclopentadiene] + $CH_3O{-}C(=O){-}C{\equiv}C{-}C(=O){-}OCH_3$

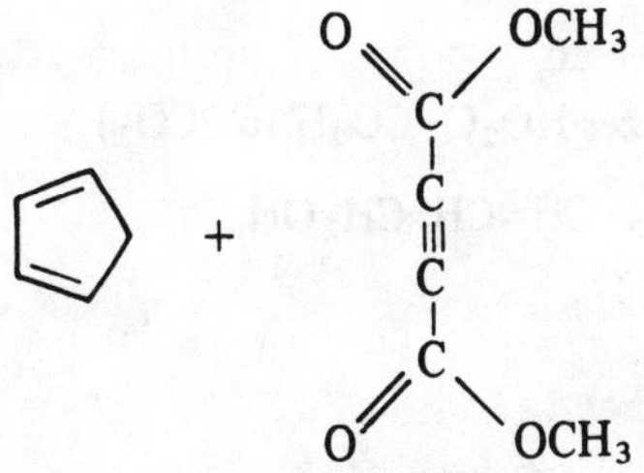

10.17

(a) $BrCH_2CH_2CH_2CH_2Br \xrightarrow[(CH_3)_3COH]{(CH_3)_3COK} CH_2{=}CH{-}CH{=}CH_2$

(b) $HOCH_2CH_2CH_2CH_2OH \xrightarrow[\text{heat}]{\text{conc. } H_2SO_4} CH_2{=}CH{-}CH{=}CH_2$

(c) $CH_2{=}CH{-}CH_2CH_2{-}OH \xrightarrow[\text{heat}]{\text{conc. } H_2SO_4} CH_2{=}CH{-}CH{=}CH_2$

(d) $CH_2{=}CH{-}CH_2CH_2{-}Cl \xrightarrow[(CH_3)_3COH]{(CH_3)_3COK} CH_2{=}CH{-}CH{=}CH_2$

(e) $CH_2{=}CH{-}CH(Cl){-}CH_3 \xrightarrow[(CH_3)_3COH]{(CH_3)_3COK} CH_2{=}CH{-}CH{=}CH_2$

(f) $CH_2{=}CH{-}CH(OH){-}CH_3 \xrightarrow[\text{heat}]{\text{conc. } H_2SO_4} CH_2{=}CH{-}CH{=}CH_2$

(g) $HC{\equiv}C{-}CH{=}CH_2 + H_2 \xrightarrow{Ni_2B\ (P\text{-}2)} CH_2{=}CH{-}CH{=}CH_2$

10.18 $CH_2{=}C(CH_3){-}C(CH_3){=}CH_2$

10.19 (a) $Cl{-}CH_2CH(Cl)CH{=}CH_2 + Cl{-}CH_2{-}CH{=}CH{-}CH_2{-}Cl$

(b) $CH_2(Cl){-}CH(Cl){-}CH(Cl){-}CH_2(Cl)$

(c) $CH_2(Br){-}CH(Br){-}CH(Br){-}CH_2(Br)$

(d) $CH_3{-}CH_2{-}CH_2{-}CH_3$

(e) No reaction

(f) $Cl{-}CH_2{-}CH(OH){-}CH{=}CH_2 + Cl{-}CH_2{-}CH{=}CH{-}CH_2{-}OH$ (+ $ClCH_2CH(Cl)CH{=}CH_2$ + $ClCH_2CH{=}CHCH_2Cl$)

(g) $4CO_2$ (*Note:* $KMnO_4$ oxidizes $HO_2C{-}CO_2H$ to $2CO_2$)

(h) $CH_3{-}CH(OH){-}CH{=}CH_2 + CH_3{-}CH{=}CH{-}CH_2OH$

10.20

(a) $CH_2{=}CH{-}CH_2{-}CH_3$ + NBS (N-bromosuccinimide: ring of H_2C, H_2C, two $C{=}O$, and NBr) $\xrightarrow{CCl_4} CH_2{=}CH{-}CH(Br){-}CH_3$ $\left(+ CH_2(Br){-}CH{=}CH{-}CH_3\right)$ $\xrightarrow[(CH_3)_3COH]{(CH_3)_3COK} CH_2{=}CH{-}CH{=}CH_2$

(NBS)

Note: In the second step both allylic halides undergo elimination of HBr to yield 1,3-butadiene and therefore separating the mixture produced in the first step is unnecessary. The $BrCH_2CH{=}CHCH_3$ undergoes a 1,4 elimination (the opposite of a 1,4 addition).

(b) $CH_2{=}CH{-}CH_2CH_2CH_3 + NBS \xrightarrow{CCl_4} CH_2{=}CH{-}\underset{}{\overset{Br}{\overset{|}{C}}}HCH_2CH_3$

$\left(+\ \underset{Br}{\underset{|}{C}}H_2CH{=}CHCH_2CH_3\right) \xrightarrow[(CH_3)_3COH]{(CH_3)_3COK} CH_2{=}CH{-}CH{=}CH{-}CH_3$

Here again both products undergo elimination of HBr to yield 1,3-pentadiene.

(c) $CH_3CH_2CH_2CH_2OH \xrightarrow[heat]{conc.\ H_2SO_4} CH_3CH_2CH{=}CH_2 \xrightarrow{[as\ in\ (a)]} CH_2{=}CH{-}CH{=}CH_2 \xrightarrow[heat]{Br_2} \underset{Br}{\underset{|}{C}}H_2{-}CH{=}CH{-}\underset{Br}{\underset{|}{C}}H_2$

(d) $CH_3{-}CH{=}CH{-}CH_3 + NBS \xrightarrow{CCl_4} CH_3{-}CH{=}CH{-}CH_2{-}Br$

$+\ CH_2{=}CH{-}CHBr{-}CH_3$

(e) cyclopentane (excess) $+ Br_2 \xrightarrow[heat]{light}$ bromocyclopentane $\xrightarrow[(CH_3)_3COH]{(CH_3)_3COK}$ cyclopentene $\xrightarrow[CCl_4]{NBS}$ 3-bromocyclopentene

(f) 3-bromocyclopentene $\xrightarrow[(CH_3)_3COH]{(CH_3)_3COK}$ 1,3-cyclopentadiene (same as 1,3-cyclopentadiene)

10.21 $R{-}\ddot{\underset{..}{O}}{-}\ddot{\underset{..}{O}}{-}R \xrightarrow[or\ heat]{light} 2R{-}\ddot{\underset{..}{O}}\cdot$

$R{-}\ddot{\underset{..}{O}}\cdot + H{-}\ddot{\underset{..}{Br}}{:} \longrightarrow R{-}\ddot{\underset{..}{O}}{-}H + \cdot\ddot{\underset{..}{Br}}{:}$

$CH_2{=}CH{-}CH{=}CH_2 + \cdot\ddot{\underset{..}{Br}}{:} \longrightarrow \left[CH_2{=}CH{-}\dot{C}H{-}\underset{Br}{\underset{|}{C}}H_2 \longleftrightarrow \dot{C}H_2{-}CH{=}CH{-}\underset{Br}{\underset{|}{C}}H_2\right]$

$\xrightarrow{HBr} CH_2{=}CH{-}\underset{H}{\underset{|}{C}}H{-}\underset{Br}{\underset{|}{C}}H_2 + \underset{H}{\underset{|}{C}}H_2{-}CH{=}CH{-}\underset{Br}{\underset{|}{C}}H_2 + \cdot\ddot{\underset{..}{Br}}{:}$

(cis and trans)

10.22 (a) $Ag(NH_3)_2OH$ gives a precipitate with 1-butyne only.

(b) 1,3-Butadiene decolorizes bromine solution; butane does not.

(c) H_2SO_4 dissolves $CH_2{=}CHCH_2CH_2OH$. Butane does not dissolve.

(d) $AgNO_3$ in C_2H_5OH gives a AgBr precipitate with $CH_2{=}CHCH_2CH_2Br$. No reaction with 1,3-butadiene.

(e) $AgNO_3$ in C_2H_5OH gives a AgBr precipitate with $BrCH_2CH{=}CHCH_2Br$ (it is an allylic bromide), but not with $CH_3\underset{Br}{\underset{|}{C}}H{=}\underset{Br}{\underset{|}{C}}HCH_3$ (a vinylic bromide).

10.23 (a) Because a highly resonance-stabilized free radical is formed:

$$CH_2{=}CH{-}\dot{C}H{-}CH{=}CH_2 \longleftrightarrow CH_2{=}CH{-}CH{=}CH{-}\dot{C}H_2 \longleftrightarrow \dot{C}H_2{-}CH{=}CH{-}CH{=}CH_2$$

(b) Because the carbanion is more stable:

$$CH_2{=}CH{-}\overset{-}{\ddot{C}}H{-}CH{=}CH_2 \longleftrightarrow CH_2{=}CH{-}CH{=}CH{-}\overset{-}{\ddot{C}}H_2 \longleftrightarrow \overset{-}{\ddot{C}}H_2{-}CH{=}CH{-}CH{=}CH_2$$

that is, we can write more reasonance structures of nearly equal energies.

10.24

$$CH_2{=}\overset{CH_3}{\overset{|}{C}}{-}CH{=}CH_2 \xrightarrow{H^+} \left[CH_3\underset{+}{\overset{CH_3}{\overset{|}{C}}}{-}CH{=}CH_2 \longleftrightarrow CH_3{-}\overset{CH_3}{\overset{|}{C}}{=}CH{-}\overset{+}{C}H_2\right] \quad \textbf{I}$$

$$\longrightarrow \left[CH_2{=}\overset{CH_3}{\overset{|}{C}}{-}\underset{+}{C}H{-}CH_3 \longleftrightarrow \overset{+}{C}H_2{-}\overset{CH_3}{\overset{|}{C}}{=}CH{-}CH_3\right] \quad \textbf{II}$$

The resonance hybrid, **I**, has the positive charge, in part, on the tertiary carbon atom; in **II**, the positive charge is on primary and secondary carbon atoms only. Therefore hybrid **I** is more stable, and will be the intermediate carbocation. 1,4 Addition to **I** gives

$$CH_3{-}\overset{CH_3}{\overset{|}{C}}{=}CH{-}CH_2Cl$$

10.25

(a) + $CH_2{=}CH{-}\overset{O}{\overset{\|}{C}}H$

(d) $CH_3{-}CH{=}CH{-}CH{=}CH_2$ + $\text{H–C(=O)–O–C(=O)–H... }$ (maleic anhydride: $HC{=}CH$ with $C{=}O$, O, $C{=}O$ ring)

(b) $CH_3{-}\overset{CH_2}{\overset{\|}{C}}{-}\underset{CH_2}{\underset{\|}{C}}{-}CH_3$ + $H(CH_3O\overset{O}{\overset{\|}{C}})C{=}C(H)\underset{O}{\underset{\|}{C}}OCH_3$

(e) + $CH_2{=}CH{-}CN$

(c) 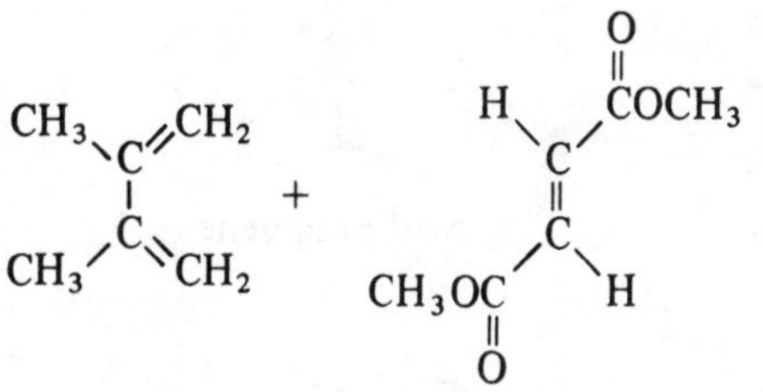 $CH_3C(=CH_2)C(=CH_2)CH_3$ + $CH_3OC(=O)CH=CHC(=O)OCH_3$ (f) 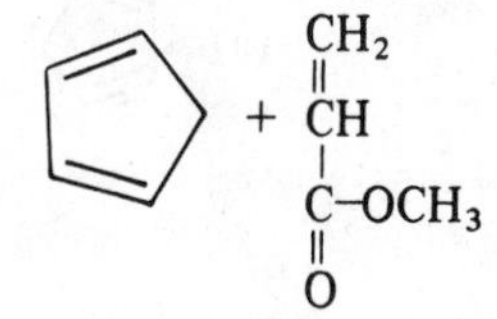 cyclopentadiene + $CH_2=CH-C(=O)-OCH_3$

10.26 Neither compound can assume the *s*-cis conformation. 1,3-Butadiyne is linear, and [cyclohexenyl]$=CH_2$ is held in an *s*-trans conformation by the requirements of the ring.

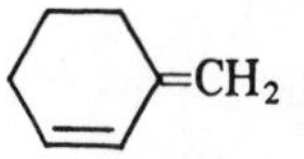

10.27

(a) 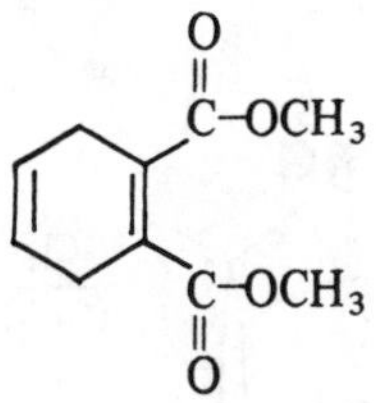 cyclohexadiene with two $C(=O)-OCH_3$ groups (b) 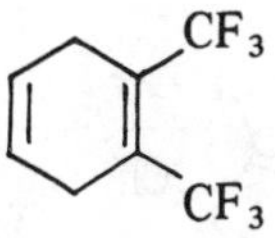 cyclohexadiene with two CF_3 groups

10.28

cyclopentadiene (CH_2) + $CH_2=CH_2$ ⟶ bicyclic product (CH_2)

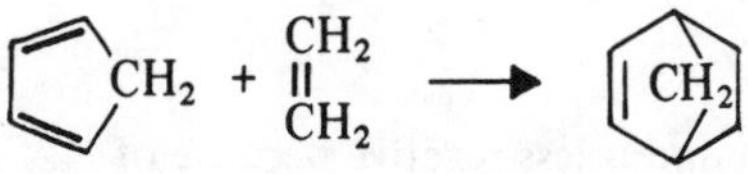

10.29 The *endo* adduct is less stable than the *exo*, but is produced at a faster rate at 25°C. At 90°C the Diels-Alder reaction becomes reversible; an equilibrium is established, and the more stable *exo* adduct predominates.

10.30

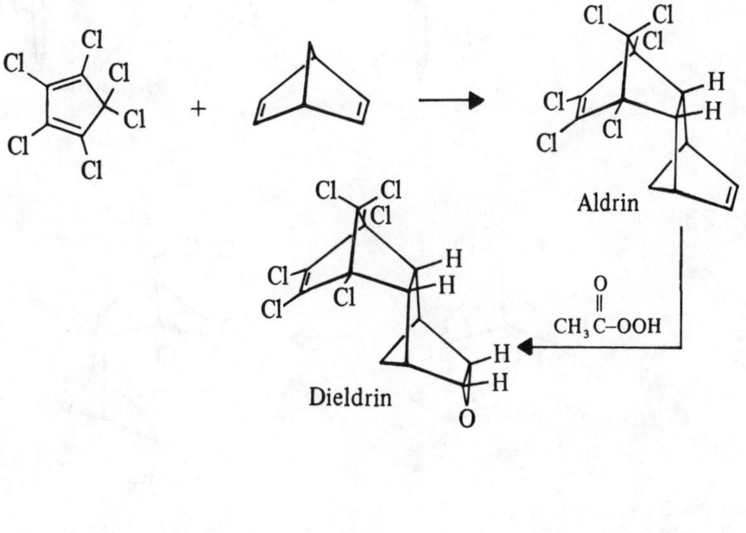

hexachlorocyclopentadiene (Cl, Cl, Cl, Cl, Cl, Cl) + norbornadiene ⟶ Aldrin (Cl, Cl, Cl, Cl, Cl, Cl, H, H)

Aldrin —$CH_3C(=O)-OOH$→ Dieldrin (Cl, Cl, Cl, Cl, Cl, Cl, H, H, H, H, O)

10.31

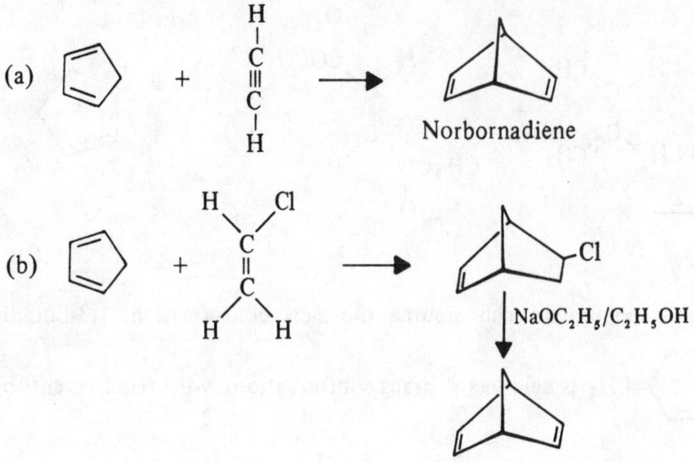

(a) + $\rightarrow$ Norbornadiene

(b) + $\rightarrow$; $NaOC_2H_5/C_2H_5OH$

10.32

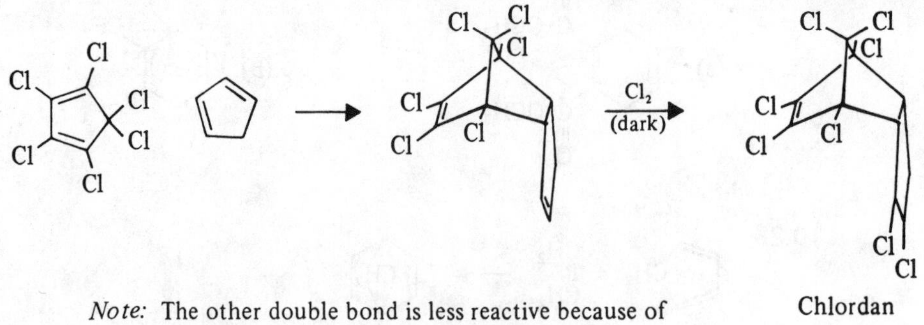

$\rightarrow$; Cl_2 (dark) $\rightarrow$ Chlordan

Note: The other double bond is less reactive because of the presence of the two chlorine substituents.

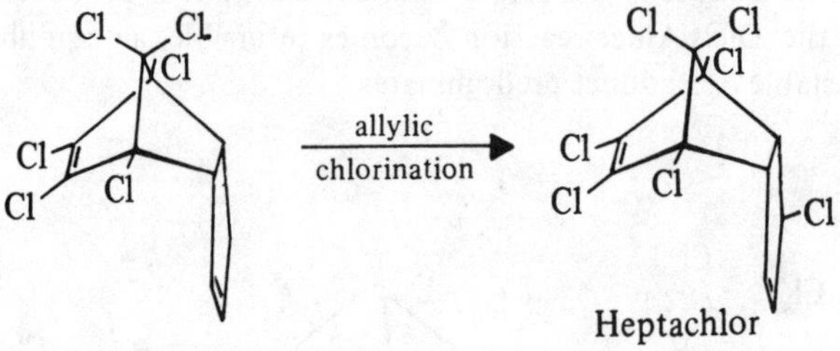

allylic chlorination $\rightarrow$ Heptachlor

10.33

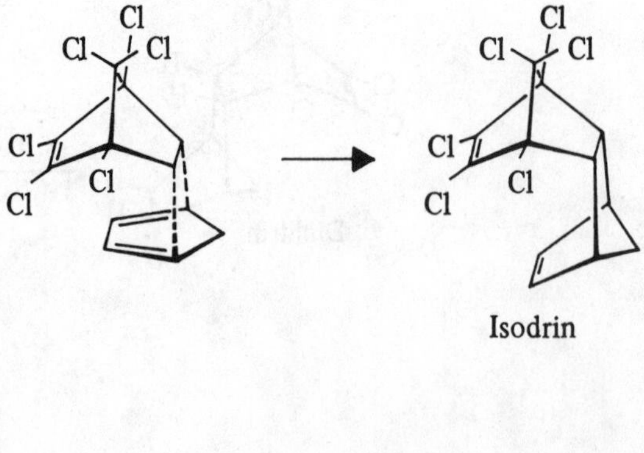

$\rightarrow$ Isodrin

10.34 Protonation of the alcohol and loss of water leads to an allylic cation that can react with a chloride ion at either C-1 or C-3.

$$CH_3CH{=}CHCH_2OH \xrightarrow{H^+} CH_3CH{=}CHCH_2{-}\overset{\displaystyle H \atop |}{O^+}{-}H \xrightarrow{-H_2O}$$

$$CH_3CH{=}CHCH_2^{\,+} \longleftrightarrow CH_3\underset{+}{C}HCH{=}CH_2 \xrightarrow{Cl^-}$$

$$CH_3CH{=}CHCH_2Cl + CH_3\underset{\displaystyle | \atop Cl}{C}HCH{=}CH_2$$

10.35 (1) $CH_2{=}CH{-}CH{=}CH_2 + Cl_2 \longrightarrow ClCH_2{-}\overset{+}{C}H{-}CH{=}CH_2$

$$\updownarrow$$

$$ClCH_2{-}CH{=}CH{-}\overset{+}{C}H_2$$

$$ClCH_2{-}\overset{\delta+}{C}H\text{---}CH\text{---}\overset{\delta+}{C}H_2$$

(2) $ClCH_2{-}\overset{\delta+}{C}H\text{---}CH\text{---}\overset{\delta+}{C}H_2 \xrightarrow[(-H^+)]{CH_3OH} ClCH_2{-}\underset{\displaystyle | \atop OCH_3}{C}H{-}CH{=}CH_2$

$$+\ ClCH_2{-}CH{=}CH{-}CH_2OCH_3$$

10.36 A six-membered ring cannot accommodate a triple bond because of the strain that would be introduced.

1,2-Dibromocyclohexane $\xrightarrow{2\ RO^-Na^+}$ ✗ cyclohexyne (CH_2, H_2C, H_2C, $C{\equiv}C$, CH_2 ring) Too highly strained

$\longrightarrow$ 1,3-cyclohexadiene

10.37 The products are $CH_3CH_2\underset{\displaystyle | \atop Br}{C}HCH{=}CH_2$ and $CH_3CH_2CH{=}CHCH_2Br$ (cis and trans). They are formed from an allylic radical in the following way:

$$\underset{\text{(from NBS)}}{Br_2} \longrightarrow 2\ Br\cdot$$

$$Br\cdot + CH_3CH_2CH_2CH{=}CH_2 \longrightarrow CH_3CH_2\dot{C}HCH{=}CH_2 \longleftrightarrow CH_3CH_2CH{=}CH\dot{C}H_2 + HBr$$

$$CH_3CH_2\overset{\delta\cdot}{C}H\text{---}CH\text{---}\overset{\delta\cdot}{C}H_2 + Br_2 \longrightarrow CH_3CH_2CH(Br)CH{=}CH_2 + CH_3CH_2CH{=}CHCH_2Br + Br\cdot$$

(cis and trans)

10.38 (a) The same carbocation (a resonance hybrid) is produced in the dissociation step:

$$CH_3C(CH_3){=}CHCH_2Cl \xrightarrow{Ag^+} \; ; \; CH_3C(CH_3)(Cl)CH{=}CH_2 \xrightarrow{Ag^+}$$

$$\underset{\text{I}}{CH_3\text{-}\overset{+}{C}(CH_3)\text{-}CH{=}CH_2} \longleftrightarrow \underset{\text{II}}{CH_3\text{-}C(CH_3){=}CH\text{-}\overset{+}{C}H_2} + AgCl$$

$$\downarrow H_2O$$

$$\underset{(85\%)}{CH_3\text{-}C(CH_3)(OH)\text{-}CH{=}CH_2} + \underset{(15\%)}{CH_3\text{-}C(CH_3){=}CH\text{-}CH_2OH}$$

(b) Structure I contributes more than II to the resonance hybrid of the carbocation (rule 8). Therefore the hybrid carbocation has a larger positive charge on the tertiary carbon atom than on the primary carbon atom. Reaction of the carbocation with water will therefore occur more frequently at the tertiary carbon atom.

10.39 (a) Propyne. (b) Base ($:B^-$) removes a proton leaving the anion whose resonance structures are shown:

$$CH_2{=}C{=}CH_2 + :B^- \rightleftarrows H:B + \underset{\text{I}}{H_2C{=}C{=}\ddot{C}^-H} \longleftrightarrow \underset{\text{II}}{H\text{-}\ddot{C}H\text{-}C{\equiv}C\text{-}H}$$

Reaction with H : B may then occur at the CH_2 carbanion. The overall reaction is

$$CH_2{=}C{=}CH_2 + :B^- \rightleftarrows [CH_2{=}C{=}\ddot{C}H \longleftrightarrow \ddot{C}H_2\text{-}C{\equiv}CH] + H:B$$

$$\rightleftarrows CH_3\text{-}C{\equiv}CH + :B^-$$

10.40 The first crystalline solid is the Diels-Alder adduct below, mp 125°C,

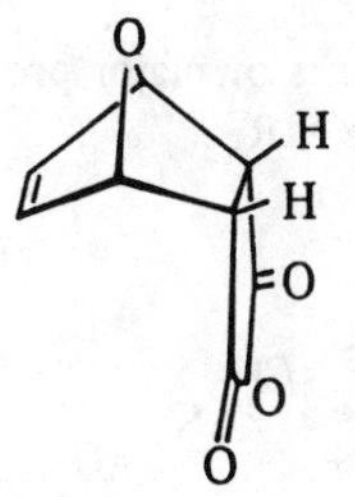

On melting, this adduct undergoes a reverse Diels-Alder reaction yielding furan (which vaporizes) and maleic anhydride, mp 56°C,

heat

Furan

Maleic anhydride (mp 56°C)

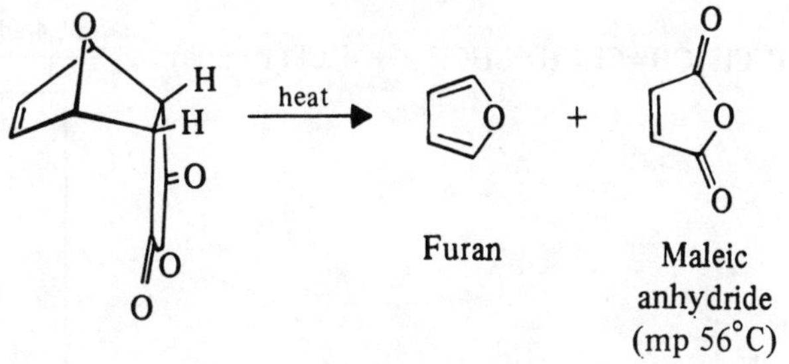

SECTION REFERENCES FOR ADDITIONAL PROBLEMS

10.17	Chapter 6	**10.29**	10.12, 10.11A
10.18	Chapter 6	**10.30**	10.12
10.19	10.11	**10.31**	10.12
10.20	10.2, 10.11	**10.32**	10.12
10.21	10.3B, 10.11	**10.33**	10.12
10.22	7.17	**10.34**	10.11
10.23	10.3B	**10.35**	10.11
10.24	10.5, 10.11	**10.36**	Chapter 7
10.25	10.12	**10.37**	10.2
10.26	10.12	**10.38**	10.11
10.27	10.12	**10.39**	10.10
10.28	10.12	**10.40**	10.12

SELF-TEST

10.1 Supply the missing reactants or major products. Show stereochemistry where applicable. If no reaction occurs write N.R.

(a) $CH_2{=}CHCH{=}CH_2$ + (CH$_3$)HC=CH(CH$_3$) [CH$_3$ and CH$_3$ on the same side; H and H on the same side] $\xrightarrow{\Delta}$ [ ] C_8H_{14}

(b) $CH_3CH{=}CHCH{=}CHCH_3$ + HCl (1 mole) $\xrightarrow{\text{(1,4 addition)}}$ [ ]

(c) $CH_3CH{=}CHCH{=}CHCH_3$ + Cl_2 (one molar equivalent) ⟶

[ ] + [ ]

(d) [ ] + [ ] $\xrightarrow{\Delta}$ [bicyclic adduct with two CN groups, each carbon also bearing H]

(e) $CH_3CH{=}CHCH_3$ + N-bromosuccinimide (N–Br) $\xrightarrow{\text{(one molar equivalent)}}$ [ ]

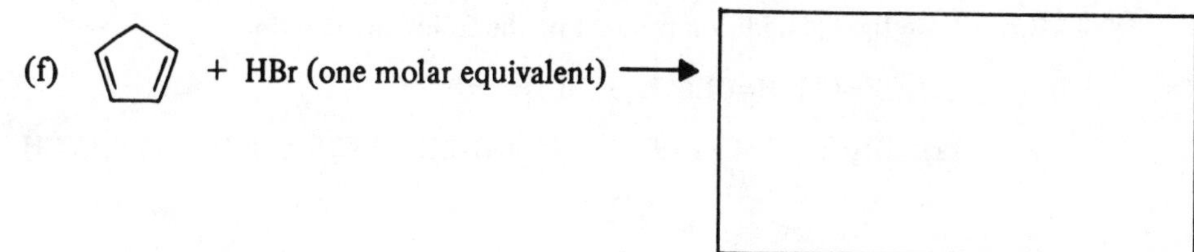

10.2 Some of the following pairs represent valid resonance structures and others do not. Place (+) in the space beside those that are valid resonance structures and (0) beside those that are not.

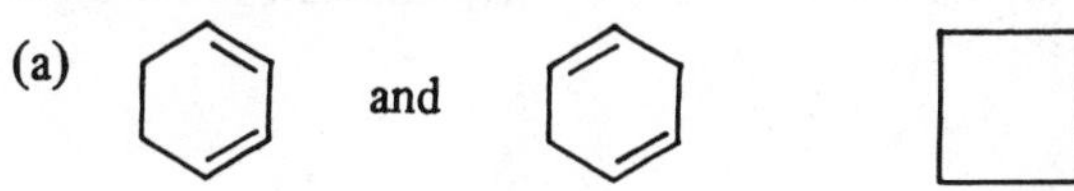

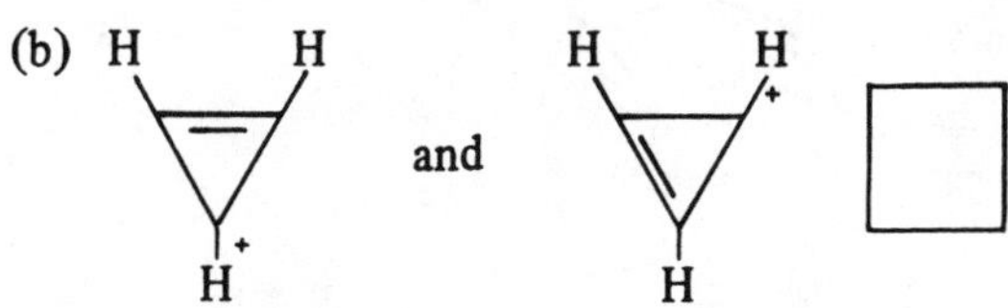

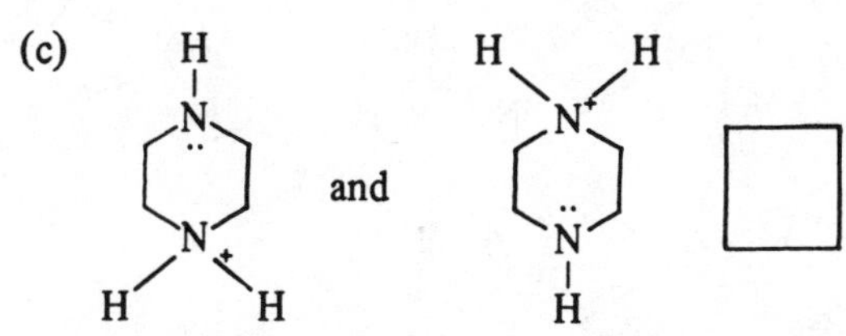

(d) $CH_2{=}CHCH{=}CH\ddot{C}H_2^-$ and $^-\ddot{C}H_2CH{=}CHCH{=}CH_2$ ☐

10.3 Which of the following compounds would absorb light at the highest wavelength?

☐

$CH_2{=}CHCH_3$	$CH_2{=}CHCH{=}CH_2$	$CH_3CH{=}CHCH_2CH{=}CH_2$
A	B	C

10.4 Which monomer can be polymerized to the polymer,

$$\left(CH_2{-}\underset{\underset{\underset{CH_3}{|}}{\underset{C=O}{|}}}{\underset{O}{|}}{CH} \right)_n$$

(a) $CH_2{=}CH{-}O{-}\overset{O}{\overset{\|}{C}}CH_3$ (b) $CH_3CH_2{-}O{-}\overset{O}{\overset{\|}{C}}CH_3$ (c) $HC{\equiv}C{-}O{-}\overset{O}{\overset{\|}{C}}CH_3$

10.5 Give the 1,4-addition product of the following reaction:

$CH_3CH{=}CHCH{=}CHCH_3 + HCl \longrightarrow ?$

(a) $CH_3CH{=}CH\underset{Cl}{\underset{|}{C}}{=}CHCH_3$ (b) $CH_3CH_2\underset{Cl}{\underset{|}{C}}HCH{=}CHCH_3$ (c) $\underset{Cl}{\underset{|}{C}}H_2CH{=}CHCH{=}CHCH_3$

(d) $CH_3CH_2CH{=}CH\underset{Cl}{\underset{|}{C}}HCH_3$ (e) $CH_3CH_2\underset{Cl}{\underset{|}{C}}H\underset{Cl}{\underset{|}{C}}HCH_2CH_3$

10.6 Which diene and dienophile could be used to synthesize the following compound?

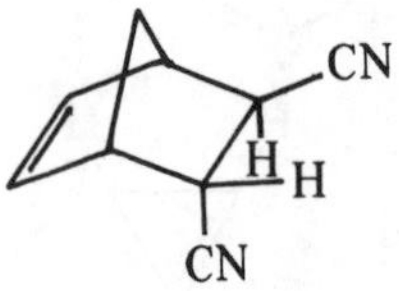

CN H H CN

(a) 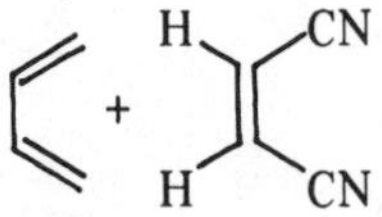 + H CN H CN (b) 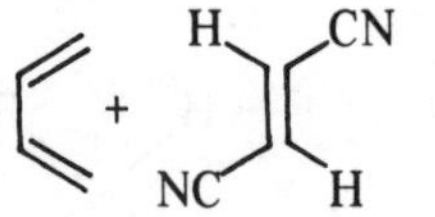 + H CN NC H (c) 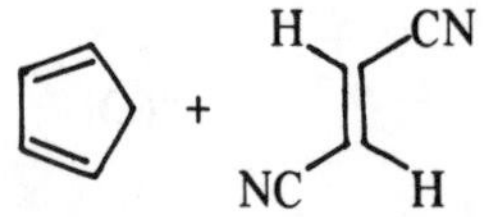 + H CN NC H

(d) 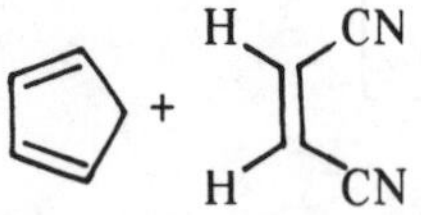 + H CN H CN (e) 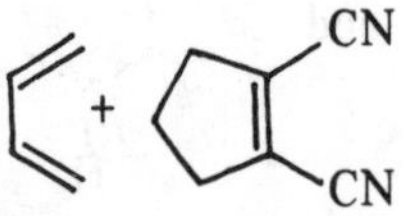 + CN CN

10.7 Which reagent(s) could be used to carry out the following reaction?

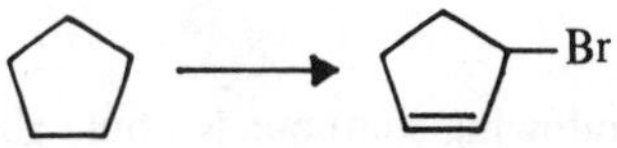

Br

(a) NBS/CCl_4 (NBS = O NBr O) (b) NBS/CCl_4, then $Br_2/h\nu$

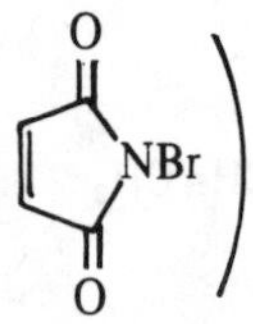

(c) $Br_2/h\nu$, then $(CH_3)_3COK/(CH_3)_3COH$, then NBS/CCl_4

(d) $(CH_3)_3COK/(CH_3)_3COH$, then NBS/CCl_4

10.8 Which of the following structures does not contribute to the hybrid for the carbocation formed when 4-chloro-2-pentene ionizes in an S_N1 reaction?

(a) $CH_3CH{=}CH\overset{+}{C}HCH_3$ (b) $CH_3\overset{+}{C}HCH{=}CHCH_3$ (c) $CH_3\overset{+}{C}HCH_2CH{=}CH_2$

(d) All of these contribute to the resonance hybrid.

10.9 Which of the folllowing resonance structures accounts at least in part for the lack of S_N2 reactivity of vinyl chloride?

(a) $CH_2{=}CH{-}\ddot{\underset{\cdot\cdot}{Cl}}:$ (b) $\overset{-}{\ddot{C}}H_2{-}CH{=}\ddot{Cl}:^{+}$ (c) Neither (d) Both

ANSWERS TO FIRST REVIEW PROBLEM SET

1

(a)

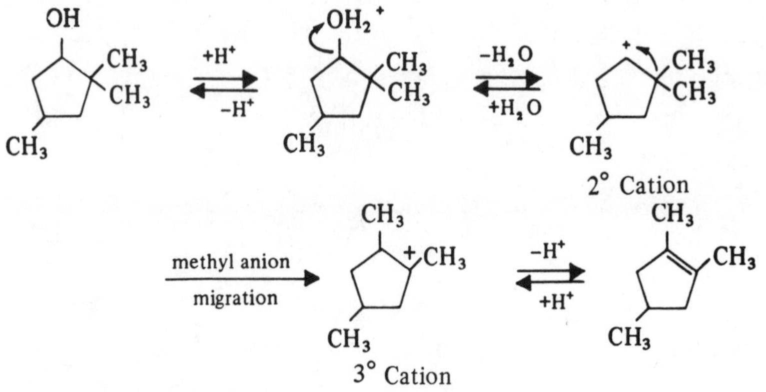

OH CH3 CH3 CH3 $\underset{-H^+}{\overset{+H^+}{\rightleftharpoons}}$ OH_2^+ CH3 CH3 CH3 $\underset{+H_2O}{\overset{-H_2O}{\rightleftharpoons}}$ CH3 CH3 CH3

2° Cation

$\xrightarrow[\text{migration}]{\text{methyl anion}}$ CH3 CH3 CH3

3° Cation

$\underset{+H^+}{\overset{-H^+}{\rightleftharpoons}}$ CH3 CH3 CH3

(b)

Br – Br ⟶ Br^+ + Br^-

then,

Br^+ Cl^- ⟶ Br Cl

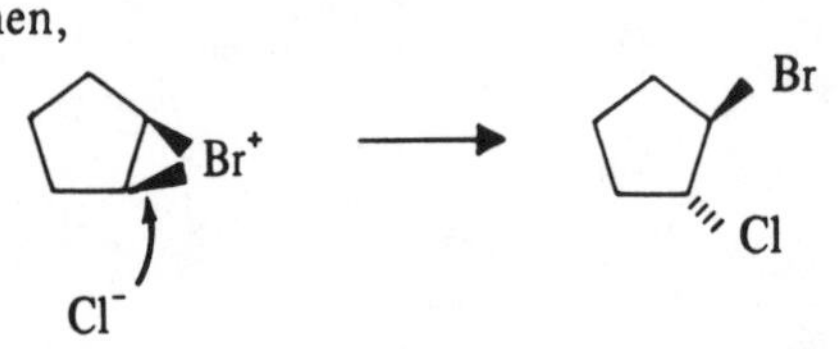

(c) The enantiomer of the product given would be formed in an equimolar amount via the following reaction:

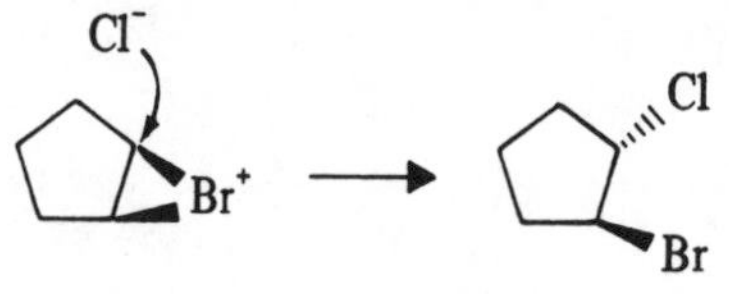

Cl^- Br^+ ⟶ Cl Br

The *trans*-1, 2-dibromocyclopentane would be formed as a racemic form via the reaction of the bromonium ion with a bromide ion:

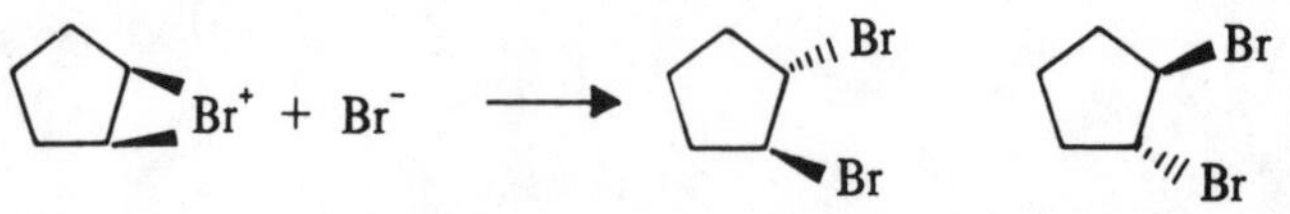

Br^+ + Br^- ⟶ Br Br   Br Br

Racemic *trans*-1,2-dibromocyclopentane

And, *trans*-2-bromocyclopentanol (the bromohydrin) would be formed (as a racemic form) via the reaction of the bromonium ion with water.

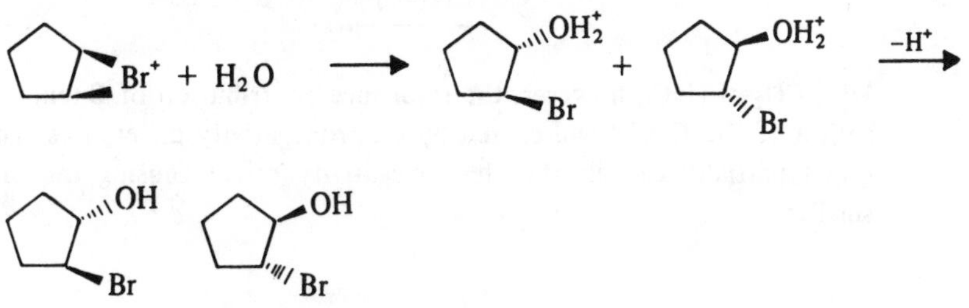

Racemic *trans*-2-bromocyclopentanol

2

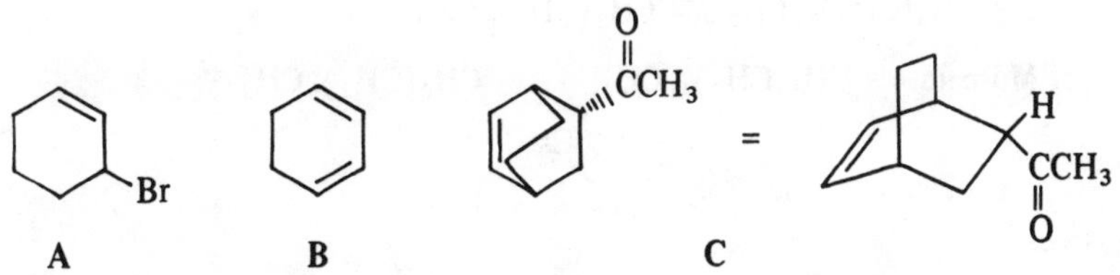

A B C

A is formed by an allylic bromination. **B** is formed by an E2 elimination. **C** is formed via a Diels-Alder reaction that yield predominantly the *endo* product. Ozonolysis of the double bond then yields the product in which all three substituents are on the same side of the cyclohexane ring.

3 All of these differences can be explained by resonance contribution to the $CH_2{=}CHCl$ molecule made by the following **A** and **B** hybrids.

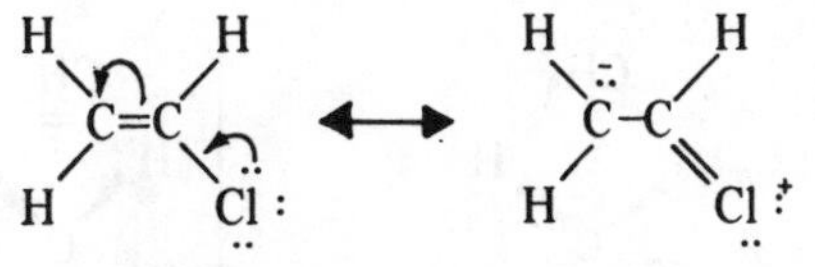

A B

(a) Because of the contribution made to the hybrid by **B**, the C–Cl bond of $CH_2{=}CH{-}Cl$ has some double bond character, and is, therefore, shorter than the "pure" single bond of $CH_3CH_2{-}Cl$.

(b) The contribution made to the hybrid by **B** imparts some single bond character to the carbon-carbon double bond of $CH_2{=}CHCl$ causing it to be longer than the "pure" double bond of $CH_2{=}CH_2$.

(c) Electronegativity differences would cause a carbon-chlorine bond to be polarized as follows:

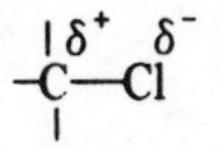

$$-\overset{|}{\underset{|}{C}}^{\delta^+}-Cl^{\delta^-}$$

And this effect accounts, almost entirely, for the dipole moment of CH_3CH_2Cl.

$$\overset{\delta^+}{CH_3CH_2}-\overset{\delta^-}{Cl} \quad \mu = 2.05\ D$$

With $CH_2{=}CH{-}Cl$, however, the resonance contribution of **B** tends to oppose the polarization of the C–Cl bond caused by electronegativity differences. That is, the resonance effect partially cancels the electronegativity effect causing the dipole moment to be smaller.

4 A = $CH_3(CH_2)_{11}CH_2C{\equiv}CH$

B = $CH_3(CH_2)_{11}CH_2C{\equiv}CNa$

C = $CH_3(CH_2)_{11}CH_2C{\equiv}CCH_2(CH_2)_6CH_3$

Muscalure = $CH_3(CH_2)_{11}CH_2(H)C{=}C(H)CH_2(CH_2)_6CH_3$ (cis: both H on the same side)

5 (E)-2, 3-Diphenyl-2-butene: $CH_3(C_6H_5)C{=}C(C_6H_5)CH_3$ (CH_3 and C_6H_5 on each carbon; C_6H_5 groups on opposite sides)

(Z)-2, 3-Diphenyl-2-butene: $CH_3(C_6H_5)C{=}C(C_6H_5)CH_3$ (C_6H_5 groups on the same side)

Because catalytic hydrogenation is a syn addition, catalytic hydrogenation of the (*Z*) isomer would yield a meso compound.

(*Z*) $C_6H_5(CH_3)C{=}C(CH_3)C_6H_5$ $\xrightarrow[Pd]{H_2}$ $C_6H_5(H)(CH_3)C{-}C(CH_3)(H)C_6H_5$ (by addition at one face) A meso compound $\equiv$ [$H(C_6H_5)(CH_3)C{-}C(CH_3)(C_6H_5)H$ (by addition at the other face)]

Syn addition of hydrogen to the (*E*) isomer would yield a racemic form:

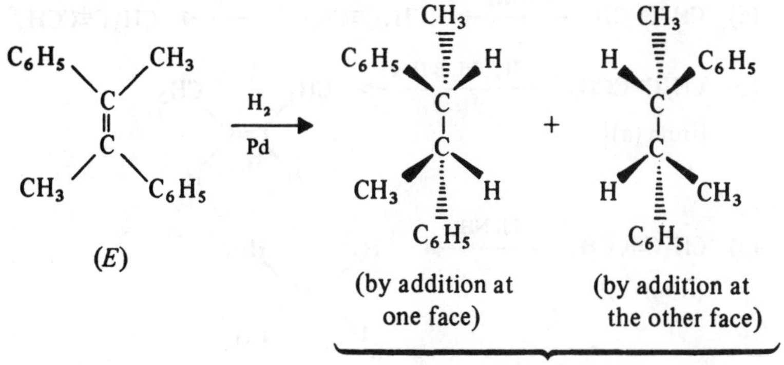

(E) $\xrightarrow[Pd]{H_2}$ (by addition at one face) + (by addition at the other face)

Enantiomers – a racemic form

6 From the molecular formula of **A** and of its hydrogenation product **B** we can conclude that **A** has two rings and a double bond. (**B** has two rings.)

From the product of strong oxidation with $KMnO_4$ and its stereochemistry (i.e., compound **C**) we can deduce the structure of **A**.

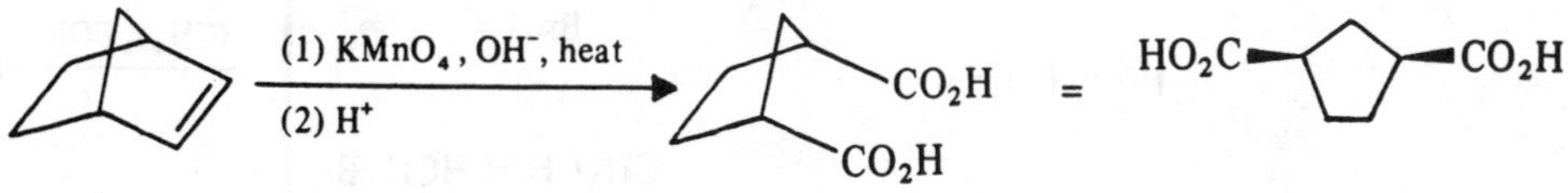

A $\xrightarrow[(2)\ H^+]{(1)\ KMnO_4,\ OH^-,\ heat}$ CO_2H, CO_2H = HO_2C–CO_2H

meso-1,3-Cyclopentanedicarboxylic acid

Compound **B** is bicycyo [2.2.1] heptane and **C** is a glycol.

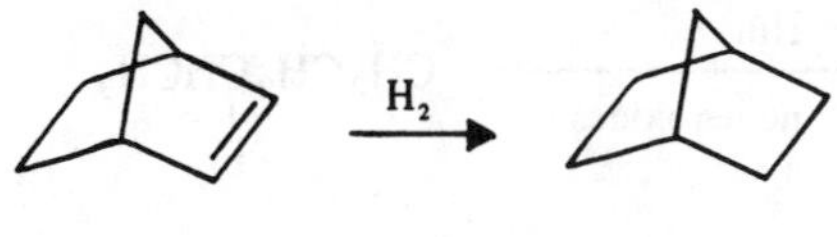

A $\xrightarrow{H_2}$ **B**

A $\xrightarrow[cold,\ dilute]{KMnO_4,\ OH^-,}$ **C** (OH, OH, H, H)

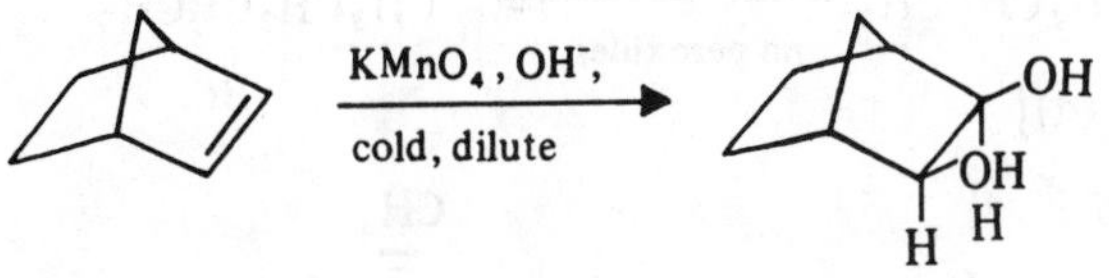

Notice that **C** is a meso compound also.

7 (a) $CH_3C{\equiv}CH \xrightarrow{NaNH_2} CH_3C{\equiv}CNa \xrightarrow{CH_3I} CH_3C{\equiv}CCH_3$

(b) $CH_3C{\equiv}CCH_3 \xrightarrow{H_2,\ Ni_2B\ (P\text{-}2)}$ (cis) $CH_3(H)C{=}C(H)CH_3$

[from (a)]

(c) $CH_3C{\equiv}CCH_3 \xrightarrow{Li,\ NH_3}$ (trans) $CH_3(H)C{=}C(H)CH_3$

[from (a)]

(d) $CH_3CH{=}CHCH_3 \xrightarrow{THF:BH_3} CH_3CH_2CH(B{<})CH_3 \xrightarrow[160°C]{heat}$

[from (b) or (c)]

$CH_3CH_2CH_2CH_2{-}B{<} \xrightarrow[160°C]{1\text{-decene}} CH_3CH_2CH{=}CH_2$

(e) $CH_3CH_2CH{=}CH_2 \xrightarrow[CCl_4]{NBS} CH_3CHBrCH{=}CH_2 + CH_3CH{=}CHCH_2Br \xrightarrow{(CH_3)_3COK}$

[from (d)]

$CH_2{=}CH{-}CH{=}CH_2$

(f) $CH_3CH_2CH{=}CH_2 \xrightarrow[ROOR]{HBr} CH_3CH_2CH_2CH_2Br$

[from (d)]

(g) $CH_3CH{=}CHCH_3 \xrightarrow[\text{no peroxides}]{HBr} CH_3CH_2CHBrCH_3$

[from (b) or (c)]

or

$CH_3CH_2CH{=}CH_2 \xrightarrow[\text{no peroxides}]{HBr} CH_3CH_2CHBrCH_3$

[from (d)]

(h) (trans) $CH_3(H)C{=}C(H)CH_3 \xrightarrow[CCl_4]{Br_2}$ (anti addition) $CH_3{-}CHBr{-}CHBr{-}CH_3$ (cf. Section 7.7A)

[from (c)]

(2*R*, 3*S*)

A meso compound

(i) $CH_3CH=CHCH_3$ (cis: H, H on same side) [from (b)] $\xrightarrow[\text{(anti addition)}]{Br_2,\ CCl_4}$ (2R,3R) + (2S,3S)

(2R,3R) (2S,3S)

A racemic form

(j) $CH_3CH=CHCH_3$ (cis) [from (b)] $\xrightarrow[\text{(2) } NaHSO_3 \text{ (syn addition)}]{\text{(1) } OsO_4}$ meso diol (cf. Section 7.7A)

or

$CH_3CH=CHCH_3$ (trans) [from (c)] $\xrightarrow[\text{(2) } H_3O^+\text{, heat (anti addition)}]{\text{(1) } RCOOH}$ meso diol (cf. Section 7.7A)

(k) $CH_3C{\equiv}C{-}CH_3 \xrightarrow[CH_3CO_2H]{HBr,\ Br^-}$ $(CH_3)(H)C=C(Br)(CH_3)$ (cf Section 7.14)

8 $CH_3CH(CH_3)CH_2CH_3 \xrightarrow{Br_2,\ h\nu,\ heat} CH_3CBr(CH_3)CH_2CH_3$ (cf. Section 9.5)

(a) $CH_3CBr(CH_3)CH_2CH_3 \xrightarrow[CH_3CH_2OH,\ heat]{CH_3CH_2ONa} CH_3C(CH_3)=CHCH_3$

(b) $CH_3C(CH_3)=CHCH_3 \xrightarrow{H_3O^+,\ H_2O} CH_3C(OH)(CH_3)CH_2CH_3$

[from (a)]

(c) $$CH_3\underset{CH_3}{C}{=}CHCH_3 \xrightarrow[(2)\ H_2O_2,\ OH^-]{(1)\ THF{:}BH_3} CH_3\underset{CH_3}{C}H\overset{OH}{C}HCH_3$$

[from (a)]

(d) $$CH_3\underset{CH_3}{C}{=}CHCH_3 \xrightarrow{THF{:}BH_3} CH_3\underset{CH_3}{C}H\overset{B-}{C}HCH_3 \xrightarrow{heat}$$

[from (a)]

$$CH_3\underset{CH_3}{C}HCH_2CH_2{-}\underset{|}{B}{-} \xrightarrow{H_2O_2,\ OH^-} CH_3\underset{CH_3}{C}HCH_2CH_2OH$$

(e) $$CH_3\underset{CH_3}{C}HCH_2CH_2{-}\underset{|}{B}{-} \xrightarrow[heat]{1\text{-decene}} CH_3\underset{CH_3}{C}HCH{=}CH_2$$

[from (d)]

(f) $$CH_3\underset{CH_3}{C}HCH{=}CH_2 \xrightarrow{Br_2} CH_3\underset{CH_3}{C}H\overset{Br}{C}HCH_2Br \xrightarrow[heat]{3NaNH_2}$$

[from (e)]

$$CH_3\underset{CH_3}{C}HC{\equiv}CNa \xrightarrow{H^+} CH_3\underset{CH_3}{C}HC{\equiv}CH$$

(g) $$CH_3\underset{CH_3}{C}HCH{=}CH_2 \xrightarrow[ROOR,\ heat]{HBr} CH_3\underset{CH_3}{C}HCH_2CH_2Br$$

[from (e)]

(h) $$CH_3\underset{CH_3}{C}HCH{=}CH_2 \xrightarrow{HCl} CH_3\underset{CH_3}{C}H\overset{Cl}{C}HCH_3$$

[from (e)]

(i) $$CH_3\underset{CH_3}{C}{=}CHCH_3 \xrightarrow{HCl} CH_3\underset{CH_3}{\overset{Cl}{C}}CH_2CH_3$$

[from (a)]

(j) $$CH_3\underset{CH_3}{C}HCH_2CH_2Br \xrightarrow[S_N2]{NaI,\ acetone} CH_3\underset{CH_3}{C}HCH_2CH_2I$$

[from (g)]

(k) $CH_3C(CH_3){=}CHCH_3 \xrightarrow[(2)\ Zn,\ H_2O]{(1)\ O_3} CH_3\overset{O}{\overset{\|}{C}}CH_3 + CH_3\overset{O}{\overset{\|}{C}}H$

[from (a)]

(l) $CH_3CH(CH_3)CH{=}CH_2 \xrightarrow[(2)\ Zn,\ H_2O]{(1)\ O_3} CH_3CH(CH_3)\overset{O}{\overset{\|}{C}}H + H\overset{O}{\overset{\|}{C}}H$

[from (e)]

(m) $CH_3CH(CH_3)C{\equiv}CH \xrightarrow[heat]{H_3O^+,\ Hg^{2+},\ H_2O} \left[CH_3CH(CH_3)C(OH){=}CH_2\right] \longrightarrow CH_3CH(CH_3)\overset{O}{\overset{\|}{C}}CH_3$

[from (f)]

9 $\underset{\textbf{A}}{CH_3C(CH_3)_2CH_2CH_3} \xrightarrow{Cl_2,\ h\nu,\ heat} \underset{\textbf{B}}{CH_3C(CH_2Cl)(CH_3)CH_2CH_3} + \underset{\textbf{C}}{CH_3C(CH_3)_2CHClCH_3} + \underset{\textbf{D}}{CH_3C(CH_3)_2CH_2CH_2Cl}$

B cannot undergo dehydrohalogenation because it has no β hydrogen, however **C** and **D** can as shown next.

$\underset{\textbf{C}}{CH_3C(CH_3)_2CHClCH_3}$ or $\underset{\textbf{D}}{CH_3C(CH_3)_2CH_2CH_2Cl} \xrightarrow[(CH_3)_3COH]{(CH_3)_3COK} \underset{\textbf{E}}{CH_3C(CH_3)_2CH{=}CH_2} \xrightarrow[Pt]{H_2} \textbf{A}$

$\underset{\textbf{E}}{CH_3C(CH_3)_2CH{=}CH_2} \xrightarrow{HCl} \left[CH_3C(CH_3)_2\overset{+}{C}HCH_3\right] + Cl^- \longrightarrow \left[CH_3\overset{+}{C}(CH_3)\text{—}CH(CH_3)CH_3\right] + Cl^- \longrightarrow$

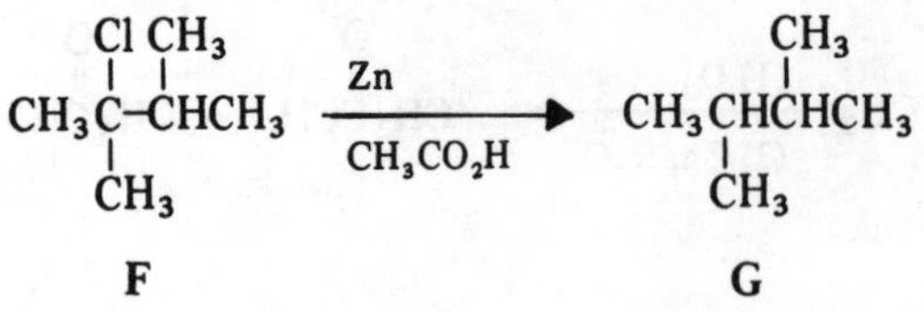

$$\underset{\textbf{F}}{CH_3C(Cl)(CH_3)CH(CH_3)CH_3} \xrightarrow[CH_3CO_2H]{Zn} \underset{\textbf{G}}{CH_3CH(CH_3)CH(CH_3)CH_3}$$

10

$$\underset{\textbf{A}}{CH_3C{\equiv}CCH_3} \xrightarrow{H_2,\ Pt} CH_3CH_2CH_2CH_3$$

$$\underset{\textbf{A}}{CH_3C{\equiv}CCH_3} \xrightarrow{Ag(NH_3)_2OH} \text{no reaction}$$

$$\textbf{A} \xrightarrow{H_2,\ Ni_2B\ (P\text{-}2)} \textbf{B}$$

B: cis-$CH_3CH{=}CHCH_3$ (CH_3, CH_3 on the same side; H, H on the same side)

$$\textbf{B} \xrightarrow[\text{(2) } NaHSO_3]{\text{(1) } OsO_4} \textbf{C}$$

(syn hydroxylation)

C: $CH_3CH(OH)CH(OH)CH_3$ (H, OH on each carbon; CH_3 groups at top and bottom)

C

(a meso compound)

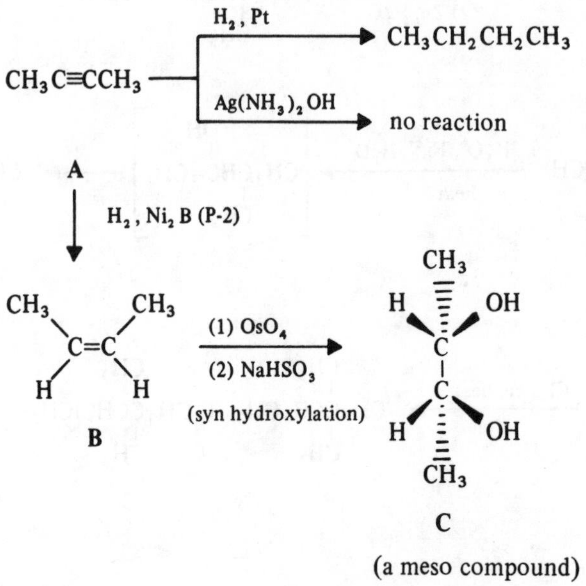

11 The eliminations are anti eliminations, requiring an anti periplanar arrangement of the bromine atoms.

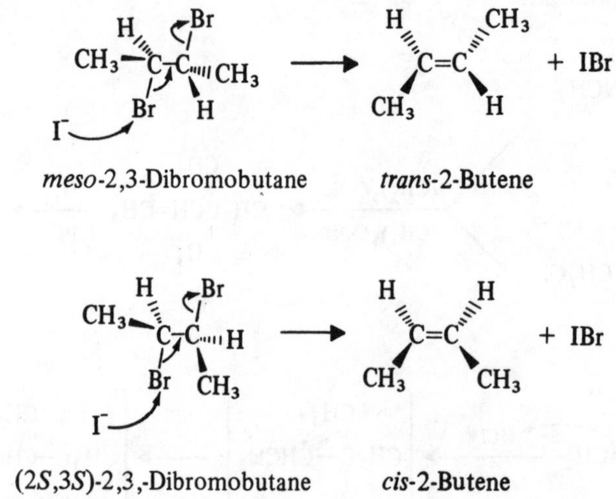

$$\underset{meso\text{-2,3-Dibromobutane}}{CH_3CHBrCHBrCH_3} \xrightarrow{I^-} \underset{trans\text{-2-Butene}}{CH_3CH{=}CHCH_3} + IBr$$

$$\underset{(2S,3S)\text{-2,3,-Dibromobutane}}{CH_3CHBrCHBrCH_3} \xrightarrow{I^-} \underset{cis\text{-2-Butene}}{CH_3CH{=}CHCH_3} + IBr$$

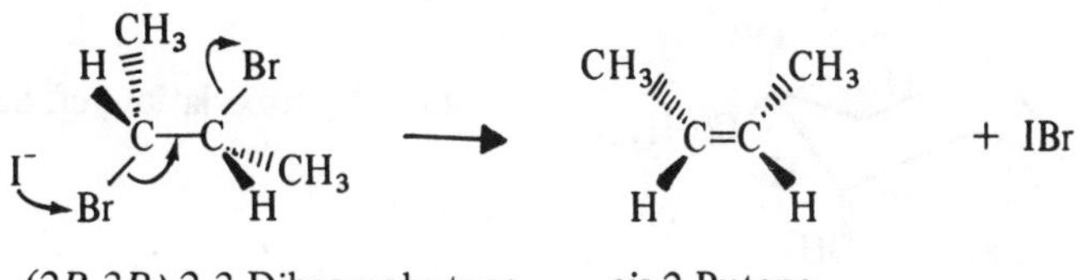

$$\text{(2}R\text{,3}R\text{)-2,3-Dibromobutane} \xrightarrow{I^-} cis\text{-2-Butene} + IBr$$

(2*R*,3*R*)-2,3-Dibromobutane   *cis*-2-Butene   + IBr

12 The eliminations are anti eliminations, requiring an anti periplanar arrangement of the –H and –Br.

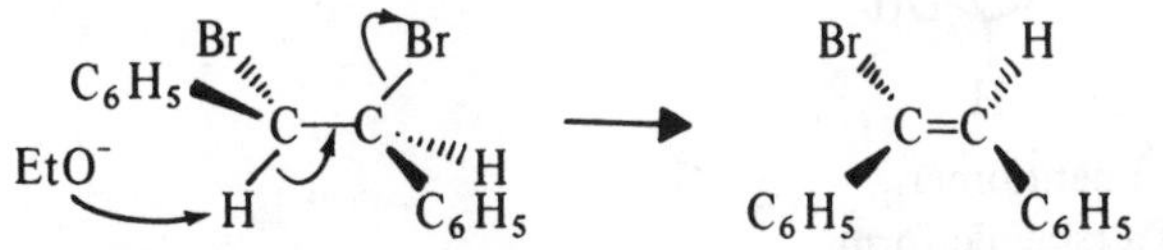

meso-1, 2-Dibromo-1, 2-diphenylethane → (*E*)-1-Bromo-1, 2-diphenylethene

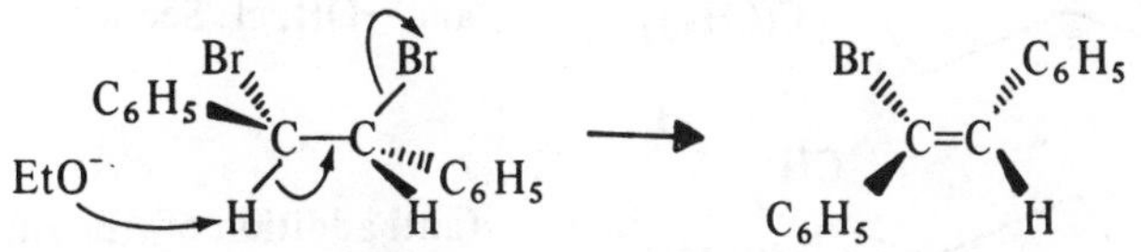

(2*R*, 3*R*) -1, 2-Dibromo-1, 2-diphenylethane → (*Z*) -1-Bromo-1, 2-diphenylethene

(2*S*,3*S*)-1,2-Dibromo-1,2-diphenylethene will also give (*Z*)-1-bromo-1,2-diphenylethene in an anti elimination.

13 In all the following structures, notice that the large *tert*-butyl group is equatorial.

(a) (bromine addition is anti, cf. Sections 7.7 and 7.7A)

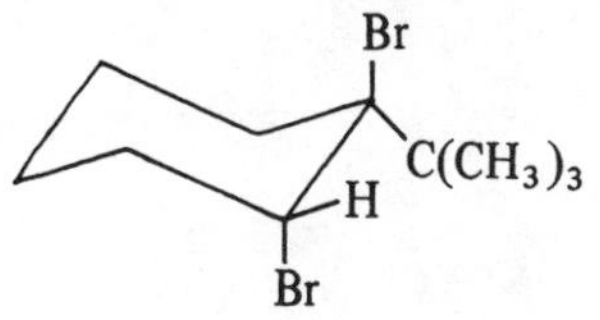

+ enantiomer
as a racemic form

(b) (syn hydroxylation, cf. Sections 7.10A and 7.7A)

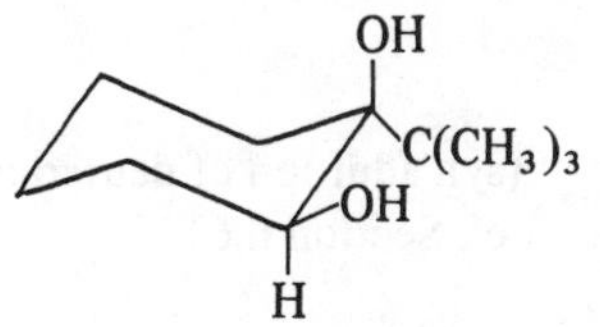

+ enantiomer
as a racemic form

(c) 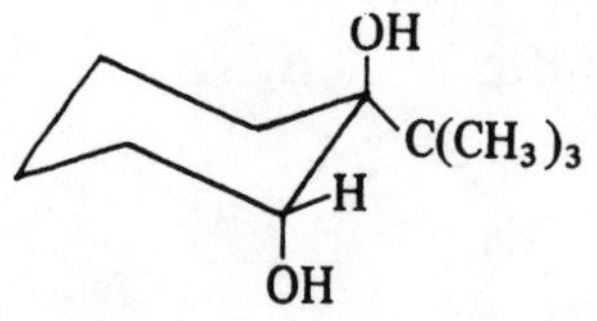 OH, $C(CH_3)_3$, H, OH

\+ enantiomer
as a racemic form

(anti hydroxylation, cf. Section 7.9)

(d) H, $C(CH_3)_3$, OH, H

\+ enantiomer
as a racemic form

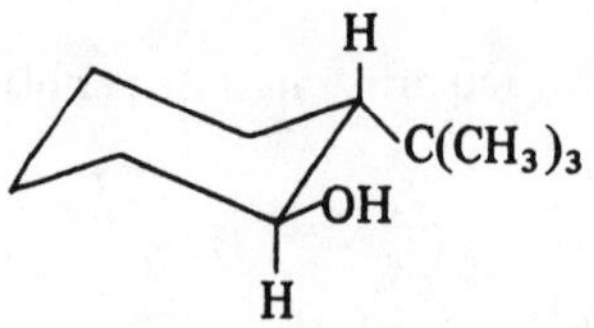

(syn and anti-Markovnikov addition of −H and −OH. cf. Section 8.7)

(e) 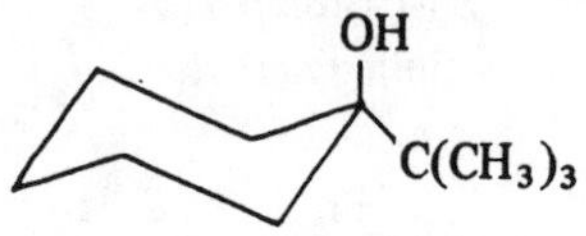 OH, $C(CH_3)_3$

(Markovnikov addition of −H and −OH, cf. Section 8.5)

(f) 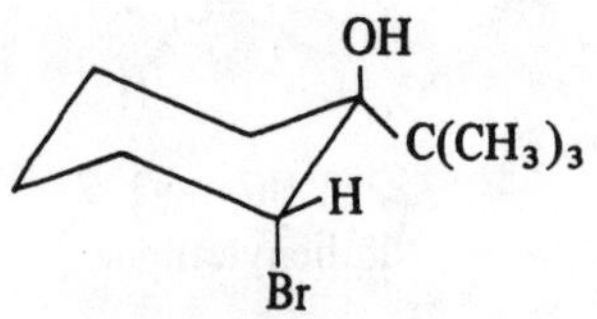 OH, $C(CH_3)_3$, H, Br

\+ enantiomer
as a racemic form

(anti addition of −Br and −OH, with −Br and −OH placement resulting from the more stable partial carbocation in the intermediate bromonium ion, cf. Section 7.8)

(g) Cl, $C(CH_3)_3$, H, I

\+ enantiomer
as a racemic form

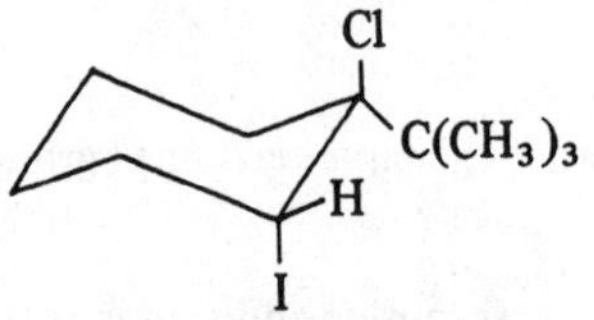

(anti addition of −I and −Cl, following Markovnikov's rule, cf. Section 7.2C)

(h) $HC(=O)(CH_2)_4C(=O)C(CH_3)_3$

(i) 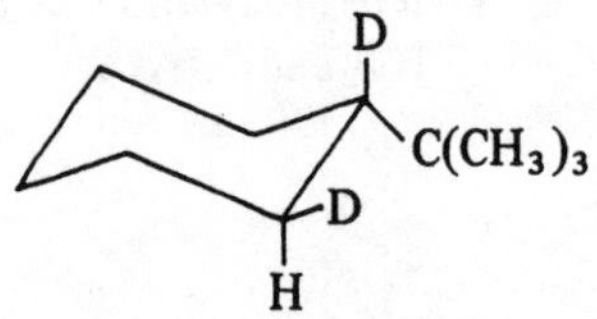 D, $C(CH_3)_3$, D, H

\+ enantiomer
as a racemic form

(syn addition of deuterium, cf. Section 6.6)

(j)

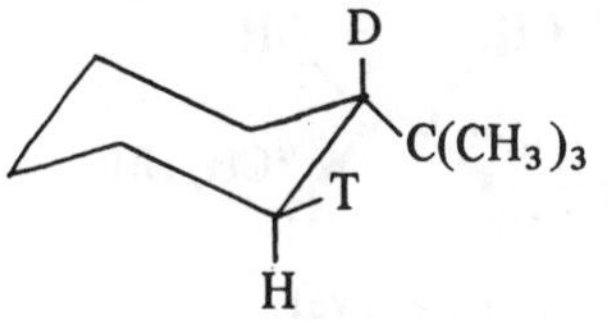

D, C(CH₃)₃, T, H

+ enantiomer
as a racemic form

(Syn, anti-Markovnikov addition of –D and –B–, with –B– being replaced by –T where it stands, cf. Section 8.7B)

14 A = $CH_3C(CH_3){=}CHCH_2CH_3$ B = $(CH_3CH(CH_3)CH(CH_2CH_3){-})_2BH$

C = $(CH_3CH(CH_3)CH_2CH_2CH_2)_2BH$ D = $CH_3CH(CH_3)CH_2CH_2CH_2OH$

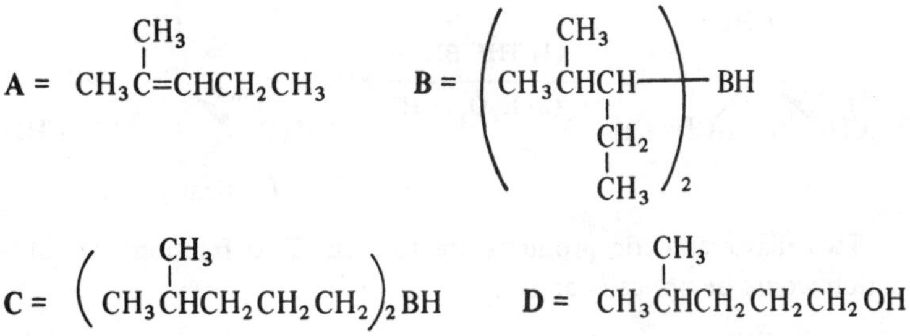

15 (a) The following products are diastereomers. They would have different boiling points and would be in separate fractions. Each fraction would be optically active.

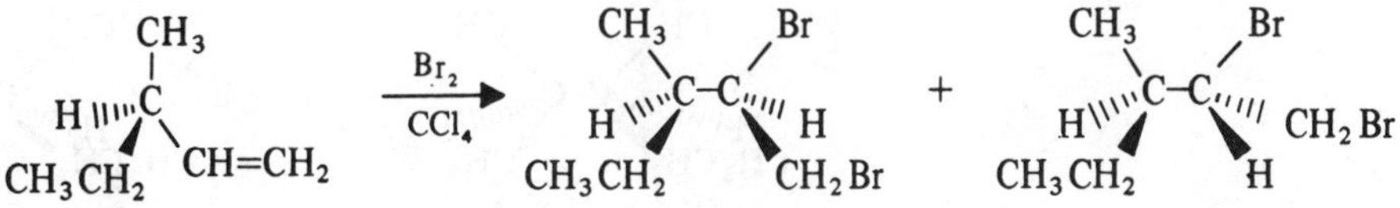

(R)-3-Methyl-1-pentene $\xrightarrow[CCl_4]{Br_2}$ (optically active) + (optically active)

Diastereomers

(b) Only one product is formed. It is achiral, and, therefore, it would not be optically active.

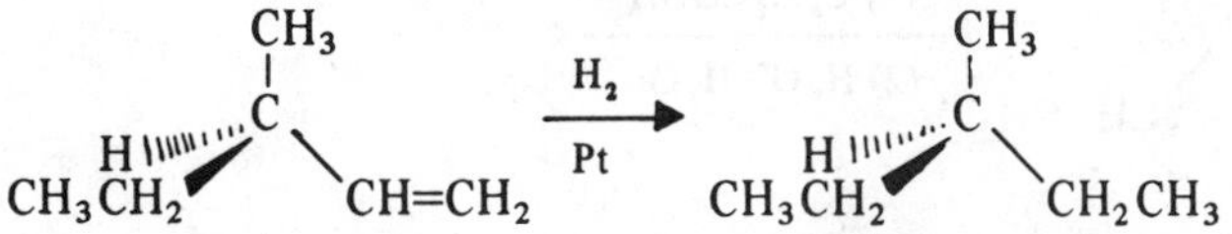

$\xrightarrow[Pt]{H_2}$ (optically inactive)

(c) Two diastereomeric products are formed. Two fractions would be obtained. Each fraction would be optically active.

$\xrightarrow[(2)\ NaHSO_3]{(1)\ OsO_4}$

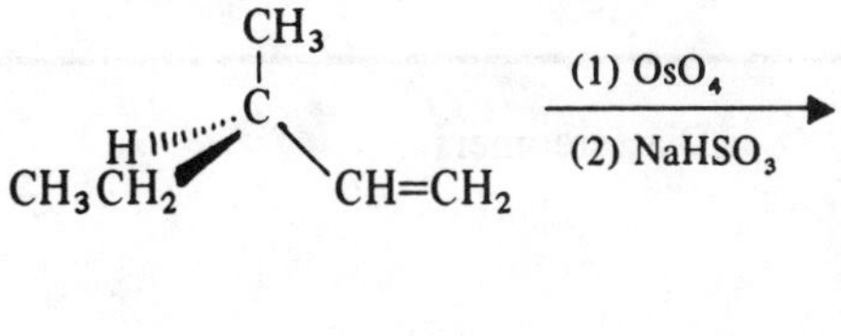

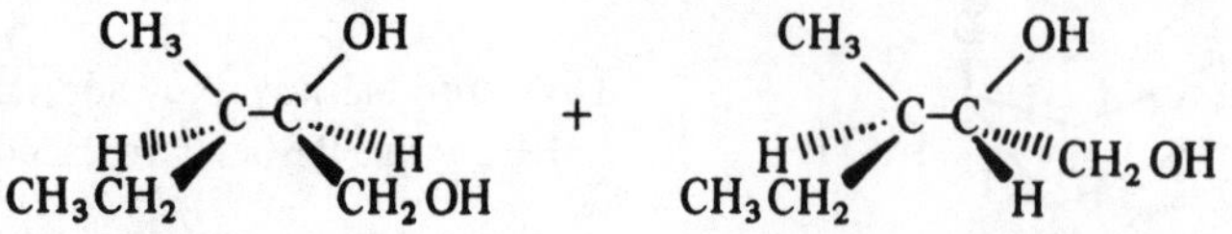

CH_3 OH / H–C–C–H / CH_3CH_2 CH_2OH + CH_3 OH / H–C–C–CH_2OH / CH_3CH_2 H

(optically active) (optically active)

Diastereomers

(d) One optically active compound is produced.

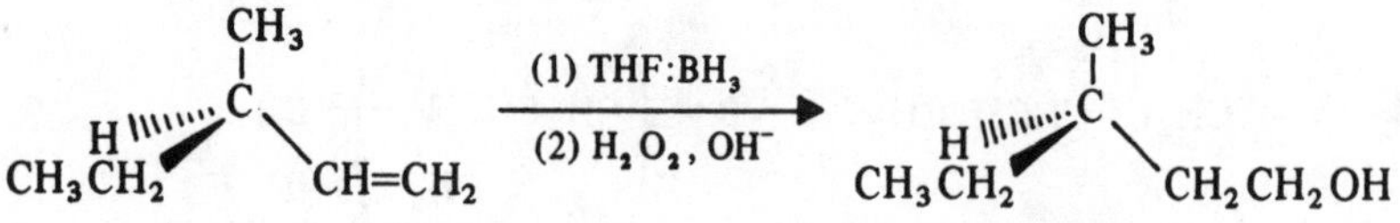

CH_3 / H–C / CH_3CH_2 $CH{=}CH_2$ $\xrightarrow[(2)\ H_2O_2,\ OH^-]{(1)\ THF{:}BH_3}$ CH_3 / H–C / CH_3CH_2 CH_2CH_2OH

(optically active)

(e) Two diastereomeric products are formed. Two fractions would be obtained. Each fraction would be optically active.

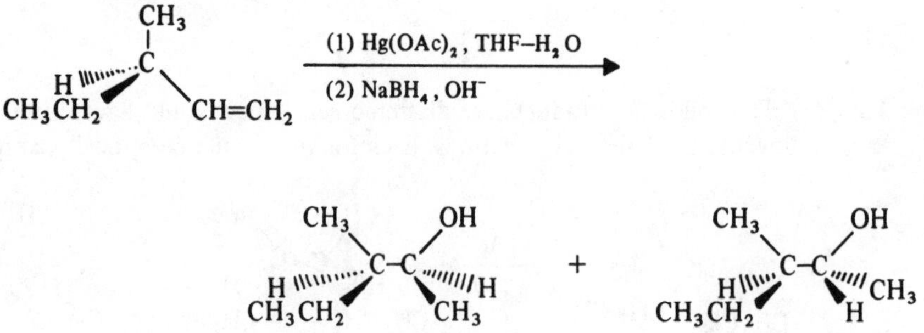

CH_3 / H–C / CH_3CH_2 $CH{=}CH_2$ $\xrightarrow[(2)\ NaBH_4,\ OH^-]{(1)\ Hg(OAc)_2,\ THF{-}H_2O}$

CH_3 OH / H–C–C–H / CH_3CH_2 CH_3 + CH_3 OH / H–C–C–CH_3 / CH_3CH_2 H

(optically active) (optically active)

Diastereomers

(f) Two diastereomeric products are formed. Two fractions would be obtained. Each fraction would be optically active.

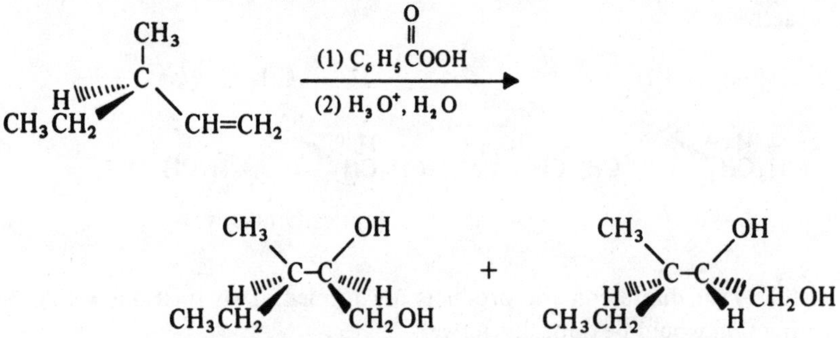

CH_3 / H–C / CH_3CH_2 $CH{=}CH_2$ $\xrightarrow[(2)\ H_3O^+,\ H_2O]{(1)\ C_6H_5C(=O)OOH}$

CH_3 OH / H–C–C–H / CH_3CH_2 CH_2OH + CH_3 OH / H–C–C–CH_2OH / CH_3CH_2 H

(optically active) (optically active)

Diastereomers

16

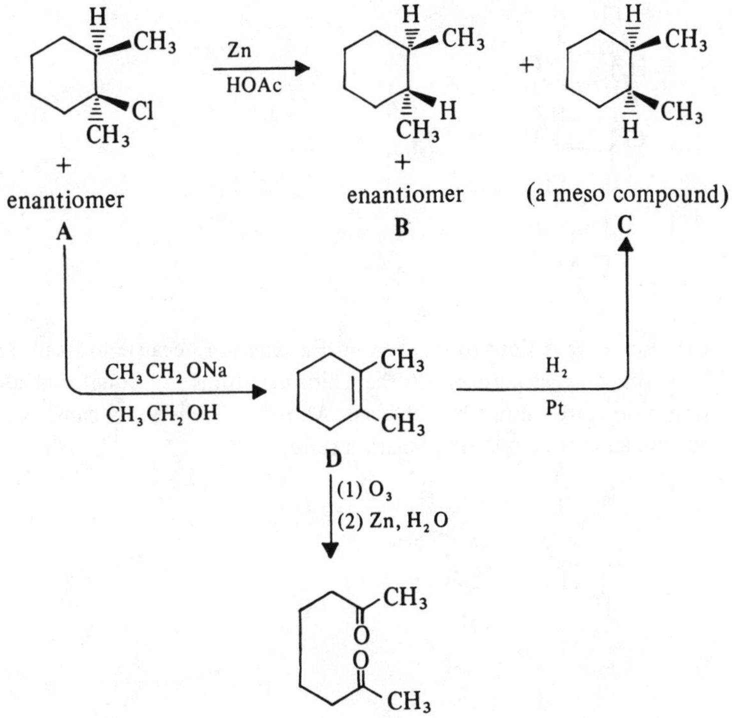

A (+ enantiomer) —Zn, HOAc→ B (+ enantiomer) + C (a meso compound)

A —CH_3CH_2ONa, CH_3CH_2OH→ D —H_2, Pt→ C

D —(1) O_3 (2) Zn, H_2O→

17

(a)

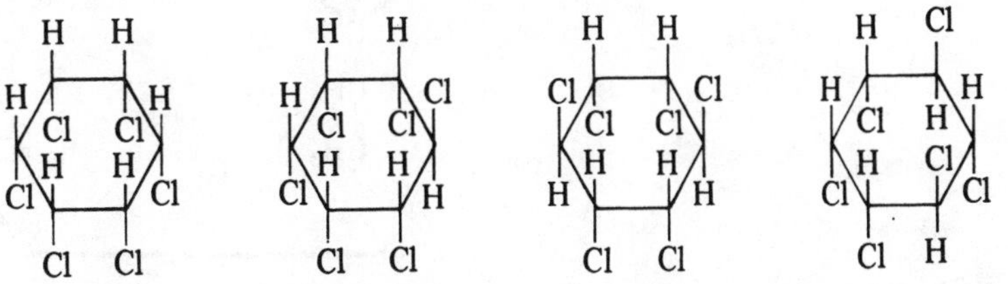

all *cis* meso **1**

1-*trans* meso **2**

1,4-*trans* meso **3**

1,3-*trans* meso **4**

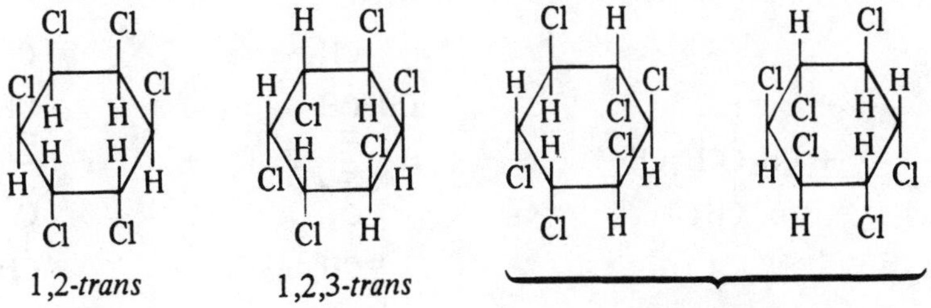

1,2-*trans* meso **5**

1,2,3-*trans* meso **6**

1,2,4-*trans* Enantiomers **7** **8**

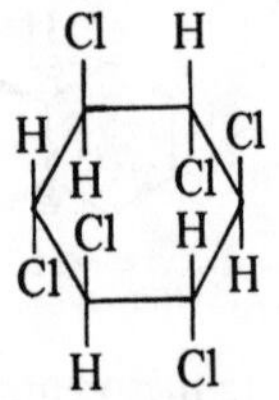

Cl H
H Cl
H Cl
Cl H
Cl H
H Cl

1,3,5-*trans*
meso
9

(b) Isomer **9** is slow to react in an E2 reaction because in its more stable conformation (see following structure) all the chlorine atoms are equatorial and an anti periplanar transition state cannot be achieved. All other isomers **1-8** can have a —Cl axial and thus achieve an anti periplanar transition state.

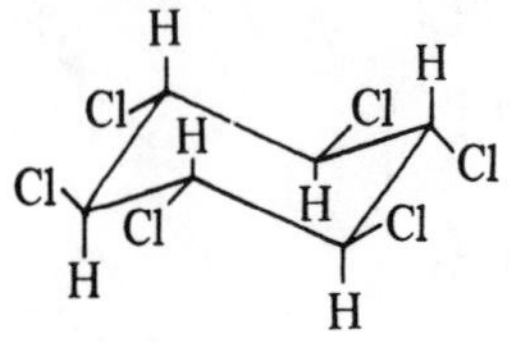

9

18 (a) $CH_3CH(CH_3)CH_2CH_3 \xrightarrow{F_2}$ **1** + **2**

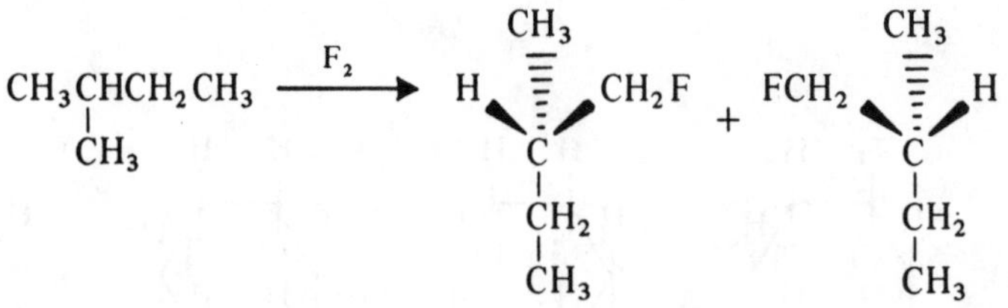

1, **2**: Enantiomers (obtained in one fraction as an optically inactive racemic form)

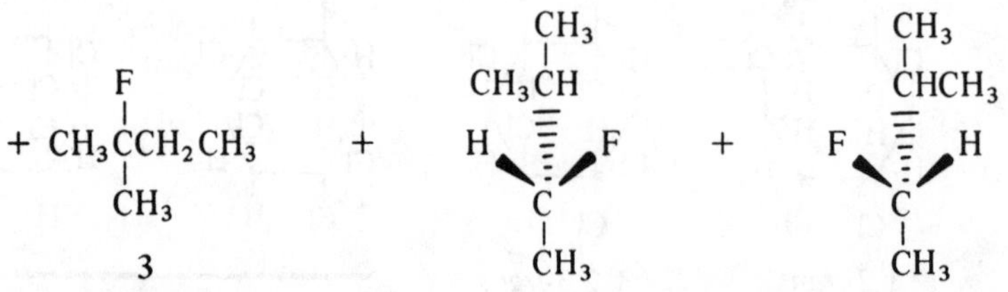

+ $CH_3CF(CH_3)CH_2CH_3$ **3** (achiral and, therefore, optically inactive) + **4** + **5**

4, **5**: Enantiomers (obtained in one fraction as an optically inactive racemic form)

$$+\ CH_3\overset{CH_3}{\overset{|}{C}}HCH_2CH_2F$$

6

(achiral and, therefore, optically inactive)

(b) Four fractions. The enantiomeric pairs would not be separated by fractional distillation because enantiomers have the same boiling points.

(c) All of the fractions would be optically inactive.

(d) The fraction containing **1** and **2** and the fraction containing **4** and **5**.

19 (R)-2-Fluorobutane $\xrightarrow{F_2}$ **1** (optically active) + **2** (achiral and, therefore, optically inactive) + **3** (optically active) + **4** meso Compound (optically inactive) + **5** (optically active)

(R)-2-Fluorobutane: CH_3, F, H, C, CH_2, CH_3

1: CH_2F, F, H, C, CH_2, CH_3 (optically active)

2: CH_3, F, F, C, CH_2, CH_3 (achiral and, therefore, optically inactive)

3: CH_3, F, H, C, C, H, F, CH_3 (optically active)

4: CH_3, F, H, C, C, F, H, CH_3 meso Compound (optically inactive)

5: CH_3, F, H, C, CH_2, CH_2F (optically active)

(b) Five. Compounds **3** and **4** are diastereomers. All others are constitutional isomers of each other.

(c) See above.

20 (R) CH_3CH_2, CH_3, C, H — cyclohexane ring — CH_2CH_3, C, CH_3, H (S); H, H on ring; meso

(R) CH_3CH_2, CH_3, C, H — cyclohexane ring with H, H — C, CH_2CH_3, CH_3, H (S)

Each of the two structures just given have a plane of symmetry (indicated by the dashed line), and, therefore, each is a meso compound. The two structures are not superposable one on the other, therefore they represent molecules of different compounds and are diastereomers.

21 Only a proton or deuteron anti to the bromine can be eliminated; that is, the two groups undergoing elimination (H and Br or D and Br) must lie in an anti periplanar arrangement. The two conformations of *erythro*-2-bromo-butane-3-*d* in which a proton or deuteron is anti periplanar to the bromine are **I** and **II**.

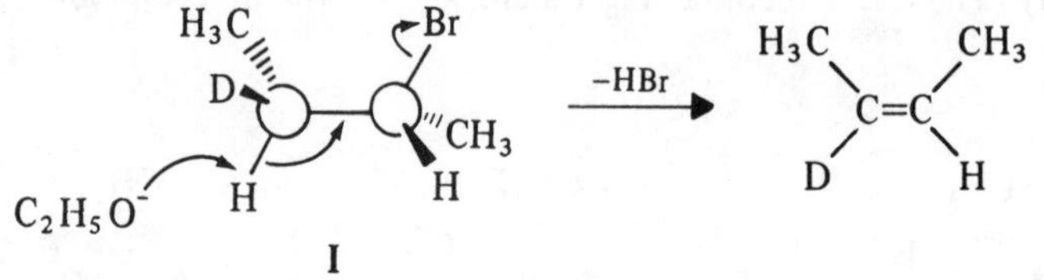

$-HBr$

I

$-DBr$

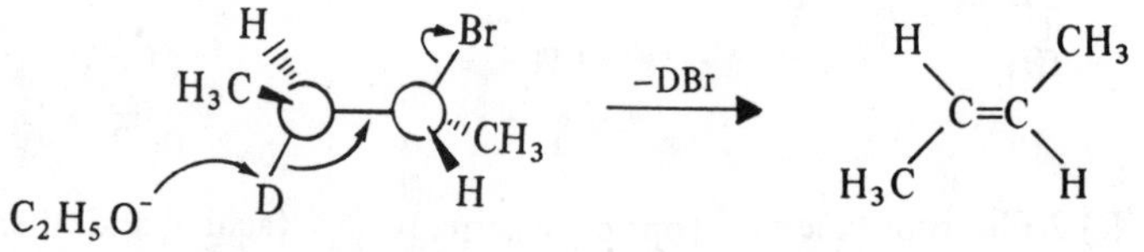

II

Conformation **I** can undergo loss of HBr to yield *cis*-2-butene-2-*d*. Conformation **II** can undergo loss of DBr to yield *trans*-2-butene.

11

AROMATIC COMPOUNDS I: THE PHENOMENON OF AROMATICITY

SOLUTIONS TO PROBLEMS

11.1 (a) None. For example, H–C≡C–CH_2CH_2–C≡C–H would yield two different monobromo products:

$$Br\text{–}C{\equiv}C\text{–}CH_2CH_2\text{–}C{\equiv}C\text{–}H \quad \text{and} \quad H\text{–}C{\equiv}C\text{–}\underset{\displaystyle Br}{\underset{|}{C}}HCH_2\text{–}C{\equiv}C\text{–}H$$

(b) None. All of these compounds should undergo addition of bromine.

11.2 Resonance structures may differ *only* in the positions of the electrons. In the two 1,3,5-cyclohexatrienes shown, the carbon atoms are in different positions; therefore they cannot be resonance structures.

11.3 (a)

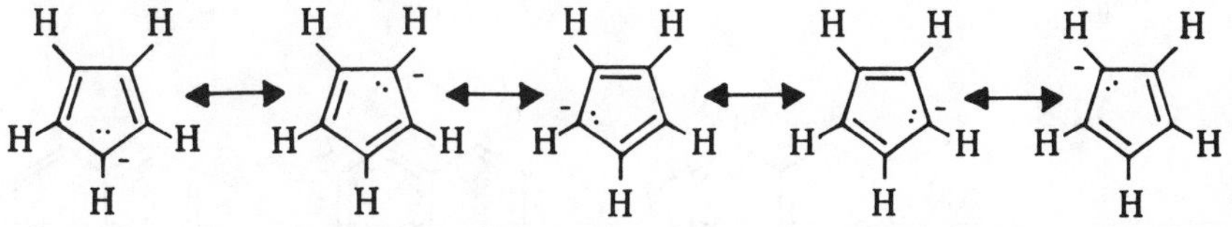

(b) Yes, all of the five resonance structures are equivalent, and all five hydrogen atoms are equivalent.

11.4 (a) ClO_4^- ⟶ ClO_4^- + $HC(C_6H_5)_3$

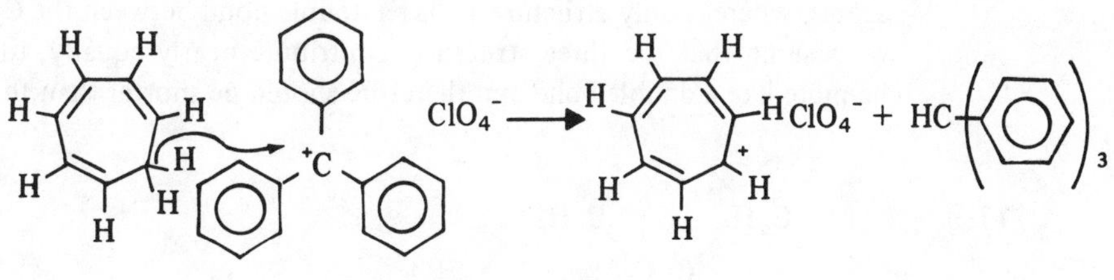

(b) Triphenylmethane [see part (a)].

(c) ClO_4^-

11.5 Tropylium bromide is ionic and has the structure, 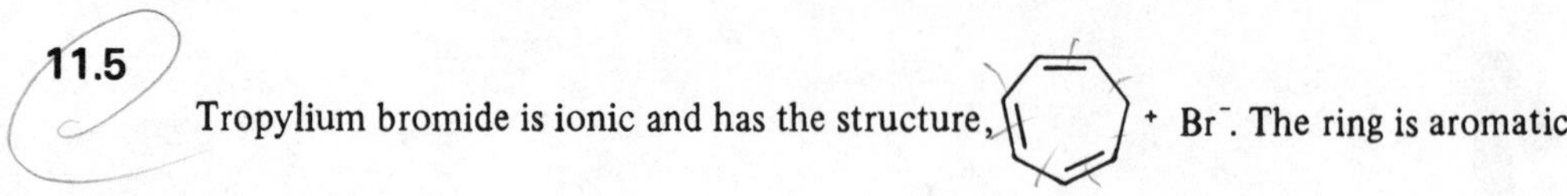 Br^-. The ring is aromatic.

11.6 (a) $^+CH(CH{=}CH_2)(CH{=}CH{-}CH{=}CH_2)$ (b) Cycloheptatrienyl cation is aromatic

11.7 (a) $^+CH(CH{=}CH_2)_2$ $\xrightarrow[\text{energy increases}]{\pi\ \text{electron}}$ cyclopentadienyl cation + H_2

The π-electron energy of cyclopentadienyl cation is higher than that of the open chain counterpart.

(b) $^+CH_2{-}CH{=}CH_2$ $\xrightarrow[\text{energy decreases}]{\pi\ \text{electron}}$ cyclopropenyl cation + H_2

The π electron energy of cyclopropenyl cation is lower than that of the open-chain counterpart.

(c) $4n+2 = 2$ when $n = 0$. Hückel's rule predicts that cyclopropenyl cation is aromatic.

(d) The π-electron energy of cyclopropenyl anion is higher than that of the open-chain counterpart.

11.8 (a) [Naphthalene resonance structures I, II, III; structure I numbered 1–8]

I II III

(b) Two of the structures (I and III) have a double bond between the $C_1{-}C_2$ carbon atoms, whereas only structure II has a double bond between the $C_2{-}C_3$ carbon atoms. If we assume that the three structures contribute nearly equally, the $C_1{-}C_2$ bond should be more like a double bond and therefore should be shorter than the $C_2{-}C_3$ bond.

11.9 C_6H_5 C_6H_5 C=C C+ :O:-

III

III is a more important contributor to the resonance hybrid of **I** than a corresponding ionic structure of **II** is to the hybrid of **II**. **III** is an important contributor to the hybrid of diphenylcyclopropenone because it resembles the aromatic cyclopropenyl cation; that is, the ring in structure **III** has 2 π electrons and is a $4n + 2$ system where $n = 0$.

11.10 (a) Different products would be obtained in each instance. By identifying the products, one can show that the product of the Birch reduction of benzene is 1,4-cyclohexadiene.

1,4-Cyclohexadiene $\xrightarrow[\text{(2) Zn, } H_2O]{\text{(1) } O_3}$ 2 $HCCH_2CH$ (with C=O on each terminal carbon: O=CH–CH₂–CH=O)

1,3-Cyclohexadiene $\xrightarrow[\text{(2) Zn, } H_2O]{\text{(1) } O_3}$ $HCCH_2CH_2CH$ + $HC{-}CH$ (each terminal carbon C=O)

(b)

1,2-Dimethyl-1,4-cyclohexadiene $\xrightarrow[\text{(2) Zn, } H_2O]{\text{(1) } O_3}$ 2 $HCCH_2CCH_3$ (O=CH–CH₂–C(=O)CH₃)

11.11 Product **X** is 1-methyl-1,3-cyclohexadiene

X $\xrightarrow[\text{(2) Zn, } H_2O]{\text{(1) } O_3}$ CH_3CCH_2CH + $HCCH_2CH$ (CH₃C(=O)CH₂CH=O + O=CHCH₂CH=O)

11.12

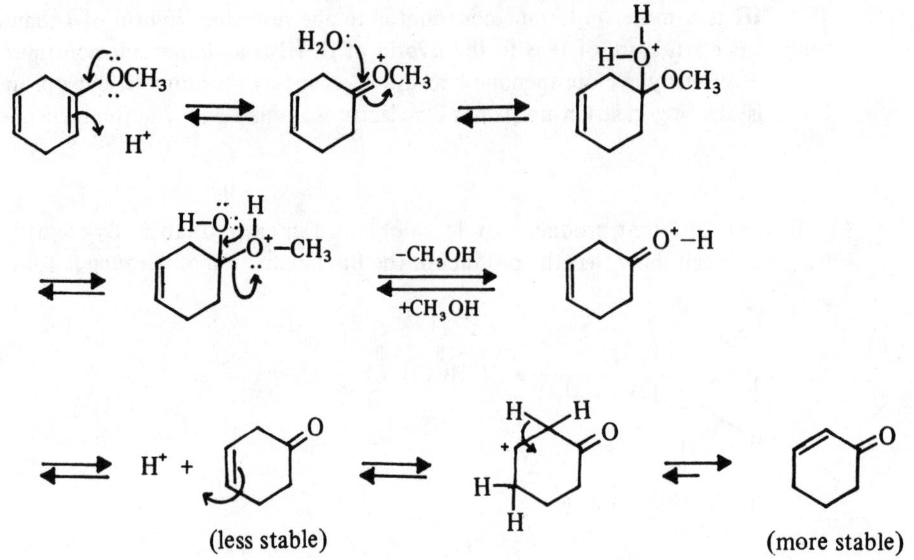

(less stable)

(more stable)

11.13 (a)

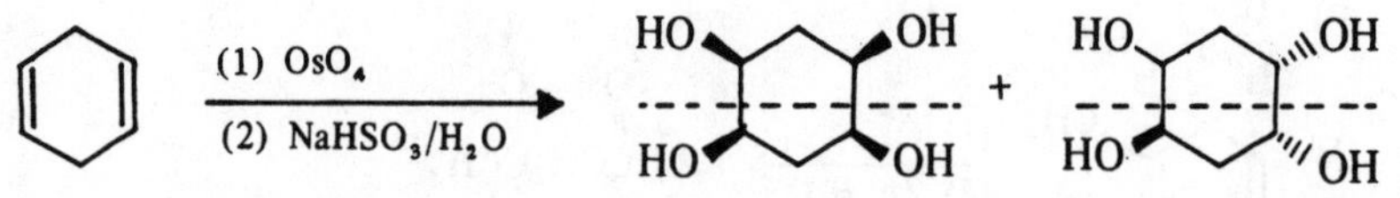

(1) OsO_4

(2) $NaHSO_3/H_2O$

(b) No, both products are achiral. (Each has at least one plane of symmetry.) See the dashed lines shown in (a).

11.14

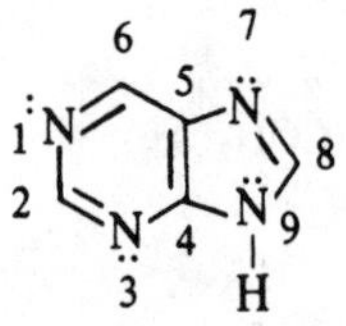

1,3,7 are pyridine-type nitrogen atoms; 9 is a pyrrole-type nitrogen atom

11.15

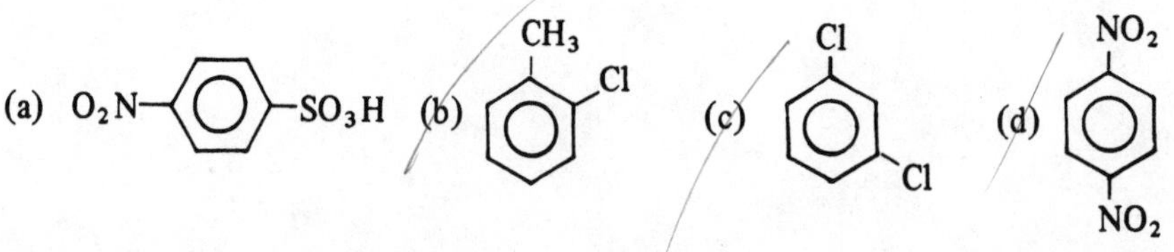

(a) $O_2N-C_6H_4-SO_3H$ (b) (c) (d)

(e) 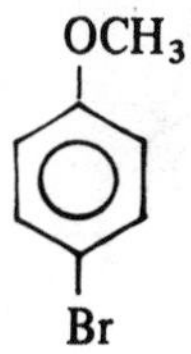 OCH_3, Br

(f) 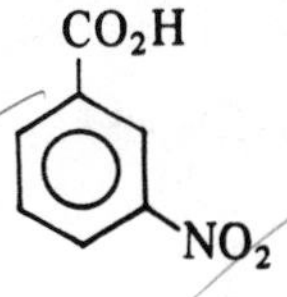 CO_2H, NO_2

(g)  OH, I

(h) CO_2H, Cl

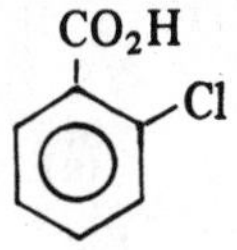

(i) Br

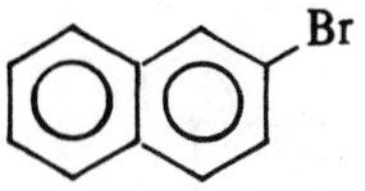

(j) 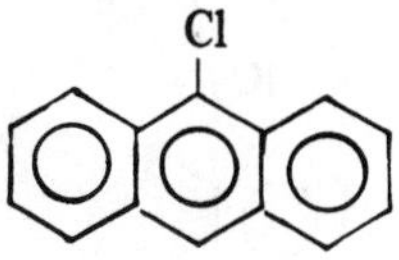 Cl

(k) 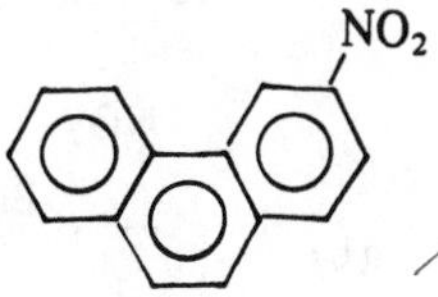 NO_2

(l) 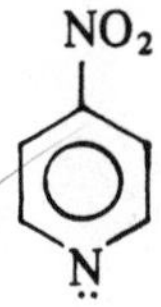 NO_2, N

(m) N, H, CH_3

(n) NO_2, Cl, Cl

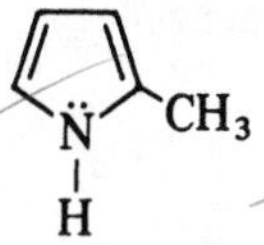

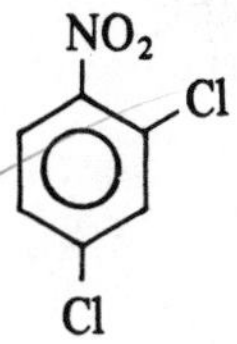

(o) 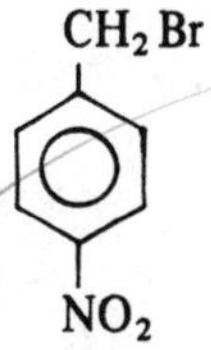 CH_2Br, NO_2

(p) 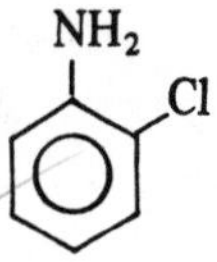 NH_2, Cl

(q) 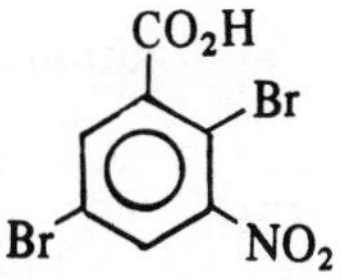 CO_2H, Br, Br, NO_2

(r) 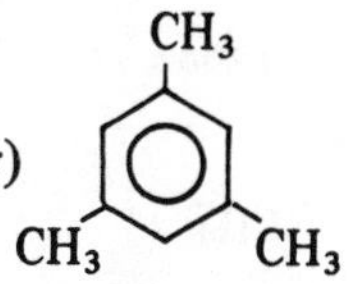 CH_3, CH_3, CH_3

(s) 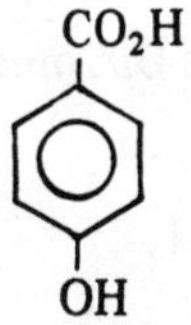 CO_2H, OH

(t) 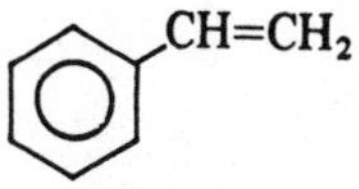 $CH{=}CH_2$

(u)

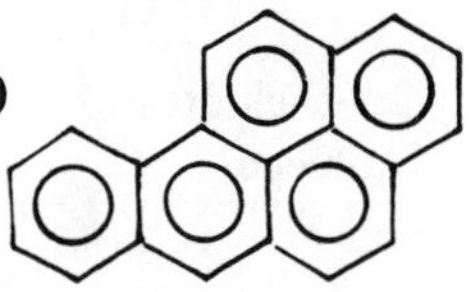

(v) 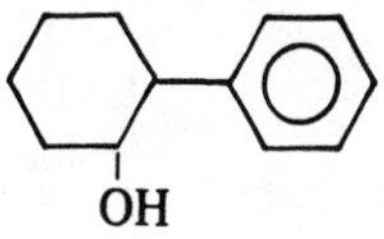 OH

(w) 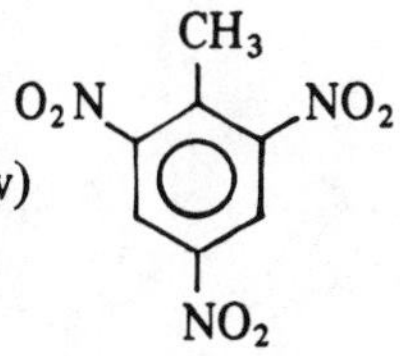 CH_3, O_2N, NO_2, NO_2

(x)

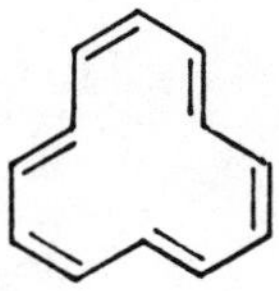

(y)

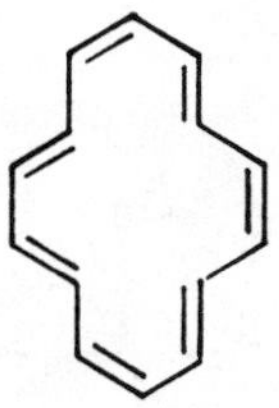

(z)

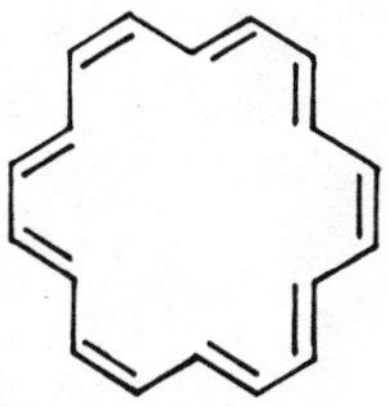

11.16

(a)

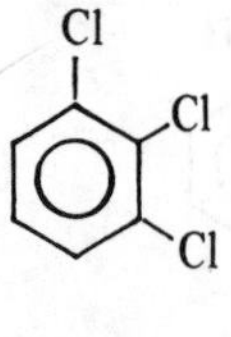
Cl
Cl
Cl

1,2,3-Trichloro-
benzene

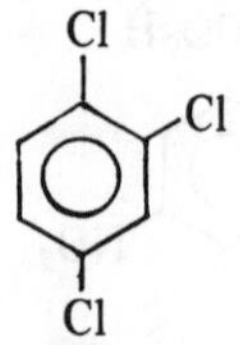
Cl
Cl
Cl

1,2,4-Trichloro-
benzene

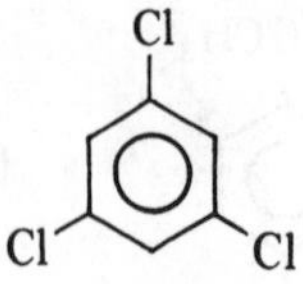
Cl
Cl
Cl

1,3,5-Trichloro-
benzene

(b)

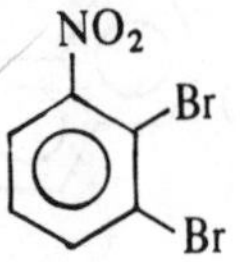
NO₂
Br
Br

2,3-Dibromo-1-
nitrobenzene

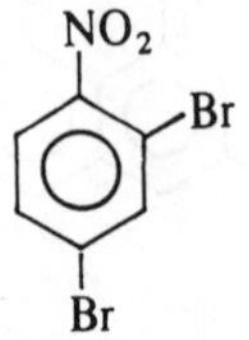
NO₂
Br
Br

2,4-Dibromo-1-
nitrobenzene

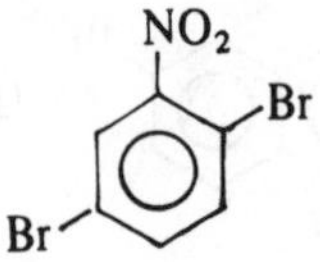
NO₂
Br
Br

1,4-Dibromo-2-
nitrobenzene

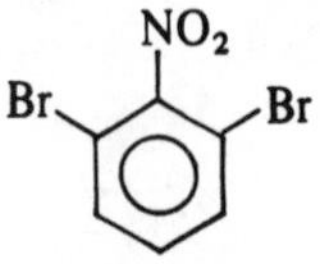
NO₂
Br
Br

1,3-Dibromo-2-
nitrobenzene

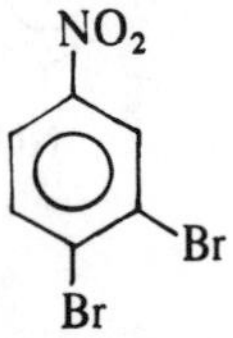
NO₂
Br
Br

1,2-Dibromo-4-nitrobenzene

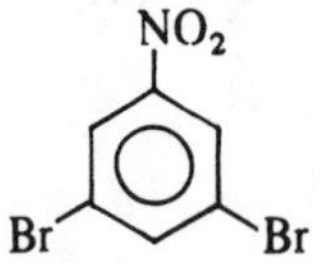
NO₂
Br
Br

3,5-Dibromo-1-nitrobenzene

(c)

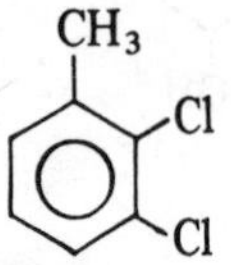
CH₃
Cl
Cl

2,3-Dichloro-
toluene

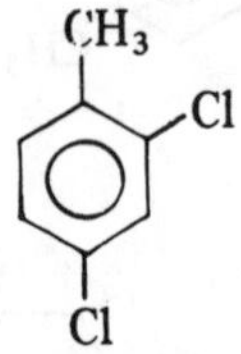
CH₃
Cl
Cl

2,4-Dichloro-
toluene

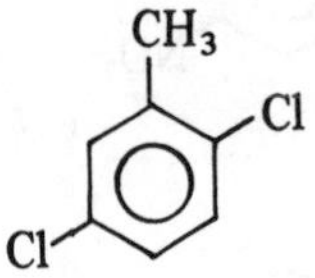
CH₃
Cl
Cl

2,5-Dichloro-
toluene

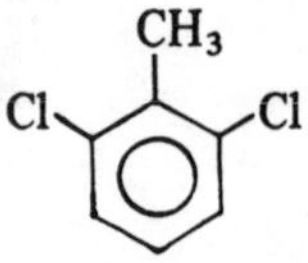
CH₃
Cl
Cl

2,6-Dichloro-
toluene

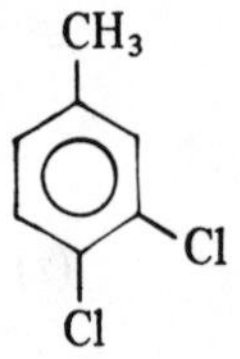
CH₃
Cl
Cl

3,4-Dichlorotoluene

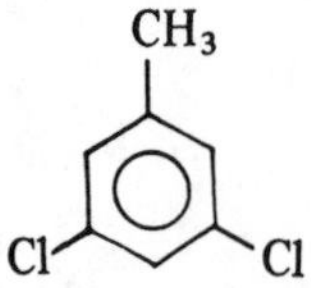
CH₃
Cl
Cl

3,5-Dichlorotoluene

(d)

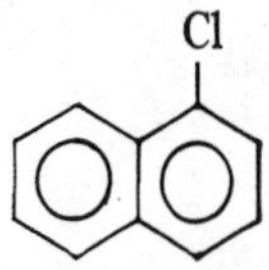
Cl

1-Chloronaphthalene

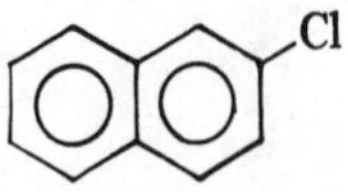
Cl

2-Chloronaphthalene

(e)

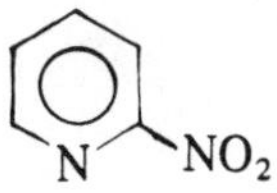

2-Nitropyridine 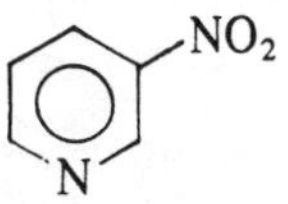 3-Nitropyridine 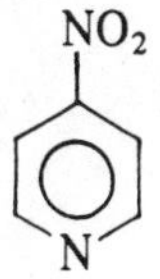 4-Nitropyridine

(f)

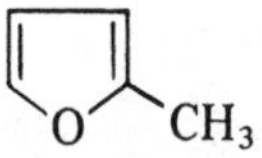

2-Methylfuran 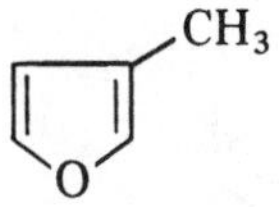 3-Methylfuran

(g)

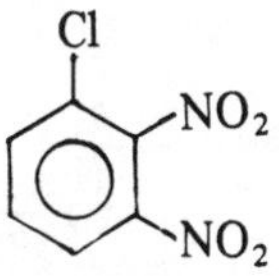

1-Chloro-2,3-dinitrobenzene 1-Chloro-2,4-dinitrobenzene 2-Chloro-1,4-dinitrobenzene 2-Chloro-1,3-dinitrobenzene

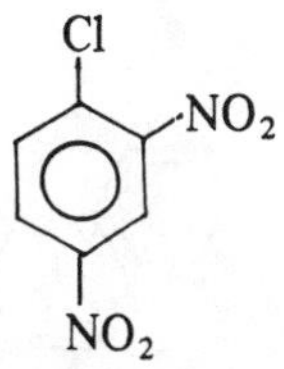

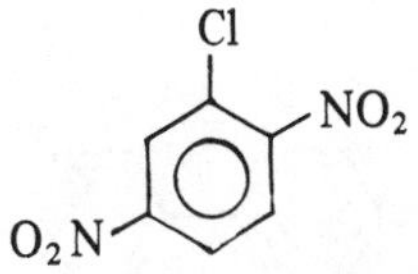

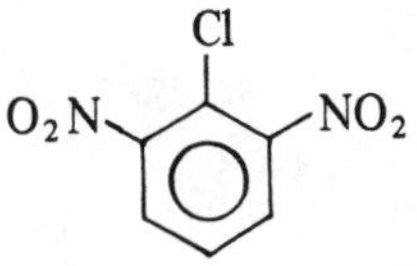

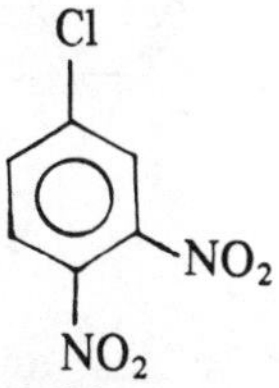

4-Chloro-1,2-dinitrobenzene 1-Chloro-3,5-dinitrobenzene

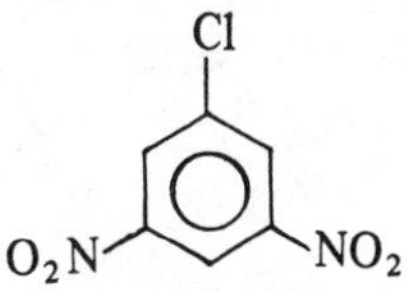

(h)

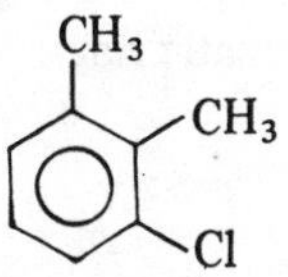

1-Chloro-2,3-dimethylbenzene 4-Chloro-1,2-dimethylbenzene 2-Chloro-1,3-dimethylbenzene 1-Chloro-2,4-dimethylbenzene

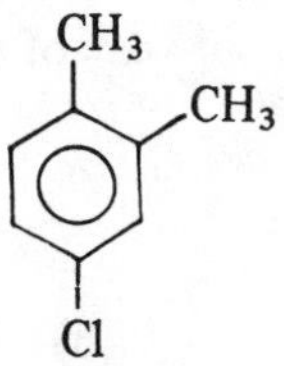

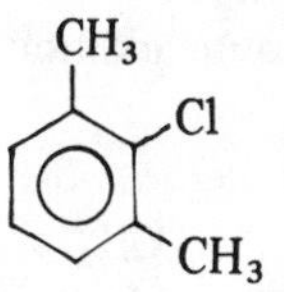

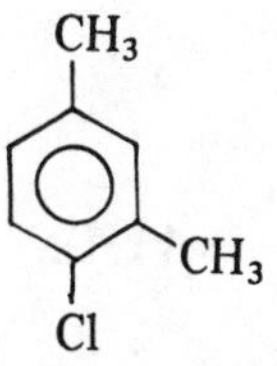

1-Chloro-3,5-dimethylbenzene

2-Chloro-1,4-dimethylbenzene

(i)

o-Cresol

m-Cresol

p-Cresol

11.17 (a)

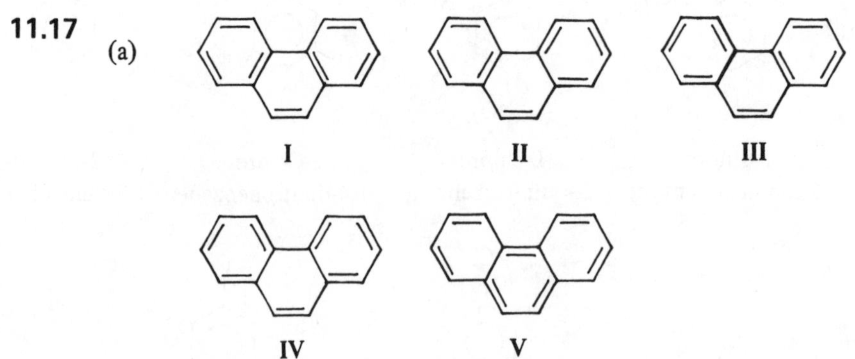

(b) The 9,10 bond should be close to that of a double bond, 1.33Å, since in four of the five contributors it is a double bond.

(c) Almost that of an actual double bond.

(d) Bromine adds to the 9,10 double bond because of its large double-bond character and because addition disrupts only one of three aromatic rings.

11.18 (a)

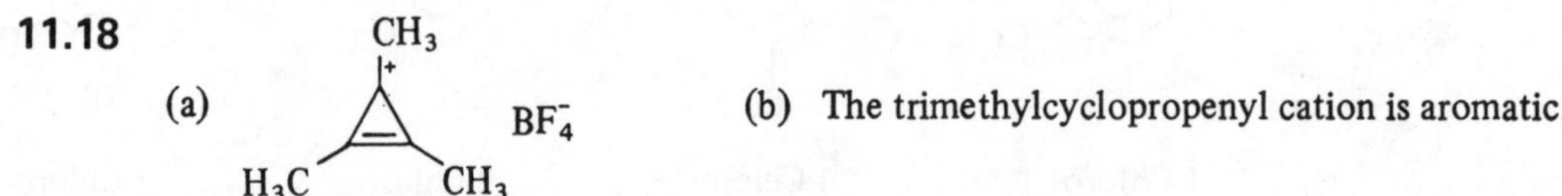

(b) The trimethylcyclopropenyl cation is aromatic.

11.19 C_6H_5 C_6H_5 Br^- OH or C_6H_5 C_6H_5 Br^- OH

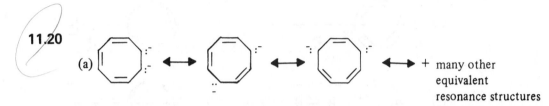

11.20 (a) ⟷ ⟷ ⟷ + many other equivalent resonance structures

(b) It has $4n + 2 = 10\,\pi$ electtrons ($n = 1$), and is therefore aromatic; that is, it illustrates Hückel's rule.

11.21 (a) Would not be aromatic; it is a monocyclic system of 12 π electrons and thus does not conform to Hückel's rule.

(b) Would not be aromatic; it is not a conjugated system.

(c) Would not be aromatic; it is an 8 π-electron monocyclic system and thus does not conform to Hückel's rule.

(d) Would not be aromatic; it is a 16 π-electron monocyclic system and thus does not conform to Hückel's rule.

(e) Would be aromatic because of resonance structures (see following structure) that consist of a cycloheptatrienyl cation and cyclopentadienyl anion.

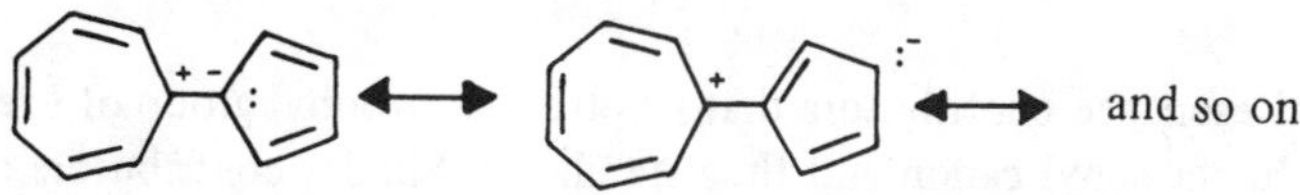

(f) Would be aromatic; it is a planar monocyclic system of 14 π electrons. (We count only two electrons of the triple bond because only two are in p orbitals that overlap with those of the double bonds on either side.)

(g) Would be aromatic; it is a planar monocyclic system of 10 π electrons.

(h) Would be aromatic; it is a nearly planar monocyclic system of 10 π electrons. (The bridging $-CH_2-$ group allows the ring system to be almost planar.)

11.22

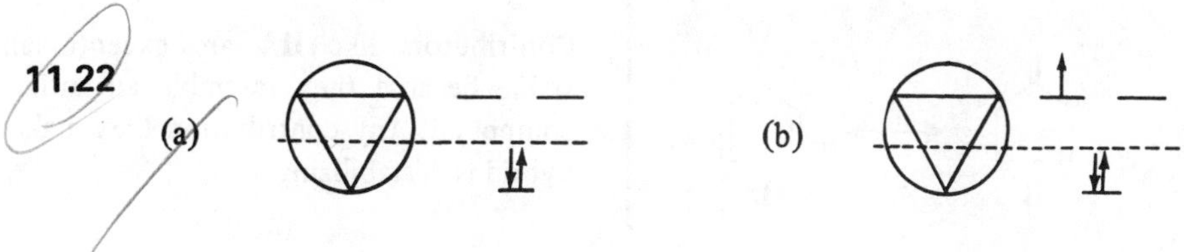

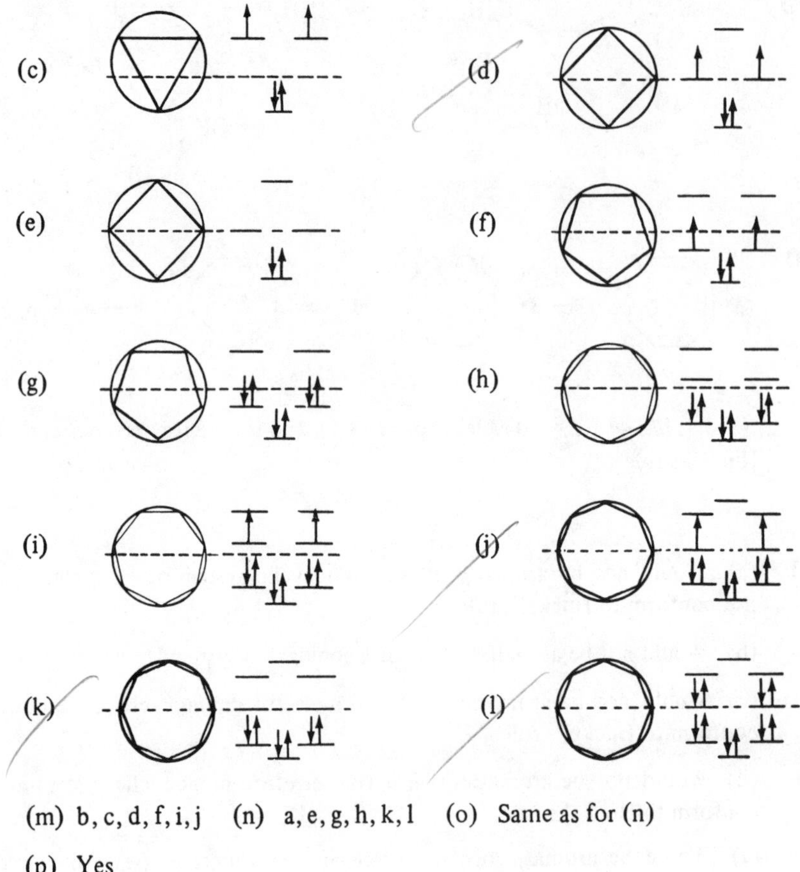

(m) b, c, d, f, i, j (n) a, e, g, h, k, l (o) Same as for (n)

(p) Yes

11.23 Resonance contributors that involve the carbonyl group of **I** resemble the *aromatic* cycloheptatrienyl cation and thus stabilize **I**. Similar contributors to the hybrid of **II** resemble the *antiaromatic* cyclopentadienyl cation (see Problem 11.7) and thus destabilize **II**.

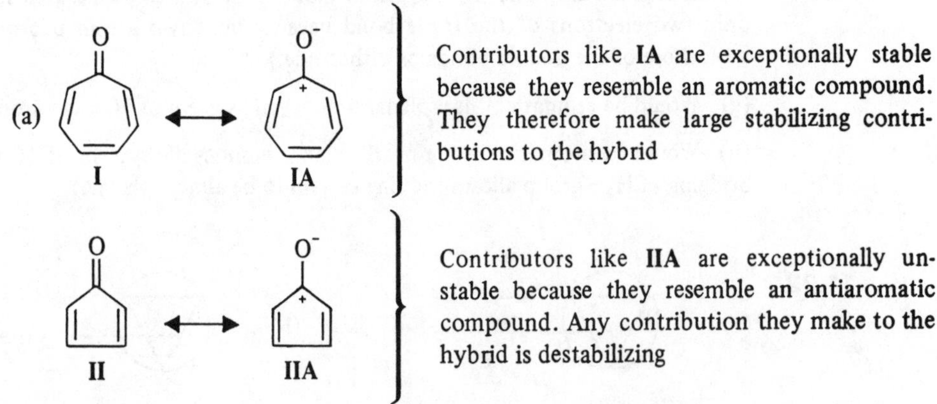

(a)
O
O⁻
I
IA
Contributors like **IA** are exceptionally stable because they resemble an aromatic compound. They therefore make large stabilizing contributions to the hybrid
O
O⁻
II
IIA
Contributors like **IIA** are exceptionally unstable because they resemble an antiaromatic compound. Any contribution they make to the hybrid is destabilizing

(b) 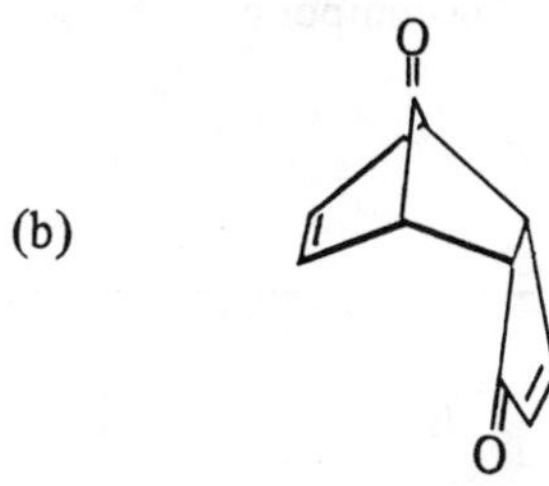

SECTION REFERENCES FOR ADDITIONAL PROBLEMS

11.15 11.8
11.16 11.8
11.17 11.7A
11.18 11.6B
11.19 11.6B
11.20 11.6B
11.21 11.6, 11.6A, 11.6B
11.22 11.6
11.23 11.6B

SELF-TEST

11.1 (a) Which of the following line formulas represent aromatic structures? (*Circle the letters that correspond to your choices.*)

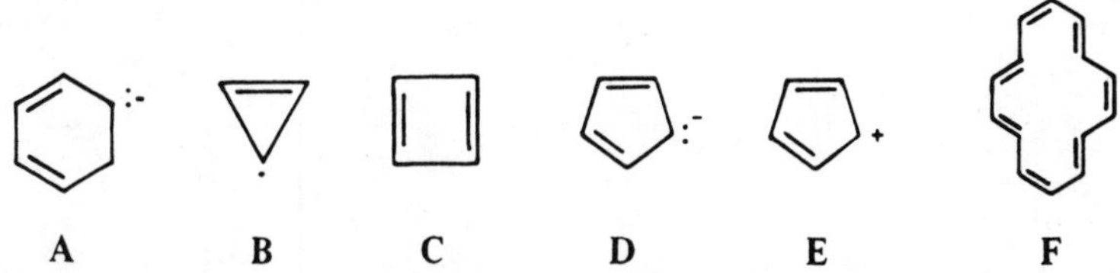

(b) Which of these formulas represent conjugated systems? (*Circle your choices below.*)

Answers: **A B C D E F**

(c) Draw all the remaining important resonance structures for

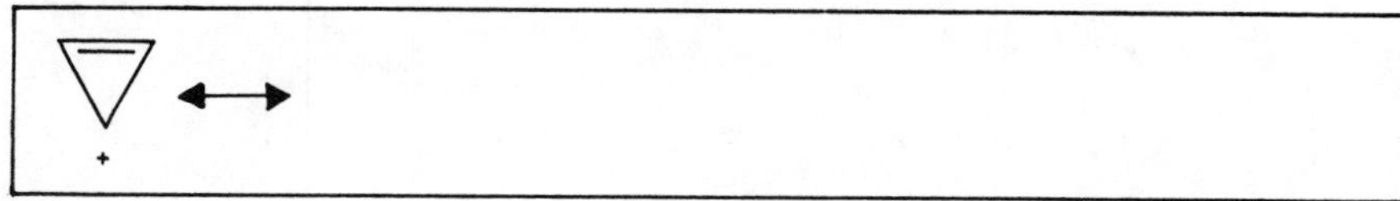

11.2 Give an acceptable name for each of the following compounds.

(a) [benzene ring with two adjacent CH_3 groups] ______________________

(b) CH_3CH_2–[benzene ring] ______________________

(c) $CH_3\overset{CH_3}{\overset{|}{C}}H$–[benzene ring]–Br ______________________

(d) [benzene ring]–$CH_2\overset{Cl}{\overset{|}{C}}HCH_2$–[benzene ring] ______________________

(e) CH_3–[benzene ring with three NO_2 groups] ______________________

11.3 Write the structural formula, **including all hydrogen atoms**, of each of the following:

(a) An aromatic seven-membered carbocyclic ring.

a.

(b) A six-membered carbocyclic ring that is not aromatic.

b.

(c) An aromatic five-membered carbocyclic ring.

c.

(d) A nonaromatic, conjugated five-membered carbocyclic ring.

d.

11.4 (a) How many isomeric trimethylbenzenes are possible?

(b) Which one undergoes ring bromination to give three different monobromotrimethylbenzenes?

(c) The name of this compound is

____

11.5 Which of the following reactions is inconsistent with the assertion that benzene is aromatic?

(a) $Br_2/CCl_4/25°C \longrightarrow$ no reaction

(b) $H_2/Pt/25°C \longrightarrow$ no reaction

(c) $Br_2/FeBr_3 \longrightarrow C_6H_5Br + HBr$

(d) $KMnO_4/H_2O/25°C \longrightarrow$ no reaction

(e) None of the above

11.6 Which is the correct name of the compound shown?

CH_3 NO_2

Cl

(a) 3-Chloro-5-nitrotoluene (b) *m*-Chloro-*m*-nitrotoluene

(c) 1-Chloro-3-nitro-5-toluene (d) *m*-Chloromethylnitrobenzene

(e) More than one of these

11.7 Which is the correct name of the compound shown?

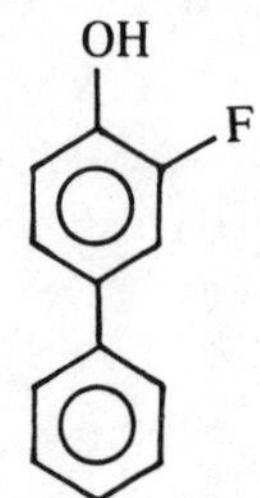

(a) 2-Fluoro-1-hydroxyphenylbenzene (b) 2-Fluoro-4-phenylphenol

(c) *m*-Fluoro-*p*-hydroxybiphenyl (d) *o*-Fluoro-*p*-phenylphenol

(e) More than one of these

11.8 Which of the following molecules or ions is not aromatic according to Hückel's rule?

(a) 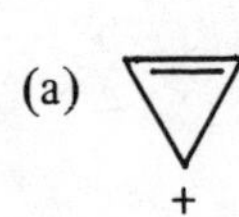

(b) 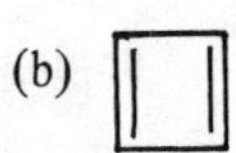

(c) 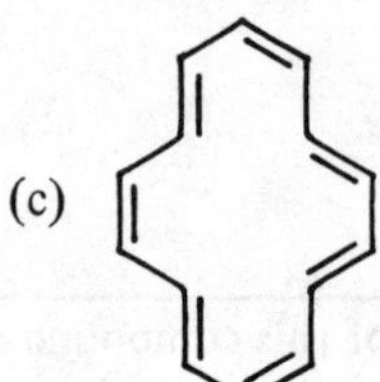

(d) 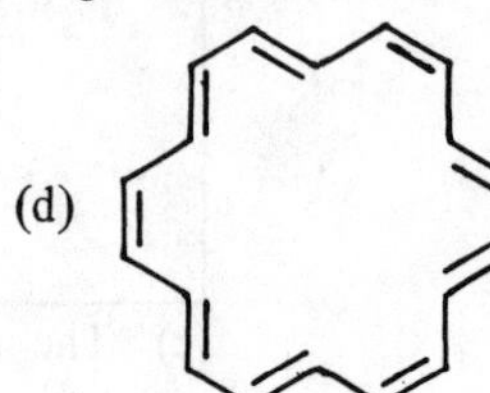

(e) All are aromatic.

11.9 Cyclopentadiene is much more acidic than cycloheptadiene. This can be explained by resonance theory.

(a) True (b) False

SUPPLEMENTARY PROBLEMS

S11.1 Select the appropriate choice (if any) in each pair of the following structures.

(a) Is the higher energy resonance structure — $C_6H_5\overset{+}{C}HCH{=}CH_2$ or $C_6H_5CH{=}CH\overset{+}{C}H_2$

(b) Is an aromatic species

(c) Is an aromatic species

(d) Forms only one tribromo derivative — Br Br or Br Br

S11.2 Which of the structures in each group is *not* a contributing resonance structure? Explain.

(a)

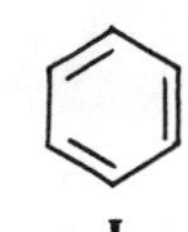

I II III IV

(b) H–C–H (I) H–C–H (II) H–C–H (III)

I II III

SOLUTIONS TO SUPPLEMENTARY PROBLEMS

S11.1 (a) Both are allylic cations conjugated with a benzene ring, however the first is 2° and the second is 1°. Therefore the 1° carbocation has higher energy.

(b) $^+$, because it has 2π electrons ($4n + 2$, where $n = 0$).

(c) Both have 6 π electrons, so both are aromatic.

(d) Br–C_6H_4–Br (p-dibromobenzene). Its only tribromo derivative is 1,2,4-tribromobenzene (Br–C_6H_3(Br)–Br).

S11.2 (a) **III**, because it has a different number of unpaired electrons.

(b) **III**, because the carbon atom has five bonds.

12 AROMATIC COMPOUNDS II: ELECTROPHILIC AROMATIC SUBSTITUTION

REACTIONS OF BENZENE

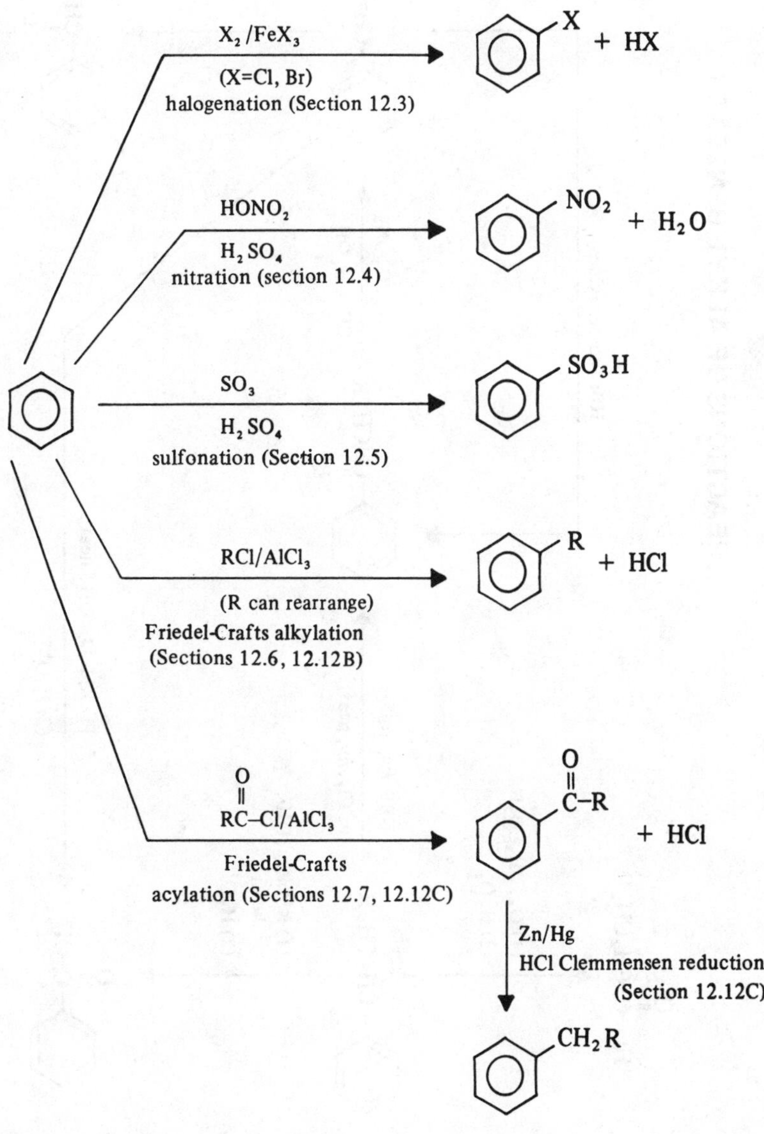

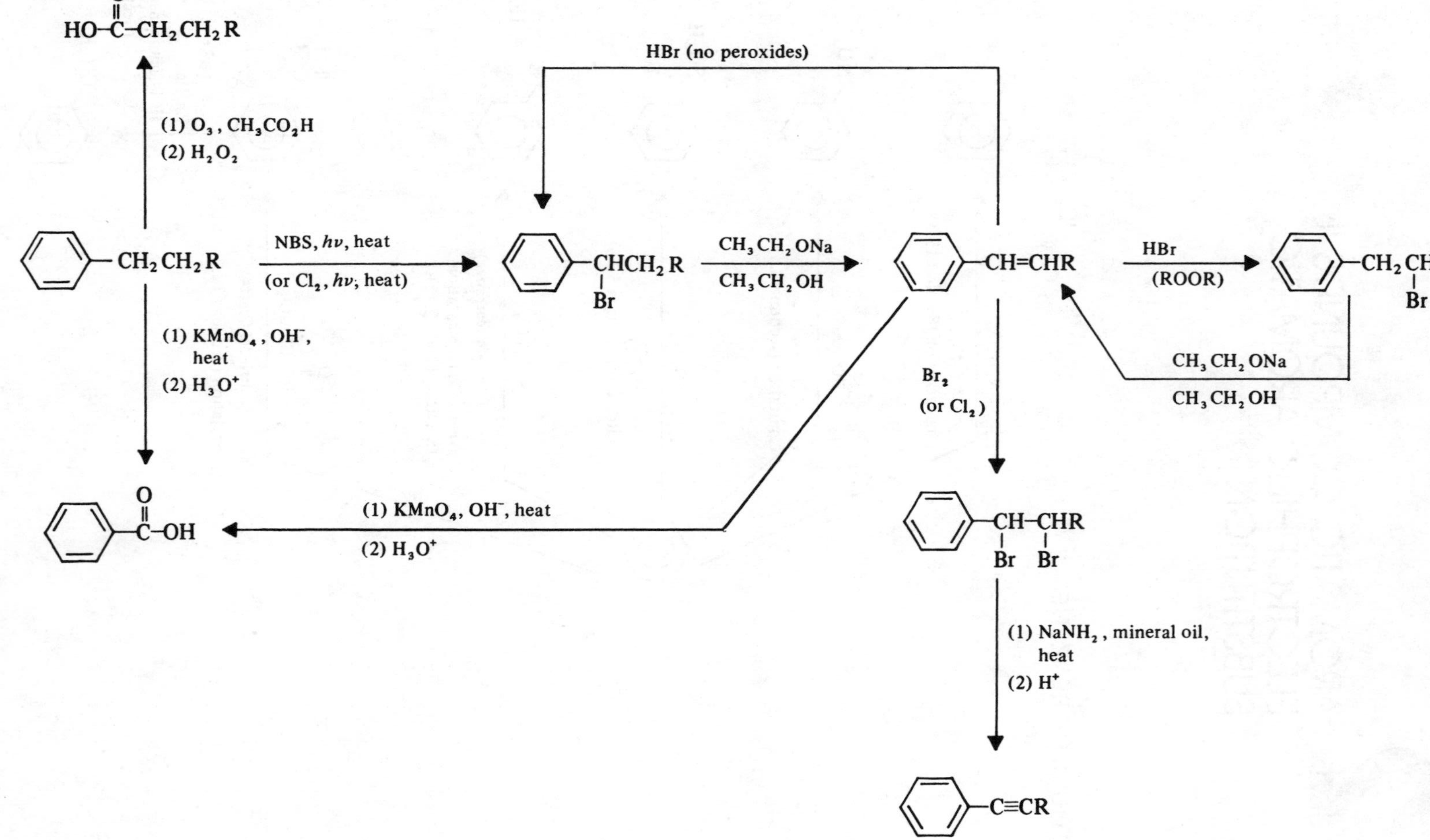

REACTIONS OF ALKYL BENZENES
HO–C(=O)–CH_2CH_2R
(1) O_3, CH_3CO_2H
(2) H_2O_2
HBr (no peroxides)
CH_2CH_2R
NBS, $h\nu$, heat
(or Cl_2, $h\nu$; heat)
$CHCH_2R$ Br
CH_3CH_2ONa
CH_3CH_2OH
CH=CHR
HBr
(ROOR)
CH_2CHR Br
CH_3CH_2ONa
CH_3CH_2OH
(1) $KMnO_4$, OH^-, heat
(2) H_3O^+
Br_2
(or Cl_2)
C(=O)–OH
(1) $KMnO_4$, OH^-, heat
(2) H_3O^+
CH–CHR Br Br
(1) $NaNH_2$, mineral oil, heat
(2) H^+
C≡CR

SOLUTIONS TO PROBLEMS

12.1

H E → E + H^+ H E → E + H^+

12.2

$H\text{–}\ddot{O}\text{–}NO_2 + H\text{–}\ddot{O}\text{–}NO_2 \longrightarrow H\text{–}\overset{+}{O}(H)\text{–}NO_2 + NO_3^-$

$H\text{–}\overset{+}{O}(H)\text{–}NO_2 + HONO_2 \longrightarrow NO_2^+ + H_3O^+ + NO_3^-$

12.3

(a) $C_6H_5SO_3H + D_2O \rightleftharpoons C_6H_5SO_3^- + D_2HO^+$

$C_6H_5SO_3^- + D_3O^+ \rightleftharpoons$ (D, SO_3^- arenium ion) $+ D_2O$

(D, SO_3^- arenium ion) $\longrightarrow C_6H_5D + :\ddot{O}=S(=\ddot{O}:)=\ddot{O}:$

(b) $:\ddot{Br}\text{–}\ddot{Br}\text{–}\bar{Fe}Br_3$ ($\delta+$ $\delta+$) is the electrophile

$C_6H_5SO_3^-$ $\delta+$ $\delta+$ $+ :\ddot{Br}\text{–}\ddot{Br}\text{–}\bar{Fe}Br_3 \longrightarrow$ (SO_3^-, $\ddot{Br}:$ arenium ion) $+ FeBr_4^-$

(Br, $S(=O)(=O)\ddot{O}:^-$ arenium ion) $\longrightarrow C_6H_5Br + SO_3$

12.4

(a) cyclohexene $\underset{}{\overset{HF}{\rightleftharpoons}}$ cyclohexyl cation

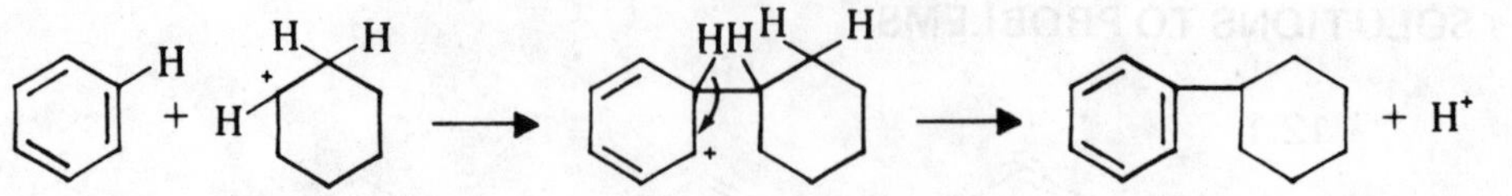

(b) + BF_3 ⇌ ⇌ + $HO\bar{B}F_3$

+ H⁺

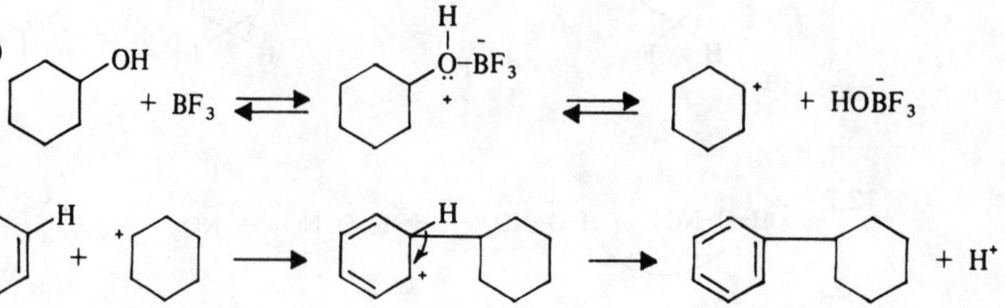

12.5

$CH_3\overset{+}{C}=\ddot{O}: + CH_3\overset{O}{\overset{\|}{C}}O\bar{A}lCl_3$

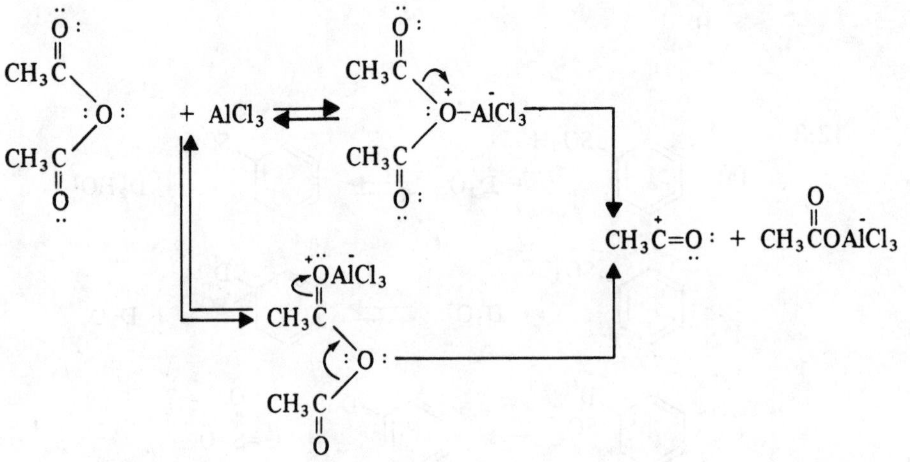

12.6 40% Ortho, 40% meta, 20% para because there are twice as many ortho and meta as para positions.

12.7

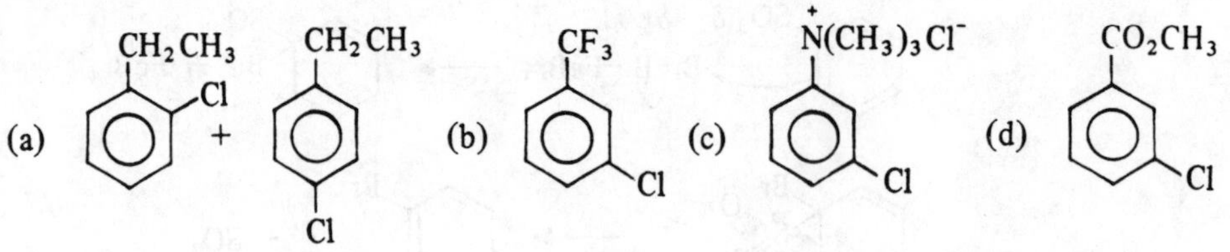

(a) CH_2CH_3, Cl + CH_2CH_3, Cl (b) CF_3, Cl (c) $\overset{+}{N}(CH_3)_3Cl^-$, Cl (d) CO_2CH_3, Cl

12.8

(a) Ortho ↔ ↔ ↔ Relatively stable

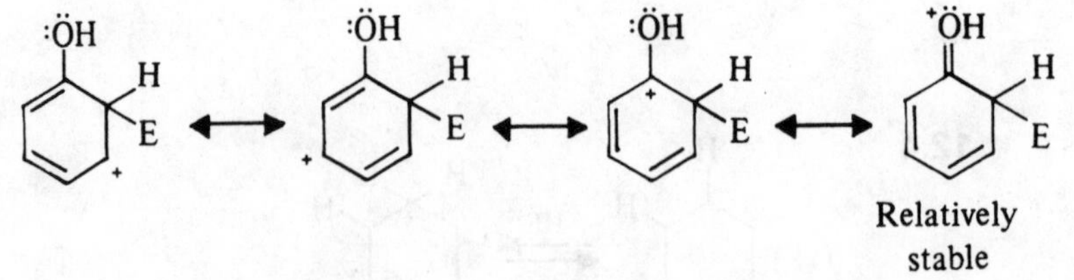

Meta

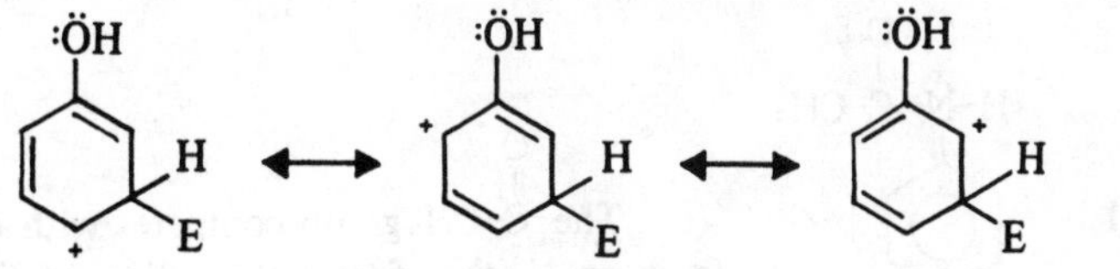

Para

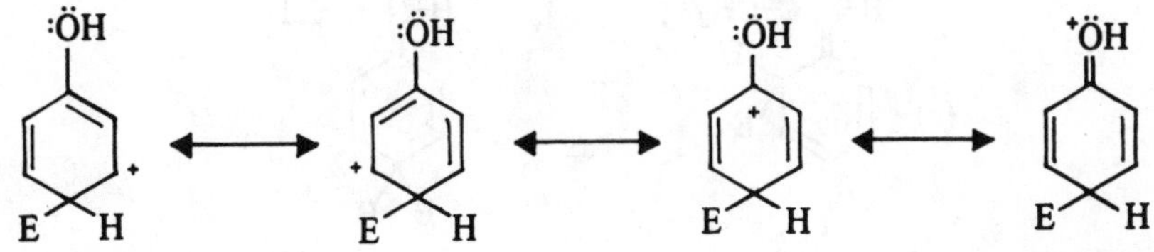

Relatively stable

(b) The electron-releasing ability of the –OH group through resonance increases the electron density of the ring, and it stabilizes the positive charge of the intermediate carbocation.

(c) An extra and relatively stable structure (just cited) contributes to the intermediate carbocation only when attack is ortho and para.

(d,e) More reactive because the extra structure (see following structure) does not have a positive charge on its oxygen as is true with phenol.

Ortho

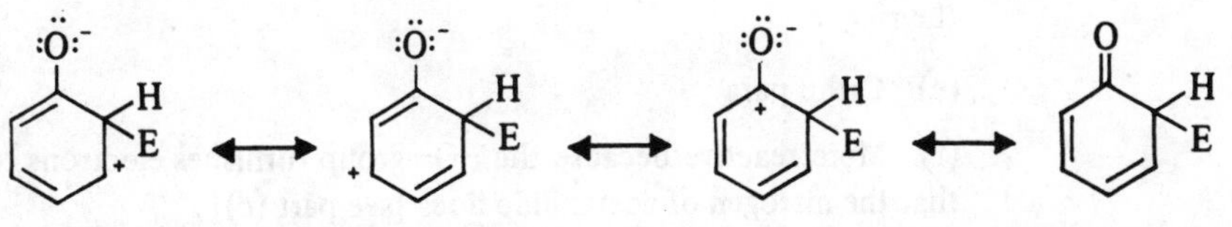

Highly stable

Para

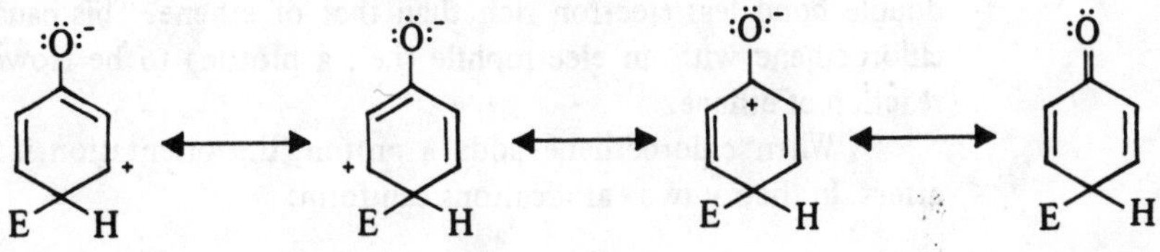

Highly stable

12.9

(a) 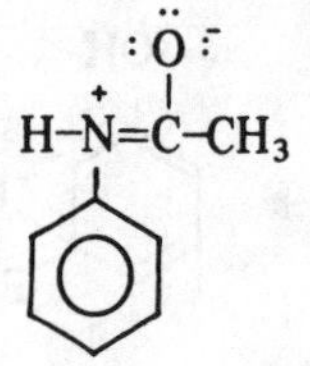 [structure: phenyl–$\overset{+}{N}$(H)=C($\ddot{O}$:$^-$)–CH_3]

(b) The–C(=O)–CH_3 group competes with the ring for the electron pair on N, therefore stabilization of the intermediate arenium ion is less effective than in aniline.

(c) 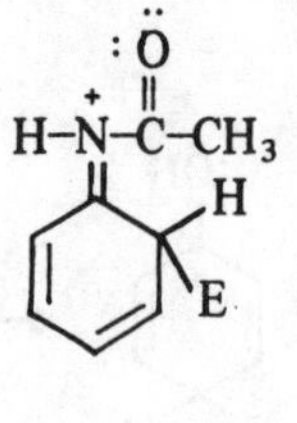 [structures A and B: H–$\overset{+}{N}$–C(=O)–CH_3 attached by a double bond to the arenium ring, with H and E on the ortho carbon (A) and on the para carbon (B)]

A

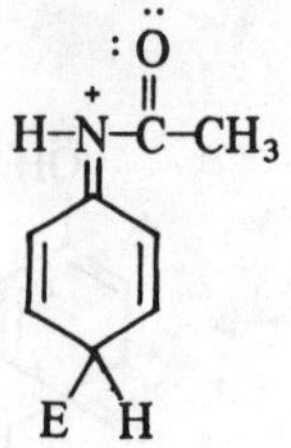

B

Yes, resonance accounts for an electron release from nitrogen to the ring, and as with aniline (Section 12.9D), extra and relatively stable structures (**A** and **B**) contribute to the arenium ions formed when attack takes place at an ortho or para carbon atom.

(d) Phenyl acetate should be less reactive than phenol because the $-COCH_3$ group competes with the ring for electrons on oxygen as shown in the following structure.

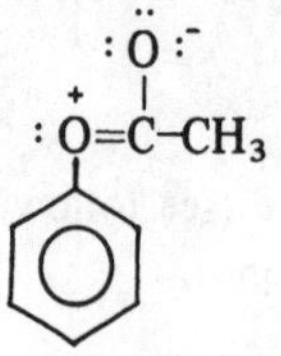

[structure: phenyl–$\overset{+}{O}$=C($\ddot{O}$:$^-$)–CH_3]

Notice that this structure also places a positive charge on the oxygen atom attached to the ring.

(e) Ortho-para.

(f) More reactive because the $-\ddot{O}-$ group furnishes electrons to the ring in the same way that the nitrogen of acetanilide does [see part (c)].

12.10 The electron-withdrawing inductive effect of the chlorine of chloroethene makes its double bond less electron rich than that of ethene. This causes the rate of reaction of chloroethene with an electrophile (i.e., a proton) to be slower than the corresponding reaction of ethene.

When chloroethene adds a proton, the orientation is governed by a resonance effect. In theory two carbocations can form:

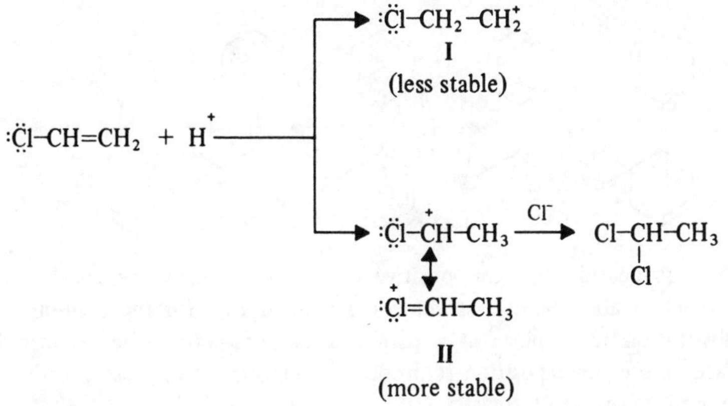

$:\ddot{Cl}-CH=CH_2 + \overset{+}{H} \longrightarrow :\ddot{Cl}-CH_2-\overset{+}{C}H_2$ **I** (less stable)

$:\ddot{Cl}-CH=CH_2 + \overset{+}{H} \longrightarrow :\ddot{Cl}-\overset{+}{C}H-CH_3 \longleftrightarrow \overset{+}{:Cl}=CH-CH_3$ **II** (more stable) $\xrightarrow{Cl^-} Cl-CH(Cl)-CH_3$

Carbocation **II** is more stable than **I** because of the resonance contribution of the extra structure just shown in which the chlorine atom donates an electron pair (see Section 12.9D).

12.11

Ortho: three resonance structures of the arenium ion (ring bearing CH_3–CH_2– with H and E at the ortho carbon); the third, with the positive charge on the carbon bearing the ethyl group, is (relatively stable)

Meta: three resonance structures of the arenium ion (ring bearing CH_3–CH_2– with H and E at the meta carbon)

Para: three resonance structures of the arenium ion (ring bearing CH_3–CH_2– with E and H at the para carbon); the second, with the positive charge on the carbon bearing the ethyl group, is (relatively stable)

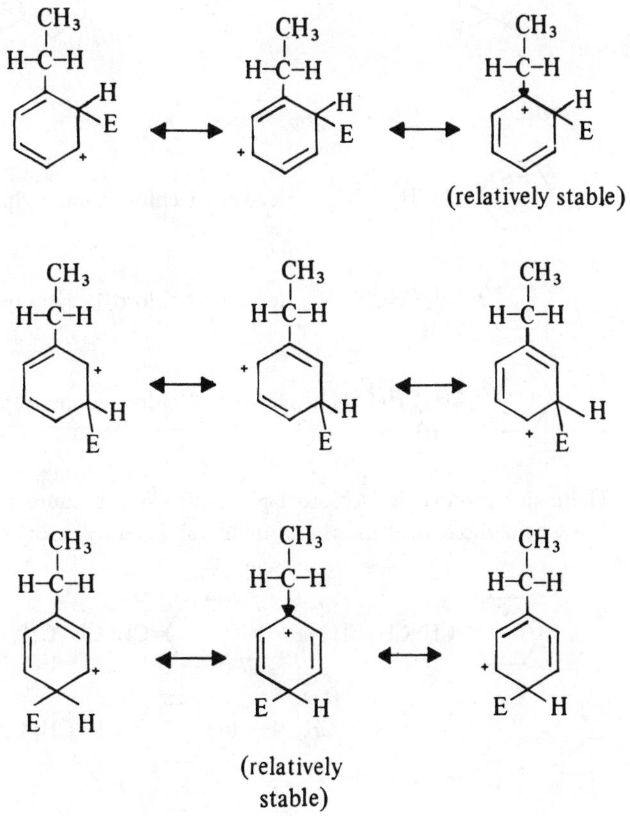

12.12

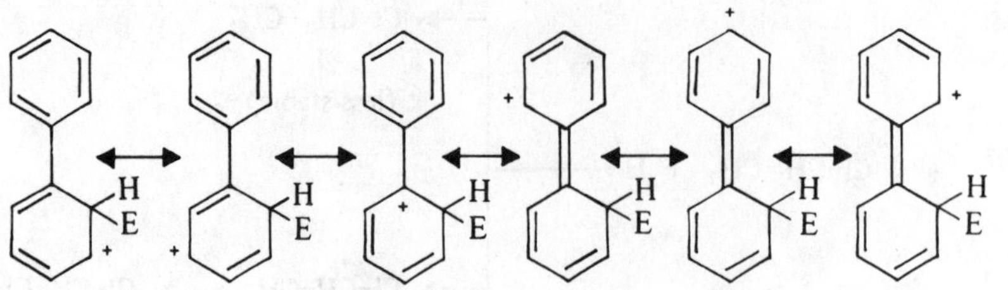

This carbocation has the positive charge delocalized over both rings and thus it is relatively stable. Similar structures can be drawn for the **arenium ions** formed when substitution takes place at a para position. However, when electrophilic attack takes place at the meta position it produces a carbocation whose positive charge cannot be delocalized over both rings:

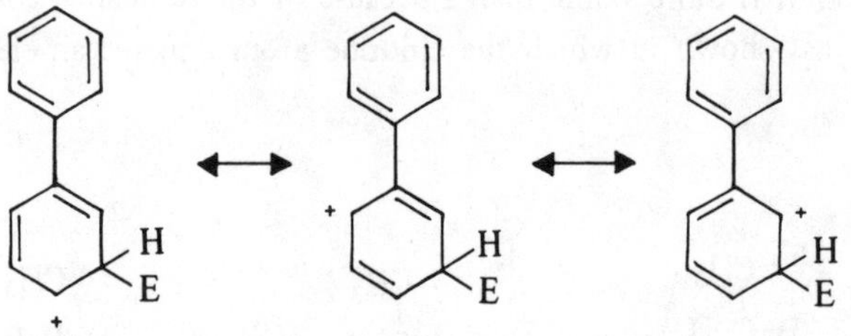

12.13

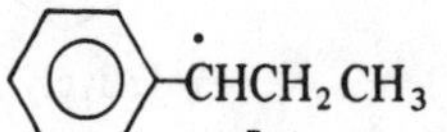

$C_6H_5\dot{C}HCH_2CH_3$ (I) leads to 1-chloro-1-phenylpropane

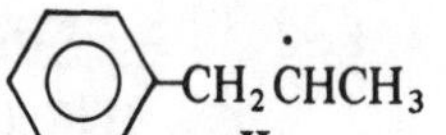

$C_6H_5CH_2\dot{C}HCH_3$ (II) leads to 2-chloro-1-phenylpropane

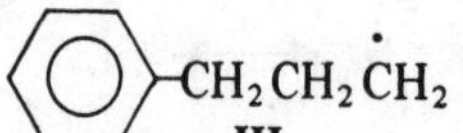

$C_6H_5CH_2CH_2\dot{C}H_2$ (III) leads to 1-chloro-3-phenylpropane

The major product is 1-chloro-1-phenylpropane because **I** is the most stable free radical. It is a benzylic radical and therefore is stabilized by resonance.

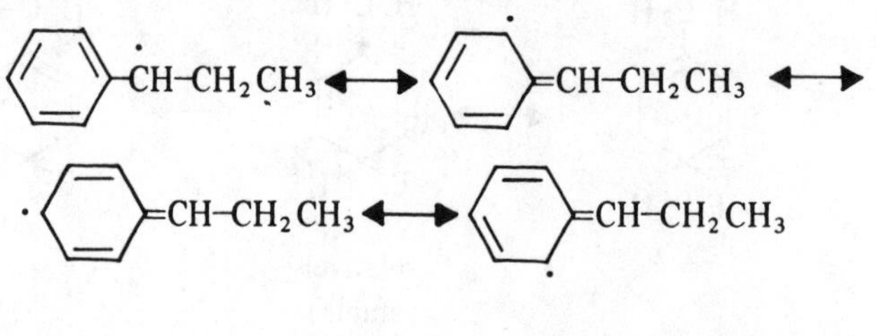

12.14

(a) 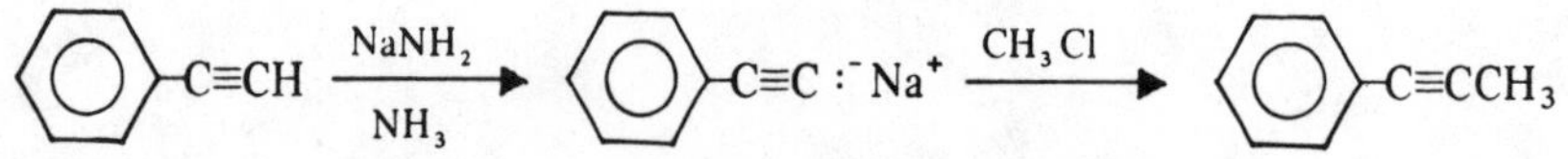 $C_6H_5C{\equiv}CH \xrightarrow[NH_3]{NaNH_2} C_6H_5C{\equiv}C{:}^-Na^+ \xrightarrow{CH_3Cl} C_6H_5C{\equiv}CCH_3$

(b) $C_6H_5C{\equiv}CH \xrightarrow[(2)\ CH_3CH_2Cl]{(1)\ NaNH_2/NH_3} C_6H_5C{\equiv}CCH_2CH_3$

(c) $C_6H_5C{\equiv}CCH_3 \xrightarrow[H_2]{Ni_2B}$ $C_6H_5C(H){=}C(H)CH_3$ (cis)

[from (a)]

(d) $C_6H_5C{\equiv}CCH_3 \xrightarrow{Li/C_2H_5NH_2}$ $C_6H_5C(H){=}C(CH_3)H$ (trans)

[from (a)]

12.15 (a) $C_6H_5\overset{+}{C}H_2 \longleftrightarrow$ (ring with + ortho, $=CH_2$) $\longleftrightarrow$ (ring with + para, $=CH_2$) $\longleftrightarrow$ (ring with + ortho, $=CH_2$)

The benzyl cation is stabilized by resonance.

12.16 (a) $C_6H_5CH(Cl)CH_2CH_3$ because the most stable carbocation intermediate is the benzylic carbocation, $C_6H_5\overset{+}{C}HCH_2CH_3$.

(b) $C_6H_5CH(OH)CH_2CH_3$ for the same reason as given in part (a).

12.17 Chlorinate the ring first and then introduce the double bond as shown here. If we were to introduce the side chain double bond first, chlorination of the ring would result in addition of chlorine to the side chain double bond.

$C_6H_5CH_2CH_2CH_3 \xrightarrow[FeCl_3]{Cl_2} p\text{-}ClC_6H_4CH_2CH_2CH_3 \xrightarrow[CCl_4]{NBS,\ h\nu} p\text{-}ClC_6H_4CHBrCH_2CH_3 \xrightarrow{(CH_3)_3COK} p\text{-}ClC_6H_4CH{=}CHCH_3$

12.18

(a) OH, NO_2 (ortho to OH), CF_3 (para to OH)

(b) CN, O_2N and SO_3H (both meta to CN)

(c) OCH_3, NO_2 (para), NO_2 (meta) + OCH_3, O_2N (ortho), NO_2 (meta)

12.19 The carbocation formed by the action of $AlCl_3$ on neopentyl chloride is primary. This carbocation rearranges to the more stable tertiary carbocation before it can react with the benzene ring:

$$CH_3C(CH_3)_2CH_2Cl + AlCl_3 \longrightarrow AlCl_4^- + CH_3C(CH_3)_2CH_2^+ \longrightarrow CH_3\overset{+}{C}(CH_3)CH_2CH_3 \xrightarrow{C_6H_6} C_6H_5C(CH_3)_2CH_2CH_3$$

12.20 $CH_3CH_2CH_2{-}OH + BF_3 \rightleftharpoons CH_3CH_2CH_2^+ + H\bar{O}BF_3$

The propyl cation can rearrange to an isopropyl cation:

$$CH_3CH_2\overset{+}{C}H_2 \xrightarrow[\text{shift}]{\text{hydride}} CH_3\overset{+}{C}HCH_3$$

Both cations can then attack the benzene ring.

$$C_6H_6 \xrightarrow{CH_3CH_2CH_2^+} [C_6H_6(H)CH_2CH_2CH_3]^+ \longrightarrow C_6H_5CH_2CH_2CH_3 + H^+$$

$$C_6H_6 \xrightarrow{CH_3\overset{+}{C}HCH_3} [C_6H_6(H)CH(CH_3)CH_3]^+ \longrightarrow C_6H_5CH(CH_3)CH_3 + H^+$$

12.21

(a) $$C_6H_6 + Cl{-}\overset{O}{\overset{\|}{C}}CH_2CH_2CH_2CH_2CH_3 \xrightarrow{AlCl_3} C_6H_5\overset{O}{\overset{\|}{C}}CH_2CH_2CH_2CH_2CH_3$$

$$\xrightarrow[\text{HCl, reflux}]{\text{Zn(Hg)}} C_6H_5CH_2CH_2CH_2CH_2CH_2CH_3$$

(b) $$C_6H_6 + Cl{-}\overset{O}{\overset{\|}{C}}{-}CH(CH_3)CH_3 \xrightarrow{AlCl_3} C_6H_5\overset{O}{\overset{\|}{C}}{-}CH(CH_3)CH_3 \xrightarrow[\text{HCl, reflux}]{\text{Zn(Hg)}} C_6H_5CH_2CH(CH_3)CH_3$$

(c) C_6H_6 + $Cl{-}C(=O){-}C_6H_5$ $\xrightarrow{AlCl_3}$ $C_6H_5{-}C(=O){-}C_6H_5$ $\xrightarrow[HCl,\ reflux]{Zn(Hg)}$ $C_6H_5{-}CH_2{-}C_6H_5$

(d) C_6H_6 + phthalic anhydride $\xrightarrow{AlCl_3}$ $C_6H_5{-}C(=O){-}C_6H_4{-}C(=O)OH$ $\xrightarrow[HCl,\ reflux]{Zn(Hg)}$

$C_6H_5{-}CH_2{-}C_6H_4{-}C(=O){-}OH$ $\xrightarrow[heat]{SOCl_2}$ $C_6H_5{-}CH_2{-}C_6H_4{-}C(=O){-}Cl$ $\xrightarrow[CS_2]{AlCl_3}$ anthrone (CH_2, H H; C=O)

12.22

(a) OCH_3-benzene + Br_2 $\xrightarrow{FeBr_3}$ OCH_3, Br (ortho) + OCH_3, Br (para)

o-Bromoanisole *p*-Bromoanisole

OCH_3-benzene + HNO_3 $\xrightarrow{H_2SO_4}$ OCH_3, NO_2 (ortho) + OCH_3, NO_2 (para)

o-Nitroanisole *p*-Nitroanisole

OCH_3-benzene + SO_3 $\xrightarrow{H_2SO_4}$ OCH_3, SO_3H (ortho) + OCH_3, SO_3H (para)

o-Methoxybenzene-sulfonic acid *p*-Methoxybenzene-sulfonic acid

Reactions are faster than the corresponding reactions of benzene.

(b) $C_6H_5CHF_2 + Br_2 \xrightarrow{FeBr_3}$ *m*-$BrC_6H_4CHF_2$

m-(Difluoromethyl)-bromobenzene

$C_6H_5CHF_2 + HNO_3 \xrightarrow{H_2SO_4}$ *m*-$O_2NC_6H_4CHF_2$

m-(Difluoromethyl)-nitrobenzene

$C_6H_5CHF_2 + SO_3 \xrightarrow{H_2SO_4}$ *m*-$HO_3SC_6H_4CHF_2$

m-(Difluoromethyl)-benzenesulfonic acid

Reactions are slower than corresponding reactions of benzene.

(c) $C_6H_5CH_2CH_3 + Br_2 \xrightarrow{FeBr_3}$ *o*-$BrC_6H_4CH_2CH_3$ + *p*-$BrC_6H_4CH_2CH_3$

o-Bromoethyl-benzene *p*-Bromoethyl-benzene

Nitration ⟶ *o*-ethylnitrobenzene and *p*-ethylnitrobenzene

Sulfonation ⟶ *o*-ethylbenzenesulfonic acid and *p*-ethylbenzenesulfonic acid

Reactions are faster than corresponding reactions of benzene.

(d) $C_6H_5NO_2 + Br_2 \xrightarrow{FeBr_3}$ *m*-$BrC_6H_4NO_2$

m-Bromonitrobenzene

Nitration ⟶ *m*-dinitrobenzene

Sulfonation ⟶ *m*-nitrobenzenesulfonic acid

Reactions are slower than corresponding reactions of benzene.

(e) 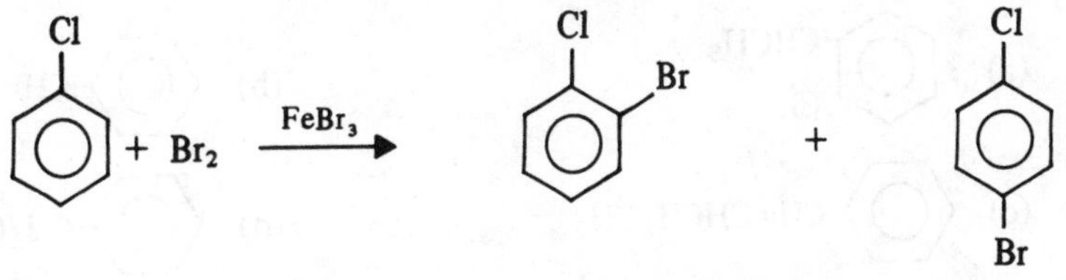 Chlorobenzene + Br_2 $\xrightarrow{FeBr_3}$ o-Bromochlorobenzene + p-Bromochlorobenzene

o-Bromochlorobenzene *p*-Bromochlorobenzene

Nitration ⟶ *o*-chloronitrobenzene + *p*-chloronitrobenzene

Sulfonation ⟶ *o*-chlorobenzenesulfonic acid + *p*-chlorobenzenesulfonic acid

Reactions are slower than corresponding reactions of benzene.

(f) 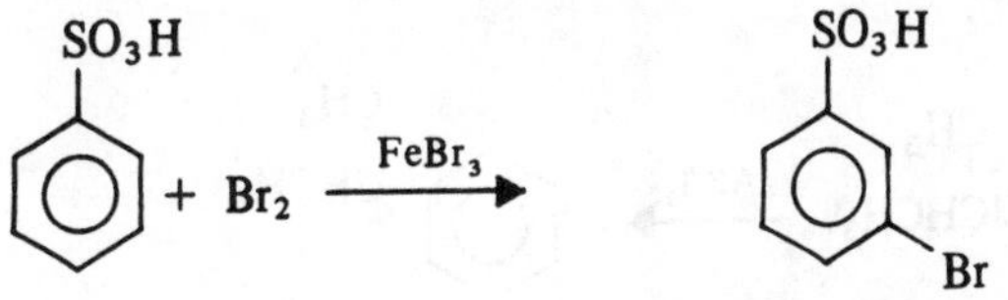 Benzenesulfonic acid (SO_3H) + Br_2 $\xrightarrow{FeBr_3}$ *m*-Bromobenzenesulfonic acid

m-Bromobenzenesulfonic acid

Nitration ⟶ *m*-nitrobenzenesulfonic acid

Sulfonation ⟶ *m*-benzenedisulfonic acid

Reactions are slower than corresponding reactions of benzene.

12.23

(a) 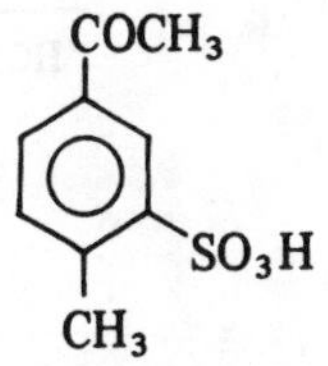 $COCH_3$, SO_3H, CH_3 (ring substituents)

(b) 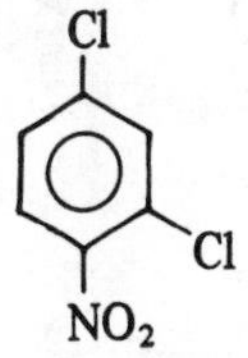 Cl, Cl, NO_2 (ring substituents)

(c) 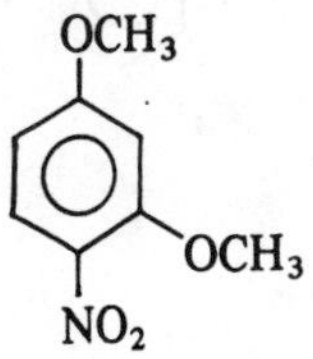 OCH_3, OCH_3, NO_2 (ring substituents)

(d) 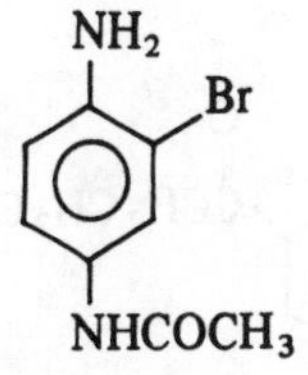 NH_2, Br, $NHCOCH_3$ (ring substituents)

(e) 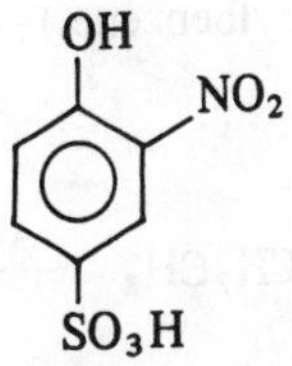 OH, NO_2, SO_3H (ring substituents)

(f) 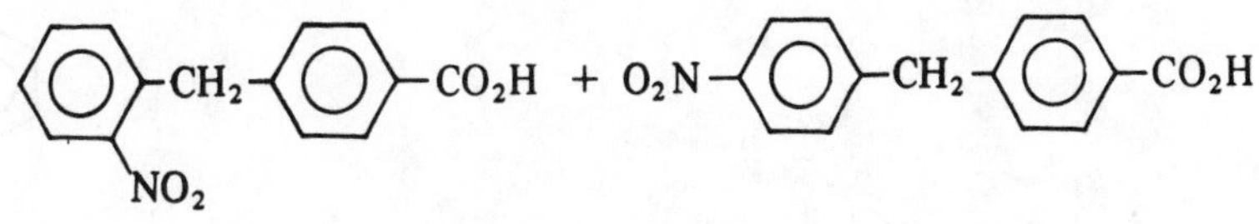 $C_6H_4(NO_2)$–CH_2–C_6H_4–CO_2H (ortho NO_2) + O_2N–C_6H_4–CH_2–C_6H_4–CO_2H

(g) 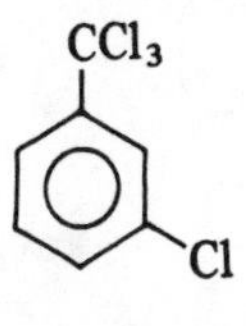 CCl_3, Cl (ring substituents, meta)

12.24

(a) 

(b) C_6H_5–$CH{=}CHCH_3$

(c) C_6H_5–$CH{=}CHCH_2CH_3$

(d) C_6H_5–$CH_2CHCH_2CH_3$ (Br on CH)

(e) C_6H_5–$CHCH_2CH_2CH_3$ (OH on CH)

(f) C_6H_5–$CH_2CH_2CH_2CH_3$

(g) C_6H_5–$\overset{O}{\overset{\|}{C}}$–OH

12.25

(a) $C_6H_6 + ClCHCH_3$ (CH_3 on CH) $\xrightarrow{AlCl_3}$ C_6H_5–$CHCH_3$ (CH_3 on CH)

(b) $C_6H_6 + ClCCH_3$ (two CH_3 on C) $\xrightarrow{AlCl_3}$ C_6H_5–CCH_3 (two CH_3 on C)

(c) C_6H_6 + Cl–$\overset{O}{\overset{\|}{C}}CH_2CH_3$ $\xrightarrow{AlCl_3}$ C_6H_5–$\overset{O}{\overset{\|}{C}}CH_2CH_3$ $\xrightarrow[\text{HCl, reflux}]{\text{Zn(Hg)}}$ C_6H_5–$CH_2CH_2CH_3$

(*Note:* The use of Cl–$CH_2CH_2CH_3$ in a Friedel-Crafts synthesis gives mainly the rearranged product, isopropylbenzene.)

(d) C_6H_6 + Cl–$\overset{O}{\overset{\|}{C}}CH_2CH_2CH_3$ $\xrightarrow{AlCl_3}$ C_6H_5–$\overset{O}{\overset{\|}{C}}CH_2CH_2CH_3$ $\xrightarrow[\text{HCl, reflux}]{\text{Zn(Hg)}}$ C_6H_5–$CH_2CH_2CH_2CH_3$

(e) C_6H_5–$C(CH_3)_3$ + Cl_2 $\xrightarrow[\text{(dark)}]{FeCl_3}$ p-Cl–C_6H_4–$C(CH_3)_3$

[from (b)]

(f) + Cl $\xrightarrow{AlCl_3}$ $\xrightarrow[h\nu]{Cl_2}$ Cl

$NaOC_2H_5/C_2H_5OH$

(g) $\xrightarrow[\text{(syn addition)}]{THF{:}BH_3}$ $\xrightarrow[OH^-]{H_2O_2}$ H HO H

H

[from (f)]

(h) + HNO_3 $\xrightarrow{H_2SO_4}$ NO_2 $\xrightarrow[H_2SO_4]{HNO_3}$ NO_2 NO_2

(i) NO_2 + Br_2 $\xrightarrow{FeBr_3}$ NO_2 Br

(j) + Br_2 $\xrightarrow{FeBr_3}$ Br $\xrightarrow[H_2SO_4]{HNO_3}$ Br NO_2 + ortho

(k) + Cl_2 $\xrightarrow{FeCl_3}$ Cl $\xrightarrow[H_2SO_4]{SO_3}$ Cl SO_3H (+ Cl SO_3H)

(separate)

H_2O/H^+, heat

(l) Chlorobenzene $\xrightarrow[H_2SO_4]{SO_3}$ (o-chlorobenzenesulfonic acid, SO_3H) (separate) + p-chlorobenzenesulfonic acid (SO_3H) $\xrightarrow[H_2SO_4]{HNO_3}$ 4-chloro-3-nitrobenzenesulfonic acid (NO_2, SO_3H) $\xrightarrow[heat]{H_2O/H^+}$ o-chloronitrobenzene (NO_2)

(ortho isomer) $\xrightarrow{H_2O/H^+, \text{heat}}$ chlorobenzene

(m) Nitrobenzene [from (h)] $\xrightarrow[H_2SO_4]{SO_3}$ m-nitrobenzenesulfonic acid (NO_2, SO_3H)

(n) $CH_3CH_2CH_2\overset{O}{\overset{\|}{C}}-Cl$ + C_6H_6 $\xrightarrow{AlCl_3}$ $CH_3CH_2CH_2\overset{O}{\overset{\|}{C}}-C_6H_5$ $\xrightarrow[HCl]{Zn(Hg)}$

$CH_3CH_2CH_2CH_2-C_6H_5$ $\xrightarrow[(2)\ H_2O_2]{(1)\ O_3,\ CH_3CO_2H}$ $CH_3CH_2CH_2CH_2CO_2H$

(o) $CH_3CH_2CH_2CH_2-C_6H_5$ [from (n)] $\xrightarrow{NBS}$ $CH_3CH_2CH_2\underset{Br}{\underset{|}{C}}H-C_6H_5$ $\xrightarrow[C_2H_5OH]{NaOC_2H_5}$

$CH_3CH_2CH{=}CH-C_6H_5$ $\xrightarrow{D_2/Ni}$ $CH_3CH_2CHDCHD-C_6H_5$ $\xrightarrow[(2)\ H_2O_2]{(1)\ O_3,\ CH_3CO_2H}$

$CH_3CH_2CHDCHDCO_2H$

(p) $CH_3CH_2CH{=}CH-C_6H_5$ [from (o)] $\xrightarrow{THF:BH_3}$ $\xrightarrow[160°C]{heat}$ $\rangle BCH_2CH_2CH_2CH_2-C_6H_5$

$\xrightarrow{CH_3CO_2D}$ $DCH_2CH_2CH_2CH_2-C_6H_5$ $\xrightarrow[(2)\ H_2O_2]{(1)\ O_3,\ CH_3CO_2H}$

$DCH_2CH_2CH_2CH_2CO_2H$

12.26

(a) $C_6H_5-CH{=}CH_2$ $\xrightarrow{Cl_2}$ $C_6H_5-CHClCH_2Cl$

(b) $C_6H_5-CH{=}CH_2$ $\xrightarrow{H_2/Ni}$ $C_6H_5-CH_2CH_3$

(c) $C_6H_5CH{=}CH_2 \xrightarrow[25°C]{KMnO_4} C_6H_5CHOHCH_2OH$

(d) $C_6H_5CH{=}CH_2 \xrightarrow[heat]{KMnO_4} C_6H_5CO_2H$

(e) $C_6H_5CH{=}CH_2 \xrightarrow[H_2SO_4]{H_2O} C_6H_5CHOHCH_3$

(f) $C_6H_5CH{=}CH_2 \xrightarrow{HBr} C_6H_5CHBrCH_3$

(g) $C_6H_5CH{=}CH_2 \xrightarrow[(2)\ H_2O_2/OH^-]{(1)\ THF:BH_3} C_6H_5CH_2CH_2OH$

(h) $C_6H_5CH{=}CH_2 \xrightarrow[(2)\ CH_3CO_2D]{(1)\ THF:BH_3} C_6H_5CH_2CH_2D$

(i) $C_6H_5CH{=}CH_2 \xrightarrow[peroxides]{HBr} C_6H_5CH_2CH_2Br$

(j) $C_6H_5CH_2CH_2Br + NaI \xrightarrow[H_2O]{acetone} C_6H_5CH_2CH_2I$

[from (i)]

(k) $C_6H_5CH_2CH_2Br + CN^- \longrightarrow C_6H_5CH_2CH_2CN$

[from (i)]

(l) $C_6H_5CH{=}CH_2 \xrightarrow[Ni]{D_2} C_6H_5CHDCH_2D$

(m) butadiene + styrene $\xrightarrow{heat}$ 4-phenylcyclohexene $\xrightarrow[Ni]{H_2}$ phenylcyclohexane

(n) $C_6H_5CH_2CH_2OH \xrightarrow{Na} C_6H_5CH_2CH_2ONa \xrightarrow{CH_3I} C_6H_5CH_2CH_2OCH_3$

[from (g)]

12.27

(a) Toluene $\xrightarrow[\text{heat}]{KMnO_4}$ $\xrightarrow{H_3O^+}$ $C_6H_5CO_2H$ $\xrightarrow[FeCl_3]{Cl_2}$ m-$ClC_6H_4CO_2H$

(b) Toluene + $Cl\text{-}\overset{O}{\overset{\|}{C}}CH_3$ $\xrightarrow{AlCl_3}$ p-$CH_3C_6H_4COCH_3$ + ortho

(c) Toluene $\xrightarrow[H_2SO_4]{HNO_3}$ p-$CH_3C_6H_4NO_2$ (+ ortho) $\xrightarrow[FeBr_3]{Br_2}$ 2-Br-4-NO_2-$C_6H_3CH_3$

(d) Toluene $\xrightarrow[FeBr_3]{Br_2}$ p-$CH_3C_6H_4Br$ (+ ortho) $\xrightarrow[\text{heat}]{KMnO_4}$ $\xrightarrow{H_3O^+}$ p-$BrC_6H_4CO_2H$

(e) Toluene $\xrightarrow[\text{light}]{Cl_2\text{ (excess)}}$ $C_6H_5CCl_3$ $\xrightarrow[FeCl_3]{Cl_2}$ m-$ClC_6H_4CCl_3$

(f) Toluene + $Cl\text{-}\overset{CH_3}{\overset{|}{C}}HCH_3$ $\xrightarrow{AlCl_3}$ p-$CH_3C_6H_4CH(CH_3)_2$ (CH_3CHCH_3) + ortho

(g) Toluene + Cl-cyclohexane $\xrightarrow{AlCl_3}$ CH_3-C_6H_4-C_6H_{11} (para) + ortho

(h) Toluene (CH_3) $\xrightarrow[H_2SO_4]{HNO_3\text{ (excess)}}$ 2,4,6-trinitrotoluene (CH_3; O_2N, NO_2, NO_2)

(i) Toluene (CH_3) $\xrightarrow[H_2SO_4]{SO_3}$ (o-$CH_3C_6H_4SO_3H$ (separate)) + p-$CH_3C_6H_4SO_3H$ $\xrightarrow[H_2SO_4]{HNO_3}$ 4-methyl-3-nitrobenzenesulfonic acid (CH_3, NO_2, SO_3H) $\xrightarrow[\text{heat}]{H_2O/H^+}$ o-nitrotoluene (CH_3, NO_2) $\xrightarrow[FeCl_3]{Cl_2}$ 4-chloro-2-nitrotoluene (CH_3, NO_2, Cl) $\xrightarrow[(2)\ H_3O^+]{(1)\ KMnO_4,\ OH^-,\ \text{heat}}$ 4-chloro-2-nitrobenzoic acid (COOH, NO_2, Cl)

The separated ortho isomer $\xrightarrow{H_2O,\ H^+,\ \text{heat}}$ toluene

(j) Toluene (CH_3) + $Cl\text{–}\overset{O}{\overset{\|}{C}}CH_2CH_2CH_3$ $\xrightarrow{AlCl_3}$ p-$CH_3C_6H_4\text{–}C(=O)\text{–}CH_2CH_2CH_3$ + ortho $\xrightarrow[HCl]{Zn(Hg)}$ p-$CH_3C_6H_4CH_2CH_2CH_2CH_3$

12.28

(a) Aniline (NH_2) $\xrightarrow{Cl\text{–}\overset{O}{\overset{\|}{C}}CH_3}$ $C_6H_5NHCOCH_3$ $\xrightarrow[FeBr_3]{Br_2}$ p-Br$C_6H_4NHCOCH_3$ + ortho $\xrightarrow{H^+,\ H_2O}$ p-Br$C_6H_4NH_2$

(b) $C_6H_5NHCOCH_3$ [from (a)] $\xrightarrow[H_2SO_4]{SO_3}$ o-$HO_3SC_6H_4NHCOCH_3$ (minor product) + p-$HO_3SC_6H_4NHCOCH_3$ (major product) $\xrightarrow[FeBr_3]{Br_2}$ 2-Br-4-SO_3H-$C_6H_3NHCOCH_3$ $\xrightarrow[heat]{H_2O/H^+}$ o-$BrC_6H_4NH_2$

(c) o-$BrC_6H_4NHCOCH_3$ [from (b)] $\xrightarrow[H_2SO_4]{HNO_3}$ 2-Br-4-NO_2-$C_6H_3NHCOCH_3$ $\xrightarrow[H_2O]{H^+}$ 2-Br-4-NO_2-$C_6H_3NH_2$

(d) p-$HO_3SC_6H_4NHCOCH_3$ [from (b)] $\xrightarrow[H_2SO_4]{HNO_3}$ 2-NO_2-4-SO_3H-$C_6H_3NHCOCH_3$ $\xrightarrow[\text{[cf. Problem 12.3 (b)]}]{Br_2, FeBr_3}$ 2-NO_2-4-Br-$C_6H_3NHCOCH_3$ $\xrightarrow[H_2O, heat]{H_3O^+,}$ 2-NO_2-4-Br-$C_6H_3NH_2$

(e) $C_6H_5NH_2$ $\xrightarrow[H_2O]{Br_2}$ 2,4,6-$Br_3C_6H_2NH_2$

12.29

(a) C_6H_5–C(=O)–$\ddot{N}$H–C_6H_5 (ring A = benzoyl ring, ring B = N-phenyl ring)

Ring **B** undergoes electrophilic substitution more readily than ring **A**

(b) Resonance structures such as the following stablize the intermediate carbocation:

C_6H_5–C(=O)–$\overset{+}{N}$(H)=C_6H_4(H)(E)

12.30

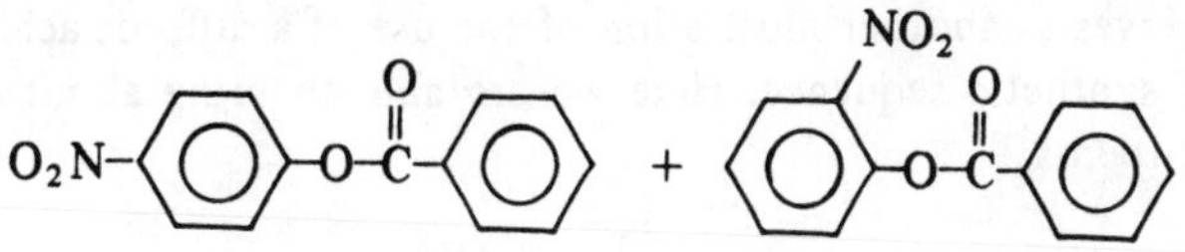

NO_2

O_2N–⟨C₆H₄⟩–O–C(=O)–⟨C₆H₅⟩ + (o-NO_2)⟨C₆H₄⟩–O–C(=O)–⟨C₆H₅⟩

See solution to Problem 12.29

12.31

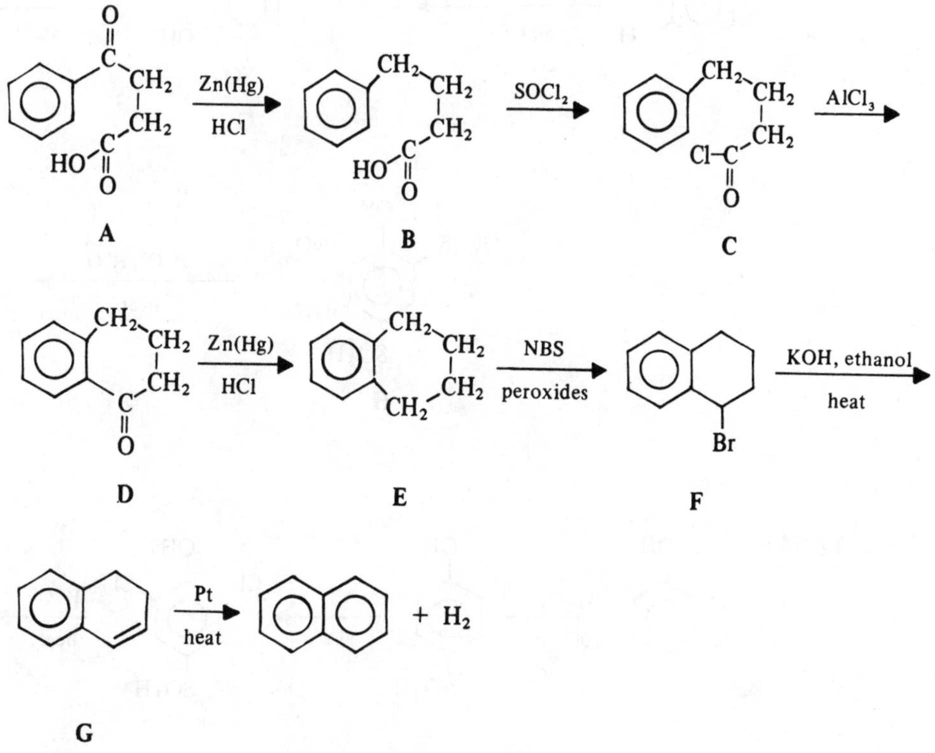

A $\xrightarrow[HCl]{Zn(Hg)}$ B $\xrightarrow{SOCl_2}$ C $\xrightarrow{AlCl_3}$

D $\xrightarrow[HCl]{Zn(Hg)}$ E $\xrightarrow[peroxides]{NBS}$ F $\xrightarrow[heat]{KOH, ethanol}$

G $\xrightarrow[heat]{Pt}$ + H_2

A B C

D E F

Br

G

12.32

$\underset{-H^+}{\overset{+H^+}{\rightleftharpoons}}$

CH_2 C=Ö: R CH_2 C^+ R OH

CH_2 C^+ R OH → CH_2 + C H R OH $\xrightarrow{-H^+}$ CH_2 C R OH

+H⁺ | −H⁺

$\xleftarrow{-H^+}$ R H H C + R $\xleftarrow{-H_2O}$ CH_2 R OH_2 +

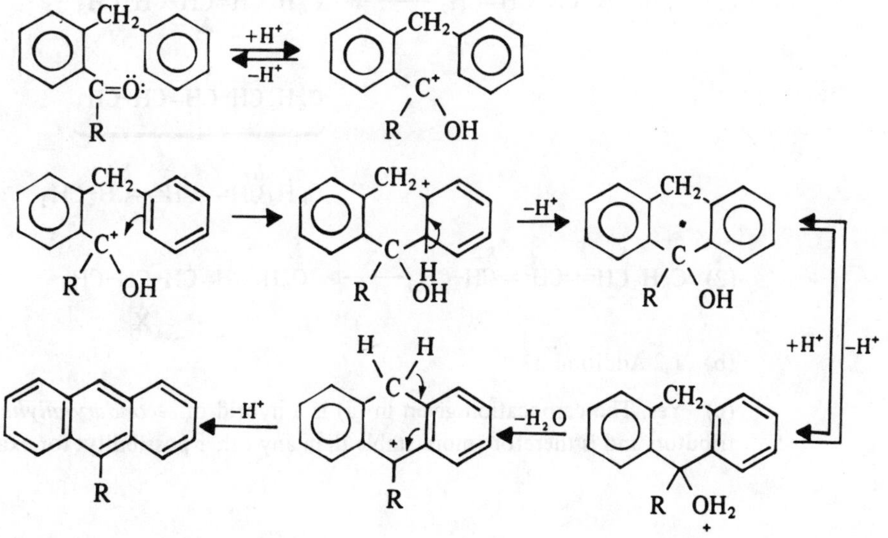

12.33 This problem serves as another illustration of the use of a sulfonic acid group as a blocking group in a synthetic sequence. Here we are able to bring about nitration between two meta substituents.

Resorcinol (1,3-dihydroxybenzene) $\xrightarrow[\text{60-65°C}]{\text{conc. } H_2SO_4}$ **G** (HO$_3$S, OH, OH, SO$_3$H substituted benzene) $\xrightarrow[\text{conc. } H_2SO_4]{\text{conc. } HNO_3}$

H (HO$_3$S, OH, NO$_2$, OH, SO$_3$H substituted benzene) $\xrightarrow[\text{heat}]{H_3O^+,\ H_2O}$ **I** (2-nitroresorcinol: OH, NO$_2$, OH)

12.34 Phenol $\xrightarrow[\text{100°C}]{H_2SO_4}$ p-HO–C$_6$H$_4$–SO$_3$H $\xrightarrow[\text{Fe}]{Cl_2}$ 3,5-dichloro-4-hydroxybenzenesulfonic acid (OH, Cl, Cl, SO$_3$H) $\xrightarrow[\text{heat}]{H_3O^+}$ 2,6-dichlorophenol (OH, Cl, Cl)

12.35 (a)

(1) $C_6H_5CH{=}CH{-}CH{=}CH_2 \xrightarrow{H^+} C_6H_5CH{=}CH{-}\overset{+}{C}H{-}CH_3$

$\updownarrow$

$C_6H_5\overset{+}{C}H{-}CH{=}CH{-}CH_3$

$C_6H_5\overset{\delta+}{C}H{\overset{\cdot\cdot\cdot}{-}}CH{\overset{\cdot\cdot\cdot}{-}}\overset{\delta+}{C}H{-}CH_3$

(2) $C_6H_5\overset{\delta+}{C}H{\overset{\cdot\cdot\cdot}{-}}CH{\overset{\cdot\cdot\cdot}{-}}\overset{\delta+}{C}H{-}CH_3 \xrightarrow{X^-} C_6H_5CH{=}CH{-}\underset{X}{\underset{|}{C}H}{-}CH_3$

(b) 1,2 Addition

(c) Yes. The carbocation given in (a) is a hybrid of *secondary allylic and benzylic* contributors and is therefore more stable than any other possibility; for example,

$$C_6H_5CH{=}CH{-}CH{=}CH_2 \xrightarrow{H^+} C_6H_5CH_2{-}\overset{+}{C}H{-}CH{=}CH_2 \longleftrightarrow C_6H_5CH_2{-}CH{=}CH{-}\overset{+}{C}H_2$$

A hybrid of allylic contributors only

(d) Since the reaction produces only *the more stable isomer*—that is, the one in which the double bond is conjugated with the benzene ring—the reaction is likely to be under equilibrium control:

$$C_6H_5\overset{\delta+}{C}H{\cdots}CH{\cdots}\overset{\delta+}{C}H{-}CH_3 + Cl^- \longrightarrow C_6H_5{-}CH{=}CH{-}CH(Cl){-}CH_3$$

More stable isomer — Actual product

$$\longrightarrow C_6H_5CH(Cl){-}CH{=}CH{-}CH_3$$

Less stable isomer — Not formed

12.36

(a) 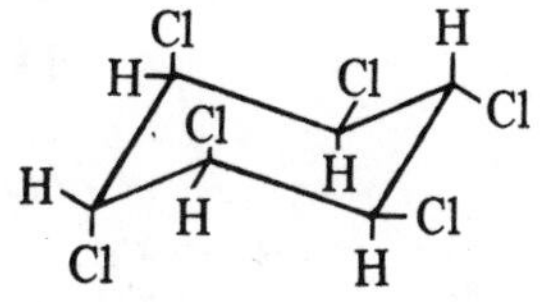 [chair structure: Cl, H, Cl, H, Cl, H, Cl, H, Cl, H, Cl, H substituents] (b) No (c) Lindane is a meso compound.

(d) 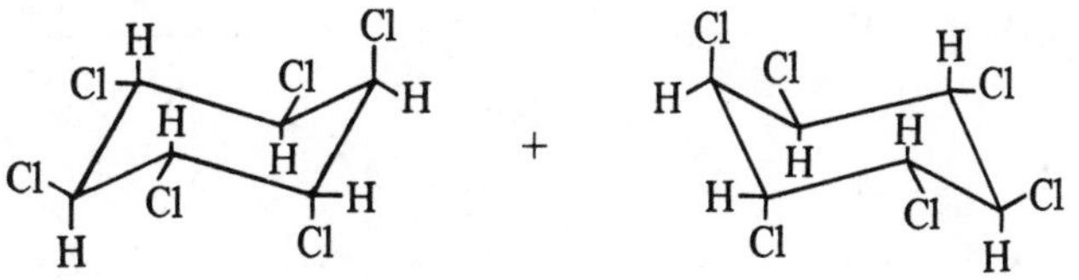 [chair structure] + [chair structure]

(see also Problem 17 of First Review Problem Set)

12.37 If we consider resonance structures for the ring that undergoes electrophilic attack, two structures are possible for the arenium ion that forms when attack takes place at the 1-position,

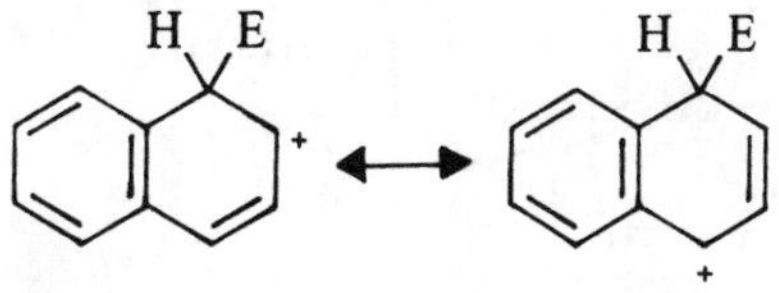

[two arenium ion resonance structures with H and E at C-1, connected by ↔]

whereas only one is possible when attack takes place at the 2 position,

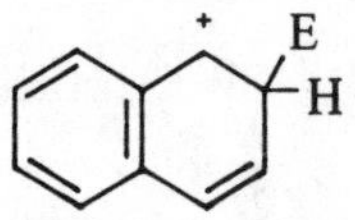

Attack at the 1 position, therefore, takes place faster.

12.38 (a) $\xrightarrow{H^+}$ $CH_2\overset{+}{O}H_2$ $\xrightarrow{-H_2O}$

$\overset{+}{C}H_2$ $\xrightarrow{\text{rearrangement}}$ $\xrightarrow{-H^+}$

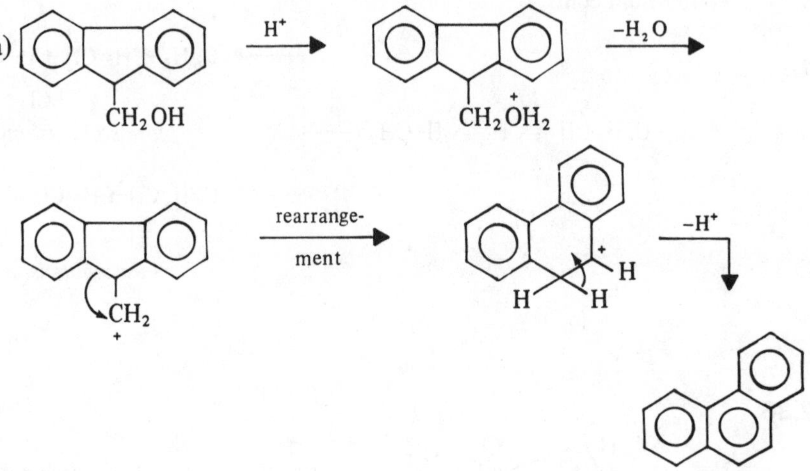

(b) $CH_3{-}C(C_6H_5){=}CH_2 \xrightarrow{H^+} CH_3{-}\overset{+}{C}(C_6H_5){-}CH_3 \xrightarrow{CH_2=C(C_6H_5)-CH_3}$

$\longrightarrow$ $\xrightarrow{-H^+}$

12.39 Benzene + $CH_3CH(CH_3)\text{—}(CH_2C(CH_3))_2\text{—}CH{=}CHCH_3$ and $CH_3CH(CH_3)\text{—}(CH_2CH(CH_3))_2\text{—}CH_2CH{=}CH_2$ $\xrightarrow[35\text{-}45°]{AlCl_3}$

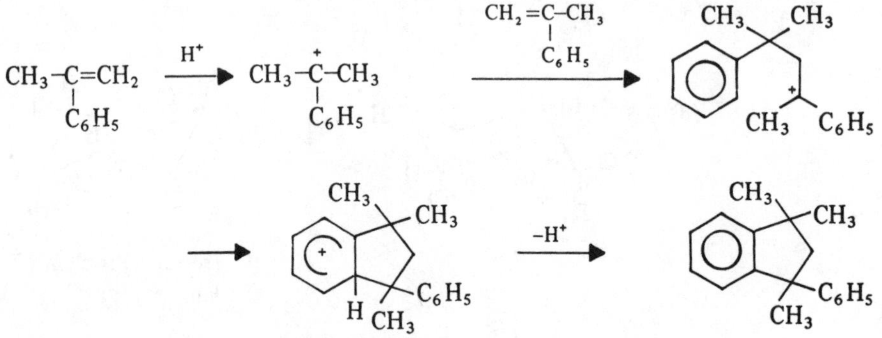

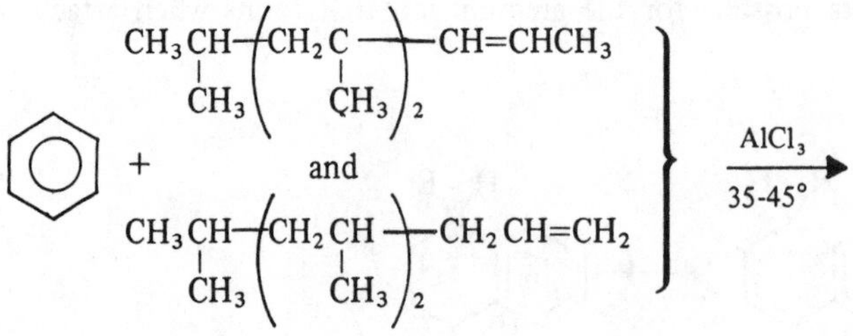

$CH_3CH(CH_3)\text{—}(CH_2CH(CH_3))_2\text{—}CH_2CH(CH_3)\text{—}C_6H_5$

and

$CH_3CH(CH_3)\text{—}(CH_2CH(CH_3))_2\text{—}CH(CH_2CH_3)\text{—}C_6H_5$

$\xrightarrow[\text{heat}]{H_2SO_4}$

$CH_3CH(CH_3)\text{—}(CH_2CH(CH_3))_2\text{—}CH_2CH(CH_3)\text{—}C_6H_4\text{—}SO_3H$

and

$CH_3CH(CH_3)\text{—}(CH_2CH(CH_3))_2\text{—}CH(CH_2CH_3)\text{—}C_6H_4\text{—}SO_3H$

$\xrightarrow{NaOH}$

$CH_3CH(CH_3)\text{—}(CH_2CH(CH_3))_2\text{—}CH_2CH(CH_3)\text{—}C_6H_4\text{—}SO_3Na$

and

$CH_3CH(CH_3)\text{—}(CH_2CH(CH_3))_2\text{—}CH(CH_2CH_3)\text{—}C_6H_4\text{—}SO_3Na$

12.40 (a) Large ortho substituents prevent the two rings from becoming coplanar and prevent rotation about the single bond that connects them. If the correct substitution patterns are present, the molecule as a whole will be chiral. Thus enantiomeric forms are possible even though the molecules do not have a stereocenter. The compound with 2-NO_2, 6-CO_2H, 2′-NO_2, 6-CO_2H is an example,

[Structures: two biphenyls, each ring bearing NO_2 and CO_2H ortho substituents (NO_2, CO_2H, NO_2, HO_2C / NO_2, CO_2H, O_2N, CO_2H)] and [mirror image]

These molecules are nonsuperposable mirror reflections and, thus, are enantiomers

(b) Yes

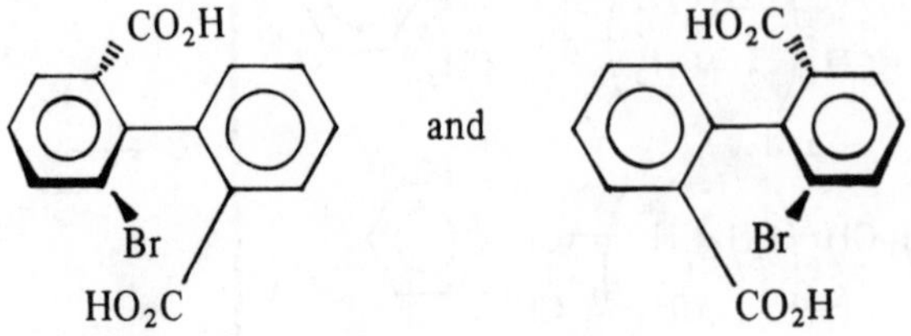

CO_2H Br HO_2C and HO_2C Br CO_2H

(c) This molecule has a plane of symmetry.

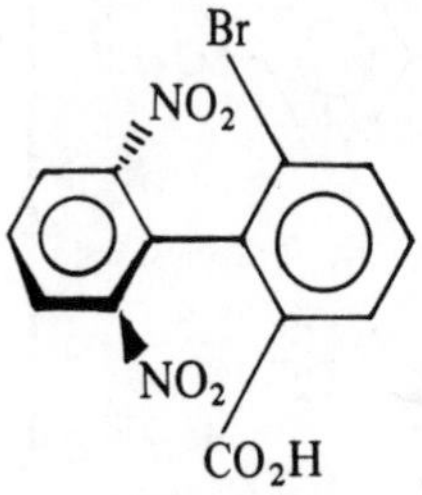

Br NO_2 NO_2 CO_2H

The plane of the page is a plane of symmetry.

12.41

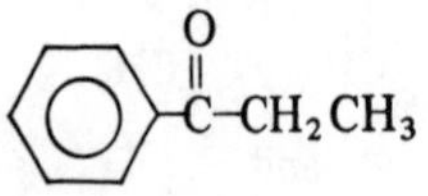

$C_6H_5C(=O)CH_2CH_3$ **A**

$C_6H_5CCl_2CH_2CH_3$ **B**

$C_6H_5C{\equiv}CCH_3$ **C**

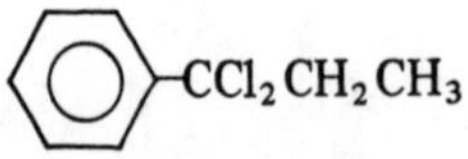

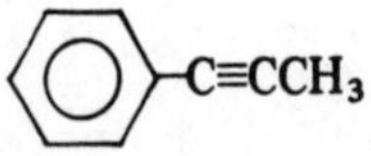

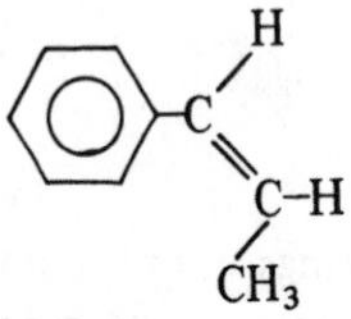

D 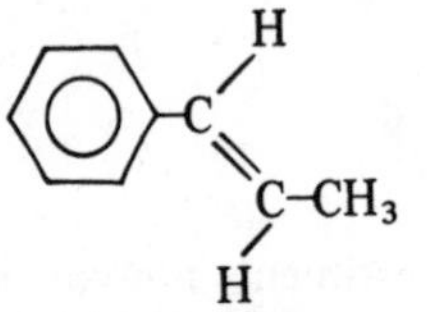 **E** 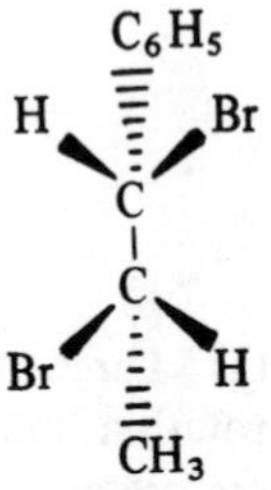 **F** 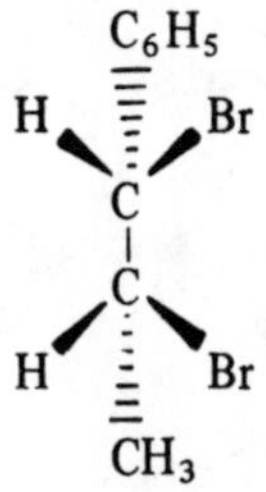 **G**

12.42

(a) [structure: $CH_3\overset{O}{\overset{\|}{C}}$, H, + arenium ion of azulene] (b) This arenium ion is especially stable because its seven-membered ring is an aromatic cation. (c) [structure: 1-acetylazulene, $CH_3-C=O$]

SECTION REFERENCES FOR ADDITIONAL PROBLEMS

12.22	12.3-12.5, 12.8	**12.31**	12.10, 12.11, 12.12C
12.23	12.8	**12.32**	12.7
12.24	12.10, 12.11	**12.33**	12.8, 12.12
12.25	12.3-12.7	**12.34**	12.12
12.26	12.11	**12.35**	12.11
12.27	12.3-12.7, 12.10	**12.36**	4.10, 7.6
12.28	12.9D	**12.37**	12.9
12.29	12.9D	**12.38**	6.13, 12.6
12.30	12.9D	**12.39**	12.5, 12.6
		12.40	4.16
		12.41	12.7, 12.10, 12.11
		12.42	12.7

SELF-TEST

12.1 Write the structural formula of the missing reactants or *major* organic products. Give more than one product *only* if they are produced in approximately equal amounts. If more than one step is needed label them (1) step 1, (2) step 2, and so on.

(a) [structure: $NH\overset{O}{\overset{\|}{C}}CH_3$ on benzene ring, para CH_3] + $HONO_2$ $\xrightarrow{H_2SO_4}$ [ ]

(b) Benzene → [ ] → $C_6H_5CCH_3$ (with C=O)

(c) C_6H_5–NH–C(=O)–C_6H_5 + SO_3 $\xrightarrow{H_2SO_4}$ [ ]

(d) C_6H_5–CH_2CH_3 + Br_2 $\xrightarrow{h\nu}$ [ ]

(e) Toluene (CH_3) → [ ] → CO_2H, NO_2 (meta-nitrobenzoic acid)

(f) Toluene (CH_3) → [ ] → O_2N, CH_3, NO_2 (major product)

(g) Benzonitrile (CN) + Br_2 $\xrightarrow{FeBr_3}$ [(C_7H_4BrN)]

(h) Draw the structural formulas of the three principal resonance structures of the intermediate (arenium ion) in reaction (g).

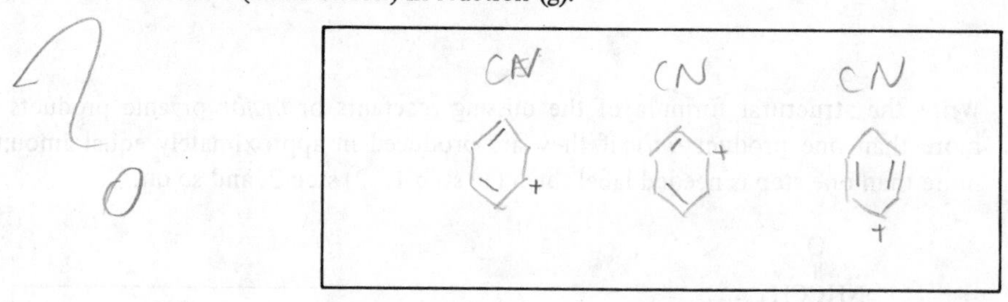

(i) $C_6H_5OCH_3$ + SO_3 $\xrightarrow{H_2SO_4}$

(j) $C_6H_5CH_3$ + $CH_3\overset{O}{\overset{\|}{C}}{-}Cl$ $\xrightarrow{AlCl_3}$

(k) O_2N–C_6H_4–CH_2–C_6H_5 $\xrightarrow[H_2SO_4]{HNO_3\text{ (one molar equivalent)}}$

(l) C_6H_5–CH_2–C_6H_5 $\xrightarrow{NBS}$

12.2 Write the structural formula for the organic ion that serves as the intermediate in the ring bromination of toluene.

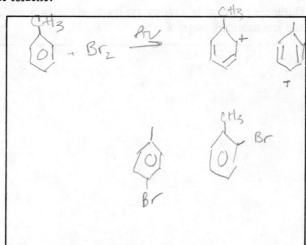

12.3 Write two additional resonance structures for the following ion:

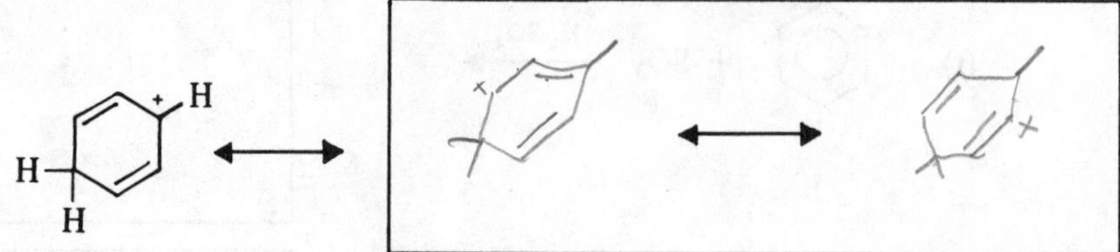

12.4 Supply the structural formulas of the missing compounds.

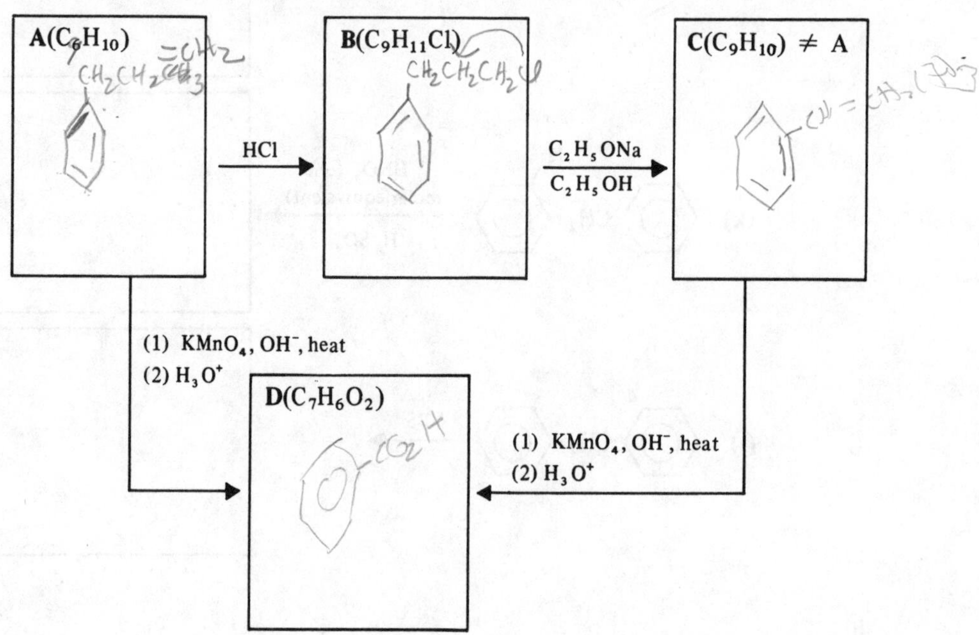

12.5 Outline a practical laboratory synthesis of *o*-nitropropylbenzene starting from benzene and any necessary reagents.

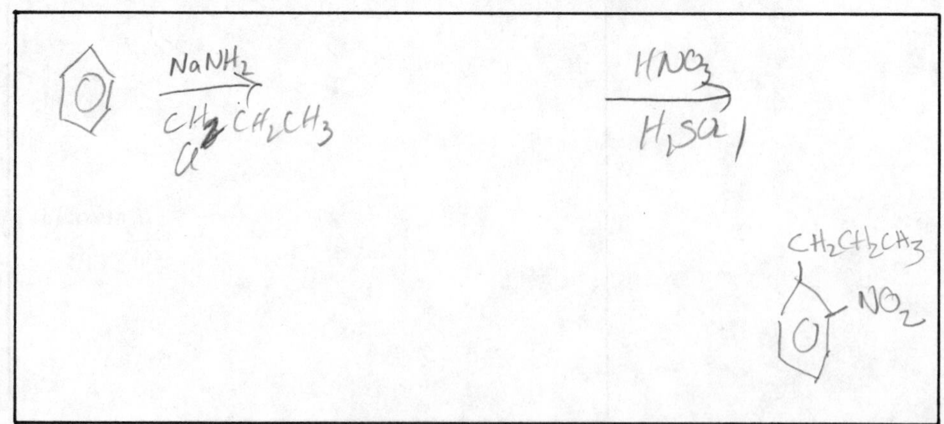

12.6 Which of the following compounds would be most reactive toward ring bromination?

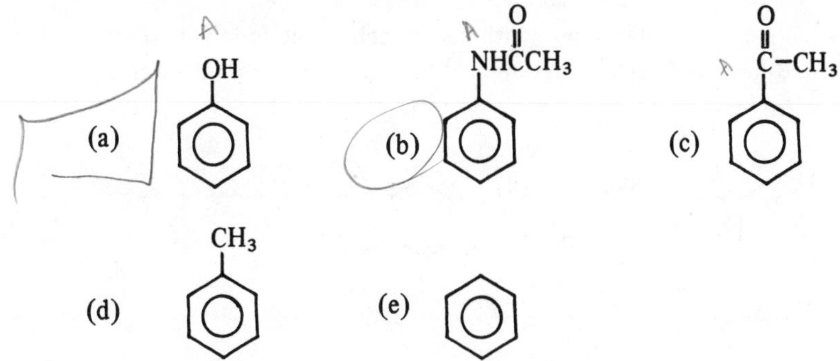

12.7 Which of the following is *not* a meta directing substituent when present on the benzene ring?

(a) $-C_6H_5$
(b) $-NO_2$
(c) $-N(CH_3)_3^+$
(d) $-C{\equiv}N$
(e) $-CO_2H$

12.8 The major product(s), **C**, of the following reaction,

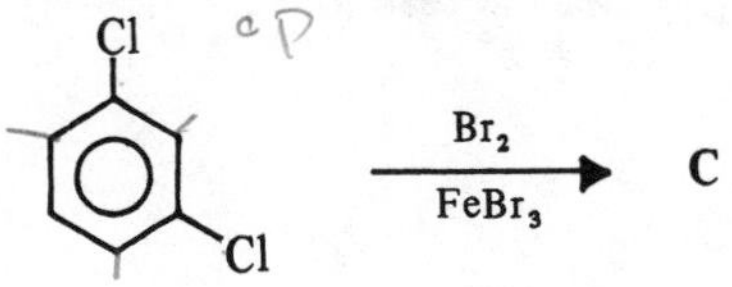

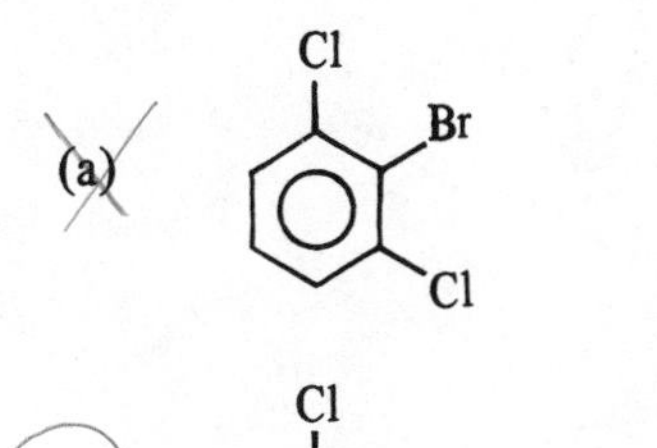

would be

(a) [1-Br-2,3-dichloro structure: Cl, Br, Cl]

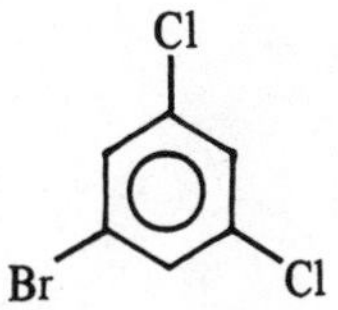

(b) 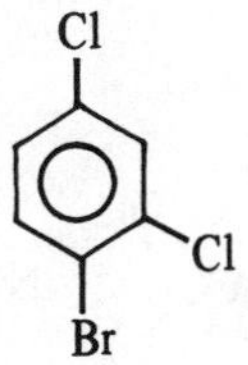 [Cl, Cl, Br structure]

(c) [Cl, Br, Cl structure]

(d) Equal amounts of (a) and (b)

(e) Equal amounts of (a) and (c)

SUPPLEMENTARY PROBLEMS

S12.1 Show all steps in a laboratory synthesis of each of the following compounds starting with phenol, toluene, and any compounds of four carbon atoms or fewer.

(a) 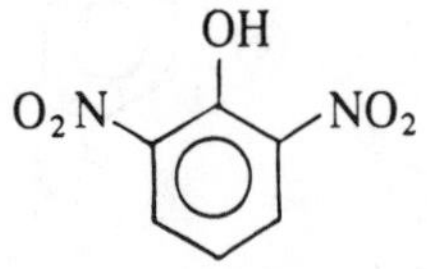 OH, O_2N, NO_2 (b) 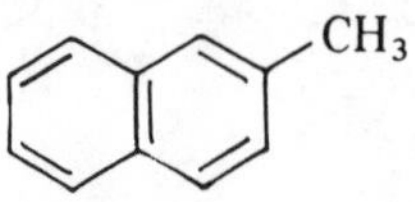 CH_3

S12.2 Supply the formulas of the unknown compounds.

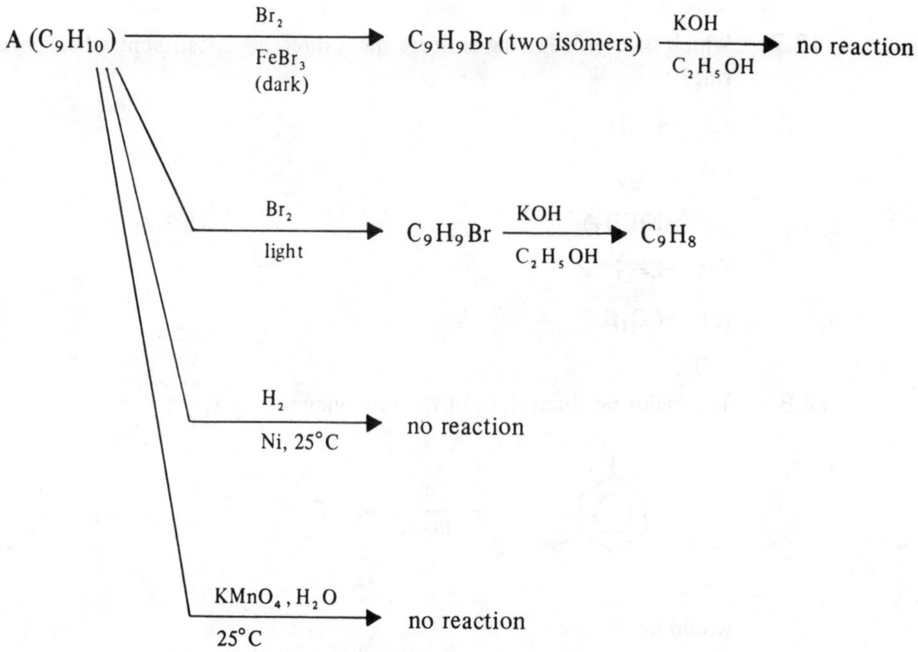

A (C_9H_{10}) —Br_2, $FeBr_3$ (dark)→ C_9H_9Br (two isomers) —KOH, C_2H_5OH→ no reaction

—Br_2, light→ C_9H_9Br —KOH, C_2H_5OH→ C_9H_8

—H_2, Ni, 25°C→ no reaction

—$KMnO_4$, H_2O, 25°C→ no reaction

SOLUTIONS TO SUPPLEMENTARY PROBLEMS

S12.1

(a) 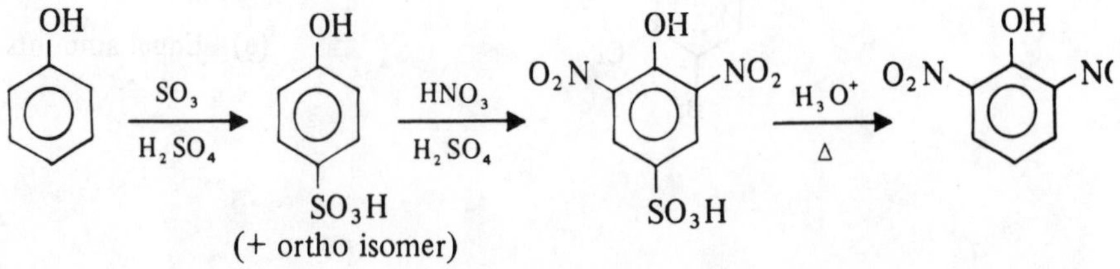 OH —SO_3, H_2SO_4→ OH, SO_3H (+ ortho isomer) —HNO_3, H_2SO_4→ OH, O_2N, NO_2, SO_3H —H_3O^+, Δ→ OH, O_2N, N(

(b)

CH3 + O O O —AlCl3→ CH3 O HO C O —Zn(Hg), HCl→ CH3 HO C O

—SOCl2→ CH3 Cl C O —AlCl3→ CH3 O —NaBH4→

CH3 OH —H2SO4, Δ→ CH3 —Pd, Δ→ CH3

S12.2 The facts that A does not add H_2 and does not react with $KMnO_4$ tell us that it has no alkene double bond. The formula C_9H_{10} suggests an aromatic ring. If there were only one alkyl substituent, it would contain: $C_9H_{10} - C_6H_5 = C_3H_5$, which would be unsaturated. The substituent must therefore be a ring. A cyclopropyl ring would have added H_2. We thus conclude that the ring is fused. The remaining reactions are shown:

A —Br_2, $FeBr_3$ (dark)→ Br + Br

A —Br_2, $h\nu$→ Br —KOH, C_2H_5OH→

13 SPECTROSCOPIC METHODS OF STRUCTURE DETERMINATION

SOLUTIONS TO PROBLEMS

13.1 The formula, C_6H_8, tells us that **A** and **B** have six hydrogen atoms less than an alkane. This unsaturation may be due to three double bonds, one triple bond and one double bond, or combinations of two double bonds and a ring, or one triple bond and a ring. Since both **A** and **B** react with 2 moles of H_2 to yield cyclohexane, they are either cyclohexyne or cyclohexadienes. The absorption maximum of 256 nm for **A** tells us that it is conjugated. **B**, with no absorption maximum beyond 200, possesses isolated double bonds. We can rule out cyclohexyne because of ring strain caused by the requirement of linearity of the $-C{\equiv}C-$ system. Therefore **A** is 1,3-cyclohexadiene; **B** is 1,4-cyclohexadiene.

13.2 All three compounds have an unbranched five-carbon chain, because the product of hydrogenation is unbranched pentane. The formula, C_5H_6, suggests that they have one double bond and one triple bond. **D**, **E**, and **F** must differ, therefore, in the way the multiple bonds are distributed in the chain. **E** and **F** have a terminal $-C{\equiv}CH$ [reaction with $Ag(NH_3)_2{}^+OH^-$]. The absorption maximum near 230 nm for **D** and **E** suggests that in these compounds, the multiple bonds are conjugated. The structures are

$CH_3-C{\equiv}C-CH{=}CH_2$ $HC{\equiv}C-CH{=}CH-CH_3$ $HC{\equiv}C-CH_2-CH{=}CH_2$

D   **E**   **F**

13.3 The methyl protons of *trans*-15,16-dimethyldihydropyrene are highly shielded by the induced field in the center of the aromatic system where the induced field opposes the applied field (Fig. 13.13).

13.4 (a) The six protons (hydrogen atoms) of ethane are equivalent:

(a) *(a)*
CH_3-CH_3

Ethane gives a single signal in its proton nmr spectrum.

(b) Propane has two different sets of equivalent protons:

(a) *(b)* *(a)*
$CH_3-CH_2-CH_3$

Propane gives two signals.

(c) The six protons of dimethyl ether are equivalent:

$$\overset{(a)}{CH_3}-O-\overset{(a)}{CH_3}$$

One signal.

(d) Three different sets of equivalent protons:

$$\overset{(a)}{CH_3}-\overset{(b)}{CH_2}-C_6H_4(\text{4 H}, (c))-\overset{(b)}{CH_2}-\overset{(a)}{CH_3}$$

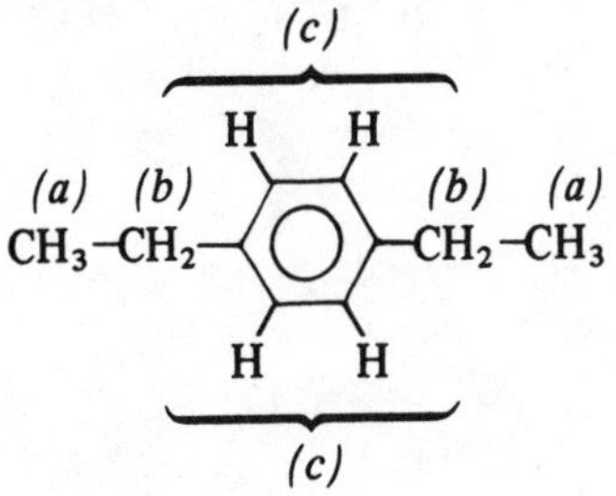

Three signals.

(e) Two different sets of equivalent protons:

$$\overset{(a)}{CH_3}-\overset{O}{\overset{\|}{C}}-O-\overset{(b)}{CH_3}$$

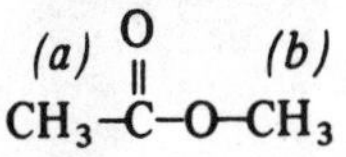

Two signals.

(f) Three different sets of equivalent protons:

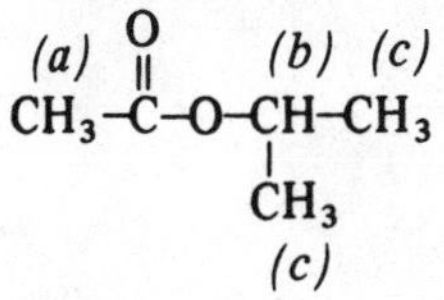

$$\overset{(a)}{CH_3}-\overset{O}{\overset{\|}{C}}-O-\overset{(b)}{CH}(\underset{(c)}{CH_3})-\overset{(c)}{CH_3}$$

Three signals.

13.5 (a) 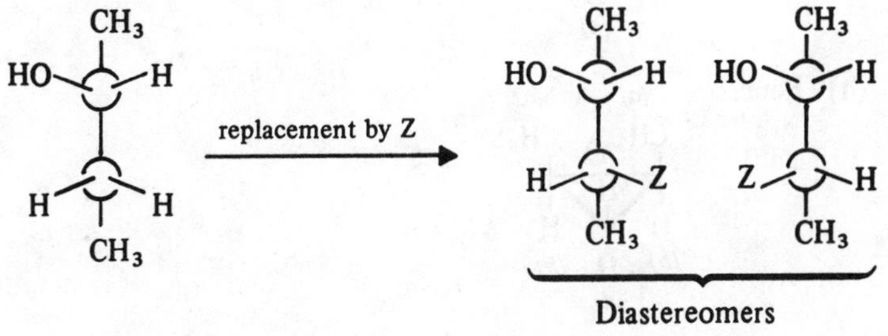 HO, H, CH_3 / H, H, CH_3 —replacement by Z→ HO, H, CH_3 / H, Z, CH_3 and HO, H, CH_3 / Z, H, CH_3

Diastereomers

(b) Six

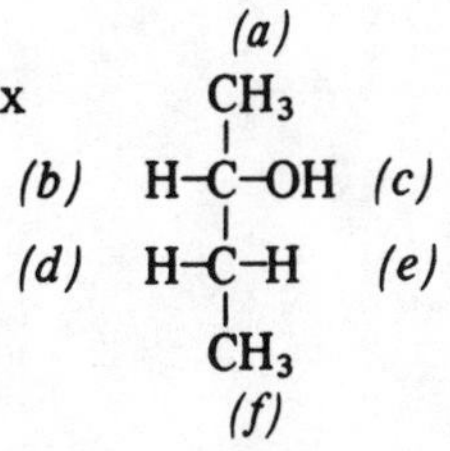

$$\begin{array}{lcl} & \overset{(a)}{CH_3} & \\ (b) & H-C-OH & (c) \\ (d) & H-C-H & (e) \\ & \underset{(f)}{CH_3} & \end{array}$$

(c) Six signals

13.6

(a) Two, $\overset{(a)}{CH_3}-\overset{(b)}{CH_2}-\overset{(b)}{CH_2}-\overset{(a)}{CH_3}$

(b) Three, $\overset{(a)}{CH_3}-\overset{(b)}{CH_2}-O-\overset{(c)}{H}$

(c) Four,

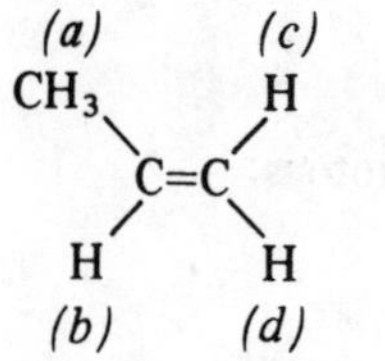

$$\begin{array}{ccc} (a) & & (c) \\ CH_3 & & H \\ & C{=}C & \\ H & & H \\ (b) & & (d) \end{array}$$

(d) Two,

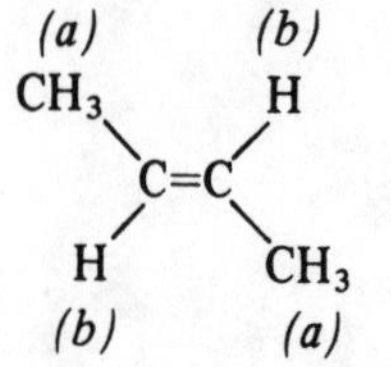

$$\begin{array}{ccc} (a) & & (b) \\ CH_3 & & H \\ & C{=}C & \\ H & & CH_3 \\ (b) & & (a) \end{array}$$

(e) Four, $\overset{(a)}{CH_3}-\overset{(b)}{CHBr}-\underset{H\,(d)}{\overset{H\,(c)}{C}}-Br$

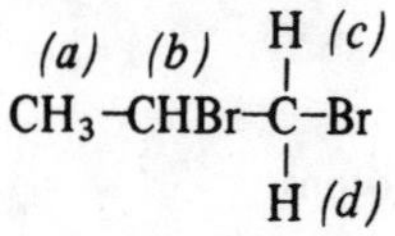

(f) Two,

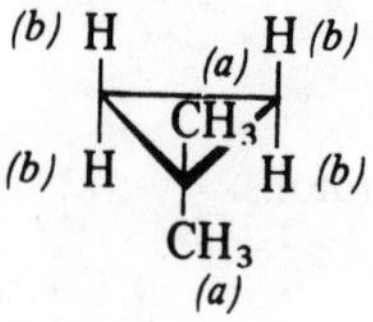

$$\begin{array}{ccc} (b)\ H & (a) & H\ (b) \\ & CH_3 & \\ (b)\ H & & H\ (b) \\ & CH_3 & \\ & (a) & \end{array}$$

(g) Three,

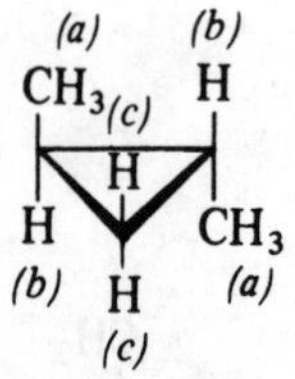

$$\begin{array}{ccc} (a) & & (b) \\ CH_3 & (c)\ H & H \\ H & & CH_3 \\ (b) & H & (a) \\ & (c) & \end{array}$$

(h) Four,

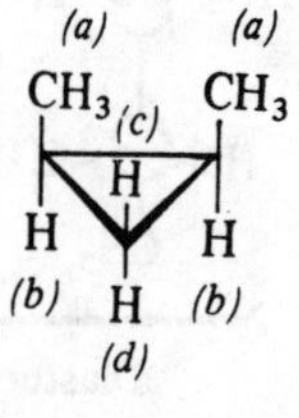

$$\begin{array}{ccc} (a) & & (a) \\ CH_3 & (c)\ H & CH_3 \\ H & & H \\ (b) & H & (b) \\ & (d) & \end{array}$$

(i) Six,

$$\begin{array}{ccc} \overset{(a)}{CH_3}-\overset{(b)}{CH_2}-\overset{(c)}{CH_2} & & H\ (e) \\ & C{=}C & \\ H & & H\ (f) \\ (d) & & \end{array}$$

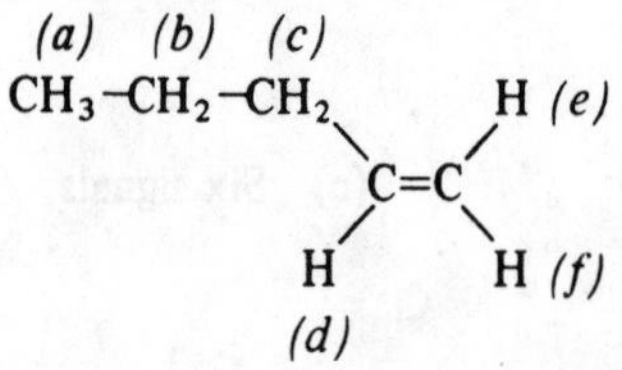

13.7 The proton nmr spectrum of $CHBr_2CHCl_2$ consists of two doublets. The doublet from the proton of the $-CHCl_2$ group should occur at lowest magnetic field strength because the greater electronegativity of chlorine reduces the electron density in the vicinity of the $-CHCl_2$ proton, and consequently, reduces its shielding relative to $-CHBr_2$.

13.8 The determining factors here are the number of chlorine atoms attached to the carbon atoms bearing protons and the deshielding that results from chlorine's electronegativity. In 1,1,2-trichloroethane and proton that gives rise to the triplet is on a carbon atom that bears two chlorines, and the signal from this proton is downfield. In 1,1,2,3,3-pentachloropropane the proton that gives rise to the triplet is on a carbon atom that bears only one chlorine; the signal from this proton is upfield.

13.9 The signal from the three equivalent protons designated *(a)* should be split into a doublet by the proton *(b)*. This doublet, because of the electronegativity of the attached chlorines, should occur downfield.

$$\overset{(a)}{(Cl_2CH)_3}-\overset{(b)}{CH}$$

The proton designated *(b)* should be split into a quartet by the three equivalent protons *(a)*. The quartet should occur upfield.

13.10

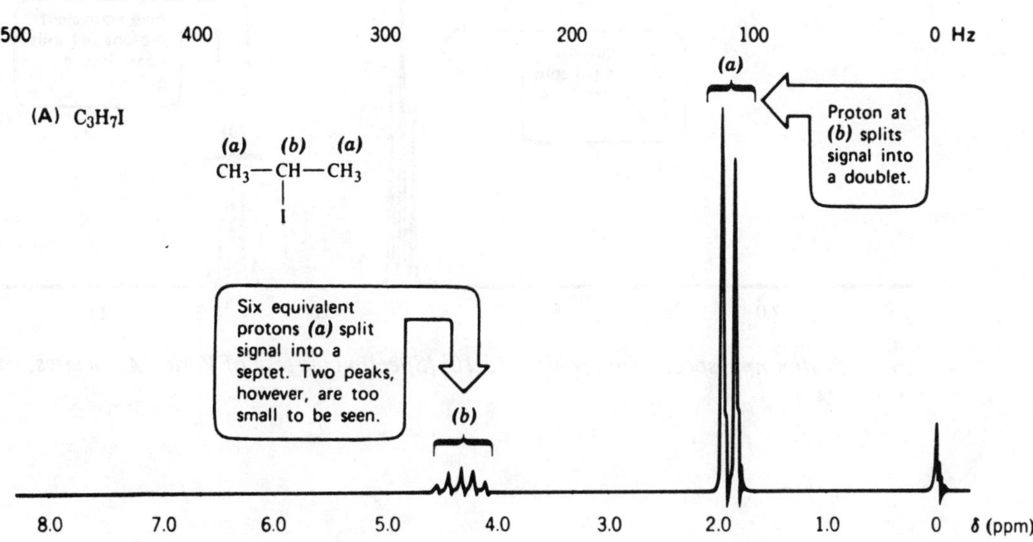

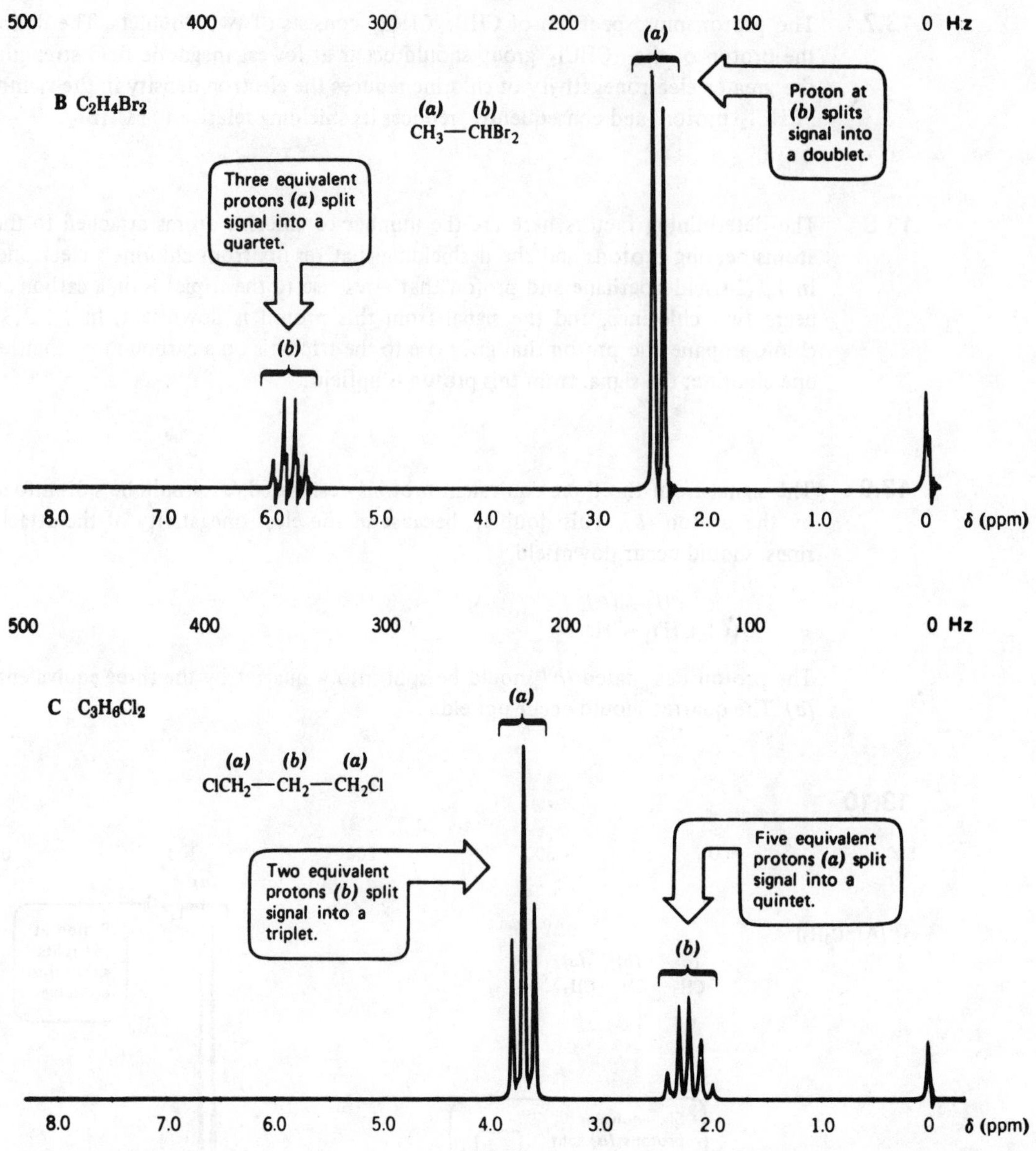

Proton nmr spectra for Problem 13.10. (Spectra courtesy of Varian Associates, Palo Alto, CA.)

13.11 (a)

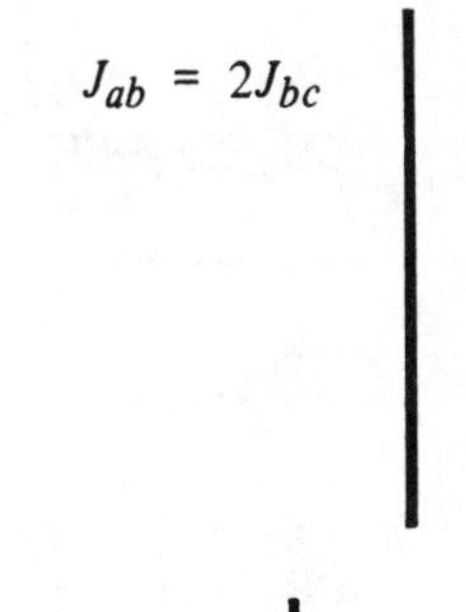

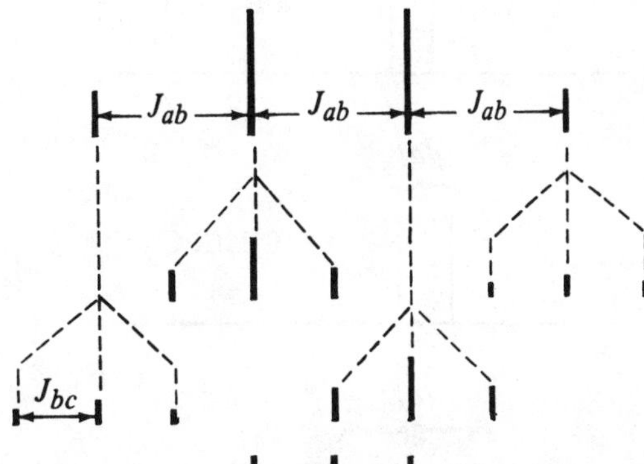

Result: (nine peaks)

(b) $J_{ab} = J_{bc}$

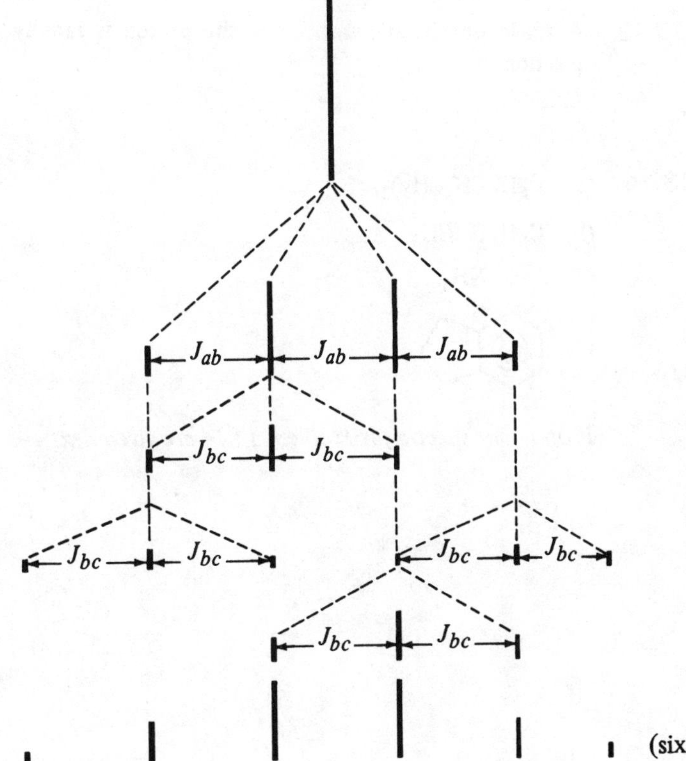

Result: (six peaks)

13.12

(a)

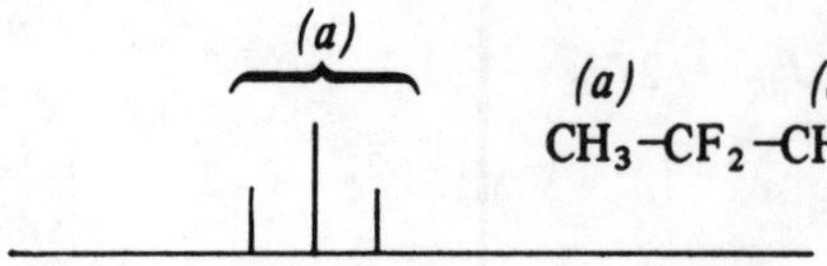

(a) (a)
$CH_3{-}CF_2{-}CH_3$

(b)

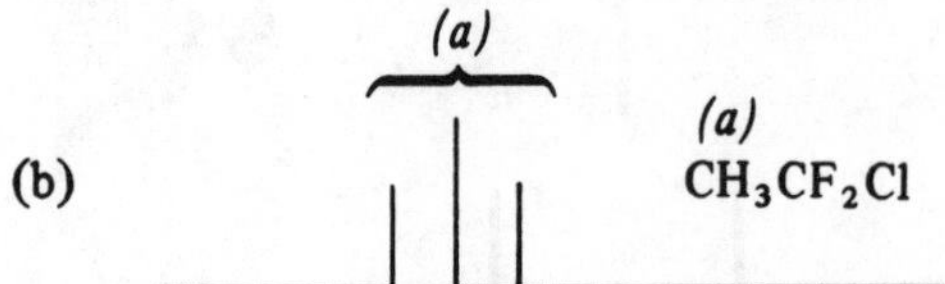

(a)
CH_3CF_2Cl

(c)

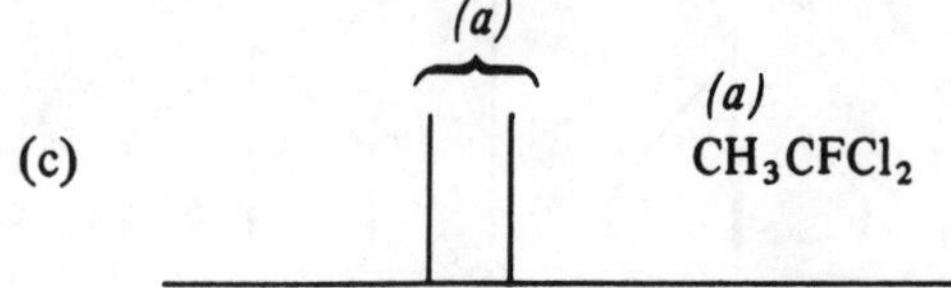

(a)
CH_3CFCl_2

(d)

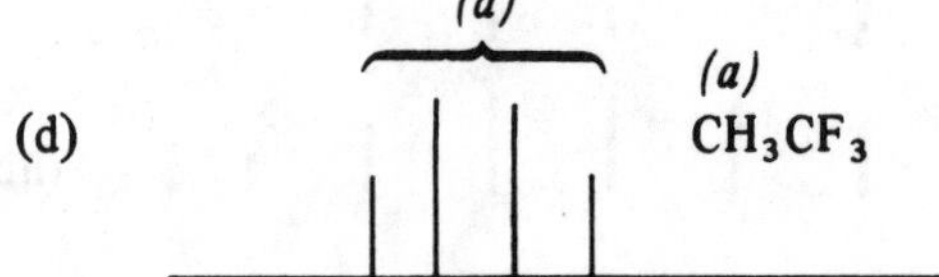

(a)
CH_3CF_3

13.13 A single unsplit signal, because the proton is rapidly shifted from axial to equatorial positions.

13.14 (a) $C_6H_5CH(CH_3)_2$

(b) $C_6H_5CH(NH_2)CH_3$

(c)

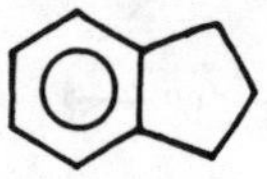

Proton nmr spectra for Problem 13.14 are given next.

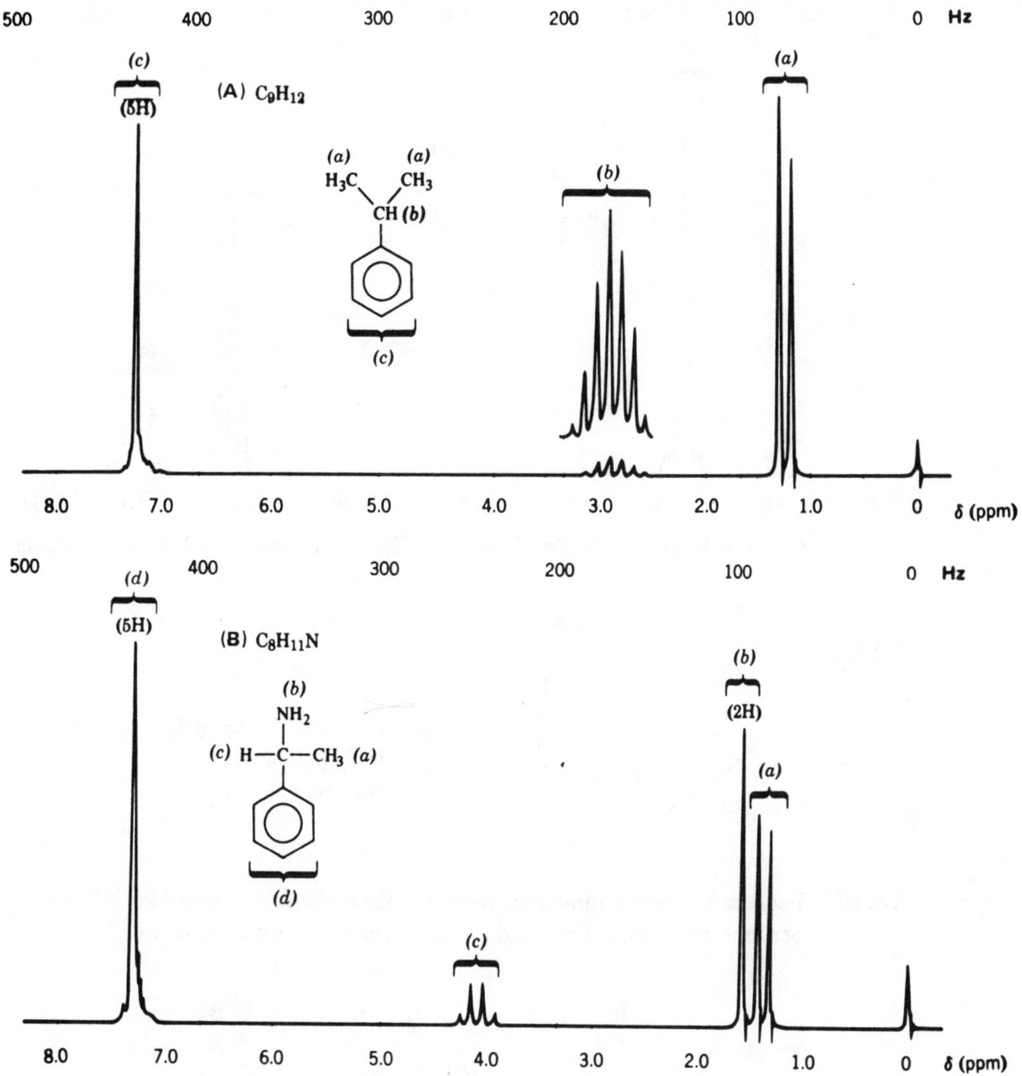

Note: There is no spin-spin coupling between the protons of the $-NH_2$ group and the proton (*c*); see Section 13.10.

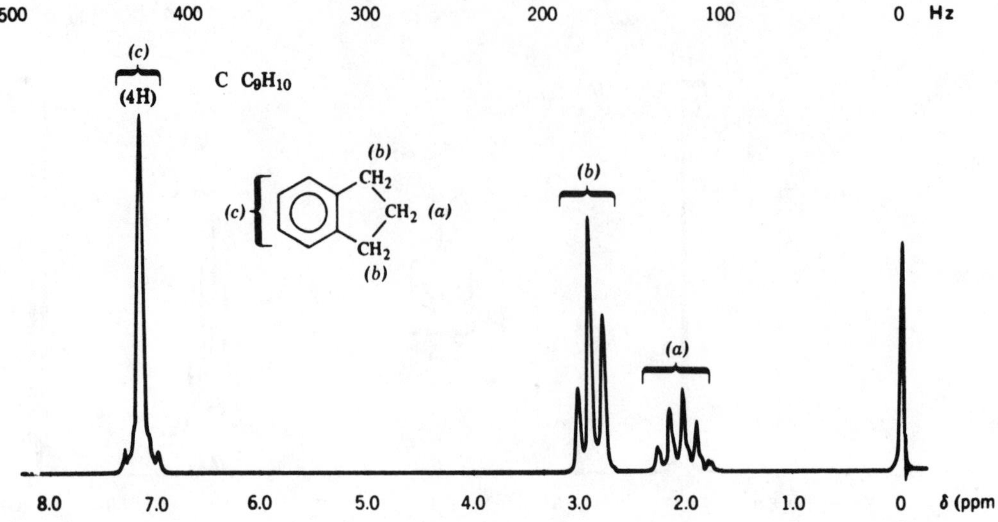

Proton nmr spectra for Problem 13.14 (Spectra courtesy of Varian Associates, Palo Alto, CA.)

13.15

$$\underset{(c)}{C_6H_5\overset{+}{C}(C_6H_5)}-\overset{(b)}{CH}(\overset{(a)}{CH_3})\overset{(a)}{CH_3}$$

(a) Doublet, δ 1.48 (6H)
(b) Multiplet, δ 4.45 (1H)
(c) Multiplet, δ 8.0 (10H)

13.16 Because of their symmetries, *p*-dibromobenzene would give two ^{13}C signals, *o*-dibromobenzene would give three, and *m*-dibromobenzene would give four.

Two signals — Three signals — Four signals

13.17 A is 1-chloro-3-methylbutane. The following are the signal assignments:

$$\overset{(d)}{ClCH_2}\overset{(c)}{CH_2}\overset{(b)}{CH}(\overset{(a)}{CH_3})_2$$

(a) δ 22*q*

(b) δ 26*d*

(c) δ 42*t*

(d) δ 43*t*

B is 2-chloro-2-methylbutane. The following are the signal assignments:

$$\underset{(a)}{CH_3}\underset{(c)}{CH_2}\overset{\overset{Cl}{|}}{\underset{(d)}{C}}\underset{(b)}{(CH_3)_2}$$

(a) δ 9*q*

(b) δ 32*q*

(c) δ 39*t*

(d) δ 71*s*

C is 1-chloropentane. The following are the signal assignments.

$$\underset{(e)}{ClCH_2}\underset{(d)}{CH_2}\underset{(c)}{CH_2}\underset{(b)}{CH_2}\underset{(a)}{CH_3}$$

(a) δ 14*q*

(b) δ 22*t*

(c) δ 29*t*

(d) δ 33*t*

(e) δ 45*t*

13.18

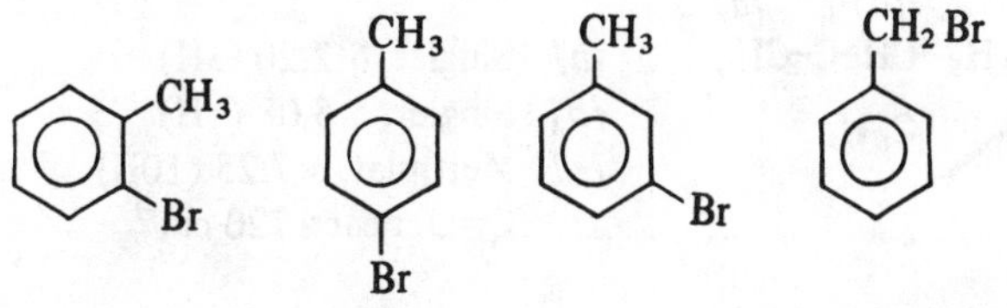

A: benzene ring with CH_3 and Br (ortho); B: benzene ring with CH_3 and Br (para); C: benzene ring with CH_3 and Br (meta); D: benzene ring with CH_2Br

A. Strong absorption at 740 cm^{-1} is characteristic of ortho substitution.

B. A very strong absorbtion peak at 795 cm^{-1} is characteristic of para substitution.

C. Strong absorbtion peaks at 680 and 760 cm^{-1} are characteristic of meta substitution.

D. Strong absorption peaks at 693 and 765 cm^{-1} are characteristic of a monosubstituted benzene ring.

13.19

(a) $CH_3-C(CH_3)_2-OH$ — (a) on each CH_3, (b) on OH

- *(a)* Singlet, δ 1.28 (9H)
- *(b)* Singlet, δ 1.35 (1H)

(b) $\overset{(a)}{CH_3}-\overset{(b)}{CH}(Br)-\overset{(a)}{CH_3}$

- *(a)* Doublet, δ 1.71 (6H)
- *(b)* Septet, δ 4.32 (1H)

(c) $\overset{(b)}{CH_3}-C(=O)-\overset{(c)}{CH_2}-\overset{(a)}{CH_3}$

- *(a)* Triplet, δ 1.05 (3H)
- *(b)* Singlet, δ 2.13 (3H)
- *(c)* Quartet, δ 2.47 (2H)
- C=O, 1720 cm^{-1}

(d) $\underset{(c)}{C_6H_5}-\overset{(b)}{CH_2}-\overset{(a)}{OH}$

- *(a)* Singlet, δ 2.43 (1H)
- *(b)* Singlet, δ 4.58 (2H)
- *(c)* Multiplet, δ 7.28 (5H)
- O–H, 3200-3600 cm^{-1}

(e) $CH_3-\overset{(b)}{CH}(CH_3)-\overset{(c)}{CH_2}Cl$ — (a) on both CH_3

- *(a)* Doublet, δ 1.04 (6H)
- *(b)* Multiplet, δ 1.95 (1H)
- *(c)* Doublet, δ 3.35 (2H)

(f) $C_6H_5-\overset{(b)}{CH}(C_6H_5)-C(=O)-\overset{(a)}{CH_3}$ — (c) on both C_6H_5

- *(a)* Singlet, δ 2.20 (3H)
- *(b)* Singlet, δ 5.08 (1H)
- *(c)* Multiplet, δ 7.25 (10H)
- C=O, near 1720 cm^{-1}

(g) $\overset{(a)}{CH_3}-\overset{(b)}{CH_2}-\overset{(c)}{CH}(Br)\overset{(d)}{CO_2H}$

- *(a)* Triplet, δ 1.08 (3H)
- *(b)* Multiplet, δ 2.07 (2H)
- *(c)* Triplet, δ 4.23 (1H)
- *(d)* Singlet, δ 10.97 (1H)
- C=O (acid) 1715 cm^{-1}

(h) $\underset{(c)}{C_6H_5}-\overset{(b)}{CH_2}-\overset{(a)}{CH_3}$

- *(a)* Triplet, δ 1.25 (3H)
- *(b)* Quartet, δ 2.68 (2H)
- *(c)* Multiplet, δ 7.23 (5H)

(i) $\overset{(a)}{CH_3}-\overset{(b)}{CH_2}-O-\overset{(c)}{CH_2}\overset{(d)}{CO_2H}$

(a) Triplet, δ 1.27 (3H)
(b) Quartet, δ 3.66 (2H)
(c) Singlet, δ 4.13 (2H)
(d) Singlet, δ 10.95 (1H)
O–H, 2500-3000 cm^{-1}
C=O (acid) 1715 cm^{-1}

(j) $\overset{(a)}{CH_3}-\overset{(b)}{CH}(NO_2)-\overset{(a)}{CH_3}$

(a) Doublet, δ 1.55 (6H)
(b) Septet, δ 4.67 (1H)

(k) $\overset{(a)}{CH_3}O-\overset{(b)}{CH_2}\overset{(b)}{CH_2}-O\overset{(a)}{CH_3}$

(a) Singlet, δ 3.25 (6H)
(b) Singlet, δ 3.45 (4H)

(l) $\overset{(b)}{CH_3}-C(=O)-\overset{(c)}{CH}(\overset{(a)}{CH_3})-\overset{(a)}{CH_3}$

(a) Doublet, δ 1.10 (6H)
(b) Singlet, δ 2.10 (3H)
(c) Septet, δ 2.50 (1H)
C=O, near 1720 cm^{-1}

(m) $\underset{(c)}{C_6H_5}-\overset{(b)}{CH}(Br)-\overset{(a)}{CH_3}$

(a) Doublet, δ 2.0 (3H)
(b) Quartet, δ 5.15 (1H)
(c) Multiplet, δ 7.35 (5H)

13.20 Compound **E** is phenylacetylene, $C_6H_5C{\equiv}CH$. We can make the following assignments in the infrared spectrum:

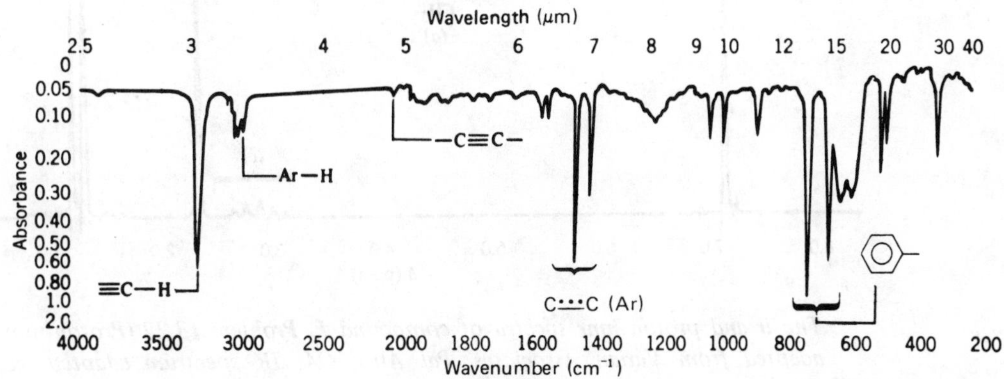

The infrared spectrum of compound **E** *(Problem 13.20). (Spectrum courtesy of Sadtler Research Laboratories Inc., Philadelphia.)*

13.21 A proton nmr signal this far upfield indicates that cyclooctatetraene is a cyclic polyene and is not aromatic

13.22 Compound F is *p*-isopropyltoluene. Assignments are shown in the following spectra.

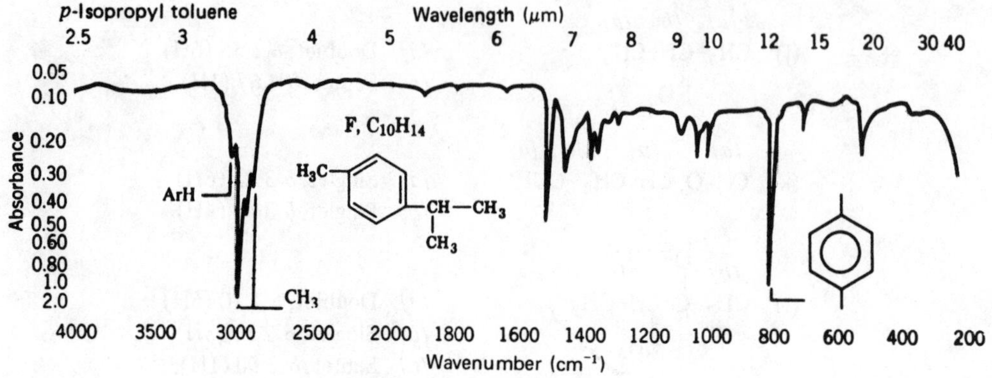

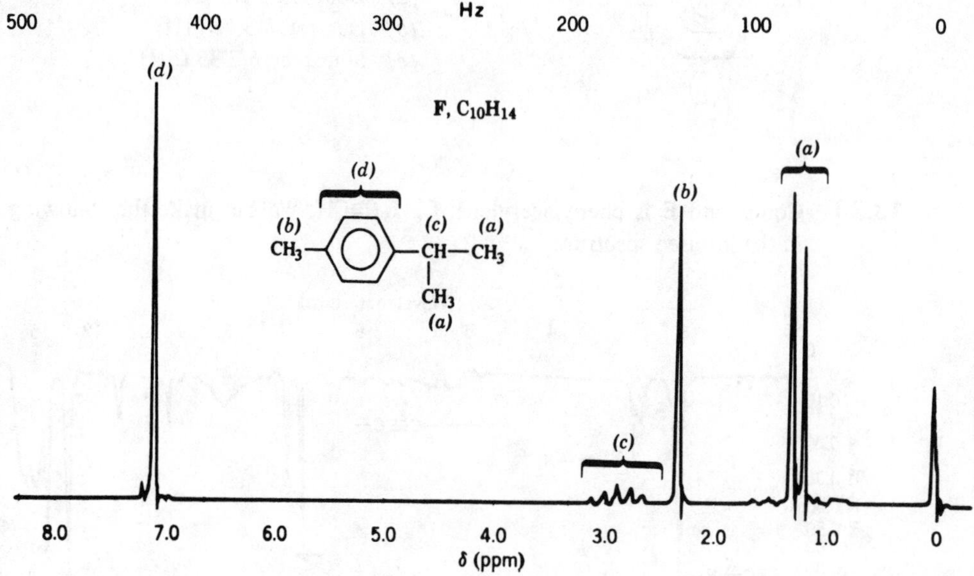

The ir and proton nmr spectra of compound F, Problem 13.22 (Proton nmr spectrum adapted from Varian Associates, Pal Alto, CA. IR spectrum adapted from Sadtlei Research Laboratories, Philadelphia.)

13.23 Compound G is 2-bromobutane. Assignments are shown in the following spectra.

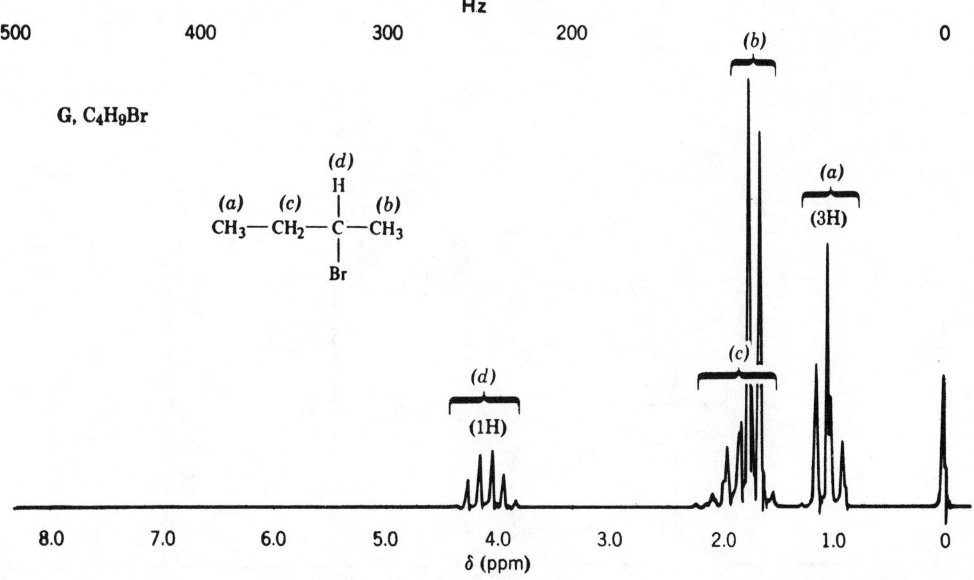

The proton nmr spectrum of compound G (Problem 13.23). (Spectrum courtesy of Varian Associates, Palo Alto, CA.)

Compound H is 2,3-dibromopropene. Assignments are shown in the following spectrum.

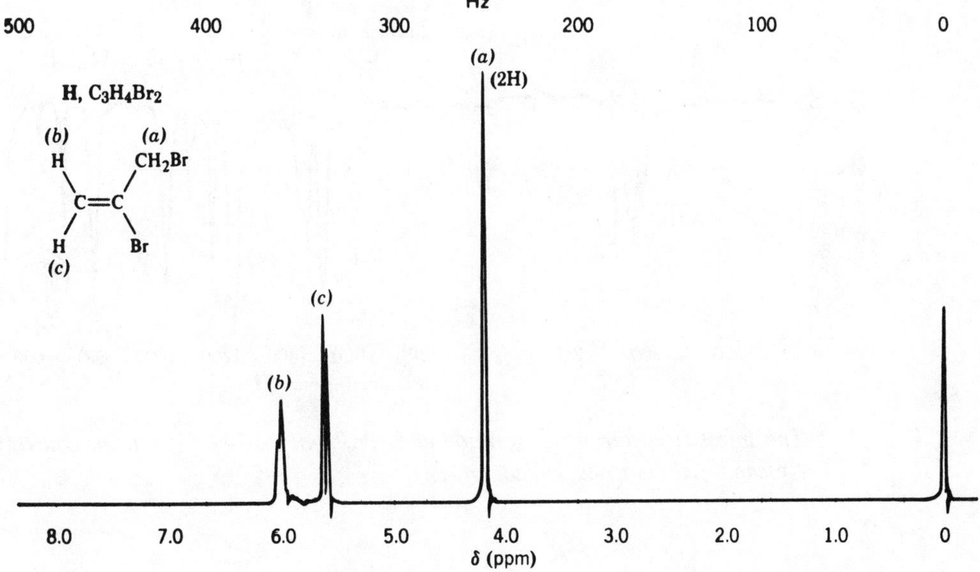

The proton nmr spectrum of compound H (Problem 13.23). (Spectrum courtesy of Varian Associates, Palo Alto, CA.)

13.24 Compound I is *p*-methoxytoluene. Assignments are shown in the spectra reproduced below.

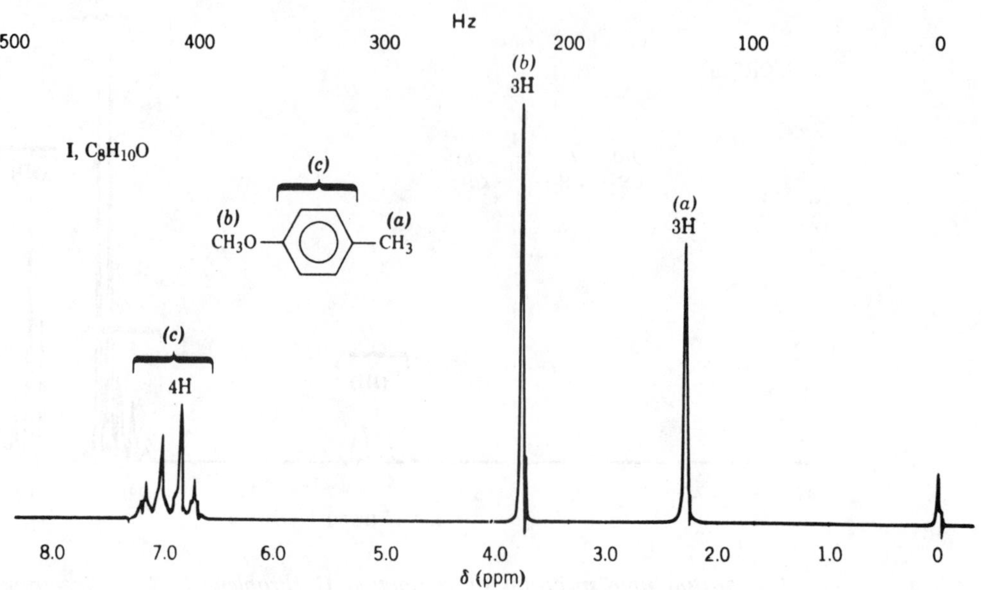

The proton nmr spectrum of compound I (Problem 13.24). (Spectrum courtesy of Varian Associates, Palo Alto, CA.)

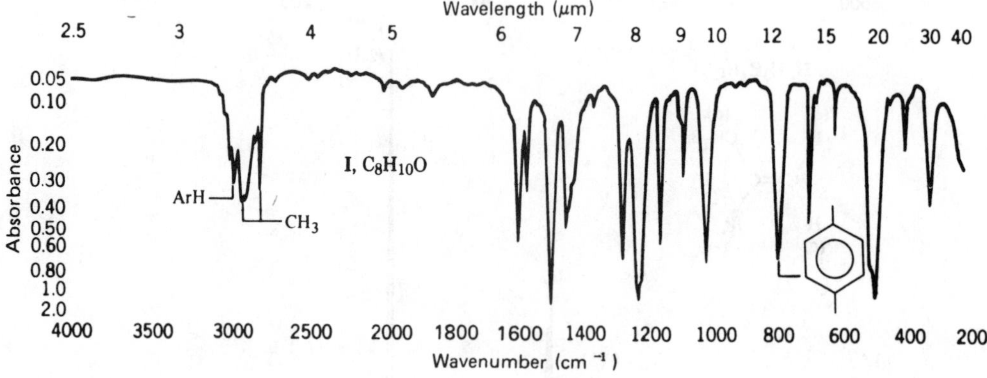

The infrared spectrum of compound I (Problem 13.24). (Spectrum courtesy of Sadtler Research Laboratories, Philadelphia.)

13.25 Compound **J** is *cis*-1,2-dichloroethene,

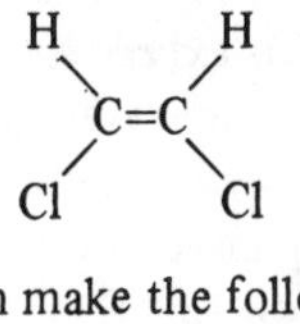

$$\begin{array}{ccc} H & & H \\ & C{=}C & \\ Cl & & Cl \end{array}$$

We can make the following infrared assignments:

3125 cm^{-1}, alkene C–H stretching
1625 cm^{-1}, C=C stretching
695 cm^{-1}, out-of-plane bending of cis double bond.
86.3, *H*–C=C

13.26 (a) Compound **K** is,

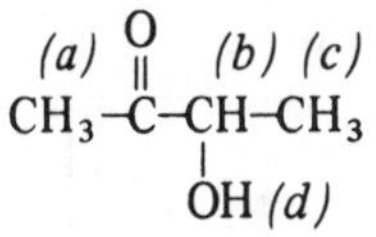

$$\overset{(a)}{CH_3}-\overset{O}{\overset{\|}{C}}-\overset{(b)}{CH}(\underset{(d)}{OH})-\overset{(c)}{CH_3}$$

(a) Singlet δ 2.15
(b) Quartet δ 4.25
(c) Doublet δ 1.35
(d) Singlet δ 3.75
C=O, 1720 cm^{-1}

(b) When the compound is dissolved in D_2O, the –OH proton (*d*) is replaced by a deuteron and thus the proton nmr absorption peak disappears.

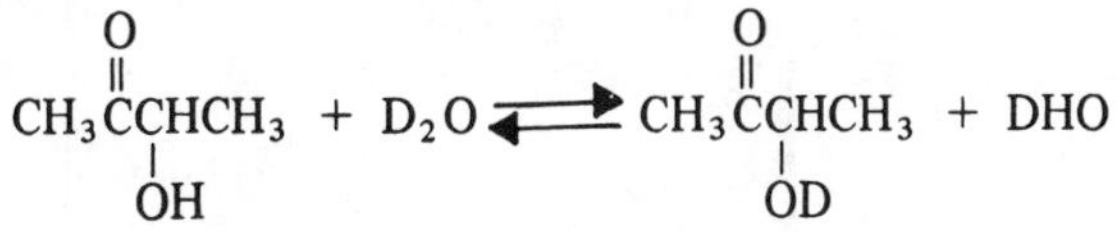

$$CH_3\overset{O}{\overset{\|}{C}}\underset{OH}{\underset{|}{C}}HCH_3 + D_2O \rightleftharpoons CH_3\overset{O}{\overset{\|}{C}}\underset{OD}{\underset{|}{C}}HCH_3 + DHO$$

13.27 Compound **L** is allylbenzene,

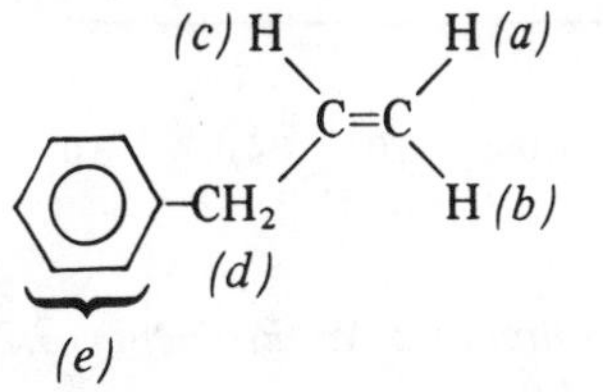

$$\underbrace{C_6H_5}_{(e)}-\underset{(d)}{CH_2}-C\overset{(c)}{H}{=}C\overset{(a)}{H}\underset{(b)}{H}$$

(d) Doublet δ 3.1 (2H)
(a) or *(b)* Multiplet δ 4.8
(a) or *(b)* Multiplet δ 5.1
(c) Multiplet δ 5.8
(e) Multiplet δ 7.1 (5H)

The following infrared assignments can be made.

3035 cm^{-1}, C–H stretching of benzene ring
3020 cm^{-1}, C–H stretching of $-CH{=}CH_2$ group
2925 cm^{-1} and 2853 cm^{-1}, C–H stretching of $-CH_2-$group
1640 cm^{-1}, C=C stretching
990 cm^{-1} and 915 cm^{-1}, C–H bendings of $-CH{=}CH_2$ group
740 cm^{-1} and 695 cm^{-1}, C–H bendings of $-C_6H_5$ group

The ultraviolet absorbance maximum at 255 nm is indicative of a benzene ring that is not conjugated with a double bond.

13.28 Run the spectrum with the spectrometer operating at a different magnetic field strength (i.e., at 30 or at 100 MHz). If the peaks are two singlets the distance between them—*when measured in hertz*—will change because chemical shifts *expressed in hertz* are proportional to the strength of the applied field (Section 13.6). If, however, the two peaks represent a doublet then the distance that separates them, expressed in hertz, will not change because this distance represents the magnitude of the coupling constant and coupling constants are independent of the applied magnetic field (Section 13.8).

13.29 Compound **M** is *m*-ethyltoluene. We can make the following assignments in the spectra.

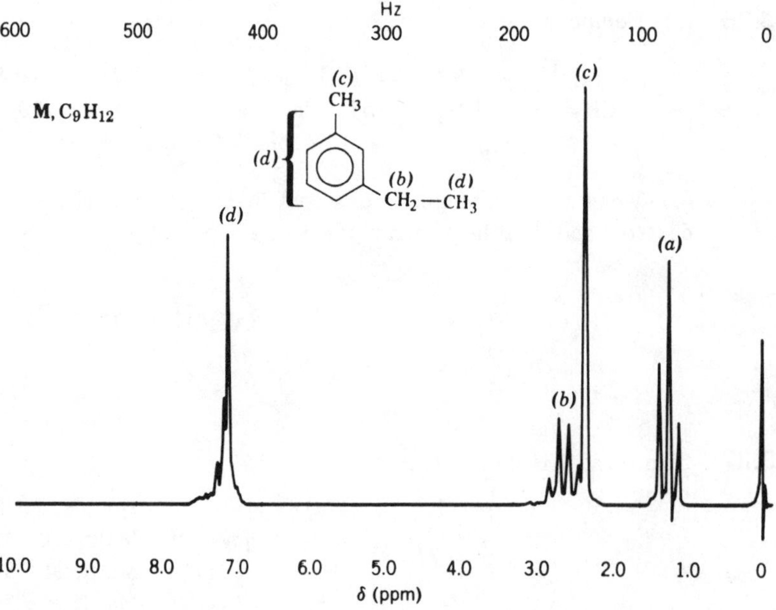

The proton nmr spectrum of compound **M**, *Problem 13.29. (Spectrum courtesy of Aldrich Chemical Co., Milwaukee, WI.)*

Meta substitution is indicated by the very strong peaks at 690 and 780 cm^{-1} in the infrared spectrum.

13.30 Compound N is $C_6H_5CH{=}CHOCH_3$. The absence of absorption peaks due to O–H or C=O stretching in the infrared spectrum of N suggests that the oxygen atom is present as part of an ether linkage. The (5H) proton nmr multiplet at δ 7.3 strongly suggests the presence of a monosubstituted benzene ring; this is confirmed by the strong peaks at ~690 and ~770 cm^{-1} in the infrared spectrum.

We can make the following assignments in the proton nmr spectrum:

(a) (b) (c) (d)
C_6H_5–CH=CH–OCH_3

(a) Multiplet δ 7.3
(c) Doublet δ 6.05
(b) Doublet δ 5.15
(d) Singlet δ 3.7

***13.31** That the proton nmr spectrum shows only one signal indicates that all 12 protons of the carbocation are equivalent, and suggests very strongly that what is being observed is the bromonium ion:

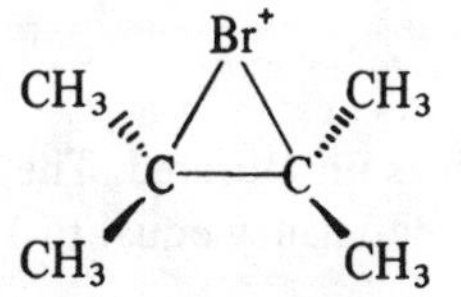

Br^+
CH_3 CH_3
C——C
CH_3 CH_3

While this experiment does not prove that bromonium ions are intermediates in alkene additions, it does show that bromonium ions are capable of existence and thus makes postulating them as intermediates more plausible.

13.32 Compound **O** is 1,4-cyclohexadiene and P is cyclohexane.

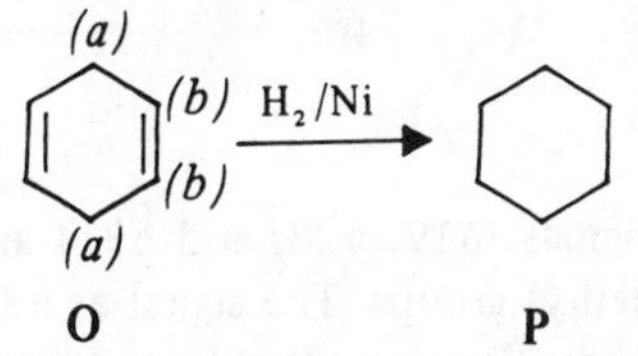

(a) (b) H_2/Ni → (b) (a)
O **P**

(a) δ 26.0*t*

(b) δ 124.5*d*

13.33 The molecular formula of **Q** (C_7H_8) indicates an index of hydrogen deficiency (Section 6.8) of four. The hydrogenation experiment suggests that **Q** contains two double bonds (or one triple bond). **Q**, therefore, must contain two rings.

The ^{13}C spectrum shows that **Q** is 2,5-bicyclo[2.2.1]heptadiene. The following reasoning shows one way to arrive at this conclusion: There is only one signal (δ 143) in the region for a doubly bonded carbon. This fact indicates that the doubly bonded carbon atoms are all equivalent. That the signal at δ 143 is a doublet in the proton off-resonance decoubled spectrum indicates that each of the doubly bonded carbon atoms bears one hydrogen atom. Because of their multiplicities in the proton off-resonance spectrum the signal at δ 75 can be assigned to a $-CH_2-$ group and the signal at δ 50 to a $\rangle$C—H group. The molecular formula tells us that the compound must contain two $\rangle$C—H groups, and since only one signal occurs in the ^{13}C spectrum, these $\rangle$C—H groups must be equivalent. Putting this all together we get the following:

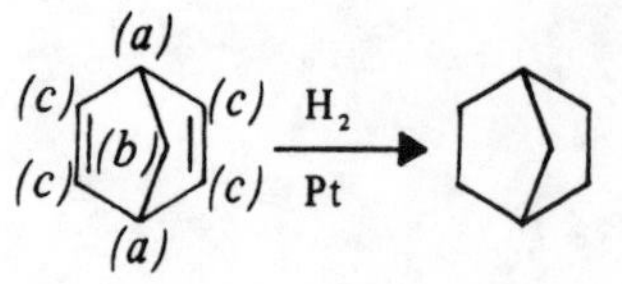

Q $\xrightarrow[\text{Pt}]{H_2}$ R

(a) δ 50d

(b) δ 75t

(c) δ 143t

13.34 That S decolorizes bromine indicates that it is unsaturated. The molecular formula of S allows us to calculate an index of hydrogen deficiency equal to 1. Therefore, we can conclude that S has one double bond.

The ^{13}C spectrum shows the doubly bonded carbon atoms at δ 130 and δ 135. In the proton off-resonance decoupled spectrum, one of these signals (δ 130) is a singlet indicating a doubly bonded carbon that bears no hydrogen atoms; the other (δ 135) is a doublet indicating that the other doubly bonded carbon bears one hydrogen atom. We can now arrive at the following partial structure.

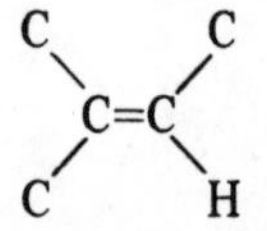

The three most upfield signals (δ 19, δ 28, and δ 31) are all quartets indicating that these signals all arise from methyl groups. The signal at δ 32 is a singlet indicating a carbon atom with no hydrogen atoms. Putting these facts together allows us to arrive at the following structure.

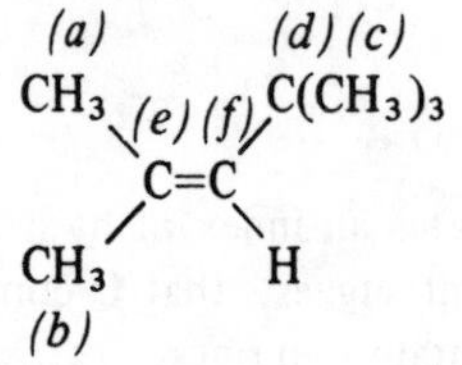

S

(a) δ 19q (d) δ 32s

(b) δ 28q (e) δ 130s

(c) δ 31q (f) δ 135d

Although the structure just given is the actual compound, other reasonable structures that one might be led to are

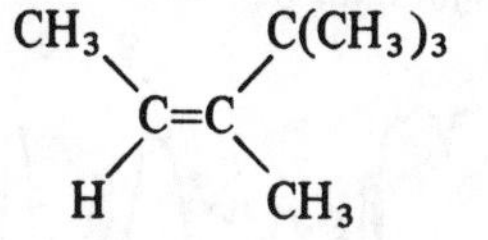

$$\underset{H}{\overset{CH_3}{}}\!\!>C=C<\!\!\underset{CH_3}{\overset{C(CH_3)_3}{}} \quad \text{and} \quad \underset{H}{\overset{CH_3}{}}\!\!>C=C<\!\!\underset{C(CH_3)_3}{\overset{CH_3}{}}$$

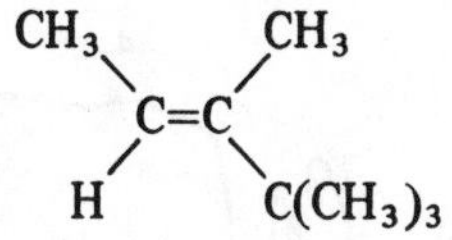

13.35 The infrared absorption band at 1745 cm^{-1} indicates the presence of a $>C=O$ group in a five-membered ring, and the signal at $\delta 218.2$ can be assigned to the carbon of this carbonyl group.

There are only two other signals in the ^{13}C spectrum; their multiplicities (triplets) in the proton off-resonance decoupled spectrum suggest two equivalent sets of two $-CH_2-$ groups each. Putting these facts together, we arrive at cyclopentanone as the structure for **T**.

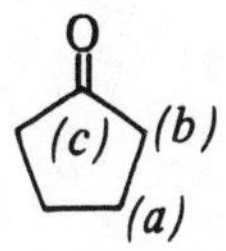

(c) (b) (a)

T

(a) δ 23.5*t*

(b) δ 38.0*t*

(c) δ 218.2*t*

13.36 Compound **X** is *meta*-xylene. The proton nmr spectrum shows only two signals. The upfield signal at $\delta 2.25$ arises from the two equivalent methyl groups. The downfield signal at $\delta 7.0$ arises from the protons of the benzene ring. Meta substitution is indicated by the strong infrared peak at 680 cm^{-1} and very strong infrared peak at 760 cm^{-1}.

13.37 The broad infrared peak at 3400 cm^{-1} indicates a hydroxy group and the two bands at 720 and 770 cm^{-1} suggest a monosubstituted benzene ring. The presence of these groups is also indicated by the peaks at δ 2.7 and δ 7.2 in the proton nmr spectrum. The proton nmr spectrum also shows a triplet at δ 0.7 indicating a $-CH_3$ group coupled with an adjacent $-CH_2-$ group. What appears at first to be a quartet at δ 1.9 actually shows further splitting. There is also a triplet at δ 4.35 (1H). Putting these pieces together in the only way possible gives us the following structure for **Y**.

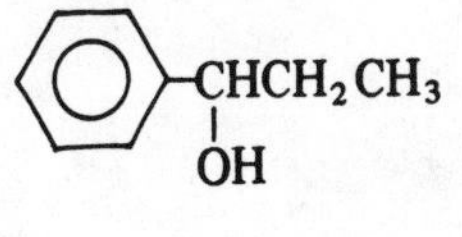

$$C_6H_5-\underset{OH}{\underset{|}{C}}HCH_2CH_3$$

Analyzed spectra are as follows:

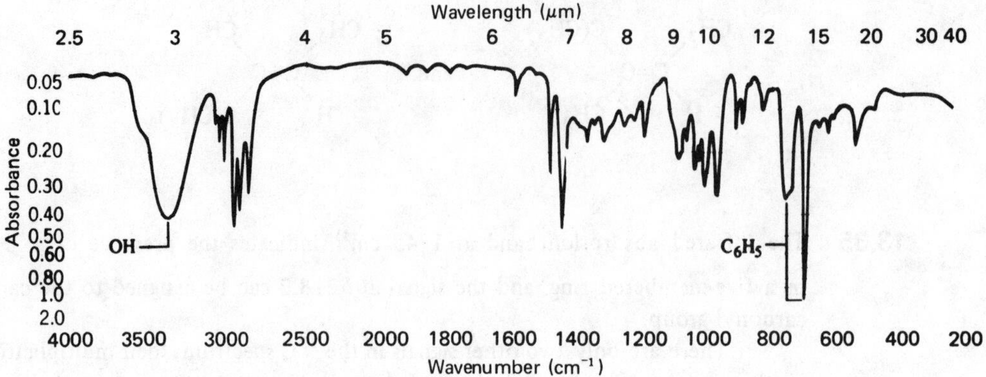

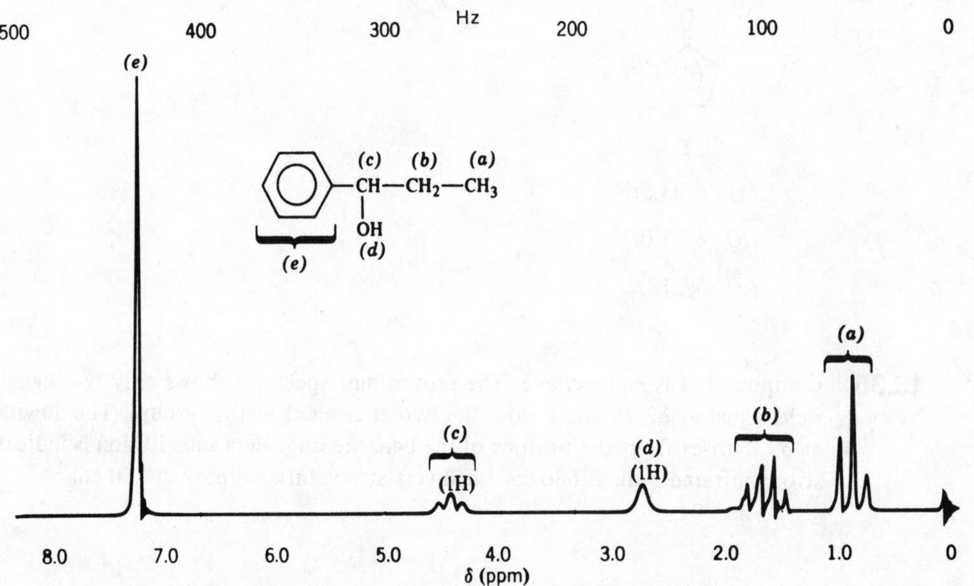

The infrared and proton nmr spectra of compound Y, Problem 13.37 (Infrared spectrum courtesy of Sadtler Research Laboratories, Philadelphia. The proton nmr spectrum courtesy of Varian Associates, Palo Alto, CA.)

13.38 Both [14]annulene and dehydro[14]annulene are aromatic as shown by the signals at δ 7.78 (10H) and at δ 8.0, (10H) respectively. [14]Annulene has four "internal" protons (δ −0.61)and dehydro[14]annulene has only two (δ 0.0).

13.39 (a) In SbF_5 the carbocations formed initially apparently undergo a complex series of rearrangements to the more stable *tert*-butyl cation.

(b) All of the cations formed initially rearrange to the more stable *tert*-pentyl cation,

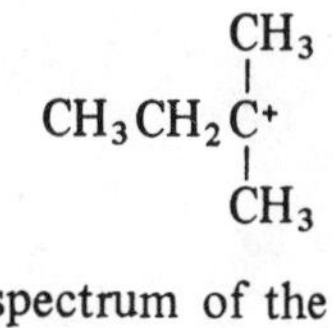

The spectrum of the *tert*-pentyl cation should consist of a singlet (6H), a quartet (2H), and a triplet (3H). The triplet should be most upfield and the quartet most downfield.

13.40 (a) Four unsplit signals,

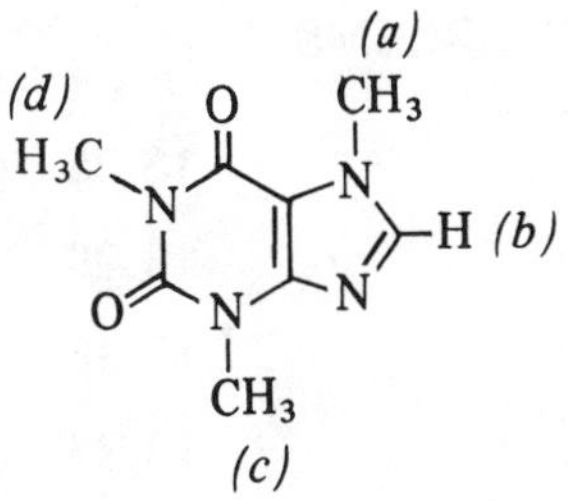

(b) Absorptions arising from: =C–H, CH_3, and C=O groups.

***13.41** In the presence of SbF_5, **I** dissociates first to the cyclic allylic cation, **II**, and then to the aromatic dication, **III**.

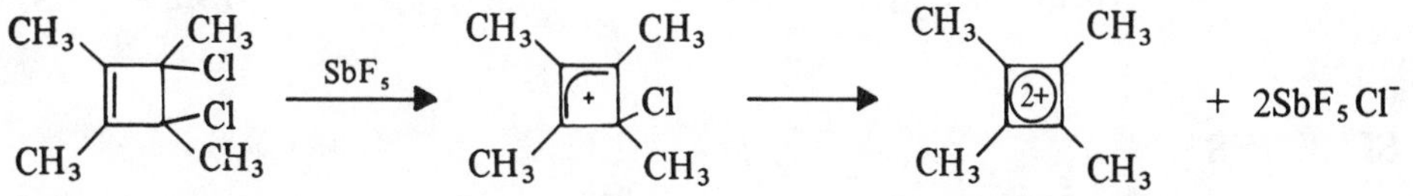

I

II
SbF_5Cl^-

III
A 2π electron aromatic dication

13.42 The vinylic protons of *p*-chlorostyrene should give a spectrum approximately like the following:

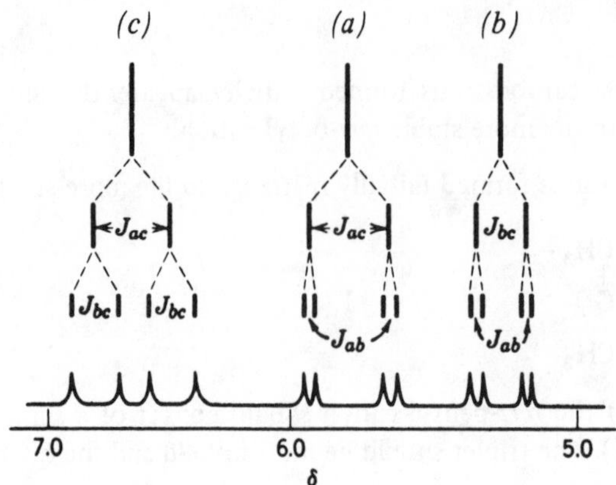

SECTION REFERENCES FOR ADDITIONAL PROBLEMS

13.19 13.7, 13.8, 13.13
13.20 13.13A, 13.13B
13.21 13.5
13.22 13.7, 13.8, 13.13A, 13.13B
13.23 13.7, 13.8
13.24 13.7, 13.8, 13.13
13.25 13.13
13.26 13.10
13.27 13.7, 13.8, 13.13
13.28 13.4
13.29 13.7, 13.8, 13.13
13.30 13.7, 13.8, 13.13
13.31 13.11
13.32 13.12
13.33 13.12
13.34 13.12
13.35 13.12, 13.13
13.36 13.8, 13.13

SELF-TEST

13.1 Propose a structure that is consistent with each set of following data.

(a) C_4H_9Br Proton nmr spectrum
singlet δ 1.7

(b) $C_4H_7Br_3$ Proton nmr spectrum
singlet δ 1.95 (3H)
singlet δ 3.9 (4H)

(c) C_8H_{16}

Proton nmr spectrum	IR spectrum
Singlet δ 1.0 (9H)	3040, 2950, 1640 cm^{-1}
Singlet δ 1.75 (3H)	and other peaks.
Singlet δ 1.9 (2H)	
Singlet δ 4.6 (1H)	
Singlet δ 4.8 (1H)	

(d) $C_9H_{10}O$

Proton nmr spectrum	IR spectrum
Singlet δ 2.0 (3H)	3100, 3000, 1720,
Singlet δ 3.75 (2H)	740, 700 cm^{-1}
Singlet δ 7.2 (5H)	and other peaks.

(e) $C_5H_7NO_2$

Proton nmr spectrum	IR spectrum
Triplet δ 1.2 (3H)	2980, 2260, 1750 cm^{-1}
Singlet δ 3.5 (2H)	and other peaks.
Quartet δ 4.2 (2H)	This compound has a nitro group.

13.2 How many proton nmr signals would the following compound give?

$$\begin{array}{c} CH_3\underset{|}{C}HCH_2Cl \\ CH_3 \end{array}$$

(a) One (b) Two (c) Three (d) Four (e) Five

13.3 How many proton nmr signals would 1,1-dichlorocyclopropane give?

(a) One (b) Two (c) Three (d) Four (e) Five

SUPPLEMENTARY PROBLEM

S13.1 Explain some advantages of ^{13}C nmr spectroscopy over proton nmr spectroscopy.

SOLUTION TO SUPPLEMENTARY PROBLEM

S13.1 In ^{13}C spectroscopy, we observe the carbon skeleton directly, and, therefore, we observe peaks for *all* carbon atoms whether they bear hydrogen atoms or not.
^{13}C chemical shifts occur over a greater range than proton nmr chemical shifts.
We do not observe spin-spin couplings between carbon nuclei in ^{13}C spectra.
In proton off-resonance decoupled spectra CH_3- groups appear as quartets, $-CH_2-$ groups as triplets, $\geqslant C-H$ groups as doublets, and $-\overset{|}{\underset{|}{C}}-$ groups as singlets.

(A disadvantage of ^{13}C spectroscopy is that we do not get a quantitative measure of the relative number of the different types of carbon atoms.)

E SPECIAL TOPIC
Mass Spectroscopy

SOLUTIONS TO PROBLEMS

E.1 The compound is methane, CH_4. The molecular ion is at *m/e* 16. (This peak happens also to be the base peak.)

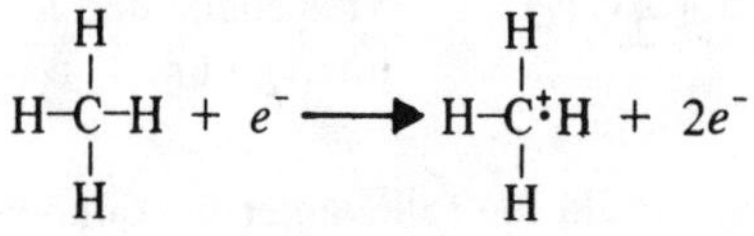

$$\begin{array}{c}\text{H}\\ |\\ \text{H—C—H}\\ |\\ \text{H}\end{array} + e^- \longrightarrow \begin{array}{c}\text{H}\\ |\\ \text{H—C}^{+\cdot}\text{H}\\ |\\ \text{H}\end{array} + 2e^-$$

m/e 16
$M^{+\cdot}$

The peaks at *m/e* 15, 14, 13, and 12 are caused by successive losses of hydrogen atoms.

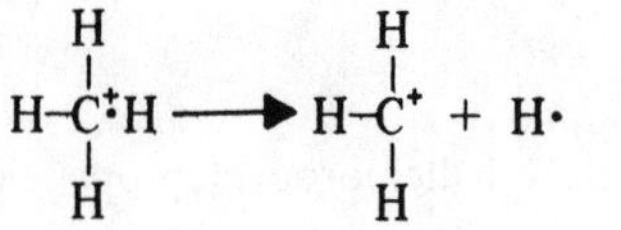

$$\begin{array}{c}\text{H}\\ |\\ \text{H—C}^{+\cdot}\text{H}\\ |\\ \text{H}\end{array} \longrightarrow \begin{array}{c}\text{H}\\ |\\ \text{H—C}^{+}\\ |\\ \text{H}\end{array} + \text{H}\cdot$$

m/e 15

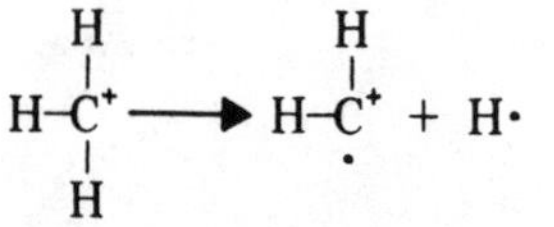

$$\begin{array}{c}\text{H}\\ |\\ \text{H—C}^{+}\\ |\\ \text{H}\end{array} \longrightarrow \begin{array}{c}\text{H}\\ |\\ \text{H—C}^{+}\\ \cdot\end{array} + \text{H}\cdot$$

m/e 14

$$\begin{array}{c}\text{H}\\ |\\ \text{H—C}^{+}\\ \cdot\end{array} \longrightarrow \text{H—C}:^{+} + \text{H}\cdot$$

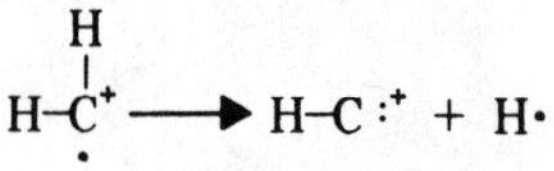

m/e 13

$$\text{H—C}:^{+} \longrightarrow \cdot\text{C}:^{+} + \text{H}\cdot$$

m/e 12

The small peak at *m/e* 17 ($M^{+\cdot} + 1$) comes mainly from methane molecules that contain ^{13}C.

$$\mathrm{H{-}^{13}CH_3} + e^- \longrightarrow \mathrm{H{-}^{13}\dot{C}^{+}H_3} + 2e^-$$

m/e 17
($M^{\ddagger}$ + 1)

E.2 The compound is water.

$$\mathrm{H{-}\ddot{O}{-}H} + e^- \longrightarrow \mathrm{H{-}\dot{O}^{+}{-}H} + 2e^-$$

m/e 18
($M^{\ddagger}$)

$$\mathrm{H{-}\dot{O}^{+}{-}H} \longrightarrow \mathrm{H{-}\ddot{O}^{+}} + \mathrm{H}\cdot$$

m/e 17

$$\mathrm{H{-}\ddot{O}^{+}} \longrightarrow \cdot\mathrm{\ddot{O}^{+}} + \mathrm{H}\cdot$$

m/e 16

The peaks at *m/e* 19 and *m/e* 20 are due (primarily) to naturally occurring oxygen isotopes.

$$\mathrm{H{-}^{17}\ddot{O}{-}H} + e^- \longrightarrow \mathrm{H{-}^{17}\dot{O}^{+}{-}H} + 2e^-$$

m/e 19
($M^{\ddagger}$ + 1)

$$\mathrm{H{-}^{18}\ddot{O}{-}H} + e^- \longrightarrow \mathrm{H{-}^{18}\dot{O}^{+}{-}H} + 2e^-$$

m/e 20
($M^{\ddagger}$ + 2)

E.3 The compound is methyl fluoride, CH_3F.

$$CH_3{-}F + e^- \longrightarrow [CH_3F]^{\ddagger} + 2e^-$$

m/e 34
($M^{\ddagger}$)

$$[CH_3F]^{\ddagger} \longrightarrow [CH_2F]^{+} + H\cdot$$

m/e 33

$$[CH_2F]^{+} \longrightarrow [CHF]^{\ddagger} + H\cdot$$

m/e 32

$$[CHF]^{\ddagger} \longrightarrow [CF]^{+} + H\cdot$$

m/e 31

$$[CH_3F]^{+\cdot} \longrightarrow [F]^+ + CH_3\cdot$$
$$m/e\ 19$$

$$[CH_3F]^{+\cdot} \longrightarrow [CH_3]^+ + F\cdot$$
$$m/e\ 15$$

$$[CH_3]^+ \longrightarrow [CH_2]^{+\cdot} + H\cdot$$
$$m/e\ 14$$

E.4 (a) $CH_3\overset{O}{\overset{\|}{C}}CH_2CH_3$ $CH_3CH_2CH_2\overset{O}{\overset{\|}{C}}H$ $CH_3\underset{CH_3}{\underset{|}{C}}H\overset{O}{\overset{\|}{C}}H$ $CH_2{=}CHCH_2OCH_3$

$CH_3CH{=}CHCH_2OH$ $CH_2{=}\overset{CH_3}{\overset{|}{C}}{-}CH_2OH$

CH₃ structures: tetrahydrofuran; 2-methyloxetane (CH₃ on ring carbon next to O); 3-methyloxetane (CH₃); ethyloxirane (CH_2CH_3); cyclopropylmethanol (CH_2OH); methoxycyclopropane ($O{-}CH_3$)

cyclobutanol (OH); 1-methylcyclopropanol (CH_3, HO); cis-2-methylcyclopropanol (CH_3, OH; H, H); trans-2-methylcyclopropanol (CH_3, H; H, OH); 2,2-dimethyloxirane (CH_3, H_3C, O); cis-2,3-dimethyloxirane (CH_3, CH_3; H, O, H); trans-2,3-dimethyloxirane (CH_3, H; H, O, CH_3)

(b) Only the first three. (The peak at 1730 cm^{-1} is due to a C=O group.)

E.5 First we recalculate the intensities of the peaks so as to base them on the $M^{+\cdot}$ peak:

m/e		INTENSITY % of $M^{+\cdot}$
86 $M^{+\cdot}$	10.0/10.0 × 100 =	100
87	0.56/10.0 × 100 =	5.6
88	0.04/10.0 × 100 =	0.4

1. Since $M^{+\cdot}$ is even, the compound must contain an even number of nitrogen atoms (i.e., 0, 2, 4, etc.)

2. The value of the $M^{+\cdot} + 1$ peak gives the number of carbon atoms

$$\text{Number of carbon atoms} = 5.6/1.1 \simeq 5$$

The compound must contain no nitrogen atoms because $C_5N_2 = (5 \times 12) + (2 \times 14) = 88$, and the molecular weight of the compound (from the $M^{+\cdot}$ peak) is only 86.

3. The very low value of the $M^{\dagger} + 2$ peak (0.4%) tells us that the compound does not contain S, Cl, or Br.

4. If the compound were composed only of C and H it would have to be C_5H_{26}:

$$H = 86 - (5 \times 12) = 26$$

But C_5H_{26} is impossible.

However, a formula with one oxygen gives a reasonable number of hydrogen atoms,

$$H = 86 - (5 \times 12) - 16 = 10$$

and thus our compound has the formula $C_5H_{10}O$.

E.6 (a) The $M^{\dagger} + 2$ peak due to $CH_3{}^{37}Cl$ (at m/e 52) should be almost one third (32.5%) as large as the $M^{\dagger}$ peak at m/e 50.

(b) The peaks due to $CH_3{}^{79}Br$ and $CH_3{}^{81}Br$ (at m/e 94 and m/e 96, respectively) should be of nearly equal intensity.

(c) That the $M^{\dagger}$ and $M^{\dagger} + 2$ peaks are of nearly equal intensity tells us that the compound contains bromine. C_3H_7Br is therefore a likely molecular formula.

$C_3 = 36$	$C_3 = 36$
$H_7 = 7$	$H_7 = 7$
$^{79}Br = \underline{79}$	$^{81}Br = \underline{81}$
$m/e = 122$	$m/e = 124$

E.7 Recalculating the intensities to base on $M^{\dagger}$

PEAK	m/e	% of BASE PEAK	% of $M^{\dagger}$
$M^{\dagger}$	73	86.1	100
$M^{\dagger} + 1$	74	3.2	3.72
$M^{\dagger} + 2$	75	0.2	0.23

These data best fit the formula C_3H_7NO.

E.8 (a) First recalculating the intensities so as to base them on the $M^{\dagger}$ peak:

m/e		INTENSITY % of $M^{\dagger}$
78 $M^{\dagger}$	$24/24 \times 100 =$	100
79	$0.8/24 \times 100 =$	3.3
80	$8/24 \times 100 =$	33

1. Since $M^{\dagger}$ is even the compound contains an even number of nitrogen atoms.

2. Number of carbon atoms = ($M^{\ddagger}$ + 1)/1.1 = 3.3/1.1 = 3.

3. The intensity of the $M^{\ddagger} + 2$ peak (33%) tells us that the compound contains one chlorine atom.

4. We use the molecular weight (from the $M^{\ddagger}$ peak) to calculate the number of hydrogen atoms.

$$H = 78 - (3 \times 12) - 35 = 7$$

Thus the formula for the compound is C_3H_7Cl.

(b) CH_3CHCH_3 (with Cl on the central C)

E.9 (a) A *tert*-butyl cation, $(CH_3)_3C^+$.

$$\left[CH_3-\underset{CH_3}{\overset{CH_3}{C}}-CH_3\right]^{\ddagger} \longrightarrow CH_3-\underset{CH_3}{\overset{CH_3}{C^+}} + CH_3\cdot$$

m/e 57

E.10 A peak at $M^{\ddagger} - 15$ involves the loss of a methyl radical and the formation of a 1° or 2° carbocation.

$$[CH_3CH_2\overset{CH_3}{CH}CH_2CH_3]^{\ddagger} \longrightarrow CH_3CH_2\overset{+}{C}HCH_2CH_3 + CH_3\cdot$$

$M^{\ddagger} - 15$

or

$$\underset{M^{\ddagger}}{[CH_3CH_2\overset{CH_3}{CH}CH_2CH_3]^{\ddagger}} \longrightarrow \underset{M^{\ddagger} - 15}{CH_3CH_2\overset{CH_3}{CH}CH_2^{+}} + CH_3\cdot$$

A peak at $M^{\ddagger} - 29$ arises from the loss of an ethyl radical and the formation of a 2° carbocation.

$$\underset{M^{\ddagger}}{[CH_3CH_2\overset{CH_3}{CH}CH_2CH_3]^{\ddagger}} \longrightarrow \underset{M^{\ddagger} - 29}{CH_3CH_2\overset{CH_3}{CH^{+}}} + CH_3CH_2\cdot$$

Since a 2° carbocation is more stable, the peak at $M^{\ddagger} - 29$ is more intense.

E.11 Both peaks arise from allylic fragmentations

$$\overset{+}{C}H_2\overset{\cdot}{-}CH-CH_2-\underset{\underset{CH_3}{|}}{C}HCH_2CH_3 \longrightarrow \overset{\cdot}{C}H_2-CH=CH_2 + \overset{+}{C}\underset{\underset{CH_3}{|}}{H}CH_2CH_3$$

Allyl radical $\quad$ *m/e* 57

$$CH_2\overset{+\cdot}{-}CH-CH_2:\underset{\underset{CH_3}{|}}{C}HCH_2CH_3 \longrightarrow \overset{+}{C}H_2-CH=CH_2 + \cdot\underset{\underset{CH_3}{|}}{C}HCH_2CH_3$$

m/e 41
Allyl cation

E.12 (a) Alcohols undergo rapid cleavage of a carbon-carbon bond next to oxygen because this leads to a resonance-stabilized cation.

$$1^\circ \text{ alcohol} \quad R:CH_2-\overset{\cdot +}{\ddot{O}}H \xrightarrow{-R\cdot} CH_2=\overset{+}{\ddot{O}}H \longleftrightarrow \overset{+}{C}H_2-\ddot{O}H$$

$$2^\circ \text{ alcohol} \quad R-\underset{}{\overset{R}{C}}H-\overset{\cdot +}{\ddot{O}}H \xrightarrow{-R\cdot} RCH=\overset{+}{\ddot{O}}H \longleftrightarrow R\overset{+}{C}H-\ddot{O}H$$

$$3^\circ \text{ alcohol} \quad R-\underset{\underset{R}{|}}{\overset{R}{C}}-\overset{\cdot +}{\ddot{O}}H \xrightarrow{-R\cdot} R-\underset{\underset{R}{|}}{C}=\overset{+}{\ddot{O}}H \longleftrightarrow R-\underset{\underset{R}{|}}{\overset{+}{C}}-\ddot{O}H$$

The cation obtained from a tertiary alcohol is the most stable (because of the electron-releasing R groups).

(b) Primary alcohols give a peak at *m/e* 31 due to $CH_2=\overset{+}{O}H$.

(c) Secondary alcohols give peaks at *m/e* 45, 59, 73, and so forth, because ions like the following are produced.

$$CH_3CH=\overset{+}{O}H \qquad CH_3CH_2CH=\overset{+}{O}H \qquad CH_3CH_2CH_2CH=\overset{+}{O}H$$

m/e 45 $\qquad$ *m/e* 59 $\qquad$ *m/e* 73

(d) Tertiary alcohols give peaks at *m/e* 59, 73, 87, and so forth, because ions like the following are produced.

$$CH_3\underset{\underset{CH_3}{|}}{C}=\overset{+}{O}H \qquad CH_3CH_2\underset{\underset{CH_3}{|}}{C}=\overset{+}{O}H \qquad CH_3CH_2CH_2\underset{\underset{CH_3}{|}}{C}=\overset{+}{O}H$$

m/e 59 $\qquad$ *m/e* 73 $\qquad$ *m/e* 87

E.13 The spectrum given in Fig. E.12 is that of isopropyl butyl ether. The main clues are the peaks at m/e 101 and m/e 73 due to the following fragmentations.

$$\left[\begin{array}{c}CH_3\\ | \\ CH_3\text{-}CH\text{-}OCH_2CH_2CH_2CH_3\end{array}\right]^{+\cdot} \xrightarrow{-CH_3\cdot} CH_3CH{=}\overset{+}{O}CH_2CH_2CH_2CH_3 \quad m/e\ 101$$

$$\left[\begin{array}{c}CH_3\\ | \\ CH_3CH\text{-}O\text{-}CH_2CH_2CH_2CH_3\end{array}\right]^{+\cdot} \xrightarrow{-CH_3CH_2CH_2\cdot} \begin{array}{c}CH_3\\ | \\ CH_3CHO^{+}{=}CH_2\end{array} \quad m/e\ 73$$

Propyl butyl ether (Fig. E.13) has no peak at m/e 101 but has a peak at m/e 87 instead.

$$[CH_3CH_2CH_2\text{-}O\text{-}CH_2CH_2CH_2CH_3]^{+\cdot} \xrightarrow{-CH_3CH_2\cdot} CH_2{=}\overset{+}{O}CH_2CH_2CH_2CH_3 \quad m/e\ 87$$

Propyl butyl ether also has a peak at m/e 73.

$$[CH_3CH_2CH_2\text{-}O\text{-}CH_2CH_2CH_2CH_3]^{+\cdot} \xrightarrow{-CH_3CH_2CH_2\cdot} CH_3CH_2CH_2\text{-}\overset{+}{O}{=}CH_2 \quad m/e\ 73$$

[Although the observation does not help us decide, it is interesting to notice that both spectra have intense peaks at m/e 43 and m/e 57 corresponding to propyl (or isopropyl) and butyl cations formed by carbon-oxygen bond cleavage.]

E.14 The compound is butanal. The peak at m/e 44 arises from a McLafferty rearrangement.

$$\left[\text{H-C}(=O)\text{-}CH_2\text{-}CH_2\text{-}CH_2\text{-H}\right]^{+\cdot} \longrightarrow \left[\text{H-C}(\text{O-H}){=}CH_2\right]^{+\cdot} + CH_2{=}CH_2$$

m/e 72 $M^{+\cdot}$ — m/e 44 $(M^{+\cdot} - 28)$

The peak at m/e 29 arises from a fragmentation producing an acylium ion.

$$\text{H-C}(=O^{+\cdot})\text{-}CH_2CH_2CH_3 \longrightarrow \text{H-C}{\equiv}\overset{+}{O} + CH_3CH_2CH_2\cdot \quad m/e\ 29$$

E.15 The ion, $CH_2{=}\overset{+}{N}H_2$, produced by the following fragmentation.

$$R\text{:}CH_2\text{-}\overset{\cdot +}{N}H_2 \xrightarrow{-R\cdot} CH_2{=}\overset{+}{N}H_2 \longleftrightarrow \overset{+}{C}H_2\text{-}\ddot{N}H_2 \quad m/e\ 30$$

E.16 Compound A is *tert*-butylamine. Our first clue is the molecular ion at *m/e* 73 (an odd-numbered mass unit) indicating the presence of an odd number of nitrogen atoms. The base peak at *m/e* 58 is our second important clue. It arises from the following fragmentation.

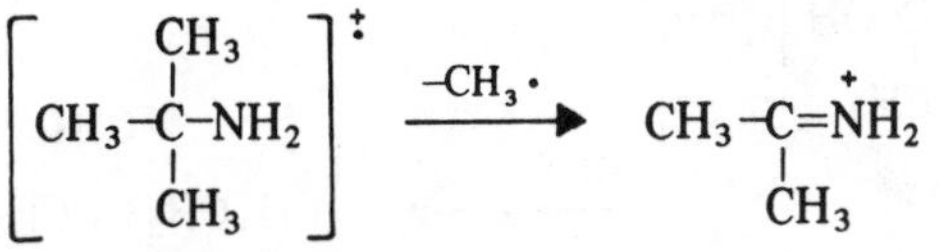

$$\left[CH_3-\underset{CH_3}{\overset{CH_3}{\underset{|}{\overset{|}{C}}}}-NH_2 \right]^{+\cdot} \xrightarrow{-CH_3\cdot} CH_3-\underset{CH_3}{\underset{|}{C}}=\overset{+}{N}H_2$$

m/e 58

The proton nmr spectrum confirms the structure

$$\overset{(a)}{(CH_3)_3}C-\overset{(b)}{NH_2}$$

(a) Singlet δ 1.2(9H)

(b) Singlet δ 1.3(2H)

E.17 The compound is 2-methyl-2-butanol. Although the molecular ion is not discernible, we are given that it is at *m/e* 88. This information gives us the molecular weight of **B** and rules out the possibility of a structure with an odd number of nitrogen atoms.

The infrared absorption (3200-3600 cm^{-1}) suggests the presence of an –OH group.

Two important peaks in the mass spectrum are the intense peaks at *m/e* 59 and *m/e* 73. These peaks correspond to fragmentation reactions that produce resonance-stabilized oxonium ions and strongly suggest that we have a tertiary alcohol [see Problem E.12, part (d)].

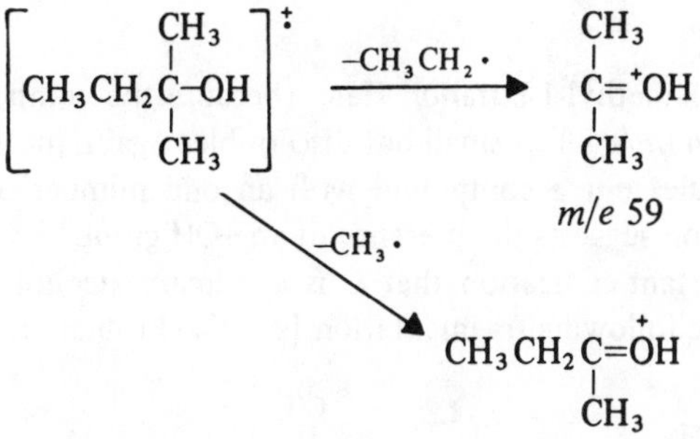

$$\left[CH_3CH_2\underset{CH_3}{\overset{CH_3}{\underset{|}{\overset{|}{C}}}}-OH \right]^{+\cdot} \xrightarrow{-CH_3CH_2\cdot} \underset{CH_3}{\overset{CH_3}{\underset{|}{\overset{|}{C}}}}=\overset{+}{O}H \quad (m/e\ 59)$$

$$\xrightarrow{-CH_3\cdot} CH_3CH_2\underset{CH_3}{\underset{|}{C}}=\overset{+}{O}H \quad (m/e\ 73)$$

The peak at *m/e* 70 corresponds to the loss of a molecule of water from the molecular ion and the peak at *m/e* 55 probably arises from a subsequent allylic cleavage

$$\left[CH_3CH_2\overset{CH_3}{\underset{CH_3}{\overset{|}{\underset{|}{C}}}}-OH\right]^{\ddagger} \xrightarrow{-H_2O} \left[CH_3CH=\overset{CH_3}{\overset{|}{C}}-CH_3\right]^{\ddagger} \quad m/e\ 70$$

$$\left[CH_3CH_2\overset{CH_3}{\underset{CH_3}{\overset{|}{\underset{|}{C}}}}-OH\right]^{\ddagger} \xrightarrow{-H_2O} \left[CH_3CH_2\overset{CH_3}{\overset{|}{C}}=CH_2\right]^{\ddagger} \quad m/e\ 70$$

$$\xrightarrow{-CH_3\cdot} \overset{+}{C}H_2-\overset{CH_3}{\overset{|}{C}}=CH_2 \quad m/e\ 55$$

The proton nmr spectrum of **B** confirms that it is 2-methyl-2-butanol

$$\overset{(a)}{CH_3}-\overset{(b)}{CH_2}-\overset{\overset{(c)}{CH_3}}{\underset{\underset{(d)}{OH}}{\overset{|}{\underset{|}{C}}}}-CH_3\ (c)$$

(a) Triplet, δ 0.9 (3H)

(b) Quartet, δ 1.6 (2H)

(c) and *(d)* Overlapping singlets, δ 1.1 (7H)

E.18 Compound **C** is 3-methyl-1-butanol. Here, (because the compound is a primary alcohol) the molecular ion (m/e 88) is small but discernible. Again, the even-numbered mass of the molecular ion rules out a compound with an odd number of nitrogen atoms and the infrared absorption suggests the presence of an $-OH$ group.

An important indication that C is a primary alcohol is the peak at m/e 31 corresponding to the following fragmentation [see also Problem E.12, part (b)].

$$\left[CH_3-\overset{CH_3}{\overset{|}{C}}HCH_2CH_2OH\right]^{\ddagger} \xrightarrow{-CH_3\overset{CH_3}{\overset{|}{C}}HCH_2\cdot} CH_2=\overset{+}{O}H$$

$$m/e\ 88 \qquad\qquad m/e\ 31$$

The peak at m/e 70 ($M^{\ddagger}$ - 18) corresponds to the loss of water from the molecular ion.

$$\left[CH_3\overset{CH_3}{\overset{|}{C}}HCH_2CH_2OH\right]^{\ddagger} \xrightarrow{-H_2O} \left[CH_3\overset{CH_3}{\overset{|}{C}}HCH=CH_2\right]^{\ddagger}$$

$$m/e\ 88 \qquad\qquad m/e\ 70$$

The peak at m/e 55 probably comes from a subsequent allylic cleavage.

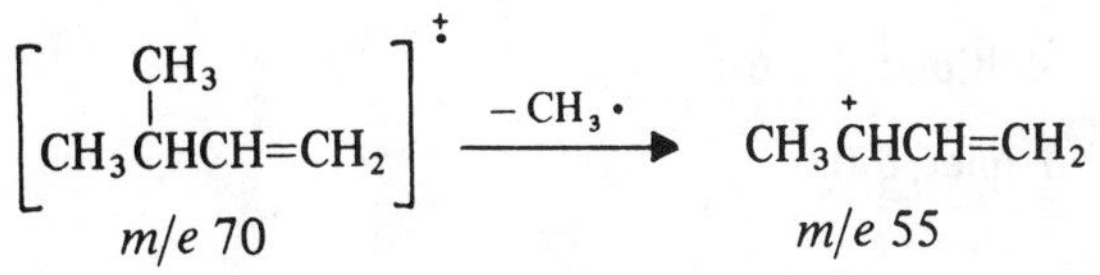

$$\left[\begin{array}{c} CH_3 \\ | \\ CH_3CHCH{=}CH_2 \end{array}\right]^{\dagger} \xrightarrow{-CH_3\cdot} CH_3\overset{+}{C}HCH{=}CH_2$$

m/e 70 → *m/e* 55

The proton nmr spectrum is consistent with this structure. We can make the following assignments.

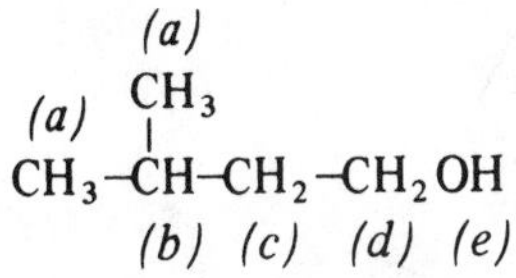

$$\overset{(a)}{\underset{(a)}{CH_3}}\ \ CH_3\text{—}\underset{(b)}{\overset{|}{C}H}\text{—}\underset{(c)}{CH_2}\text{—}\underset{(d)}{CH_2}\underset{(e)}{OH}$$

(a) Doublet, δ 0.9

(b) and *(c)* Multiplet δ 1.5

(d) Triplet δ 3.7

(e) Singlet δ 2.2

E.19 The compound is 2-pentanone. The infrared absorption at 1710 cm^{-1} strongly indicates the presence of a carbonyl group. In the mass spectrum the molecular ion peak at *m/e* 86 gives the molecular weight and rules out structures with an odd number of nitrogens. A possible formula is $C_5H_{10}O$. (See Problem E.5.)

The peaks at *m/e* 71 and *m/e* 43 correspond to $M^{\dagger} - 15$ and $M^{\dagger} - 43$. Fragmentations of 2-pentanone would produce acylium ions with these mass numbers.

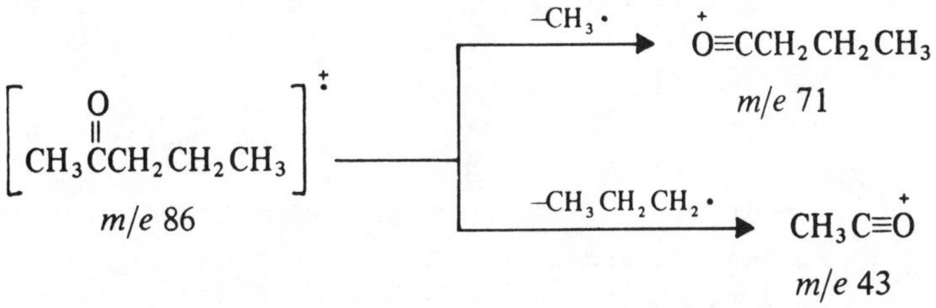

$$\left[CH_3\overset{O}{\overset{\|}{C}}CH_2CH_2CH_3\right]^{\dagger} \xrightarrow{-CH_3\cdot} \overset{+}{O}{\equiv}CCH_2CH_2CH_3$$

m/e 86 → *m/e* 71

$$\left[CH_3\overset{O}{\overset{\|}{C}}CH_2CH_2CH_3\right]^{\dagger} \xrightarrow{-CH_3CH_2CH_2\cdot} CH_3C{\equiv}\overset{+}{O}$$

m/e 86 → *m/e* 43

The peak at *m/e* 58 ($M^{\dagger} - 28$) comes from a McLafferty rearrangement.

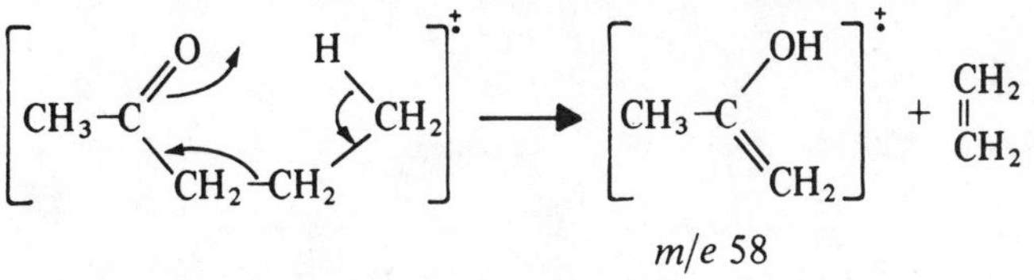

$$\left[CH_3\text{—}C(=O)\text{—}CH_2\text{—}CH_2\text{—}CH_2\text{—}H\right]^{\dagger} \longrightarrow \left[CH_3\text{—}C(OH){=}CH_2\right]^{\dagger} + CH_2{=}CH_2$$

m/e 58

The proton nmr spectrum confirms our structure.

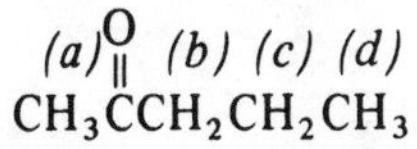

$$\underset{(a)}{CH_3}\overset{O}{\overset{\|}{C}}\underset{(b)}{CH_2}\underset{(c)}{CH_2}\underset{(d)}{CH_3}$$

(a) Singlet, δ 2.2

(b) Triplet, δ 2.4

(c) Multiplet, δ 1.6

(d) Triplet, δ 0.9

E.20 The compound is bromobenzene. That the compound contains bromine is indicated by the $M^{\ddagger}$ and $M^{\ddagger} + 2$ peaks of nearly equal intensity at *m/e* 156 and *m/e* 158. The peak at *m/e* 77 (the base peak) strongly suggests the presence of a benzene ring.

$$C_6H_5Br^{+\cdot} \xrightarrow{-Br\cdot} C_6H_5^{+}$$

m/e 77

Putting these facts together with the molecular weight (156) leads us to only one logical conclusion.

SECTION REFERENCES FOR ADDITIONAL PROBLEMS

E.14	E.4C	**E.18**	E.4B, E.4C
E.15	E.4B	**E.19**	E.4B, E.4C
E.16	E.4B	**E.20**	E.3A
E.17	E.4B, E.4C		

14 PHENOLS AND ARYL HALIDES. NUCLEOPHILIC AROMATIC SUBSTITUTION

SUMMARY OF PHENOLS

SO_3, H_2SO_4 ; SO_3H ; (1) NaOH (2) dry ; SO_3Na ; NaOH 350°C ; ONa ; H_3O^+

Cl ; NaOH 350°C, pressure ; ONa ; H_3O^+ ; OH

$CH_2{=}CHCH_3$, H_3PO_4, heat, pressure ; $CH(CH_3)_2$; O_2 ; CH_3, $CH_3{-}C{-}O{-}O{-}H$; H_3O^+ heat

NaOH ; ONa ; $(CH_3)_2SO_4$; OCH_3 ; CO_2 ; OH, CO_2Na

$NaHCO_3$; no evolution of CO_2

RCOCl or $(RCO)_2O$; O-C(=O)-R

Br_2, H_2O ; OH, Br, Br, Br

dil. HNO_3 ; OH, NO_2 + OH, NO_2

H_2SO_4, 25°C ; OH, SO_3H

H_2SO_4, 100°C ; OH, SO_3H

SOLUTIONS TO PROBLEMS

14.1 In structures 2-4, the carbon-oxygen bond is a double bond. Thus we would expect the carbon-oxygen bond of a phenol to be much stronger than that of an alcohol.

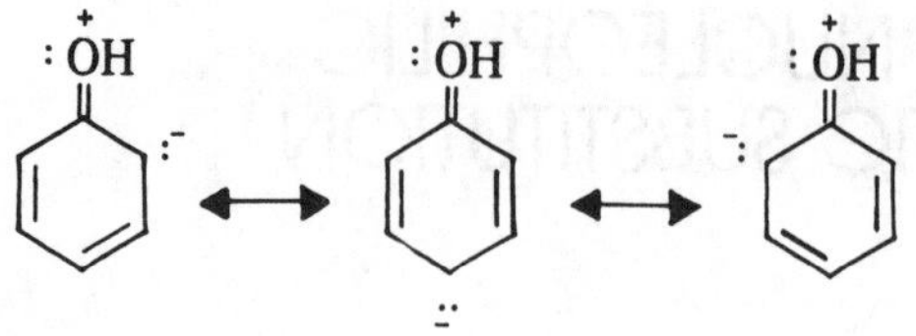

2 3 4

The strength of the carbon-oxygen bond is one factor that helps explain the low reactivity of phenol toward conc. HBr. (Another factor is the high energy associated with phenol protonated on oxygen. Structures comparable to 2, 3, and 4 would not contribute appreciably to such a hybrid, because the oxygen would bear a double positive charge.)

14.2 An electron-releasing group (i.e., $-CH_3$) destabilizes the phenoxide anion by intensifying its negative charge. This effect makes the substituted phenol less acidic than phenol itself.

OH

$+ H_2O \rightleftharpoons$

O^-

$+ H_3O^+$

CH_3 CH_3

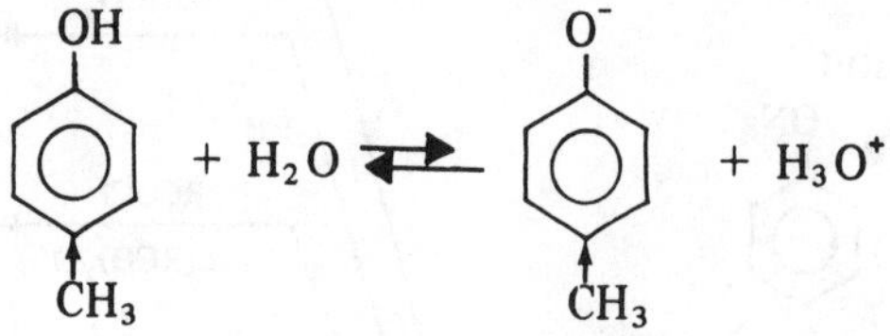

Electron-releasing $-CH_3$ destabilizes the anion more than the acid – K_a is smaller than for phenol.

An electron-withdrawing group such as chlorine can stabilize the phenoxide ion by dispersing its negative charge through an inductive effect. This effect makes the substituted phenol more acidic than phenol itself.

OH

$+ H_2O \rightleftharpoons$

O^-

$+ H_3O^+$

Cl Cl δ-

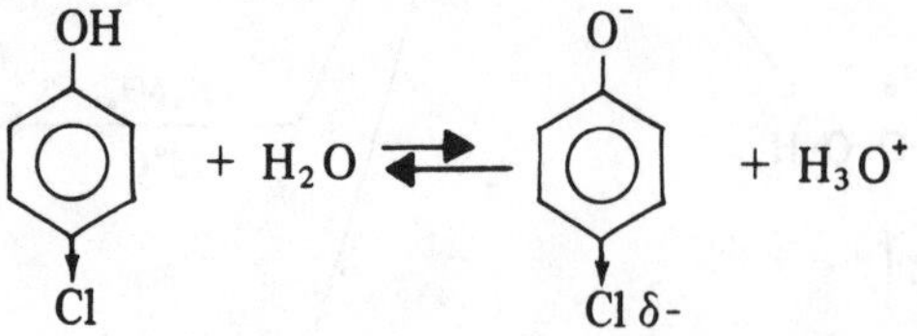

Electron-withdrawing chlorine stabilizes the anion by dispersing its negative charge. K_a is larger than for phenol.

Nitro groups are electron withdrawing by their inductive and resonance effects. The resonance effect is especially important in stabilizing the phenoxide anion. In the 2,4,6-

trinitrophenoxide anion, for example, structures, **B**, **C**, and **D** contribute to the resonance hybrid and stabilize it by dispersing the negative charge. These contributions explain why 2,4,6-trinitro phenol (picric acid) is so exceptionally acidic.

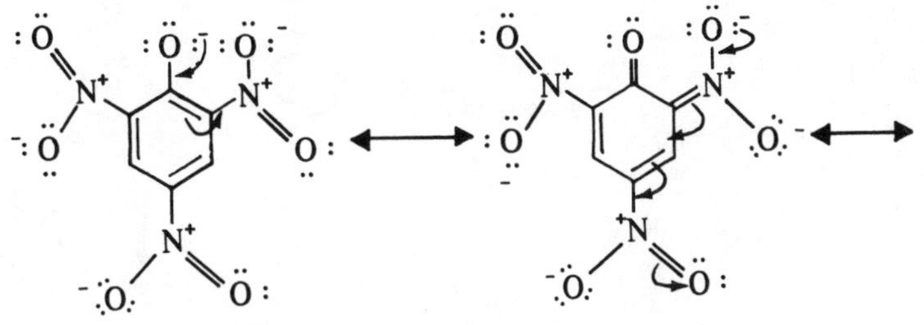

A

B

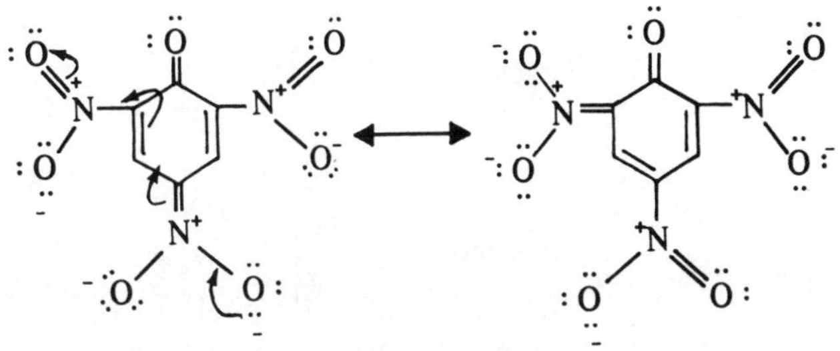

C

D

14.3 (d), (e), (f) All of these are stronger acids than H_2CO_3 (see Table 14.2), thus they would all be converted to their soluble sodium salts when treated with aqueous $NaHCO_3$. With 2,4-dinitrophenol, for example, the following reaction would take place.

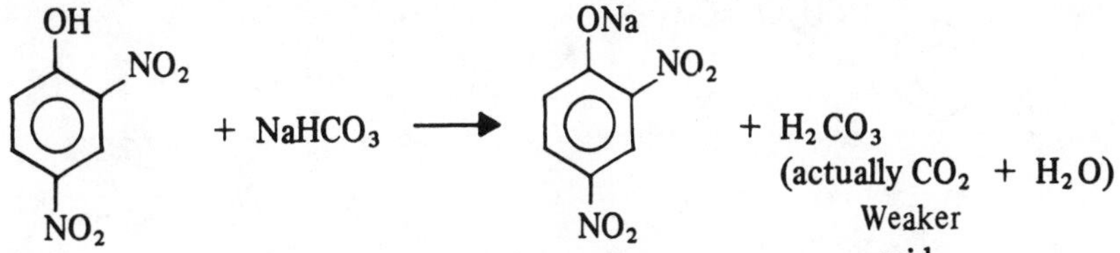

2,4-dinitrophenol (OH, NO_2, NO_2)	+ $NaHCO_3$ ⟶	sodium 2,4-dinitrophenoxide (ONa, NO_2, NO_2)	+ H_2CO_3 (actually CO_2 + H_2O)
Stronger acid (K_a = 1.1 × 10^{-4})		Water soluble	Weaker acid (K_a = 4.3 × 10^{-7})

14.4 (a) The para-sulfonated phenol, because it is the major product at the higher temperature—when the reaction is under equilibrium control.

(b) For ortho sulfonation, because it is the major reaction pathway at the lower temperature—when the reaction is under rate control.

14.5 If the mechanism involved dissociation into an allyl cation and a phenoxide ion, then recombination would lead to two products: one in which the labeled carbon atom is bonded to the ring and one in which an unlabeled carbon atom is bonded to the ring.

$C_6H_5-O-CH_2-CH{=}^{14}CH_2 \xrightarrow{\text{(dissociation)}} C_6H_5-O^- + \overset{\delta+}{C}H_2 \cdots CH \cdots {}^{14}\overset{\delta+}{C}H_2$

(recombination)

$o\text{-}HOC_6H_4-CH_2-CH{=}^{14}CH_2$ + $o\text{-}HOC_6H_4-{}^{14}CH_2-CH{=}CH_2$

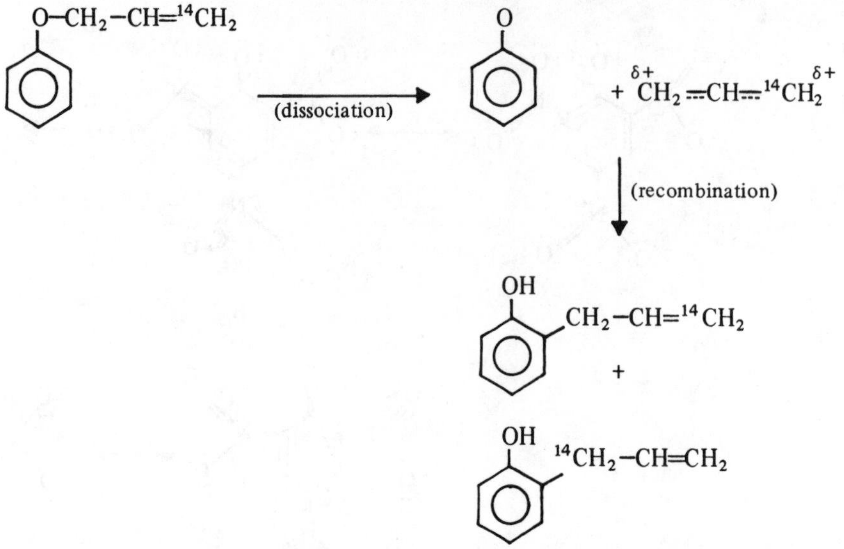

The fact that all of the product has the labeled carbon atom bonded to the ring eliminates this mechanism from consideration.

14.6 $C_6H_5OH \xrightarrow{NaOH} C_6H_5O^-Na^+ \xrightarrow[(S_N2)]{CH_2{=}CHCH_2-Br} C_6H_5OCH_2CH{=}CH_2 + NaBr$

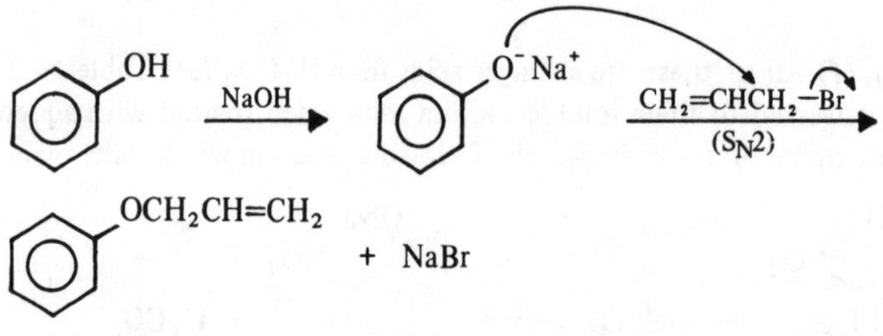

14.7 (a) (b) (c) CH_2

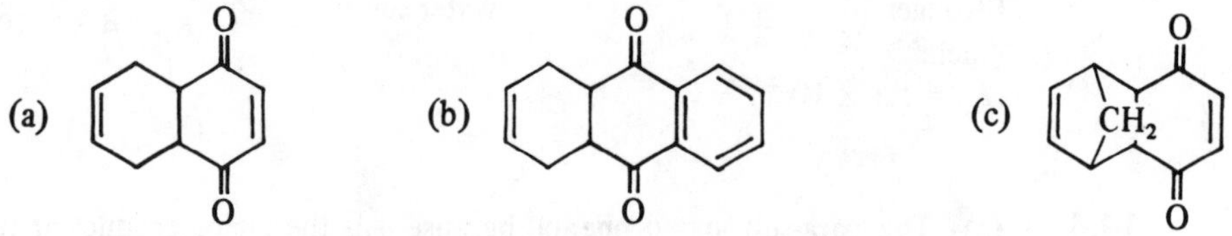

14.8 + → →

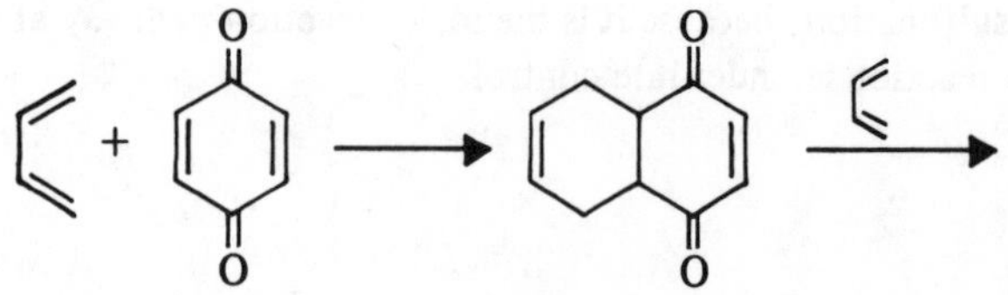

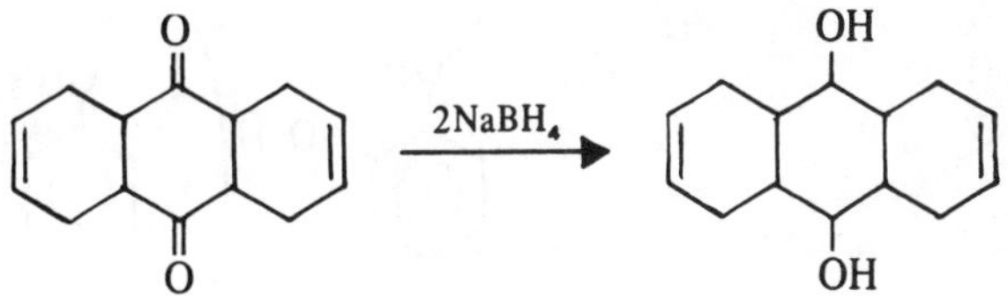

$\xrightarrow{2NaBH_4}$

14.9

(a) 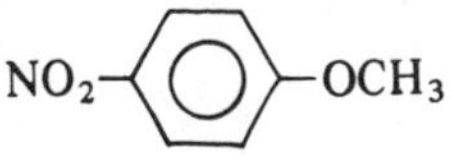 NO_2–C_6H_4–OCH_3 (b) 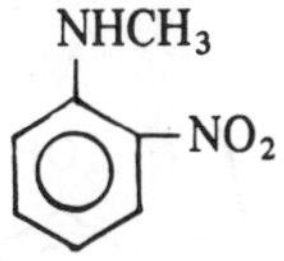 $NHCH_3$, NO_2 (c) NHC_6H_5, NO_2, NO_2

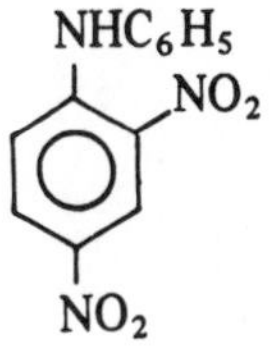

14.10 (a) 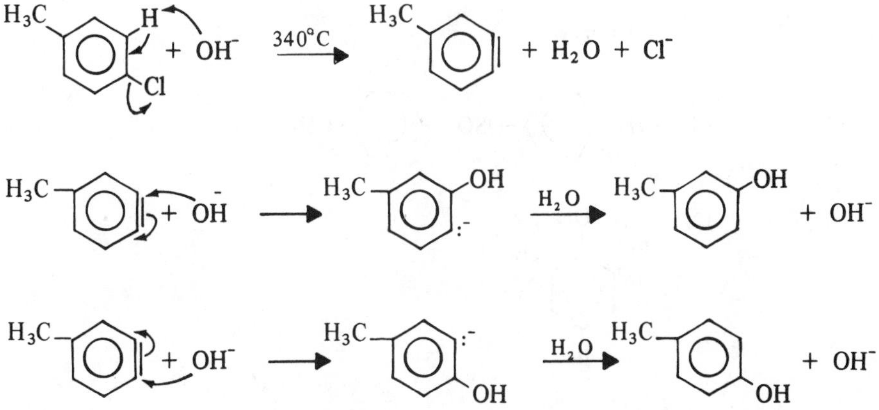 H_3C–$C_6H_3(H)(Cl)$ + OH^- $\xrightarrow{340°C}$ H_3C–benzyne + H_2O + Cl^-

H_3C–benzyne + $\bar{O}H$ ⟶ H_3C–$C_6H_3(OH)^-$ $\xrightarrow{H_2O}$ H_3C–C_6H_4–OH + OH^-

H_3C–benzyne + OH^- ⟶ H_3C–$C_6H_3(OH)^-$ $\xrightarrow{H_2O}$ H_3C–C_6H_4–OH + OH^-

(b) It suggests that the Dow Process also occurs by an elimination-addition mechanism.

14.11 Since there are no hydrogen atoms ortho to halogen, elimination cannot take place. (Reaction by a bimoleuclar displacement is not possible either, because the substrate lacks strong electron-withdrawing groups.) Thus the absence of a reaction must be due to the inability of 2-bromo-3-methylanisole to form a benzyne intermediate.

14.12

(a) 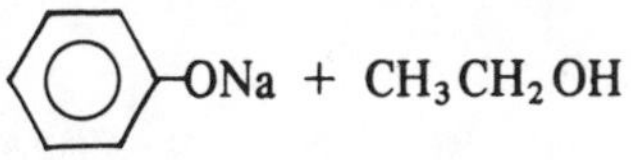 C_6H_5ONa + CH_3CH_2OH (b) C_6H_5ONa + H_2O

(c) C_6H_5OH + NaCl (d) 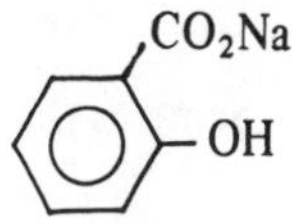 $C_6H_4(CO_2Na)(OH)$

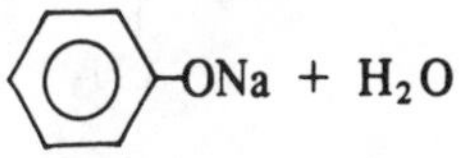

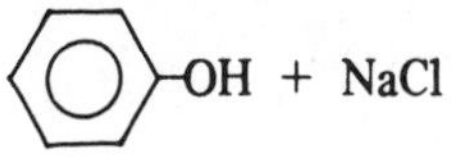

14.13 (a) p-bromophenol: OH, Br (para)

(major)

(b) OH, SO_3H (ortho) (+ OH, SO_3H (para))

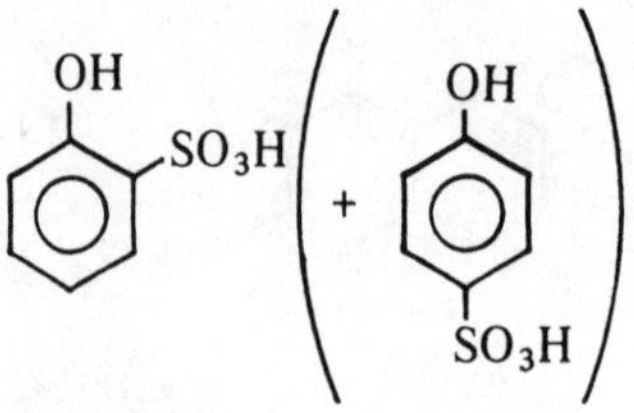

(major)

(c) OH, SO_3H (para) (+ OH, SO_3H (ortho))

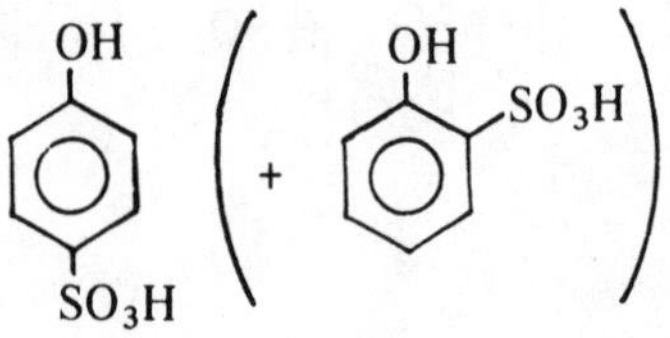

(major)

(d) CH_3–C_6H_4–OSO_2–C_6H_4–CH_3

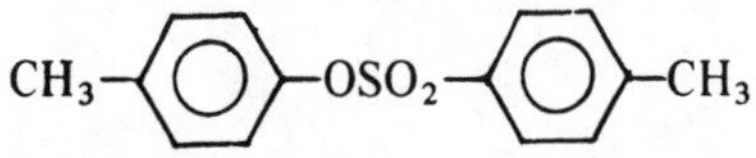

(e) OH, Br, Br, Br (2,4,6-tribromophenol)

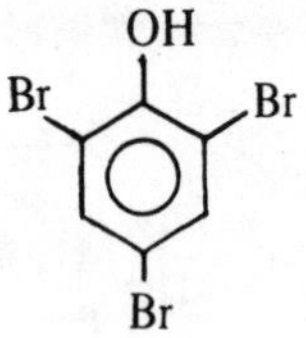

(f) C_6H_4(–$\overset{O}{\overset{\|}{C}}OC_6H_5$)(–$\underset{\underset{O}{\|}}{C}OH$)

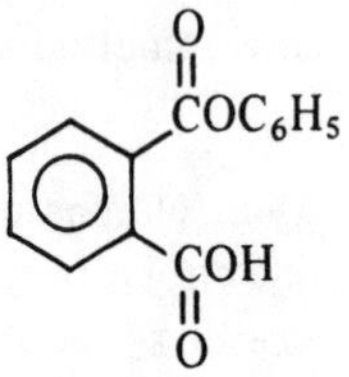

(g) OH, Br, Br, CH_3 (2,6-dibromo-4-methylphenol)

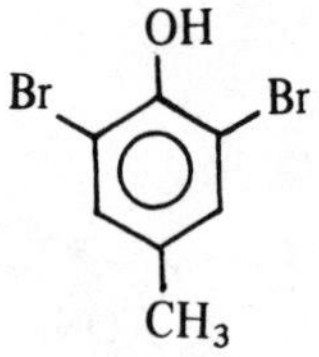

(h) C_6H_5–O–$\overset{O}{\overset{\|}{C}}$–$C_6H_5$

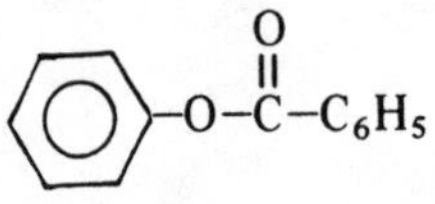

(i) Same as (h)

(j) C_6H_5–ONa

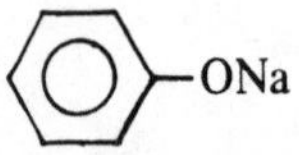

(k) C_6H_5–OCH_3

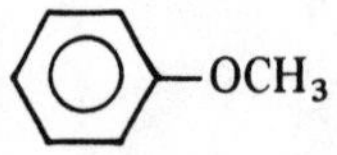

(l) Same as (k)

(m) C_6H_5–$OCH_2C_6H_5$

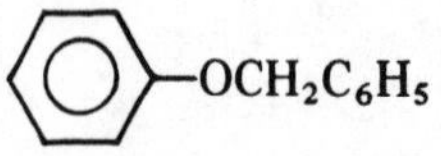

14.14 (a) *p*-Cresol is soluble in aqueous NaOH; benzyl alcohol is not.

(b) Phenol is soluble in aqueous NaOH; cyclohexane is not.

(c) Cyclohexene will decolorize Br_2/CCl_4 solution; cyclohexanol will not.

(d) Allyl phenyl ether will decolorize Br_2/CCl_4 solution; phenyl propyl ether will not.

(e) *p*-Cresol is soluble in aqueous NaOH; anisole is not.

(f) Picric acid is soluble in aqueous $NaHCO_3$: 2,4,6-trimethylphenol is not (cf. Problem 14.3)

14.15

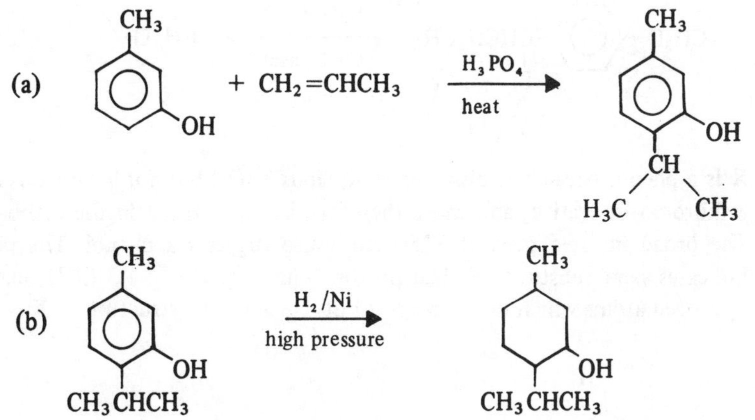

(a) m-Cresol (CH_3, OH) + $CH_2{=}CHCH_3$ $\xrightarrow[\text{heat}]{H_3PO_4}$ product (CH_3, OH, $CH(CH_3)_2$: H$_3$C–CH–CH$_3$)

(b) (CH_3, OH, CH_3CHCH_3 on benzene ring) $\xrightarrow[\text{high pressure}]{H_2/Ni}$ (CH_3, OH, CH_3CHCH_3 on cyclohexane ring)

14.16 The position ortho to the isopropyl group is sterically more hindered than the position ortho to the methyl group.

14.17

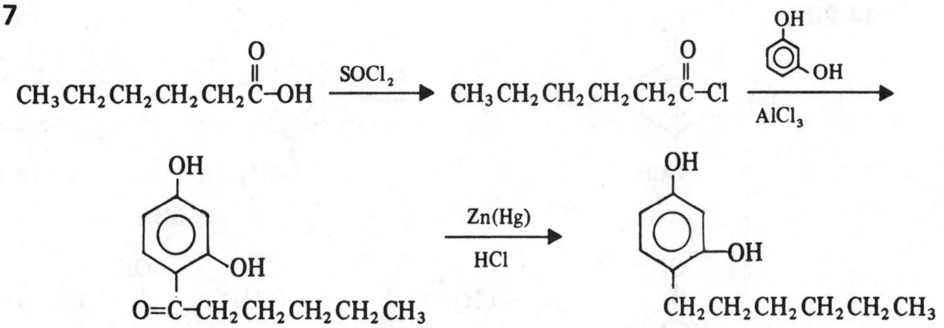

$$CH_3CH_2CH_2CH_2CH_2\overset{O}{\overset{\|}{C}}{-}OH \xrightarrow{SOCl_2} CH_3CH_2CH_2CH_2CH_2\overset{O}{\overset{\|}{C}}{-}Cl \xrightarrow[AlCl_3]{\text{resorcinol (OH, OH)}}$$

(OH, OH, $O{=}C{-}CH_2CH_2CH_2CH_2CH_3$ on benzene ring) $\xrightarrow[HCl]{Zn(Hg)}$ (OH, OH, $CH_2CH_2CH_2CH_2CH_2CH_3$ on benzene ring)

14.18

$$CH_3CH_2\overset{O}{\overset{\|}{C}}{-}OH \xrightarrow{SOCl_2} CH_3CH_2\overset{O}{\overset{\|}{C}}{-}Cl \xrightarrow[AlCl_3]{C_6H_5OCH_3} p\text{-}CH_3O{-}C_6H_4{-}\overset{O}{\overset{\|}{C}}CH_2CH_3$$

$$CH_3O{-}C_6H_4{-}\overset{O}{\overset{\|}{C}}CH_2CH_3 \xrightarrow[HCl]{Zn(Hg)} CH_3O{-}C_6H_4{-}CH_2CH_2CH_3 \xrightarrow[h\nu,\ heat]{NBS,\ CCl_4}$$

$$CH_3O{-}C_6H_4{-}\underset{Br}{\underset{|}{C}}HCH_2CH_3 \xrightarrow[EtOH,\ heat]{NaOEt} CH_3O{-}C_6H_4{-}CH{=}CH{-}CH_3$$

14.19 **X** is a phenol because it dissolves in aqueous NaOH but not in aqueous $NaHCO_3$. It gives a dibromo derivative, and must therefore be substituted in the ortho or para position. The broad infrared peak at 3250 cm^{-1} also suggests a phenol. The peak at 830 cm^{-1} indicates para substitution. The proton nmr singlet at δ 1.3 (9H) suggests nine methyl hydrogen atoms which must be a *tert*-butyl group. The structure of **X** is

$$HO{-}C_6H_4{-}C(CH_3)_3 \quad (\text{para})$$

14.20

(a)
$$p\text{-}HO{-}C_6H_4{-}OCH_3 + CH_2{=}\overset{CH_3}{\overset{|}{C}}{-}CH_3 \xrightarrow{H^+} \underset{\textbf{BHA}}{2\text{-}C(CH_3)_3\text{-}4\text{-}CH_3O{-}C_6H_3{-}OH + 3\text{-}C(CH_3)_3\text{-}4\text{-}CH_3O{-}C_6H_3{-}OH}$$

(b)
$$p\text{-}HO{-}C_6H_4{-}CH_3 + 2\,CH_2{=}\overset{CH_3}{\overset{|}{C}}{-}CH_3 \xrightarrow{H^+} \underset{\textbf{BHT}}{2,6\text{-}[(CH_3)_3C]_2\text{-}4\text{-}CH_3{-}C_6H_2{-}OH}$$

Notice that both reactions are Friedel-Crafts alkylations.

14.21

$$\text{phenol} \xrightarrow[25^\circ C]{H_2SO_4} \text{2-hydroxybenzenesulfonic acid } (SO_3H) \xrightarrow{2Cl_2} \text{3,5-dichloro-2-hydroxybenzenesulfonic acid} \xrightarrow[\substack{H_2O \\ \text{(steam)}}]{H_3O^+}$$

$$\text{2,4-dichlorophenol} \xrightarrow[ClCH_2CO_2H]{NaOH} \text{2,4-dichlorophenoxide } (O^-Na^+) + Cl\text{—}CH_2CO_2Na \longrightarrow$$

$$\text{2,4-}Cl_2C_6H_3OCH_2CO_2Na \xrightarrow{H_3O^+} \text{2,4-}Cl_2C_6H_3OCH_2CO_2H$$

2,4-D

14.22 The broad IR peak at 3200-3600 cm^{-1} suggests a hydroxyl group. The two proton nmr peaks at $\delta 1.7$ and $\delta 1.8$ are not a doublet because their separation is not equal to other splittings; therefore these peaks are singlets. Reaction with Br_2/CCl_4 suggests an alkene. If we put these bits of information together, we conclude that Z is 3-methyl-2-buten-1-ol.

$$\overset{(a)}{CH_3}(\overset{(b)}{CH_3})C{=}C(\overset{(c)}{H})\underset{(d)\,(e)}{CH_2OH}$$

The analyzed spectrum is

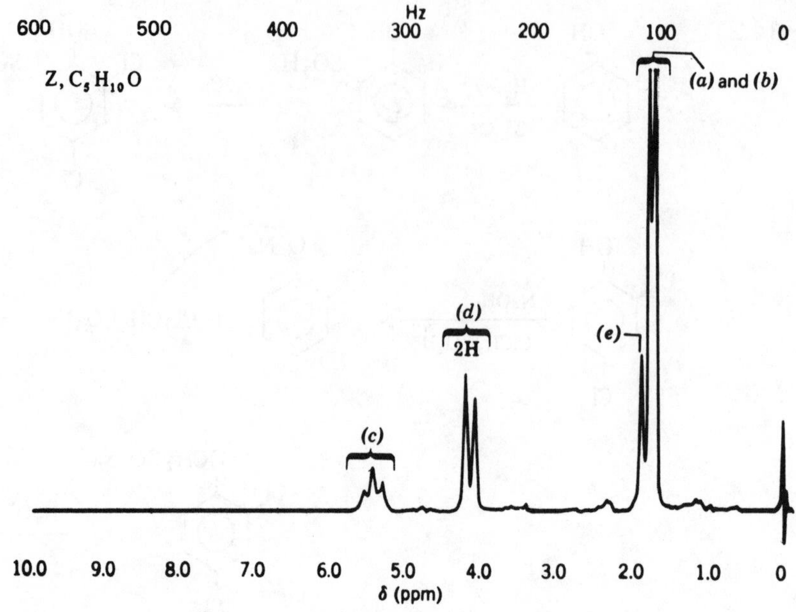

The proton nmr spectrum of compound Z, Problem 14.22. (Spectrum courtesy of Aldrich Chemical Co., Milwaukee, WI.)

SECTION REFERENCES FOR ADDITIONAL PROBLEMS

14.12 14.5, 14.6, 14.8

14.13 14.5, 14.6, 14.8

14.14 7.17, 14.5

14.15 14.3

14.16 12.12, 14.3

14.17 12.7, 12.12, 14.8

14.18 12.7, 12.12, 14.5

14.19 13.6, 13.10, 14.5, 14.8

14.20 12.6, 14.5

14.21 12.3, 14.6, 14.8

14.22 13.6, 13.10, 14.8

SELF-TEST

Complete the following reactions

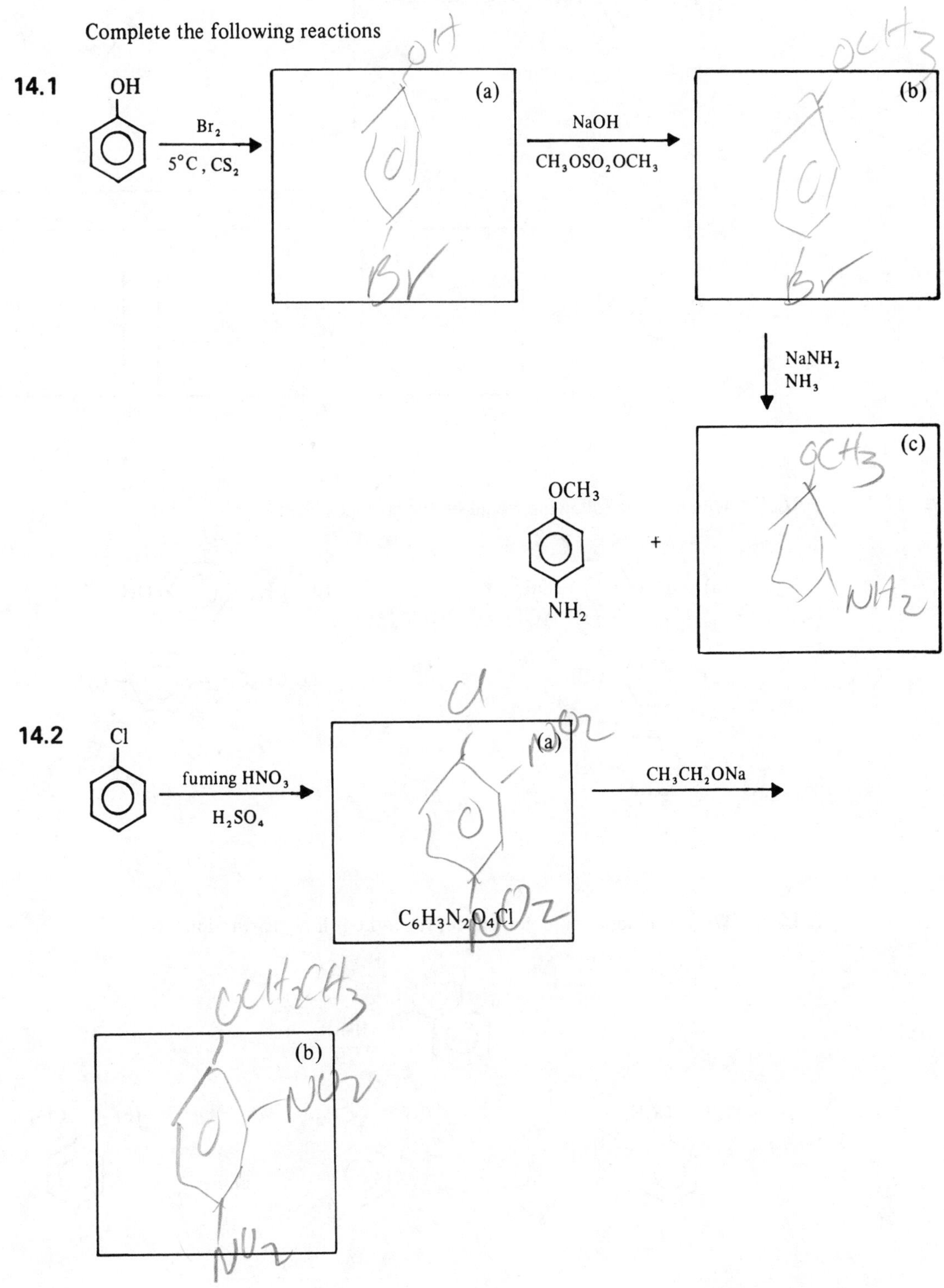

14.3

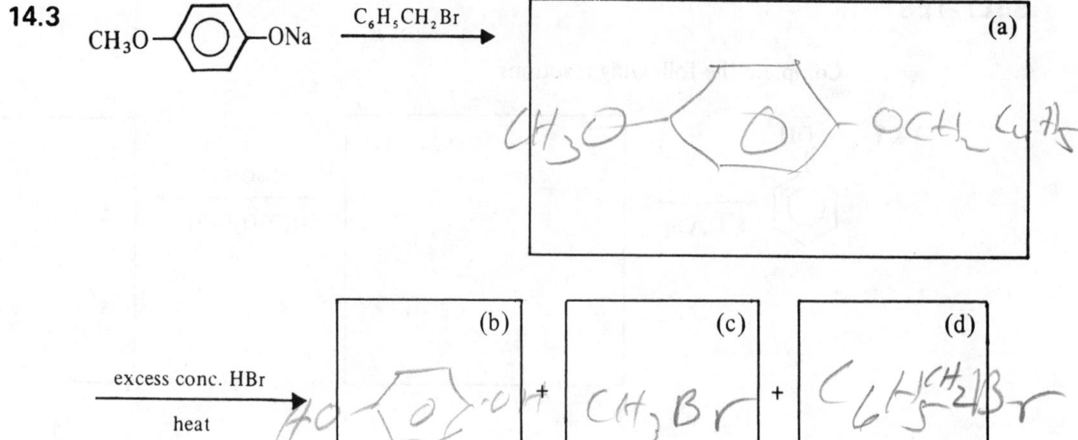

excess conc. HBr
heat

(b) + (c) + (d)

14.4 Which of the following would be the strongest acid?

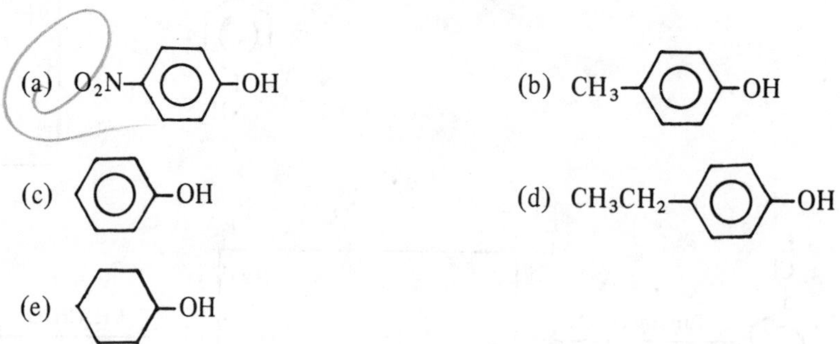

14.5 What products would you expect from the following reaction?

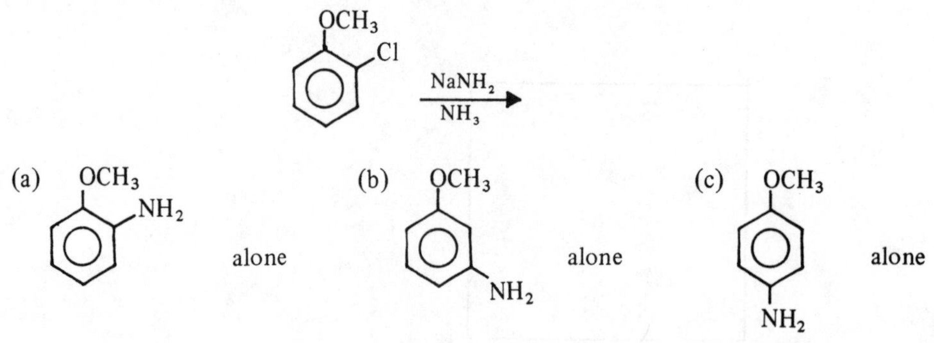

(d) More than one of the above

(e) All of the above

14.6 Which of the reagents listed here would serve as the basis for a simple chemical test to distinguish between

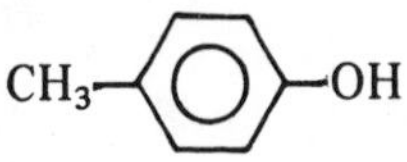

$$CH_3-C_6H_4-OH \text{ and } (CH_3)_3CHCH_2OH?$$

(a) $Ag(NH_3)_2OH$

(b) $NaOH/H_2O$

(c) Dilute HCl

(d) Cold conc. H_2SO_4

(e) None of the above

14.7 Indicate the correct product, if any, of the following reaction.

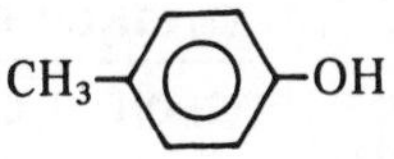

$$CH_3-C_6H_4-OH + HBr \longrightarrow ?$$

(a) 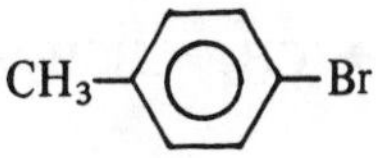 $CH_3-C_6H_4-Br$

(b) 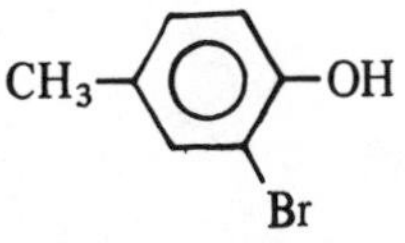 $CH_3-C_6H_3(Br)-OH$ (Br ortho to OH)

(c) $CH_3-C_6H_3(Br)-OH$ (Br ortho to OH)

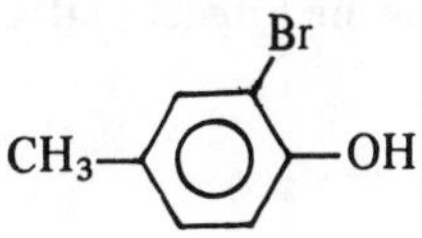

(d) 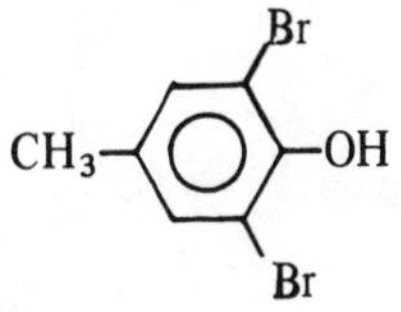 $CH_3-C_6H_2(Br)_2-OH$ (both Br ortho to OH)

(e) There is no reaction

F SPECIAL TOPIC
Thiols, Thioethers, and Thiophenols

SOLUTIONS TO PROBLEMS

F.1 (a) $C_6H_5-CH_2-\overset{+}{S}=C(NH_2)_2\ Br^-$ (b) $C_6H_5-CH_2SH$

(c) $C_6H_5-CH_2-S-S-CH_2-C_6H_5$ (d) $C_6H_5-CH_2-S^-\ Na^+$

(e) $C_6H_5-CH_2-S-CH_2-C_6H_5$

F.2

$$CH_2{=}CHCH_2Br + S{=}C(NH_2)_2 \xrightarrow[(2)\ OH^-,\ H_2O]{(1)\ CH_3CH_2OH} CH_2{=}CHCH_2SH$$

$$\xrightarrow{H_2O_2} CH_2{=}CHCH_2-S-S-CH_2CH{=}CH_2$$

F.3

$$CH_2{=}CHCH_2OH \xrightarrow{Br_2} CH_2BrCHBrCH_2OH \xrightarrow{NaSH} \underset{SH}{CH_2}-\underset{SH}{CH}-CH_2OH$$

F.4 (a) $ClCH_2CH_2\overset{O}{\overset{\|}{C}}(CH_2)_4CO_2C_2H_5$ (this step is the Friedel-Crafts acylation of an alkene)

(b) $SOCl_2$

(c) $2C_6H_5CH_2SH$ and KOH

(d) H_3O^+

(e) Ring structure: H_2C (top) bonded to H_2C (left) and to $CH(CH_2)_4CO_2H$ (right); H_2C bears $S-H$ and $CH(CH_2)_4CO_2H$ bears $S-H$ (two SH groups).

F.5 $H_2\ddot{S}: + H_2C\text{—}CH_2$ (with O bridging, epoxide) $\longrightarrow$ $H\ddot{S}\text{—}CH_2CH_2OH$ $\xrightarrow{H_2C\text{—}CH_2 \text{ (epoxide, O)}}$

$$HOCH_2CH_2SCH_2CH_2OH \xrightarrow[ZnCl_2]{HCl} ClCH_2CH_2SCH_2CH_2Cl$$

$(C_4H_{10}SO_2)$ — Mustard gas

15 ORGANIC OXIDATION AND REDUCTION REACTIONS. ORGANOMETALLIC COMPOUNDS

SOLUTIONS TO PROBLEMS

15.1 (a) H–C(H)(H)–O–H

3 H = 3(−1)
1 O = +1
Total = −2 = oxidation state of C

H–C(=O)–O–H

1 H = −1
3 O = +3
Total = +2 = oxidation state of C

H–C(=O)–H

2 H = −2
2 O = +2
Total = 0 = oxidation state of C

(b)

CH_4	CH_3OH	HC(=O)H	HC(=O)OH	CO_2
−4	−2	0	+2	+4

(c) A change from −2 to 0

(d) An oxidation, since the oxidation state increases

(e) A reduction from +6 to +3

15.2 (a) H_3C–CH_2–OH

For the CH_3 carbon:
3 H = −3
1 C = 0
Total = −3

For the CH_2OH carbon:
2 H = −2
1 C = 0
1 O = +1
Total = −1

H_3C–C(=O)–H

For the CH_3 carbon:
3 H = −3
1 C = 0
Total = −3

For the CHO carbon:
1 H = −1
1 C = 0
2 O = +2
Total = +1

(b) Only the carbon atom of the $-CH_2OH$ group of ethanol undergoes a change in oxidation state. The oxidation state of the carbon atom in the CH_3- group remains unchanged.

(c)

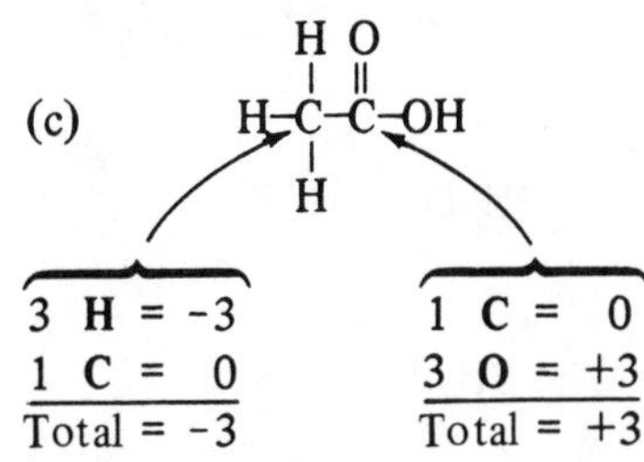

The oxygen-bearing carbon atom increases its oxidation state from +1 (in acetaldehyde) to +3 (in acetic acid)

15.3 (a) If we consider the hydrogenation of ethene as an example, we find that the oxidation state of carbon decreases. Thus, because the reaction involves the *addition* of *hydrogen,* it is both an *addition reaction* and a *reduction.*

$$H_2C{=}CH_2 + H_2 \xrightarrow{Ni} H_3C{-}CH_3$$

2 H = -2	3 H = -3
2 C = 0	1 C = 0
Total = -2	Total = -3

(b) The hydrogenation of acetaldehyde is not only an addition reaction, it is also a *reduction* because the carbon atom of the C=O group goes from a + 1 to a - 1 oxidation state. The reverse reaction (the *dehydrogenation* of ethyl alcohol) is not only an *elimination* reaction, it is also an *oxidation.*

Ion-Electron Half-Reaction Method for Balancing Organic Oxidation Reduction Equations

Only two simple rules are needed:

Rule 1 Electrons (e^-) together with protons (H^+) are arbitrarily considered the reducing agents in the half-reaction for the reduction of the oxidizing agent. Ion charges are balanced by *adding electrons to the left-hand side.* (If the reaction is run in neutral or basic solution, add an equal number of OH^- ions to both sides of the balanced half-reaction to neutralize the H^+, and show the resulting $H^+ + OH^-$ as H_2O.

Rule 2 Water (H_2O) is arbitrarily taken as the formal source of oxygen for the oxidation of the organic compound, producing *product, protons,* and *electrons* on the right-hand side. (Again, use OH^- to neutralize H^+ in the *balanced* half-reaction in neutral or basic media.)

EXAMPLE 1

Write a balanced equation for the oxidation of RCH_2OH to RCO_2H by $Cr_2O_7^{2-}$ in acid solution.

Reduction half-reaction:

$$Cr_2O_7^{2-} + H^+ + e^- \longrightarrow 2Cr^{3+} + 7H_2O$$

Balancing atoms and charges:

$$Cr_2O_7^{2-} + 14H^+ + 6e^- = 2Cr^{3+} + 7H_2O$$

Oxidation half-reaction:

$$RCH_2OH + H_2O = RCO_2H + 4H^+ + 4e^-$$

The least common multiple of a 6-electron uptake in the reduction step and a 4-electron loss in the oxidation step is 12, so we multiply the first half-reaction by 2 and the second by 3, and add:

$$3RCH_2OH + 3H_2O + 2Cr_2O_7^{2-} + 28H^+ = 3RCO_2H + 12H^+ + 4Cr^{3+} + 14H_2O$$

Canceling common terms we get:

$$3RCH_2OH + 2Cr_2O_7^{2-} + 16H^+ = 3RCO_2H + 4Cr^{3+} + 11H_2O$$

This shows that the oxidation of 3 moles of a primary alcohol to a carboxylic acid requires 2 moles of dichromate.

EXAMPLE 2

Write a balanced equation for the oxidation of styrene to benzoate ion and carbonate ion by MnO_4^- in alkaline solution.

Reduction:

$$MnO_4^- + 4H^+ + 3e^- = MnO_2 + 2H_2O \quad \text{(in acid)}$$

Since this reaction is carried out in basic solution, we must add 4 OH^- to neutralize the $4H^+$ on the left side, and, of course, 4 OH^- to the right side to maintain a balanced equation.

$$MnO_4^- + 4H^+ + 4\,OH^- + 3e^- = MnO_2 + 2H_2O + 4\,OH^-$$

or, $$MnO_4^- + 2H_2O + 3e^- = MnO_2 + 4\,OH^-$$

Oxidation:

$$ArCH{=}CH_2 + 5H_2O = ArCO_2^- + CO_3^{2-} + 13H^+ + 10e^-$$

We add 13 OH^- to each side to neutralize the H^+ on the right side,

$$ArCH{=}CH_2 + 5H_2O + 13\,OH^- = ArCO_2^- + CO_3^{2-} + 13H_2O + 10e^-$$

The least common multiple is 30, so we multiply the reduction half-reaction by 10 and the oxidation half-reaction by 3 and add:

$$3ArCH{=}CH_2 + 39\,OH^- + 10MnO_4^- + 20H_2O = 3ArCO_2^- + 3CO_3^{2-} + 24H_2O + 10MnO_2 + 40\,OH^-$$

Canceling:

$$3ArCH{=}CH_2 + 10MnO_4^- = 3ArCO_2^- + 3CO_3^{2-} + 4H_2O + 10MnO_2 + OH^-$$

SAMPLE PROBLEMS

Using the ion-electron half-reaction method, write balanced equations for the following oxidation reactions.

(a) $\text{Cyclohexene} + MnO_4^- + H^+ \xrightarrow{\text{(hot)}} HO_2C(CH_2)_4CO_2H + Mn^{2+} + H_2O$

(b) $\text{Cyclopentene} + MnO_4^- + H_2O \xrightarrow{\text{(cold)}}$ *cis*-1,2-cyclopentanediol $+ MnO_2 + OH^-$

(c) $\text{Cyclopentanol} + HNO_3 \xrightarrow{\text{(hot)}} HO_2C(CH_2)_3CO_2H + NO_2 + H_2O$

(d) $\text{1,2,3-Cyclohexanetriol} + HIO_4 \xrightarrow{\text{(cold)}} HO_2C(CH_2)_3CHO + HCO_2H + HIO_3$

SOLUTIONS TO SAMPLE PROBLEMS

(a) Reduction:

$$MnO_4^- + 8H^+ + 5e^- = Mn^{2+} + 4H_2O$$

Oxidation:

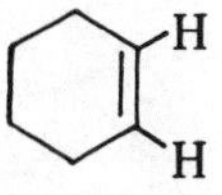

$$(\text{cyclohexene, H, H}) + 4H_2O = (CO_2H,\ CO_2H) + 8H^+ + 8e^-$$

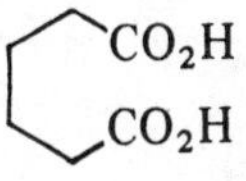

The least common multiple is 40:

$$8MnO_4^- + 64H^+ + 40\,e^- = 8Mn^{2+} + 32\,H_2O$$

$$5\,(\text{cyclohexene, H, H}) + 20\,H_2O = 5\,(CO_2H,\ CO_2H) + 40\,H^+ + 40e^-$$

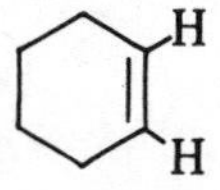

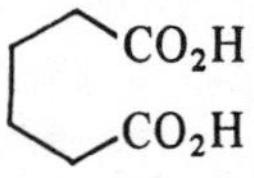

Adding and canceling:

$$5\,(\text{cyclohexene, H, H}) + 8MnO_4^- + 24H^+ = 5\,(CO_2H,\ CO_2H) + 8Mn^{2+} + 12H_2O$$

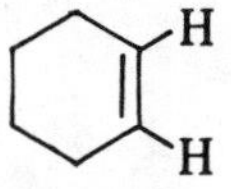

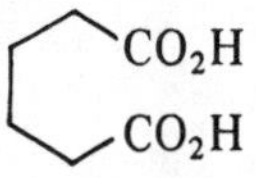

(b) Reduction:

$$MnO_4^- + 2H_2O + 3e^- = MnO_2 + 4\ OH^-$$

Oxidation:

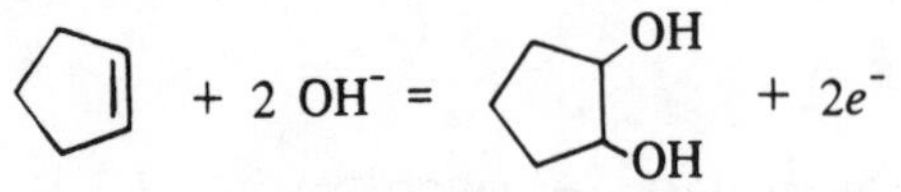

$$+ 2\ OH^- = \quad \begin{matrix} OH \\ OH \end{matrix} \quad + 2e^-$$

The least common multiple is 6:

$$2MnO_4^- + 4H_2O + 6e^- = 2MnO_2 + 8\ OH^-$$

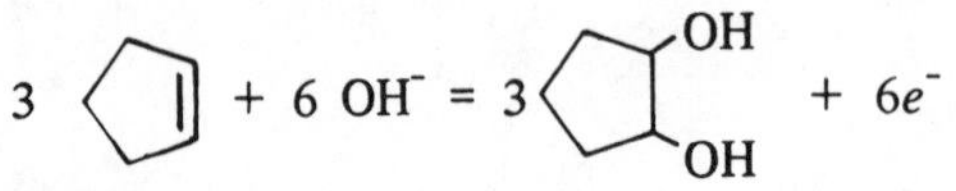

$$3 \quad + 6\ OH^- = 3 \quad \begin{matrix} OH \\ OH \end{matrix} \quad + 6e^-$$

Adding and canceling:

$$3 \quad + 2MnO_4^- + 4H_2O = 3 \quad \begin{matrix} OH \\ OH \end{matrix} \quad + 2MnO_2 + 2\ OH^-$$

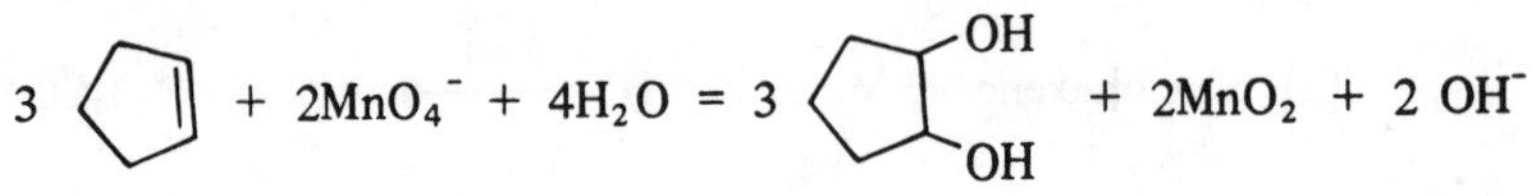

(c) Reduction:

$$HNO_3 + H^+ + e^- = NO_2 + H_2O$$

Oxidation:

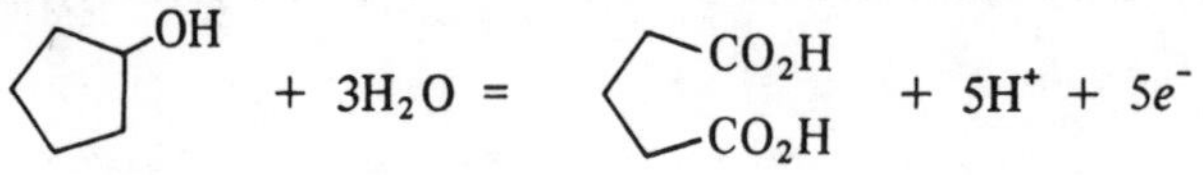

$$OH \quad + 3H_2O = \quad \begin{matrix} CO_2H \\ CO_2H \end{matrix} \quad + 5H^+ + 5e^-$$

The least common multiple is 5:

$$5HNO_3 + 5H^+ + 5e^- = 5NO_2 + 5H_2O$$

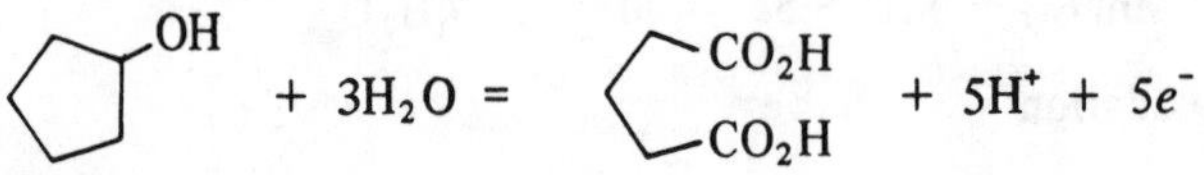

$$OH \quad + 3H_2O = \quad \begin{matrix} CO_2H \\ CO_2H \end{matrix} \quad + 5H^+ + 5e^-$$

Adding and canceling:

$$OH \quad + 5HNO_3 = \quad \begin{matrix} CO_2H \\ CO_2H \end{matrix} \quad + 5NO_2 + 2H_2O$$

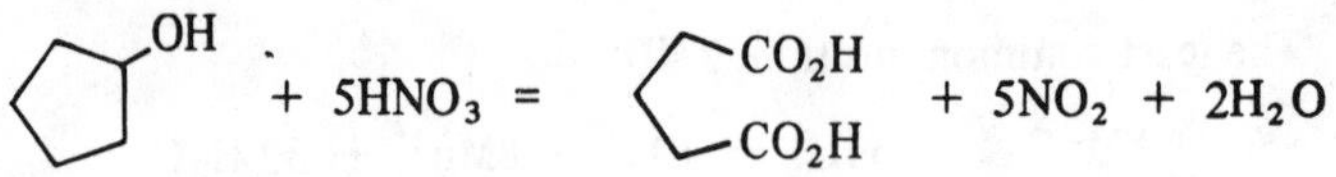

(d) Reduction:

$$HIO_4 + 2H^+ + 2e^- = HIO_3 + H_2O$$

Oxidation:

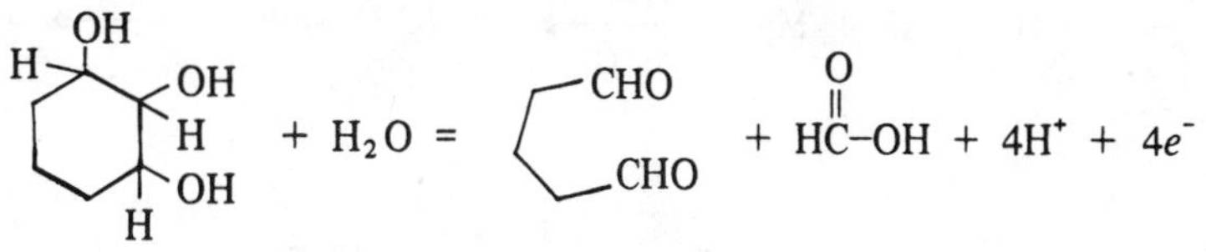

OH, H, OH, H, OH, H (cyclohexanetriol) $+ H_2O =$ CHO, CHO $+ HC(=O)-OH + 4H^+ + 4e^-$

The least common multiple is 4:

$$2HIO_4 + 4H^+ + 4e^- = 2HIO_3 + 2H_2O$$

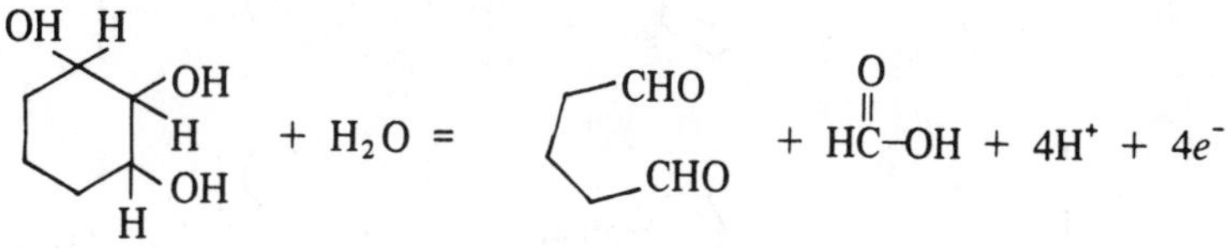

OH H, OH, H, OH, H (cyclohexanetriol) $+ H_2O =$ CHO, CHO $+ HC(=O)-OH + 4H^+ + 4e^-$

Adding and canceling:

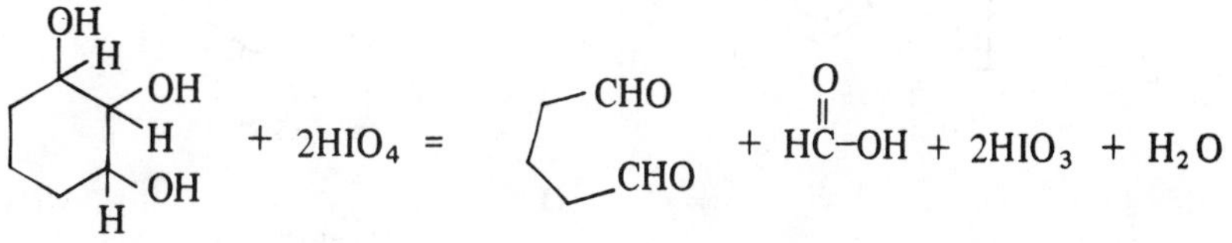

OH, H, OH, H, OH, H (cyclohexanetriol) $+ 2HIO_4 =$ CHO, CHO $+ HC(=O)-OH + 2HIO_3 + H_2O$

15.4 (a) $LiAlH_4$ (b) $NaBH_4$ ($LiAlH_4$ would reduce both carbonyl groups.)

(c) $LiAlH_4$

15.5 (a) 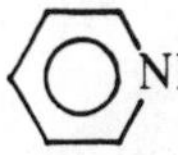 $C_5H_5NH^+ CrO_3Cl^-$ (PCC)/CH_2Cl_2

(b) $KMnO_4$, OH^-, H_2O, heat

(c) H_2CrO_4/acetone

(d) (1) O_3 (2) Zn, H_2O

15.6 (a) $\overset{\delta-}{CH_3CH_2CH_2CH_2}:\overset{\delta+}{Li} + H:\ddot{O}H \longrightarrow CH_3CH_2CH_2CH_2{-}H + Li^{+}\ {}^{-}:\ddot{O}H$

(stronger base) (stronger acid) (weaker acid) (weaker base)

(b) $\overset{\delta-}{CH_3CH_2CH_2CH_2}:\overset{\delta+}{Li} + H:\ddot{O}CH_2CH_3 \longrightarrow CH_3CH_2CH_2CH_2{-}H + Li^{+}\ {}^{-}:\ddot{O}CH_2CH_3$

(stronger base) (stronger acid) (weaker acid) (weaker base)

15.7

$$(CH_3)_3C{-}Br + Mg \xrightarrow[35^\circ C]{\text{ether}} (CH_3)_3C{-}MgBr \xrightarrow{D_2O} (CH_3)_3C{-}D$$

15.8

$$C_6H_5MgBr + Cl{-}\overset{}{C}(=O){-}C_6H_5 \longrightarrow \left[(C_6H_5)_2C(Cl){-}O{-}MgBr\right] \xrightarrow{-MgBrCl}$$

$$\left[(C_6H_5)_2C{=}O\right] \xrightarrow{C_6H_5MgBr} (C_6H_5)_3C{-}OMgBr \xrightarrow{H_3O^+} (C_6H_5)_3C{-}OH$$

15.9 (a) (1) $CH_3MgBr + CH_3\overset{O}{\overset{\|}{C}}CH_3 \xrightarrow[(2)\ H_3O^+]{(1)\ \text{ether}} CH_3{-}\underset{CH_3}{\overset{CH_3}{C}}{-}OH$

(2) $2CH_3MgBr + CH_3\overset{O}{\overset{\|}{C}}{-}OC_2H_5 \xrightarrow[(2)\ H_3O^+]{(1)\ \text{ether}} CH_3{-}\underset{CH_3}{\overset{CH_3}{C}}{-}OH$

(b) (1) $CH_3MgBr + CH_3CH_2CH_2\overset{O}{\overset{\|}{C}}H \xrightarrow[(2)\ H_3O^+]{(1)\ \text{ether}} CH_3CH_2CH_2\overset{OH}{C}HCH_3$

(2) $CH_3CH_2CH_2MgBr + CH_3\overset{O}{\overset{\|}{C}}H \xrightarrow[(2)\ H_3O^+]{(1)\ \text{ether}} CH_3CH_2CH_2\overset{OH}{C}HCH_3$

(c) (1) $C_6H_5MgBr + CH_3\overset{O}{\overset{\|}{C}}CH_2CH_3 \xrightarrow[(2)\ H_3O^+]{(1)\ \text{ether}} C_6H_5\underset{OH}{\overset{CH_3}{C}}CH_2CH_3$

(2) $CH_3MgBr + C_6H_5\overset{O}{\overset{\|}{C}}CH_2CH_3 \xrightarrow[(2)\ H_3O^+]{(1)\ \text{ether}} C_6H_5\underset{OH}{\overset{CH_3}{C}}CH_2CH_3$

(3) $CH_3CH_2MgBr + C_6H_5\overset{O}{\overset{\|}{C}}CH_3 \xrightarrow[(2)\ H_3O^+]{(1)\ \text{ether}} C_6H_5\underset{OH}{\overset{CH_3}{C}}CH_2CH_3$

(d) (1) $CH_3CH_2CH_2CH_2MgBr + H_2C\overset{O}{-}CH_2 \xrightarrow[(2)\ H_3O^+]{(1)\ ether} CH_3CH_2CH_2CH_2CH_2CH_2OH$

(2) $CH_3CH_2CH_2CH_2CH_2MgBr + CH_2O \xrightarrow[(2)\ H_3O^+]{} CH_3CH_2CH_2CH_2CH_2CH_2OH$

15.10 (a) $C_6H_6 \xrightarrow[CH_3CH_2Br]{AlCl_3} C_6H_5CH_2CH_3 \xrightarrow[CCl_4,\ h\nu]{NBS}$

$CH_3CH_2OH \xrightarrow{PBr_3} CH_3CH_2Br$

$C_6H_5CH(Br)CH_3 \xrightarrow[ether]{Mg} C_6H_5CH(MgBr)CH_3 \xrightarrow[(2)\ H_3O^+]{(1)\ HCHO}$

$C_6H_5CH(CH_3)CH_2OH \xrightarrow[CH_2Cl_2]{PCC} C_6H_5CH(CH_3)CHO$

(b) $CH_3CH_2CH_2OH \xrightarrow[CH_2Cl_2]{PCC} CH_3CH_2CHO \xrightarrow{C_6H_5MgBr,\ ether}$

$CH_3CH_2CH(OMgBr)C_6H_5 \xrightarrow[H_2O]{H_3O^+} CH_3CH_2CH(OH)C_6H_5$

(c) $C_6H_5MgBr \xrightarrow[ether\ (2)\ H_3O^+]{(1)\ HCHO} C_6H_5CH_2OH \xrightarrow[CH_2Cl_2]{PCC} C_6H_5CHO$

(d) $CH_3CH_2COOCH_3 \xrightarrow[\text{[from part (a)]}]{2C_6H_5MgBr,\ ether} C_6H_5C(OMgBr)(C_6H_5)CH_2CH_3$

$\xrightarrow[H_2O]{H_3O^+} C_6H_5C(OH)(C_6H_5)CH_2CH_3$

(e)
$$\underset{\underset{CH_3}{|}}{CH_3CHCH_2OH} \xrightarrow[CH_2Cl_2]{PCC} \underset{\underset{CH_3}{|}}{CH_3CH\overset{\overset{O}{\|}}{C}H} \xrightarrow{C_6H_5MgBr,\ ether}$$

$$CH_3\underset{\underset{CH_3}{|}}{C}H\overset{\overset{OMgBr}{|}}{C}HC_6H_5 \xrightarrow[H_2O]{H_3O^+} CH_3\underset{\underset{CH_3}{|}}{C}H\overset{\overset{OH}{|}}{C}HC_6H_5$$

15.11

(a)
$$CH_3CH_2Br \xrightarrow[\text{ether}\ (-LiBr)]{Li} CH_3CH_2Li \xrightarrow[(-LiI)]{CuI} (CH_3CH_2)_2CuLi$$

$$(CH_3CH_2)_2CuLi + CH_3I \longrightarrow CH_3CH_2CH_3 + CH_3CH_2Cu + LiI$$

(b)
$$(CH_3CH_2)_2CuLi + CH_3CH_2I \longrightarrow CH_3CH_2CH_2CH_3 + CH_3CH_2Cu + LiI$$

[from (a)]

(c)
$$CH_3\overset{\overset{CH_3}{|}}{C}HCH_2Br \xrightarrow[\text{ether}\ (-LiBr)]{Li} CH_3\overset{\overset{CH_3}{|}}{C}HCH_2Li \xrightarrow[(-LiI)]{CuI} (CH_3\overset{\overset{CH_3}{|}}{C}HCH_2)_2CuLi$$

$$\xrightarrow{CH_3I} CH_3\overset{\overset{CH_3}{|}}{C}HCH_2CH_3 + CH_3\overset{\overset{CH_3}{|}}{C}HCH_2Cu + LiI$$

(d)
$$CH_3\overset{\overset{CH_3}{|}}{C}HCH_2CH_2I \xrightarrow[\text{ether}\ (-LiI)]{Li} CH_3\overset{\overset{CH_3}{|}}{C}HCH_2CH_2Li \xrightarrow[(-LiI)]{CuI}$$

$$(CH_3\overset{\overset{CH_3}{|}}{C}HCH_2CH_2)_2CuLi \xrightarrow{CH_3\overset{\overset{CH_3}{|}}{C}HCH_2CH_2I} CH_3\overset{\overset{CH_3}{|}}{C}HCH_2CH_2CH_2CH_2\overset{\overset{CH_3}{|}}{C}HCH_3$$

$$+\ CH_3\overset{\overset{CH_3}{|}}{C}HCH_2CH_2Cu + LiI$$

Other syntheses are possible in each part except (a).

15.12

(a)
$$CH_3CH_2CH_2Br \xrightarrow[\text{ether}\ (-LiBr)]{Li} CH_3CH_2CH_2Li \xrightarrow[(-LiI)]{CuI}$$

$$(CH_3CH_2CH_2)_2CuLi \xrightarrow{CH_3CH_2CH_2Br} CH_3CH_2CH_2CH_2CH_2CH_3$$

$$+\ CH_3CH_2CH_2Cu + LiBr$$

(b) $CH_3CH_2CH_2CH_2Br \xrightarrow[\text{ether (-LiBr)}]{Li} CH_3CH_2CH_2CH_2Li \xrightarrow[(-LiI)]{CuI}$

$(CH_3CH_2CH_2CH_2)_2CuLi \xrightarrow{CH_3CH_2Br} CH_3CH_2CH_2CH_2CH_2CH_3 + CH_3CH_2CH_2CH_2Cu + LiBr$

(c) $CH_3CH_2CH_2CH_2CH_2Br \xrightarrow[\text{ether (-LiBr)}]{Li} CH_3CH_2CH_2CH_2CH_2Li \xrightarrow[(-LiI)]{CuI}$

$(CH_3CH_2CH_2CH_2CH_2)_2CuLi \xrightarrow{CH_3Br} CH_3CH_2CH_2CH_2CH_2CH_3 + CH_3CH_2CH_2CH_2CH_2Cu + LiBr$

(d) $CH_3CH_2CH_2CH_2CH_2CH_2Br \xrightarrow[Zn]{H^+} CH_3CH_2CH_2CH_2CH_2CH_3 + ZnBr_2$

(e) $CH_3CH_2CH{=}CHCH_2CH_3 \xrightarrow[C_2H_5OH\ (25°C,\ 50\ atm)]{Ni,\ H_2} CH_3CH_2CH_2CH_2CH_2CH_3$

15.13 (a) $(CH_3)_2CHCH_2OH + (CH_3)_2C{=}CH_2$ (b) $(CH_3)_2CHCH_2CN$

(c) $CH_2{=}C(CH_3)_2$ (d) $CH_3CH(CH_3)CH_2OCH_3 + (CH_3)_2C{=}CH_2$

(e) $(CH_3)_2CHCH_2{-}C(OH)(CH_3){-}CH_3$ (f) $(CH_3)_2CHCH_2CH(OH)CH_3$

(g) $(CH_3)_2CHCH_2C(OH)(CH_3)CH_2CH(CH_3)_2$ (h) $(CH_3)_2CHCH_2CH_2CH_2OH$

(i) $(CH_3)_2CHCH_2CH_2OH$ (j) $(CH_3)_2CHCH_3$

(k) $(CH_3)_2CHCH_3 + CH_3C{\equiv}CLi$

15.14 (a) CH_3CH_3 (b) CH_3CH_2D (c) $C_6H_5CH(OH)CH_2CH_3$

(d) $C_6H_5{-}C(OH)(CH_2CH_3){-}C_6H_5$ (e) $C_6H_5{-}C(OH)(CH_2CH_3){-}CH_2CH_3$ (f) $C_6H_5{-}C(OH)(CH_3){-}CH_2CH_3$

(g) $CH_3CH_3 + CH_3CH_2C{\equiv}C{-}CH(OH)CH_3$ (h) CH_3CH_3 + cyclopentadienyl–MgBr

(i) $(CH_3CH_2)_2Hg + 2MgBrCl$ (j) $(CH_3CH_2)_2Cd$ (k) $(CH_3CH_2)_3P$

15.15 (a) $(CH_3)_2CHCH(OH)CH_2CH_2CH_3$ (b) $(CH_3)_2CHC(OH)(CH_3)CH_2CH_2CH_3$

(c) $CH_3CH_2CH_3$ + $CH_3CH_2CH_2C{\equiv}C{-}C(OH)(CH_3){-}CH_3$ (d) $CH_3CH_2CH_3$

(e) $CH_3CH_2CH_2CH_2CH{=}CH_2$ (f) $CH_3CH_2CH_2{-}$cyclopentyl

(g) $CH_3CH_2CH_2(H)C{=}C(H)CH_3$ (cis) (h) $CH_3CH_2CH_2CH_3$

(i) $CH_3CH_2CH_2D$ (j) $(CH_3CH_2CH_2)_4Si$ (k) $(CH_3CH_2CH_2)_2Zn$

15.16 (a) (1) $CH_3CH_2MgBr + (CH_3)_2C{=}O \xrightarrow[(2)\ H_3O^+]{(1)\ \text{ether}} CH_3CH_2C(CH_3)_2{-}OH$

(2) $CH_3MgBr + CH_3CH_2C(CH_3){=}O \xrightarrow[(2)\ H_3O^+]{(1)\ \text{ether}} CH_3CH_2C(CH_3)_2{-}OH$

(3) $CH_3CH_2C(=O){-}OCH_3 + 2CH_3MgBr \xrightarrow[(2)\ H_3O^+]{(1)\ \text{ether}} CH_3CH_2C(CH_3)_2{-}OH$

(b) (1) $CH_3CH_2MgBr + C_6H_5C(=O){-}CH_2CH_3 \xrightarrow[(2)\ H_3O^+]{(1)\ \text{ether}} C_6H_5C(OH)(CH_2CH_3){-}CH_2CH_3$

(2) $C_6H_5MgBr + CH_3CH_2C(=O)CH_2CH_3 \xrightarrow[(2)\ H_3O^+]{(1)\ \text{ether}} C_6H_5C(OH)(CH_2CH_3){-}CH_2CH_3$

(3) $C_6H_5C(=O){-}OCH_3 + 2CH_3CH_2MgBr \xrightarrow[(2)\ H_3O^+]{(1)\ \text{ether}} C_6H_5C(OH)(CH_2CH_3){-}CH_2CH_3$

(c) cyclohexanone ($C_6H_{10}{=}O$) + $C_6H_5MgBr \xrightarrow[(2)\ H_3O^+]{(1)\ \text{ether}}$ 1-phenylcyclohexanol (ring C bearing OH and C_6H_5)

(d) $C_5H_9MgBr + H_2C\overset{O}{-}CH_2 \xrightarrow[(2)\ H_3O^+]{(1)\ ether} C_5H_9CH_2CH_2OH$ (cyclopentyl)

(e) (1) $C_4H_7MgBr + CH_3\overset{O}{\overset{\|}{C}}H \xrightarrow[(2)\ H_3O^+]{(1)\ ether} C_4H_7\underset{OH}{\underset{|}{C}}HCH_3$ (cyclobutyl)

(2) $C_4H_7\overset{O}{\overset{\|}{C}}H + CH_3MgBr \xrightarrow[(2)\ H_3O^+]{(1)\ ether} C_4H_7\underset{OH}{\underset{|}{C}}HCH_3$ (cyclobutyl)

15.17 (a) $3(CH_3)_2CHOH + PBr_3 \longrightarrow (CH_3)_2CHBr + H_3PO_3$

$(CH_3)_2CHBr + Mg \xrightarrow{ether} (CH_3)_2CHMgBr$

$(CH_3)_2CHMgBr + CH_3\overset{O}{\overset{\|}{C}}H \xrightarrow{(2)\ H_3O^+} (CH_3)_2CH\overset{OH}{\overset{|}{C}}HCH_3$

(b) $(CH_3)_2CHMgBr + H\overset{O}{\overset{\|}{C}}H \xrightarrow{(2)\ H_3O^+} (CH_3)_2CHCH_2OH$
[from (a)]

(c) $(CH_3)_2CHMgBr + CH_2\overset{O}{-}CH_2 \xrightarrow{(2)\ H_3O^+} (CH_3)_2CHCH_2CH_2OH$
[from (a)]

$(CH_3)_2CHCH_2CH_2OH \xrightarrow{SOCl_2} (CH_3)_2CHCH_2CH_2Cl$

(d) $(CH_3)_2CHMgBr + H\overset{O}{\overset{\|}{C}}CH(CH_3)_2 \xrightarrow{(2)\ H_3O^+} (CH_3)_2CH\overset{OH}{\overset{|}{C}}HCH(CH_3)_2$
[from (a)]

(e) $(CH_3)_2CHMgBr + D_2O \longrightarrow (CH_3)_2CHD$
[from (a)]

(f) $(CH_3)_2CHBr + Li \longrightarrow (CH_3)_2CHLi \xrightarrow{CuI} [(CH_3)_2CH)]_2CuLi$
[from (a)]

$[(CH_3)_2CH)]_2CuLi \xrightarrow{C_6H_{11}Br} (CH_3)_2CH\text{–}C_6H_{11}$ (cyclohexyl)

15.18 Ionization of $(C_6H_5)_2CHCl$ yields a carbocation that is highly resonance stabilized:

$$(C_6H_5)_2CHCl \longrightarrow C_6H_5\overset{+}{C}H{-}C_6H_5 + Cl^-$$

$$\updownarrow$$

$$C_6H_5{-}CH{=}C_6H_5^{+}$$

$$\updownarrow$$

$$C_6H_5{-}CH{=}C_6H_5^{+}$$

$$\updownarrow$$

and so on

15.19 The Grignard reagent that forms early in the reaction reacts with the allyl halide in an S_N2 reaction:

$$RCH{=}CH\overset{\delta-}{C}H_2{:}\overset{\delta+}{Mg}X + RCH{=}CHCH_2{-}X \longrightarrow RCH{=}CHCH_2{-}CH_2CH{=}CHR + MgX_2$$

15.20 (a) cyclopentyl bromide (Br) $+ (CH_3)_2CuLi \xrightarrow[\text{ether}]{0°C}$ methylcyclopentane (CH_3) $+ CH_3Cu + LiBr$

(b) 3-bromocyclopentene (Br) $+ (CH_3)_2CuLi \xrightarrow[\text{ether}]{0°C}$ 3-methylcyclopentene (CH_3) $+ CH_3Cu + LiBr$

(c) $CH_2{=}CH{-}CH_2Br + (CH_3CH_2)_2CuLi \xrightarrow[\text{ether}]{0°C} CH_2{=}CH{-}CH_2{-}CH_2{-}CH_3 + CH_3CH_2Cu + LiBr$

(d) $(CH_3)(H)C{=}C(CH_3)(I) + (CH_3CH_2CH_2CH_2)_2CuLi \xrightarrow[\text{ether}]{0°C}$

$(CH_3)(H)C{=}C(CH_3)(CH_2CH_2CH_2CH_3) + CH_3CH_2CH_2CH_2Cu + LiI$

15.21 (a) $C_6H_6 + CH_3CO_2^-\,Li^+$ (b) $C_6H_6 + CH_3O^-\,Li^+$

(c) $CH_4 + MgBrNH_2$ (d) $(CH_3)_4Si + 4MgBrCl$

(e) $(C_6H_5)_3P + 3MgBrCl$ (f) $(CH_3CH_2)_2Cd + 2MgBrCl$

(g) $C_6H_5CH_2OH + Mg^{2+}$

15.22 (a) Allyl bromide decolorizes Br_2/CCl_4 solution; propyl bromide does not.

(b) Benzyl bromide gives an AgBr precipitate with $AgNO_3$ in alcohol; *p*-bromotoluene does not.

(c) Benzyl chloride gives an AgCl precipitate with $AgNO_3$ in alcohol; or vinyl chloride decolorizes Br_2/CCl_4 solution.

(d) Phenyllithium (a small amount) reacts vigorously with water to give benzene and a strongly basic aqueous solution (LiOH). Diphenylmercury does not react in this way.

(e) Bromocyclohexane gives a AgBr precipitate with $AgNO_3$ in alcohol; bromobenzene does not.

15.23

(a) (1) $$C_6H_5CH{=}CH_2 + H_2O \xrightarrow[\text{heat}]{H^+} C_6H_5CH(OH)CH_3$$

(2) $$C_6H_5CH{=}CH_2 \xrightarrow[(2)\ NaBH_4,\ OH^-]{(1)\ Hg(OAc)_2,\ THF,\ H_2O} C_6H_5CH(OH)CH_3$$

(b) $$C_6H_5CH{=}CH_2 \xrightarrow{THF:BH_3} (C_6H_5CH_2CH_2)_3B \xrightarrow[OH^-,\ H_2O]{H_2O_2} C_6H_5CH_2CH_2OH$$

(c) $$C_6H_5CH_2CH_2OH \xrightarrow{Na} C_6H_5CH_2CH_2ONa \xrightarrow{CH_3Br} C_6H_5CH_2CH_2OCH_3$$

[from (b)]

(d) $C_6H_5CH(CH_3)OH \xrightarrow{Na} C_6H_5CH(CH_3)O^- \xrightarrow{CH_3CH_2Br} C_6H_5CH(CH_3)-O-CH_2CH_3$

(e) $C_6H_5CH_2CO_2H \xrightarrow[(2)\ H_2O]{(1)\ LiAlH_4,\ ether} C_6H_5CH_2CH_2OH$

(f) $C_6H_5C(=O)CH_3 \xrightarrow[H_2O]{NaBH_4} C_6H_5CH(OH)CH_3$

(g) $C_6H_5CH_3 \xrightarrow[CCl_4,\ light]{NBS} C_6H_5CH_2Br \xrightarrow[ether]{Mg} C_6H_5CH_2MgBr \xrightarrow[(2)\ H_3O^+]{(1)\ CH_2O} C_6H_5CH_2CH_2OH$

(h) $C_6H_6 \xrightarrow[FeBr_3]{Br_2} C_6H_5Br \xrightarrow[ether]{Mg} C_6H_5MgBr \xrightarrow[(2)\ H_3O^+]{(1)\ H_2C\overset{O}{—}CH_2} C_6H_5CH_2CH_2OH$

(i) $C_6H_5CH_2CO_2CH_3 \xrightarrow[ether]{LiAlH_4} C_6H_5CH_2CH_2OH$

15.24

(a) $CH_3CH_2CH_2CH_2OH \xrightarrow[heat]{PBr_3} CH_3CH_2CH_2CH_2Br \xrightarrow{(CH_3)_3COK/(CH_3)_3COH} CH_3CH_2CH{=}CH_2$

(b) $CH_3CH_2CH{=}CH_2$ [from (a)] $\xrightarrow[(2)\ NaBH_4,\ OH^-]{(1)\ Hg(OAc)_2,\ THF\text{-}H_2O} CH_3CH_2CH(OH)CH_3$

(c) $CH_3CH_2CH(OH)CH_3 \xrightarrow{H_2CrO_4} CH_3CH_2C(=O)CH_3$

(d) $$CH_3CH_2CH_2CH_2OH \xrightarrow{PBr_3} CH_3CH_2CH_2CH_2Br$$

(e) $$CH_3CH_2CH{=}CH_2 + HBr \xrightarrow[\text{(no peroxides)}]{} CH_3CH_2\underset{}{\overset{Br}{\overset{|}{C}}}HCH_3$$

(f) $$CH_3CH_2CH_2CH_2Br \xrightarrow[\text{ether}]{Mg} CH_3CH_2CH_2CH_2MgBr \xrightarrow[(2)\ H_3O^+]{(1)\ CH_2O} CH_3CH_2CH_2CH_2CH_2OH$$

(g) $$\underset{\text{[from (f)]}}{CH_3CH_2CH_2CH_2MgBr} \xrightarrow[(2)\ H_3O^+]{(1)\ H_2C\overset{O}{—}CH_2} CH_3CH_2CH_2CH_2CH_2CH_2OH$$

$$\xrightarrow{PBr_3} CH_3CH_2CH_2CH_2CH_2CH_2Br \xrightarrow{(CH_3)_3COK} CH_3CH_2CH_2CH_2CH{=}CH_2$$

(h) $$\underset{\text{[from (f)]}}{CH_3CH_2CH_2CH_2MgBr} + CH_3\overset{O}{\overset{\|}{C}}CH_2CH_3 \xrightarrow[(2)\ H_3O^+]{(1)\ \text{ether}} CH_3CH_2CH_2CH_2\overset{OH}{\overset{|}{\underset{CH_3}{\underset{|}{C}}}}CH_2CH_3$$

(i) $$CH_3CH_2CH_2CH_2OH \xrightarrow{PCC/CH_2Cl_2} CH_3CH_2CH_2\overset{O}{\overset{\|}{C}}H$$

(j) $$\underset{\text{[from (f)]}}{CH_3CH_2CH_2CH_2MgBr} + \underset{\text{[from (i)]}}{CH_3CH_2CH_2\overset{O}{\overset{\|}{C}}H} \xrightarrow[(2)\ H_3O^+]{(1)\ \text{ether}} CH_3CH_2CH_2CH_2\overset{OH}{\overset{|}{C}}HCH_2CH_2CH_3$$

(k) $$\underset{\text{[from (e)]}}{CH_3CH_2\overset{Br}{\overset{|}{C}}HCH_3} \xrightarrow[\text{ether}]{Mg} CH_3CH_2\overset{CH_3}{\overset{|}{C}}HMgBr \xrightarrow[(2)\ H_3O^+]{(1)\ CH_3CH_2CH_2\overset{O}{\overset{\|}{C}}H\ \text{[from (i)]}} CH_3CH_2\overset{CH_3}{\overset{|}{C}}H—\overset{OH}{\overset{|}{C}}HCH_2CH_2CH_3$$

(l) $$\underset{\text{[from (f)]}}{CH_3CH_2CH_2CH_2CH_2OH} + \xrightarrow[(2)\ H_3O^+]{(1)\ KMnO_4,\ OH^-} CH_3CH_2CH_2CH_2COOH$$

(m) $$\underset{\text{[from (b)]}}{CH_3CH_2\overset{CH_3}{\overset{|}{C}}HOH} \xrightarrow{Na} CH_3CH_2\overset{CH_3}{\overset{|}{C}}HONa \xrightarrow{CH_3CH_2CH_2CH_2Br} CH_3CH_2\overset{CH_3}{\overset{|}{C}}H—O—CH_2CH_2CH_2CH_3$$

or $CH_3CH_2CH{=}CH_2 + Hg(OAc)_2 \xrightarrow[CH_3CH_2CH_2CH_2OH]{THF} CH_3CH_2CH(OCH_2CH_2CH_2CH_3){-}CH_2{-}HgOAc$

$\xrightarrow[OH^-]{NaBH_4} CH_3CH_2CH(CH_3){-}O{-}CH_2CH_2CH_2CH_3$

(n) (1) $2CH_3CH_2CH_2CH_2OH \xrightarrow[140^\circ C]{H_2SO_4} (CH_3CH_2CH_2CH_2)_2O$

(2) $CH_3CH_2CH_2CH_2OH + Na \longrightarrow CH_3CH_2CH_2CH_2ONa \xrightarrow{CH_3CH_2CH_2CH_2Br} (CH_3CH_2CH_2CH_2)_2O$

(o) $CH_3CH_2CH_2CH_2Br + 2Li \longrightarrow CH_3CH_2CH_2CH_2Li + LiBr$
[from (a)]

(p) $CH_3CH_2CH_2CH_2Li \xrightarrow{CuI} (CH_3CH_2CH_2CH_2)_2CuLi \xrightarrow{CH_3CH_2CH_2CH_2Br} CH_3CH_2CH_2CH_2CH_2CH_2CH_2CH_3$
[from (o)]

15.25 That A gives a positive test with chromic oxide in aqueous sulfuric acid indicates that A is a 1° or 2° alcohol.

The ^{13}C spectrum shows the presence of $\overset{|}{\underset{|}{-CH-}}$ and $-CH_2-$ groups only (doublet and triplets in the proton off-resonance decoupled spectrum).

These facts when coupled with the molecular formula indicate that A is cyclohexanol. The following peak assignments can be made:

[Cyclohexanol structure with carbons labelled (a), (b), (c), (d); OH on carbon (d)]

(a) δ 24*t*

(b) δ 26*t*

(c) δ36*t*

(d) δ 70*d*

SECTION REFERENCES FOR ADDITIONAL PROBLEMS

15.13	5.18, 15.5-15.7	**15.20**	15.9
15.14	15.6–15.8	**15.21**	15.6, 15.8
15.15	15.6–15.9	**15.22**	7.17, 10.7, 14.10, 15.6, 15.8
15.16	15.5–15.7	**15.23**	7.5, 8.5, 8.6, 15.2, 15.7
15.17	12.6, 15.2, 15.5-15.7	**15.24**	6.13, 7.5, 8.14, 15.3, 15.7
15.18	10.6, 10.7	**15.25**	13.12, 15.4E
15.19	10.7		

SELF-TEST

15.1 Give the structural formula of the missing reactants or major organic product in each of the following reactions. Write N.R. if no reaction occurs. If two steps are needed, label them (1), (2), and so on.

(a) 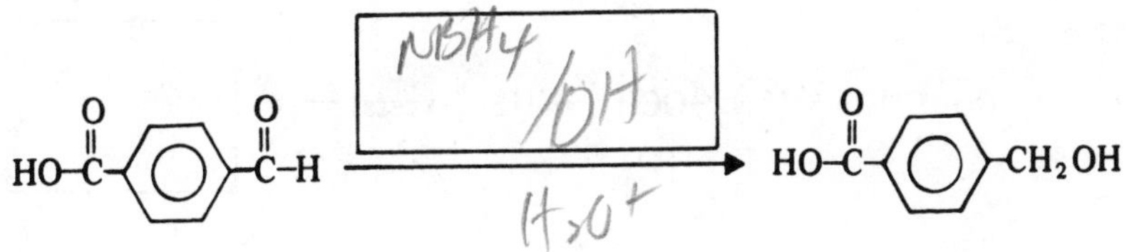 $HO-\overset{O}{\overset{\|}{C}}-C_6H_4-\overset{O}{\overset{\|}{C}}-H \xrightarrow{[\quad]} HO-\overset{O}{\overset{\|}{C}}-C_6H_4-CH_2OH$

(b) $C_6H_5-C(CH_3)_2-OH + Li \xrightarrow{\text{ether}} [\quad]$

(c) $C_6H_5MgBr \xrightarrow{H_2C\overset{O}{—}CH_2} \xrightarrow{H_3O^+} [\quad]$

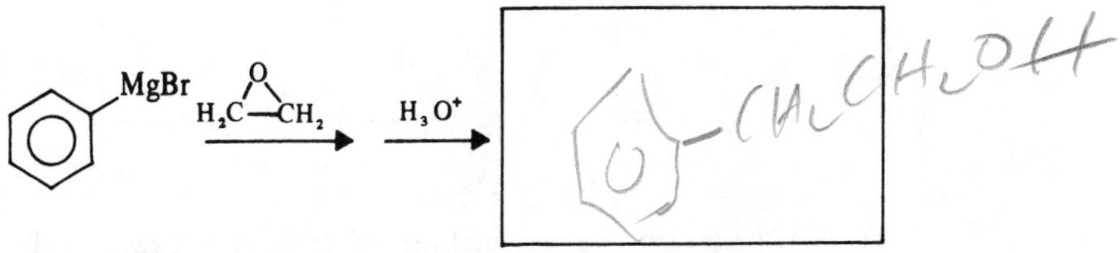

(d) 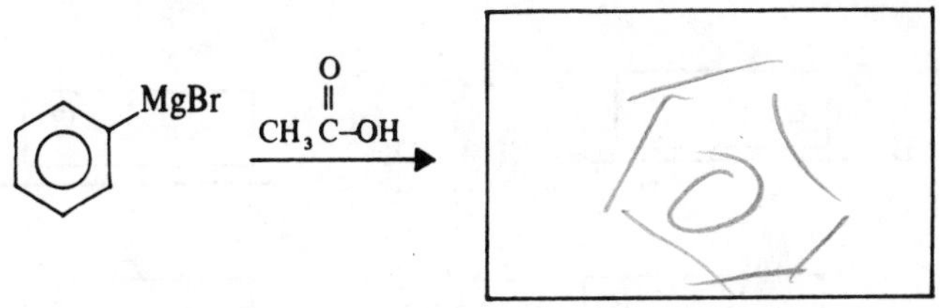 $C_6H_5MgBr \xrightarrow{CH_3\overset{O}{\overset{\|}{C}}-OH} [\quad]$

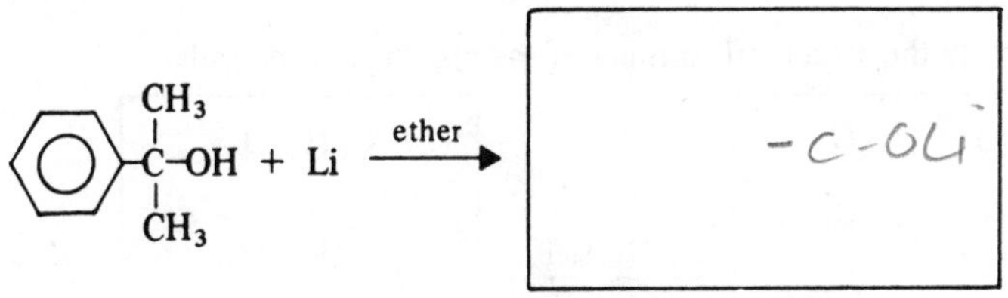

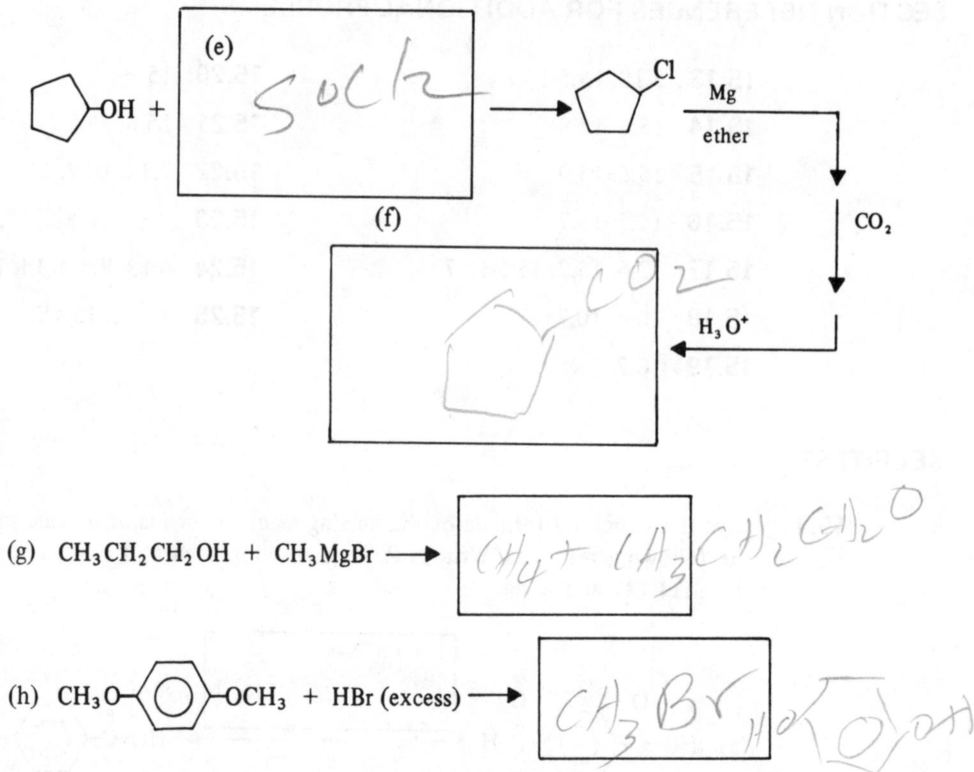

15.2 Supply the structural formula of the missing compounds.

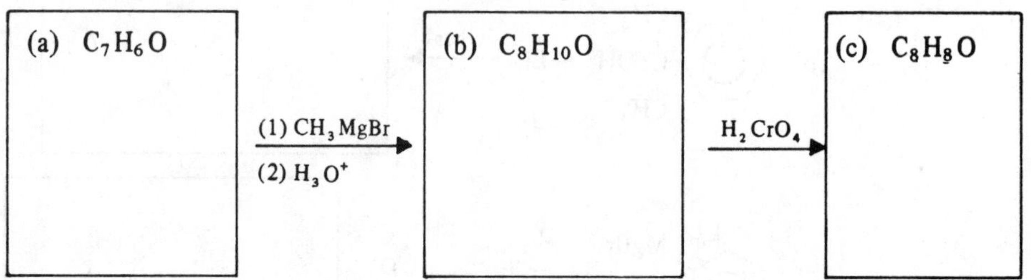

15.3 Provide the reagent (or reagents) needed to carry out each of the following transformations

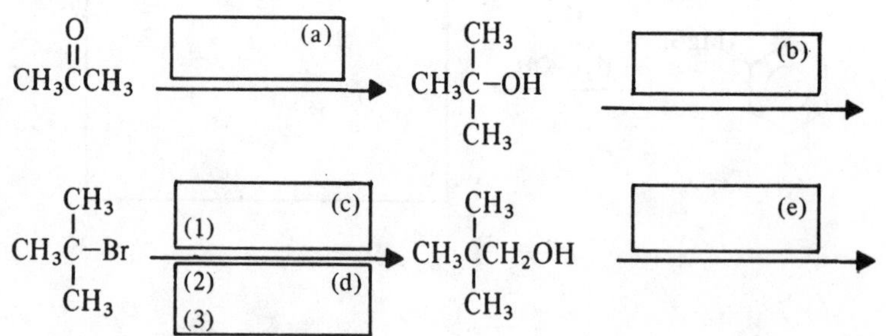

$$\underset{\displaystyle CH_3}{\overset{\displaystyle CH_3}{\overset{|}{\underset{|}{CH_3C}}}}-CHO \xrightarrow[(2)]{(1) \qquad (f)} \underset{\displaystyle CH_3}{\overset{\displaystyle CH_3}{\overset{|}{\underset{|}{CH_3C}}}}-\underset{\displaystyle OH}{\underset{|}{CH}}C\equiv CH$$

15.4 Which of the following synthetic procedures could be employed to transform ethanol into $CH_3CH_2CH_2OH$?

(a) Ethanol + HBr, then Mg/ether, then H_3O^+

(b) Ethanol + HBr, then Mg/ether, then $\overset{\displaystyle O}{\overset{\|}{HCH}}$, then H_3O^+

(c) Ethanol + H_2SO_4/140°C

(d) Ethanol + Na, then $\overset{\displaystyle O}{\overset{\|}{HCH}}$, then H_3O^+

(e) Ethanol + H_2SO_4/180°C, then $H_2C\overset{O}{-}CH_2$ (ethylene oxide)

15.5 The principal product(s) formed when *1 mole* of methyl magnesium iodide reacts with 1 mole of $CH_3\underset{\displaystyle O}{\underset{\|}{C}}CH_2CH_2OH$.

(a) $CH_4 + CH_3\underset{\displaystyle O}{\underset{\|}{C}}CH_2CH_2OMgI$

(b) $CH_3\overset{\displaystyle OMgI}{\overset{|}{\underset{\displaystyle CH_3}{\underset{|}{C}}}}CH_2CH_2OH$

(c) $CH_3\underset{\displaystyle O}{\underset{\|}{C}}CH_2CH_2OCH_3$

(d) $CH_3\overset{\displaystyle CH_3}{\overset{|}{\underset{\displaystyle OH}{\underset{|}{C}}}}CH_2CH_2OCH_3$

(e) None of the above

SPECIAL TOPIC
Transition Metal Organic Compounds

SOLUTIONS TO PROBLEMS

G.1

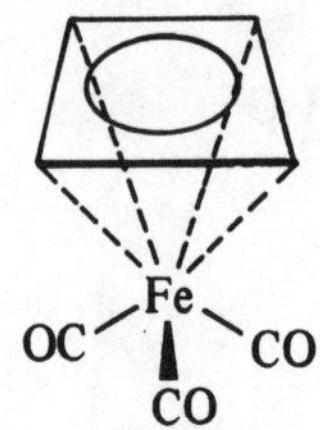

Cyclobutadiene iron tricarbonyl

d^n = Total number of valence electrons (both s and d electrons) of elemental iron − oxidation state of the metal in the complex

$d^n = 8 - 0 = 8$

Total number of valence electrons of iron in the complex = d^n + electrons donated by ligands

= 8 + 3(CO) + cyclobutadiene

= 8 + 3(2) + 4 = 18

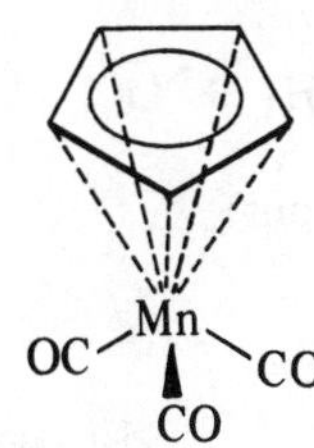

Cyclopentadienylmanganese tricarbonyl

$d^n = 7 - 1 = 6$

Total number of valence electrons of Mn in complex $= 6 + 3(CO) + Cp$

$= 6 + 3(2) + 6 = 18$

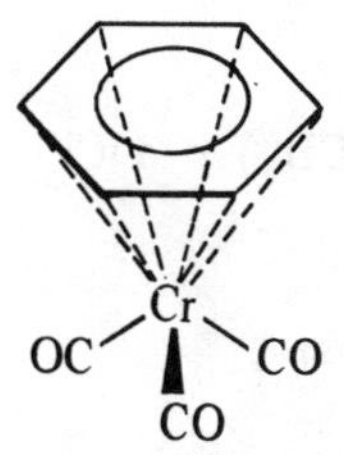

Benzene chromium tricarbonyl

$d^n = 6 - 0 = 6$

Total number of valence electrons of Cr in complex $= 6 + 3(CO) + \text{benzene}$

$= 6 + 3(2) + 6 = 18$

G.2 A syn addition of D_2 to the trans alkene would produce the following racemic form.

H, CO_2Et C=C H, EtO_2C $+ D_2 \xrightarrow{RhPh_3Cl}$ [H, CO_2Et, D; H, EtO_2C, D] + [D, CO_2Et, H; D, H, CO_2Et]

G.3 $(Ph_3P)_3RhCl + CH_3Li \xrightarrow[\text{exchange}]{\text{ligand}} (Ph_3P)_3RhCH_3 + LiCl$

(16 electrons) Rh^{I} → (16 electrons) Rh^{I}

Oxidative addition (with C_6H_5I):

$(Ph_3P)_3Rh(CH_3)I$ (with C_6H_5 on Rh) (18 electrons) Rh^{III}

$\xrightarrow[\text{elimination}]{\text{reductive}}$ $C_6H_5CH_3$ + $(Ph_3P)_3RhI$ (16 electrons) Rh^{I}

G.4 $(Ph_3P)_2Rh(CO)Cl + CH_3Li \xrightarrow{(a)} (Ph_3P)_2Rh(CO)(CH_3) + LiCl$

1 (16 electrons) Rh^{I} → **2** (16 electrons) Rh^{I}

$\xrightarrow[(b)]{C_6H_5\overset{O}{\overset{\|}{C}}Cl}$ $(Ph_3P)_2Rh(CO)(COC_6H_5)(CH_3)Cl$ **3** (18 electrons) Rh^{III}

$\xrightarrow{(c)}$ $(Ph_3P)_2Rh(CO)Cl + C_6H_5\overset{O}{\overset{\|}{C}}CH_3$ (16 electrons) Rh^{I}

(a) Is a ligand exchange

(b) Is an oxidative addition

(c) Is a reductive elimination

G.5 1. $(Ph_3P)_3Rh(CO)H \xrightarrow{-Ph_3P} (Ph_3P)_2Rh(CO)H$ **Ligand dissociation**

(18 electrons, Rh^{I}) → (16 electrons, Rh^{I})

2. $(Ph_3P)_2Rh(CO)H$ (16 electrons, Rh^{I}) $\xrightarrow{CH_3OC(=O)C{\equiv}CC(=O)OCH_3}$ OC, PPh_3, Ph_3P, H on $Rh \leftarrow C(CO_2CH_3){\equiv}C(CO_2CH_3)$ (18 electrons, Rh^{I}) **Ligand association**

3. $Rh(CO)(PPh_3)_2H \leftarrow C(CO_2CH_3){\equiv}C(CO_2CH_3)$ (18 electrons, Rh^{I}) $\longrightarrow$ $(OC)(Ph_3P)_2Rh{-}C(CO_2CH_3){=}CH(CO_2CH_3)$ (16 electrons, Rh^{I}) **Insertion**

4. $(OC)(Ph_3P)_2Rh{-}C(CO_2CH_3){=}CH(CO_2CH_3)$ (16 electrons, Rh^{I}) $\xrightarrow{CH_3{-}I}$ $(OC)(Ph_3P)_2(CH_3)(I)Rh{-}C(CO_2CH_3){=}CH(CO_2CH_3)$ (18 electrons, Rh^{III}) **Oxidative addition**

5. $(OC)(Ph_3P)_2(CH_3)(I)Rh{-}C(CO_2CH_3){=}CH(CO_2CH_3)$ (18 electrons, Rh^{III}) $\longrightarrow$ $RhI(CO)(PPh_3)_2$ (16 electrons, Rh^{I}) + $CH_3(CO_2CH_3)C{=}C(H)(CO_2CH_3)$ **Reductive Elimination**

G.6 $(CH_3)_2CuLi$ + $C_6H_5{-}I$ $\longrightarrow$ $[CH_3{-}Cu(I)(CH_3){-}C_6H_5]^-$ Li^+ **Oxidative addition**

$[CH_3{-}Cu(I)(CH_3){-}C_6H_5]^-$ Li^+ $\longrightarrow$ $C_6H_5{-}CH_3$ + CH_3Cu + LiI **Reductive elimination**

G.7 $L = Ph_3P$

1. $L_3RhCl + C_6H_5\overset{O}{\overset{\|}{C}}\text{–}H \longrightarrow$ $L_3Rh(Cl)(H)(COC_6H_5)$ **Oxidative addition**

(16 electrons, Rh^{I}) (18 electrons, Rh^{III})

2. $L_3Rh(Cl)(H)(COC_6H_5) \underset{+L}{\overset{-L}{\rightleftharpoons}} L_2Rh(Cl)(H)\text{–}COC_6H_5$ **Ligand dissociation**

(18 electrons, Rh^{III}) (16 electrons, Rh^{III})

3. $L_2Rh(Cl)(H)\text{–}COC_6H_5 \rightleftharpoons L_2Rh(Cl)(CO)(H)(C_6H_5)$ **Deinsertion**

(16 electrons, Rh^{III}) (18 electrons, Rh^{III})

4. $L_2Rh(Cl)(CO)(H)(C_6H_5) \longrightarrow L_2Rh(CO)Cl + C_6H_5\text{–}H$ **Reductive elimination**

(18 electrons, Rh^{III}) (16 electrons, Rh^{I})

H SPECIAL TOPIC
Organic Halides and Organometallic Compounds in the Environment

SOLUTIONS TO PROBLEMS

H.1

$$Cl_3C\text{-}\overset{O}{\overset{\|}{C}}H + H_2SO_4 \rightleftharpoons \left[Cl_3C\text{-}\overset{\overset{+}{O}H}{\overset{\|}{C}}H \longleftrightarrow Cl_3C\text{-}\overset{OH}{\overset{|}{\underset{+}{C}}}H \right] + HSO_4^-$$

$$Cl_3C\text{-}\overset{OH}{\overset{|}{\underset{+}{C}}}H + C_6H_5Cl \longrightarrow Cl_3C\text{-}\overset{OH}{\overset{|}{\underset{H}{\underset{|}{C}}}}\text{-}(C_6H_5Cl)^+ \xrightarrow{-H^+} Cl_3C\text{-}\overset{OH}{\overset{|}{\underset{H}{\underset{|}{C}}}}\text{-}C_6H_4\text{-}Cl$$

$$Cl_3C\text{-}\overset{OH}{\overset{|}{C}}H\text{-}C_6H_4\text{-}Cl + H_2SO_4 \rightleftharpoons Cl_3C\text{-}\overset{^+OH_2}{\overset{|}{C}}H\text{-}C_6H_4\text{-}Cl + HSO_4^- \rightleftharpoons Cl_3C\text{-}{^+C}H\text{-}C_6H_4\text{-}Cl + H_2O$$

$$Cl\text{-}C_6H_5 + \overset{+}{C}H(CCl_3)\text{-}C_6H_4\text{-}Cl \longrightarrow Cl\text{-}(C_6H_5)^+\text{-}CH(CCl_3)\text{-}C_6H_4\text{-}Cl \xrightarrow{-H^+}$$

$$Cl\text{-}C_6H_4\text{-}\underset{CCl_3}{\underset{|}{C}}H\text{-}C_6H_4\text{-}Cl$$

H.2 An elimination reaction.

H.3

(a)

$$C_6H_2Cl_4 \text{ (1,2,4,5-tetrachlorobenzene)} + OH^- \rightleftharpoons$$

$$\left[\text{(anionic adduct with OH and Cl on same carbon, Cl's at 1,2,4,5)} \longleftrightarrow \text{(resonance form)} \longleftrightarrow \text{and so on} \right]$$

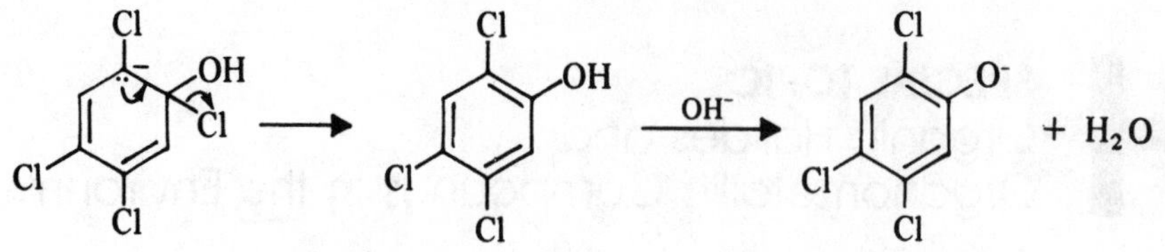

(b)

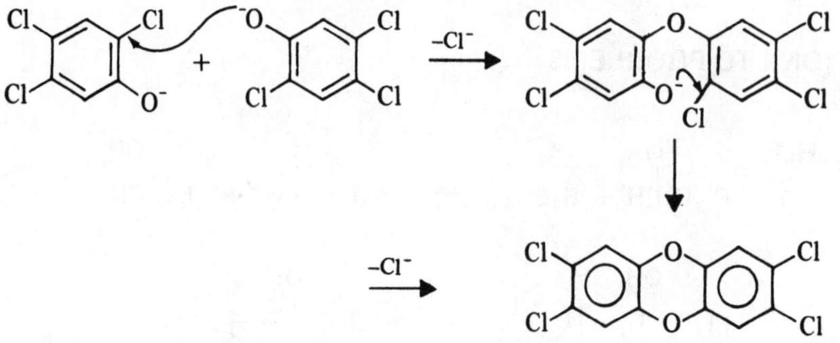

$-Cl^-$

$-Cl^-$

H.4 An S_N2 reaction:

$+ Cl^-$

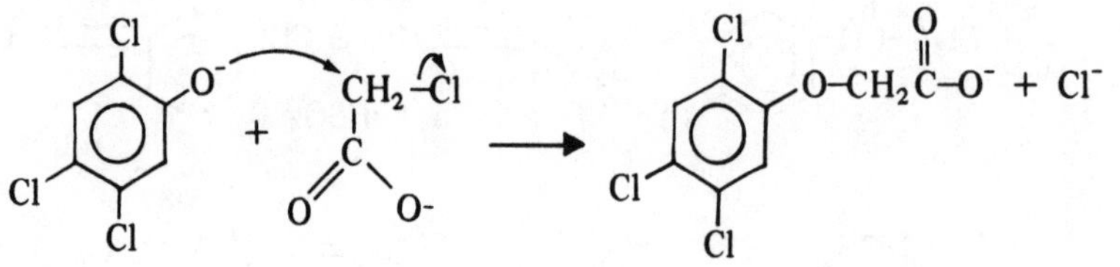

16 ALDEHYDES AND KETONES I. NUCLEOPHILIC ADDITIONS TO THE CARBONYL GROUP

PREPARATION AND REACTIONS OF ALDEHYDES

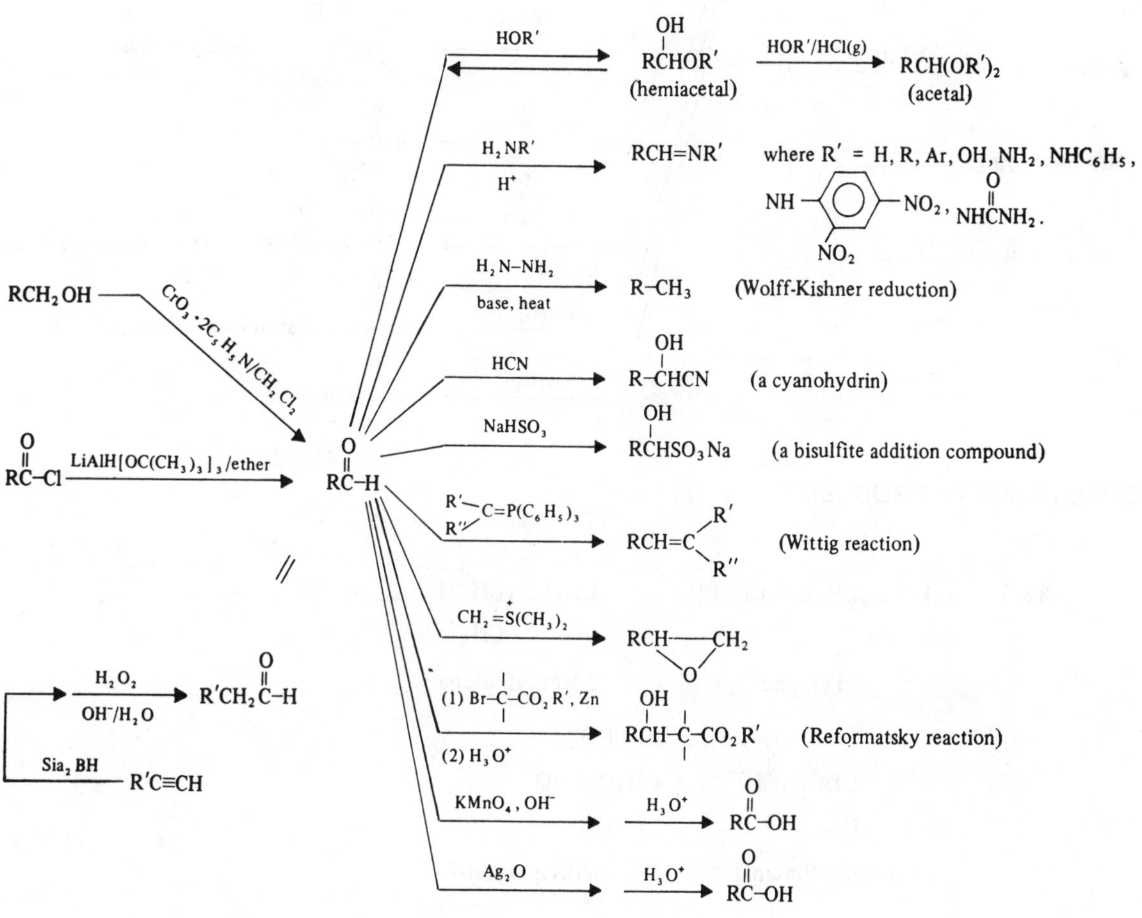

PREPARATION AND REACTIONS OF KETONES

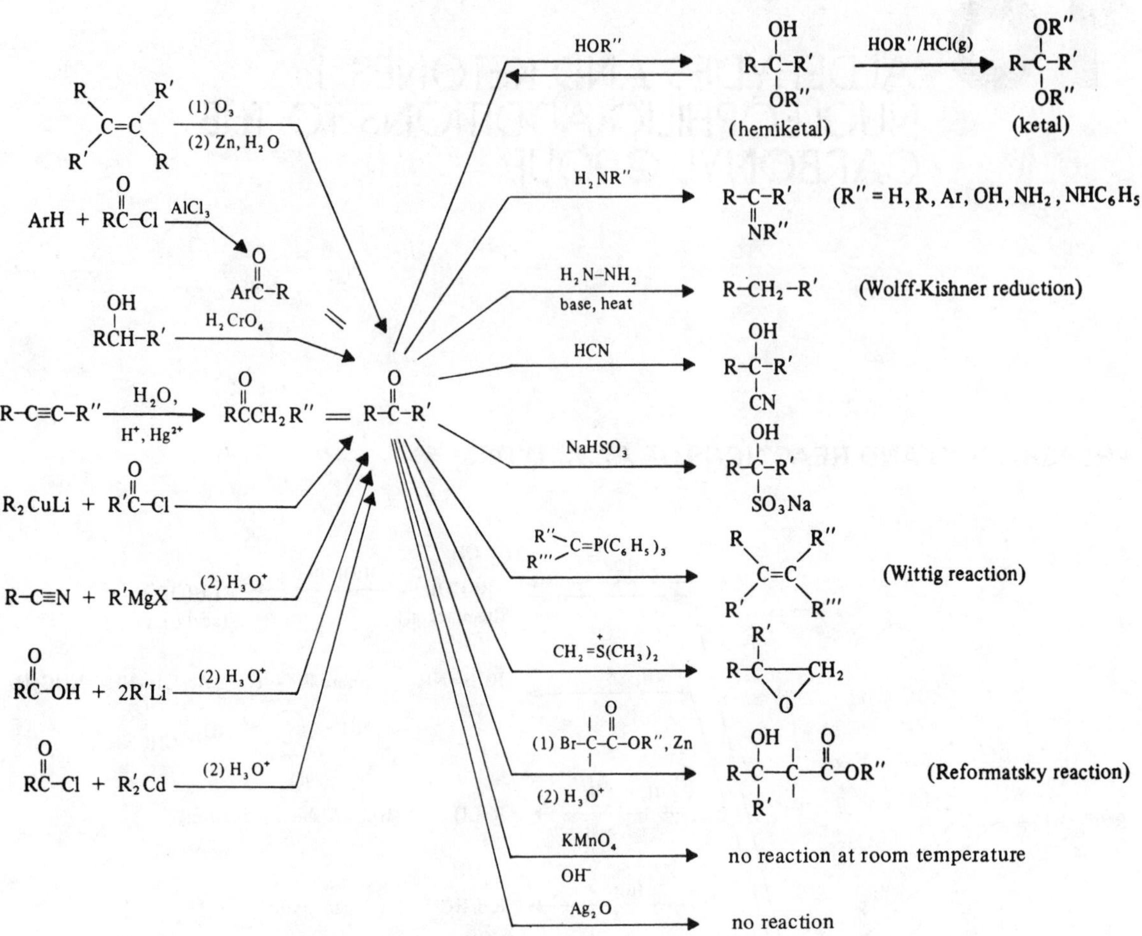

SOLUTIONS TO PROBLEMS

16.1 (a) $CH_3CH_2CH_2CH_2CHO$ — Pentanal

$CH_3CH_2CH(CH_3)CHO$ — 2-Methylbutanal

$CH_3CH(CH_3)CH_2CHO$ — 3-Methylbutanal

$CH_3C(CH_3)_2CHO$ — 2,2-Dimethylpropanal

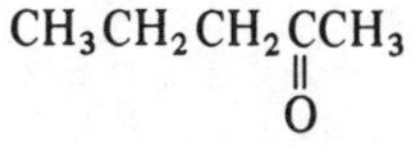

$CH_3CH_2CH_2\overset{O}{\overset{\|}{C}}CH_3$ $\quad$ $CH_3CH_2\overset{O}{\overset{\|}{C}}CH_2CH_3$

2-Pentanone 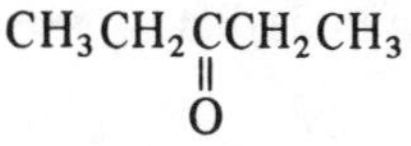 $\quad$ 3-Pentanone

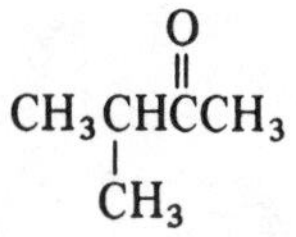

$CH_3\underset{CH_3}{\underset{|}{C}}H\overset{O}{\overset{\|}{C}}CH_3$

3-Methyl-2-butanone

(b) and (c)

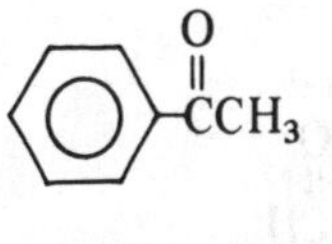

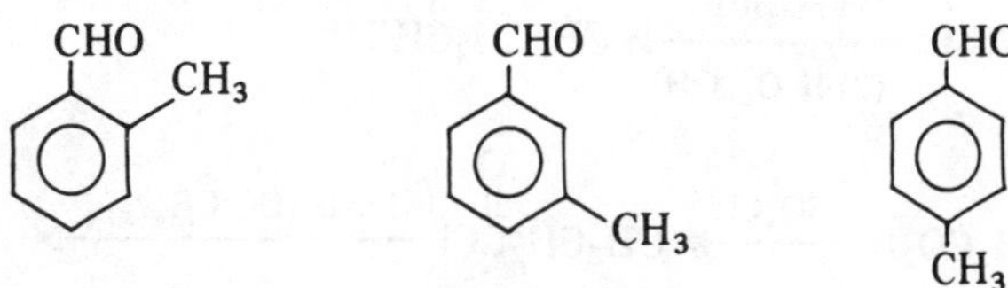

Acetophenone or methyl phenyl ketone 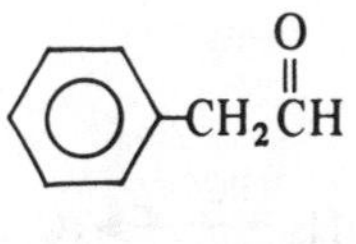 $\quad$ Phenylethanal or phenylacetaldehyde

2-Methylbenzaldehyde (o-tolualdehyde) $\quad$ 3-Methylbenzaldehyde (m-tolualdehyde) $\quad$ 4-Methylbenzaldehyde (p-tolualdehyde)

16.2 (a) 1-Pentanol, because its molecules form hydrogen bonds to each other.

(b) 2-Pentanol, because its molecules form hydrogen bonds to each other.

(c) Pentanal, because its molecules are more polar.

(d) 2-Phenylethanol, because its molecules form hydrogen bonds to each other.

(e) Benzyl alcohol because its molecules form hydrogen bonds to each other.

16.3 Two molar equivalents of $CH_3\overset{CH_3}{\overset{|}{C}}H\underset{CH_3}{\underset{|}{C}}HOH$ from the two Sia groups (i.e., $CH_3\overset{CH_3}{\overset{|}{\underset{CH_3}{\underset{|}{C}}}}CH-$ groups) of $CH_3(CH_2)_3CH{=}CH{-}BSia_2$.

16.4 $CH_3CH_2C{\equiv}CCH_2CH_3$ + Sia_2BH ⟶
3-Hexyne

$(CH_3CH_2)(H)C{=}C(CH_2CH_3)(BSia_2)$ $\xrightarrow{CH_3CO_2H}$

$(CH_3CH_2)(H)C{=}C(CH_2CH_3)(H)$
cis-3-Hexene

16.5 (a) $CH_3CH_2CH_2OH \xrightarrow[CH_2Cl_2]{PCC} CH_3CH_2\overset{O}{\overset{\|}{C}}H$

(b) $CH_3C{\equiv}CH \xrightarrow[(2)\ H_2O_2,\ OH^-]{(1)\ Sia_2BH} CH_3CH_2\overset{O}{\overset{\|}{C}}H$

(c) $CH_3CH_2CO_2H \xrightarrow{SO_2Cl_2} CH_3CH_2\overset{O}{\overset{\|}{C}}Cl \xrightarrow[ether]{LiAlH[OC(CH_3)_3]_3}$

$CH_3CH_2\overset{O}{\overset{\|}{C}}H$

16.6 (a) $C_6H_6 \xrightarrow{Br_2,\ Fe} C_6H_5Br \xrightarrow[ether]{Mg} C_6H_5MgBr \xrightarrow[(2)\ H^+]{(1)\ HCHO}$

$C_6H_5CH_2OH \xrightarrow[CH_2Cl_2]{PCC} C_6H_5CHO$

(b) $C_6H_5CH_3 \xrightarrow[(2)\ H^+]{(1)\ KMnO_4,\ OH^-,\ heat} C_6H_5CO_2H \xrightarrow{SOCl_2}$

$C_6H_5COCl \xrightarrow[ether]{LiAlH[OC(CH_3)_3]_3} C_6H_5CHO$

(c) $CH_3CH_2Br \xrightarrow{HC{\equiv}CNa} CH_3CH_2C{\equiv}CH \xrightarrow[(2)\ H_2O_2,\ OH^-]{(1)\ Sia_2BH} CH_3CH_2CH_2\overset{O}{\overset{\|}{C}}H$

(d) $CH_3C{\equiv}CCH_3 \xrightarrow[H_2O]{H_3O^+,\ Hg^{2+}} CH_3\overset{O}{\overset{\|}{C}}CH_2CH_3$

(e) $C_6H_5CH(OH)CH_3 \xrightarrow[H_2SO_4]{CrO_3} C_6H_5C(=O)CH_3$

(f) $C_6H_6 \xrightarrow[AlCl_3]{CH_3COCl} C_6H_5C(=O)CH_3$

(g) $C_6H_5C(=O)Cl \xrightarrow{(CH_3)_2CuLi} C_6H_5C(=O)CH_3$

(h) $C_6H_5C(=O)OH \xrightarrow{SOCl_2} C_6H_5C(=O)Cl \xrightarrow{(CH_3)_2CuLi} C_6H_5C(=O)CH_3$

(i) $C_6H_5CH_2Br \xrightarrow{CN^-} C_6H_5CH_2CN \xrightarrow[ether]{CH_3CH_2MgBr} C_6H_5CH_2C(=NMgBr)CH_2CH_3$

$\xrightarrow{H_3O^+} C_6H_5CH_2C(=O)CH_2CH_3$

(j) $C_6H_5CH_2CN \xrightarrow[(2)\ H_2O]{(1)\ i\text{-}Bu_2AlH} C_6H_5CH_2CH(=O)$

(k) $CH_3(CH_2)_4CO_2CH_3 \xrightarrow[(2)\ H_2O]{(1)\ i\text{-}Bu_2AlH} CH_3(CH_2)_4CH(=O)$

16.7 (a) The nucleophile is the negatively charged carbon of the Grignard reagent *acting as a carbanion.*

(b) The magnesium portion of the Grignard reagent acts as a Lewis acid and accepts an electron pair of the carbonyl oxygen. This acid-base interaction makes the carbonyl carbon even more positive and, therefore, even more susceptible to nucleophilic attack.

$$\overset{\delta^+}{-\overset{|}{C}}=\overset{\delta^-}{\ddot{O}}:\cdots\overset{\delta^+}{Mg}-X,\ \overset{\delta^-}{R} \longrightarrow -\underset{R}{\overset{|}{C}}-\ddot{O}-MgX$$

(c) The product that forms initially (above) is a magnesium derivative of an alcohol.

(d) On addition of water, the organic product that forms is an alcohol.

16.8 The nucleophile is a hydride ion.

16.9

$$H_2C{=}\ddot{O}: + H_2\ddot{O}: \rightleftharpoons H_2C(\ddot{O}:^-)(\overset{+}{O}H_2) \rightleftharpoons H_2C(OH)(OH)$$

16.10 **Acid-Catalyzed Reaction**

$$CH_3-\overset{O}{\overset{\|}{C}}-CH_3 \underset{-H^+}{\overset{+H^+}{\rightleftharpoons}} CH_3-\overset{+OH}{\overset{\|}{C}}-CH_3 \underset{-H_2{}^{18}O}{\overset{+H_2{}^{18}O}{\rightleftharpoons}} CH_3-\overset{OH}{\underset{H_2{}^{18}O^+}{C}}-CH_3 \underset{+H^+}{\overset{-H^+}{\rightleftharpoons}}$$

$$CH_3-\overset{OH}{\underset{^{18}OH}{C}}-CH_3 \underset{-H^+}{\overset{+H^+}{\rightleftharpoons}} CH_3-\overset{+OH_2}{\underset{^{18}OH}{C}}-CH_3 \underset{+H_2O}{\overset{-H_2O}{\rightleftharpoons}} CH_3-\underset{^{18}OH^+}{\underset{\|}{C}}-CH_3 \underset{+H^+}{\overset{-H^+}{\rightleftharpoons}} CH_3-\underset{^{18}O}{\underset{\|}{C}}-CH_3$$

Base-Catalyzed Reaction

$$OH^- + H_2{}^{18}O \rightleftharpoons H_2O + {}^{18}OH^-$$

$$CH_3\overset{O}{\overset{\|}{C}}CH_3 + {}^{18}OH^- \rightleftharpoons CH_3\overset{O^-}{\underset{^{18}OH}{C}}CH_3 \underset{OH^-}{\overset{H_2O}{\rightleftharpoons}} CH_3\overset{OH}{\underset{^{18}OH}{C}}CH_3 \overset{OH^-}{\rightleftharpoons}$$

$$CH_3\overset{OH}{\underset{^{18}O^-}{C}}CH_3 \underset{+OH^-}{\overset{-OH^-}{\rightleftharpoons}} CH_3\underset{^{18}O}{\underset{\|}{C}}CH_3$$

16.11

$$C_6H_5-\underset{H}{C}{=}\ddot{O}: \underset{-H^+}{\overset{+H^+}{\rightleftharpoons}} C_6H_5-\underset{H}{C}{=}\overset{+}{O}H \underset{-HO-CH_3}{\overset{H\ddot{O}-CH_3}{\rightleftharpoons}} C_6H_5-\underset{H}{C}(\overset{+}{O}(H)CH_3)(\ddot{O}H) \underset{+H^+}{\overset{-H^+}{\rightleftharpoons}}$$

$$\underset{\text{(hemiacetal)}}{C_6H_5-\underset{H}{C}(\ddot{O}CH_3)(\ddot{O}H)} \underset{-H^+}{\overset{+H^+}{\rightleftharpoons}} C_6H_5-\underset{H}{C}(\ddot{O}CH_3)(\overset{+}{O}H_2) \underset{+H_2O}{\overset{-H_2O}{\rightleftharpoons}} C_6H_5-\underset{H}{C}{=}\overset{+}{O}CH_3$$

$$C_6H_5-\underset{H}{C}{=}\overset{+}{O}CH_3 \underset{-H\ddot{O}CH_3}{\overset{+H\ddot{O}CH_3}{\rightleftharpoons}} C_6H_5-\underset{H}{C}(\overset{+}{O}(H)CH_3)(\ddot{O}CH_3) \underset{+H^+}{\overset{-H^+}{\rightleftharpoons}} \underset{\text{(acetal)}}{C_6H_5-\underset{H}{C}(\ddot{O}CH_3)(\ddot{O}CH_3)}$$

16.12

$$(CH_3)_2C{=}O \underset{-H^+}{\overset{+H^+}{\rightleftharpoons}} (CH_3)_2C{=}\overset{+}{O}H \underset{-HOCH_2CH_2OH}{\overset{+HOCH_2CH_2OH}{\rightleftharpoons}} (CH_3)_2C(O{-}H)(\overset{+}{O}(H)CH_2CH_2OH) \underset{+H^+}{\overset{-H^+}{\rightleftharpoons}}$$

$$(CH_3)_2C(O{-}H)(OCH_2CH_2OH) \underset{-H^+}{\overset{+H^+}{\rightleftharpoons}} (CH_3)_2C(\overset{+}{O}H_2)(\ddot{O}CH_2CH_2OH) \underset{+H_2O}{\overset{-H_2O}{\rightleftharpoons}}$$

$$(CH_3)_2C{=}\overset{+}{O}CH_2CH_2\ddot{O}H \rightleftharpoons (CH_3)_2C(\overset{+}{O}(H){-}CH_2)(O{-}CH_2) \underset{+H^+}{\overset{-H^+}{\rightleftharpoons}} (CH_3)_2C(O{-}CH_2)(O{-}CH_2)$$

16.13 HO–cyclopentane–CH_2OH (1,3-disubstituted cyclopentane)

16.14

(a) A (ethyl 3-oxocyclopentanecarboxylate, O= ring –$CO_2C_2H_5$) $\xrightarrow[H^+]{HOCH_2CH_2OH}$ ethylene ketal of A (–$CO_2C_2H_5$)

$\xrightarrow{2\ CH_3MgI}$ ketal –$C(OMgI)(CH_3)-CH_3$ $\xrightarrow[H_2O]{H_3O^+}$ C (O= ring –$C(OH)(CH_3)-CH_3$)

(b) Addition would take place at the ketone group as well as at the ester group. The product (after hydrolysis) would be,

HO, CH_3 on one ring carbon of cyclopentane; –$C(OH)(CH_3)-CH_3$ on the 3-position

16.15

(a) dihydropyran $\xrightarrow{+H^+}$ oxocarbenium ion ($\overset{+}{O}$=) $\xrightarrow{+ROH}$ tetrahydropyranyl–$\overset{+}{O}(H)-R$ $\xrightarrow{-H^+}$ tetrahydropyranyl–O–R

(b) Tetrahydropyranyl ethers are acetals; thus they are stable in aqueous base and hydrolyze readily in aqueous acid.

$$\text{THP–O–R} \xrightarrow{H^+} \text{THP–}\overset{+}{O}(H)\text{–R} \xrightarrow{-ROH} \text{(oxocarbenium ion, }\overset{+}{O}\text{)} \xrightarrow{H_2O}$$

$$\text{THP–}\overset{+}{O}H_2 \xrightarrow{-H^+} \text{THP–OH} \rightleftarrows HOCH_2CH_2CH_2CH_2\overset{O}{\overset{\|}{C}}H$$

5-Hydroxybutanal

(c) $$HOCH_2CH_2CH_2CH_2Cl \xrightarrow[H^+]{\text{dihydropyran}} \text{THP–}OCH_2CH_2CH_2CH_2Cl$$

$$\xrightarrow[\text{ether}]{Mg} \text{THP–}OCH_2CH_2CH_2CH_2MgCl \xrightarrow{CH_3\overset{O}{\overset{\|}{C}}CH_3}$$

$$\text{THP–}OCH_2CH_2CH_2CH_2C(CH_3)_2OMgCl \xrightarrow[H_2O]{H^+} HOCH_2CH_2CH_2CH_2C(CH_3)_2OH$$

$$(+\ HOCH_2CH_2CH_2CH_2\overset{O}{\overset{\|}{C}}H)$$

16.16 (a) $$\text{Cyclohexanone} + HSCH_2CH_2SH \xrightarrow{BF_3} \text{cyclohexane spiro-1,3-dithiolane } (S\text{–}CH_2,\ S\text{–}CH_2)$$

$$\xrightarrow[\text{Ni } (H_2)]{\text{Raney}} \text{Cyclohexane} + CH_3CH_3 + NiS$$

(b) $$C_6H_5\overset{O}{\overset{\|}{C}}H + HSCH_2CH_2SH \xrightarrow{BF_3} C_6H_5CH(S\text{–}CH_2)(S\text{–}CH_2) \xrightarrow[\text{Ni } (H_2)]{\text{Raney}}$$

$$C_6H_5CH_3 + NiS + CH_3CH_3$$

16.17 (a) $CH_3\overset{O}{\overset{\|}{C}}H \xrightarrow{HCN} CH_3\overset{OH}{\overset{|}{C}}HCN \xrightarrow[\text{reflux}]{HCl, H_2O} CH_3\overset{OH}{\overset{|}{C}}HCO_2H$ Lactic acid

(b) A racemic form

16.18 (a) $CH_3I \xrightarrow[(2)\ RLi]{(1)\ (C_6H_5)_3P} \overset{..}{:}\bar{C}H_2-\overset{+}{P}(C_6H_5)_3 \xrightarrow{C_6H_5\overset{O}{\overset{\|}{C}}CH_3} C_6H_5\underset{CH_3}{\underset{|}{C}}=CH_2$

(b) $CH_3CH_2Br \xrightarrow[(2)\ RLi]{(1)\ (C_6H_5)_3P} CH_3\overset{..-}{C}H-\overset{+}{P}(C_6H_5)_3 \xrightarrow{C_6H_5\overset{O}{\overset{\|}{C}}CH_3} C_6H_5\underset{CH_3}{\underset{|}{C}}=CHCH_3$

(c) $\overset{..}{:}\bar{C}H_2-\overset{+}{P}(C_6H_5)_3 \xrightarrow{CH_3\overset{O}{\overset{\|}{C}}CH_3} (CH_3)_2C=CH_2$

[from part (a)]

(d) $\overset{..}{:}\bar{C}H_2-\overset{+}{P}(C_6H_5)_3 \xrightarrow{\text{cyclopentanone}}$ methylenecyclopentane (CH_2=cyclopentane)

[from part (a)]

(e) $CH_3CH_2CH_2Br \xrightarrow[(2)\ RLi]{(1)\ (C_6H_5)_3P} CH_3CH_2\overset{..-}{C}H-\overset{+}{P}(C_6H_5)_3$

$\xrightarrow{CH_3\overset{O}{\overset{\|}{C}}CH_2CH_3} CH_3CH_2CH=\overset{CH_3}{\overset{|}{C}}CH_2CH_3$

(f) $CH_2=CHCH_2Br \xrightarrow[(2)\ RLi]{(1)\ (C_6H_5)_3P} CH_2=CH\overset{..-}{C}H-\overset{+}{P}(C_6H_5)_3$

$\xrightarrow{C_6H_5\overset{O}{\overset{\|}{C}}H} C_6H_5CH=CHCH=CH_2$

(g) $C_6H_5CH_2Br \xrightarrow[(2)\ RLi]{(1)\ (C_6H_5)_3P} C_6H_5\overset{..-}{C}H-\overset{+}{P}(C_6H_5)_3 \xrightarrow{C_6H_5\overset{O}{\overset{\|}{C}}H}$

$C_6H_5CH=CHC_6H_5$

16.19

$$(C_6H_5)_3P: + C_6H_5CH\!-\!CHCH_3 \text{ (epoxide, O bridging)} \longrightarrow$$

$$C_6H_5CH(\overset{+}{P}(C_6H_5)_3)\!-\!CH(O^-)CH_3 \longrightarrow C_6H_5CH\!-\!CHCH_3 \text{ (ring with } (C_6H_5)_3P\!-\!O)$$

$$\longrightarrow C_6H_5CH{=}CHCH_3 + (C_6H_5)_3P{=}O$$

16.20 (a) $CH_3OCH_2Br + (C_6H_5)_3P \xrightarrow{(2)\ RLi} CH_3OCH{=}P(C_6H_5)_3$

(b) Hydrolysis of the ether yields a hemiacetal that then goes on to form an aldehyde:

$$CH_3O\!-\!C_6H_4\!-\!C(CH_3){=}CHOCH_3 \underset{-H^+}{\overset{+H^+}{\rightleftharpoons}} CH_3O\!-\!C_6H_4\!-\!CH(CH_3)\!-\!CH{=}\overset{+}{\ddot{O}}CH_3 \underset{-H_2O}{\overset{+H_2O}{\rightleftharpoons}}$$

$$CH_3O\!-\!C_6H_4\!-\!CH(CH_3)\!-\!CH(\overset{+}{O}H_2)\!-\!OCH_3 \underset{+H^+}{\overset{-H^+}{\rightleftharpoons}} CH_3O\!-\!C_6H_4\!-\!CH(CH_3)\!-\!CH(OH)\!-\!OCH_3 \underset{+CH_3OH}{\overset{-CH_3OH}{\rightleftharpoons}}$$

(hemiacetal)

$$CH_3O\!-\!C_6H_4\!-\!CH(CH_3)\!-\!C(=O)\!-\!H$$

(c) $\text{cyclohexanone} + CH_3OCH{=}P(C_6H_5)_3 \longrightarrow$

$$C_6H_{10}{=}CHOCH_3 \underset{-H^+}{\overset{+H^+}{\rightleftharpoons}} C_6H_{10}(H)\!-\!CH{=}\overset{+}{O}CH_3 \underset{-H_2O}{\overset{+H_2O}{\rightleftharpoons}}$$

$$C_6H_{11}\!-\!CH(\overset{+}{O}H_2)\!-\!OCH_3 \underset{+H^+}{\overset{-H^+}{\rightleftharpoons}} C_6H_{11}\!-\!CH(OH)\!-\!OCH_3 \underset{+CH_3OH}{\overset{-CH_3OH}{\rightleftharpoons}} C_6H_{11}\!-\!CH{=}O$$

16.21

(a) $\text{cyclohexanone}{=}O + CH_2{=}S(CH_3)_2 \longrightarrow \text{spiro-epoxide } (C_6H_{10}\!<\!O\!>\!CH_2) + CH_3SCH_3$

(b) $(CH_3)_2C{=}O + CH_2{=}S(CH_3)_2 \longrightarrow (CH_3)_2C\overset{O}{—}CH_2 + CH_3SCH_3$

16.22 (a) $(CH_3)_2C{=}O + BrCH_2CO_2CH_2CH_3 \xrightarrow[\text{benzene}]{\text{Zn}} (CH_3)_2\overset{OZn}{\overset{|}{C}}CH_2CO_2CH_2CH_3$

$\xrightarrow{H_3O^+} (CH_3)_2\overset{OH}{\overset{|}{C}}CH_2CO_2CH_2CH_3$

(b) cyclohexanone $+ \underset{CH_3}{BrCHCO_2CH_2CH_3} \xrightarrow[(2)\ H_3O^+]{(1)\ \text{Zn, benzene}}$ 1-hydroxycyclohexyl$-\underset{CH_3}{CHCO_2CH_2CH_3}$ (OH on ring carbon)

(c) $CH_3CH_2\overset{O}{\overset{\|}{C}}H + BrCH_2CO_2CH_2CH_3 \xrightarrow[(2)\ H_3O^+,\ \text{heat}]{(1)\ \text{Zn, benzene}} CH_3CH_2CH{=}CHCO_2CH_2CH_3$

$\xrightarrow[\text{Pt}]{H_2} CH_3CH_2CH_2CH_2CO_2CH_2CH_3$

16.23

$C_6H_5\overset{O}{\overset{\|}{C}}{-}H + :\overset{H}{\overset{|}{O}}{-}O{-}\overset{O}{\overset{\|}{C}}{-}R \rightleftharpoons C_6H_5\underset{H}{\overset{OH}{C}}{-}O{-}O{-}\overset{O}{\overset{\|}{C}}{-}R \overset{H^+}{\rightleftharpoons} C_6H_5\underset{H}{\overset{OH}{C}}{-}O{-}O{-}\overset{^+OH}{\overset{\|}{C}}{-}R$

$\xrightarrow{-R\overset{O}{\overset{\|}{C}}{-}OH} C_6H_5\underset{H}{\overset{O-H}{C}}{-}\ddot{O}:^+ \xrightarrow{H:^-\ \text{migration}} C_6H_5\overset{O}{\overset{\|}{C}}{-}OH + H^+$

16.24 The product is a lactone, formed as follows:

cyclopentanone $+ \overset{H}{\overset{|}{O}}{-}O{-}\overset{O}{\overset{\|}{C}}{-}R \rightleftharpoons$ 1-HO-1-$(O{-}O{-}\overset{O}{\overset{\|}{C}}{-}R)$cyclopentane $\overset{H^+}{\rightleftharpoons}$ 1-HO-1-$(O{-}O{-}\overset{^+OH}{\overset{\|}{C}}{-}R)$cyclopentane

$\xrightarrow{-R\overset{O}{\overset{\|}{C}}{-}OH}$ 1-HO-1-O^+cyclopentane $\xrightarrow{-H^+}$ six-membered ring lactone ($C{=}O$ adjacent to ring O)

(a lactone)

16.25 $CH_3\overset{O}{\overset{\|}{C}}-O-\underset{CH_3}{\underset{|}{C}}HCH_3$. The isopropyl group has a greater migratory aptitude than the methyl group. The mechanism is as follows:

$$CH_3-\underset{(CH_3)_2CH}{\underset{|}{\overset{O}{\overset{\|}{C}}}} + :\overset{H}{\overset{|}{O}}-O-\overset{O}{\overset{\|}{C}}-R \rightleftharpoons CH_3-\underset{(CH_3)_2CH}{\underset{|}{\overset{OH}{\overset{|}{C}}}}-O-O-\overset{O}{\overset{\|}{C}}-R \overset{H^+}{\rightleftharpoons}$$

$$CH_3-\underset{(CH_3)_2CH}{\underset{|}{\overset{OH}{\overset{|}{C}}}}-O-O-\overset{^+OH}{\overset{\|}{C}}-R \xrightarrow{-R\overset{O}{\overset{\|}{C}}-OH} CH_3-\underset{(CH_3)_2CH}{\underset{|}{\overset{O-H}{\overset{|}{C}}}}-\ddot{O}:^+ \xrightarrow[\text{anion migration}]{\text{isopropyl}} CH_3\overset{O}{\overset{\|}{C}}-OCH(CH_3)_2 + H^+$$

16.26

(a)	$HCHO$	Methanal
(b)	CH_3CHO	Ethanal
(c)	$C_6H_5CH_2CHO$	Phenylethanal
(d)	CH_3COCH_3	Propanone
(e)	$CH_3COCH_2CH_3$	Butanone
(f)	$CH_3COC_6H_5$	Methyl phenyl ketone
(g)	$C_6H_5COC_6H_5$	Diphenyl ketone
(h)	benzene ring with CHO and OH (ortho)	2-Hydroxybenzaldehyde
(i)	benzene ring with CH_3O, HO, CHO	4-Hydroxy-3-methoxybenzaldehyde
(j)	$CH_3CH_2COCH_2CH_3$	3-Pentanone
(k)	$CH_3CH_2COCH(CH_3)_2$	2-Methyl-3-pentanone
(l)	$(CH_3)_2CHCOCH(CH_3)_2$	2,4-Dimethyl-3-Pentanone
(m)	$CH_3(CH_2)_3CO(CH_2)_2CH_3$	5-Nonanone
(n)	$CH_3(CH_2)_2CO(CH_2)_2CH_3$	4-Heptanone
(o)	$C_6H_5CH{=}CHCHO$	3-Phenyl-2-propenal

16.27

(a) $CH_3CH_2CH_2OH$

(b) $CH_3CH_2CHOHC_6H_5$

(c) $CH_3CH_2CH_2OH$

(d) $CH_3CH_2\overset{\overset{\displaystyle O}{\|}}{C}\text{–}O^-$

(e) $CH_3CH_2CH{=}CH_2$

(f) $CH_3CH_2CH_2OH$

(g) 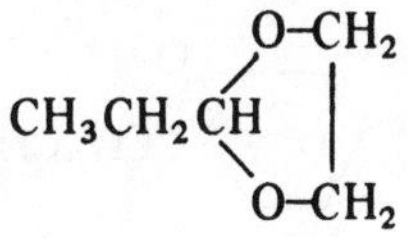 CH_3CH_2CH (cyclic acetal: CH bonded to two O atoms of an $O–CH_2–CH_2–O$ ring)

(h) $CH_3CH_2CH{=}CHCH_3$

(i) $CH_3CH_2CHOHCH_2CO_2C_2H_5$

(j) $CH_3CH_2CO_2^-\ NH_4^+ + Ag\downarrow$

(k) $CH_3CH_2CH{=}NOH$

(l) $CH_3CH_2CH{=}NNHCONH_2$

(m) $CH_3CH_2CH{=}NNHC_6H_5$

(n) $CH_3CH_2CO_2H$

(o) 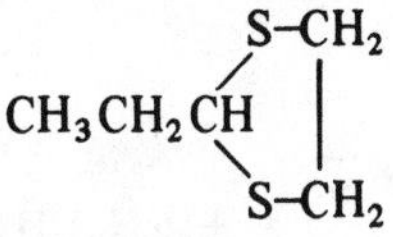 CH_3CH_2CH (cyclic thioacetal: CH bonded to two S atoms of an $S–CH_2–CH_2–S$ ring)

(p) $CH_3CH_2CH_3 + CH_3CH_3 + NiS$

16.28

(a) $CH_3CHOHCH_3$

(b) $C_6H_5\underset{\underset{\displaystyle CH_3}{|}}{C}OHCH_3$

(c) $CH_3CHOHCH_3$

(d) No reaction

(e) $CH_3\overset{\overset{\displaystyle CH_3}{|}}{C}{=}CH_2$

(f) $CH_3CHOHCH_3$

(g) 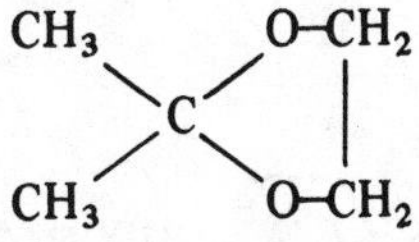 $(CH_3)_2C$ (cyclic ketal: C bonded to two O atoms of an $O–CH_2–CH_2–O$ ring)

(h) $CH_3CH{=}C(CH_3)_2$

(i) $CH_3\overset{\overset{\displaystyle OH}{|}}{\underset{\underset{\displaystyle CH_3}{|}}{C}}CH_2CO_2C_2H_5$

(j) No reaction

(k) $CH_3\underset{\underset{\displaystyle CH_3}{|}}{C}{=}NOH$

(l) $CH_3\underset{\underset{\displaystyle CH_3}{|}}{C}{=}NNHCONH_2$

(m) $CH_3\underset{\underset{\displaystyle CH_3}{|}}{C}{=}NNHC_6H_5$

(n) No reaction

(o) 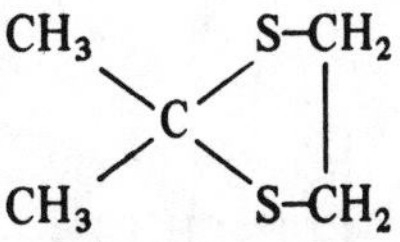 $(CH_3)_2C$ (cyclic thioketal: C bonded to two S atoms of an $S–CH_2–CH_2–S$ ring)

(p) $CH_3CH_2CH_3 + CH_3CH_3 + NiS$

16.29

(a) 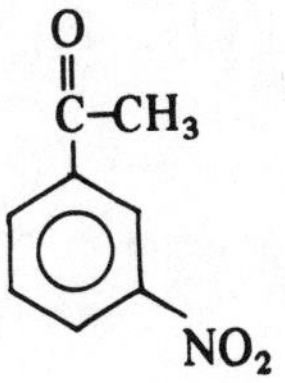 Benzene ring bearing $\overset{\overset{\displaystyle O}{\|}}{C}\text{–}CH_3$ and, meta to it, NO_2

(b) 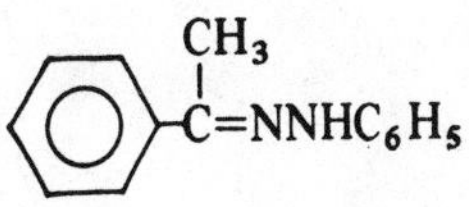 $C_6H_5\overset{\overset{\displaystyle CH_3}{|}}{C}{=}NNHC_6H_5$

(c) $C_6H_5C(=CH_2)CH_3$

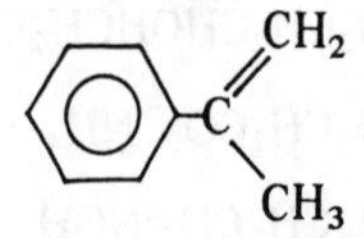

(d) $C_6H_5CH(OH)CH_3$

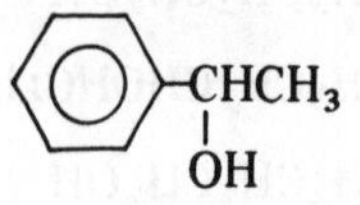

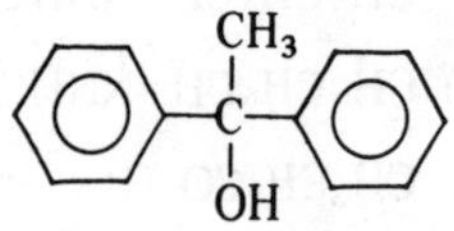

(e) $(C_6H_5)_2C(OH)CH_3$

16.30 (a) $C_6H_6 + CH_3CH_2CH_2COCl \xrightarrow{AlCl_3} C_6H_5COCH_2CH_2CH_3$

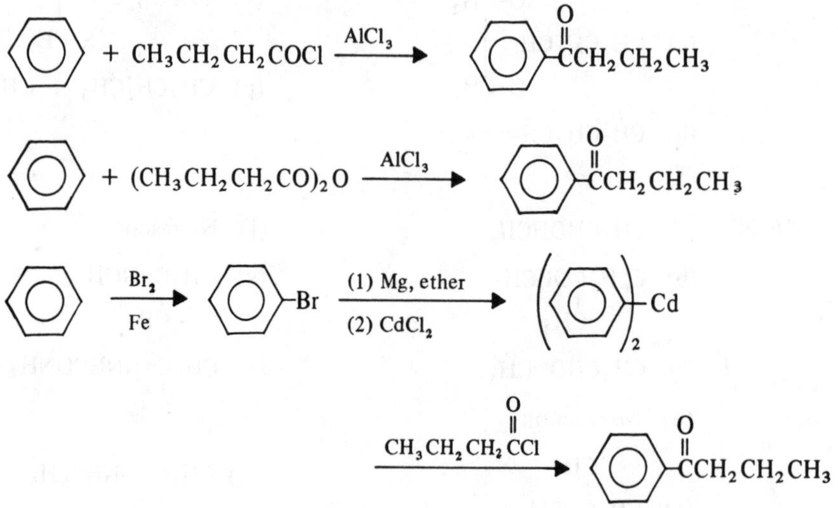

$C_6H_6 + (CH_3CH_2CH_2CO)_2O \xrightarrow{AlCl_3} C_6H_5COCH_2CH_2CH_3$

$C_6H_6 \xrightarrow[Fe]{Br_2} C_6H_5Br \xrightarrow[(2)\ CdCl_2]{(1)\ Mg,\ ether} (C_6H_5)_2Cd \xrightarrow{CH_3CH_2CH_2COCl} C_6H_5COCH_2CH_2CH_3$

(b)

$C_6H_5COCH_2CH_2CH_3 \xrightarrow[HCl]{Zn(Hg)} C_6H_5CH_2CH_2CH_2CH_3$

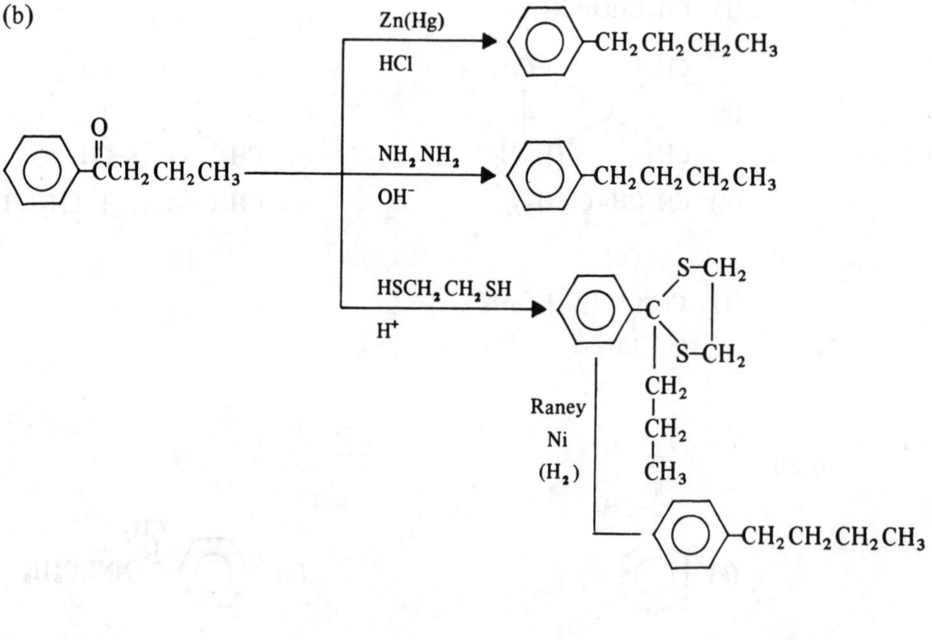

$C_6H_5COCH_2CH_2CH_3 \xrightarrow[OH^-]{NH_2NH_2} C_6H_5CH_2CH_2CH_2CH_3$

$C_6H_5COCH_2CH_2CH_3 \xrightarrow[H^+]{HSCH_2CH_2SH} C_6H_5C(SCH_2CH_2S)CH_2CH_2CH_3 \xrightarrow{Raney\ Ni\ (H_2)} C_6H_5CH_2CH_2CH_2CH_3$

16.31 (a) $C_6H_5CHO \xrightarrow{NaBH_4} C_6H_5CH_2OH$

(b) $C_6H_5CHO \xrightarrow[NH_3]{Ag(NH_3)_2^+} \xrightarrow{H_3O^+} C_6H_5CO_2H$

(c) $C_6H_5CO_2H \xrightarrow{SOCl_2} C_6H_5COCl$
[from (b)]

(d) $C_6H_5CHO + C_6H_5MgBr \xrightarrow{ether} C_6H_5CH(OMgBr)C_6H_5$

$\xrightarrow{H_3O^+} C_6H_5CH(OH)C_6H_5 \xrightarrow{H_2CrO_4} C_6H_5C(=O)C_6H_5$

or

$C_6H_5C(=O)Cl + C_6H_6 \xrightarrow[(2)\ H_3O^+]{(1)\ AlCl_3} C_6H_5C(=O)C_6H_5$
[from (c)]

(e) $C_6H_5C(=O)Cl + (CH_3)_2CuLi \longrightarrow C_6H_5C(=O)CH_3 + CH_3Cu + LiCl$
[from (c)]

(f) $C_6H_5C(=O)H \xrightarrow[(2)\ H_3O^+]{(1)\ CH_3MgI} C_6H_5CH(OH)CH_3$

(g) $C_6H_5C(=O)H \xrightarrow[(2)\ H_3O^+]{(1)\ (CH_3)_2CHCH_2MgBr} C_6H_5CH(OH)CH_2CH(CH_3)CH_3$

(h) $C_6H_5CH_2OH \xrightarrow{PBr_3} C_6H_5CH_2Br$
[from (a)]

(i) $C_6H_5CH_2Br \xrightarrow[CH_3CO_2H]{Zn} C_6H_5CH_3$

[from (h)]

or

$C_6H_5CHO \xrightarrow[BF_3]{HSCH_2CH_2SH} C_6H_5CH(SCH_2)_2$ (S–CH₂ / S–CH₂ ring) $\xrightarrow[(H_2)]{\text{Raney Ni}} C_6H_5CH_3$

(j) $C_6H_5CHO \xrightarrow{CH_3OH,\ H^+} C_6H_5CH(OCH_3)_2$

(k) $C_6H_5CHO \xrightarrow[H_3{}^{18}O^+]{H_2{}^{18}O} C_6H_5CH{=}{}^{18}O$ (See Problem 16.10 for the mechanism)

(l) $C_6H_5CHO \xrightarrow[(2)\ H_3O^+]{(1)\ NaBD_4} C_6H_5CHDOH$

(m) $C_6H_5CHO \xrightarrow{HCN} C_6H_5CH(OH)CN$

(a cyanohydrin)

(n) $C_6H_5CHO \xrightarrow{NH_2OH} C_6H_5CH{=}NOH$

(an oxime)

(o) $C_6H_5CHO + H_2NNHC_6H_5 \xrightarrow[CH_3CO_2H]{H_3O^+} C_6H_5CH{=}NNHC_6H_5$

(a phenylhydrazone)

(p) $C_6H_5CHO + H_2NNHCONH_2 \longrightarrow C_6H_5CH{=}NNHCONH_2$

(a semicarbazone)

(q) $C_6H_5CHO + (C_6H_5)_3\overset{+}{P}{-}\overset{-}{\ddot{C}}HCH{=}CH_2 \longrightarrow C_6H_5CH{=}CHCH{=}CH_2$

(a Wittig reagent)

(r) $C_6H_5CHO + NaHSO_3 \longrightarrow C_6H_5CH(OH)SO_3Na$

16.32

(a) $$C_6H_6 + CH_3CH_2\overset{O}{\overset{\|}{C}}Cl \xrightarrow{AlCl_3} C_6H_5\overset{O}{\overset{\|}{C}}CH_2CH_3$$

(b) $$C_6H_5\overset{O}{\overset{\|}{C}}{-}Cl + (CH_3CH_2)_2CuLi \longrightarrow C_6H_5\overset{O}{\overset{\|}{C}}CH_2CH_3$$

(c) $$C_6H_5C{\equiv}N + CH_3CH_2Li \xrightarrow{(2)\ H_3O^+} C_6H_5\overset{O}{\overset{\|}{C}}{-}CH_2CH_3$$

(d) $$C_6H_5CHO + CH_3CH_2MgBr \xrightarrow{(2)\ H_3O^+} C_6H_5\overset{OH}{\overset{|}{C}}HCH_2CH_3 \xrightarrow{H_2CrO_4} C_6H_5\overset{O}{\overset{\|}{C}}CH_2CH_3$$

(e) $$C_6H_5\overset{O}{\overset{\|}{C}}{-}OH + 2CH_3CH_2Li \xrightarrow{(2)\ H_2O} C_6H_5\overset{O}{\overset{\|}{C}}CH_2CH_3$$

16.33

(a) $$C_6H_5CH_2OH \xrightarrow[CH_2Cl_2]{PCC} C_6H_5\overset{O}{\overset{\|}{C}}H$$

(b) $$C_6H_5\overset{O}{\overset{\|}{C}}{-}OH \xrightarrow{SOCl_2} C_6H_5\overset{O}{\overset{\|}{C}}{-}Cl \xrightarrow{LiAlH[OC(CH_3)_3]_3} C_6H_5\overset{O}{\overset{\|}{C}}H$$

(c) $$C_6H_5C{\equiv}CH \xrightarrow[(2)\ H_3O^+]{(1)\ KMnO_4,\ OH^-} C_6H_5CO_2H \xrightarrow{[\text{as in (b)}]} C_6H_5\overset{O}{\overset{\|}{C}}H$$

(d) $$C_6H_5CH{=}CH_2 \xrightarrow[(2)\ H_3O^+]{(1)\ KMnO_4,\ OH^-} C_6H_5CO_2H \xrightarrow{[\text{as in (b)}]} C_6H_5\overset{O}{\overset{\|}{C}}H$$

(e) $$C_6H_5\overset{O}{\overset{\|}{C}}{-}OCH_3 \xrightarrow[(2)\ H_2O]{(1)\ (i\text{-Bu})_2AlH,\ \text{hexane},\ -78^\circ C} C_6H_5\overset{O}{\overset{\|}{C}}H$$

(f) $$C_6H_5{-}C{\equiv}N \xrightarrow[(2)\ H_2O]{(1)\ (i\text{-Bu})_2AlH,\ \text{hexane},\ -78^\circ C} C_6H_5\overset{O}{\overset{\|}{C}}H$$

16.34

$$CH_3\overset{\overset{\ddot{O}:}{\|}}{C}{-}H + :CN^- \underset{}{\overset{slow}{\rightleftharpoons}} CH_3\overset{\overset{:\ddot{O}:^-}{|}}{\underset{\underset{CN}{|}}{C}}{-}H \underset{fast}{\overset{H^+}{\rightleftharpoons}} CH_3\overset{\overset{OH}{|}}{\underset{\underset{CN}{|}}{C}}{-}H$$

The first step is the slow step. If the second step were slower, then the rate of the reaction would depend also on the concentration of hydrogen ion.

16.35

Cyclohexanol $\xrightarrow[acetone]{H_2CrO_4}$ cyclohexanone (**A**) $\xrightarrow[(2)\ H_3O^+]{(1)\ CH_3MgI}$ 1-methylcyclohexanol (OH, CH_3) (**B**) $\xrightarrow[heat]{H^+}$

1-methylcyclohexene (CH_3) (**C**) $\xrightarrow[(2)\ Zn,\ H_2O]{(1)\ O_3}$ $O{=}CHCH_2CH_2CH_2CH_2\overset{O}{\overset{\|}{C}}CH_3$ (**D**) $\xrightarrow[(2)\ H^+]{(1)\ Ag_2O,\ OH^-}$

$HO{-}\overset{O}{\overset{\|}{C}}CH_2CH_2CH_2CH_2\overset{O}{\overset{\|}{C}}CH_3$ (**E**)

16.36

Cyclopentanone (=O) $\xrightarrow[(2)\ H_3O^+]{(1)\ NaC{\equiv}CH}$ cyclopentane with OH and $C{\equiv}CH$ (**F**) $\xrightarrow[(2)\ H_2O_2,\ OH^-]{(1)\ Sia_2BH}$ cyclopentane with OH and $CH_2\overset{O}{\overset{\|}{C}}{-}H$ (**G**)

$\xrightarrow{CH_3OH,\ H^+}$ cyclopentane with OH and $CH_2CH(OCH_3)_2$ (**H**) $\xrightarrow[(2)\ CH_3I]{(1)\ NaH}$ cyclopentane with OCH_3 and $CH_2CH(OCH_3)_2$ (**I**)

$\xrightarrow[H_2O]{H^+}$ cyclopentane with OCH_3 and CH_2CHO (**J**)

16.37

$CH_3CH_2\overset{O}{\overset{\|}{C}}H \xrightarrow[Zn]{BrCH_2CO_2Et,} \xrightarrow{H_3O^+} CH_3CH_2\overset{OH}{\overset{|}{C}}HCH_2CO_2Et$ (**K**) $\xrightarrow[heat]{H^+,}$

$CH_3CH_2CH{=}CHCO_2Et$ (**L**) $\xrightarrow{H_2,\ Pt}$ $CH_3CH_2CH_2CH_2CO_2Et$ (**M**) $\xrightarrow[(2)\ H_2O]{(1)\ LiAlH_4}$

$CH_3CH_2CH_2CH_2CH_2OH$ (**N**) $\xrightarrow[CH_2Cl_2]{PCC}$ $CH_3CH_2CH_2CH_2\overset{O}{\overset{\|}{C}}H$

16.38

$$\text{3,4-(methylenedioxy)benzaldehyde} \xrightarrow[\text{warm}]{H_3O^+} \text{3,4-dihydroxybenzaldehyde (CHO, OH, OH)} + H\!-\!\overset{O}{\overset{\|}{C}}\!-\!H$$

The compound $C_7H_6O_3$ is 3,4-dihydroxybenzaldehyde. The reaction involves hydrolysis of the acetal of formaldehyde.

16.39

(a)

$$C_6H_5CH_2Br \xrightarrow[\text{ether}]{Mg} C_6H_5CH_2MgBr \xrightarrow[(2)\ H_3O^+]{(1)\ CH_3CHO} C_6H_5CH_2\overset{OH}{\overset{|}{C}}HCH_3$$

(b)

$$\underset{\text{[from (a)]}}{C_6H_5CH_2MgBr} \xrightarrow[(2)\ H_3O^+]{(1)\ H_2\overset{O}{C{-}C}H_2} C_6H_5CH_2CH_2CH_2OH$$

$$\xrightarrow[CH_2Cl_2]{PCC} C_6H_5CH_2CH_2CHO$$

(c)

$$\underset{\text{[from (a)]}}{C_6H_5CH_2Br} \xrightarrow[(2)\ C_6H_5Li]{(1)\ (C_6H_5)_3P} \underset{\text{(a Wittig reagent)}}{C_6H_5\ddot{C}H{-}\overset{+}{P}(C_6H_5)_3} \xrightarrow{C_6H_5CH{=}CH\overset{O}{\overset{\|}{C}}H}$$

$$C_6H_5CH{=}CH{-}CH{=}CH{-}C_6H_5$$

(d)

$$\underset{\text{[from (a)]}}{C_6H_5CH_2MgBr} + H{-}\overset{O}{\overset{\|}{C}}CH_2CH_3 \xrightarrow{(2)\ H_3O^+} C_6H_5CH_2\overset{OH}{\overset{|}{C}}HCH_2CH_3$$

$$\xrightarrow{H_2CrO_4} C_6H_5CH_2\overset{O}{\overset{\|}{C}}CH_2CH_3$$

16.40

$$BrCH_2CH_2CH_2\overset{O}{\overset{\|}{C}}{-}H \xrightarrow{HO\frown\frown OH,\ H^+} \underset{\mathbf{A}}{BrCH_2CH_2CH_2CH\langle O\text{-}CH_2CH_2\text{-}O\rangle} \xrightarrow{Mg,\ ether}$$

$$\underset{\mathbf{B}}{BrMgCH_2CH_2CH_2CH\langle O\text{-}CH_2CH_2\text{-}O\rangle} \xrightarrow[(2)\ H_3O^+,\ H_2O]{(1)\ CH_3CHO} \underset{\mathbf{C}}{CH_3\overset{OH}{\overset{|}{C}}HCH_2CH_2CH_2\overset{O}{\overset{\|}{C}}{-}H}$$

$$\rightleftharpoons \text{(2-methyl-6-hydroxytetrahydropyran: ring with } CH_3 \text{, O, OH)} \xrightarrow[H^+]{CH_3OH} \text{(ring with } CH_3\text{, O, } OCH_3\text{)}$$

(a hemiacetal) **D** (an acetal)

16.41 (a) $(CH_3)_2SO_4$, NaOH or CH_3I, NaOH

(b) PCC (c) Zn, $Br\overset{CH_3}{\overset{|}{C}}HCO_2Et$, then H_3O^+

(d) $LiAlH_4$

16.42

$$CH_2{=}CHCH_2OH \xrightarrow[CH_2Cl_2]{PCC} \underset{\mathbf{A}}{CH_2{=}CH\overset{O}{\overset{\|}{C}}H} \xrightarrow{CH_3OH,\ H^+}$$

$$\underset{\mathbf{B}}{CH_2{=}CH{-}CH(OCH_3)_2} \xrightarrow[\text{cold, dilute}]{KMnO_4,\ OH^-} \underset{\mathbf{C}}{\underset{OH\ \ OH}{CH_2CHCH(OCH_3)_2}} \xrightarrow[H_2O]{H_3O^+} \underset{\text{Glyceraldehyde}}{\underset{OH\ \ OH}{CH_2CH\overset{O}{\overset{\|}{C}}H}}$$

The product would be racemic as no chiral reagents were used.

16.43

$$CH_3\overset{O}{\overset{\|}{C}}{-}C(C_6H_5)(H)(CH_2CH_3) \xrightarrow{NaBH_4} \underset{(R)\qquad (R)}{HO(H)C(CH_3){-}C(C_6H_5)(H)(CH_2CH_3)} + \underset{(S)\qquad (R)}{H(OH)C(CH_3){-}C(C_6H_5)(H)(CH_2CH_3)}$$

(R)-3-Phenyl-2-pentanone

Diastereomers

16.44 $BrCH_2(CH_2)_7CH_2Br \xrightarrow[(2)\ RLi]{(1)\ (C_6H_5)_3P} (C_6H_5)_3\overset{+}{P}-\ddot{C}H(CH_2)_7\ddot{C}H-\overset{+}{P}(C_6H_5)_3$

A

$\xrightarrow{CH_3(CH_2)_{11}\overset{O}{\overset{\|}{C}}CH_3} CH_3(CH_2)_{11}\overset{CH_3}{\overset{|}{C}}=CH(CH_2)_7CH=\overset{CH_3}{\overset{|}{C}}(CH_2)_{11}CH_3 \xrightarrow{H_2,\ Pt}$

B

$CH_3(CH_2)_{11}\overset{CH_3}{\overset{|}{C}H}(CH_2)_9\overset{CH_3}{\overset{|}{C}H}(CH_2)_{11}CH_3$

C

16.45 (a) $Ag(NH_3)_2{}^+OH^-$ (positive test with benzaldehyde)

(b) $Ag(NH_3)_2{}^+OH^-$ (positive test with hexanal)

(c) Concentrated H_2SO_4 (2-hexanone is soluble)

(d) CrO_3 in H_2SO_4 (positive test with 2-hexanol)

(e) Br_2 in CCl_4 (decolorization with $C_6H_5CH{=}CHCOC_6H_5$)

(f) $Ag(NH_3)_2{}^+OH^-$ (positive test with pentanal)

(g) Br_2 in CCl_4 (immediate decolorization occurs with enol form)

(h) $Ag(NH_3)_2{}^+OH^-$ (positive test with cyclic hemiacetal)

16.46 Compound **W** is

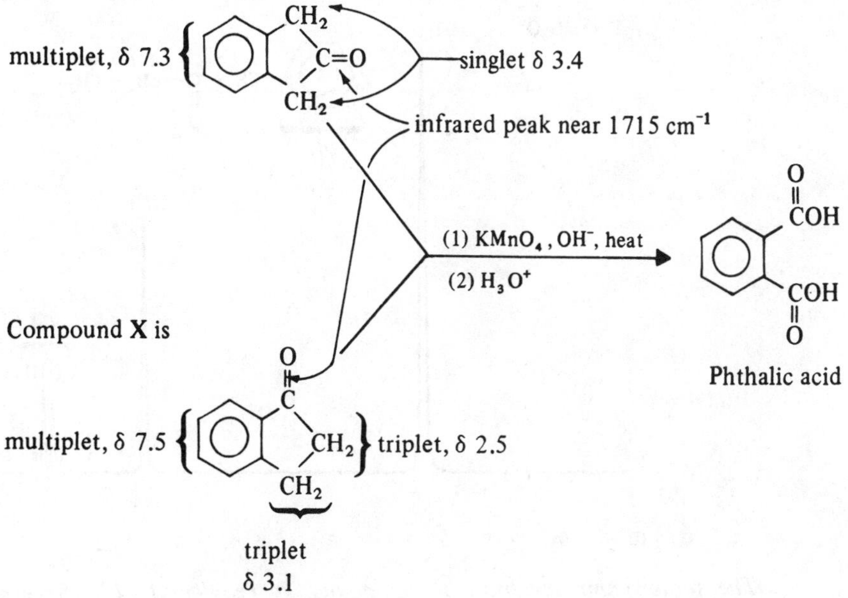

multiplet, δ 7.3 — benzene ring; CH_2, $C{=}O$, CH_2 — singlet δ 3.4; infrared peak near 1715 cm^{-1}

$\xrightarrow[(2)\ H_3O^+]{(1)\ KMnO_4,\ OH^-,\ heat}$ Phthalic acid (benzene ring with two $\overset{O}{\overset{\|}{C}}OH$ groups)

Compound **X** is

multiplet, δ 7.5 — benzene ring; $\overset{O}{\overset{\|}{C}}$; CH_2 triplet, δ 2.5; CH_2 triplet δ 3.1

16.47 Each proton nmr spectra (Figs. 16.2 and 16.3) has a five hydrogen peak near δ 7.1, suggesting that **Y** and **Z** each have a C_6H_5- group. The infrared spectrum of each compound show a strong peak near 1705 cm^{-1}. This absorption indicates that each compound has a C=O group not adjacent to the phenyl group. We have, therefore, the following pieces,

C_6H_5- and $-\overset{\overset{\large O}{\|}}{C}-$

If we subtract the atoms of these pieces from the molecular formula,

$$\begin{array}{l} C_{10}H_{12}O \\ \underline{-C_7H_5O} \quad (C_6H_5 + C{=}O) \end{array}$$

We are left with, C_3H_7

In the proton nmr spectrum of **Y** we see an ethyl group [triplet, δ 1.0 (3H) and quartet, δ 2.3 (2H)] and an unsplit $-CH_2-$ group [singlet, δ 3.7 (2H)]. This means that **Y** must be,

$C_6H_5-CH_2\overset{\overset{\large O}{\|}}{C}CH_2CH_3$

1-Phenyl-2-butanone

In the proton nmr spectrum of **Z**, we see an unsplit $-CH_3$ group [singlet, δ 2.0 (3H)] and a multiplet (actually two superimposed triplets) at δ 2.8. This means **Z** must be,

$C_6H_5-CH_2CH_2\overset{\overset{\large O}{\|}}{C}CH_3$

4-Phenyl-2-butanone

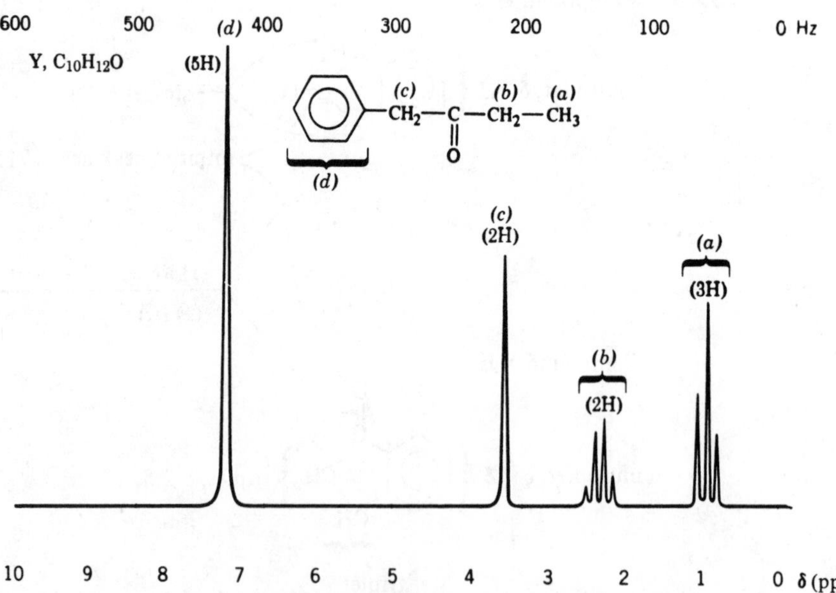

The proton nmr spectrum of compound Y, Problem 16.47. (Spectrum courtesy of Aldrich Chemical Co., Milwaukee, WI.)

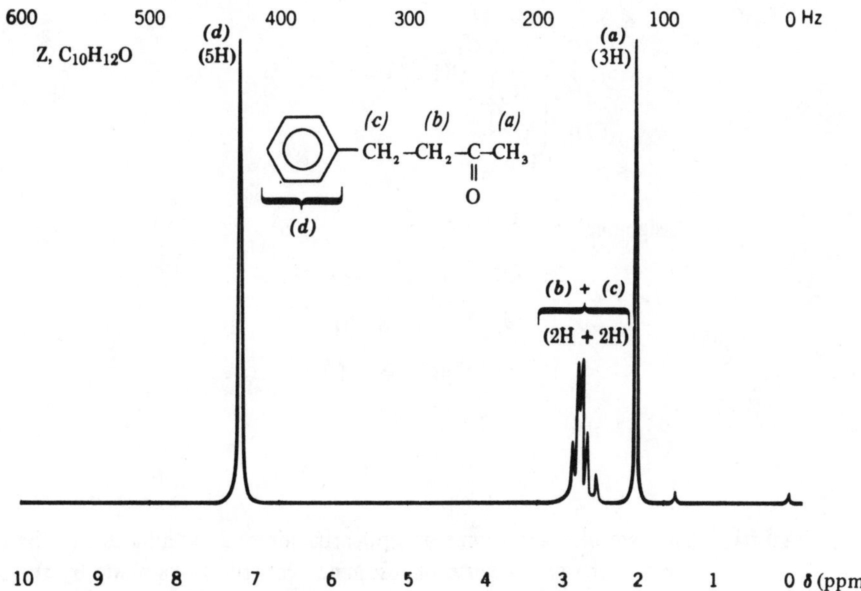

The proton nmr spectrum of compound Z, Problem 16.47. (Spectrum courtesy of Aldrich Chemical Co., Milwaukee, WI.)

16.48 That compound **A** forms a phenylhydrazone, gives a negative Tollens' test, and gives an infrared band near 1710 cm^{-1} indicates that **A** is a ketone. The ^{13}C spectrum of **A** contains only four signals indicating that **A** has a high degree of symmetry. The splitting patterns of the proton off-resonance decoupled spectrum enable us to conclude that **A** is diisobutyl ketone:

$$\overset{(a)}{(CH_3)_2}\overset{(b)}{CH}\overset{(c)}{CH_2}\underset{(d)}{\overset{O}{\overset{\|}{C}}}CH_2CH(CH_3)_2$$

Assignments:

(a) Quartet δ 22.6

(b) Doublet δ 24.4

(c) Triplet δ 52.3

(d) Singlet δ 210.0

16.49 That the ^{13}C spectrum of **B** contains only three signals indicates that **B** has a highly symmetrical structure. The splitting patterns of the proton off-resonance decoupled spectrum indicate the presence of equivalent methyl groups (quartet at δ 18.8), equivalent $-\overset{|}{\underset{|}{C}}-$ groups (singlet at δ 70.4), and equivalent $\rangle C{=}O$ groups (singlet at δ 215.0). These features allow only one possible structure for **B**:

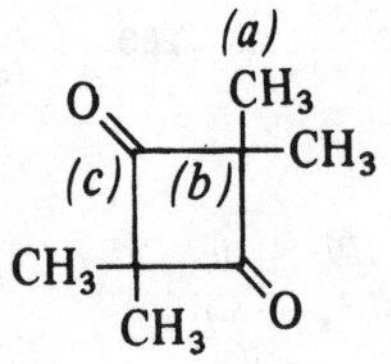

(a) CH_3, (b), (c) labelled on the structure: 2,2,4,4-tetramethylcyclobutane-1,3-dione

Assignments:

(a) Quartet δ 18.8

(b) Singlet δ 70.4

(c) Singlet δ 210.0

16.50 The two nitrogen atoms of semicarbazide that are adjacent to the C=O group bear partial positive charges because of resonance contributions made by the second and third structures below,

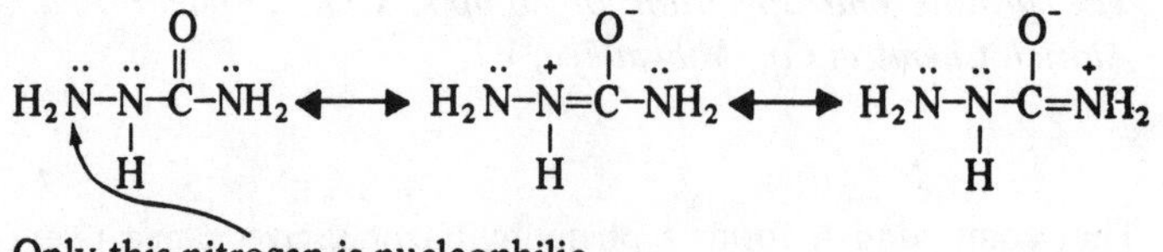

$$H_2\ddot{N}-\ddot{N}(H)-C(=O)-\ddot{N}H_2 \longleftrightarrow H_2\ddot{N}-\overset{+}{N}(H)=C(O^-)-\ddot{N}H_2 \longleftrightarrow H_2\ddot{N}-\ddot{N}(H)-C(O^-)=\overset{+}{N}H_2$$

Only this nitrogen is nucleophilic.

16.51

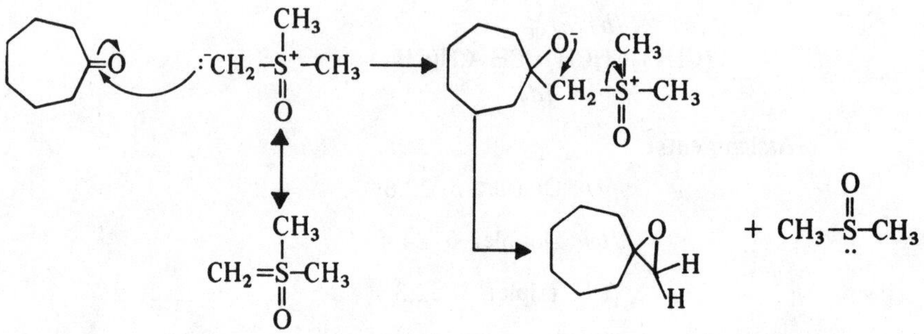

$$\text{cycloheptanone} + :\overset{-}{C}H_2-\overset{+}{S}(CH_3)(=O)-CH_3 \longrightarrow \text{cycloheptane-C}(O^-)(CH_2-\overset{+}{S}(CH_3)(=O)-CH_3)$$

$$:\overset{-}{C}H_2-\overset{+}{S}(CH_3)(=O)-CH_3 \longleftrightarrow CH_2=S(CH_3)(=O)-CH_3$$

$$\longrightarrow \text{cycloheptane spiro-epoxide (}CH_2\text{ with H, H)} + CH_3-\ddot{S}(=O)-CH_3$$

SECTION REFERENCES FOR ADDITIONAL PROBLEMS

16.26 16.2
16.27 16.6-16.14
16.28 16.6-16.14
16.29 16.6-16.14
16.30 12.12C 16.5, 16.8C
16.31 16.6-16.14
16.32 16.5
16.33 16.4
16.34 16.9
16.35 16.5-16.14
16.36 16.6-16.14
16.37 16.4, 16.11
16.38 16.7A
16.39 16.4, 16.14
16.40 16.17A
16.41 16.4, 16.11, 16.14
16.42 16.4, 16.7
16.43 16.6, 16.14
16.44 16.10A
16.45 16.13
16.46 16.13
16.47 16.13
16.48 16.13
16.49 16.13
16.50 16.8

SELF-TEST

16.1 Give an acceptable name for

$$\begin{array}{c} \quad OH \qquad\qquad\quad O \\ \quad | \qquad\qquad\qquad \| \\ CH_3CHCH_2CHCH_2CH \\ \qquad\qquad | \\ \qquad\qquad CH_3 \end{array}$$

16.2 Which of the following compounds has the highest boiling point?

(a) Propanal (b) Butanal (c) Butanone (d) 1-Butanol

16.3 Give a simple chemical test that would serve to distinguish between the compounds in each of the following pairs.

(a)

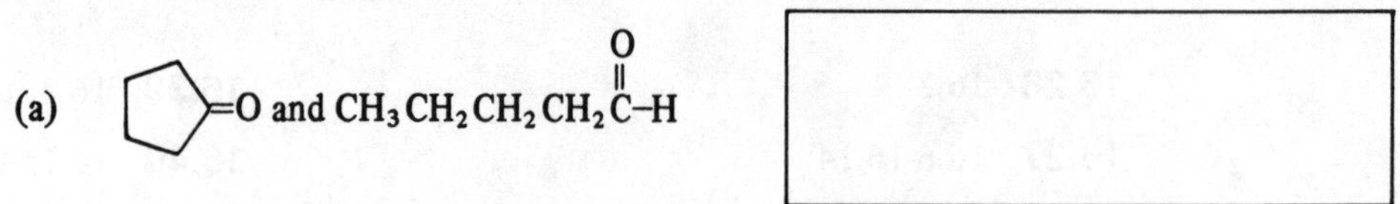

(b) $CH_3O-C_6H_4-\overset{O}{\overset{\|}{C}}-H$ and $C_6H_5-\overset{O}{\overset{\|}{C}}-OCH_3$

16.4 Give the structural formula of the missing reactant or major organic product. Write N.R. if no reaction occurs.

(a) [ketone] + [ ] ⟶ [CH_2 / CH_2 alkene product]

(b) [cyclohexanone] + [ ] ⟶ $\xrightarrow[\text{heat}]{\text{conc. } H_2SO_4}$ [cyclohexene-1-carboxylic acid]

(c)

(d) $CH_3CHO + CH_3OH \xrightarrow{HCl(g)}$

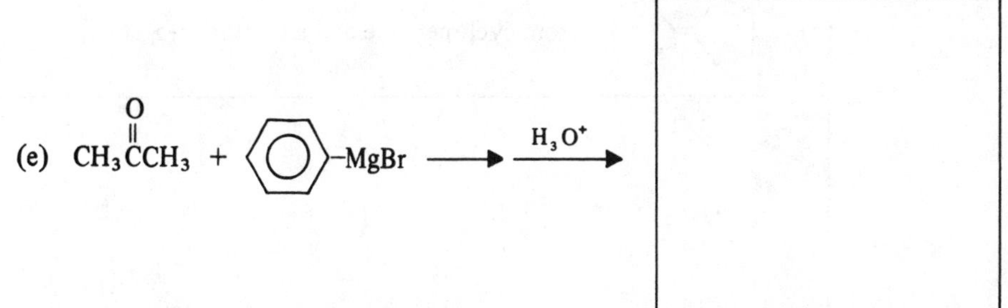

(f) $CH_3CH_2\overset{O}{\overset{\|}{C}}C_6H_5 \xrightarrow{R\overset{O}{\overset{\|}{C}}OOH}$ [ ]

(g) $CH_3\overset{O}{\overset{\|}{C}}CH_3 \xrightarrow{[\quad]} CH_3\underset{CH_3}{\underset{|}{\overset{OH}{\overset{|}{C}}}}CH_2\overset{O}{\overset{\|}{C}}OCH_3$

16.5 Write equations for a reasonable laboratory synthesis of

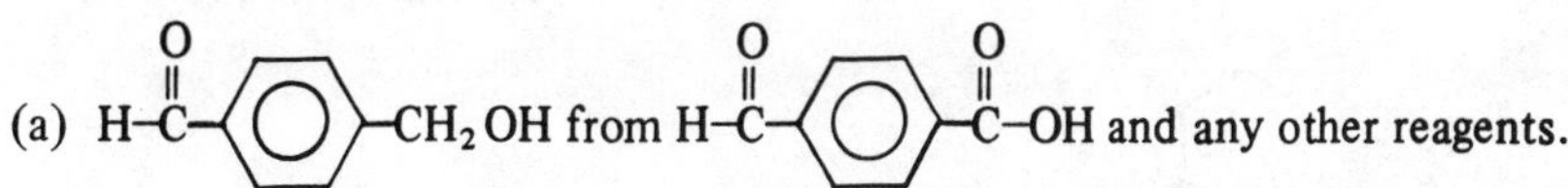

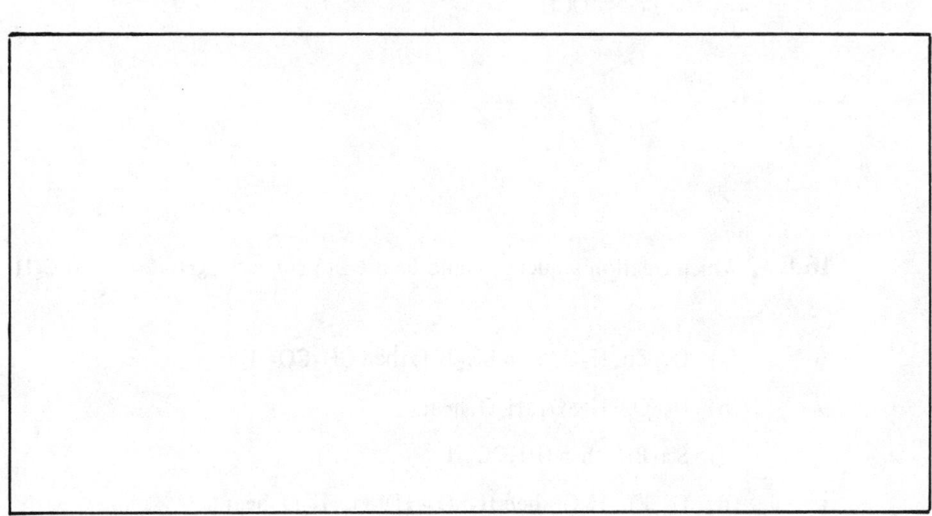

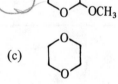

(b) from cyclopentene and any other reagents.

16.6 Which Wittig reagent could be used to synthesize $C_6H_5CH{=}CHCH_2CH_3$? (Assume any other needed reagents are available.)

(a) $C_6H_5\overset{-}{\ddot{C}}H\overset{+}{P}(C_6H_5)_3$

(b) $C_6H_5CH{=}CH\overset{-}{\ddot{C}}H\overset{+}{P}(C_6H_5)_3$

(c) $CH_3CH_2\overset{-}{\ddot{C}}H\overset{+}{P}(C_6H_5)_3$

(d) More than one of the above

(e) None of the above

16.7 Which compound is an acetal?

(a) $C_6H_5\underset{\underset{OH}{|}}{C}HOCH_3$

(b) 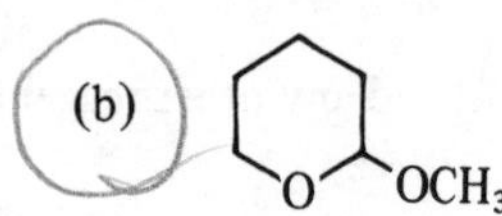

(c)

(d) More than one of the above

(e) All of the above

16.8 Which reaction sequence could be used to convert $C_6H_{13}C{\equiv}CH$ to $C_6H_{13}CH_2\underset{\underset{O}{\|}}{C}H$?

(a) O_3, Zn, H_2O, then Sia_2BH, then CH_3CO_2H

(b) H_2SO_4, $HgSO_4$, H_2O, heat

(c) Sia_2BH, then CH_3CO_2H

(d) O_3, Zn, H_2O, then H_2SO_4, $HgSO_4$, H_2O, heat

(e) Sia_2BH, then H_2O_2, OH^-/H_2O

SUPPLEMENTARY PROBLEMS

S16.1 What major reaction type occurs readily with aldehydes but not as readily with ketones?

S16.2 Many reactions of the carbonyl group of aldehydes and ketones are catalyzed by *both* acids and bases. Explain.

SOLUTIONS TO SUPPLEMENTARY PROBLEMS

S16.1 Oxidation. Ketones do not undergo oxidation with common oxidizing agents such as $KMnO_4$, H_2CrO_4, and Ag_2O. However, ketones may be conveniently oxidized in the Baeyer-Villager oxidation.

S16.2 The carbonyl group is polar:

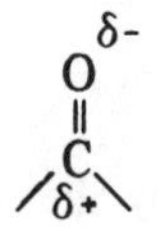

$$\overset{\delta-}{O}=\underset{\delta+}{C}<$$

Reactions occur through a nucleophilic attack at the positive carbon atom. Acid catalysts convert the carbonyl group to the cation,

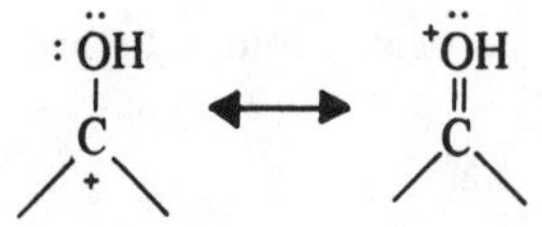

$$:\ddot{O}H-\overset{+}{C}< \longleftrightarrow {}^{+}\ddot{O}H=C<$$

which is more reactive than the neutral carbonyl group because it has an even greater positive charge on carbon than the neutral carbonyl group.

Base catalysts often increase the nucleophilicity of the nucleophile by removing a proton; for example,

$$\underset{\text{(nucleophile)}}{H-C\equiv N:} \xrightarrow{\text{base}} \underset{\text{(better nucleophile)}}{^{-}:C\equiv N:} + \text{base-H}^+$$

17 ALDEHYDES AND KETONES II: REACTIONS AT THE α CARBON. ALDOL CONDENSATION

SOLUTIONS TO PROBLEMS

17.1

2,4-Cyclohexadien-1-one (keto form) ⇌ Phenol (enol form)

The enol form is aromatic, and it is therefore stabilized by the resonance energy of the benzene ring.

17.2 No. 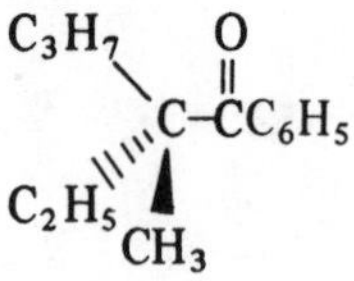 $(C_3H_7)(C_2H_5)(CH_3)C{-}\overset{O}{\overset{\|}{C}}C_6H_5$ does not have a hydrogen atom attached to its α carbon atom (which is a stereocenter) and thus enol formation involving the stereocenter is not possible. With $(C_2H_5)(CH_3)(H)CCH_2\overset{O}{\overset{\|}{C}}C_6H_5$ the stereocenter is a β carbon and thus enol formation does not affect it.

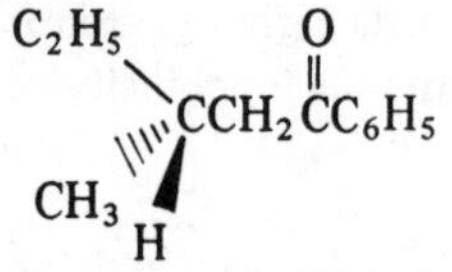

17.3 **In OD^-/D_2O**

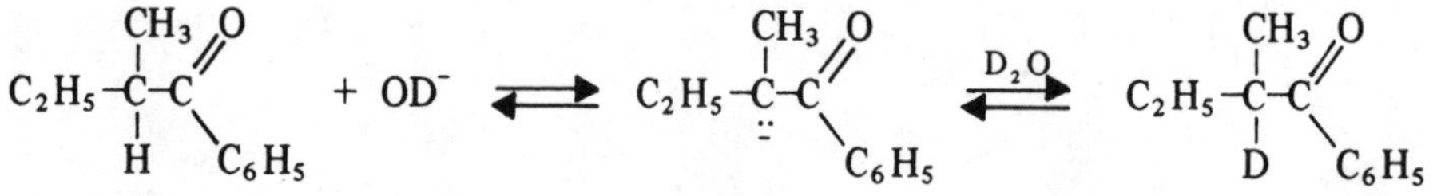

$$C_2H_5{-}C(CH_3)(H){-}C(=O)C_6H_5 + OD^- \rightleftharpoons C_2H_5{-}\ddot{C}^-(CH_3){-}C(=O)C_6H_5 \xrightleftharpoons{D_2O} C_2H_5{-}C(CH_3)(D){-}C(=O)C_6H_5$$

In D_3O^+/D_2O

$$C_2H_5{-}C(CH_3)(H){-}C(=O)C_6H_5 + D_3O^+ \xrightleftharpoons{+D^+} C_2H_5{-}C(CH_3)(H){-}C(=\overset{+}{O}{-}D)C_6H_5 \xrightleftharpoons{-H^+}$$

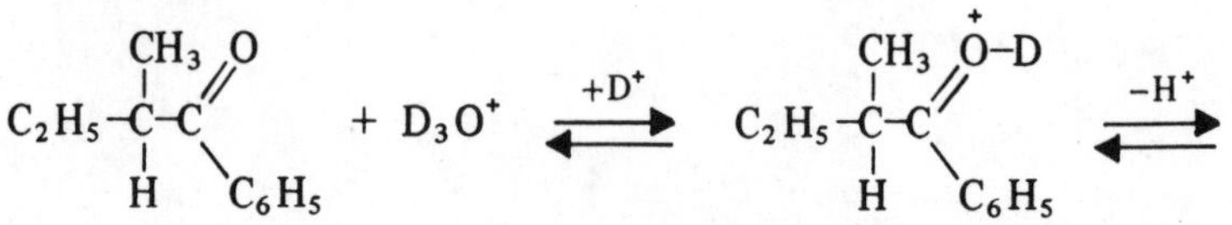

$$(CH_3)(C_2H_5)C{=}C(OD)(C_6H_5) \underset{}{\overset{+D^+}{\rightleftharpoons}} C_2H_5\text{-}C(CH_3)(D)\text{-}C({=}\overset{+}{O}\text{-}D)C_6H_5 \overset{-D^+}{\rightleftharpoons} C_2H_5\text{-}C(CH_3)(D)\text{-}C({=}O)C_6H_5$$

17.4 The reaction is said to be "base promoted" because base is consumed as the reaction takes place. A catalyst is, by definition, not consumed.

17.5 (a) The slow step in base-catalyzed racemization is the same as that in base-promoted halogenation–*the formation of an enolate ion.* (Formation of an enolate ion from *sec*-butyl phenyl ketone leads to racemization because the enolate ion is achiral. When it accepts a proton it yields a racemic form.) The slow step in acid-catalyzed racemization is also the same as that in acid-catalyzed halogenation–*the formation of an enol.* (The enol, like the enolate ion, is achiral and tautomerizes to yield a racemic form of the ketone.)

(b) According to the mechanism given, the slow step for acid-catalyzed iodination (formation of the enol) is the same as that for acid-catalyzed bromination. Thus we would expect both reactions to occur at the same rate.

(c) Again, the slow step for both reactions (formation of the enolate ion) is the same, and consequently, both reactions take place at the same rate.

17.6

(a) Acetone, $CH_3\overset{O}{\overset{\|}{C}}CH_3$

(b) Acetophenone, $C_6H_5\overset{O}{\overset{\|}{C}}CH_3$

(d) 2-Pentanone, $CH_3CH_2CH_2\overset{O}{\overset{\|}{C}}CH_3$

(f) 1-Phenylethanol, $C_6H_5\overset{OH}{\overset{|}{C}}HCH_3$

(h) 2-Butanol, $CH_3CH_2\overset{OH}{\overset{|}{C}}HCH_3$

(i) Methyl 2-naphthyl ketone, $C_{10}H_7\overset{O}{\overset{\|}{C}}CH_3$ (2-naphthyl)

17.7

(a) $\overset{\beta}{C}H_3\overset{\alpha}{C}H_2\overset{O}{\overset{\|}{C}}H + OH^- \rightleftharpoons CH_3\overset{..}{C}H\overset{O}{\overset{\|}{C}}H + H_2O$

$$CH_3CH_2\overset{O}{\overset{\|}{C}}H + {}^{-}\!:\!\underset{CH_3}{\underset{|}{C}}H\overset{O}{\overset{\|}{C}}H \rightleftharpoons CH_3CH_2\overset{O^-}{\overset{|}{C}}H\underset{CH_3}{\underset{|}{C}}H\overset{O}{\overset{\|}{C}}H$$

$$CH_3CH_2\overset{O^-}{\overset{|}{C}}H\underset{CH_3}{\underset{|}{C}}H\overset{O}{\overset{\|}{C}}H + HOH \rightleftharpoons CH_3CH_2\overset{OH}{\overset{|}{C}}H\underset{CH_3}{\underset{|}{C}}H\overset{O}{\overset{\|}{C}}H + OH^-$$

(b) For $CH_3CH_2\overset{OH}{\overset{|}{C}}HCH_2CH_2\overset{O}{\overset{\|}{C}}H$ to form, a hydroxide ion would have to remove a β proton in the first step. This does not happen because the anion that would be produced, that is, $\bar{:}CH_2CH_2CHO$, cannot be stabilized by resonance.

(c) $CH_3CH_2CH{=}\underset{CH_3}{\underset{|}{C}}\overset{O}{\overset{\|}{C}}H$

17.8

$$CH_3CHO \xrightarrow[5°C]{10\%NaOH} \underset{(aldol)}{CH_3\overset{OH}{\overset{|}{C}}HCH_2CHO} \xrightarrow{heat} CH_3CH{=}CHCHO \xrightarrow{H_2,\,Ni} CH_3CH_2CH_2CH_2OH$$

17.9

(a) $$2CH_3CH_2CH_2CHO \xrightarrow[H_2O]{OH^-} CH_3CH_2CH_2\overset{OH}{\overset{|}{C}}H\underset{\underset{CH_3}{|}}{\underset{CH_2}{\underset{|}{C}}}HCHO$$

(b) $$\text{Product of (a)} \xrightarrow[-H_2O]{H^+} CH_3CH_2CH_2CH{=}\underset{\underset{CH_3}{|}}{\underset{CH_2}{\underset{|}{C}}}CHO \xrightarrow{LiAlH_4} CH_3CH_2CH_2CH{=}\underset{\underset{CH_3}{|}}{\underset{CH_2}{\underset{|}{C}}}CH_2OH$$

(c) $$\text{Product of (b)} \xrightarrow[Pt]{H_2,} CH_3CH_2CH_2CH_2\underset{\underset{CH_3}{|}}{\underset{CH_2}{\underset{|}{C}}}HCH_2OH$$

(d) Product of (a) $\xrightarrow{NaBH_4}$ $CH_3CH_2CH_2CH(OH)CH(CH_2CH_3)CH_2OH$

$$\begin{array}{c}\quad\quad\quad\quad\quad\quad OH\\ \quad\quad\quad\quad\quad\quad |\\ CH_3CH_2CH_2CHCHCH_2OH\\ \quad\quad\quad\quad\quad\quad\quad |\\ \quad\quad\quad\quad\quad\quad\quad CH_2\\ \quad\quad\quad\quad\quad\quad\quad |\\ \quad\quad\quad\quad\quad\quad\quad CH_3\end{array}$$

17.10

(a) 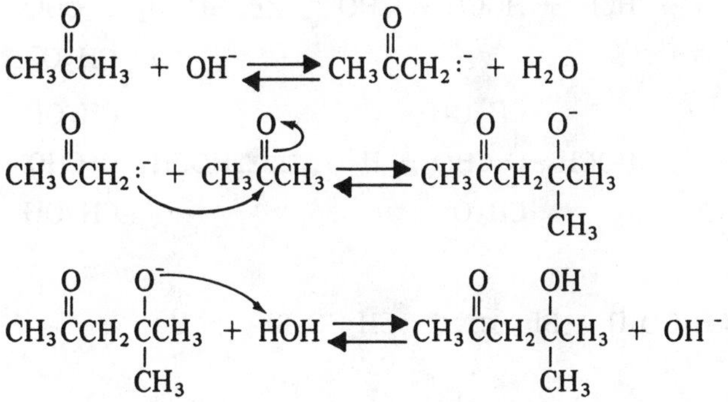 $CH_3\overset{O}{\overset{\|}{C}}CH_3 + OH^- \rightleftharpoons CH_3\overset{O}{\overset{\|}{C}}CH_2{:}^- + H_2O$

$$CH_3\overset{O}{\overset{\|}{C}}CH_2{:}^- + CH_3\overset{O}{\overset{\|}{C}}CH_3 \rightleftharpoons CH_3\overset{O}{\overset{\|}{C}}CH_2\overset{O^-}{\overset{|}{\underset{CH_3}{\underset{|}{C}}}}CH_3$$

$$CH_3\overset{O}{\overset{\|}{C}}CH_2\overset{O^-}{\overset{|}{\underset{CH_3}{\underset{|}{C}}}}CH_3 + HOH \rightleftharpoons CH_3\overset{O}{\overset{\|}{C}}CH_2\overset{OH}{\overset{|}{\underset{CH_3}{\underset{|}{C}}}}CH_3 + OH^-$$

(b) $CH_3\overset{O}{\overset{\|}{C}}CH{=}\underset{CH_3}{\underset{|}{C}}CH_3$

17.11

$$C_6H_5\overset{O}{\overset{\|}{C}}H + CH_3CHO \xrightarrow[5°C]{10\%NaOH} \xrightarrow{heat} C_6H_5CH{=}CHCHO$$

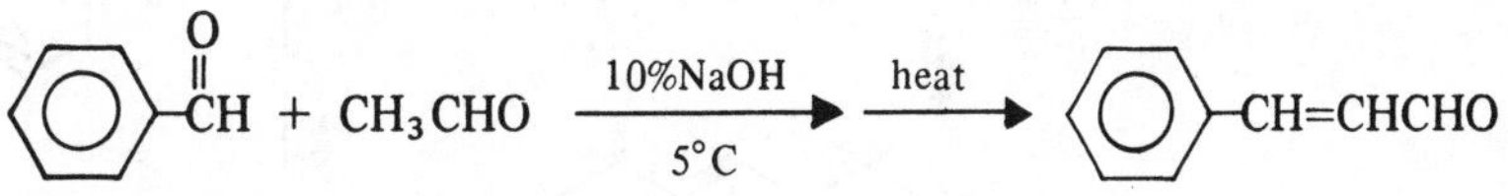

17.12 Three successive aldol additions occur.

First Aldol Addition

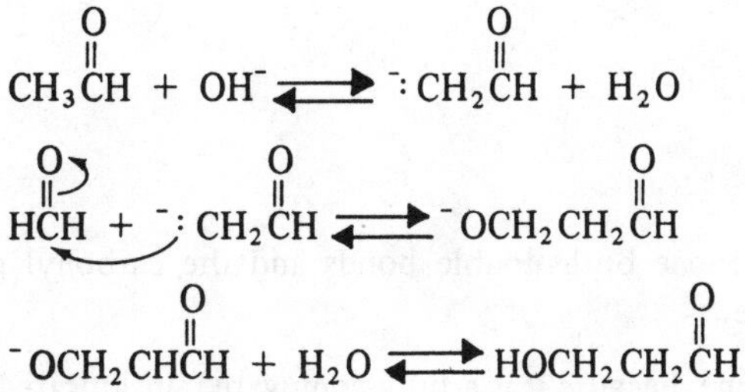

$$CH_3\overset{O}{\overset{\|}{C}}H + OH^- \rightleftharpoons {}^-{:}CH_2\overset{O}{\overset{\|}{C}}H + H_2O$$

$$H\overset{O}{\overset{\|}{C}}H + {}^-{:}CH_2\overset{O}{\overset{\|}{C}}H \rightleftharpoons {}^-OCH_2CH_2\overset{O}{\overset{\|}{C}}H$$

$${}^-OCH_2CH\overset{O}{\overset{\|}{C}}H + H_2O \rightleftharpoons HOCH_2CH_2\overset{O}{\overset{\|}{C}}H$$

Second Aldol Addition

$$HOCH_2CH_2\overset{O}{\overset{\|}{C}}H + OH^- \rightleftharpoons HOCH_2\overset{..}{C}H\overset{O}{\overset{\|}{C}}H + H_2O$$

$$H\overset{O}{\overset{\|}{C}}H + HOCH_2\overset{..}{C}H\overset{O}{\overset{\|}{C}}H \rightleftharpoons HOCH_2\overset{CH_2O^-}{\overset{|}{C}}HCHO$$

$$HOCH_2\overset{CH_2O^-}{\overset{|}{C}}HCHO + H_2O \rightleftharpoons HOCH_2\overset{CH_2OH}{\overset{|}{C}}HCHO + OH^-$$

Third Aldol Addition

$$HOCH_2CH(CH_2OH){-}CHO + OH^- \rightleftharpoons HOCH_2\ddot{C}^-(CH_2OH){-}CHO$$

$$HCH{=}O + HOCH_2\ddot{C}^-(CH_2OH){-}CHO \rightleftharpoons HOCH_2{-}C(CH_2OH)(CH_2O^-){-}CHO$$

$$HOCH_2{-}C(CH_2OH)(CH_2O^-){-}CHO + H_2O \rightleftharpoons HOCH_2{-}C(CH_2OH)(CH_2OH){-}CHO + OH^-$$

17.13 (a) $CH_3CO_2H + BF_3 \rightleftharpoons CH_3CO_2BF_3^- + H^+$

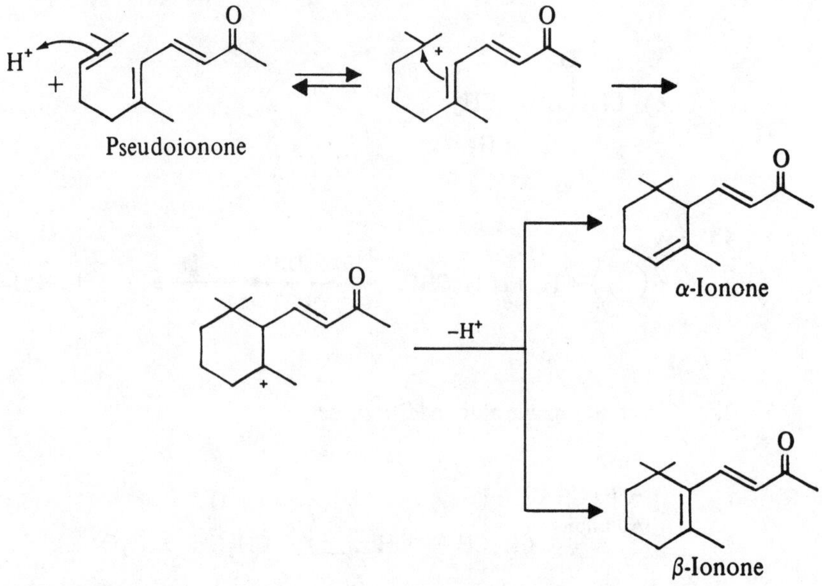

(b) In β-ionone both double bonds and the carbonyl group are conjugated, thus it is more stable.

(c) β-Ionone, because it is a fully conjugated unsaturated system.

17.14

(a)

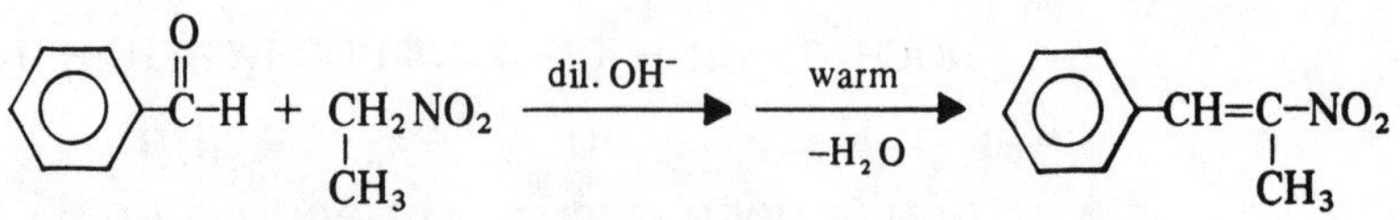

(b) $HCH{=}O + CH_3NO_2 \xrightarrow{\text{dil. } OH^-} HOCH_2CH_2NO_2$

(c) cyclohexanone + $CH_3CH_2NO_2$ $\xrightarrow{\text{dil. OH}^-}$ $\xrightarrow[-H_2O]{\text{warm}}$ cyclohexylidene$=C(CH_3)-NO_2$

17.15

(a) $^-{:}CH_2-C{\equiv}N{:} \longleftrightarrow CH_2{=}C{=}\ddot{N}{:}^-$

(b) $CH_3-C{\equiv}N{:} \underset{}{\overset{EtO^-}{\rightleftharpoons}} [{:}\overset{-}{C}H_2-C{\equiv}N{:} \longleftrightarrow CH_2{=}C{=}\ddot{\overset{-}{N}}{:}] + EtOH$

$C_6H_5-CH{=}O + {:}\overset{-}{C}H_2-C{\equiv}N{:} \longrightarrow C_6H_5-CH(O^-)-CH_2-C{\equiv}N{:} \xrightarrow{EtOH}$

$C_6H_5-CH(OH)-CH_2-C{\equiv}N{:} \xrightarrow[-H_2O]{\text{heat}} C_6H_5-CH{=}CH-CN$

17.16

$CH_3\overset{O}{\overset{\|}{C}}(CH_2)_4\overset{O}{\overset{\|}{C}}H \xrightarrow{OH^-} {}^-{:}CH_2\overset{O}{\overset{\|}{C}}(CH_2)_4\overset{O}{\overset{\|}{C}}H \longrightarrow$ cyclic $(CH_2)_4$, $CH(O^-)$, CH_2, $C{=}O$ $\xrightarrow{H_2O}$

cyclic $(CH_2)_4$, $CH(OH)$, CH_2, $C{=}O$ $\xrightarrow{-H_2O}$ cyclohept-2-enone

$CH_3\overset{O}{\overset{\|}{C}}CH_2CH_2CH_2CH_2\overset{O}{\overset{\|}{C}}H \xrightarrow{OH^-} CH_3\overset{O}{\overset{\|}{C}}CH_2CH_2CH_2\ddot{\overset{-}{C}}H\overset{O}{\overset{\|}{C}}H \longrightarrow$

cyclic $CH_3-C(O^-)$, H_2C, H_2C, CH_2, $\overset{O}{\overset{\|}{C}H}CH$ $\xrightarrow{H_2O}$ 1-methyl-1-hydroxycyclopentane-2-$\overset{O}{\overset{\|}{C}}H$ $\xrightarrow{-H_2O}$ 2-methylcyclopent-1-ene-1-$\overset{O}{\overset{\|}{C}}H$

17.17 (a) $CH_3-C(=O)-CH_2-CH_2-C(=O)-CH_3$ (b) $HC(=O)-CH_2-CH_2-CH_2-C(=O)-CH_3$ (c) $CH_3-C(=O)-CH_2-CH_2-CH_2-CH_2-C(=O)-CH_3$

Notice that starting compounds are drawn so as to indicate which atoms are involved in the cyclization reaction.

17.18

$$CH_3\overset{O}{\overset{\|}{C}}CH{=}\overset{CH_3}{\overset{|}{C}}CH_3 + HCl \rightleftharpoons CH_3\overset{^+OH}{\overset{\|}{C}}CH{=}\overset{CH_3}{\overset{|}{C}}CH_3 + Cl^- \rightleftharpoons CH_2{=}\overset{OH}{\overset{|}{C}}{-}CH{=}\overset{CH_3}{\overset{|}{C}}{-}CH_3 + HCl$$

(shown in text)

$$CH_3\overset{O}{\overset{\|}{C}}CH_3 + HCl \rightleftharpoons CH_3\overset{^+OH}{\overset{\|}{C}}CH_3 + Cl^-$$

$$H\overset{+}{O}{=}C(CH_3)_2 + CH_2{=}\overset{:\ddot{O}H}{\overset{|}{C}}{-}CH{=}\overset{CH_3}{\overset{|}{C}}{-}CH_3 \rightleftharpoons HO{-}\overset{CH_3}{\underset{CH_3}{C}}{-}CH_2{-}\overset{^+OH}{\overset{\|}{C}}{-}CH{=}\overset{CH_3}{\overset{|}{C}}CH_3$$

$$\rightleftharpoons H_2\overset{+}{O}{-}\overset{CH_3}{\underset{CH_3}{C}}{-}\underset{H}{CH}{-}\overset{O}{\overset{\|}{C}}{-}CH{=}\overset{CH_3}{\overset{|}{C}}{-}CH_3 \xrightarrow[-H^+]{-H_2O} CH_3\overset{CH_3}{\overset{|}{C}}{=}CH\overset{O}{\overset{\|}{C}}CH{=}\overset{CH_3}{\overset{|}{C}}CH_3$$

2,6-Dimethyl-2,5-hepta-dien-4-one

17.19 Drawing the molecules as they will appear in the final product helps to visualize the necessary steps:

Mesitylene

The two molecules that lead to mesitylene are shown as follows:

$CH_3-C(=O)-CH=C(CH_3)-CH_3$ and $CH_3-C(=O)-CH_3$

This molecule (4-methyl-3-penten-2-one) is formed by an acid-catalyzed condensation between two molecules of acetone as shown in the text

The mechanism is,

$$CH_3COCH_3 + H^+ \rightleftharpoons CH_3C(=\overset{+}{O}H)CH_3 \quad ; \quad (CH_3)C(=O)CH=C(CH_3)CH_3 \overset{H^+}{\rightleftharpoons} CH_3C(=\overset{+}{O}H)CH=C(CH_3)CH_2\text{-}H$$

$$\overset{-H^+}{\rightleftharpoons} HO\text{-}C(CH_3)=CH\text{-}C(CH_3)=CH_2$$

$$CH_3C(=\overset{+}{O}H)CH_3 + HO\text{-}C(CH_3)=CH\text{-}C(CH_3)=CH_2 \rightleftharpoons \text{cyclic adduct } HO\text{-}C(CH_3)_2\text{-}CH_2\text{-}C(CH_3)=CH\text{-}C(CH_3)=\overset{+}{O}H \xrightarrow{-H_2O}$$

$$\rightleftharpoons^{-H^+} \longrightarrow \text{cyclic } HO\text{-}C(CH_3) \text{ with } H\text{-}CH,\ C(CH_3)=CH,\ CH=C\text{-}CH_3$$

$$\xrightarrow{H^+} H_2\overset{+}{O}\text{-}C(CH_3) \text{ intermediate} \xrightarrow[-H^+]{-H_2O} \text{1,3,5-trimethylbenzene (HC, CH, C-CH}_3\text{, C-H)}$$

17.20

$$\text{2-methylcyclohexane-1,3-dione (CH}_3\text{, H)} + OH^- \rightleftharpoons \text{carbanion}$$

$$\text{carbanion} + CH_2=CH\text{-}\overset{O}{\overset{\|}{C}}\text{-}CH_3 \rightleftharpoons \text{2-methyl-2-(}CH_2\text{-}CH{=}{=}{=}C(O^-)CH_3\text{)cyclohexane-1,3-dione} \overset{+H^+}{\underset{-H^+}{\rightleftharpoons}}$$

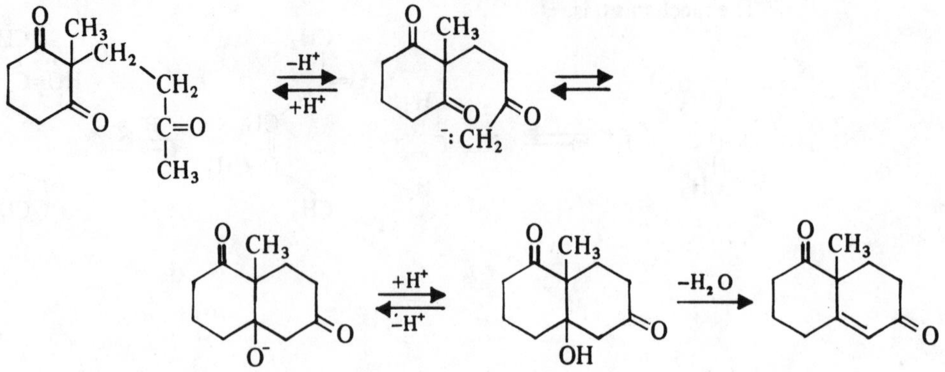

(b) 2-Methyl-1,3-cyclohexanedione is more acidic because its enolate ion is stabilized by an additional resonance structure.

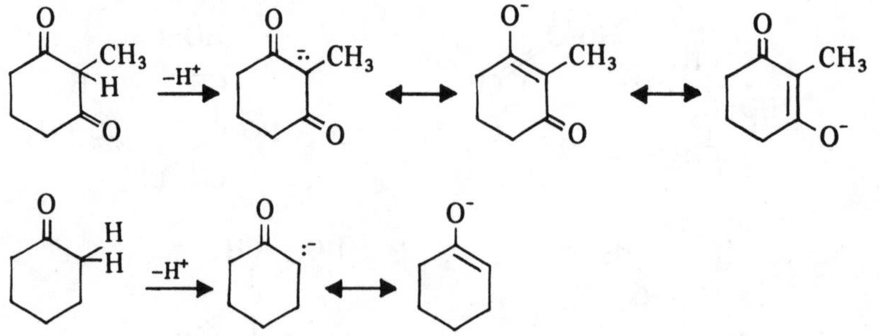

17.21

(a) $C_6H_5\overset{O}{\overset{\|}{C}}CH_3 \underset{+H^+}{\overset{-H^+}{\rightleftharpoons}} C_6H_5\overset{O}{\overset{\|}{C}}CH_2:^-$

$C_6H_5\overset{O}{\overset{\|}{C}}CH_2:^- + C_6H_5CH{=}CH\overset{O}{\overset{\|}{C}}C_6H_5 \rightleftharpoons$

$C_6H_5CH(CH_2COC_6H_5){-}CH{=}C(O^-)C_6H_5 \underset{-H^+}{\overset{+H^+}{\rightleftharpoons}} C_6H_5CH(CH_2COC_6H_5)CH_2\overset{O}{\overset{\|}{C}}C_6H_5$

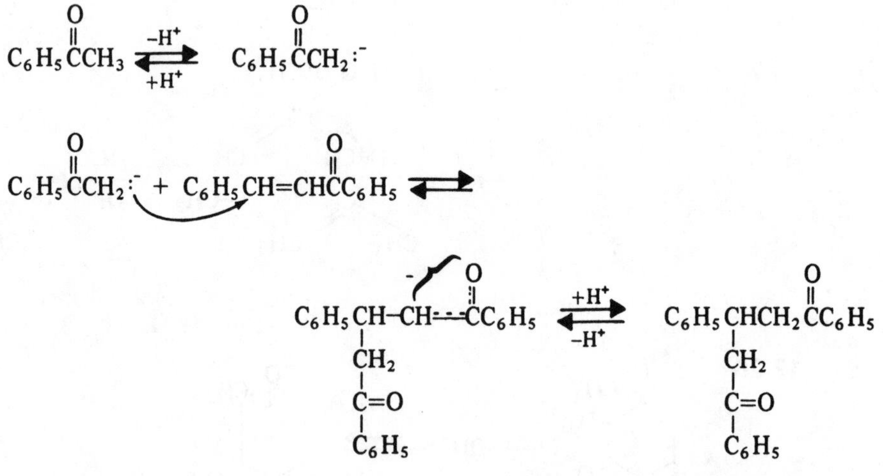

(b) $\underset{+H^+}{\overset{-H^+}{\rightleftharpoons}}$

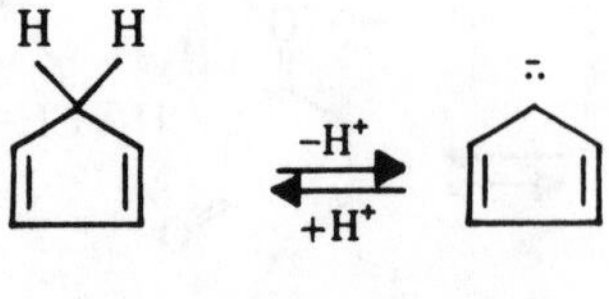

$$\text{cyclopentadienide} + C_6H_5CH{=}CHCC_6H_5 \ (\text{C=O}) \rightleftharpoons$$

$$C_6H_5CH\text{–}CH{=}CC_6H_5\ (\text{O}^-,\ \text{cyclopentadienyl on CH}) \underset{-H^+}{\overset{+H^+}{\rightleftharpoons}} C_6H_5CHCH_2CC_6H_5\ (\text{C=O, cyclopentadienyl on CH})$$

17.22

$$H_2\ddot{N}\text{–}\ddot{N}H_2 + CH_2{=}CH\text{–}CH{=}O \xrightarrow[\text{addition}]{\text{conjugate}} CH_2\text{–}C(=O)\text{–}H,\ CH_2\text{–}NH\text{–}\ddot{N}H_2 \rightleftharpoons \text{(cyclic } O^-,\ N^+\text{–H, N–H)}$$

$$\rightleftharpoons \text{(cyclic OH, N–H, N–H)} \xrightarrow{-H_2O} \text{(cyclic C=N, N–H)}$$

17.23

(a) $CH_3CH_2CH(OH)CH(CH_3)CHO$ (b) $C_6H_5\text{–}CH{=}C(CH_3)\text{–}CHO$ (c) $CH_3CH_2CH(OH)CN$

(d) $CH_3CH_2CH_2OH$ (e) CH_3CH_2CH (cyclic acetal: $O\text{–}CH_2\text{–}CH_2\text{–}O$) (f) $CH_3CH_2\overset{O}{\overset{\|}{C}}\text{–}OH$

(g) $CH_3CH_2CH(OH)CH_3$ (h) $CH_3CH_2\overset{O}{\overset{\|}{C}}\text{–}OH$ (i) $CH_3CH_2CH{=}NOH$

(j) $CH_3CH_2CH{=}CHC_6H_5$ (k) $CH_3CH_2CH(OH)\text{–}C_6H_5$ (l) $CH_3CH_2CH(OH)C{\equiv}CH$

(m) $CH_3CH_2CH_3$ (n) $CH_3CH_2CH(OH)\text{–}CH(CH_2CH_3)CO_2Et$

17.24

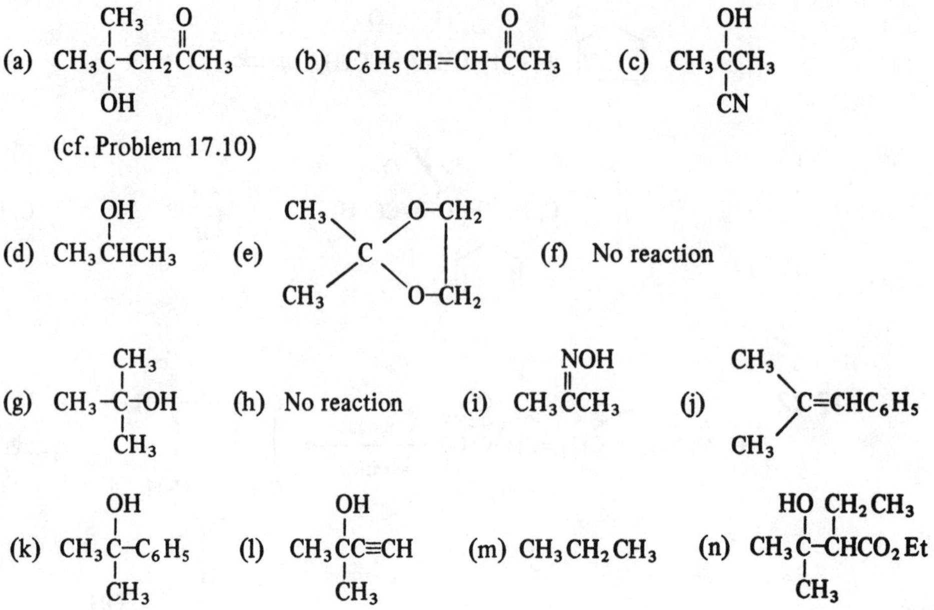

(a) $CH_3C(CH_3)(OH)-CH_2C(=O)CH_3$ (cf. Problem 17.10) (b) $C_6H_5CH=CH-C(=O)CH_3$ (c) $CH_3C(OH)(CN)CH_3$

(d) $CH_3CH(OH)CH_3$ (e) $(CH_3)_2C(O-CH_2)(O-CH_2)$ (cyclic) (f) No reaction

(g) $CH_3-C(CH_3)_2-OH$ (h) No reaction (i) $CH_3C(=NOH)CH_3$ (j) $(CH_3)_2C=CHC_6H_5$

(k) $CH_3C(OH)(CH_3)-C_6H_5$ (l) $CH_3C(OH)(CH_3)C\equiv CH$ (m) $CH_3CH_2CH_3$ (n) $CH_3C(OH)(CH_3)-CH(CH_2CH_3)CO_2Et$

17.25

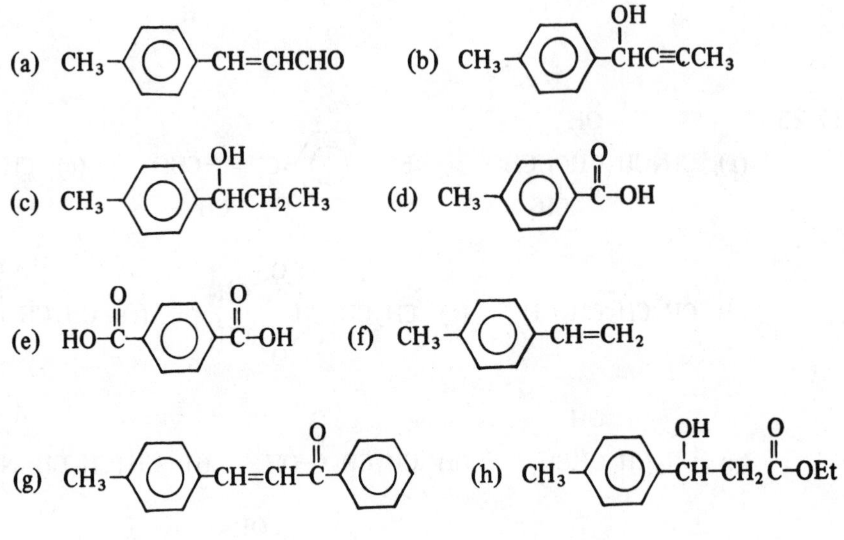

(a) $CH_3-C_6H_4-CH=CHCHO$ (b) $CH_3-C_6H_4-CH(OH)C\equiv CCH_3$

(c) $CH_3-C_6H_4-CH(OH)CH_2CH_3$ (d) $CH_3-C_6H_4-C(=O)-OH$

(e) $HO-C(=O)-C_6H_4-C(=O)-OH$ (f) $CH_3-C_6H_4-CH=CH_2$

(g) $CH_3-C_6H_4-CH=CH-C(=O)-C_6H_5$ (h) $CH_3-C_6H_4-CH(OH)-CH_2C(=O)-OEt$

17.26

(a) $C_6H_5CHO + CH_3-C(=O)-C(CH_3)_3 \xrightarrow{\text{dil. OH}^-} C_6H_5CH=CH-C(=O)-C(CH_3)_3$

(b) C_6H_5CHO + cyclopentanone $\xrightarrow{\text{dil. OH}^-}$ $C_6H_5CH=$(2-cyclopentanone) (2-benzylidenecyclopentanone)

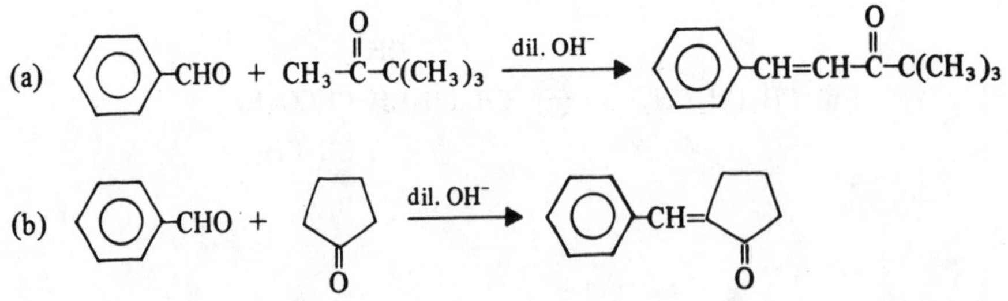

(c) $C_6H_5CHO + CH_3CH_2NO_2 \xrightarrow{\text{dil. OH}^-} C_6H_5CH{=}C(CH_3){-}NO_2 \xrightarrow{H_2,\ Pt} C_6H_5CH_2CH(NH_2)CH_3$

(d) $CH_3C(=O)CH_2CH_2CH_2C(=O)CH_3 \xrightarrow{\text{dil. OH}^-}$ 1-acetyl-2-methylcyclopentene (CH_3, $C(=O)CH_3$ on the ring double bond) $\xrightarrow{LiAlH_4}$ 1-(1-hydroxyethyl)-2-methylcyclopentene (CH_3, $CH(OH)CH_3$)

(e) $CH_3O{-}C_6H_4{-}CHO + CH_3CN \xrightarrow{\text{base}} CH_3O{-}C_6H_4{-}CH{=}CHCN$

(f) $2CH_3CH_2CH_2CH_2\overset{O}{\overset{\|}{C}}H \xrightarrow[5^\circ C]{\text{dil. OH}^-} CH_3CH_2CH_2CH_2CH(OH){-}CH(CHO)(CH_2)_2CH_3 \xrightarrow{\text{heat}}$

$CH_3(CH_2)_3CH{=}C(CHO)(CH_2)_2CH_3 \xrightarrow{LiAlH_4} CH_3(CH_2)_3CH{=}C(CH_2OH)(CH_2)_2CH_3$

(g) 1-indanone $+ \ HC(=O){-}C_6H_5 \xrightarrow{\text{dil. OH}^-}$ 2-benzylidene-1-indanone ($=CH{-}C_6H_5$)

$\xrightarrow[BF_3]{HSCH_2CH_2SH}$ dithiolane (S, S spiro at C-1) of 2-benzylidene-indan ($=CH{-}C_6H_5$) $\xrightarrow[H_2]{\text{Raney Ni,}}$ 2-benzylideneindan ($=CH{-}C_6H_5$)

17.27

$C_6H_5\overset{O}{\overset{\|}{C}}CH_2CH_3 \xrightarrow{OH^-} C_6H_5\overset{O}{\overset{\|}{C}}{-}\ddot{C}^{-}HCH_3 \xrightarrow{CH_2{=}C(CH_3){-}C(=O){-}CH_3}$

$CH_3CH(C(=O)C_6H_5){-}CH_2{-}C(CH_3){=}C(O^-){-}CH_3$ (HOH) $\longrightarrow CH_3CH(C(=O)C_6H_5){-}CH_2{-}CH(CH_3){-}C(=O){-}CH_3 \xrightarrow{OH^-}$

CH_3, CH, CH_2, CH, CH_3, C(=O), $CH_2^{:-}$, C(=O)C_6H_5 $\longrightarrow$ CH_3, CH, CH_2, CH, CH_3, C(=O), CH_2, C(O^-)C_6H_5 $\xrightarrow{H_2O}$

CH_3, OH, C_6H_5, CH_3, O $\xrightarrow{-H_2O}$ CH_3, C_6H_5, CH_3, O

17.28

$$HC{\equiv}CH \xrightarrow[(2)\ CH_3COCH_3 \quad (3)\ H^+]{(1)\ NaNH_2} \underset{\mathbf{A}}{HC{\equiv}C{-}C(CH_3)_2{-}OH} \xrightarrow[H_2O]{Hg^{2+},\ H_3O^+,} \underset{\mathbf{B}}{CH_3\overset{O}{\overset{\|}{C}}{-}C(CH_3)_2{-}OH} \xrightarrow[OH^-]{C_6H_5CHO}$$

$$\underset{\mathbf{C}}{C_6H_5{-}CH{=}CH{-}\overset{O}{\overset{\|}{C}}{-}C(CH_3)_2{-}OH}$$

17.29 (a) The conjugate base is a hybrid of the following structures:

$$\ddot{:}CH_2{-}CH{=}CH{-}\overset{O}{\overset{\|}{C}}H \longleftrightarrow CH_2{=}CH{-}\ddot{C}H{-}\overset{O}{\overset{\|}{C}}H \longleftrightarrow CH_2{=}CH{-}CH{=}\overset{O^-}{\overset{|}{C}}H$$

This structure is especially stable because the negative charge is on the oxygen atom

(b) $CH_3CH{=}CHCHO \underset{+H^+}{\overset{-H^+}{\rightleftharpoons}} \overset{-}{:}CH_2CH{=}CHCHO$

$C_6H_5CH{=}CH\overset{O}{\overset{\|}{C}}H + \overset{-}{:}CH_2CH{=}CHCHO \rightleftharpoons$

$C_6H_5CH{=}CH\overset{O^-}{\overset{|}{C}}H{-}CH_2CH{=}CHCHO \underset{-H^+}{\overset{+H^+}{\rightleftharpoons}} C_6H_5CH{=}CH\overset{OH}{\overset{|}{C}}H{-}CH_2CH{=}CHCHO$

$\xrightarrow{-H_2O} C_6H_5CH{=}CHCH{=}CHCH{=}CHCHO$

17.30

(a) (1) O_3 (2) Zn, H_2O → [o-phenylene with CH=O and $CH_2CH_2CH=O$] base (aldol condensation) → [indene-CHO]

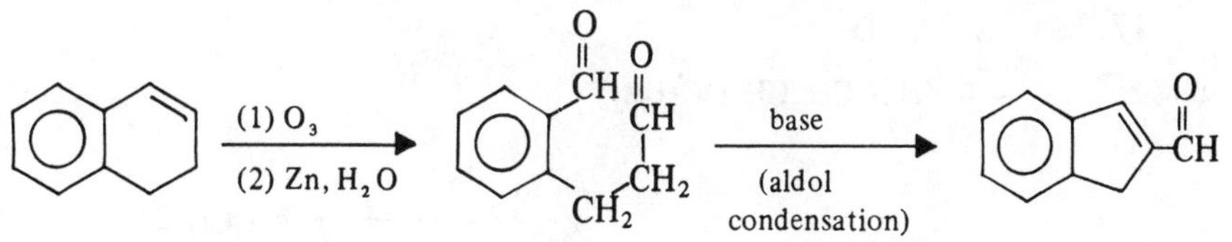

(b) (1) O_3 (2) Zn, H_2O → [cyclodecane-1,6-dione] base (aldol condensation) → [bicyclic enone]

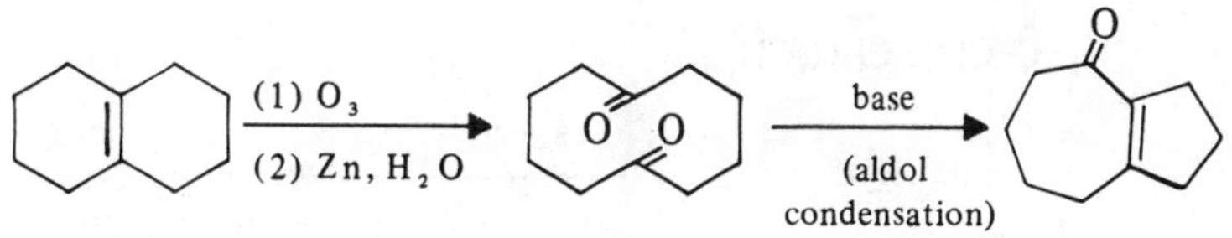

(c) [CH_3-substituted bicyclic alkene] (1) OsO_4 (2) Na_2SO_3 → [diol, OH, OH] HIO_4 (Section 21.4D) → [CH_3, CHO, CHO] base (aldol condensation) → [CH_3, CHO] + [CH_3, CHO]

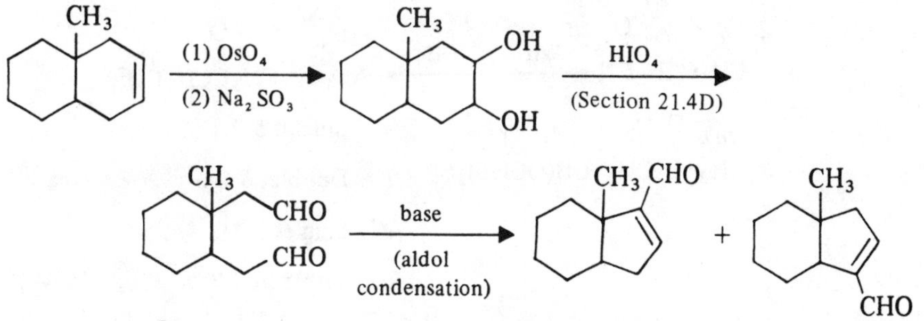

(d) [pyrazolium with $CH_2\overset{O}{\overset{\|}{C}}C_6H_5$ on each N] base (aldol condensation) → [C_6H_5, C=O, C_6H_5]

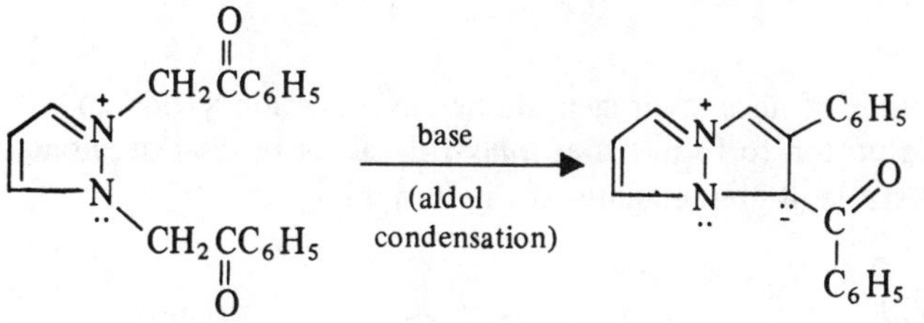

17.31 (a) In simple addition the carbonyl peak (1665-1780-cm^{-1} region) does not appear in the product; in conjugate addition it does.

(b) As the reaction takes place, the long-wavelength absorption arising from the conjugated system should disappear. One could follow the rate of the reaction by following the rate at which this absorption peak disappears.

17.32 (a) Compound **U** is phenyl ethyl ketone: (b) Compound **V** is benzyl methyl ketone:

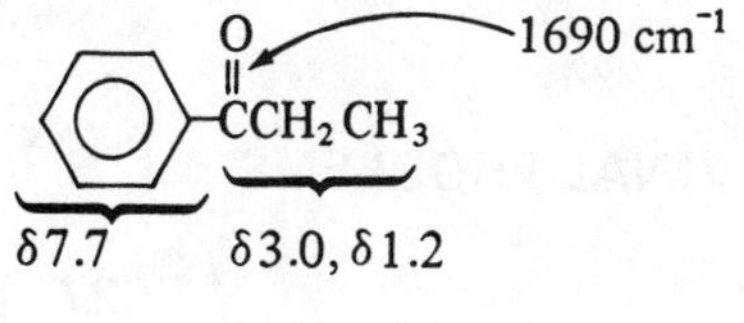

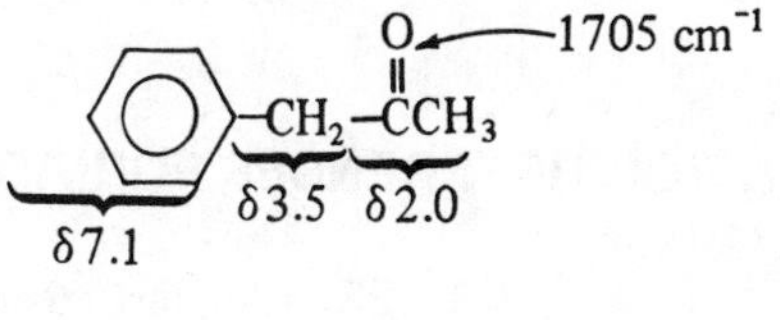

(a) $C_6H_5\overset{O}{\overset{\|}{C}}CH_2CH_3$; 1690 cm^{-1}; δ7.7; δ3.0, δ1.2

(b) $C_6H_5-CH_2-\overset{O}{\overset{\|}{C}}CH_3$; 1705 cm^{-1}; δ7.1; δ3.5; δ2.0

17.33 A is $CH_3\overset{\displaystyle O}{\overset{\|}{C}}CH_2CH(OCH_3)_2$

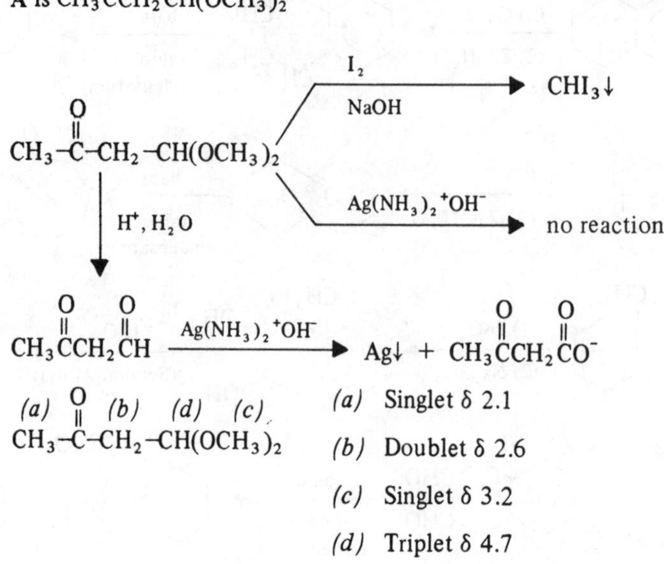

17.34 Abstraction of an α hydrogen at the ring junction yields an enolate ion that can then accept a proton to form either *trans*-1-decalone or *cis*-1-decalone. Since *trans*-1-decalone is more stable, it predominates at equilibrium.

$\xrightarrow[(-H^+)]{B:^-}$ $\longleftrightarrow$ $\xrightarrow{+H^+}$ +

(95%)
trans-1-Decalone
(more stable)

(5%)
cis-1-Decalone
(less stable)

SECTION REFERENCES FOR ADDITIONAL PROBLEMS

17.23	Chapters 16 and 17	**17.27**	17.9B
17.24	Chapters 16 and 17	**17.28**	16.11, 17.6
17.25	16.10-16.12, 17.6, 17.10	**17.29**	17.1, 17.5
17.26	16.7, 16.9, 17.5, 17.6	**17.30**	7.13, 17.6

17.31 13.13, 16.13, 17.9

17.32 16.13, 17.4

17.33 13.13, 16.12, 16.13, 17.4

17.34 3.13, 3.14, 17.1–17.3

SELF-TEST

Supply formulas for the missing reagents and intermediates in the following syntheses.

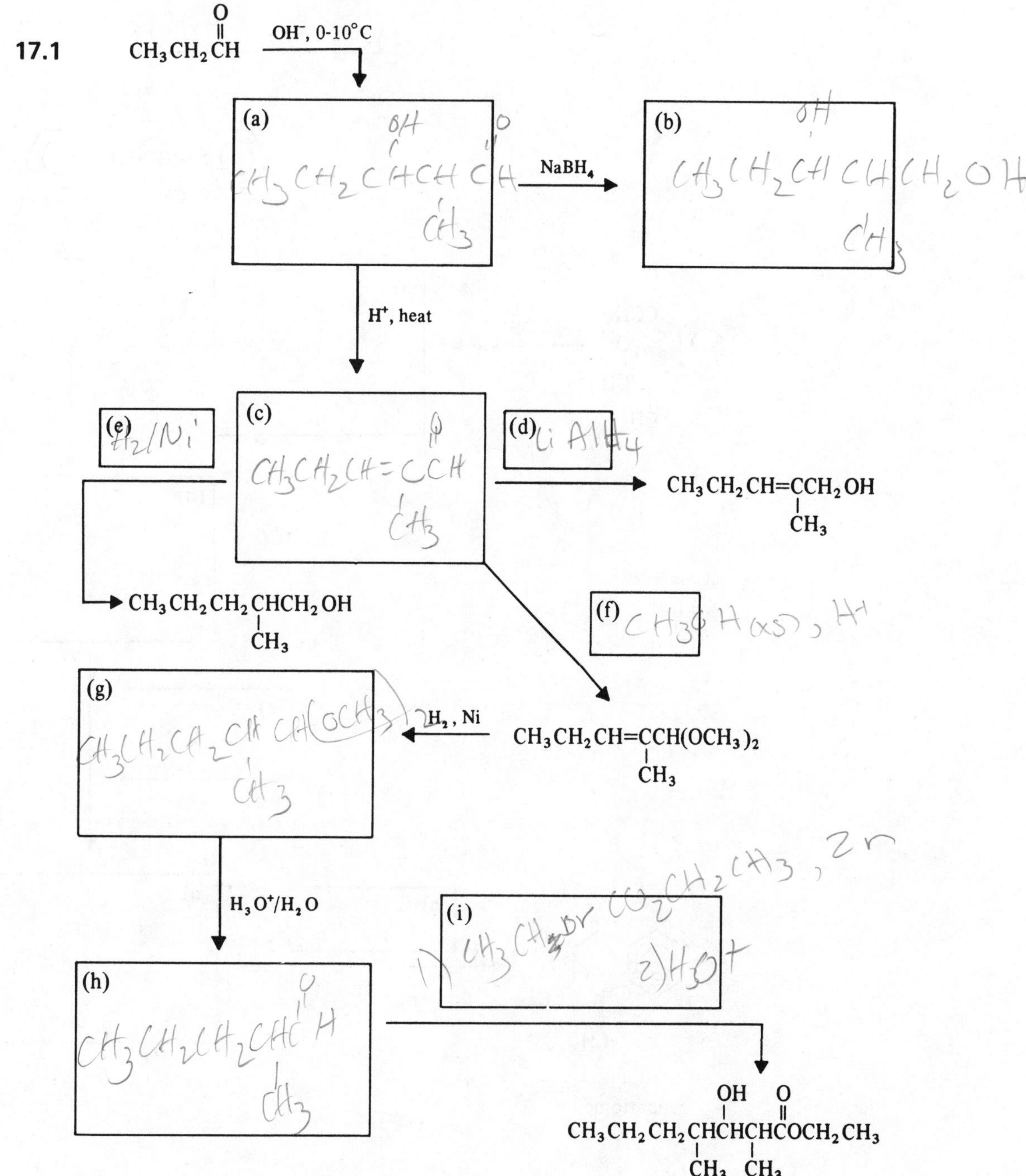

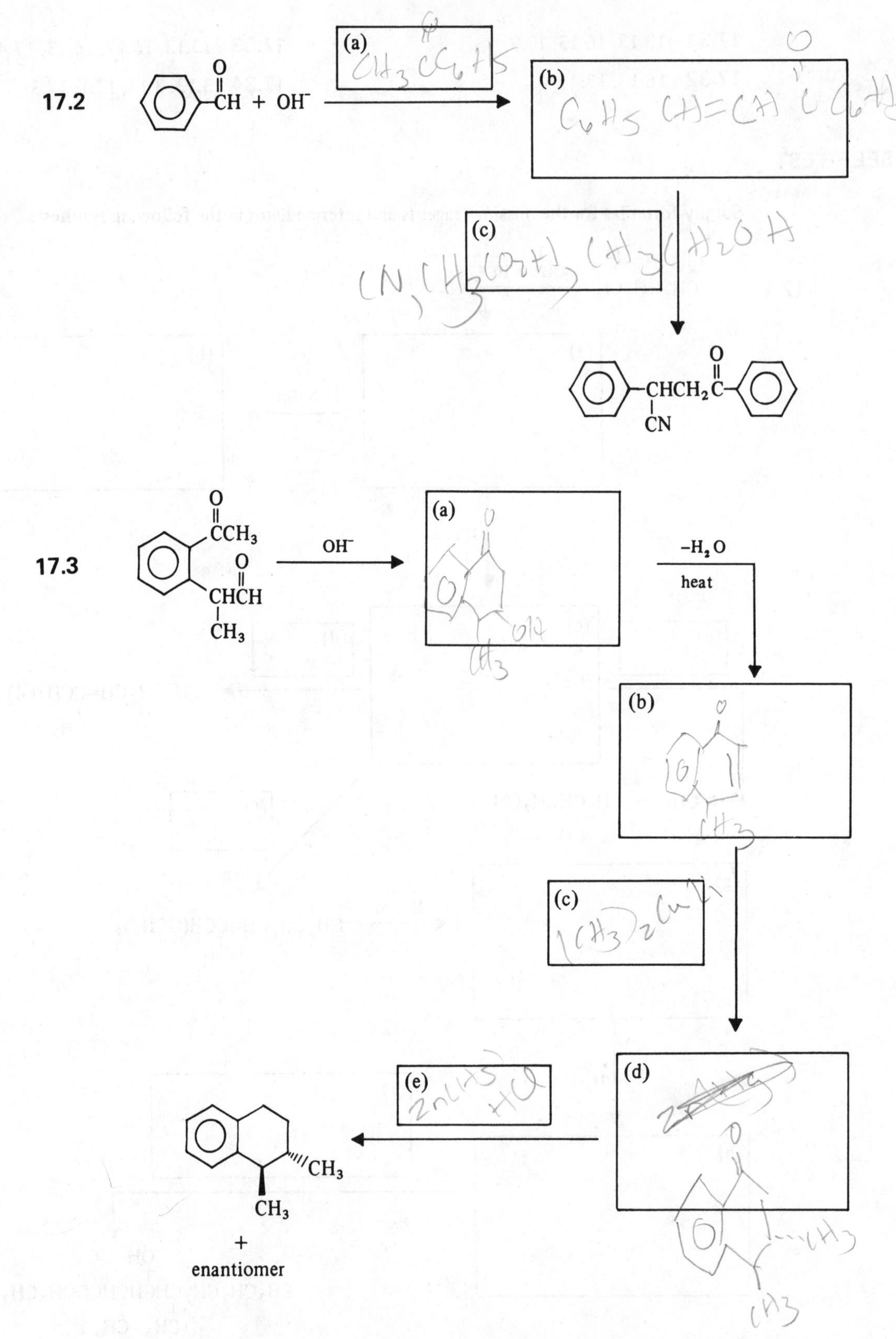
17.2
CH + OH⁻
(a)
(b)
(c)
CHCH₂C
CN
17.3
CCH₃
CHCH
CH₃
OH⁻
(a)
−H₂O
heat
(b)
(c)
(d)
(e)
CH₃
CH₃
+
enantiomer

17.4 Which would be formed in the following reaction?

$$CH_3\overset{O}{\overset{\|}{C}}H + CH_3CH_2\overset{O}{\overset{\|}{C}}H \xrightarrow[25°C]{OH^-} ?$$

(a) $CH_3\overset{OH}{\overset{|}{C}}HCH_2\overset{O}{\overset{\|}{C}}H$ (b) $CH_3CH_2\overset{OH}{\overset{|}{C}}H\underset{CH_3}{\underset{|}{C}}H\overset{O}{\overset{\|}{C}}H$ (c) $CH_3\overset{OH}{\overset{|}{C}}H\underset{CH_3}{\underset{|}{C}}H\overset{O}{\overset{\|}{C}}H$

(d) $CH_3CH_2\overset{OH}{\overset{|}{C}}HCH_2\overset{O}{\overset{\|}{C}}H$ (e) All of these will be formed

17.5 What would be the major product of the following reaction?

$$C_6H_5\overset{O}{\overset{\|}{C}}\underset{CH_3}{\underset{|}{C}}HCH_3 + Br_2 + OH^- \longrightarrow ?$$

(a) $C_6H_5\overset{O}{\overset{\|}{C}}\underset{CH_3}{\underset{|}{C}}BrCH_3$ (b) $C_6H_5\overset{O}{\overset{\|}{C}}\underset{CH_3}{\underset{|}{C}}HCH_2Br$ (c) $C_6H_5\overset{O}{\overset{\|}{C}}\underset{CH_3}{\underset{|}{C}}HCH_2OH$

(d) $C_6H_5CBr_2\overset{CH_3}{\overset{|}{C}}HCH_3$ (e) None of the above

SPECIAL TOPIC
Lithium Enolates in Organic Synthesis

SOLUTIONS TO PROBLEMS

I.1

Enolate from cyclohexanone $\xrightarrow{C_6H_5CH_2Br}$ O-Alkylated product ($OCH_2C_6H_5$) + C-Alkylated product ($CH_2C_6H_5$)

O-Alkylated product

C-Alkylated product

I.2

(a) 2-Methylcyclopentanone $\xrightarrow{LDA}$ Kinetic enolate (O^-Li^+) $\xrightarrow[(2)\ H_2O]{(1)\ CH_3CHO}$ product bearing CH_3, O and $CHCH_3$ (OH)

Kinetic enolate

(b) $CH_3CH_2\overset{O}{\overset{\|}{C}}CH_3 \xrightarrow{LDA} CH_3CH_2\overset{O^-Li^+}{\overset{|}{C}}{=}CH_2 \xrightarrow[(2)\ H_2O]{(1)\ C_6H_5CHO} CH_3CH_2\overset{O}{\overset{\|}{C}}CH_2\overset{OH}{\overset{|}{C}}HC_6H_5$

Kinetic enolate

(c) $CH_3\underset{CH_3}{\underset{|}{C}}H\overset{O}{\overset{\|}{C}}CH_3 \xrightarrow{LDA} CH_3\underset{CH_3}{\underset{|}{C}}H\overset{O^-Li^+}{\overset{|}{C}}{=}CH_2 \xrightarrow[(2)\ H_2O]{(1)\ CH_3CH_2CHO} CH_3\underset{CH_3}{\underset{|}{C}}H\overset{O}{\overset{\|}{C}}CH_2\overset{OH}{\overset{|}{C}}HCH_2CH_3$

Kinetic enolate

(d) $CH_3CH{=}CH\overset{O}{\overset{\|}{C}}CH_3 \xrightarrow{LDA} CH_3CH{=}CH\overset{O^-Li^+}{\overset{|}{C}}{=}CH_2 \xrightarrow{(1)\ CH_3CHO} CH_3CH{=}CH\overset{O}{\overset{\|}{C}}CH_2\overset{OH}{\overset{|}{C}}HCH_3$

I.3

(a) $CH_3C(CH_3){=}CHC(=O)CH_3 \xrightarrow{LDA} CH_3C(CH_3){=}CHC(O^-Li^+){=}CH_2 \xrightarrow[(2)\ H_2O]{(1)\ CH_3\text{-}C_6H_8\text{-}C(=O)CH_3}$

CH3 OH

CH3

O

α-Bisabolanone

(b) $CH_3C(CH_3){=}CHC(O^-Li^+){=}CH_2 \xrightarrow[(2)\ H_2O]{(1)\ CH_3\text{-}C_6H_4\text{-}C(=O)H}$

[prepared as in (a)]

H OH

CH3

O

Ocimenone

I.4

Step 1 $\text{cyclohexenyl-}O\text{-}Si(CH_3)_3 + \bar{F}\overset{+}{N}(C_4H_9)_4 \longrightarrow \text{cyclohexenyl-}\bar{O}\ \overset{+}{N}(C_4H_9)_4 + (CH_3)_3SiF$

Step 2 $\text{cyclohexenyl-}\bar{O}\ \overset{+}{N}(C_4H_9)_4 + C_6H_5CH(=O) \longrightarrow$ 2-[$CH(\bar{O}\ \overset{+}{N}(C_4H_9)_4)C_6H_5$]cyclohexanone

Step 3 2-[$CH(\bar{O}\ \overset{+}{N}(C_4H_9)_4)C_6H_5$]cyclohexanone $\xrightarrow{H_2O}$ 2-[$CH(OH)C_6H_5$]cyclohexanone $+ (C_4H_9)_4\overset{+}{N}\ \overset{-}{O}H$

I.5

CH_3-cyclohexanone $\xrightarrow{\text{LDA}}$ [O^- Li^+ enolate] $\xrightarrow{C_6H_5Se\text{-}Br}$ 2-methyl-6-(SeC_6H_5)cyclohexanone

Kinetic enolate

$\xrightarrow{H_2O_2}$ [selenoxide intermediate: CH_3, O, Se, C_6H_5, O^-, H] $\longrightarrow$ 6-methylcyclohex-2-enone (CH_3) + C_6H_5SeOH

18 CARBOXYLIC ACIDS AND THEIR DERIVATIVES: NUCLEOPHILIC SUBSTITUTION AT ACYL CARBON

REACTIONS OF CARBOXYLIC ACIDS AND THEIR DERIVATIVES

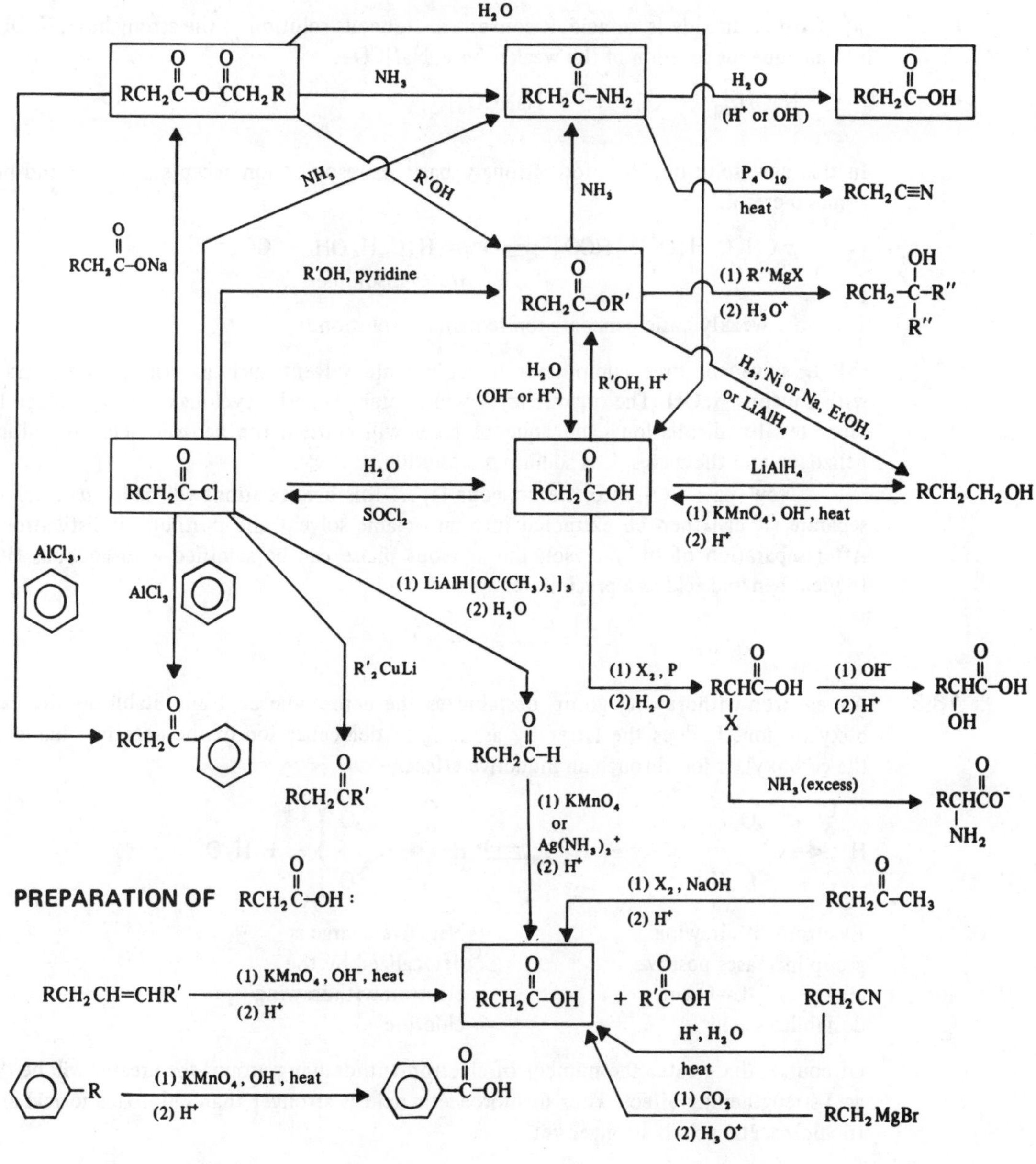

SOLUTIONS TO PROBLEMS

18.1 (a) 2-Methylbutanoic acid

(b) 3-Pentenoic acid

(c) Sodium 4-bromobutanoate

(d) 5-Phenylpentanoic acid

(e) 3-Methyl-3-pentenoic acid

18.2 (a) Carbon dioxide is an acid; it converts an aqueous solution of the strong base, NaOH, into an aqueous solution of the weaker base, $NaHCO_3$.

$$NaOH_{(aq)} + CO_2 \longrightarrow NaHCO_{3\,(aq)}$$

In this new solution, the more strongly basic *p*-cresoxide ion accepts a proton and becomes *p*-cresol,

$$p\text{-}CH_3C_6H_4O^- + HCO_3^- \rightleftharpoons \underset{\text{Water-insoluble}}{p\text{-}CH_3C_6H_4OH} + CO_3^{2-}$$

The more weakly basic benzoate ion remains in solution.

(b) Dissolve all three compounds in an organic solvent such as ether, then extract with aqueous NaOH. The organic layer will contain 1-methylcyclohexanol, which can be separated by distillation. The aqueous layer will contain the benzoic acid, as sodium benzoate, and the *p*-cresol, as sodium *p*-cresoxide.

Now pass CO_2 into the aqueous layer; this acidification will cause *p*-cresol to separate (it can then be extracted into an organic solvent and purified by distillation). After separation of the *p*-cresol, the aqueous phase can be acidified with aqueous HCl to yield benzoic acid as a precipitate.

18.3 An electron-withdrawing group destabilizes the carboxylic acid and stabilizes the carboxylate ion. It does the latter by assisting in delocalization of the negative charge of the carboxylate ion through an inductive effect.

$$ClCH_2COOH + H_2O \rightleftharpoons [ClCH_2CO_2]^- + H_3O^+$$

Electron-withdrawing group increases positive charge on >C=O and destabilizes acid

Negative charge is delocalized by the electron-withdrawing chlorine

Of course, the greater the number of electron-withdrawing groups the greater will be the acid-strengthening effect. Thus dichloroacetic acid is stronger than chloroacetic acid and trichloroacetic acid is stronger yet.

18.4 (a) CH_2FCO_2H (F– is more electronegative than H–)

(b) CH_2FCO_2H (F– is more electronegative than Cl–)

(c) CH_2ClCO_2H (Cl– is more electronegative than Br–)

(d) $CH_3CHClCH_2CO_2H$ (Cl– is closer to $-CO_2H$)

(e) $CH_3CH_2CHClCO_2H$ (Cl– is closer to $-CO_2H$)

(f) $(CH_3)_3\overset{+}{N}-C_6H_4-CO_2H$ (para) [$(CH_3)_3\overset{+}{N}-$ is more electronegative than H–]

(g) $CF_3-C_6H_4-CO_2H$ (para) (CF_3- is more electronegative than CH_3-)

18.5 (a) The carboxyl group is an electron-withdrawing group; thus in a dicarboxylic acid such as those in Table 18.3, one carboxyl group increases the acidity of the other.

(b) As the distance between the carboxyl groups increases the acid-strengthening, inductive effect decreases.

18.6 (a) Heptanedioic acid

(b) Methyl butanoate

(c) 2-Chlorobutanoic acid

(d) Propanoic anhydride

(e) Butanoyl chloride

(f) Propanamide

(g) *N*-Methylbutanamide

(h) 4-Phenylbutanoyl chloride

18.7 (a) $CH_3CH_2\overset{O}{\overset{\|}{C}}-OCH_3$

(b) $O_2N-C_6H_4-\overset{O}{\overset{\|}{C}}-OCH_2CH_3$ (para)

(c) $CH_3O-\overset{O}{\overset{\|}{C}}CH_2\overset{O}{\overset{\|}{C}}-OCH_3$

(d) $C_6H_5-\overset{O}{\overset{\|}{C}}-N(CH_3)_2$

(e) $C_6H_4(\overset{O}{\overset{\|}{C}}-OCH_3)_2$ (ortho)

(f) $\mathrm{H}(\overset{O}{\overset{\|}{C}}-OCH_2CH_2CH_3)C=C(\overset{O}{\overset{\|}{C}}-OCH_2CH_2CH_3)\mathrm{H}$ (both H on one carbon: $H_2C=C(\overset{O}{\overset{\|}{C}}-OCH_2CH_2CH_3)_2$)

(g) $H-\overset{O}{\overset{\|}{C}}-N(CH_3)_2$

(h) $CH_3\underset{Br}{\underset{|}{C}}H\overset{O}{\overset{\|}{C}}-Br$

18.8 These syntheses are easy to see if we work backward.

(a) $$C_6H_5CH_2CO_2H \xleftarrow[(2)\ H^+]{(1)\ CO_2} C_6H_5CH_2MgBr \xleftarrow{Mg,\ ether} C_6H_5CH_2Br$$

(b) $$CH_3CH_2CH_2C(CH_3)_2CO_2H \xleftarrow[(2)\ H^+]{(1)\ CO_2} CH_3CH_2CH_2C(CH_3)_2MgBr \xleftarrow{Mg,\ ether} CH_3CH_2CH_2C(CH_3)_2Br$$

(c) $$CH_2{=}CHCH_2CO_2H \xleftarrow[(2)\ H^+]{(1)\ CO_2} CH_2{=}CHCH_2MgBr \xleftarrow{Mg,\ ether} CH_2{=}CHCH_2Br$$

(d) $$CH_3\text{-}C_6H_4\text{-}CO_2H \xleftarrow[(2)\ H^+]{(1)\ CO_2} CH_3\text{-}C_6H_4\text{-}MgBr \xleftarrow{Mg,\ ether} CH_3\text{-}C_6H_4\text{-}Br$$

(e) $$CH_3CH_2CH_2CH_2CH_2CO_2H \xleftarrow[(2)\ H^+]{(1)\ CO_2} CH_3CH_2CH_2CH_2CH_2MgBr \xleftarrow{Mg,\ ether} CH_3CH_2CH_2CH_2CH_2Br$$

18.9

(a) $$C_6H_5CH_2CO_2H \xleftarrow[(2)\ H^+,\ H_2O,\ heat]{(1)\ CN^-} C_6H_5CH_2Br$$

$$CH_2{=}CHCH_2CO_2H \xleftarrow[(2)\ H^+,\ H_2O,\ heat]{(1)\ CN^-} CH_2{=}CHCH_2Br$$

$$CH_3CH_2CH_2CH_2CH_2CO_2H \xleftarrow[(2)\ H^+,\ H_2O,\ heat]{(1)\ CN^-} CH_3CH_2CH_2CH_2CH_2Br$$

(b) A nitrile synthesis. Preparation of a Grignard reagent from $HOCH_2CH_2CH_2CH_2Br$ would not be possible because of the presence of the acidic hydroxyl group.

18.10 Since maleic acid is a cis dicarboxylic acid, dehydration occurs readily:

$$\text{Maleic acid (cis-HOOC-CH=CH-COOH)} \xrightarrow{200°C} \text{Maleic anhydride} + H_2O$$

Maleic acid — Maleic anhydride

Being a trans dicarboxylic acid, fumaric acid must undergo isomerization to maleic acid first. This isomerization requires a higher temperature.

$$\text{Fumaric acid (trans-HOC(=O)-CH=CH-C(=O)OH)} \xrightarrow{\text{heat}} \text{cis-HOC(=O)-CH=CH-C(=O)OH} \xrightarrow{-H_2O} \text{Maleic anhydride}$$

Fumaric acid

18.11 The labeled oxygen atom should appear in the carboxyl group of the acid. (Follow the reverse steps of the mechanism in Section 18.7A of the text using $H_2{}^{18}O$.)

18.12

$$CH_2{=}CHC(=O)OCH_3 \underset{-H^+}{\overset{+H^+}{\rightleftharpoons}} CH_2{=}CHC(=\overset{+}{O}H)OCH_3 \underset{-C_4H_9OH}{\overset{+C_4H_9OH}{\rightleftharpoons}}$$

$$CH_2{=}CHC(OH)(OCH_3)\overset{+}{O}(H)C_4H_9 \rightleftharpoons CH_2{=}CHC(OH)(OC_4H_9)\overset{+}{O}(H)CH_3 \underset{+CH_3OH}{\overset{-CH_3OH}{\rightleftharpoons}}$$

$$CH_2{=}CHC(=\overset{+}{O}H)OC_4H_9 \underset{+H^+}{\overset{-H^+}{\rightleftharpoons}} CH_2{=}CHC(=O)OC_4H_9$$

18.13 (a)

(1) H_3C–(cyclopentyl)–OH + $C_6H_5SO_2Cl$ ⟶ H_3C–(cyclopentyl)–$OSO_2C_6H_5$ (A) $\xrightarrow[\text{(inversion)}]{OH^-,\ \text{heat}}$ H_3C–(cyclopentyl)–OH (B) + $C_6H_5SO_3^-$

(2) H_3C–(cyclopentyl)–OH + $C_6H_5\overset{O}{\overset{\|}{C}}$–Cl ⟶ H_3C–(cyclopentyl)–O–$\overset{\|}{\underset{O}{C}}C_6H_5$ (C) $\xrightarrow[\text{(retention)}]{OH^-,\ \text{heat}}$ H_3C–(cyclopentyl)–OH (D) + $C_6H_5CO_2^-$

(3) H, C_6H_{13}, CH_3 (stereocenter)–Br + $CH_3\overset{O}{\overset{\|}{C}}O^-Na^+$ $\xrightarrow{\text{(inversion)}}$ $CH_3\overset{O}{\overset{\|}{C}}O$–(stereocenter: H, C_6H_{13}, CH_3) (E) $\xrightarrow[\text{(retention)}]{OH^-,\ \text{heat}}$ HO–(stereocenter: H, C_6H_{13}, CH_3) (F)

(4) H, C_6H_{13}, CH_3 (stereocenter)–Br $\xrightarrow[\text{(inversion)}]{OH^-,\ \text{heat}}$ HO–(stereocenter: H, C_6H_{13}, CH_3) F

(b) Method (3) should give a higher yield of **F** than method (4). Since the hydroxide ion is a strong base and since the alkyl halide is secondary, method (4) is likely to be accompanied by considerable elimination. Method (3), on the other hand, employs a weaker base, acetate ion, in the S_N2 step and is less likely to be complicated by elimination. Hydrolysis of the ester **E** that results should also proceed in high yield.

18.14 (a) Steric hindrance presented by the di-ortho methyl groups of methyl mesitoate prevents formation of the tetrahedral intermediate that must accompany attack at the acyl carbon.

(b) Carry out hydrolysis with labeled $^{18}OH^-$ in labeled $H_2{}^{18}O$. The label should appear in the methanol.

18.15

(a) $C_6H_5\overset{O}{\overset{\|}{C}}N(CH_2CH_3)_2 \xrightarrow[H_2O]{OH^-} C_6H_5CO_2^- + (CH_3CH_2)_2NH$

$C_6H_5\overset{O}{\overset{\|}{C}}N(CH_2CH_3)_2 \xrightarrow[H_2O]{H^+} C_6H_5CO_2H + (CH_3CH_2)_2\overset{+}{N}H_2$

(b) 2-piperidinone (six-membered ring with C=O and N–H)

$\xrightarrow[H_2O]{OH^-} {}^-O\overset{O}{\overset{\|}{C}}CH_2CH_2CH_2CH_2NH_2$

$\xrightarrow[H_2O]{H^+} HO\overset{O}{\overset{\|}{C}}CH_2CH_2CH_2CH_2\overset{+}{N}H_3$

(c) $HO_2C\underset{CH_3}{CH}-NH\overset{O}{\overset{\|}{C}}\underset{CH_2C_6H_5}{CH}NH_2$

$\xrightarrow[H_2O]{OH^-} {}^-O_2C\underset{CH_3}{CH}NH_2 + {}^-O_2C\underset{CH_2C_6H_5}{CH}NH_2$

$\xrightarrow[H_2O]{H^+} HO_2C\underset{CH_3}{CH}\overset{+}{N}H_3 \quad HO_2C\underset{CH_2C_6H_5}{CH}\overset{+}{N}H_3$

18.16 (a) $(CH_3)_3CCO_2H \xrightarrow{SOCl_2} (CH_3)_3CCOCl \xrightarrow{NH_3}$

$(CH_3)_3CCONH_2 \xrightarrow[heat]{P_4O_{10}} (CH_3)_3CC{\equiv}N$

(b) An elimination reaction would take place because CN^- is a strong base.

$CN^- + H-CH_2-\underset{CH_3}{\overset{CH_3}{C}}-Br \longrightarrow HCN + CH_2{=}C(CH_3)_2 + Br^-$

18.17

(a) $C_6H_5CH_2OH + O{=}C{=}N{-}C_6H_5 \longrightarrow C_6H_5CH_2{-}O{-}\overset{O}{\overset{\|}{C}}{-}\overset{H}{N}{-}C_6H_5$

(b) $Cl{-}\overset{O}{\overset{\|}{C}}{-}Cl + 4CH_3NH_2 \longrightarrow CH_3\underset{H}{N}{-}\overset{O}{\overset{\|}{C}}{-}\underset{H}{N}CH_3 + 2CH_3\overset{+}{N}H_3 + 2Cl^-$

(c) $C_6H_5CH_2O\overset{O}{\overset{\|}{C}}Cl + H_3\overset{+}{N}CH_2CO_2^- \xrightarrow{OH^-} C_6H_5CH_2O\overset{O}{\overset{\|}{C}}\underset{H}{\underset{|}{N}}CH_2CO_2H + HCl$

(d) $C_6H_5CH_2O\overset{O}{\overset{\|}{C}}NHCH_2CO_2H \xrightarrow{H_2, Pd} H_3\overset{+}{N}CH_2CO_2^- + CO_2 + C_6H_5CH_3$

(e) $C_6H_5CH_2O\overset{O}{\overset{\|}{C}}NHCH_2CO_2H \xrightarrow{HBr, CH_3CO_2H} H_3\overset{+}{N}CH_2CO_2H + CO_2 + C_6H_5CH_2Br$

(f) $H_2N\overset{O}{\overset{\|}{C}}NH_2 \xrightarrow{OH^-, H_2O, heat} 2\,NH_3 + CO_3^{-2}$

18.18 (a) By a Kolbe electrolysis of hexanoic acid:

(1) $CH_3(CH_2)_4\overset{O}{\overset{\|}{C}}O^- \xrightarrow[-e^-]{anode} CH_3(CH_2)_4\overset{O}{\overset{\|}{C}}O\cdot$

(2) $CH_3(CH_2)_4\overset{O}{\overset{\|}{C}}O\cdot \longrightarrow CH_3(CH_2)_3CH_2\cdot + CO_2$

(3) $2CH_3(CH_2)_3CH_2\cdot \longrightarrow CH_3(CH_2)_8CH_3$

(b) By decarboxylation of a β-keto acid:

$$CH_3(CH_2)_3\overset{O}{\overset{\|}{C}}CH_2\overset{O}{\overset{\|}{C}}OH \xrightarrow{100\text{-}150^\circ C} CH_3(CH_2)_3\overset{O}{\overset{\|}{C}}CH_3 + CO_2$$

(c) By decarboxylation of a substituted malonic acid

$$CH_3CH_2\underset{CH_3}{\underset{|}{\overset{CO_2H}{\overset{|}{C}}}}\!-\!CO_2H \xrightarrow{100\text{-}150^\circ C} CH_3CH_2\underset{CH_3}{\underset{|}{C}}HCO_2H + CO_2$$

(d) By a Hunsdiecker reaction

$$C_6H_5CH_2CO_2Ag + Br_2 \xrightarrow[heat]{CCl_4} C_6H_5CH_2Br + CO_2 + AgBr$$

(e) By decarboxylation of a β-keto acid

$$CH_3CH_2\overset{O}{\overset{\|}{C}}CH_2\overset{O}{\overset{\|}{C}}OH \xrightarrow{100\text{-}150°C} CH_3CH_2\overset{O}{\overset{\|}{C}}CH_3 + CO_2$$

(f) By a Hundieker reaction followed by treatment with zinc and acid.

$$C_6H_{11}CO_2H \xrightarrow{AgOH} C_6H_{11}CO_2Ag \xrightarrow[\text{heat}]{Br_2/CCl_4} C_6H_{11}Br$$

$$\xrightarrow[H^+]{Zn,} C_6H_{12} + ZnBr_2$$

(g) By decarboxylation of a β-keto acid

$$\text{2-oxocyclohexane-}\overset{O}{\overset{\|}{C}}OH \xrightarrow{100\text{-}150°C} \text{cyclohexanone} + CO_2$$

(h) By decarboxylation of a substituted malonic acid.

$$CH_3CH_2CH_2CH(CO_2H)_2 \xrightarrow{100\text{-}150°C} CH_3CH_2CH_2CH_2CO_2H + CO_2$$

18.19 (a) The oxygen-oxygen bond of the diacyl peroxide has a low homolytic bond dissociation energy ($DH° \simeq 35$ kcal/mole). This allows the following reaction to occur at a moderate temperature.

$$R\text{–}\overset{O}{\overset{\|}{C}}\text{–}O\text{–}O\text{–}\overset{O}{\overset{\|}{C}}R \longrightarrow 2R\text{–}\overset{O}{\overset{\|}{C}}\text{–}O\cdot \qquad \Delta H° \simeq 35 \text{ kcal/mole}$$

(b) By decarboxylation of the carboxylate radical produced in part (a).

$$R\text{–}\overset{O}{\overset{\|}{C}}\text{–}O\cdot \longrightarrow R\cdot + CO_2$$

(c) **Chain initiation**

Step 1 $R\text{–}\overset{O}{\overset{\|}{C}}\text{–}O\text{–}O\text{–}\overset{O}{\overset{\|}{C}}\text{–}R \xrightarrow{\text{heat}} 2R\text{–}\overset{O}{\overset{\|}{C}}\text{–}O\cdot$

Step 2 $R\text{–}\overset{O}{\overset{\|}{C}}\text{–}O\cdot \longrightarrow R\cdot + CO_2$

Chain Propagation

Step 3 $R\cdot + CH_2{=}CH_2 \longrightarrow RCH_2CH_2\cdot$

Step 4 $RCH_2CH_2\cdot + CH_2{=}CH_2 \longrightarrow RCH_2CH_2CH_2CH_2\cdot$

Step 3, 4, 3, 4, and so on.

18.20

(a) $CH_3(CH_2)_4CO_2H$

(b) $CH_3(CH_2)_4CONH_2$

(c) $CH_3(CH_2)_4CONHC_2H_5$

(d) $CH_3(CH_2)_4CON(C_2H_5)_2$

(e) $CH_3CH_2CH{=}CHCH_2CO_2H$

(f) $CH_3CH{=}CHCH_2CH(CH_3)CO_2H$

(g) $HO_2CCH_2CH_2CH_2CH_2CO_2H$

(h) o-$C_6H_4(CO_2H)_2$ (benzene ring with CO_2H groups at adjacent positions)

(i) m-$C_6H_4(CO_2H)_2$ (benzene ring with CO_2H groups at 1,3-positions)

(j) p-$C_6H_4(CO_2H)_2$ (benzene ring with CO_2H groups at 1,4-positions)

(k) $C_2H_5O_2C{-}CO_2C_2H_5$

(l) $C_2H_5O_2C(CH_2)_4CO_2C_2H_5$

(m) $CH_3CH_2CO_2CH_2CH(CH_3)_2$

(n) Naphthalene ring with CO_2H at the 2-position

(o) $HO_2C(H)C{=}C(H)CO_2H$ (cis: both CO_2H on the same side)

(p) $HO_2CCHOHCH_2CO_2H$

(q) $HO_2C(H)C{=}C(H)CO_2H$ (trans: CO_2H groups on opposite sides)

(r) $HO_2CCH_2CH_2CO_2H$

(s) Cyclic imide: $H_2C{-}C(=O){-}NH{-}C(=O){-}CH_2$ (ring)

(t) $HO_2CCH_2CO_2H$

(u) $C_2H_5O_2CCH_2CO_2C_2H_5$

18.21

(a) Benzoic acid

(b) Benzoyl chloride

(c) Benzamide

(d) Benzoic anhydride

(e) Benzyl benzoate

(f) Phenyl benzoate

(g) Isopropyl acetate

(h) *N,N*-Dimethylacetamide

(i) Acetonitrile

(j) Maleic anhydride

(k) Phthalic anhydride

(l) Phthalimide

(m) Glyceryl tripalmitate

(n) α-Ketosuccinic acid

(o) Methyl salicylate

18.22

(a) $C_6H_5Br \xrightarrow{\text{Mg, ether}} C_6H_5MgBr \xrightarrow[\text{(2) } H_3O^+]{\text{(1) } CO_2} C_6H_5CO_2H$

(b) $C_6H_5CH_3 \xrightarrow[\text{(2) } H_3O^+]{\text{(1) } KMnO_4\text{, } OH^-\text{, heat}} C_6H_5CO_2H$

(c) $C_6H_5CN \xrightarrow[\text{heat}]{H_3O^+\text{, } H_2O} C_6H_5CO_2H + NH_4^+$

(d) $C_6H_5\overset{O}{\overset{\|}{C}}CH_3 \xrightarrow[-CHCl_3]{Cl_2\text{, } OH^-} C_6H_5CO_2^- \xrightarrow{H_3O^+} C_6H_5CO_2H$

(e) $C_6H_5CHO \xrightarrow{Ag(NH_3)_2^+\ OH^-} C_6H_5CO_2^- \xrightarrow{H_3O^+} C_6H_5CO_2H$

(f) $C_6H_5CH{=}CH_2 \xrightarrow[\text{(2) } H_3O^+]{\text{(1) } KMnO_4\text{, } OH^-\text{, heat}} C_6H_5CO_2H$

(g) $C_6H_5CH_2OH \xrightarrow[\text{(2) } H_3O^+]{\text{(1) } KMnO_4\text{, } OH^-\text{, heat}} C_6H_5CO_2H$

18.23

(a) $C_6H_5CH_2CHO \xrightarrow{Ag(NH_3)_2^+\ OH^-} C_6H_5CH_2CO_2^- \xrightarrow{H_3O^+} C_6H_5CH_2CO_2H$

(b) $C_6H_5CH_2Br \xrightarrow[\text{(2) } CO_2]{\text{(1) Mg, ether}} C_6H_5CH_2CO_2MgBr \xrightarrow{H_3O^+} C_6H_5CH_2CO_2H$

$C_6H_5CH_2Br \xrightarrow{CN^-} C_6H_5CH_2CN \xrightarrow[\text{heat}]{H_3O^+\text{, } H_2O} C_6H_5CH_2CO_2H$

18.24

(a) $CH_3CH_2CH_2CH_2CH_2OH \xrightarrow[\text{(2) } H_3O^+]{\text{(1) } KMnO_4\text{, } OH^-\text{, heat}} CH_3CH_2CH_2CH_2CO_2H$

(b) $CH_3CH_2CH_2CH_2Br \xrightarrow[(2)\ CO_2]{(1)\ Mg,\ ether} CH_3CH_2CH_2CH_2CO_2MgBr \xrightarrow{H_3O^+} CH_3CH_2CH_2CH_2CO_2H$

$CH_3CH_2CH_2CH_2Br \xrightarrow{CN^-} CH_3CH_2CH_2CH_2CN \xrightarrow[heat]{H_3O^+,\ H_2O,} CH_3CH_2CH_2CH_2CO_2H$

(c) $CH_3(CH_2)_3CH{=}CH(CH_2)_3CH_3 \xrightarrow[(2)\ H_3O^+]{(1)\ KMnO_4,\ OH^-,\ heat} 2CH_3(CH_2)_3CO_2H$

(d) $CH_3CH_2CH_2CH_2CHO \xrightarrow[(2)\ H_3O^+]{(1)\ Ag(NH_3)_2^+\ OH^-} CH_3CH_2CH_2CH_2CO_2H$

18.25
(a) $CH_3CO_2H + HCl$
(b) $CH_3CO_2H + AgCl$
(c) $CH_3CO_2CH_2(CH_2)_2CH_3$
(d) CH_3CONH_2
(e) o-methylacetophenone ($C_6H_4(CH_3)(COCH_3)$, 1,2) + $CH_3-C_6H_4-\overset{O}{\overset{\|}{C}}CH_3$ (para)
(f) CH_3CHO
(g) CH_3COCH_3
(h) CH_3CO_2Na
(i) $CH_3CONHCH_3$
(j) $CH_3CONHC_6H_5$
(k) $CH_3CON(CH_3)_2$
(l) $CH_3CO_2CH_2CH_3$
(m) $(CH_3CO)_2O$
(n) $(CH_3CO)_2O$
(o) $CH_3CO_2C_6H_5$

18.26
(a) $CH_3CONH_2 + CH_3CO_2NH_4$
(b) $2CH_3CO_2H$
(c) $CH_3CO_2CH_2CH_2CH_3 + CH_3CO_2H$
(d) $C_6H_5COCH_3 + CH_3CO_2H$
(e) $CH_3CONHCH_2CH_3 + CH_3CO_2^-CH_3CH_2NH_3^+$
(f) $CH_3CON(CH_2CH_3)_2 + CH_3CO_2^-(CH_3CH_2)_2NH_2^+$

18.27
(a) $CONH_2-CH_2-CH_2-CO_2^-NH_4^+$
(b) $CO_2H-CH_2-CH_2-CO_2H$

(c) $CH_3CH_2CH_2O_2C-CH_2-CH_2-CO_2H$

(d) $C_6H_5-C(=O)-CH_2-CH_2-CO_2H$

(e) $CH_3CH_2NHCO-CH_2-CH_2-CO_2^- \ CH_3CH_2\overset{+}{N}H_3$

(f) $(CH_3CH_2)_2NCO-CH_2-CH_2-CO_2^-(CH_3CH_2)_2NH_2^+$

18.28

(a) Succinic anhydride + benzene $\xrightarrow{AlCl_3}$ $C_6H_5-C(=O)-CH_2-CH_2-CO_2H$ $\xrightarrow[\text{HCl, heat}]{\text{Zn(Hg)}}$

$C_6H_5-CH_2-CH_2-CH_2-CO_2H$ $\xrightarrow{SOCl_2}$ $C_6H_5-CH_2-CH_2-CH_2-C(=O)-Cl$ $\xrightarrow{AlCl_3}$ 1-tetralone

(b) 1-tetralone $\xrightarrow[\text{(2) } H_3O^+]{\text{(1) } CH_3MgBr}$ 1-methyl-1-tetralol (OH, CH_3) $\xrightarrow[-H_2O]{H^+\text{, heat}}$ 4-methyl-1,2-dihydronaphthalene (CH_3)

(c) 4-methyl-1,2-dihydronaphthalene (CH_3) $\xrightarrow[-H_2]{\text{Pd, heat}}$ 1-methylnaphthalene (CH_3)

(d) Phthalic anhydride + benzene $\xrightarrow{AlCl_3}$ 2-benzoylbenzoic acid ($C_6H_5-C(=O)-C_6H_4-COOH$) $\xrightarrow[\text{heat}]{\text{Zn(Hg), HCl}}$

2-benzylbenzoic acid ($C_6H_5-CH_2-C_6H_4-COOH$) $\xrightarrow[\text{(2) } AlCl_3]{\text{(1) } SOCl_2}$ anthrone (O)

(e) 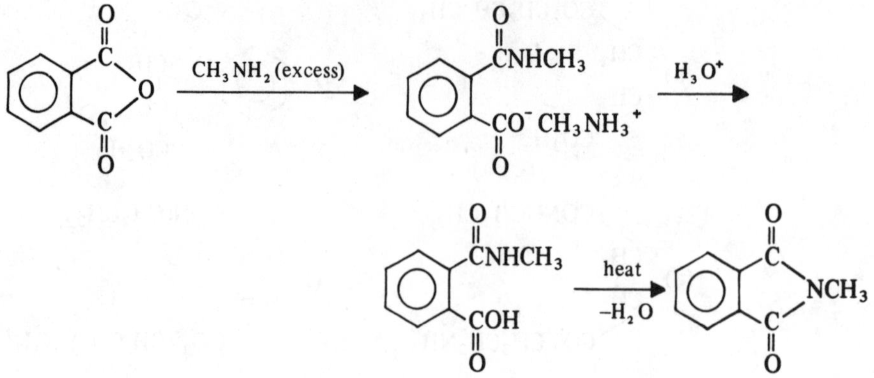 Phthalic anhydride $\xrightarrow{CH_3NH_2\ (excess)}$ $C_6H_4(CONHCH_3)(CO_2^-\ CH_3NH_3^+)$ $\xrightarrow{H_3O^+}$

$C_6H_4(CONHCH_3)(COOH)$ $\xrightarrow[-H_2O]{heat}$ N-methylphthalimide ($C_6H_4(CO)_2NCH_3$)

(f) 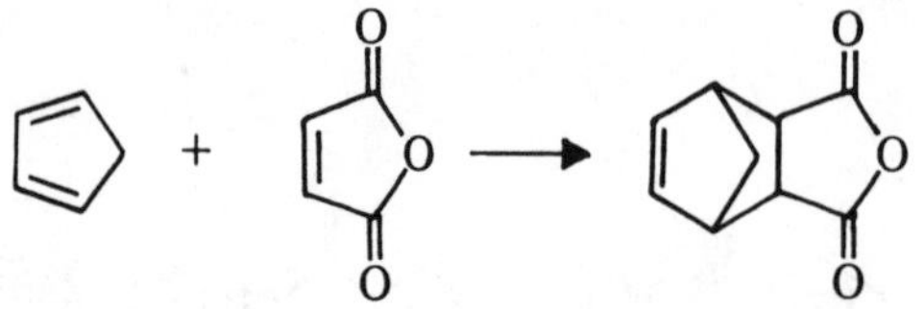 cyclopentadiene + maleic anhydride $\longrightarrow$ bicyclic anhydride adduct

(g) 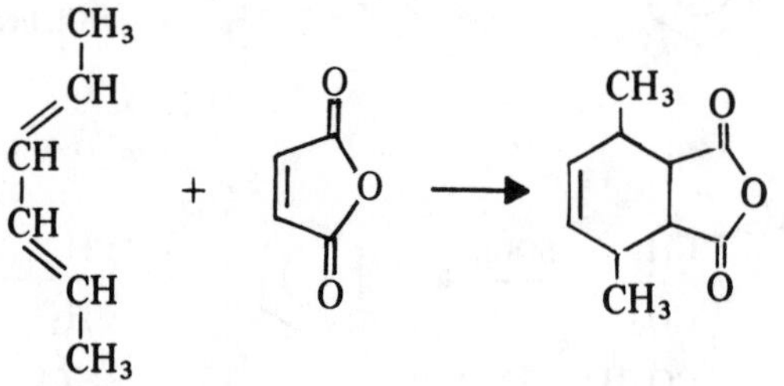 $CH_3CH{=}CH{-}CH{=}CHCH_3$ + maleic anhydride $\longrightarrow$ 3,6-dimethylcyclohex-4-ene-1,2-dicarboxylic anhydride (with two CH_3 groups)

18.29 (a) $CH_3CH_2CO_2H + CH_3CH_2OH$

(b) $CH_3CH_2CO_2^- + CH_3CH_2OH$

(c) $CH_3CH_2CO_2(CH_2)_7CH_3 + CH_3CH_2OH$

(d) $CH_3CH_2CONHCH_3 + CH_3CH_2OH$

(e) $CH_3CH_2CH_2OH + CH_3CH_2OH$

(f) $CH_3CH_2C(C_6H_5)_2OH + CH_3CH_2OH$

18.30 (a) $CH_3CH_2CO_2H + NH_4^+$

(b) $CH_3CH_2CO_2^- + NH_3$

(c) CH_3CH_2CN

18.31 (a) Benzoic acid dissolves in aqueous $NaHCO_3$. Methyl benzoate does not.

(b) Benzoyl chloride gives a precipitate (AgCl) when treated with alcoholic $AgNO_3$. Benzoic acid does not.

(c) Benzoic acid dissolves in aqueous $NaHCO_3$. Benzamide does not.

(d) Benzoic acid dissolves in aqueous $NaHCO_3$. *p*-Cresol does not.

(e) Refluxing benzamide with aqueous NaOH liberates NH_3, which can be detected in the vapors with moist red litmus paper. Ethyl benzoate does not liberate NH_3.

(f) Cinnamic acid, because it has a double bond, decolorizes Br_2 in CCl_4. Benzoic acid does not.

(g) Benzoyl chloride gives a precipitate (AgCl) when treated with alcoholic $AgNO_3$. Ethyl benzoate does not.

(h) 2-Chlorobutanoic acid gives a precipitate (AgCl) when treated with alcoholic silver nitrate. Butanoic acid does not.

18.32

(a)
$$\begin{array}{lll} H_2C-CH_2 & & \\ | & & C=O \\ H_2C-O & & \end{array}$$

(b) $CH_3CH{=}CHCO_2H$

(c)
$$\begin{array}{ccc} & O & \\ & \| & \\ & C-O & \\ CH_3CH_2CH & & CHCH_2CH_3 \\ & O-C & \\ & \| & \\ & O & \end{array}$$

(d)
$$\begin{array}{lll} & & O \\ & CH_2-C & \\ H_2C & & O \\ & CH_2-C & \\ & & O \end{array}$$

(e)
$$\begin{array}{llll} & & O & \\ & CH_2-C & & \\ H_2C & & & NH \\ & CH_2-CH & & \\ & & & CH_3 \end{array}$$

(f)
$$\begin{array}{ll} & O \\ & \| \\ & C \\ C_6H_4 & \quad NH \\ & \quad | \\ & CH_2 \end{array}$$

18.33

(a) (R)-(−)-2-butanol [HO, H, CH_3, CH_2CH_3] $\xrightarrow[\text{(retention)}]{\text{TsCl, pyridine}}$ **A** [TsO, H, CH_3, CH_2CH_3] $\xrightarrow[\text{(inversion)}]{CN^-}$ **B** [H, CN, CH_3, CH_2CH_3]

$\xrightarrow[\text{(retention)}]{H_2SO_4,\ H_2O}$ (+) **C** [H, CO_2H, CH_3, CH_2CH_3] $\xrightarrow[\text{(retention)}]{LiAlH_4}$ (−) **D** [H, CH_2OH, CH_3, CH_2CH_3]

(b) (R)-(−)-2-butanol [HO, H, CH_3, CH_2CH_3] $\xrightarrow[\text{pyridine (inversion)}]{PBr}$ **E** [H, Br, CH_3, CH_2CH_3] $\xrightarrow[\text{(inversion)}]{CN^-}$ **F** [NC, H, CH_3, CH_2CH_3]

$\xrightarrow[\text{(retention)}]{H_2SO_4,\ H_2O}$ (−) **C** [HO_2C, H, CH_3, CH_2CH_3] $\xrightarrow{LiAlH_4}$ (+) **D** [$HOCH_2$, H, CH_3, CH_2CH_3]

(c) **A** [TsO, H, CH_3, CH_2CH_3] $\xrightarrow[\text{(inversion)}]{CH_3CO_2^-}$ **G** [H, $OCCH_3$ (C=O), CH_3, CH_2CH_3] $\xrightarrow[\text{(retention)}]{OH^-}$ (+) **H** [H, OH, CH_3, CH_2CH_3]

(S)-(+)-2-butanol

(d) (−) **D** [H, CH_2OH, CH_3, CH_2CH_3] $\xrightarrow[\text{(retention)}]{PBr_3}$ **J** [H, CH_2Br, CH_3, CH_2CH_3] $\xrightarrow[\text{ether (retention)}]{Mg}$

$$\underset{\mathbf{K}}{H\text{—}C(CH_3)(CH_2CH_3)\text{—}CH_2MgBr} \xrightarrow[\text{(2) } H^+ \text{ (retention)}]{\text{(1) } CO_2} \underset{\mathbf{L}}{H\text{—}C(CH_3)(CH_2CH_3)\text{—}CH_2CO_2H}$$

(e) (R)-(+)-Glyceraldehyde [CHO; H, OH; CH_2OH] $\xrightarrow{HCN}$ M [CN; H, OH; H, OH; CH_2OH] + N [CN; HO, H; H, OH; CH_2OH]

(f) M $\xrightarrow[\text{H}_2\text{O, heat}]{H_2SO_4}$ P [CO_2H; H, OH; H, OH; CH_2OH] $\xrightarrow[HNO_3]{[O]}$ *meso*-Tartaric acid [CO_2H; H, OH; H, OH; CO_2H]

(g) N $\xrightarrow[\text{H}_2\text{O, heat}]{H_2SO_4}$ Q [CO_2H; HO, H; H, OH; CH_2OH] $\xrightarrow[HNO_3]{[O]}$ (−)-Tartaric acid [CO_2H; HO, H; H, OH; CO_2H]

18.34

(a) $CH_3\overset{CH_3}{\overset{|}{C}}HCHO + H\overset{O}{\overset{\|}{C}}H \xrightarrow[H_2O]{K_2CO_3} \underset{\mathbf{A}}{CH_3\overset{CH_3}{\overset{|}{C}}(CH_2OH)CHO} \xrightarrow{HCN}$

$$\underset{(\pm)\text{-}\mathbf{B}}{CH_3C(CH_3)(CH_2OH)\text{—}CH(OH)CN} \xrightarrow[\text{heat}]{H_3O^+} \left[\underset{(\pm)\text{-}\mathbf{C}}{CH_3C(CH_3)(CH_2OH)\text{—}CH(OH)CO_2H}\right] \xrightarrow{-H_2O}$$

$$\underset{(\pm)\text{-D}}{\text{lactone: } CH_3-\overset{CH_3}{\underset{H_2C}{C}}-\overset{OH}{\underset{C=O}{CH}}, \; H_2C-O-C} \xrightarrow{H_2NCH_2CH_2\overset{O}{\overset{\|}{C}}OH} (\pm)\text{-pantothenic acid}$$

$$(\pm)\text{-D} \xrightarrow{H_2NCH_2CH_2\overset{O}{\overset{\|}{C}}NHCH_2CH_2SH} (\pm)\text{-pantetheine}$$

(b) $(CH_3)_2\overset{CH_2OH}{\overset{|}{C}}-\overset{OH}{\underset{H}{C}}-\overset{O}{\overset{\|}{C}}-NHCH_2CH_2\overset{O}{\overset{\|}{C}}-NHCH_2CH_2SH$

(c) $\xrightarrow[\text{heat}]{OH^-,\ H_2O} (CH_3)_2\overset{CH_2OH}{\overset{|}{C}}-\overset{OH}{\underset{H}{C}}-CO_2^- + H_2NCH_2CH_2CO_2^- + H_2NCH_2CH_2S^-$

18.35

$$\underset{\text{Phenacetin}}{CH_3CH_2O-C_6H_4-NH-\overset{O}{\overset{\|}{C}}-CH_3} \xrightarrow[\substack{H_2O \\ \text{reflux}}]{OH^-} \underset{\text{Phenetidine}}{CH_3CH_2O-C_6H_4-NH_2} + CH_3CO_2^-$$

An interpretation of the spectral data for phenacetin is given in Fig. 18.2. (See following figure.)

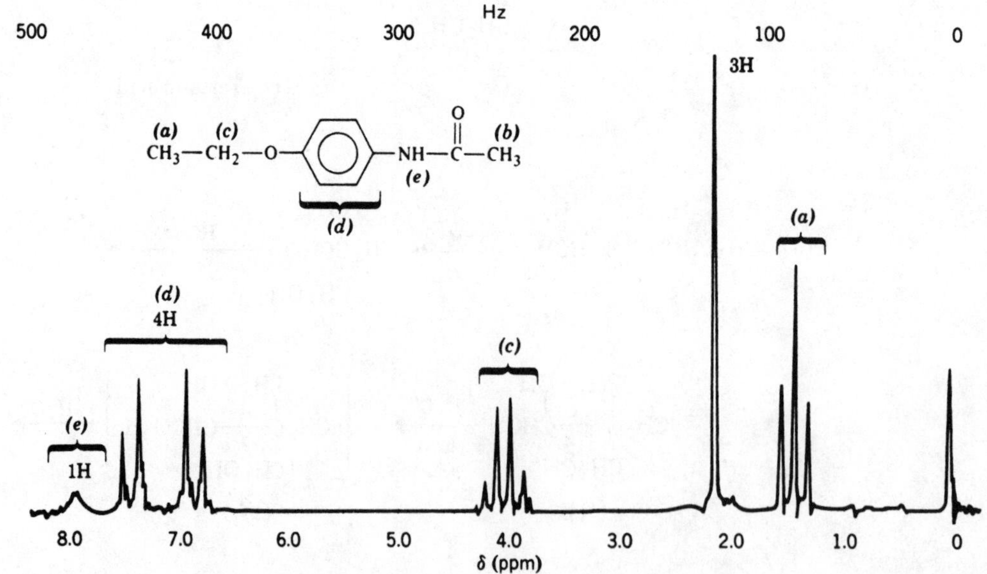

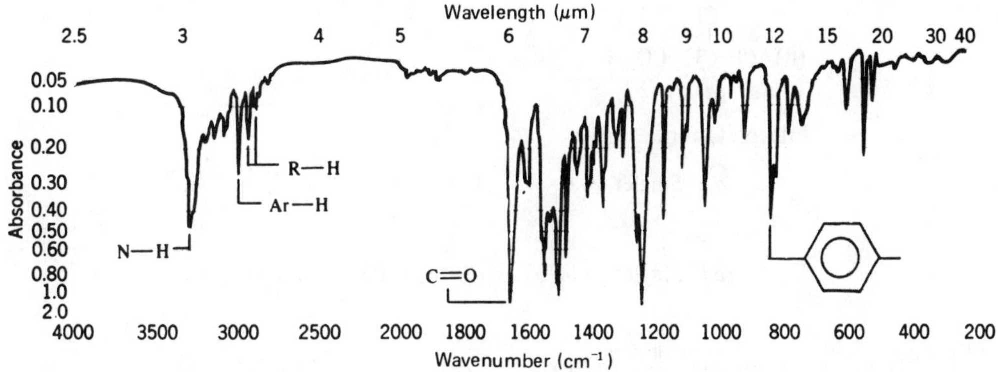

The proton nmr and infrared spectra of phenacetin. (The proton nmr spectrum courtesy of Varian Associates, Palo Alto, CA. IR spectrum courtesy of Sadtler Research Laboratories, Philadelphia.

18.36

(a) $\underset{(a)}{CH_3}\underset{(c)}{CH_2}-O-\overset{O}{\overset{\|}{C}}-\underset{(b)}{CH_2}\underset{(b)}{CH_2}-\overset{O}{\overset{\|}{C}}-O-\underset{(c)}{CH_2}\underset{(a)}{CH_3}$

Interpretation:

(a) Triplet δ1.2 (6H) $\quad 2-\overset{O}{\overset{\|}{C}}-O-$, 1740 cm^{-1}

(b) Singlet δ2.5 (4H)

(c) Quartet δ4.1 (4H)

(b) $\underset{(d)}{C_6H_5}-\overset{O}{\overset{\|}{C}}-O-\underset{(c)}{CH_2}-\underset{(b)}{\overset{CH_3\,(a)}{\overset{|}{CH}}}-\underset{(a)}{CH_3}$

Interpretation:

(a) Doublet δ1.0 (6H) $\quad -\overset{O}{\overset{\|}{C}}-$, 1720 cm^{-1} (ester)

(b) Multiplet δ2.1 (1H)

(c) Doublet δ4.1 (2H)

(d) Multiplet δ7.8 (5H)

(c) $\underset{(d)}{C_6H_5}-\underset{(b)}{CH_2}-\overset{O}{\overset{\|}{C}}-O-\underset{(c)}{CH_2}\underset{(a)}{CH_3}$

Interpretation:

(a) Triplet δ1.2 (3H) $\quad -\overset{O}{\overset{\|}{C}}-$, 1740 cm^{-1} (ester)

(b) Singlet δ3.5 (2H)

(c) Quartet δ4.1 (2H)

(d) Multiplet δ7.3 (5H)

(d) $Cl-\overset{\displaystyle Cl}{\overset{|}{C}H}-CO_2H$
(a) *(b)*

Interpretation:

(a) Singlet δ6.0 $-OH$, 2500-2700 cm^{-1}

(b) Singlet δ11.70 $-\overset{O}{\overset{\|}{C}}-$, 1705 cm^{-1} (acid)

(e) $Cl-CH_2-\overset{O}{\overset{\|}{C}}-OCH_2CH_3$
(b) *(c)* *(a)*

Interpretation:

(a) Triplet δ1.3 $-\overset{O}{\overset{\|}{C}}-$, 1745 cm^{-1} (ester)

(b) Singlet δ4.0

(c) Quartet δ4.2

18.37

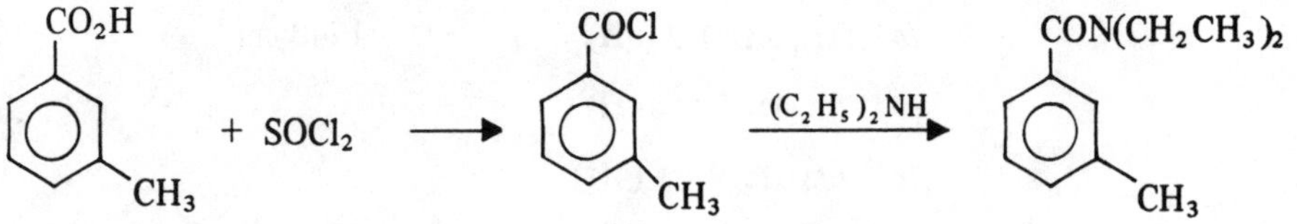

$$m\text{-}CH_3C_6H_4CO_2H + SOCl_2 \longrightarrow m\text{-}CH_3C_6H_4COCl \xrightarrow{(C_2H_5)_2NH} m\text{-}CH_3C_6H_4CON(CH_2CH_3)_2$$

18.38 Alkyl groups are electron releasing; they help disperse the positive charge of an alkylammonium salt and thereby help to stabilize it.

$$R\ddot{N}H_2 + H_3O^+ \longrightarrow R{\rightarrow}NH_3^+ + H_2O$$

Stabilized by electron-releasing alkyl group

Alkylamines, consequently, are somewhat stronger bases than ammonia.

Amides, on the other hand, have acyl groups, $R-\overset{O}{\overset{\|}{C}}-$, attached to nitrogen, and acyl groups are electron withdrawing. They are especially electron withdrawing because of resonance contributions of the kind shown here,

$$R-\overset{:\ddot{O}}{\overset{\|}{C}}-\ddot{N}H_2 \longleftrightarrow R-\overset{:\ddot{O}:^-}{\overset{|}{C}}=\overset{+}{N}H_2$$

This kind of resonance also *stabilizes* the amide. The tendency of the acyl group to be electron withdrawing, however, *destabilizes* the conjugate acid of an amide and reactions such as the following do not take place to an appreciable extent.

$$\mathrm{R\overset{O}{\overset{\|}{C}}-\ddot{N}H_2 + H_3O^+ \rightleftharpoons R\overset{O}{\overset{\|}{C}}-NH_3^+ + H_2O}$$

Stabilized by resonance — *Destabilized by electron-withdrawing acyl group*

18.39 (a) The conjugate base of an amide is stabilized by resonance.

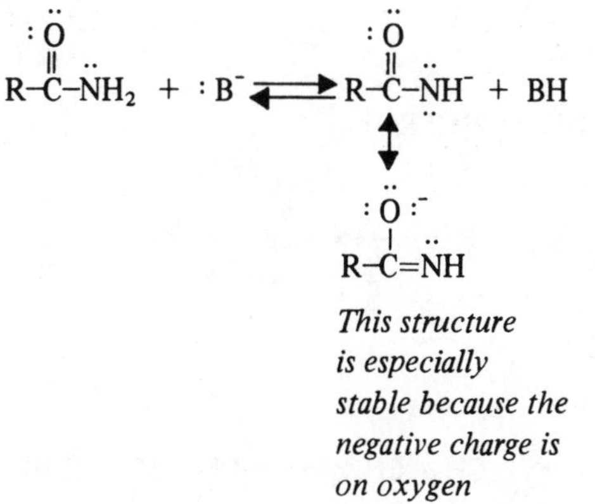

This structure is especially stable because the negative charge is on oxygen

(b) The conjugate base of an imide is stabilized by an additional resonance structure,

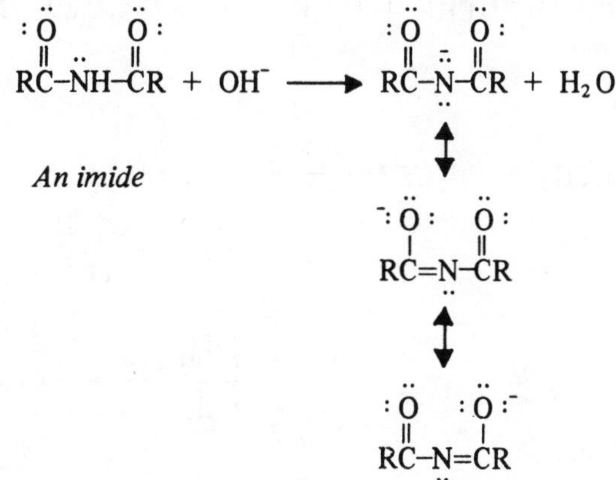

18.40 That compound X does not dissolve in aqueous sodium bicarbonate indicates that X is not a carboxylic acid. That X has an infrared absorption peak at 1740 cm^{-1} indicates the presence of a carbonyl group, probably that of an ester (Table 18.5). That the molecular formula of X ($C_7H_{12}O_4$) contains four oxygen atoms suggests that X is a diester.

The ^{13}C spectrum shows only four signals indicating a high degree of symmetry for X. The single signal at δ 166.7 is that of an ester carbonyl carbon, indicating that both ester groups of X are equivalent.

Putting these observations together with the proton off-resonance decoupled spectrum and the molecular formula leads us to the conclusion that X is diethyl malonate. The assignments are

$$\underset{(a)}{CH_3}\underset{(c)}{CH_2}O\underset{(d)}{\overset{O}{\overset{\|}{C}}}\underset{(b)}{CH_2}\overset{O}{\overset{\|}{C}}O\underset{(c)}{CH_2}\underset{(a)}{CH_3}$$

(a) Quartet δ 14.2

(b) Triplet δ 41.6

(c) Triplet δ 61.3

(d) Singlet δ 166.7

18.41 (a) **Chain Initiation**

Step 1 $RO{-}OR \longrightarrow 2\,RO\cdot$

Step 2 $CH_3\overset{O}{\overset{\|}{C}}SH + RO\cdot \longrightarrow R\overset{O}{\overset{\|}{C}}S\cdot + ROH$

Chain Propagation

Step 3 $CH_3\overset{O}{\overset{\|}{C}}S\cdot + CH_2{=}CHR \longrightarrow CH_3\overset{O}{\overset{\|}{C}}SCH_2\underset{\cdot}{C}HR$

Step 4 $CH_3\overset{O}{\overset{\|}{C}}SCH_2\dot{C}HR + CH_3\overset{O}{\overset{\|}{C}}SH \longrightarrow CH_3\overset{O}{\overset{\|}{C}}SCH_2CH_2R + CH_3\overset{O}{\overset{\|}{C}}S\cdot$

(b) $CH_3\overset{CH_3}{\overset{|}{C}}{=}CHCH_3 + CH_3\overset{O}{\overset{\|}{C}}SH \xrightarrow{ROOR} CH_3\overset{CH_3}{\overset{|}{C}}H\underset{\underset{\underset{O}{\|}}{SCCH_3}}{\underset{|}{C}}HCH_3$

$$\xrightarrow[(2)\ H_3O^+]{(1)\ OH^-,\ heat} CH_3\overset{CH_3}{\overset{|}{C}}H\underset{SH}{\underset{|}{C}}HCH_3 + CH_3\overset{O}{\overset{\|}{C}}OH$$

18.42 *cis*-4-Hydroxycyclohexanecarboxylic acid can assume a boat conformation that permits lactone formation.

[Chair form with HOC(=O) and OH ⇌ boat form with COH(=O) and HO; $\xrightarrow{-H_2O}$ bicyclic lactone with C(=O)—O bridge]

Neither of the chair conformations nor the boat form of *trans*-4-hydroxycyclohexanecarboxylic acid places the $-OH$ group and the $-CO_2H$ group close enough together to permit lactonization.

18.43 CHO / H—◯—OH / CH_2OH $\xrightarrow[\text{oxidation (see p. 1008)}]{Br_2,\ H_2O}$ CO_2H / H—◯—OH / CH_2OH $\xrightarrow{PBr_3}$ CO_2H / H—◯—OH / CH_2Br $\xrightarrow{NaCN}$

(*R*)-(+)-Glyceraldehyde; (R)-(−)-Glyceric acid; (R)-(−)-3-Bromo-2-hydroxypropanoic acid

CO_2H / H—◯—OH / CH_2 / CN $\xrightarrow[\text{heat}]{H_3O^+}$ CO_2H / H—◯—OH / CH_2 / CO_2H

(R)-($C_4H_5NO_3$); (*R*)-(+)-Malic acid

18.44 CHO / H—◯—OH / CH_2OH $\xrightarrow{HCN}$ CN / H—◯—OH / H—◯—OH / CH_2OH + CN / HO—◯—H / H—◯—OH / CH_2OH [cf. Problem 18.33(e)]

(*R*)-(+)-Glyceraldehyde; **M**; **N**

N $\xrightarrow[H_2O]{H_2SO_4}$ CO_2H / HO—◯—H / H—◯—OH / CH_2OH $\xrightarrow[HNO_3]{[O]}$ CO_2H / HO—◯—H / H—◯—OH / CO_2H $\xrightarrow{PBr_3}$ CO_2H / HO—◯—H / Br—◯—H / CO_2H

(−)-Tartaric acid

[cf. Problem 18.33 (g)]

$\xrightarrow[H^+]{Zn,}$ CO_2H / HO—◯—H / CH_2 / CO_2H

(b) Replacement of either alcoholic –OH by a reaction that proceeds with inversion produces the same stereoisomer.

CO_2H / HO–(1)–H / H–(2)–OH / CO_2H $\xrightarrow[\text{(inversion at C-2)}]{PBr_3}$ CO_2H / HO–C–H / Br–C–H / CO_2H $\equiv$ CO_2H / H–C–Br / H–C–OH / CO_2H $\xleftarrow[\text{(inversion at C-1)}]{PBr_3}$ CO_2H / HO–(1)–H / H–(2)–OH / CO_2H

(c) Two. The stereoisomer given in (b) and the one given next. below.

CO_2H / HO–(1)–H / H–(2)–OH / CO_2H $\xrightarrow[\text{(retention at C-2)}]{PBr_3}$ CO_2H / HO–C–H / H–C–Br / CO_2H $\equiv$ CO_2H / Br–C–H / H–C–OH / CO_2H $\xleftarrow[\text{(retention at C-1)}]{PBr_3}$ CO_2H / HO–(1)–H / H–(2)–OH / CO_2H

(d) It would have made no difference because treating either isomer (or both together) with zinc and acid produces (–)-malic acid.

CO_2H / HO–C–H / Br–C–H / CO_2H $\xrightarrow[H^+]{Zn,}$ CO_2H / HO–C–H / CH_2 / CO_2H $\equiv$ CO_2H / HO–C–H / CH_2 / CO_2H $\xleftarrow[H^+]{Zn,}$ CO_2H / HO–C–H / H–C–Br / CO_2H

(–)-Malic acid

18.45 (a) $CH_3O_2C–C{\equiv}C–CO_2CH_3$. This is a Diels-Alder reaction.

(b) H_2, Pd. The disubstituted double bond is less hindered than the tetrasubstituted double bond and hence is more reactive.

(c) $CH_2{=}CH–CH{=}CH_2$. Another Diels-Alder reaction.

(d) $LiAlH_4$

(e) $CH_3S(=O)_2–Cl$ and pyridine

(f) $CH_3CH_2S^-$

(g) OsO_4

(h) Raney Ni

(i) Base. This is an aldol condensation.

(j) C_6H_5Li (or C_6H_5MgBr) followed by H_3O^+

(k) H_3O^+. This is an acid-catalyzed rearrangement of an allylic alcohol.

(l) $CH_3\overset{O}{\overset{\|}{C}}Cl$, pyridine

(m) O_3 followed by oxidation.

(n) Heat

18.46

(a) Furan + Dimethylmaleic anhydride $\xrightarrow{\text{heat}}$ (Diels-Alder adduct) $\xrightarrow[\text{Pt}]{H_2,}$ Cantharidin

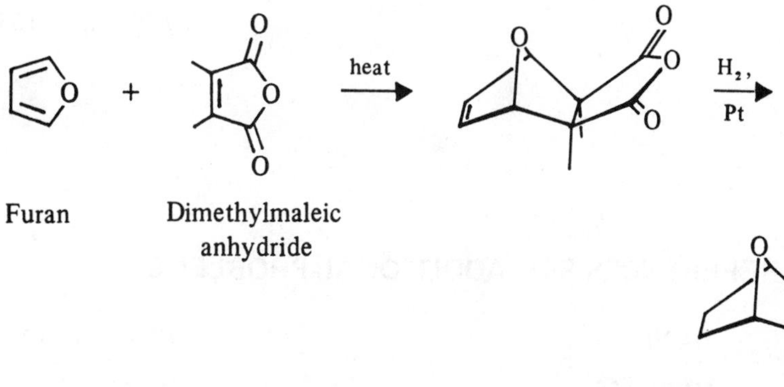

(b) Cantharidin apparently undergoes dehydrogenation to the Diels-Alder adduct shown here and then the adduct spontaneously decomposes through a reverse Diels-Alder reaction to furan and dimethylmaleic anhydride. These results suggest that the attempted Diels-Alder synthesis fails because the position of equilibrium favors reactants rather than products.

18.47 The very low hydrogen content of the molecular formula of **Y** ($C_8H_4O_3$) indicates that **Y** is highly unsaturated. That **Y** dissolves slowly in warm aqueous $NaHCO_3$ suggests that **Y** is a carboxylic acid anhydride that hydrolyses, and dissolves because it forms a carboxylate salt:

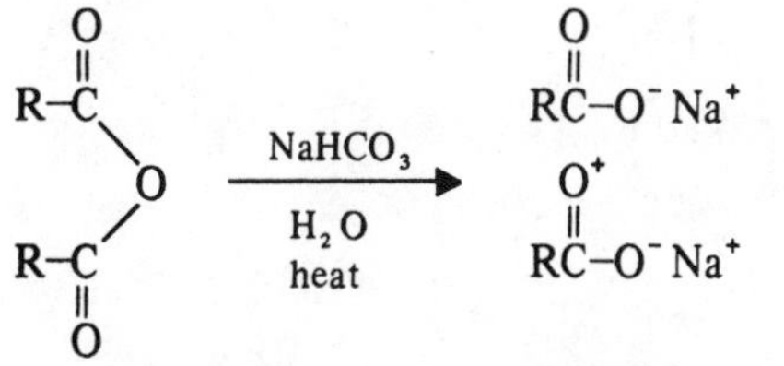

$$(RCO)_2O \text{ (insoluble)} \xrightarrow[\text{H}_2\text{O, heat}]{\text{NaHCO}_3} RCO_2^-Na^+ + RCO_2^-Na^+ \text{ (soluble)}$$

The infrared absorption peaks at 1779 and 1854 cm^{-1} are consistent with those of an aromatic carboxylic anhydride (Table 18.5).

That only four signals appear in the ^{13}C spectrum of **Y** indicates a high degree of symmetry for **Y**. Three of the signals occur in the aromatic region (δ120–δ140) and one signal is downfield (δ163.1).

These signals and their splitting patterns in the proton off-resonance decoupled spectrum lead us to conclude that **Y** is phthalic anhydride. The assignments are

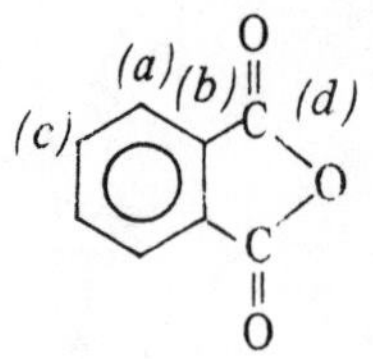

(a) Doublet δ 124.3

(b) Singlet δ 131.1

(c) Doublet δ 136.1

(d) Singlet δ 163.1

SECTION REFERENCES FOR ADDITIONAL PROBLEMS

18.20 18.2
18.21 18.2
18.22 18.3
18.23 18.3
18.24 18.3
18.25 18.13B
18.26 18.13C
18.27 18.13C
18.28 12.12, 14.12
18.29 18.13D
18.30 18.13E
18.31 18.12
18.32 18.7, 18.13A
18.33 18.3, 8.12, 8.13
18.34 16.9, 17.6, 18.3, 18.7C, 18.8
18.35 18.2 I, 18.7B
18.36 18.2 I
18.37 18.8
18.38 18.2C
18.39 18.2C
18.40 18.2C, 18.2I

SELF-TEST

18.1 Give an acceptable name for each of the following compounds.

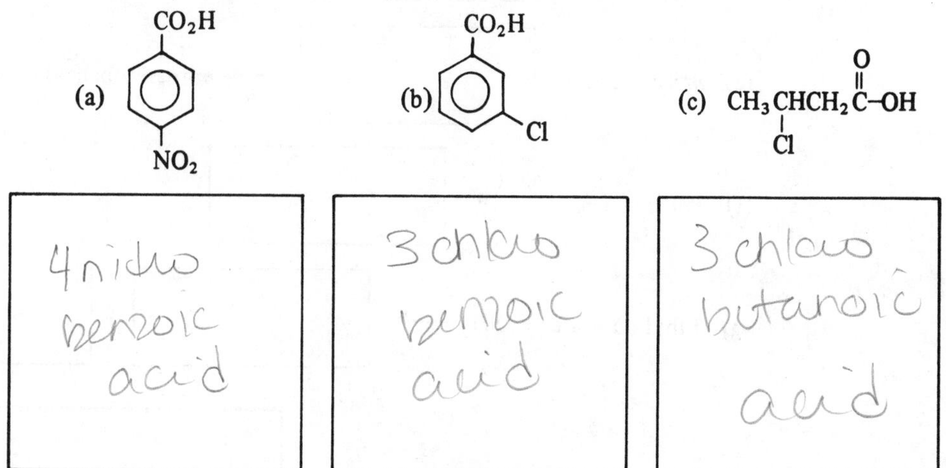

18.2 Use the letters of the appropriate compounds below to answer the following questions.

(a) The least reactive carboxylic acid derivative toward nucleophilic substitution at the acyl carbon is D

(b) The most readily hydrolyzed compound is B

(c) Besides RCO_2H itself, the only carboxylic acid derivative that can be prepared directly from all of the others is D

A. CH_3CO_2H

B. CH_3COCl

C. $CH_3CO_2C_2H_5$

D. CH_3CONH_2

E. 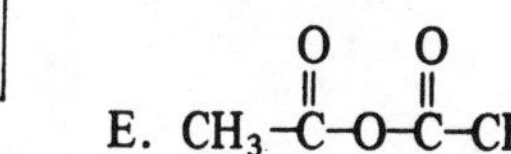

18.3 Supply the structural formula of the missing reactant, reagent, or major product. Show stereochemistry where appropriate. More than one step may be needed.

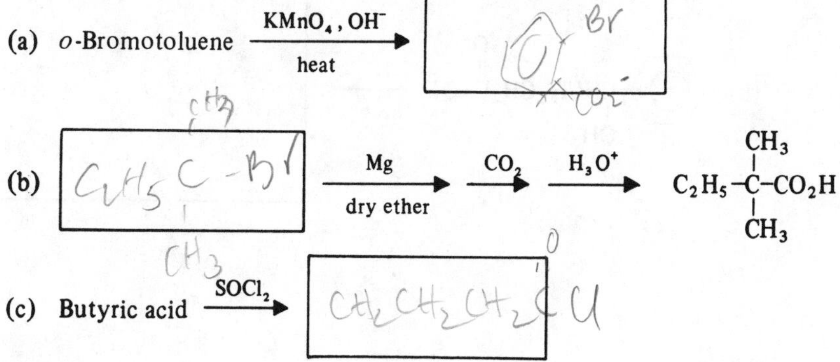

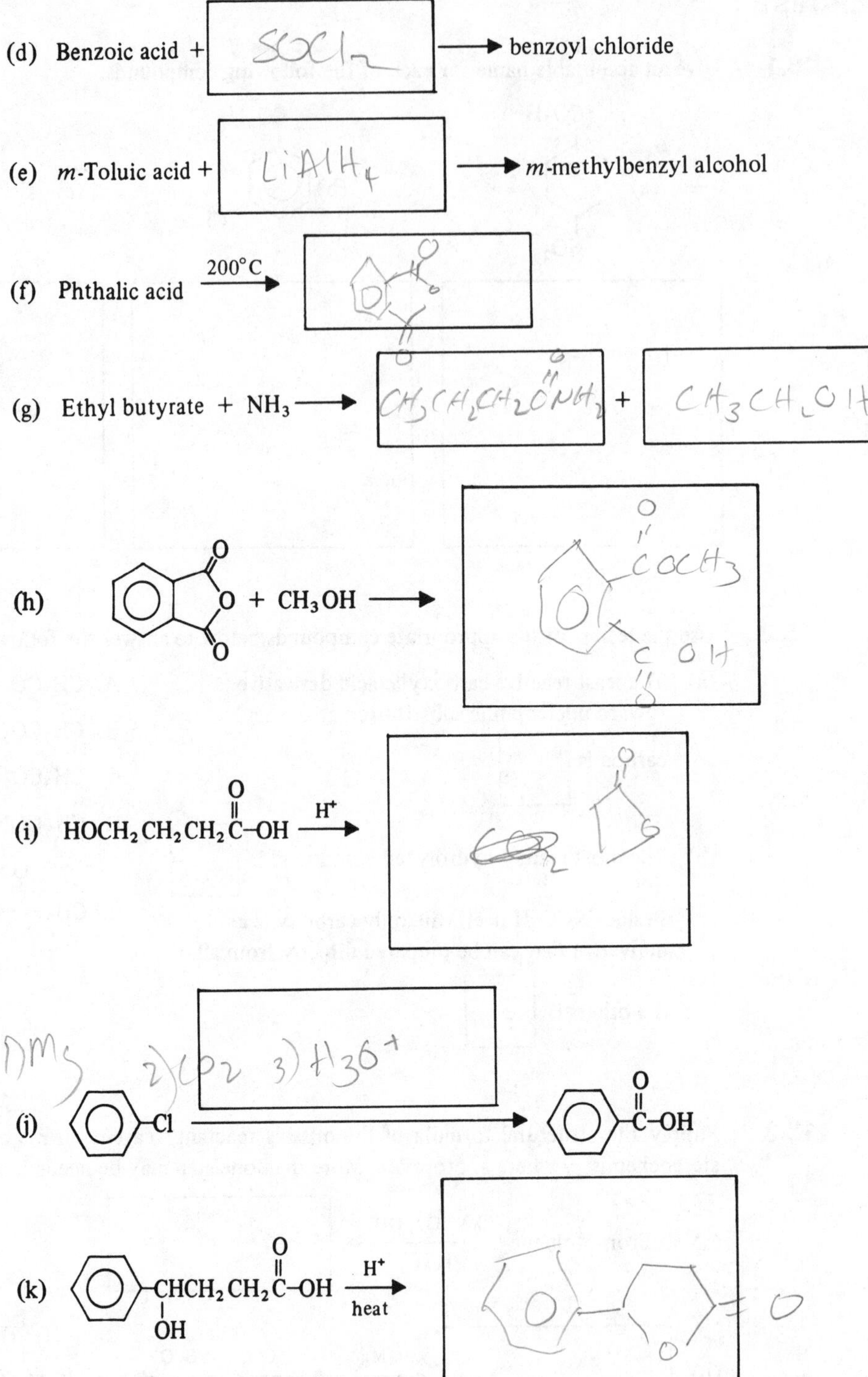
(d) Benzoic acid + → benzoyl chloride
(e) m-Toluic acid + → m-methylbenzyl alcohol
(f) Phthalic acid 200°C →
(g) Ethyl butyrate + NH3 → +
(h) O + CH3OH →
(i) HOCH2CH2CH2C-OH H+ →
(j) Cl → C-OH
(k) CHCH2CH2C-OH OH H+ heat →

(l) Phthalic anhydride + NH_3(excess) $\xrightarrow[\text{(2) } H_3O^+]{\text{(1) } H_2O\text{, warm}}$

(m) Phthalic anhydride + benzene $\xrightarrow{AlCl_3}$

(n) + $(CH_3)_2NH$ $\xrightarrow{25°C}$ C_6H_5–C(=O)–$N(CH_3)_2$

18.4 What reagent would distinguish between the compounds in each of the following pairs?

(a) CH_3O–C_6H_4–C(=O)–OH and HO–C_6H_4–C(=O)–OCH_3

(b) C_6H_5–C(=O)–NH_2 and C_6H_5–C(=O)–OH

18.5 Which of the following would be the strongest acid?

(a) Benzoic acid (b) 4-Nitrobenzoic acid (c) 4-Methylbenzoic acid

(d) 4-Methoxybenzoic acid (e) 4-Ethylbenzoic acid

18.6 Which of the following would yield (*S*)-2-butanol?

(a) (*R*)-2-Bromobutane + $CH_3CO_2^- Na^+$ $\longrightarrow$ product $\xrightarrow[\text{heat}]{OH^-, H_2O}$

(b) (*R*)-2-Bromobutane $\xrightarrow[\text{heat}]{OH^-, H_2O}$

(c) (*S*)-2-Butyl acetate $\xrightarrow[\text{heat}]{OH^-, H_2O}$

(d) All of the above

(e) None of the above

18.7 Which reagent would serve as the basis for a simple chemical test to distinguish between hexanoic acid and hexanamide?

(a) Cold dilute NaOH

(b) Cold dilute $NaHCO_3$

(c) Cold conc. H_2SO_4

(d) More than one of these

(e) None of these

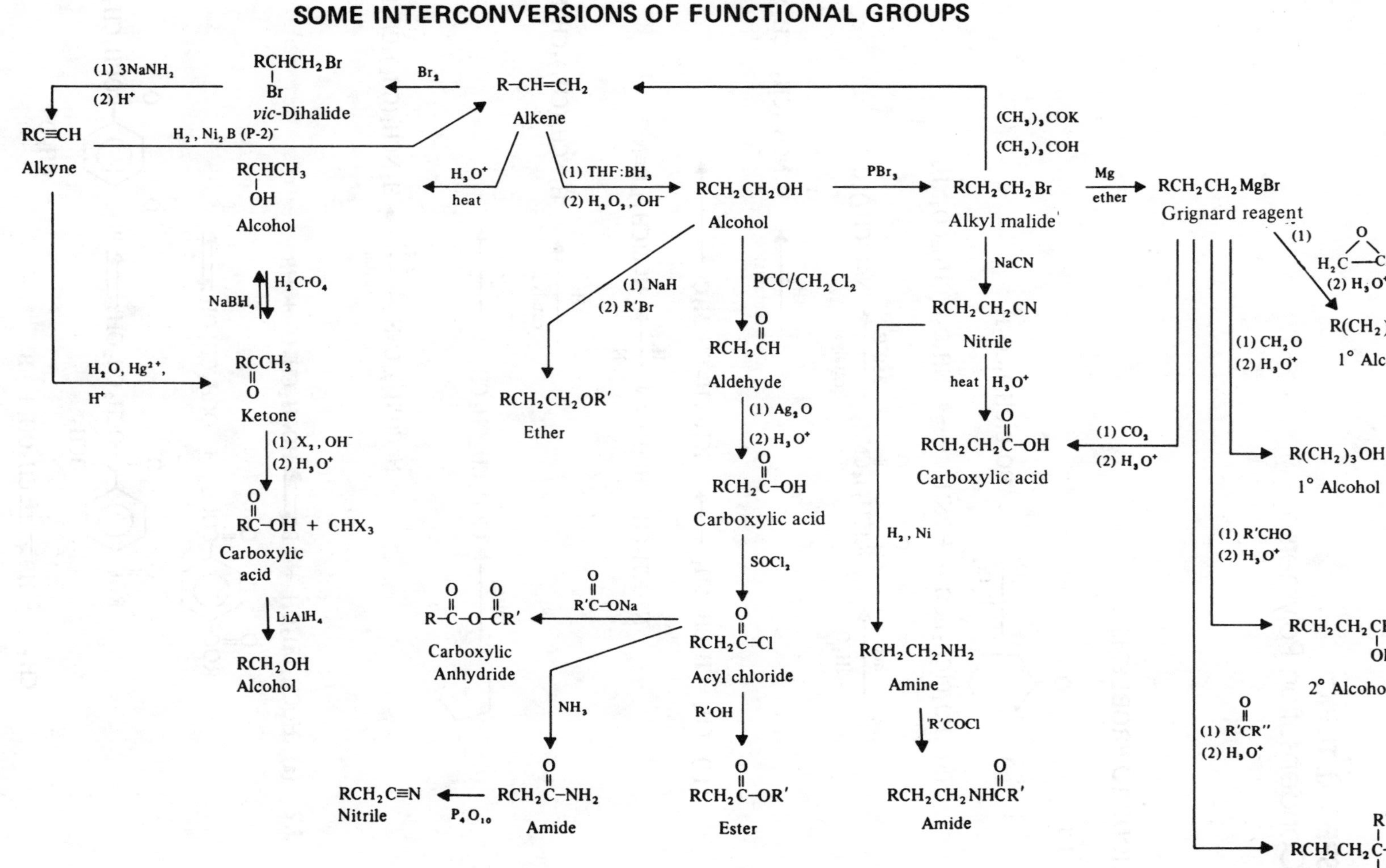
SOME INTERCONVERSIONS OF FUNCTIONAL GROUPS
(1) 3NaNH2
(2) H+
RCHCH2Br
Br
vic-Dihalide
Br2
R–CH=CH2
Alkene
RC≡CH
Alkyne
H2, Ni2B (P-2)
(CH3)3COK
(CH3)3COH
RCHCH3
OH
Alcohol
H3O+
heat
(1) THF:BH3
(2) H2O2, OH–
RCH2CH2OH
Alcohol
PBr3
RCH2CH2Br
Alkyl malide
Mg
ether
RCH2CH2MgBr
Grignard reagent
(1) H2C—CH2 (O)
(2) H3O+
R(CH2)4OH
1° Alcohol
H2CrO4
NaBH4
(1) NaH
(2) R'Br
PCC/CH2Cl2
NaCN
RCH2CH
Aldehyde
RCH2CH2CN
Nitrile
(1) CH2O
(2) H3O+
H2O, Hg2+,
H+
RCCH3
O
Ketone
RCH2CH2OR'
Ether
(1) Ag2O
(2) H3O+
heat H3O+
RCH2CH2C–OH
Carboxylic acid
(1) CO2
(2) H3O+
R(CH2)3OH
1° Alcohol
(1) X2, OH–
(2) H3O+
RCH2C–OH
Carboxylic acid
RC–OH + CHX3
Carboxylic
acid
H2, Ni
(1) R'CHO
(2) H3O+
SOCl2
R'C–ONa
R–C–O–CR'
Carboxylic
Anhydride
LiAlH4
RCH2OH
Alcohol
RCH2C–Cl
Acyl chloride
RCH2CH2NH2
Amine
RCH2CH2CHR'
OH
2° Alcohol
NH3
R'OH
R'COCl
(1) R'CR''
(2) H3O+
RCH2C≡N
Nitrile
P4O10
RCH2C–NH2
Amide
RCH2C–OR'
Ester
RCH2CH2NHCR'
Amide
R'
RCH2CH2C–R''
OH
3° Alcohol

J SPECIAL TOPIC
Condensation Polymers

SOLUTIONS TO PROBLEMS

J.1

(a) Cyclohexanone $\xrightarrow[H_2Cr_2O_7]{[O]}$ $HO_2C(CH_2)_4CO_2H$

(b) $HO_2C(CH_2)_4CO_2H + 2NH_3 \longrightarrow NH_4O_2C(CH_2)_4CO_2NH_4$

$$\xrightarrow[-2H_2O]{heat} H_2N\overset{O}{\overset{\|}{C}}(CH_2)_4\overset{O}{\overset{\|}{C}}NH_2 \xrightarrow[catalyst]{350°C} N{\equiv}C(CH_2)_4C{\equiv}N$$

$$\xrightarrow[catalyst]{4H_2} H_2NCH_2(CH_2)_4CH_2NH_2$$

(c) $CH_2{=}CH{-}CH{=}CH_2 \xrightarrow{Cl_2} ClCH_2CH{=}CHCH_2Cl \xrightarrow{2NaCN}$

$$N{\equiv}CCH_2CH{=}CHCH_2C{\equiv}N \xrightarrow[Ni]{H_2,} N{\equiv}C(CH_2)_4C{\equiv}N$$

$$\xrightarrow[catalyst]{4H_2} H_2NCH_2(CH_2)_4CH_2NH_2$$

(d) Tetrahydrofuran $\xrightarrow{2HCl} ClCH_2CH_2CH_2CH_2Cl \xrightarrow{2NaCN}$

$$N{\equiv}C(CH_2)_4C{\equiv}N \xrightarrow[catalyst]{4H_2} H_2NCH_2(CH_2)_4CH_2NH_2$$

J.2 (a) $HOCH_2CH_2OH + {}^{-}\!:B \rightleftharpoons HOCH_2CH_2O^- + HB$

$$RO\overset{O}{\overset{\|}{C}}{-}C_6H_4{-}\overset{O}{\overset{\|}{C}}OCH_3 + {}^{-}OCH_2CH_2OH \rightleftharpoons$$

$$RO\overset{O}{\overset{\|}{C}}{-}C_6H_4{-}\underset{OCH_3}{\underset{|}{\overset{O^-}{\overset{|}{C}}}}{-}OCH_2CH_2OH \rightleftharpoons RO\overset{O}{\overset{\|}{C}}{-}C_6H_4{-}\overset{O}{\overset{\|}{C}}OCH_2CH_2OH + CH_3O^-$$

$$CH_3O^- + HB \rightleftharpoons CH_3OH + :B^-$$

R = CH_3- or $HOCH_2CH_2-$

(b) $ROC(=O)-C_6H_4-C(=O)-OCH_3 \underset{-H^+}{\overset{+H^+}{\rightleftharpoons}} ROC(=O)-C_6H_4-C(=\overset{+}{O}H)-OCH_3$

$\underset{-HOCH_2CH_2OH}{\overset{+HOCH_2CH_2OH}{\rightleftharpoons}} ROC(=O)-C_6H_4-C(OH)(OCH_3)-\overset{+}{O}(H)-CH_2CH_2OH \rightleftharpoons$

$ROC(=O)-C_6H_4-C(OH)(\overset{+}{O}(H)CH_3)-OCH_2CH_2OH \underset{+CH_3OH}{\overset{-CH_3OH}{\rightleftharpoons}} ROC(=O)-C_6H_4-C(=\overset{+}{O}H)OCH_2CH_2OH$

$\underset{+H^+}{\overset{-H^+}{\rightleftharpoons}} ROC(=O)-C_6H_4-C(=O)-OCH_2CH_2OH$

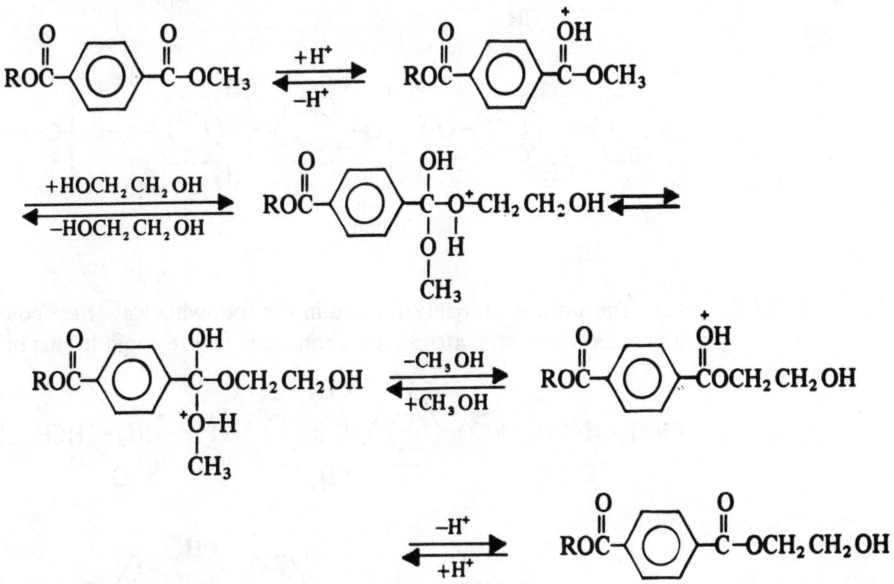

R = CH_3- or $HOCH_2CH_2-$

J.3

(a) $CH_3OC(=O)-C_6H_4-C(=O)OCH_3 + HOCH_2-C_6H_{10}-CH_2OH$

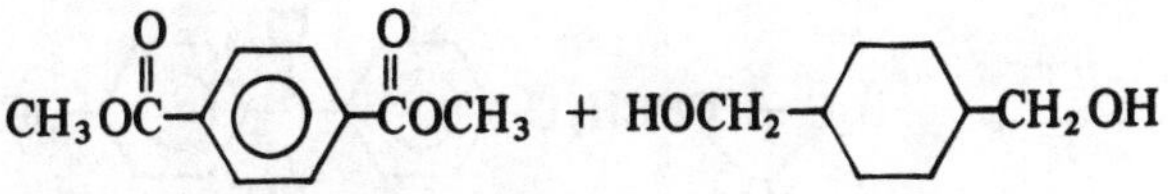

(b) By high-pressure catalytic hydrogenation

J.4

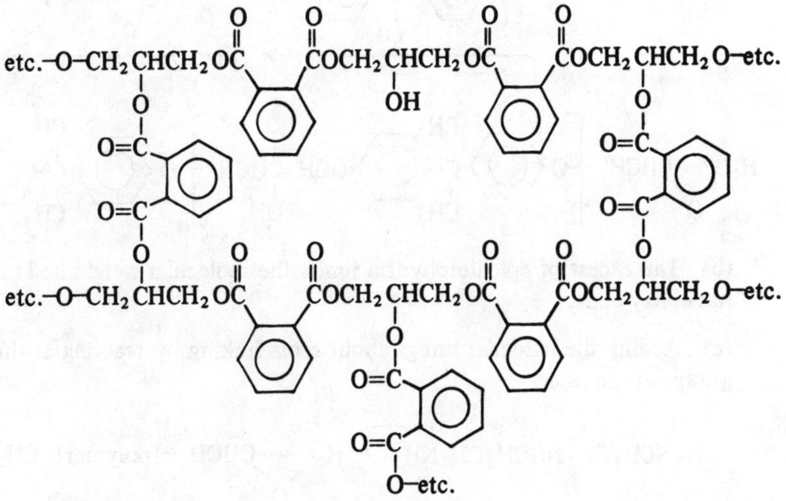

etc.$-O-CH_2CHCH_2OC(=O)-C_6H_4-C(=O)OCH_2CHCH_2OC(=O)-C_6H_4-C(=O)OCH_2CHCH_2O-$etc.
(first CH bears $-O-C(=O)-C_6H_4-C(=O)-O-$ linked to the CH of the lower chain; middle CH bears OH; last CH bears $-O-C(=O)-C_6H_4-C(=O)-O-$ linked to the lower chain)

etc.$-O-CH_2CHCH_2OC(=O)-C_6H_4-C(=O)OCH_2CHCH_2OC(=O)-C_6H_4-C(=O)OCH_2CHCH_2O-$etc.
(middle CH bears $-O-C(=O)-C_6H_4-C(=O)-O-$etc.)

J.5

$$HO-C_6H_4-C(CH_3)_2-C_6H_4-OH + Cl-\overset{O}{\overset{\|}{C}}-Cl \xrightarrow[-HCl]{\text{pyridine}}$$

$$-C_6H_4-C(CH_3)_2-C_6H_4-O-\overset{O}{\overset{\|}{C}}\left[-O-C_6H_4-C(CH_3)_2-C_6H_4-O-\overset{O}{\overset{\|}{C}}-\right]_n O-\text{etc.}$$

Lexan

J.6 (a) The resin is probably formed in the following way. Base converts the bisphenol A to a phenoxide ion that attacks a carbon atom of the epoxide ring of epichlorohydrin:

$$ClCH_2CH\overset{O}{-}CH_2 + {}^-O-C_6H_4-C(CH_3)_2-C_6H_4-O^- + CH_2\overset{O}{-}CHCH_2Cl \longrightarrow$$

$$Cl-CH_2-\underset{O^-}{CH}-CH_2-O-C_6H_4-C(CH_3)_2-C_6H_4-OCH_2-\underset{O^-}{CH}-CH_2-Cl$$

$$\xrightarrow{-2Cl^-} H_2C\overset{O}{-}CHCH_2O-C_6H_4-C(CH_3)_2-C_6H_4-OCH_2CH\overset{O}{-}CH_2$$

$$\xrightarrow{{}^-O-C_6H_4-C(CH_3)_2-C_6H_4-O^-} \text{then} \xrightarrow{H_2C\overset{O}{-}CHCH_2Cl}$$

$$H_2C\overset{O}{-}CHCH_2\left[-O-C_6H_4-C(CH_3)_2-C_6H_4-OCH_2\underset{OH}{CH}CH_2-\right]_n O-C_6H_4-C(CH_3)_2-C_6H_4-OCH_2CH\overset{O}{-}CH_2$$

(b) The excess of epichlorohydrin limits the molecular weight and insures that the resin has epoxy ends.

(c) Adding the hardener brings about cross linking by reacting at the terminal epoxide groups of the resin:

$$H_2NCH_2CH_2NHCH_2CH_2\ddot{N}H_2 + H_2C\overset{O}{-}CHCH_2-[\text{polymer}]-CH_2CH\overset{O}{-}CH_2 \longrightarrow$$

$$-CH_2-\underset{OH}{CH}CH_2-\underset{H}{N}CH_2CH_2-\underset{\substack{| \\ CH_2 \\ | \\ CHOH \\ | \\ CH_2 \\ | \\ [polymer] \\ | \\ CH_2 \\ | \\ CHOH \\ | \\ CH_2 \\ |}}{N}-CH_2CH_2-\underset{H}{N}-CH_2\underset{OH}{CH}CH_2\,[polymer]\,CH_2\underset{OH}{CH}CH_2-etc.$$

$$N(H)-CH_2CH_2\underset{H}{N}-CH_2CH_2-\underset{H}{N}-CH_2\underset{OH}{CH}CH_2\,[polymer]\,CH_2\overset{etc.}{\underset{OH}{CH}}CH_2$$

J.7

(a) $\left[\text{-}C_6H_4\text{-}NH\overset{O}{\overset{\|}{C}}OCH_2CH_2O\overset{O}{\overset{\|}{C}}(CH_2)_6\overset{O}{\overset{\|}{C}}OCH_2CH_2O\overset{O}{\overset{\|}{C}}NH-\right]_n$ (1,3-disubstituted benzene ring)

(b) To ensure that the polyester chain has $-CH_2OH$ end groups.

J.8 Because the para position is occupied by a methyl group, cross linking does not occur and the resulting polymer remains thermoplastic. (See Section J.4.)

J.9

$$H-\overset{O}{\overset{\|}{C}}-H \xrightarrow{H^+} H-\overset{O^+-H}{\overset{\|}{C}}-H \xrightarrow{\text{phenol (C}_6H_5OH)} \text{(cyclohexadienyl cation: }{}^+OH\text{, H, C(H)(H)-O-H)} \xrightarrow{-H^+}$$

$$\text{o-}HOC_6H_4CH_2OH \xrightarrow[\text{(as before)}]{H-\overset{O^+-H}{\overset{\|}{C}}-H} \text{2,6-}(HOCH_2)_2C_6H_3OH \xrightarrow[\text{(as before)}]{H-\overset{O^+-H}{\overset{\|}{C}}-H}$$

$$\text{2,4,6-}(HOCH_2)_3C_6H_2OH \xrightarrow{H^+} \text{(OH; HOCH}_2\text{; CH}_2OH_2^+\text{; CH}_2OH) \xrightarrow{-H_2O}$$

OH HOCH₂ CH₂⁺ + phenol (OH) ⟶ HOCH₂ OH CH₂ OH (phenol ring) CH₂OH, then $H-\overset{\overset{+}{O}-H}{\overset{\|}{C}}-H$ ⟶ etc.

$\xrightarrow{\text{phenol (OH)}}$ bakelite

19 AMINES

PREPARATION AND REACTIONS OF AMINES

A. Preparation

1. Preparation via nucleophilic substitution reactions.

$$RCH_2Br \xrightarrow{NaN_3} RCH_2{-}N_3 \xrightarrow[\text{or } LiAlH_4]{Na/alcohol} RCH_2NH_2$$

$$RCH_2Br \xrightarrow{\text{potassium phthalimide}} RCH_2{-}N(\text{phthalimide}) \xrightarrow{NH_2NH_2} RCH_2NH_2 + \text{phthalhydrazide}$$

$$RCH_2Br \xrightarrow{NH_3} RCH_2NH_2 + (RCH_2)_2NH + (RCH_2)_3N \quad \text{(poor method)}$$

2. Preparation through reduction of nitro compounds.

$$C_6H_6 \xrightarrow[H_2SO_4]{HNO_3} C_6H_5NO_2 \xrightarrow[HCl]{Fe} C_6H_5NH_2$$

3. Preparation via reductive amination.

$$RR'C{=}O + NH_3 \xrightarrow{H_2,\,Ni} R{-}CHNH_2{-}R'$$

4. Preparation of amines through reduction of amides, oximes, and nitriles.

$$R{-}CH_2Br \xrightarrow{CN^-} RCH_2CN \xrightarrow{H_2,Ni} RCH_2CH_2NH_2$$

$$RR'C{=}O \xrightarrow{NH_2OH} RR'C{=}NOH \xrightarrow{Na/C_2H_5OH} R{-}CHNH_2{-}R'$$

$$R{-}NH_2 + R'\overset{O}{\overset{\|}{C}}Cl \longrightarrow R{-}NH\overset{O}{\overset{\|}{C}}R' \xrightarrow[(2)\ H_2O]{(1)\ LiAlH_4} RNHCH_2R'$$

5. Preparation through the Hofmann rearrangement of amides

$$R{-}\overset{O}{\overset{\|}{C}}OH \xrightarrow{SOCl_2} R\overset{O}{\overset{\|}{C}}Cl \xrightarrow{NH_3} R\overset{O}{\overset{\|}{C}}NH_2 \xrightarrow[(NaOBr)]{Br_2/NaOH} RNH_2 + CO_3^{2-}$$

B. Reactions of Amines

1. As a base or a nucleophile.

As a base

$$\gt N{:} + H{-}\underset{H}{O}{-}H^+ \rightleftharpoons -\overset{|}{\underset{|}{N}}{}^+{-}H + H_2O$$

As a nucleophile in alkylation

$$\gt N{:} + RCH_2{-}X \longrightarrow -\overset{|}{\underset{|}{N}}{}^+{-}CH_2R + X^-$$

As a nucleophile in acylation

$$\gt N(H){:} + R{-}\overset{O}{\overset{\|}{C}}{-}Cl \xrightarrow[(-HCl)]{} -\overset{|}{\ddot{N}}{-}\overset{O}{\overset{\|}{C}}{-}R$$

2. With nitrous acid

$$\underset{1^\circ\ aliphatic}{R{-}NH_2} \xrightarrow[HX]{HONO} \underset{(unstable)}{R{-}N_2^+X^-} \xrightarrow{-N_2} R^+ \longrightarrow \text{alkenes, alcohols, and so on}$$

$$\underset{1^\circ\ aromatic}{ArNH_2} \xrightarrow[0\text{-}5^\circ C]{HONO,HX} ArN_2^+X^-$$

- $\xrightarrow{CuCl}$ $ArCl + N_2$
- $\xrightarrow{CuBr}$ $ArBr + N_2$
- $\xrightarrow{CuCN}$ $ArCN + N_2$
- $\xrightarrow{KI}$ $ArI + N_2$
- $\xrightarrow{HBF_4}$ $ArN_2^+BF_4^- \xrightarrow{heat} ArF + N_2 + BF_3$
- $\xrightarrow[heat]{H_3O^+}$ $ArOH + N_2$
- $\xrightarrow{H_3PO_2}$ $ArH + N_2$
- $\xrightarrow{C_6H_5{-}OH}$ $Ar{-}N{=}N{-}C_6H_4{-}OH$
- $\xrightarrow{C_6H_5{-}NR_2}$ $Ar{-}N{=}N{-}C_6H_4{-}NR_2$

$$R_2NH \xrightarrow{HONO} R_2N{-}N{=}O$$

2° aliphatic

$$ArNHR \xrightarrow{HONO} ArN(N{=}O){-}R$$

2° aromatic

$$R_3N \xrightarrow{HX,\ NaNO_2} R_3NH^+X^- + R_3\overset{+}{N}{-}N{=}O\ X^-$$

3° aliphatic

$$R_2N{-}C_6H_5 \xrightarrow{HONO} R_2N{-}C_6H_4{-}N{=}O$$

3° aromatic

3. With sulfonyl chlorides

$$R{-}NH_2 + ArSO_2Cl \xrightarrow[(-HCl)]{} RNHSO_2Ar \underset{H^+}{\overset{OH^-}{\rightleftharpoons}} [RNSO_2Ar]^- + H_2O$$

1° amine

$$R_2NH + ArSO_2Cl \xrightarrow[(-HCl)]{} R_2NSO_2Ar$$

2° amine

4. The Hofmann elimination

$$HO^- + -\overset{H}{\underset{|}{\overset{|}{C}}}-\underset{\underset{+}{N(CH_3)_3}}{\overset{|}{\underset{|}{C}}}- \xrightarrow{heat} >C{=}C< + (CH_3)_3N + H_2O$$

SOLUTIONS TO PROBLEMS

19.1 Dissolve both compounds in ether and extract with aqueous HCl. This procedure gives an ether layer that contains cyclohexane and an aqueous layer that contains hexylaminium chloride. Cyclohexane may then be recovered from the ether layer by distillation. Hexylamine may be recovered from the aqueous layer by adding aqueous NaOH (to convert hexylaminium chloride to hexylamine) and then by ether extraction and distillation

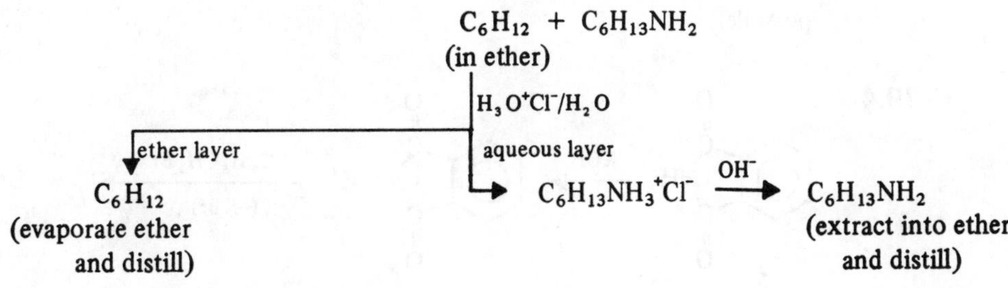

19.2 We begin by dissolving the mixture in a water-immiscible organic solvent such as CH_2Cl_2 or ether. Then, extractions with aqueous acids and bases allow us to separate the components. (We separate *p*-cresol from benzoic acid by taking advantage of benzoic acid's solubility in the more weakly basic aqueous $NaHCO_3$, whereas *p*-cresol requires the more strongly basic, aqueous NaOH.)

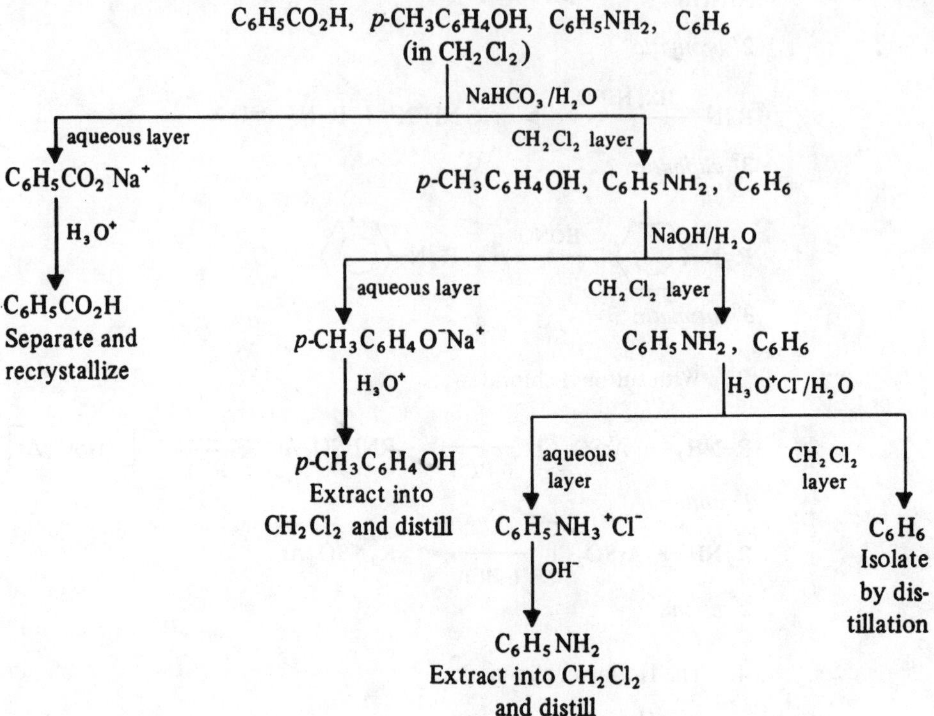

19.3 (a) Neglecting Kekulé forms of the ring, we can write the following resonance structures for the phthalimide anion.

(b) Phthalimide is more acidic than benzamide because its anion is stabilized by resonace to a greater extent than the anion of benzamide. (Benzamide has only one carbonyl group attached to the nitrogen atom and thus fewer resonance contributors are possible.)

19.4

$\xrightarrow{KOH}$ $:N:^-K^+$ $\xrightarrow[(-KBr)]{C_6H_5CH_2Br}$

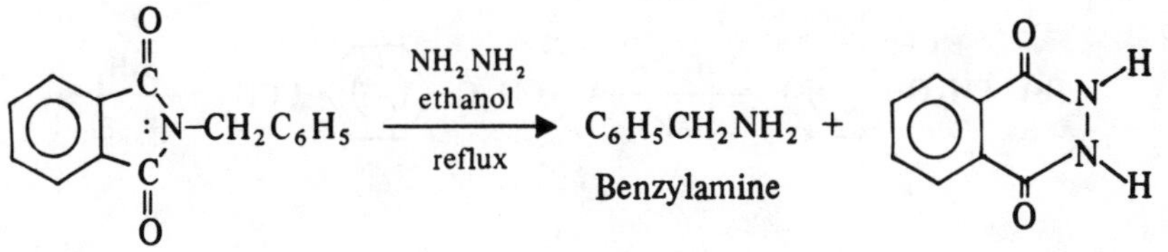

19.5 (a) $CH_3(CH_2)_3CHO + NH_3 \xrightarrow{H_2,\ Ni} CH_3(CH_2)_3CH_2NH_2$

(b) $C_6H_5C(=O)CH_3 + NH_3 \xrightarrow{H_2,\ Ni} C_6H_5CH(NH_2)CH_3$

(c) $CH_3(CH_2)_4CHO + C_6H_5NH_2 \xrightarrow[CH_3OH]{LiBH_3CN} CH_3(CH_2)_4CH_2NHC_6H_5$

19.6 The reaction of a secondary halide with ammonia would inevitably be accompanied by considerable elimination thus decreasing the yield.

$$RCH(R')\text{–}X + \underset{(excess)}{NH_3} \xrightarrow{substitution} RCH(R')NH_2$$

$$RCH(R')\text{–}X + \underset{(excess)}{NH_3} \xrightarrow{elimination} \text{alkene}$$

19.7 (a) $C_6H_5CO_2H \xrightarrow{SOCl_2} C_6H_5COCl \xrightarrow{CH_3CH_2NH_2}$

$C_6H_5CONHCH_2CH_3 \xrightarrow{LiAlH_4} C_6H_5CH_2NHCH_2CH_3$

(b) $CH_3CH_2CH_2CH_2CH_2Br \xrightarrow{NaCN} CH_3CH_2CH_2CH_2CH_2CN$

$\xrightarrow{LiAlH_4} CH_3CH_2CH_2CH_2CH_2CH_2NH_2$

(c) $CH_3CH_2CO_2H \xrightarrow{SOCl_2} CH_3CH_2COCl \xrightarrow{(CH_3CH_2CH_2)_2NH}$

$CH_3CH_2CON(CH_2CH_2CH_3)_2 \xrightarrow{LiAlH_4} (CH_3CH_2CH_2)_3N$

(d) $CH_3C(=O)CH_2CH_3 \xrightarrow{NH_2OH} CH_3C(=NOH)CH_2CH_3 \xrightarrow{Na/C_2H_5OH} CH_3CH(NH_2)CH_2CH_3$

19.8 (a)

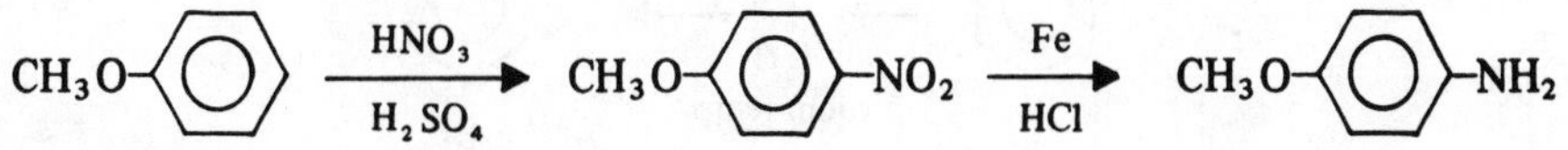

(b) 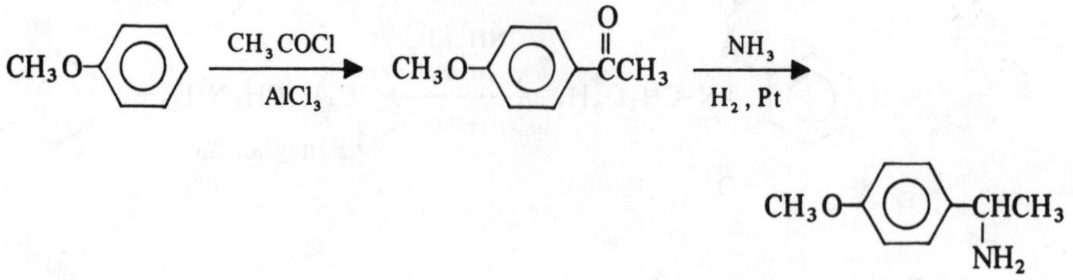 $CH_3O-C_6H_4 \xrightarrow[AlCl_3]{CH_3COCl} CH_3O-C_6H_4-\overset{O}{\overset{\|}{C}}CH_3 \xrightarrow[H_2, Pt]{NH_3} CH_3O-C_6H_4-\underset{NH_2}{\underset{|}{C}}HCH_3$

(c) $C_6H_5-CH_3$ (excess) $\xrightarrow{Cl_2,\ h\nu} C_6H_5-CH_2Cl \xrightarrow{(CH_3)_3N} C_6H_5-CH_2\overset{+}{N}(CH_3)_3Cl^-$

(d) 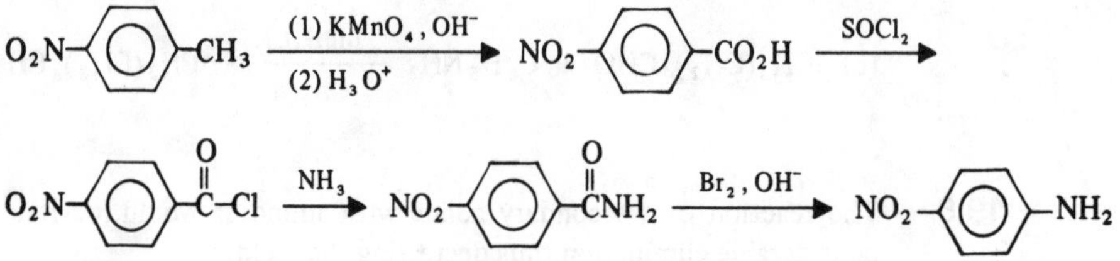 $O_2N-C_6H_4-CH_3 \xrightarrow[(2)\ H_3O^+]{(1)\ KMnO_4,\ OH^-} NO_2-C_6H_4-CO_2H \xrightarrow{SOCl_2}$

$O_2N-C_6H_4-\overset{O}{\overset{\|}{C}}-Cl \xrightarrow{NH_3} NO_2-C_6H_4-\overset{O}{\overset{\|}{C}}NH_2 \xrightarrow{Br_2,\ OH^-} NO_2-C_6H_4-NH_2$

(e) $CH_3-C_6H_5 + NBS \xrightarrow{ROOR} C_6H_5-CH_2Br \xrightarrow{KCN}$

$C_6H_5-CH_2CN \xrightarrow{LiAlH_4} C_6H_5-CH_2CH_2NH_2$

19.9 An amine acting as a base.

$$CH_3CH_2\ddot{N}H_2 + H_3O^+ \rightleftharpoons CH_3CH_2NH_3^+ + H_2O$$

An amine acting as a nucleophile in an alkylation reaction.

$$(CH_3CH_2)_3N: + CH_3-I \longrightarrow (CH_3CH_2)_3\overset{+}{N}-CH_3\ I^-$$

An amine acting as a nucleophile in an acylation reaction.

$$\underset{\text{(excess)}}{(CH_3)_2\ddot{N}H} + CH_3C(=O)Cl \longrightarrow (CH_3)_2N\overset{O}{\overset{\|}{C}}CH_3 + (CH_3)_2NH_2Cl$$

An amino group acting as an activating group and as an ortho-para director in electrophilic aromatic substitution.

$$C_6H_5NH_2 \xrightarrow[\text{room temp.}]{Br_2,\ H_2O} \text{2,4,6-}Br_3C_6H_2NH_2$$

19.10 (a, b) $^{-}O{-}N{=}O + H_3O^+ \rightleftharpoons HO{-}N{=}O + H_2O$

$$HO{-}N{=}O + H_3O^+ \rightleftharpoons HO^{+}(H){-}N{=}O + H_2O$$

$$HO^{+}(H){-}N{=}O \rightleftharpoons H_2O + \overset{+}{N}{=}O$$

$$C_6H_5N(CH_3)_2 + \overset{+}{N}O \longrightarrow [\,\overset{+}{N}(CH_3)_2{=}C_6H_5(H)(NO)\,] \xrightarrow{-H^+} p\text{-}ON{-}C_6H_4{-}N(CH_3)_2$$

(c) The $\overset{+}{N}O$ ion is a weak electrophile. For it to react with an aromatic ring, the ring must have a powerful activating group such as $-OH$ or $-NR_2$.

19.11

(a) benzene $\xrightarrow[\text{H}_2\text{SO}_4,\ \text{heat}]{\text{fuming HNO}_3}$ m-dinitrobenzene (NO_2, NO_2) $\xrightarrow[\text{NH}_3,\ \text{C}_2\text{H}_5\text{OH}]{\text{H}_2\text{S}}$ m-nitroaniline (NO_2, NH_2)

(b) m-nitroaniline (NO_2, NH_2) [as in (a)] $\xrightarrow[\text{H}_2\text{O}\ (0\text{-}5°\text{C})]{\text{HCl/NaNO}_2}$ (NO_2, $N_2^+Cl^-$) $\xrightarrow{\text{CuCl}}$ (NO_2, Cl) $\xrightarrow[\text{heat}]{\text{Fe, HCl}}$ (NH_2, Cl)

(c) nitrobenzene (NO_2) (by nitration of benzene) $\xrightarrow[\text{FeBr}_3]{\text{Br}_2}$ (NO_2, Br) $\xrightarrow[(2)\ \text{OH}^-]{(1)\ \text{Fe, HCl, heat}}$ (NH_2, Br)

(d) nitrobenzene (NO_2) [as in (c)] $\xrightarrow[(2)\ \text{OH}^-]{(1)\ \text{Fe, HCl, heat}}$ aniline (NH_2) $\xrightarrow{(\text{CH}_3\text{CO})_2\text{O}}$

$C_6H_5NHCOCH_3 \xrightarrow[H_2SO_4]{conc.} p\text{-}HO_3SC_6H_4NHCOCH_3 \xrightarrow{HNO_3}$ (NHCOCH$_3$, NO$_2$ ortho, SO$_3$H para) $\xrightarrow[heat]{57\%\ H_2SO_4}$ o-nitroaniline (NH$_2$, NO$_2$)

(e) $C_6H_5NHCOCH_3$ [from part (d)] $\xrightarrow[H_2SO_4]{HNO_3}$ $p\text{-}O_2NC_6H_4NHCOCH_3$ (plus a trace of ortho) $\xrightarrow[H_2O]{H^+}$ $p\text{-}O_2NC_6H_4NH_2$

19.12 *p*-Toluidine $\xrightarrow[H_2O]{Br_2}$ 2,6-dibromo-4-methylaniline (CH$_3$, Br, Br, NH$_2$) $\xrightarrow[H_2O,\ 0\text{-}5°C]{H_2SO_4,\ NaNO_2}$ (CH$_3$, Br, Br, N_2^+) $\xrightarrow[H_2O]{H_3PO_2}$ 3,5-Dibromotoluene + N_2

19.13 (a) Toluene $\xrightarrow[H_2SO_4]{HNO_3}$ *p*-Nitrotoluene (+ *o*-nitrotoluene) $\xrightarrow[(2)\ OH^-]{(1)\ Fe,\ HCl}$

$p\text{-}CH_3C_6H_4NH_2 \xrightarrow{(CH_3CO)_2O} p\text{-}CH_3C_6H_4NHCOCH_3 \xrightarrow[H_2SO_4]{HNO_3}$ (CH$_3$, NO$_2$, NHCOCH$_3$) $\xrightarrow[H_2O]{H^+}$

(CH$_3$, NO$_2$, NH$_2$) $\xrightarrow[0\text{-}5°C]{HONO}$ (CH$_3$, NO$_2$, N_2^+) $\xrightarrow[(-N_2)]{H_3PO_2}$ m-nitrotoluene (CH$_3$, NO$_2$)

$\xrightarrow[\text{(2) } OH^-]{\text{(1) Fe, HCl}}$ 3-methylaniline (CH_3, NH_2)

3-methylaniline (CH_3, NH_2) $\xrightarrow{HONO}$ 3-methylbenzenediazonium (CH_3, N_2^+)

(b) $\xrightarrow{CuCl}$ 3-chlorotoluene (CH_3, Cl)

(c) $\xrightarrow{CuBr}$ 3-bromotoluene (CH_3, Br)

(d) $\xrightarrow{KI}$ 3-iodotoluene (CH_3, I)

(e) $\xrightarrow{CuCN}$ 3-methylbenzonitrile (CH_3, CN) (f) $\xrightarrow[H_2O]{H^+}$ 3-methylbenzoic acid (CH_3, CO_2H)

19.14 4-nitroaniline (NH_2, NO_2) $\xrightarrow[FeBr_3]{Br_2}$ 2,6-dibromo-4-nitroaniline (Br, NH_2, Br, NO_2) $\xrightarrow[0\text{-}5^\circ C]{H_2SO_4/NaNO_2}$ 2,6-dibromo-4-nitrobenzenediazonium (Br, N_2^+, Br, NO_2) $\xrightarrow{CuBr}$

1,2,3-tribromo-5-nitrobenzene (Br, Br, Br, NO_2) $\xrightarrow{H_2,\ Pt}$ 3,4,5-tribromoaniline (Br, Br, Br, NH_2) $\xrightarrow[0\text{-}5^\circ C]{H_2SO_4/NaHO_2}$

$\longrightarrow$ 3,4,5-tribromobenzenediazonium (Br, Br, Br, N_2^+) $\xrightarrow{H_3PO_2}$ 1,2,3-tribromobenzene (Br, Br, Br)

19.15

$$HO_3S\text{-}C_6H_4\text{-}NH_2 \xrightarrow[0\text{-}5°C]{H_2SO_4/NaNO_2} HO_3S\text{-}C_6H_4\text{-}N_2^+$$

$$\xrightarrow[\text{NaOH, pH 8-10}]{\text{2-naphthol (OH)}} \text{1-}(N{=}N\text{-}C_6H_4\text{-}SO_3Na)\text{-2-naphthol (OH)}$$

Orange II

19.16

$$C_6H_6 \xrightarrow[H_2SO_4]{HNO_3} C_6H_5NO_2 \xrightarrow[HCl]{Fe} C_6H_5NH_2 \xrightarrow[0\text{-}5°C]{H_2SO_4/NaNO_2}$$

$$C_6H_5N_2^+ \xrightarrow[\text{pH 5-7}]{C_6H_5N(CH_3)_2} C_6H_5\text{-}N{=}N\text{-}C_6H_4\text{-}N(CH_3)_2$$

19.17

$$CH_3CH_2O\text{-}C_6H_4\text{-}NH_2 \xrightarrow{H_2SO_4/NaNO_2} CH_3CH_2O\text{-}C_6H_4\text{-}N_2^+ \xrightarrow[OH^-]{C_6H_5\text{-}OH}$$

$$\underset{\textbf{A}}{CH_3CH_2O\text{-}C_6H_4\text{-}N{=}N\text{-}C_6H_4\text{-}OH} \xrightarrow[CH_3CH_2Br]{NaOH}$$

$$\underset{\textbf{B}}{CH_3CH_2O\text{-}C_6H_4\text{-}N{=}N\text{-}C_6H_4\text{-}OCH_2CH_3} \xrightarrow{SnCl_2} \underset{\textbf{C}}{2CH_3CH_2O\text{-}C_6H_4\text{-}NH_2}$$

$$\xrightarrow{(CH_3\overset{O}{\overset{\|}{C}})_2O} 2CH_3CH_2O\text{-}C_6H_4\text{-}NH\overset{O}{\overset{\|}{C}}CH_3$$

Phenacetin

19.18 (1) That **A** reacts with benzenesulfonyl chloride in aqueous KOH to give a clear solution, which on acidification yields a precipitate, shows that **A** is a primary amine.

(2) That diazotization of **A** followed by treatment with 2-naphthol gives an intensely colored precipitate shows that **A** is a primary aromatic amine; that is, **A** is a substituted aniline.

(3) Consideration of the molecular formula of A leads us to conclude that A is a toluidine.

$$\frac{C_7H_9N - C_6H_6N}{CH_3} = -C_6H_4-NH_2$$

But is A *o*-toluidine, *m*-toluidine, or *p*-toluidine?

(4) This question is answered by the infrared data. A single absorption peak in the 680-840 cm^{-1} region at 815 cm^{-1} is indicative of a para substituted benzene. Thus A is *p*-toluidine.

p-$CH_3C_6H_4NH_2$

A

19.19 First convert the sulfonamide to its anion, then alkylate the anion with an alkyl halide, then remove the $-SO_2C_6H_5$ group by hydrolysis. For example,

$$R\text{-}NH\text{-}SO_2C_6H_5 \xrightarrow{OH^-} R\text{-}\ddot{N}^-\text{-}SO_2C_6H_5 \xrightarrow{R'\text{-}CH_2X}$$

$$R\text{-}N(CH_2R')\text{-}SO_2C_6H_5 \xrightarrow[heat]{H_3O^+} \xrightarrow{OH^-} R\text{-}N(CH_2R')\text{-}H + C_6H_5SO_3$$

19.20 (a)

$$\underset{\text{Aniline}}{C_6H_5NH_2} \xrightarrow{(CH_3CO)_2O} C_6H_5NHCOCH_3 \xrightarrow{HOSO_2Cl}$$

$$p\text{-}CH_3CONHC_6H_4SO_2Cl \xrightarrow{H_2N\text{-(2-thiazolyl)}} p\text{-}CH_3CONHC_6H_4SO_2NH\text{-(2-thiazolyl)} \xrightarrow[heat]{dil.\ HCl}$$

(b) Sulfathiazole (p-aminobenzenesulfonamide N-linked to 2-thiazolyl: NH_2–C_6H_4–SO_2NH–thiazole) + succinic anhydride → Succinylsulfathiazole ($HO_2CCH_2CH_2CONH$–C_6H_4–SO_2NH–thiazole)

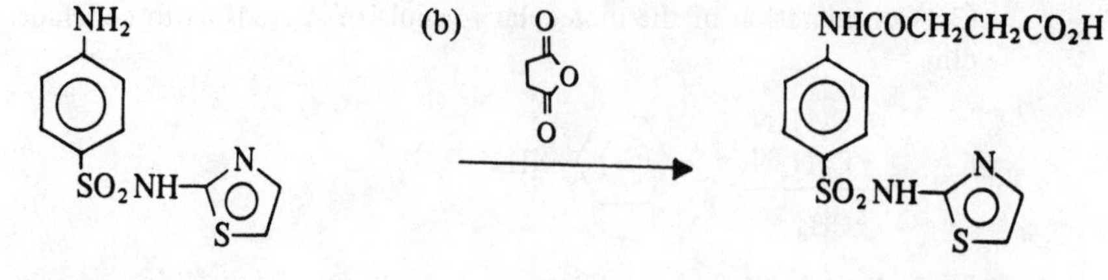

Sulfathiazole

Succinylsulfathiazole

19.21 (a) $C_6H_5CH_2NHCH_3$

(b) $\left(\underset{CH_3}{CH_3CH}\right)_3N$

(c) $C_6H_5N(CH_3)CH_2CH_3$

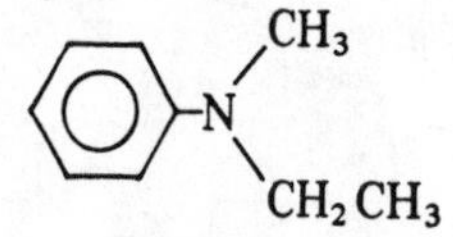

(d) 3-methylaniline (CH_3 and NH_2 meta on benzene ring)

(e) 2-methylpyrrole (N–H)

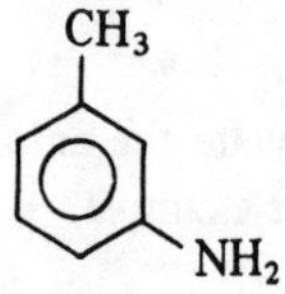

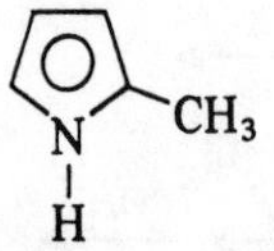

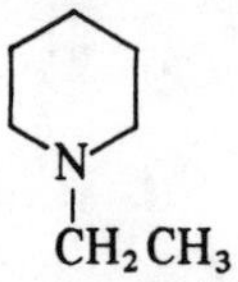

(f) N-ethylpiperidine (N–CH_2CH_3)

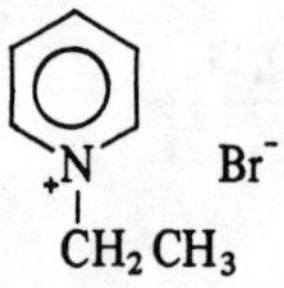

(g) N-ethylpyridinium bromide ($\overset{+}{N}$–CH_2CH_3 Br^-)

(h) 3-pyridinecarboxylic acid (CO_2H)

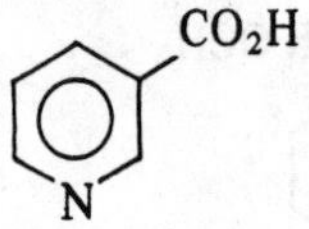

(i) 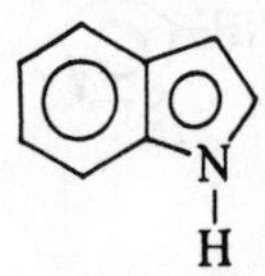 indole (N–H)

(j) 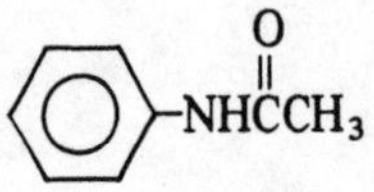 $C_6H_5NH\overset{O}{\overset{\|}{C}}CH_3$

(k) $CH_3-\overset{H}{\underset{CH_3}{N^+}}-H\ Cl^-$

(l) 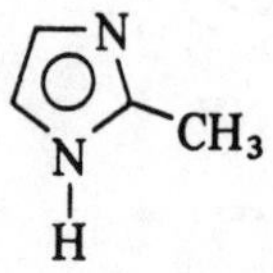 2-methylimidazole (CH_3, N–H)

(m) $H_2NCH_2CH_2CH_2OH$

(n) $(CH_3CH_2CH_2)_4N^+\ Cl^-$

(o)  pyrrolidine (N–H)

(p) 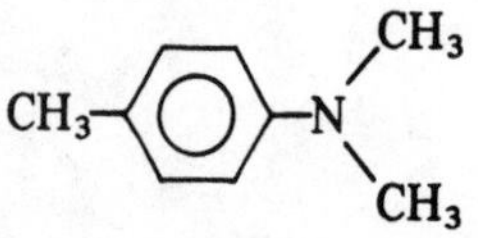 CH_3–C_6H_4–$N(CH_3)_2$

(q) 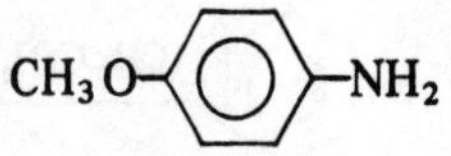 CH_3O–C_6H_4–NH_2

(r) $(CH_3)_4N^+\ OH^-$

(s) 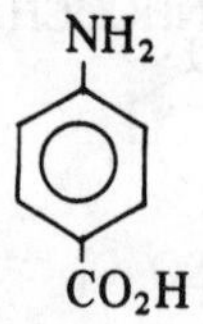 NH_2–C_6H_4–CO_2H (para)

(t) 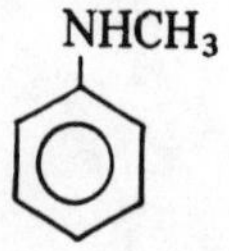 $C_6H_5NHCH_3$

19.22

(a) Propylamine

(b) *N*-methylaniline

(c) Isopropyltrimethylammonium iodide

(d) *o*-Toluidine

(e) *o*-Anisidine (or *o*-methoxyaniline)

(f) Pyrazole

(g) 2-Aminopyrimidine

(h) Benzylaminium chloride

(i) *N,N*-Dipropylaniline

(j) Benzenesulfonamide

(k) Methylaminium acetate

(l) 3-Aminopropanol

(m) Purine

(n) *N*-Methylpyrrole

19.23

(a) $C_6H_5C{\equiv}N + LiAlH_4 \longrightarrow C_6H_5CH_2NH_2$

(b) $C_6H_5\overset{O}{\overset{\|}{C}}{-}NH_2 + LiAlH_4 \longrightarrow C_6H_5CH_2NH_2$

(c) $C_6H_5CH_2Br + \underset{(excess)}{NH_3} \longrightarrow C_6H_5CH_2NH_2$

$C_6H_5CH_2Br$ + potassium phthalimide (NK) $\longrightarrow C_6H_5CH_2{-}N$(phthalimide)

$\xrightarrow{NH_2NH_2} C_6H_5CH_2NH_2$ + phthalhydrazide (cyclic N–H, N–H)

(d) $C_6H_5CH_2OTs + \underset{(excess)}{NH_3} \longrightarrow C_6H_5CH_2NH_2$

(e) $C_6H_5CHO + NH_3 \xrightarrow{H_2,\ Ni} C_6H_5CH_2NH_2$

(f) $C_6H_5CH_2NO_2 + 3H_2 \xrightarrow{Pt} C_6H_5CH_2NH_2$

(g) $C_6H_5CH_2\overset{O}{\overset{\|}{C}}NH_2 \xrightarrow{Br_2,\ OH^-} C_6H_5CH_2NH_2 + CO_3^{2-}$

19.24

(a) Benzene $\xrightarrow[H_2SO_4]{HNO_3}$ nitrobenzene (NO_2) $\xrightarrow[(2)\ OH^-]{(1)\ Fe,\ HCl,\ heat}$ aniline (NH_2)

(b) Bromobenzene (Br) $\xrightarrow[liq\ NH_3]{NaNH_2}$ aniline (NH_2)

(c) Benzamide ($CONH_2$) $\xrightarrow{Br_2,\ OH^-}$ aniline (NH_2)

19.25

(a) $CH_3(CH_2)_2CH_2OH \xrightarrow{PBr_3} CH_3(CH_2)_2CH_2Br \xrightarrow{\text{potassium phthalimide (NK)}}$

N-butylphthalimide ($NCH_2(CH_2)_2CH_3$) $\xrightarrow{NH_2NH_2} CH_3(CH_2)_2CH_2NH_2$ + phthalhydrazide

(b) $CH_3(CH_2)_2CH_2Br \xrightarrow{NaCN} CH_3(CH_2)_3CN \xrightarrow{LiAlH_4} CH_3(CH_2)_3CH_2NH_2$
[from part (a)]

(c) $CH_3(CH_2)_2CH_2OH \xrightarrow[(2)\ H_3O^+]{(1)\ KMnO_4,\ OH^-} CH_3CH_2CH_2CO_2H$

$\xrightarrow[(2)\ NH_3]{(1)\ SOCl_2} CH_3CH_2CH_2CONH_2 \xrightarrow{Br_2,\ OH^-} CH_3CH_2CH_2NH_2$

(d) $CH_3CH_2CH_2CH_2OH \xrightarrow[CH_2Cl_2]{PCC} CH_3CH_2CH_2CHO \xrightarrow[H_2,\ Ni]{CH_3NH_2}$
$CH_3CH_2CH_2CH_2NHCH_3$

19.26

(a) Aniline (NH_2) $\xrightarrow{(CH_3CO)_2O}$ acetanilide ($NHCOCH_3$)

(b) $C_6H_5NH_2$ + phthalic anhydride $\xrightarrow{\text{heat}}$ N-phenylphthalimide

(c) $C_6H_5NHCOCH_3$ [from part (a)] $\xrightarrow[H_2SO_4]{HNO_3}$ p-$O_2NC_6H_4NHCOCH_3$ $\xrightarrow[(2)\ OH^-]{(1)\ H^+,\ H_2O}$ p-$O_2NC_6H_4NH_2$

(d) $C_6H_5NHCOCH_3$ [from part (a)] $\xrightarrow{HOSO_2Cl}$ p-$ClO_2SC_6H_4NHCOCH_3$ $\xrightarrow[(2)\ H_3O^+,\ heat]{(1)\ NH_3}$ p-$H_2NO_2SC_6H_4NH_2$

(e) $C_6H_5NH_2$ $\xrightarrow[base]{2CH_3I}$ $C_6H_5N(CH_3)_2$

(f) $C_6H_5NH_2$ $\xrightarrow[0\text{-}5°C]{HONO}$ $C_6H_5N_2^+\ X^-$ $\xrightarrow{HBF_4}$ $C_6H_5N_2^+\ BF_4^-$ $\xrightarrow{heat}$ C_6H_5F

(g) $C_6H_5N_2^+\ X^-$ [from part (f)] $\xrightarrow{CuCl}$ C_6H_5Cl

(h) $C_6H_5N_2^+\ X^-$ [from part (f)] $\xrightarrow{CuBr}$ C_6H_5Br

(i) $C_6H_5N_2^+\ X^-$ [from part (f)] $\xrightarrow{KI}$ C_6H_5I

(j) $C_6H_5N_2^+\ X^-$ [from part (f)] $\xrightarrow{CuCN}$ C_6H_5CN

(k) C_6H_5CN [from part (j)] $\xrightarrow[\text{heat}]{H_3O^+, H_2O}$ $C_6H_5CO_2H$

(l) $C_6H_5N_2^+\ X^-$ [from part (f)] $\xrightarrow[\text{heat}]{H_3O^+, H_2O}$ C_6H_5OH

(m) $C_6H_5N_2^+\ X^-$ [from part (f)] $\xrightarrow[H_2O]{H_3PO_2}$ C_6H_6

(n) $C_6H_5N_2^+\ X^-$ [from part (f)] + C_6H_5OH [from part (l)] $\xrightarrow[\text{(pH 8-10)}]{OH^-}$ $C_6H_5\text{–N=N–}C_6H_4\text{–}OH$

(o) $C_6H_5N_2^+\ X^-$ [from part (f)] + $C_6H_5N(CH_3)_2$ [from part (e)] $\xrightarrow[\text{(pH 5-7)}]{H_3O^+}$ $C_6H_5\text{–N=N–}C_6H_4\text{–}N(CH_3)_2$

19.27

(a) $CH_3CH_2CH_2NH_2 \xrightarrow[NaNO_2/HCl]{HONO} [CH_3CH_2CH_2N_2^+] \xrightarrow{-N_2}$

$[CH_3CH_2CH_2^+] \xrightarrow[\text{shift}]{\text{hydride}} [CH_3\overset{+}{C}HCH_3]$

$[CH_3CH_2CH_2^+] \xrightarrow{Cl^-} CH_3CH_2CH_2Cl$

$[CH_3CH_2CH_2^+] \xrightarrow{H_2O} CH_3CH_2CH_2OH$

$[CH_3CH_2CH_2^+] \xrightarrow{-H^+} CH_3CH{=}CH_2$

$[CH_3\overset{+}{C}HCH_3] \xrightarrow{-H^+} CH_3CH{=}CH_2$

$[CH_3\overset{+}{C}HCH_3] \xrightarrow{H_2O} CH_3CH(OH)CH_3$

$[CH_3\overset{+}{C}HCH_3] \xrightarrow{Cl^-} CH_3CH(Cl)CH_3$

(b) $(CH_3CH_2CH_2)_2NH \xrightarrow[NaNO_2/HCl]{HONO} (CH_3CH_2CH_2)_2N{-}N{=}O$

(c) $C_6H_5N(H)CH_2CH_2CH_3 \xrightarrow[NaNO_2/HCl]{HONO} C_6H_5N(N{=}O)CH_2CH_2CH_3$

(d) $C_6H_5N(CH_2CH_2CH_3)_2 \xrightarrow[NaNO_2/HCl]{HONO} O{=}N\text{-}C_6H_4\text{-}N(CH_2CH_2CH_3)_2$

(e) $CH_3CH_2CH_2\text{-}C_6H_4\text{-}NH_2 \xrightarrow[NaNO_2/HCl]{HONO,\ 0\text{-}5^\circ C} CH_3CH_2CH_2\text{-}C_6H_4\text{-}N_2^+\ Cl^-$

19.28

(a) $CH_3CH_2CH_2NH_2 + C_6H_5SO_2Cl \xrightarrow[H_2O]{KOH} CH_3CH_2CH_2\overset{-}{N}SO_2C_6H_5\ K^+$

Clear solution

$\xrightarrow{H_3O^+} CH_3CH_2CH_2NHSO_2C_6H_5$

Precipitate

(b) $(CH_3CH_2CH_2)_2NH + C_6H_5SO_2Cl \xrightarrow[H_2O]{KOH} (CH_3CH_2CH_2)_2NSO_2C_6H_5$

Precipitate

$\xrightarrow{H_3O^+}$ no reaction (precipitate remains)

(c) $C_6H_5N(H)CH_2CH_2CH_3 + C_6H_5SO_2Cl \xrightarrow[H_2O]{KOH} C_6H_5N(SO_2C_6H_5)CH_2CH_2CH_3$

Precipitate

$\xrightarrow{H_3O^+}$ no reaction (precipitate remains)

(d) $C_6H_5N(CH_2CH_2CH_3)_2 + C_6H_5SO_2Cl \xrightarrow[H_2O]{KOH}$ no reaction (3° amine is insoluble)

$\xrightarrow{H_3O^+} C_6H_5\overset{+}{N}H(CH_2CH_2CH_3)_2$

3° Amine dissolves

(e) $C_3H_7-C_6H_4-NH_2 + C_6H_5SO_2Cl \xrightarrow[H_2O]{KOH} C_3H_7-C_6H_4-\bar{N}SO_2C_6H_5\ K^+$ Clear solution

$\xrightarrow{H_3O^+} C_3H_7-C_6H_4-NHSO_2C_6H_5$ Precipitate

19.29 (a) piperidine N–H $\xrightarrow[NaNO_2/HCl]{HONO}$ piperidine N–N=O

(b) piperidine N–H + $C_6H_5SO_2Cl \xrightarrow[H_2O]{KOH}$ piperidine N–$SO_2C_6H_5$

19.30 (a) $2CH_3CH_2NH_2 + C_6H_5COCl \longrightarrow CH_3CH_2NHCOC_6H_5 + CH_3CH_2NH_3^+Cl^-$

(b) $2CH_3NH_2 + (CH_3\overset{O}{\overset{\|}{C}})_2O \longrightarrow CH_3NH\overset{O}{\overset{\|}{C}}CH_3 + CH_3\overset{+}{N}H_3\ \ CH_3\overset{O}{\overset{\|}{C}}O^-$

(c) succinic anhydride (cyclic $H_2C-C(=O)-O-C(=O)-CH_2$) $+ 2CH_3NH_2 \longrightarrow H_2C(-\overset{O}{\overset{\|}{C}}-NHCH_3)-H_2C(-\overset{O}{\overset{\|}{C}}O^-CH_3NH_3^+)$

(d) [product of (c)] $\xrightarrow{heat}$ N-methylsuccinimide (cyclic $H_2C-C(=O)-N(CH_3)-C(=O)-CH_2$) $+ H_2O + CH_3NH_2$

(e) pyrrolidine (N–H) + phthalic anhydride $\longrightarrow$ C_6H_4 bearing $\overset{O}{\overset{\|}{C}}$–N(pyrrolidine) and $\underset{\|}{\underset{O}{C}}OH$ (ortho)

(f) pyrrole (N–H) + $(CH_3CO)_2O \longrightarrow$ N-acetylpyrrole (N–C(=O)CH$_3$) + CH_3CO_2H

(g) $2\ C_6H_5\text{-}NH_2 + CH_3CH_2\overset{O}{\overset{\|}{C}}Cl \longrightarrow C_6H_5\text{-}NH\overset{O}{\overset{\|}{C}}CH_2CH_3 + C_6H_5\text{-}NH_3^+\ Cl^-$

(h) $CH_3CH_2\text{-}\overset{CH_2CH_3}{\underset{CH_2CH_3}{N^+}}\text{-}CH_2CH_3\ OH^- \xrightarrow{\text{heat}} CH_2{=}CH_2 + (CH_3CH_2)_3N + H_2O$

(i) 1,3-dinitrobenzene (NO_2, NO_2) $+ H_2S \xrightarrow[C_2H_5OH]{NH_3}$ 3-nitroaniline (NO_2, NH_2)

(j) 4-methylaniline (CH_3, NH_2) $+ Br_2$ (excess) $\xrightarrow{H_2O}$ 2,6-dibromo-4-methylaniline (CH_3, Br, NH_2, Br)

19.31

(a) toluene (CH_3) $\xrightarrow[H_2SO_4]{HNO_3}$ o-nitrotoluene (CH_3, NO_2) + p-nitrotoluene (CH_3, NO_2)

Separate isomers

o-nitrotoluene (CH_3, NO_2) $\xrightarrow[(2)\ OH^-]{(1)\ Fe,\ HCl,\ heat}$ o-toluidine (CH_3, NH_2) $\xrightarrow{HONO}$

o-methylbenzenediazonium (CH_3, N_2^+) X^- $\xrightarrow[heat]{H_3O^+}$ o-cresol (CH_3, OH)

(b) m-toluidine (CH_3, NH_2) $\xrightarrow[(2)\ H_3O^+,\ heat]{(1)\ HCl/NaNO_2}$ m-cresol (CH_3, OH)

[from Problem 19.13(a)]

(c) p-nitrotoluene (4-CH_3-C_6H_4-NO_2) $\xrightarrow[\text{(2) } OH^-]{\text{(1) Fe, HCl, heat}}$ p-toluidine (4-CH_3-C_6H_4-NH_2) $\xrightarrow[\text{(2) } H_3O^+\text{, heat}]{\text{(1) } H_2SO_4/NaNO_2}$ p-cresol (4-CH_3-C_6H_4-OH)

[from part (a)]

(d) m-phenylenediamine (NH_2, NH_2) $\xrightarrow{H_2SO_4/NaNO_2}$ bis-diazonium salt (N_2^+ X^-, N_2^+ X^-) $\xrightarrow{CuCl}$ m-dichlorobenzene (Cl, Cl)

[by reduction of *m*-dinitrobenzene, cf. Problem 19.11(a)]

(e) bis-diazonium salt (N_2^+ X^-, N_2^+ X^-) $\xrightarrow{CuCN}$ m-dicyanobenzene (CN, CN)

[cf. part (d)]

(f) m-nitroaniline (NO_2, NH_2) $\xrightarrow[\text{(2) KI}]{\text{(1) } H_2SO_4/NaNO_2}$ m-iodonitrobenzene (NO_2, I) $\xrightarrow[\text{(2) } OH^-]{\text{(1) Fe, HCl, heat}}$ m-iodoaniline (NH_2, I)

[from Problem 19.11(a)]

$\xrightarrow[\text{(2) } H_3O^+\text{, heat}]{\text{(1) HONO}}$ m-iodophenol (OH, I)

(g) m-nitroaniline (NO_2, NH_2) $\xrightarrow[\text{(2) CuBr}]{\text{(1) } H_2SO_4/NaNO_2}$ m-bromonitrobenzene (NO_2, Br) $\xrightarrow[\text{(2) } OH^-]{\text{(1) Fe, HCl, heat}}$

[from Problem 19.11(a)]

m-bromoaniline (NH_2, Br) $\xrightarrow[\text{(2) CuCN}]{\text{(1) } H_2SO_4/NaNO_2}$ m-bromobenzonitrile (CN, Br)

(h) p-nitroaniline (NO_2, NH_2) $\xrightarrow[FeBr_3]{Br_2}$ 2,6-dibromo-4-nitroaniline (NO_2, Br, NH_2, Br) $\xrightarrow{H_2SO_4/NaNO_2}$ diazonium salt (NO_2, Br, N_2^+, Br) $\xrightarrow{H_3PO_2}$ 3,5-dibromonitrobenzene (NO_2, Br, Br)

[from Problem 19.11(e)]

(i) 3,5-dibromonitrobenzene [from part (h)] $\xrightarrow[\text{(2) } OH^-]{\text{(1) Fe, HCl, heat}}$ 3,5-dibromoaniline

(j) 2,6-dibromo-4-nitroaniline [from part (h)] $\xrightarrow[\text{(2) CuBr}]{\text{(1) } H_2SO_4/NaNO_2}$ 3,4,5-tribromonitrobenzene $\xrightarrow[\text{(2) } OH^-]{\text{(1) Fe, HCl, heat}}$ 3,4,5-tribromoaniline $\xrightarrow[\text{(2) } H_3O^+\text{, heat}]{\text{(1) } H_2SO_4/NaNO_2}$ 3,4,5-tribromophenol

(k) 3,4,5-tribromoaniline [from part (j)] $\xrightarrow[\text{(2) CuCN}]{\text{(1) } H_2SO_4/NaNO_2}$ 3,4,5-tribromobenzonitrile

(l) 2,6-dibromo-4-nitroaniline [from part (h)] $\xrightarrow[\text{(2) CuCN}]{\text{(1) } H_2SO_4/NaNO_2}$ 2,6-dibromo-4-nitrobenzonitrile $\xrightarrow[\text{heat}]{H_3O^+}$ 2,6-dibromo-4-nitrobenzoic acid (CO_2H) $\xrightarrow{H_2\text{, Pt}}$

4-amino-2,6-dibromobenzoic acid $\xrightarrow[\text{(2) } H_3PO_2]{\text{(1) } H_2SO_4/NaNO_2}$ 2,6-dibromobenzoic acid

(m) 2,6-dibromo-4-nitroaniline [from part (h)] $\xrightarrow[\text{(2) KI}]{\text{(1) } H_2SO_4/NaNO_2}$ 3,5-dibromo-4-iodonitrobenzene $\xrightarrow[\text{(2) } OH^-]{\text{(1) Fe, HCl, heat}}$ 3,5-dibromo-4-iodoaniline $\xrightarrow[\text{(2) } H_3PO_2]{\text{(1) } H_2SO_4/NaNO_2}$ 1,3-dibromo-2-iodobenzene

(n) CH_3 (toluene) $\xrightarrow[H_2SO_4]{\text{fuming } HNO_3}$ CH_3, NO_2, NO_2 $\xrightarrow[\text{(cf. Section 19.5B)}]{H_2S,\ NH_3}$

CH_3, NO_2, NH_2 $\xrightarrow[(2)\ CuBr]{(1)\ H_2SO_4/NaNO_2}$ CH_3, NO_2, Br

(o) CH_3, NO_2, NH_2 $\xrightarrow[(2)\ H_3O^+,\ \text{heat}]{(1)\ H_2SO_4/NaNO_2}$ CH_3, NO_2, OH

[from part (n)]

(p) CH_3, NO_2, NH_2 $\xrightarrow[(2)\ CuBr]{(1)\ H_2SO_4/NaNO_2}$ CH_3, NO_2, Br $\xrightarrow[(2)\ OH^-]{(1)\ Fe,\ HCl,\ \text{heat}}$

[from part (n)]

CH_3, NH_2, Br $\xrightarrow[(2)\ CuCN]{(1)\ H_2SO_4/NaNO_2}$ CH_3, CN, Br

(q) $CH_3-C_6H_4-NH_2$ $\xrightarrow{H_2SO_4/NaNO_2}$ $CH_3-C_6H_4-N_2^+$ X^-

[from part (c)]

$\downarrow$ C_6H_5-OH, pH 8-10

$CH_3-C_6H_4-N{=}N-C_6H_4-OH$

(r) $CH_3-C_6H_4-N_2^+$ X^- $\xrightarrow[\text{pH 8-10}]{CH_3-C_6H_4-OH\ \text{[from part (c)]}}$ $CH_3-C_6H_4-N{=}N-$ (ring with OH, CH_3)

[from part (q)]

19.32 (a) Benzylamine dissolves in dilute HCl at room temperature,

$$C_6H_5CH_2NH_2 + H_3O^+ + Cl^- \xrightarrow{25°C} C_6H_5CH_2\overset{+}{N}H_3Cl^-$$

benzamide does not dissolve:

$$C_6H_5CONH_2 + H_3O^+ + Cl^- \xrightarrow{25°C} \text{no reaction}$$

(b) Allylamine reacts with (and decolorizes) bromine in carbon tetrachloride instantly,

$$CH_2{=}CHCH_2NH_2 + Br_2 \xrightarrow{CCl_4} \underset{Br}{CH_2}\underset{Br}{CH}CH_2NH_2$$

propylamine does not:

$$CH_3CH_2CH_2NH_2 + Br_2 \xrightarrow{CCl_4} \text{no reaction if the mixture is not heated or irradiated}$$

(c) The Hinsberg test:

$$CH_3\text{-}C_6H_4\text{-}NH_2 + C_6H_5SO_2Cl \xrightarrow[H_2O]{KOH} \underset{\text{Soluble}}{CH_3\text{-}C_6H_4\text{-}\overset{-}{N}SO_2C_6H_5\ K^+} \xrightarrow{H_3O^+} \underset{\text{Precipitate}}{CH_3\text{-}C_6H_4\text{-}NHSO_2C_6H_5}$$

$$C_6H_5\text{-}NHCH_3 + C_6H_5SO_2Cl \xrightarrow[H_2O]{KOH} \underset{\text{Precipitate}}{C_6H_5\text{-}N(CH_3)SO_2C_6H_5} \xrightarrow{H_3O^+} \text{precipitate remains}$$

(d) The Hinsberg test:

$$C_6H_{11}\text{-}NH_2 + C_6H_5SO_2Cl \xrightarrow[H_2O]{KOH} \underset{\text{Soluble}}{C_6H_{11}\text{-}\overset{-}{N}SO_2C_6H_5\ K^+} \xrightarrow{H_3O^+} \underset{\text{Precipitate}}{C_6H_{11}\text{-}NHSO_2C_6H_5}$$

$$\text{(piperidine) }N\text{-}H + C_6H_5SO_2Cl \xrightarrow[H_2O]{KOH} \underset{\text{Precipitate}}{N\text{-}SO_2C_6H_5} \xrightarrow{H_3O^+} \text{precipitate remains}$$

(e) Pyridine dissolves in dilute HCl,

$$C_5H_5N + H_3O^+ + Cl^- \longrightarrow C_5H_5\overset{+}{N}{-}H \quad Cl^-$$

benzene does not:

$$C_6H_6 + H_3O^+ + Cl^- \longrightarrow \text{no reaction}$$

(f) Aniline reacts with nitrous acid at 0-5°C to give a stable diazonium salt that couples with 2-naphthol yielding an intensely colored azo compound.

$$C_6H_5{-}NH_2 \xrightarrow[0\text{-}5°C]{H_2SO_4/NaNO_2} C_6H_5{-}N_2^+ \xrightarrow{\text{2-naphthol}} C_6H_5{-}N{=}N{-}C_{10}H_6OH$$

Cyclohexylamine reacts with nitrous acid at 0-5°C to yield a highly unstable diazonium salt–one that decomposes so rapidly that the addition of 2-naphthol gives no azo compound.

$$C_6H_{11}{-}NH_2 \xrightarrow[0\text{-}5°C]{H_2SO_4/NaNO_2} \left[C_6H_{11}{-}N_2^+\right] \xrightarrow{-N_2} \left[C_6H_{11}^+\right] \longrightarrow$$

$$\text{alkenes, alcohols, and so on} \xrightarrow{\text{2-naphthol}} \text{no reaction}$$

(g) The Hinsberg test:

$$(C_2H_5)_3N + C_6H_5SO_2Cl \xrightarrow[H_2O]{KOH} \text{no reaction} \xrightarrow{H_3O^+} (C_2H_5)_3\overset{+}{N}H$$
Soluble

$$(C_2H_5)_2NH + C_6H_5SO_2Cl \xrightarrow[H_2O]{KOH} (C_2H_5)_2NSO_2C_6H_5 \xrightarrow{H_3O^+} \text{precipitate remains}$$
Precipitate

(h) Tripropylaminium chloride reacts with aqueous NaOH to give a water insoluble tertiary amine.

$$(CH_3CH_2CH_2)_3\overset{+}{N}H\ Cl^- \xrightarrow[H_2O]{NaOH} (CH_3CH_2CH_2)_3N$$
Water soluble → Water insoluble

Tetrapropylammonium chloride does not react with aqueous NaOH (at room temperature) and the tetrapropylammonium ion remains in solution.

$$(CH_3CH_2CH_2)_4N^+Cl^- \xrightarrow[H_2O]{NaOH} (CH_3CH_2CH_2)_4N^+\ [Cl^- \text{ or } OH^-]$$
Water soluble → Water soluble

(i) Tetrapropylammonium chloride dissolves in water to give a neutral solution. Tetrapropylammonium hydroxide dissolves in water to give a strongly basic solution.

19.33 Follow the procedure outlined in the answer to Problem 19.2. Toluene will show the same solubility behavior as benzene.

19.34

$$\text{(succinic anhydride: } H_2C-C(=O) / H_2C-C(=O) \text{ bridged by O)} + NH_3 \xrightarrow[\text{(2) } H^+]{} H_2N\overset{O}{\overset{\|}{C}}CH_2CH_2\overset{O}{\overset{\|}{C}}OH$$

$$\xrightarrow[-CO_3^{2-}]{Br,\ OH^-} H_2NCH_2CH_2CO_2^- \xrightarrow{H^+} H_3\overset{+}{N}CH_2CH_2CO_2^-$$

19.35

(a) $$HOCH_2(CH_2)_8CH_2OH \xrightarrow{2PBr_3} BrCH_2(CH_2)_8CH_2Br \xrightarrow{2\ (CH_3)_3N}$$

$$(CH_3)_3\overset{+}{N}CH_2(CH_2)_8CH_2\overset{+}{N}(CH_3)_3\ 2Br^-$$

(b) $$HO_2CCH_2CH_2CO_2H + 2\ BrCH_2CH_2OH \xrightarrow{H^+}$$

$$BrCH_2CH_2O_2CCH_2CH_2CO_2CH_2CH_2Br \xrightarrow{2\ (CH_3)_3N}$$

$$(CH_3)_3\overset{+}{N}CH_2CH_2O_2CCH_2CH_2CO_2CH_2CH_2\overset{+}{N}(CH_3)_3\ 2\ Br^-$$

(c) $$(CH_3)_3N + H_2C\overset{O}{—}CH_2 \longrightarrow (CH_3)_3\overset{+}{N}CH_2CH_2O^- \xrightarrow{CH_3\overset{O}{\overset{\|}{C}}Cl}$$

$$(CH_3)_3\overset{+}{N}CH_2CH_2O\overset{O}{\overset{\|}{C}}CH_3\ Cl^-$$

19.36

$$\text{(2-amino-4,5-diamino-6-hydroxypyrimidine: } H_2N,\ NH_2,\ NH_2,\ OH) + \underset{O=\ C-CH_2Br}{CHBr_2} \xrightarrow[-H_2O]{} \text{(pyrimidine: } H_2N,\ OH,\ \ddot{N}H_2-\underset{|}{\overset{Br}{CH}}-Br,\ N=C-CH_2Br)$$

$\xrightarrow{(-HBr)}$ H, H_2N, N, N, N, Br, N, N, CH_2Br, OH $\xrightarrow[(-HBr)]{HCO_3^-}$

H_2N, N, N, N, N, OH, CH_2, Br + H_2N–C_6H_4–CONHCH(CO_2H)$CH_2CH_2CO_2H$ $\xrightarrow{-HBr}$ folic acid

19.37 The results of the Hinsberg test indicate that compound **W** is a tertiary amine. The proton nmr spectrum provides evidence for the following:

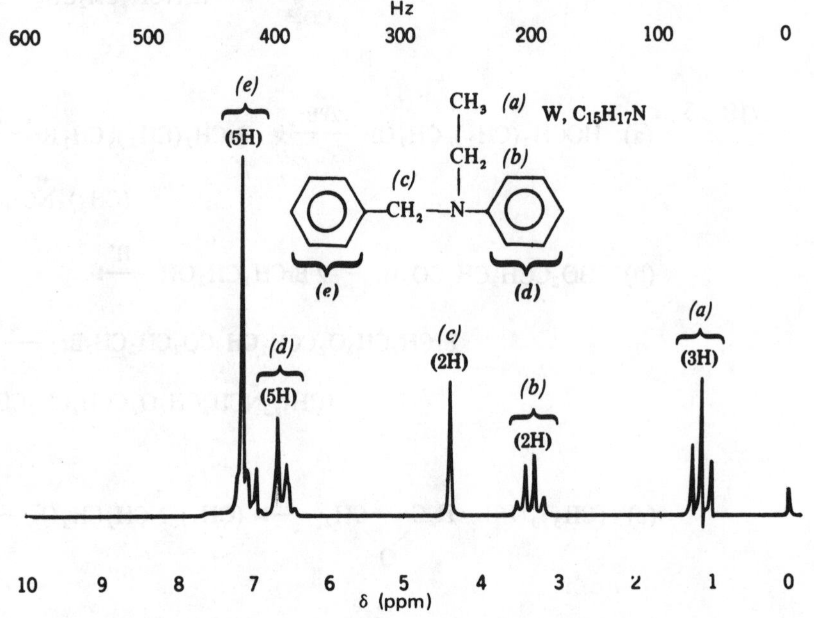

The proton nmr spectrum of W (Problem 19.37). (Courtesy of Aldrich Chemical Company, Inc., Milwaukee, WI.)

Thus **W** is *N*-benzyl-*N*-ethylaniline.

19.38 Compound **X** is benzyl bromide, $C_6H_5CH_2Br$. This is the only structure consistent with the proton nmr and infrared data. (The monosubstituted benzene ring is strongly indicated by the (5H), δ7.3 proton nmr absorption and is confirmed by the peaks at 690 and 770 cm^{-1} in the infrared spectrum.)

Compound **Y**, therefore must be phenylacetonitrile, ($C_6H_5CH_2CN$) and **Z** must be 2-phenylethylamine, $C_6H_5CH_2CH_2NH_2$.

$$\underset{\substack{\mathbf{X} \\ C_7H_7Br}}{C_6H_5CH_2Br} \xrightarrow{NaCN} \underset{\substack{\mathbf{Y} \\ C_8H_7N}}{C_6H_5CH_2CN} \xrightarrow{LiAlH_4} \underset{\substack{\mathbf{Z} \\ C_8H_{11}N}}{C_6H_5CH_2CH_2NH_2}$$

Interpretations of the infrared and proton nmr spectra of **Z** are given in Fig. 19.5 to follow.

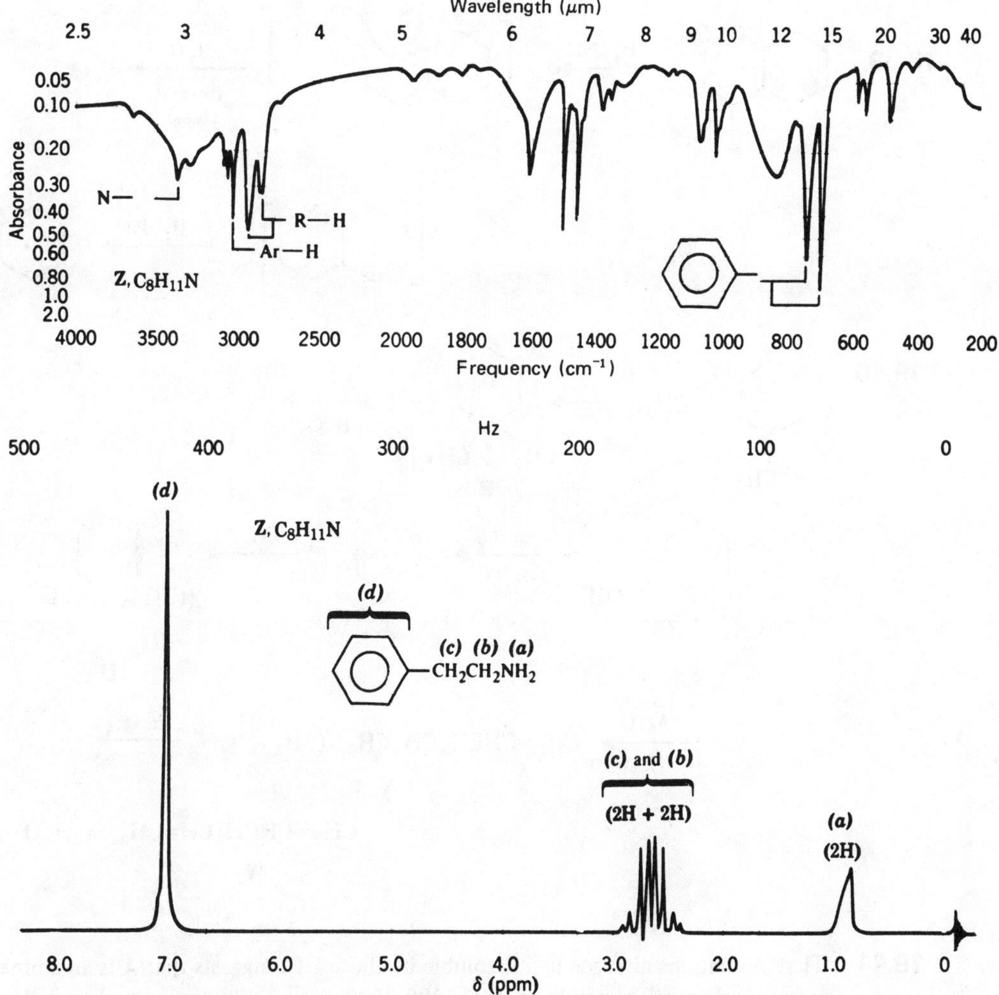

Infrared and proton nmr spectra for compound Z, Problem 19.38. (Courtesy of Sadtler Research Laboratories, Inc., Philadelphia.)

(1) Two different C_6H_5-groups (one absorbing at δ7.2 and one at δ6.7).
(2) A CH_3CH_2-group (the quartet at δ3.3 and the triplet at δ1.2).
(3) An unsplit $-CH_2-$group (the singlet at δ4.4).

There is only one reasonable way to put all of this together.

$C_6H_5-CH_2-N(CH_2CH_3)-C_6H_5$

Thus W is *N*-benzyl-*N*-ethylaniline.

19.39 2-(2-cyanoethyl)cyclohexanone $\xrightarrow{H_2,\ Pd}$ [2-(3-aminopropyl)cyclohexanone] $\xrightarrow[\text{(several steps)}]{-H_2O}$ [cyclic imine] $\xrightarrow{H_2,\ Pd}$ decahydroquinoline

19.40 N-methylpiperidine $+ CH_3I \longrightarrow$ N,N-dimethylpiperidinium I^- (**R**) $\xrightarrow[H_2O]{Ag_2O}$

N,N-dimethylpiperidinium OH^- (**S**) $\xrightarrow[-H_2O]{heat}$ $CH_2=CHCH_2CH_2CH_2N(CH_3)_2$ (**T**) $\xrightarrow{CH_3I}$ $CH_2=CHCH_2CH_2CH_2\overset{+}{N}(CH_3)_3\ I^-$ (**U**)

$\xrightarrow[H_2O]{Ag_2O}$ $CH_2=CHCH_2CH_2CH_2\overset{+}{N}(CH_3)_3\ OH^-$ (**V**) $\xrightarrow{heat}$

$CH_2=CHCH_2CH=CH_2 + H_2O + (CH_3)_3N$ (**W**)

19.41 That A contains nitrogen and is soluble in dilute HCl suggests that A is an amine. The two infrared absorption bands in the 3300-3500 cm^{-1} region suggest that A is a primary amine. The ^{13}C spectrum shows only two signals in the upfield aliphatic region. There are four signals downfield in the aromatic region. The splitting patterns of the aliphatic peaks in the proton off-resonance spectrum suggest an ethyl group *or two equivalent ethyl groups*. Assuming the latter, and assuming that **A** is a primary amine, we can conclude from the molecular formula and from the splitting patterns of the aromatic signals that **A** is 2,6-diethylaniline. The assignments are:

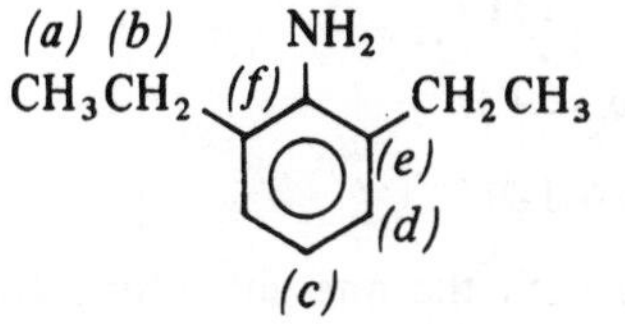

(a) (b) NH_2
CH_3CH_2 (f) CH_2CH_3
(e)
(d)
(c)

(a)	Quartet δ 12.9	*(d)*	Doublet δ 125.9
(b)	Triplet δ 24.2	*(e)*	Singlet δ 127.4
(c)	Doublet δ 118.1	*(f)*	Singlet δ 141.5

(An equally plausible answer would be that **A** is 3,5-diethylaniline.)

19.42 That **B** dissolves in dilute HCl suggests that **B** is an amine. That the infrared spectrum of **B** lacks bands in the 3300-3500 cm^{-1} region suggests that **B** is a tertiary amine. The upfield signals in the ^{13}C spectrum, and the splitting patterns in the proton off-resonance decoupled spectrum suggest two equivalent ethyl groups (as was also true of **A** in the preceding problem). The splitting of the downfield peaks (in the aromatic region) is consistent with a monosubstituted benzene ring. Putting all of these observations together with the molecular formula leads us to conclude that **B** is *N,N*-diethylaniline. The assignments are

(b) (a)
$N(CH_2CH_3)_2$
(f)
(c)
(e)
(d)

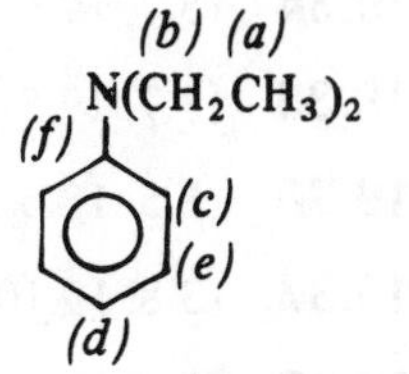

(a) Quartet δ 12.5

(b) Triplet δ 44.2

(c) Doublet δ 112.0

(d) Doublet δ 115.5

(e) Doublet δ 128.1

(f) Singlet δ 147.8

19.43 That **C** gives a positive Tollens' test indicates the presence of an aldehyde group; the solubility of **C** in aqueous HCl suggests that **C** is also an amine. The absence of bands in the 3300-3500 cm^{-1} region of the infrared spectrum of **C** suggests that **C** is a tertiary amine. The signal at δ 189.7 in the ^{13}C spectrum can be assigned to the aldehyde group. The signal at δ 39.7 is the only one in the aliphatic region and its splitting (a quartet in the proton off-resonance decoupled spectrum) is consistent with a methyl group or with two equivalent methyl groups. The remaining signals are in the aromatic region. If we assume that **C** has a benzene ring containing a $-\overset{O}{\overset{\|}{C}}H$ group and a $-N(CH_3)_2$ group then the aromatic signals and their splittings are consistent with **C** being *p*-(*N,N*-dimethylamino)-benzaldehyde. The assignments are

$N(CH_3)_2$ (a), ring carbons (e), (b), (d), (c), $C=O$ with H (f)

(a) Quartet δ 39.7
(b) Doublet δ 110.8
(c) Singlet δ 124.9
(d) Doublet δ 131.6
(e) Singlet δ 154.1
(f) Doublet δ 189.7

You should now compare this spectrum with the one given for *p*-(*N,N*-diethylamino)-benzaldehyde given in Fig. 13.27 and the analysis of that spectrum given in Section 13.10.

SECTION REFERENCES FOR ADDITIONAL PROBLEMS

19.21	12.12, 17.2, 19.1	**19.30**	12.8, 18.8, 19.5, 19.13
19.22	19.1	**19.31**	12.8, 19.7-19.9
19.23	19.5	**19.32**	18.8, 19.9, 19.11, 19.12
19.24	12.4, 14.14, 19.5	**19.33**	17.13, 19.3, 19.12
19.25	14.7, 15.8, 18.5, 18.8, 19.5	**19.34**	18.7, 19.6
19.26	12.12, 18.11, 19.6-19.11	**19.35**	18.7, 19.5
19.27	19.7	**19.36**	18.5, 18.6
19.28	19.10	**19.37**	13.8, 19.10, 19.12
19.29	19.7, 19.10	**19.38**	13.8, 13.12, 19.5, 19.12
		19.39	19.6

SELF-TEST

19.1 Circle the stronger base in each of the following pairs.

(a) C_6H_5–$NHCOCH_3$ and C_6H_5–$NHCH_2CH_3$

(b) C_6H_5–NH_2 and C_6H_5–CH_2NH_2

(c) CH_3O–C_6H_4–NH_2 and O_2N–C_6H_4–NH_2

(d) 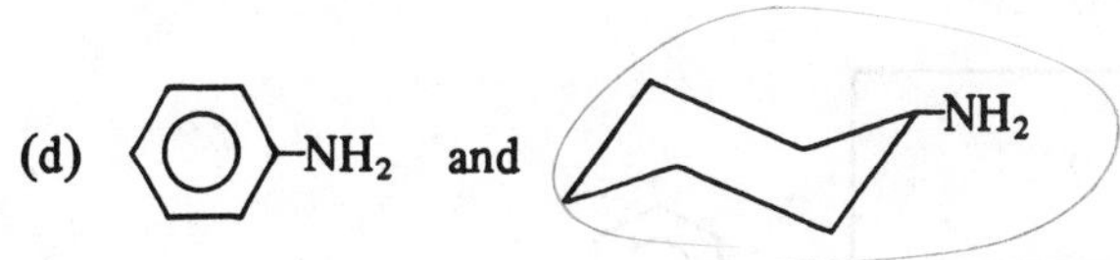

19.2 Arrange the following compounds in order of increasing basicity. Place a *1* beside the most basic, and a *4* beside the least basic. Use *2* and *3* for the remaining compounds accordingly. *All four numbers must be correct for this question to be marked correct.*

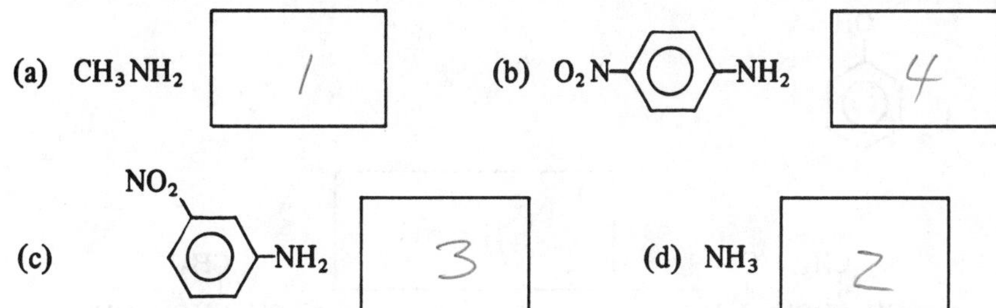

19.3 Supply the formulas of the missing reactants, reagents, and products in the following reaction sequences.

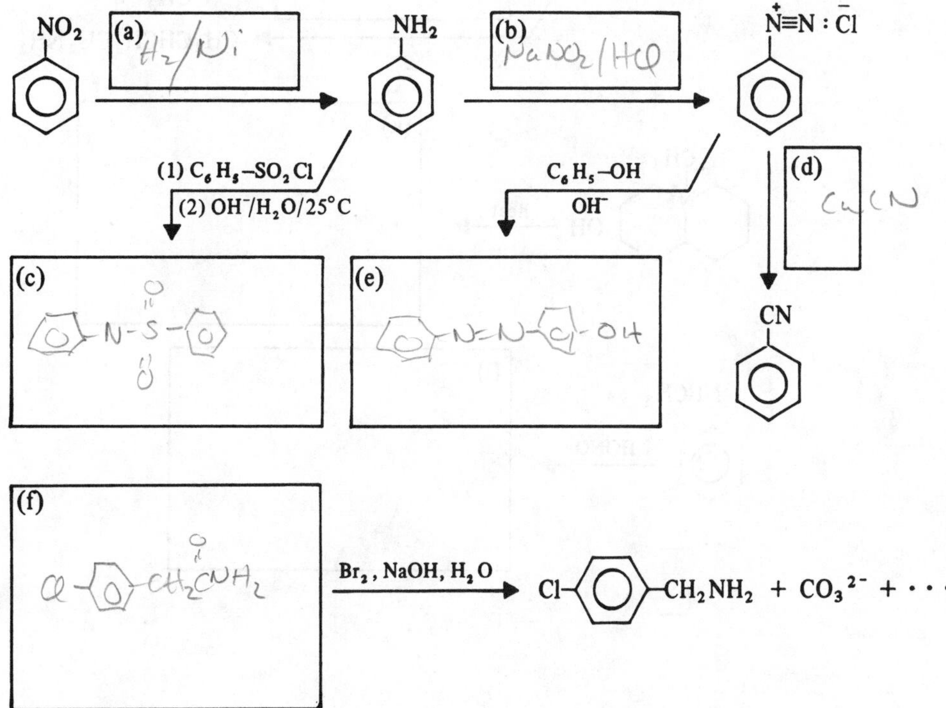

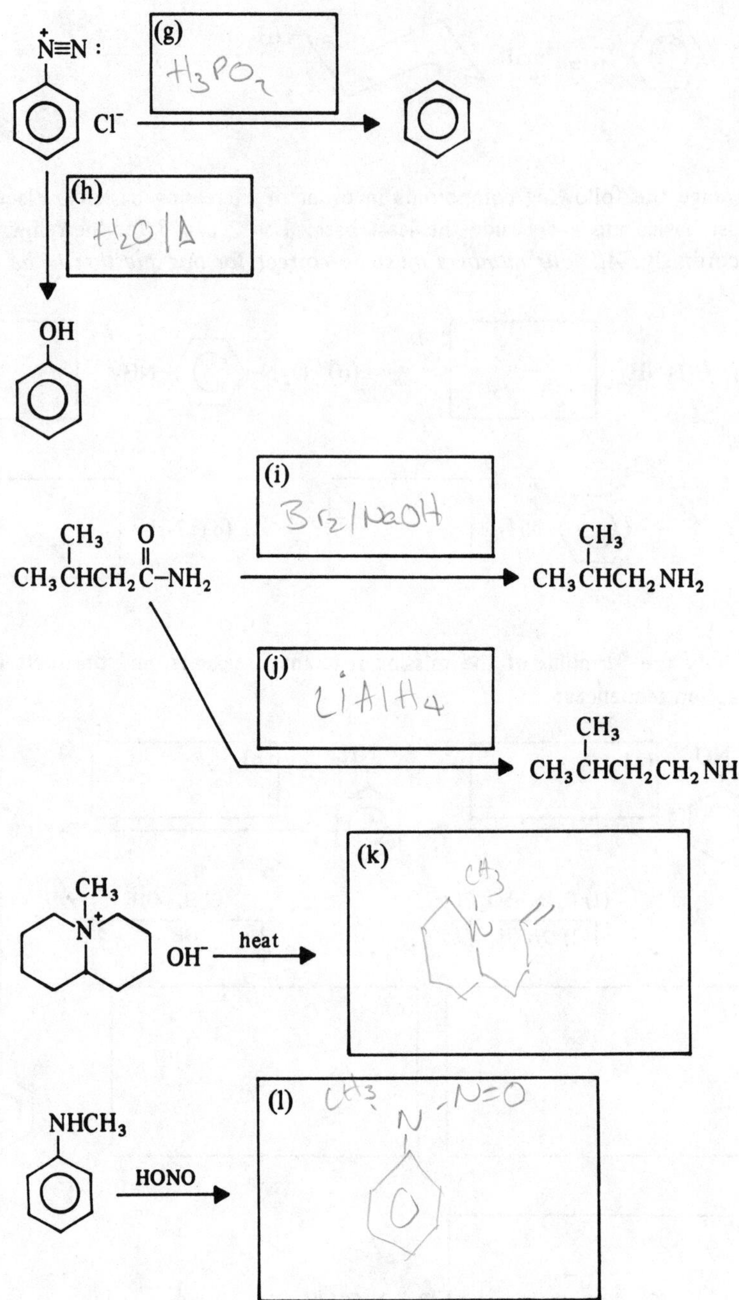

(g)
H3PO2
Cl⁻
(h)
H2O | Δ
OH
(i)
Br2/NaOH
CH3CHCH2C–NH2
CH3CHCH2NH2
(j)
LiAlH4
CH3CHCH2CH2NH2
(k)
heat
OH⁻
(l)
NHCH3
HONO
N–N=O

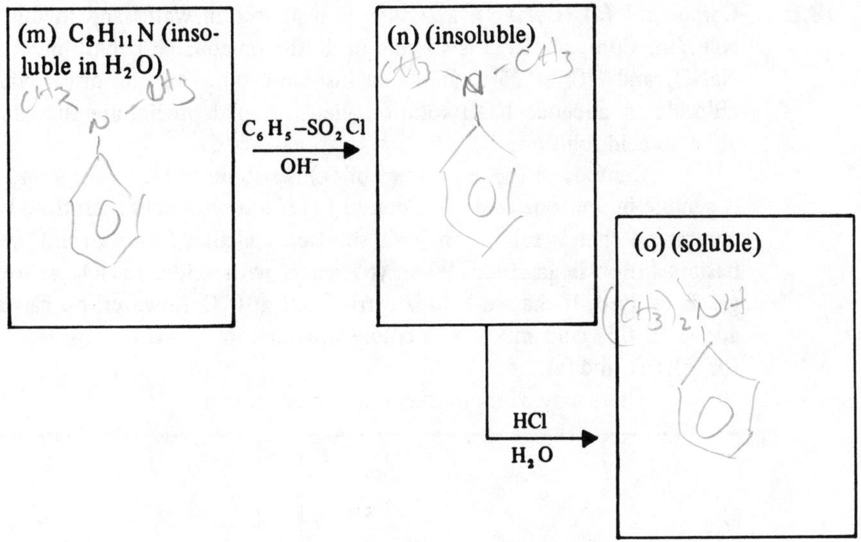

19.4 Write equations for a practical laboratory synthesis of each of the following.

(a) 3-Aminopropanoic acid from succinic anhydride (structure of succinic anhydride)

(b) *m*-Nitrotoluene from *p*-nitrotoluene

19.5 Compound (a) (C_7H_9N), a liquid, is insoluble in water and in dilute aqueous NaOH solution. Compound (a) is soluble in dilute aqueous HCl solution. Reaction of (a) with $NaNO_2$ and HCl at 25°C yields an insoluble oil. Reaction of (a) with benzenesulfonyl chloride in aqueous KOH solution yields a solid precipitate that does not dissolve in aqueous acid solution.

Compound (b), an isomer of (a) is also insoluble in water and aqueous base, and is soluble in aqueous acid. Compound (b) reacts with benzenesulfonyl chloride to yield a compound that is soluble in basic solution and that forms a solid precipitate when the basic solution is acidified. When (b) reacts with acidic $NaNO_2$ at room temperature, a gas is evolved. If the reaction is carried out at 0°C, however, no gas is evolved. If (a) is added to this cold mixture, a colored precipitate forms (c) ($C_{14}H_{15}N_3$). Give formulas for (a), (b), and (c).

There may be more than one correct answer.

(a)

CH_3NH

(b)

NH_2

or o, m

CH_3

(c)

CH_3–(O)–N=N–(O)–$NHCH_3$

19.6 Complete the following reaction sequence by drawing the correct formulas in the blocks provided

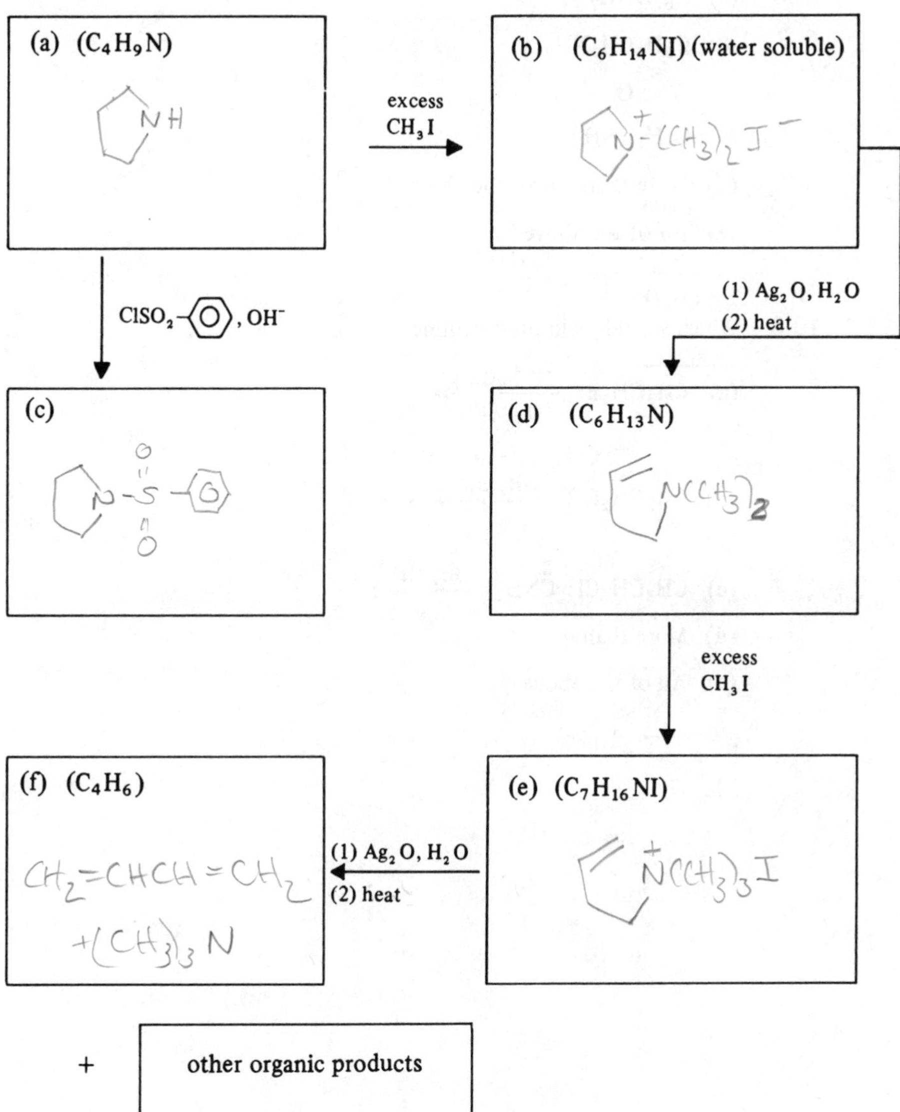

19.7 Which of the following would be soluble in dilute aqueous HCl?

(a) $C_6H_5NH_2$

(b) $C_6H_5CH_2NH_2$

(c) $C_6H_5\overset{O}{\overset{\|}{C}}NH_2$

(d) More than one of the above

(e) All of the above

19.8 Which would yield propylamine?

(a) $CH_3CH_2Br \xrightarrow[(2)\ LiAlH_4]{(1)\ NaCN}$

(b) $CH_3CH_2\overset{O}{\overset{\|}{C}}H \xrightarrow{NH_3, H_2/Ni}$

(c) $CH_3CH_2CH_2\overset{O}{\overset{\|}{C}}NH_2 \xrightarrow{Br_2/OH^-}$

(d) More than one of these

(e) All of the above

K SPECIAL TOPIC
Reactions and Synthesis of Heterocyclic Amines

SOLUTIONS TO PROBLEMS

K.1 (a) 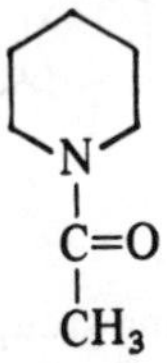 N-acetylpiperidine: piperidine ring with N–C(=O)–CH_3 (b) N-methylpyridinium ring, N^+–CH_3 I^-

(c) benzene ring bearing –C(=O)–N(pyrrolidine) and –C(=O)–OH (ortho) (d) 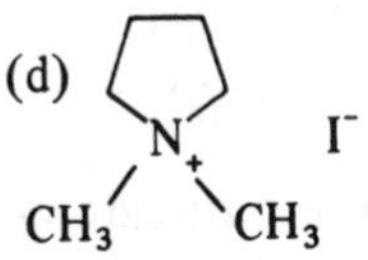 pyrrolidine ring with N^+(CH_3)(CH_3) I^-

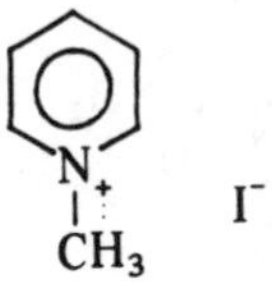

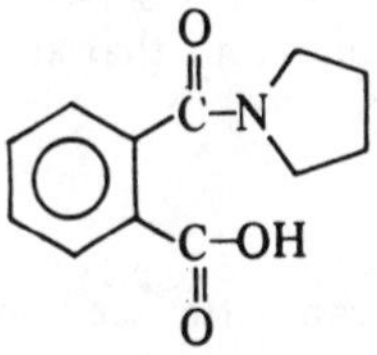

(e) $CH_2{=}CHCH_2CH_2N{-}CH_3$ with CH_3 on N

K.2 (a) The cyclopentadienyl anion.

(b) The pyrrole anion is a resonance hybrid of the following structures:

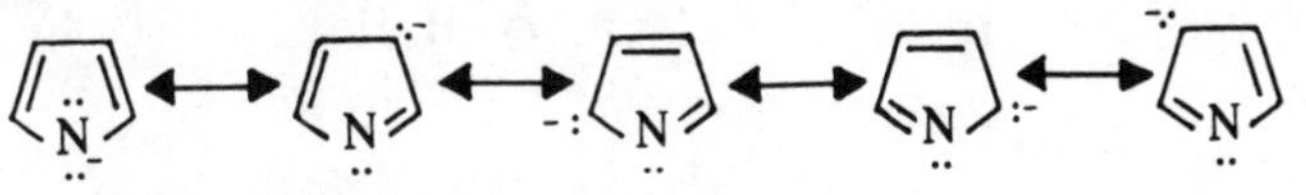

[five pyrrole anion resonance structures linked by ↔]

The imidazole anion is a hybrid of these:

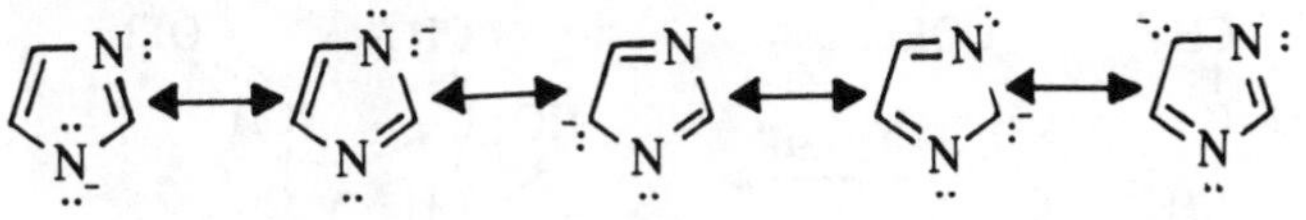

[five imidazole anion resonance structures linked by ↔]

K.3 A mechanism involving a "pyridyne" intermediate would involve a net loss (of 50%) of the deuterium label.

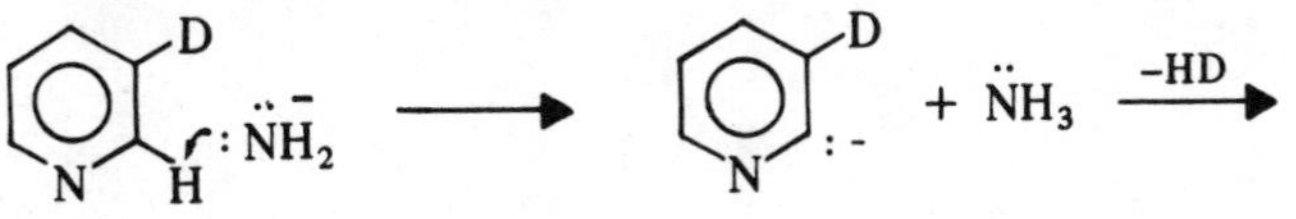

[3-deuteriopyridine + $:\ddot{N}H_2^-$ (removing H at C-2) ⟶ 3-deuterio-2-pyridyl anion + $\ddot{N}H_3$ $\xrightarrow{-HD}$]

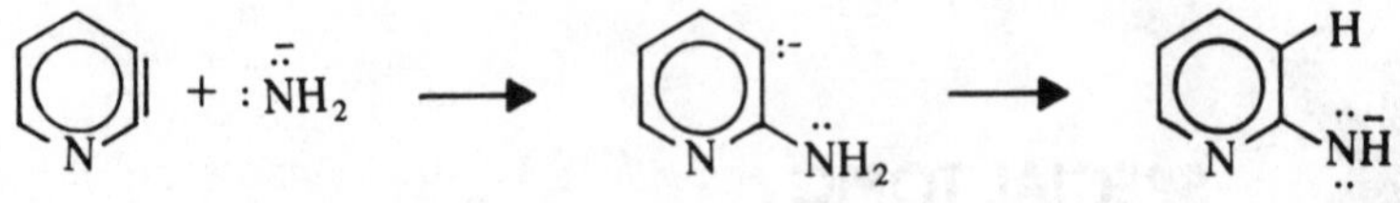

$$\text{pyridine} + :\ddot{N}H_2^- \longrightarrow \text{(carbanion)} \longrightarrow \text{(anion)}$$

2-Pyridyne

Since in the actual experiment there was no loss of deuterium this mechanism was disallowed.

The mechanism given in Section K.4 would not be expected to result in a loss of deuterium, thus it is consistent with the labeling experiment.

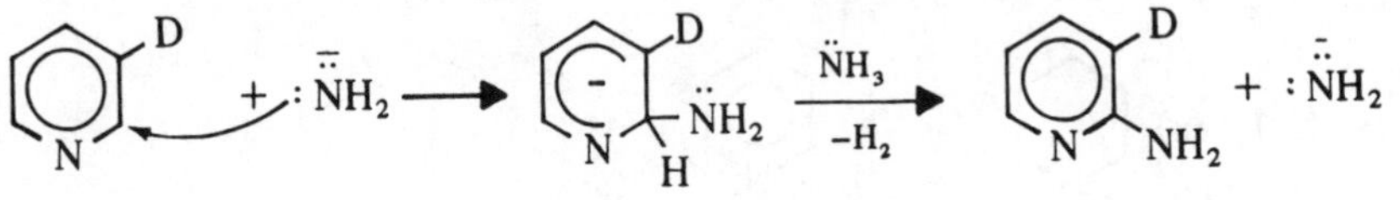

$$\xrightarrow{\ddot{N}H_3,\ -H_2} \quad + :\ddot{N}H_2^-$$

K.4 When pyridine undergoes nucleophilic substitution, the leaving group is a hydride ion–an ion that is a strong base and, consequently, a poor leaving group. With 2-halopyridines, on the other hand, the leaving groups are halide ions–ions that are weak bases and thus good leaving groups.

K.5 If we write the reactants in the following way we can better see how the reaction occurs.

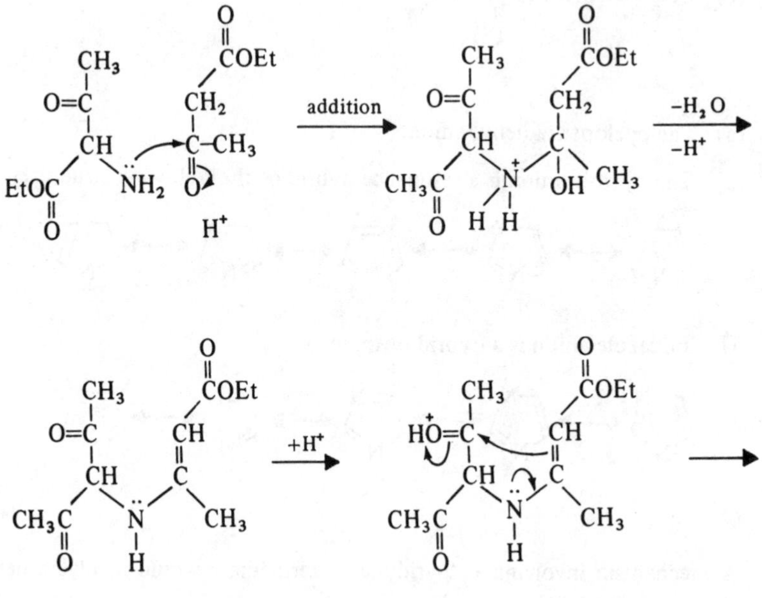

$$\xrightarrow{\text{addition}} \qquad \xrightarrow[-H^+]{-H_2O}$$

$$\xrightarrow{+H^+} \qquad \longrightarrow$$

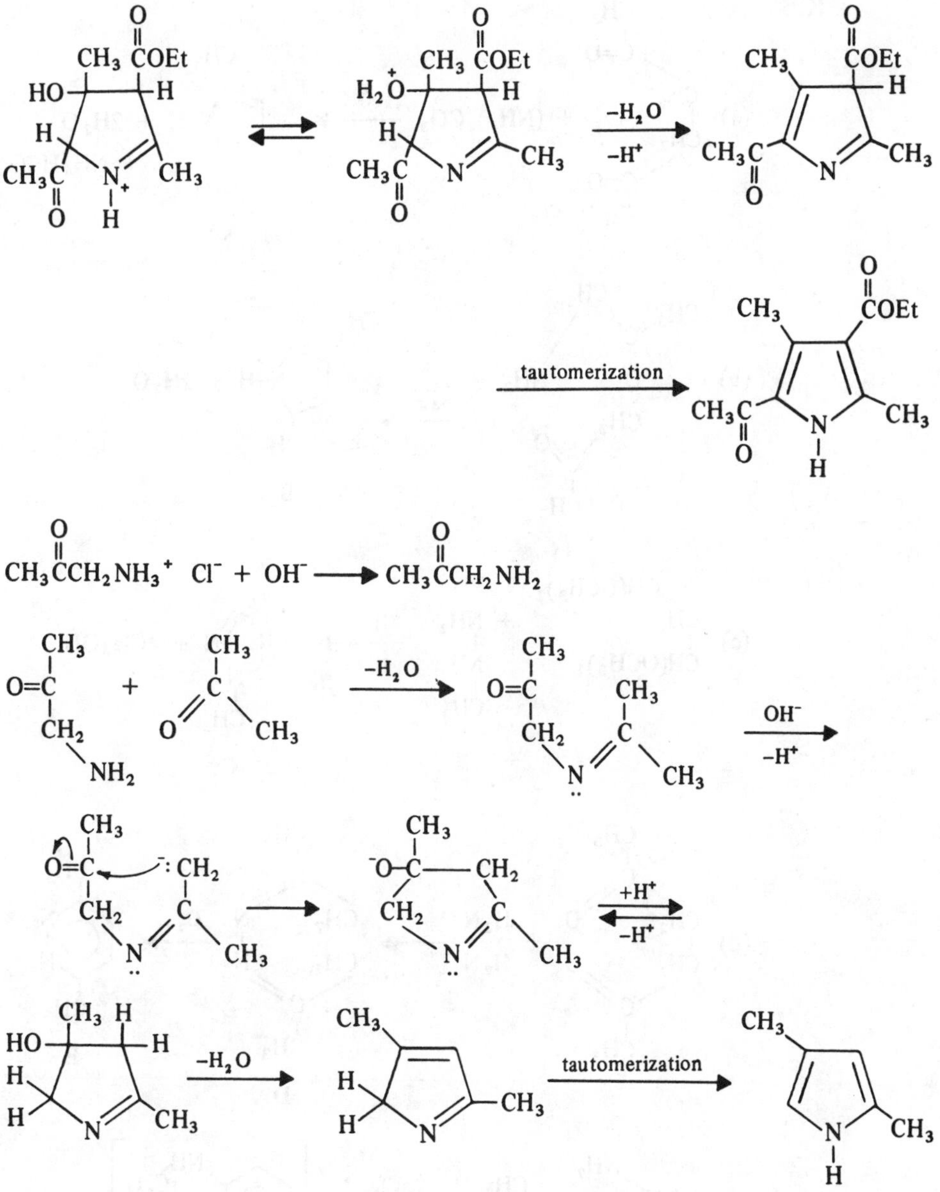

$$CH_3CCH_2NH_3^+\ Cl^- + OH^- \longrightarrow CH_3CCH_2NH_2$$

(with C=O on the carbonyl carbon in each structure)

$-H_2O$, $-H^+$

tautomerization

$-H_2O$

OH^-, $-H^+$

$+H^+$ / $-H^+$

$-H_2O$

tautomerization

K.6

(a) $CH_3COCH_2CH_2COCH_3 + (NH_4)_2CO_3 \xrightarrow{100°C}$ A (2,5-dimethylpyrrole, N–H) $+ 2H_2O + NH_4HCO_3$

(b) $CH_3COCH_2NH_2 + CH_3COCH_3 \xrightarrow{\text{base}}$ B (2,4-dimethylpyrrole, N–H) $+ 2H_2O$

(c) $CH(OCH_3)_2CH_2CH(OCH_3)_2 + NH_2NHCH_3 \xrightarrow[H_2O]{H^+}$ C (1-methylpyrazole, N–CH_3) $+ 4CH_3OH$

(d) $CH_3COCH_2CH_2COCH_3 + H_2N{-}NH_2 \longrightarrow$ D (3,6-dimethyl-4,5-dihydropyridazine) $\xrightarrow{O_2}$ E (3,6-dimethylpyridazine)

(e) $C_6H_5NH_2 + CH_2{=}CH{-}COCH_3 \xrightarrow[FeCl_3]{ZnCl_2}$ [$C_6H_5NHCH_2CH_2COCH_3$] $\longrightarrow$ F (4-methylquinoline) $+ H_2O$

(f) [3-pyridyl]-CH(I)-CH$_2$-CH$_2$-CH$_2$-NH-CH$_3$ $\xrightarrow{\text{heat}}$ [N-methylpyrrolidinium, N^+(H)(CH$_3$)] I^- $\xrightarrow{OH^-}$

[3-pyridyl-N-methylpyrrolidine] $\xrightarrow[(2)\ H^+]{(1)\ KMnO_4,\ OH^-}$ [3-pyridyl-CO_2H]

G
Nicotine

H
Nicotinic acid

20 SYNTHESIS AND REACTIONS OF β-DICARBONYL COMPOUNDS: MORE CHEMISTRY OF ENOLATE IONS

SUMMARY OF ACETOACETIC ESTER AND MALONIC ESTER SYNTHESES

A. Acetoacetic Ester Synthesis

$$CH_3\overset{O}{\overset{\|}{C}}CH_2\overset{O}{\overset{\|}{C}}OEt \xrightarrow[(2)\ RX]{(1)\ NaOEt} CH_3\overset{O}{\overset{\|}{C}}\underset{R}{\underset{|}{C}H}\overset{O}{\overset{\|}{C}}OEt \xrightarrow[(2)\ R'X]{(1)\ (CH_3)_3COK}$$

$$CH_3\overset{O}{\overset{\|}{C}}-\underset{R}{\underset{|}{\overset{R'}{\overset{|}{C}}}}-\overset{O}{\overset{\|}{C}}OEt \xrightarrow[(2)\ H_3O^+]{(1)\ OH^-/H_2O} CH_3\overset{O}{\overset{\|}{C}}-\underset{R}{\underset{|}{\overset{R'}{\overset{|}{C}}}}-\overset{O}{\overset{\|}{C}}OH \xrightarrow[-CO_2]{heat} CH_3\overset{O}{\overset{\|}{C}}\underset{R}{\underset{|}{C}}HR'$$

B. Malonic Ester Synthesis

$$EtO\overset{O}{\overset{\|}{C}}CH_2\overset{O}{\overset{\|}{C}}OEt \xrightarrow[(2)\ RX]{(1)\ NaOEt} EtO\overset{O}{\overset{\|}{C}}\underset{R}{\underset{|}{C}H}\overset{O}{\overset{\|}{C}}OEt \xrightarrow[(2)\ R'X]{(1)\ (CH_3)_3COK}$$

$$EtO\overset{O}{\overset{\|}{C}}-\underset{R}{\underset{|}{\overset{R'}{\overset{|}{C}}}}-\overset{O}{\overset{\|}{C}}OEt \xrightarrow[(2)\ H_3O^+]{(1)\ OH^-/H_2O} HO\overset{O}{\overset{\|}{C}}-\underset{R}{\underset{|}{\overset{R'}{\overset{|}{C}}}}-\overset{O}{\overset{\|}{C}}OH \xrightarrow[-CO_2]{heat} HO\overset{O}{\overset{\|}{C}}\underset{R}{\underset{|}{C}}HR'$$

SOLUTIONS TO PROBLEMS

20.1

(a) Step 1

$$CH_3\underset{H}{\underset{|}{C}H}-\overset{O}{\overset{\|}{C}}OC_2H_5 \quad {}^-OC_2H_5 \rightleftharpoons CH_3\underset{..}{\overset{-}{C}}H-\overset{O}{\overset{\|}{C}}OC_2H_5 + C_2H_5OH$$

$$\updownarrow$$

$$CH_3CH=\overset{O^-}{\overset{|}{C}}OC_2H_5$$

Step 2 $CH_3CH_2C(=O)OC_2H_5 + {:}\overline{C}HCOC_2H_5(CH_3) \rightleftharpoons CH_3CH_2C(O^-)(OC_2H_5)-CH(CH_3)-COC_2H_5$

$\rightleftharpoons C_2H_5O^- + CH_3CH_2\overset{O}{\overset{\|}{C}}-CH(CH_3)-\overset{O}{\overset{\|}{C}}OC_2H_5$

Step 3 $CH_3CH_2\overset{O}{\overset{\|}{C}}-\overset{H}{\overset{|}{C}}(CH_3)-\overset{O}{\overset{\|}{C}}OC_2H_5 + {}^-OC_2H_5 \rightleftharpoons [CH_3CH_2\overset{O}{C}\cdots C(CH_3)\cdots\overset{O}{C}OC_2H_5]^- + C_2H_5OH$

(b) $CH_3CH_2\overset{O}{\overset{\|}{C}}\underset{CH_3}{\underset{|}{C}H}\overset{O}{\overset{\|}{C}}OC_2H_5 + CH_3CH_2\overset{OH}{\overset{|}{C}}=\underset{CH_3}{\underset{|}{C}}\overset{O}{\overset{\|}{C}}OC_2H_5$

20.2

(a) $C_2H_5O\overset{O}{\overset{\|}{C}}CH_2CH_2CH_2CH_2\overset{O}{\overset{\|}{C}}OC_2H_5 \underset{+H^+}{\overset{-H^+}{\rightleftharpoons}}$ [C_2H_5O–C(=O)–CH_2–CH_2–CH_2–${:}\overline{C}HCO_2C_2H_5$, intramolecular attack]

$\rightleftharpoons$ [cyclopentane with $C(O^-)(OC_2H_5)$ adjacent to $CH(COC_2H_5)$] $\rightleftharpoons$ [2-oxocyclopentane with H and $\overset{O}{\overset{\|}{C}}OC_2H_5$] $+ C_2H_5O^-$

$\rightleftharpoons$ [enolate of 2-(ethoxycarbonyl)cyclopentanone]$^-$ $+ C_2H_5OH$ $\xrightarrow{H^+}$ [2-oxocyclopentane with $\overset{O}{\overset{\|}{C}}OC_2H_5$] + enol form

(b) [2-oxocyclohexane with $\overset{O}{\overset{\|}{C}}OC_2H_5$]

(c) To undergo a Dieckmann condensation, diethyl glutarate would have to form a highly strained four-membered ring.

20.3

$$CH_3\overset{O}{\overset{\|}{C}}OC_2H_5 + C_2H_5O^- \rightleftharpoons {}^-:CH_2\overset{O}{\overset{\|}{C}}OC_2H_5 + C_2H_5OH$$

$$C_6H_5\overset{O}{\overset{\|}{C}}OC_2H_5 + {}^-:CH_2\overset{O}{\overset{\|}{C}}OC_2H_5 \rightleftharpoons C_6H_5\underset{OC_2H_5}{\underset{|}{\overset{O^-}{\overset{|}{C}}}}-CH_2\overset{O}{\overset{\|}{C}}OC_2H_5$$

$$\rightleftharpoons C_6H_5\overset{O}{\overset{\|}{C}}CH_2\overset{O}{\overset{\|}{C}}OC_2H_5 + C_2H_5O^- \rightleftharpoons \overbrace{C_6H_5\overset{O}{\overset{\|}{C}}\text{---}CH\text{---}\overset{O}{\overset{\|}{C}}OC_2H_5}^{-} + C_2H_5OH$$

$$\xrightarrow{H^+} C_6H_5\overset{O}{\overset{\|}{C}}CH_2\overset{O}{\overset{\|}{C}}OC_2H_5$$

$$C_6H_5CH_2\overset{O}{\overset{\|}{C}}OC_2H_5 + C_2H_5O^- \rightleftharpoons C_6H_5\overset{..}{\overset{-}{C}}H\overset{O}{\overset{\|}{C}}OC_2H_5 + C_2H_5OH$$

$$C_6H_5\overset{..}{\overset{-}{C}}H\overset{O}{\overset{\|}{C}}OC_2H_5 + C_2H_5O\overset{O}{\overset{\|}{C}}OC_2H_5 \rightleftharpoons C_6H_5\underset{\underset{\underset{O}{\|}}{COC_2H_5}}{\underset{|}{CH}}-\overset{O^-}{\overset{|}{C}}(OC_2H_5)_2$$

$$\rightleftharpoons C_6H_5\underset{\underset{\underset{O}{\|}}{COC_2H_5}}{\underset{|}{\overset{\overset{\overset{O}{\|}}{COC_2H_5}}{\overset{|}{CH}}}} + C_2H_5O^- \rightleftharpoons C_6H_5\underset{\underset{\underset{O}{\|}}{COC_2H_5}}{\underset{|}{\overset{\overset{\overset{O}{\|}}{COC_2H_5}}{\overset{|}{C}}}}:^- + C_2H_5OH$$

Resonance stabilized

$$\xrightarrow{H^+} C_6H_5\underset{\underset{\underset{O}{\|}}{COC_2H_5}}{\underset{|}{\overset{\overset{\overset{O}{\|}}{COC_2H_5}}{\overset{|}{CH}}}}$$

20.4

(a) $$CH_3CH_2\overset{O}{\overset{\|}{C}}OC_2H_5 + C_2H_5O\overset{O}{\overset{\|}{C}}-\overset{O}{\overset{\|}{C}}OC_2H_5 \xrightarrow[(2)\ H^+]{(1)\ NaOCH_2CH_3} CH_3\underset{\underset{\overset{\|\ \ \|}{O\ \ O}}{C-COC_2H_5}}{\underset{|}{CH}}\overset{O}{\overset{\|}{C}}OC_2H_5$$

(b) $CH_3\overset{O}{\overset{\|}{C}}OC_2H_5 + H\overset{O}{\overset{\|}{C}}OC_2H_5 \xrightarrow[(2)\ H^+]{(1)\ NaOCH_2CH_3} H\overset{O}{\overset{\|}{C}}CH_2\overset{O}{\overset{\|}{C}}OC_2H_5$

20.5

(a) cyclohexanone $+ H\overset{O}{\overset{\|}{C}}OC_2H_5 \xrightarrow[(2)\ H^+]{(1)\ NaOC_2H_5}$ 2-formylcyclohexanone ($\overset{O}{\overset{\|}{C}}H$)

(b) $CH_3CH_2\overset{O}{\overset{\|}{C}}CH_2CH_2CH_2\overset{O}{\overset{\|}{C}}OC_2H_5 \xrightarrow[(2)\ H^+]{(1)\ NaOC_2H_5}$ 2-methylcyclohexane-1,3-dione (CH_3)

(c) $C_2H_5O_2CCH_2C(CH_3)_2CH_2CO_2C_2H_5 + C_2H_5O\overset{O}{\overset{\|}{C}}-\overset{O}{\overset{\|}{C}}OC_2H_5 \xrightarrow[(2)\ H^+]{(1)\ NaOC_2H_5}$

$C_2H_5O-\overset{O}{\overset{\|}{C}}-\overset{O}{\overset{\|}{C}}-CH(CO_2C_2H_5)-C(CH_3)_2-CH_2CO_2C_2H_5 \xrightarrow[(2)\ H^+]{(1)\ NaOC_2H_5}$ 1,2-dioxocyclopentane ring bearing two $CO_2C_2H_5$ groups and $C(CH_3)_2$

20.6

$CH_3\overset{O}{\overset{\|}{C}}CH_2CH_2CH_2CH_2\overset{O}{\overset{\|}{C}}OC_2H_5 + {}^-OC_2H_5 \underset{-C_2H_5OH}{\rightleftharpoons} C_2H_5O\overset{O}{\overset{\|}{C}}-CH_2-CH_2-CH_2-\overset{-}{:}CH-\overset{O}{\overset{\|}{C}}-CH_3$

$\rightleftharpoons$ (cyclic tetrahedral intermediate: C_2H_5O, O^-, H, $\overset{O}{\overset{\|}{C}}-CH_3$) $\rightleftharpoons$ 2-acetylcyclopentanone ($\overset{O}{\overset{\|}{C}}-CH_3$, H) $+ C_2H_5O^-$

$\rightleftharpoons$ (resonance-stabilized enolate anion, CH_3) $+ C_2H_5OH \xrightarrow{H^+}$ 2-acetylcyclopentanone ($\overset{O}{\overset{\|}{C}}CH_3$)

20.7 The partially negative oxygen atom of sodioacetoacetic ester acts as the nucleophile.

$$CH_3\overset{O}{\overset{\|}{C}}-\ddot{C}H^{-}-\overset{O}{\overset{\|}{C}}-OC_2H_5 \longleftrightarrow CH_3\overset{O^-}{\overset{|}{C}}=CH-\overset{O}{\overset{\|}{C}}-OC_2H_5$$

20.8 Again, working backward,

(a)
$$CH_3\overset{O}{\overset{\|}{C}}CH_2CH_2CH_3 \xleftarrow[-CO_2]{\text{heat}} CH_3\overset{O}{\overset{\|}{C}}\underset{\underset{CH_3}{|}}{\underset{CH_2}{\underset{|}{C}H}}-\overset{O}{\overset{\|}{C}}OH \xleftarrow[(2)\ H_3O^+]{(1)\ \text{dil. NaOH, heat}}$$

$$CH_3\overset{O}{\overset{\|}{C}}\underset{\underset{CH_3}{|}}{\underset{CH_2}{\underset{|}{C}H}}\overset{O}{\overset{\|}{C}}OC_2H_5 \xleftarrow[(2)\ CH_3CH_2Br]{(1)\ NaOC_2H_5} CH_3\overset{O}{\overset{\|}{C}}CH_2\overset{O}{\overset{\|}{C}}OC_2H_5$$

(b)
$$CH_3\overset{O}{\overset{\|}{C}}\underset{\underset{\underset{CH_3}{|}}{CH_2}}{\underset{|}{\underset{CH_2}{\underset{|}{C}H}}}CH_2CH_2CH_3 \xleftarrow[-CO_2]{\text{heat}} CH_3\overset{O}{\overset{\|}{C}}-\overset{CH_3CH_2CH_2}{\underset{CH_2CH_2CH_3}{C}}-CO_2H \xleftarrow[(2)\ H_3O^+]{(1)\ \text{dil. NaOH, heat}}$$

$$CH_3\overset{O}{\overset{\|}{C}}-\overset{CH_3CH_2CH_2}{\underset{CH_2CH_2CH_3}{C}}-CO_2C_2H_5 \xleftarrow[(2)\ CH_3CH_2CH_2Br]{(1)\ (CH_3)_3COK} CH_3\overset{O}{\overset{\|}{C}}-\underset{CH_2CH_2CH_3}{CH}\overset{O}{\overset{\|}{C}}OC_2H_5$$

$$\xleftarrow[(2)\ CH_3CH_2CH_2Br]{(1)\ NaOC_2H_5} CH_3\overset{O}{\overset{\|}{C}}CH_2\overset{O}{\overset{\|}{C}}OC_2H_5$$

(c) $CH_3\overset{O}{\overset{\|}{C}}CH_2CH_2C_6H_5 \xleftarrow[-CO_2]{\text{heat}} CH_3\overset{O}{\overset{\|}{C}}\underset{\underset{C_6H_5}{|}}{\underset{CH_2}{|}}{CH}\overset{O}{\overset{\|}{C}}OH \xleftarrow[(2)\ H_3O^+]{(1)\ NaOH,\ heat} CH_3\overset{O}{\overset{\|}{C}}\underset{\underset{C_6H_5}{|}}{\underset{CH_2}{|}}{CH}\overset{O}{\overset{\|}{C}}OC_2H_5$

$\xleftarrow[(2)\ C_6H_5CH_2Br]{(1)\ NaOC_2H_5} CH_3\overset{O}{\overset{\|}{C}}CH_2\overset{O}{\overset{\|}{C}}OC_2H_5$

20.9 (a) Reactivity is the same as with any second order reaction. With primary halides substitution is highly favored, with secondary halides elimination competes with substitution, and with tertiary halides elimination is the exclusive course of reaction.

(b) Acetoacetic ester and 2-methylpropene.

(c) Bromobenzene is unreactive toward nucleophilic substitution (cf. Section 14.11 of the text).

20.10

$CH_3CH_2CH_2\overset{O}{\overset{\|}{C}}OC_2H_5 \xrightarrow[(2)\ H^+]{(1)\ NaOC_2H_5} CH_3CH_2CH_2\overset{O}{\overset{\|}{C}}\underset{\underset{CH_3}{|}}{\underset{CH_2}{|}}{CH}\overset{O}{\overset{\|}{C}}OC_2H_5 \xrightarrow[(2)\ H_3O^+]{(1)\ NaOH,\ H_2O,\ heat}$

$CH_3CH_2CH_2\overset{O}{\overset{\|}{C}}\underset{\underset{CH_3}{|}}{\underset{CH_2}{|}}{CH}\overset{O}{\overset{\|}{C}}OH \xrightarrow[-CO_2]{\text{heat}} CH_3CH_2CH_2\overset{O}{\overset{\|}{C}}CH_2CH_2CH_3$

20.11 The carboxyl group that is lost most readily is the one that is β to the keto group (cf. Section 18.11 of the text).

20.12

$CH_3\overset{O}{\overset{\|}{C}}CH_2CH_2\overset{O}{\overset{\|}{C}}C_6H_5 \xleftarrow[-CO_2]{\text{heat}} CH_3\overset{O}{\overset{\|}{C}}\underset{\underset{\underset{C_6H_5}{|}}{\underset{C=O}{|}}}{\underset{CH_2}{|}}{CH}\overset{O}{\overset{\|}{C}}OH \xleftarrow[(2)\ H_3O^+]{(1)\ OH^-,\ H_2O,\ heat}$

$CH_3\overset{O}{\overset{\|}{C}}\underset{\underset{\underset{C_6H_5}{|}}{\underset{C=O}{|}}}{\underset{CH_2}{|}}{CH}\overset{O}{\overset{\|}{C}}OC_2H_5 \xleftarrow[(2)\ C_6H_5COCH_2Br]{(1)\ NaOC_2H_5} CH_3\overset{O}{\overset{\|}{C}}CH_2\overset{O}{\overset{\|}{C}}OC_2H_5$

20.13

$$CH_3\overset{O}{\overset{\|}{C}}CH_2\overset{O}{\overset{\|}{C}}C_6H_5 \xleftarrow[-CO_2]{\text{heat}} CH_3\overset{O}{\overset{\|}{C}}\underset{\underset{C_6H_5}{|}}{\underset{C=O}{\underset{|}{C}}}H\overset{O}{\overset{\|}{C}}OH \xleftarrow[(2)\ H_3O^+]{(1)\ OH^-,\ H_2O,\ \text{heat}}$$

$$CH_3\overset{O}{\overset{\|}{C}}\underset{\underset{C_6H_5}{|}}{\underset{C=O}{\underset{|}{C}}}H\overset{O}{\overset{\|}{C}}OC_2H_5 \xleftarrow[(2)\ C_6H_5COCl]{(1)\ NaH} CH_3\overset{O}{\overset{\|}{C}}CH_2\overset{O}{\overset{\|}{C}}OC_2H_5$$

20.14 (a) One molar equivalent of $NaNH_2$ converts acetoacetic ester to its anion,

$$CH_3\overset{O}{\overset{\|}{C}}CH_2\overset{O}{\overset{\|}{C}}OEt + NH_2^- \longrightarrow CH_3\overset{O}{\overset{\|}{C}}-\overset{..-}{C}H-\overset{O}{\overset{\|}{C}}OEt + NH_3$$

and one molar equivalent of $NaNH_2$ converts bromobenzene to benzyne (cf. Section 14.11B):

$$NH_2^- + C_6H_5Br \longrightarrow \text{benzyne} + NH_3 + Br^-$$

Then the anion of acetoacetic ester adds to the benzyne as it forms in the mixture.

$$\text{benzyne} + {}^-\!:CH(COEt)(CCH_3) \longrightarrow C_6H_4^{-}\text{-}CH(\overset{O}{\overset{\|}{C}}-OEt)(\overset{O}{\overset{\|}{C}}CH_3) \longrightarrow$$

$$C_6H_5\overset{..}{C}(\overset{O}{\overset{\|}{C}}OEt)(\overset{O}{\overset{\|}{C}}CH_3)\ {}^-Na^+ \xrightarrow[NH_4Cl]{\text{then}} C_6H_5CH(\overset{O}{\overset{\|}{C}}OEt)(\overset{O}{\overset{\|}{C}}CH_3)$$

This is the end product of the addition

(b) 1-phenyl-2-propanone, as follows:

$$C_6H_5CH(\overset{O}{\overset{\|}{C}}OEt)(\overset{O}{\overset{\|}{C}}CH_3) \xrightarrow[(2)\ H_3O^+]{(1)\ OH^-,\ H_2O,\ \text{heat}} C_6H_5CH(\overset{O}{\overset{\|}{C}}OH)(\overset{O}{\overset{\|}{C}}CH_3)$$

$$\xrightarrow[-CO_2]{\text{heat}} C_6H_5CH_2\overset{O}{\overset{\|}{C}}CH_3$$

(c) By treating bromobenzene with diethyl malonate and two molar equivalents of $NaNH_2$ to form diethyl phenylmalonate.

$$C_6H_5Br + CH_2(COEt)_2 \xrightarrow{2NaNH_2} EtO\overset{O}{\overset{\|}{C}}-\underset{C_6H_5}{CH}-\overset{O}{\overset{\|}{C}}OEt$$

[The mechanism for this reaction is analogous to that given in part (a).]

Then hydrolyis and decarboxylation will convert diethyl phenylmalonate to phenylacetic acid

$$EtO\overset{O}{\overset{\|}{C}}-\underset{C_6H_5}{CH}-\overset{O}{\overset{\|}{C}}OEt \xrightarrow[(2)\ H_3O^+]{(1)\ OH^-,\ H_2O,\ heat} HO\overset{O}{\overset{\|}{C}}-\underset{C_6H_5}{CH}-\overset{O}{\overset{\|}{C}}OH \xrightarrow{\text{heat}} C_6H_5-CH_2CO_2H + CO_2$$

20.15 Here we alkylate the dianion,

$$CH_3-\overset{O}{\overset{\|}{C}}-CH_2-\overset{O}{\overset{\|}{C}}OC_2H_5 \xrightarrow[\text{liq. }NH_3]{2KNH_2} \bar{:}CH_2-\overset{O}{\overset{\|}{C}}-\bar{\ddot{C}}H-\overset{O}{\overset{\|}{C}}OC_2H_5$$

$$\xrightarrow[(2)\ NH_4Cl]{(1)\ C_6H_5CH_2Cl} C_6H_5CH_2CH_2\overset{O}{\overset{\|}{C}}CH_2\overset{O}{\overset{\|}{C}}OC_2H_5$$

20.16 Working backward,

(a) $$CH_3CH_2CH_2CH_2CO_2H \xleftarrow[-CO_2]{\text{heat}} CH_3CH_2CH_2CH(CO_2H)_2 \xleftarrow[(2)\ H_3O^+]{(1)\ OH^-,\ H_2O,\ heat}$$

$$CH_3CH_2CH_2CH(CO_2C_2H_5)_2 \xleftarrow[CH_3CH_2CH_2Br]{NaOC_2H_5} CH_2(CO_2C_2H_5)_2$$

(b) $CH_3CH_2CH_2\underset{\displaystyle CH_3}{\underset{|}{C}}HCO_2H \xleftarrow[-CO_2]{heat} (CH_3CH_2CH_2)(CH_3)C(CO_2H)_2 \xleftarrow[(2)\ H_3O^+]{(1)\ OH^-,\ H_2O,\ heat}$

$(CH_3CH_2CH_2)(CH_3)C(CO_2C_2H_5)_2 \xleftarrow[(CH_3)_3COK]{CH_3I} CH_3CH_2CH_2CH(CO_2C_2H_5)_2$

$\xleftarrow[NaOC_2H_5]{CH_3CH_2CH_2Br} CH_2(CO_2C_2H_5)_2$

(c) $CH_3\underset{\displaystyle CH_3}{\underset{|}{C}}HCH_2CH_2CO_2H \xleftarrow[-CO_2]{heat} CH_3\underset{\displaystyle CH_3}{\underset{|}{C}}HCH_2CH(CO_2H)_2 \xleftarrow[(2)\ H_3O^+]{(1)\ OH^-,\ H_2O,\ heat}$

$CH_3\underset{\displaystyle CH_3}{\underset{|}{C}}HCH_2CH(CO_2C_2H_5)_2 \xleftarrow[CH_3\underset{CH_3}{CH}CH_2Br]{NaOC_2H_5} CH_2(CO_2C_2H_5)_2$

20.17

(a) Formaldehyde, $H-\overset{O}{\overset{\|}{C}}-H$

(b) 1,3-dithiane (S, S; H, H) $\xrightarrow[-C_4H_{10}]{C_4H_9Li}$ 2-lithio-1,3-dithiane (S, S; H, Li^+) $\xrightarrow[-LiBr]{C_6H_5CH_2Br}$ 2-benzyl-1,3-dithiane (S, S; H, $CH_2C_6H_5$)

$\xrightarrow[-HSCH_2CH_2CH_2SH]{HgCl_2,\ CH_3OH,\ H_2O} C_6H_5CH_2\overset{O}{\overset{\|}{C}}H$

(c) $C_6H_5\overset{O}{\overset{\|}{C}}H + HSCH_2CH_2CH_2SH \xrightarrow{H^+}$ 1,3-dithiane (S, S; C_6H_5, H) $\xrightarrow[(2)\ CH_3I]{(1)\ C_4H_9Li}$

1,3-dithiane (S, S; C_6H_5, CH_3) $\xrightarrow{HgCl_2,\ CH_3OH,\ H_2O} C_6H_5\overset{O}{\overset{\|}{C}}CH_3 + HSCH_2CH_2CH_2SH$

20.18 By treating the thioketal with Raney nickel.

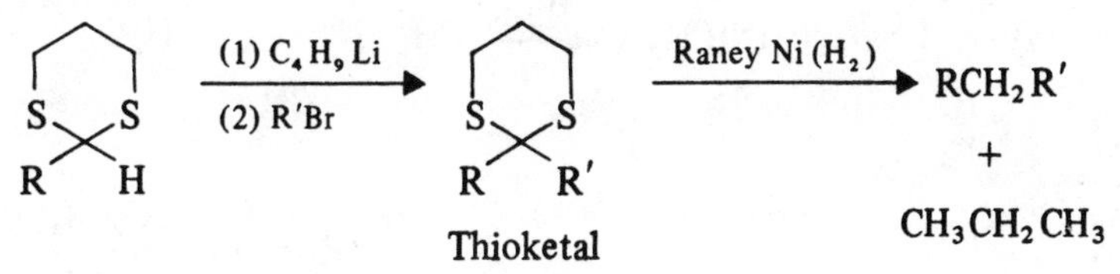

$$\text{(1,3-dithiane, R, H)} \xrightarrow[\text{(2) R'Br}]{\text{(1) } C_4H_9Li} \underset{\text{Thioketal}}{\text{(1,3-dithiane, R, R')}} \xrightarrow{\text{Raney Ni } (H_2)} RCH_2R' + CH_3CH_2CH_3$$

20.19

(a)

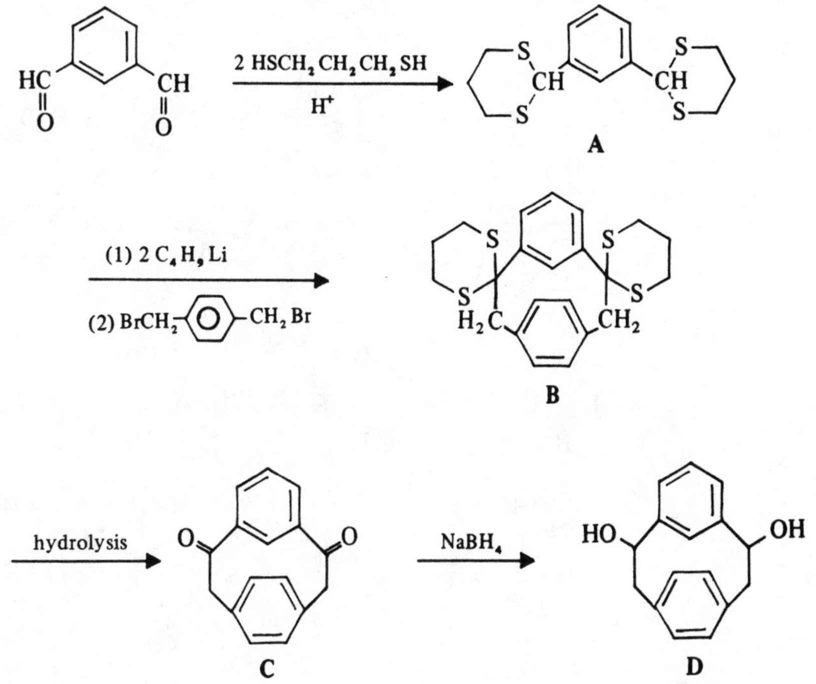

$$\text{HC(=O)–}C_6H_4\text{–CH(=O)} \xrightarrow[H^+]{2\ HSCH_2CH_2CH_2SH} \mathbf{A}$$

$$\xrightarrow[\text{(2) } BrCH_2\text{–}C_6H_4\text{–}CH_2Br]{\text{(1) } 2\ C_4H_9Li} \mathbf{B}$$

$$\xrightarrow{\text{hydrolysis}} \mathbf{C} \xrightarrow{NaBH_4} \mathbf{D}$$

(b)

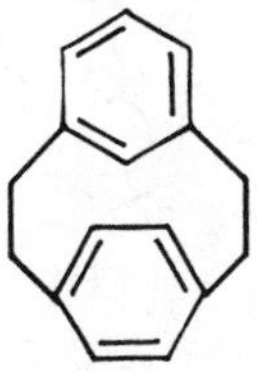

20.20

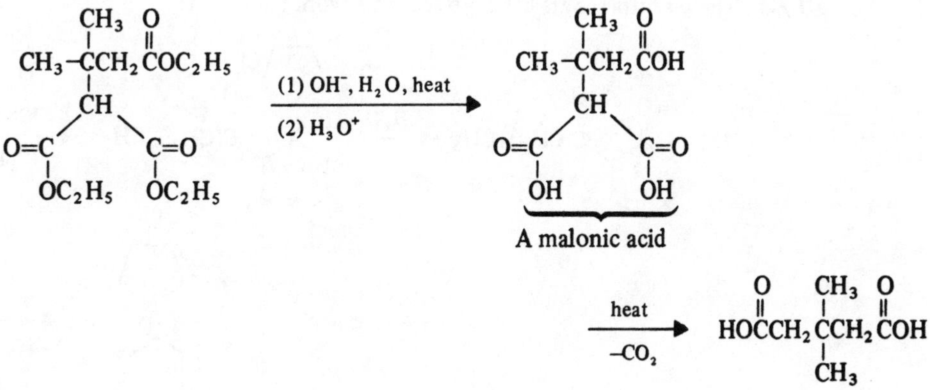

$$\begin{array}{c} CH_3\ \ O \\ CH_3\text{–}\overset{|}{C}CH_2\overset{\|}{C}OC_2H_5 \\ CH \\ O{=}C \quad C{=}O \\ OC_2H_5 \quad OC_2H_5 \end{array} \xrightarrow[\text{(2) } H_3O^+]{\text{(1) } OH^-,\ H_2O,\ \text{heat}} \underset{\text{A malonic acid}}{\begin{array}{c} CH_3\ \ O \\ CH_3\text{–}\overset{|}{C}CH_2\overset{\|}{C}OH \\ CH \\ O{=}C \quad C{=}O \\ OH \quad OH \end{array}}$$

$$\xrightarrow[-CO_2]{\text{heat}} \begin{array}{c} O \quad CH_3 \quad O \\ HO\overset{\|}{C}CH_2\overset{|}{C}CH_2\overset{\|}{C}OH \\ CH_3 \end{array}$$

20.21

(a) $H_2C{=}O + HN(CH_3)_2 \underset{}{\overset{H^+}{\rightleftharpoons}} CH_2{=}\overset{+}{N}(CH_3)_2 + H_2O$

Cyclohexanone ⇌ enol (O–H) + $CH_2{=}\overset{+}{N}(CH_3)_2$ $\xrightarrow[-H^+]{}$ 2-($CH_2N(CH_3)_2$)cyclohexanone

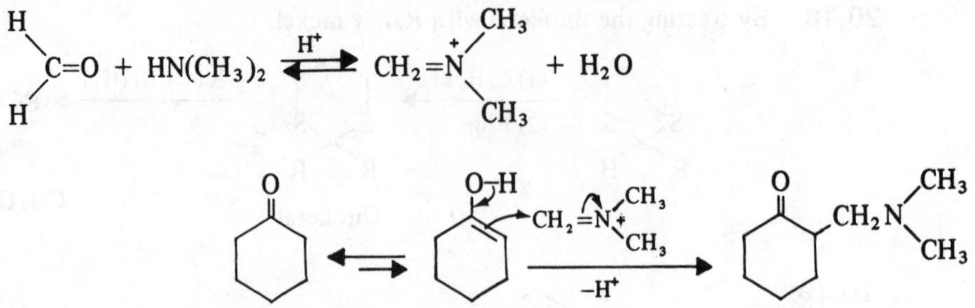

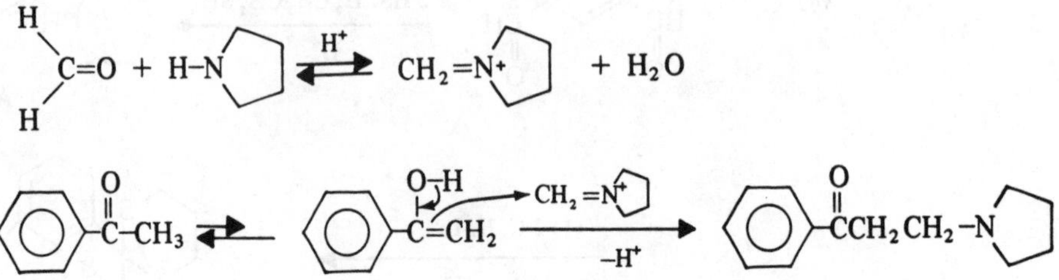

(b) $H_2C{=}O + H{-}N$(pyrrolidine) $\overset{H^+}{\rightleftharpoons}$ $CH_2{=}N^+$(pyrrolidinium) $+ H_2O$

$C_6H_5\overset{O}{\overset{\|}{C}}{-}CH_3 \rightleftharpoons C_6H_5C(OH){=}CH_2$ + $CH_2{=}N^+$(pyrrolidinium) $\xrightarrow[-H^+]{}$ $C_6H_5\overset{O}{\overset{\|}{C}}CH_2CH_2{-}N$(pyrrolidine)

(c) $H_2C{=}O + HN(CH_3)_2 \overset{H^+}{\rightleftharpoons} CH_2{=}\overset{+}{N}(CH_3)_2 + H_2O$

p-cresol (:Ö–H, CH_3) + $CH_2{=}\overset{+}{N}(CH_3)_2$ ⟶ intermediate ($\overset{+}{O}$–H, H, $CH_2N(CH_3)_2$, CH_3) $\xrightarrow{-H^+}$

2-($CH_2N(CH_3)_2$)-4-methylphenol (OH, CH_3) $\xrightarrow[\text{of similar steps}]{\text{repetition}}$ 2,6-bis($CH_2N(CH_3)_2$)-4-methylphenol: $(CH_3)_2NCH_2$, OH, $CH_2N(CH_3)_2$, CH_3

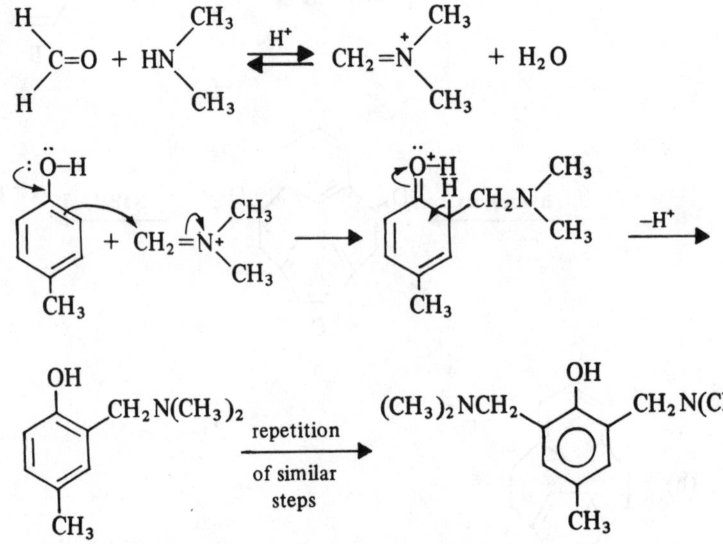

20.22 These syntheses are easier to see if we work backward.

(a) 2-($\overset{O}{\overset{\|}{C}}(CH_2)_4CH_3$)cyclohexanone $\xleftarrow{H_2O}$ iminium salt (pyrrolidinium Cl^-, $\overset{O}{\overset{\|}{C}}(CH_2)_4CH_3$) $\xleftarrow[(C_2H_5)_3N]{ClC(O)(CH_2)_4CH_3}$

1-(1-cyclohexenyl)pyrrolidine $\xleftarrow[H^+, -H_2O]{\text{pyrrolidine (N–H)}}$ cyclohexanone

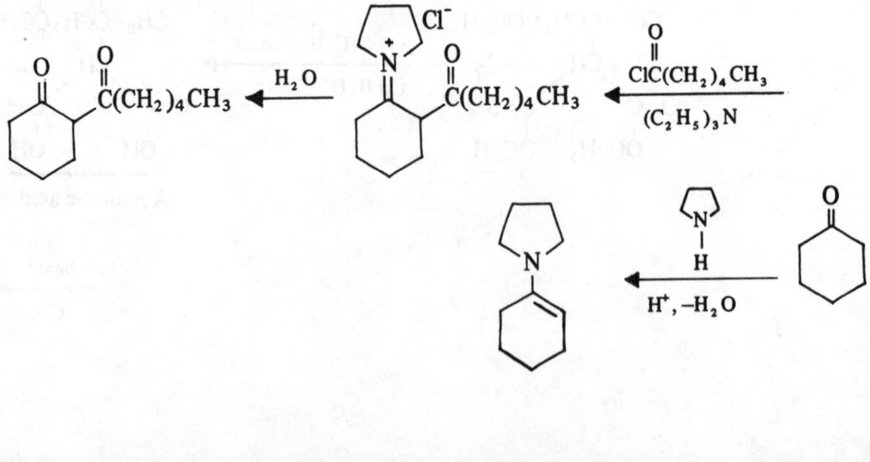

(b) 2-($CH_2CH{=}CHCH_3$)cyclopentanone $\xleftarrow{H_2O}$ iminium salt (Br^-) with $CH_2CH{=}CHCH_3$ $\xleftarrow{BrCH_2CH{=}CHCH_3}$ 1-pyrrolidinylcyclopentene $\xleftarrow[H^+, -H_2O]{\text{pyrrolidine (N–H)}}$ cyclopentanone

(c) 2-($CH_2\overset{O}{\overset{\|}{C}}CH_3$)cyclopentanone $\xleftarrow[(2)\ H_2O]{(1)\ BrCH_2\overset{O}{\overset{\|}{C}}CH_3}$ 1-pyrrolidinylcyclopentene $\xleftarrow[H^+, -H_2O]{\text{pyrrolidine (N–H)}}$ cyclopentanone

(d) 1-($CH_2\overset{O}{\overset{\|}{C}}OC_2H_5$)-2-tetralone $\xleftarrow[(2)\ H_2O]{(1)\ BrCH_2\overset{O}{\overset{\|}{C}}OC_2H_5}$ enamine $\xleftarrow[H^+, -H_2O]{\text{pyrrolidine (N–H)}}$ 2-tetralone

20.23 $C_6H_5CH_3 \xrightarrow[CCl_4]{NBS}$ $C_6H_5CH_2Br$ (**A**) $\xrightarrow[(2)\ CO_2,\ \text{then}\ H^+]{(1)\ Mg,\ Et_2O}$ $C_6H_5CH_2CO_2H$ (**B**)

$\xrightarrow{SOCl_2}$ $C_6H_5CH_2\overset{O}{\overset{\|}{C}}Cl$ (**C**) $\xrightarrow{EtOH}$ $C_6H_5CH_2\overset{O}{\overset{\|}{C}}OEt$ (**D**) $\xrightarrow[NaOEt]{EtO\overset{O}{\overset{\|}{C}}OEt}$

$C_6H_5CH(\overset{O}{\overset{\|}{C}}OEt)_2$ (**E**) $\xrightarrow[CH_3CH_2Br]{KOC(CH_3)_3}$ $(CH_3CH_2)(C_6H_5)C(\overset{O}{\overset{\|}{C}}OEt)_2$ (**F**) $\xrightarrow[NaOEt]{H_2N\overset{O}{\overset{\|}{C}}NH_2}$

Phenobarbital (5-ethyl-5-phenyl barbituric acid: CH_3CH_2, C_6H_5, N–H, =O)

20.24 $CH_2(CO_2Et)_2 \xrightarrow[(2)\ CH_3CH_2Br]{(1)\ NaOEt} CH_3CH_2CH(CO_2Et)_2 \xrightarrow[(2)\ CH_3CH_2Br]{(1)\ KOC(CH_3)_3} (CH_3CH_2)_2C(COEt)_2$ (each COEt is C=O)

$\xrightarrow[NaOEt]{H_2NCNH_2\ (C{=}O)}$ 5,5-diethyl barbituric acid (CH_3CH_2, CH_3CH_2 on C-5; ring C=O, N–H, C=O, N–H)

Veronal

$CH_2(CO_2Et)_2 \xrightarrow[(2)\ CH_3(CH_2)_2CHBrCH_3]{(1)\ NaOEt} CH_3(CH_2)_2CH(CH_3)–CH(CO_2Et)_2 \xrightarrow[(2)\ CH_2{=}CHCH_2Br]{(1)\ KOC(CH_3)_3}$

$(CH_2{=}CHCH_2)(CH_3(CH_2)_2CH(CH_3))C(COEt)_2$ (each COEt is C=O) $\xrightarrow[NaOEt]{H_2NCNH_2\ (C{=}O)}$ 5-allyl-5-(1-methylbutyl) barbituric acid ($CH_2{=}CHCH_2$, $CH_3(CH_2)_2CH(CH_3)$ on C-5; ring C=O, N–H, C=O, N–H)

Seconal

20.25

(a) $CH_3CH_2CH_2\overset{O}{\overset{\|}{C}}\underset{\substack{|\\CH_2\\|\\CH_3}}{CH}\overset{O}{\overset{\|}{C}}OC_2H_5 \xleftarrow[(2)\ H^+]{(1)\ NaOC_2H_5} CH_3CH_2CH_2\overset{O}{\overset{\|}{C}}OC_2H_5$

(b) $CH_3CH_2CH_2\overset{O}{\overset{\|}{C}}CH_2CH_2CH_3 \xleftarrow[-CO_2]{heat} CH_3CH_2CH_2\overset{O}{\overset{\|}{C}}\underset{\substack{|\\CH_2\\|\\CH_3}}{CH}\overset{O}{\overset{\|}{C}}OH$

$\xleftarrow[(2)\ H_3O^+]{(1)\ OH^-,\ H_2O,\ heat}$ product of (a)

(c) $C_6H_5CH(CH_3)CO_2H \xleftarrow[-CO_2]{\text{heat}} (CH_3)(C_6H_5)C(CO_2H)_2 \xleftarrow[(2)\ H_3O^+]{(1)\ OH^-,\ H_2O,\ \text{heat}}$

$(CH_3)(C_6H_5)C(CO_2C_2H_5)_2 \xleftarrow[CH_3I]{NaOC_2H_5} C_6H_5\text{—}CH(CO_2C_2H_5)_2$

$\xleftarrow[NaOC_2H_5,\ (2)\ H^+]{(1)\ C_2H_5OCOC_2H_5\ (C{=}O)} C_6H_5CH_2\overset{O}{\overset{\|}{C}}OC_2H_5$

(d) $CH_3CH_2CH(\overset{O}{\overset{\|}{C}}OC_2H_5)\text{—}\overset{O}{\overset{\|}{C}}\text{—}\overset{O}{\overset{\|}{C}}OC_2H_5 \xleftarrow[NaOC_2H_5,\ (2)\ H^+]{(1)\ C_2H_5O\overset{O}{\overset{\|}{C}}\text{—}\overset{O}{\overset{\|}{C}}OC_2H_5} CH_3CH_2CH_2\overset{O}{\overset{\|}{C}}OC_2H_5$

(e) $CH_3CH_2CH_2\overset{O}{\overset{\|}{C}}\text{—}\overset{O}{\overset{\|}{C}}OC_2H_5 \xleftarrow[C_2H_5OH]{H^+} CH_3CH_2CH_2\overset{O}{\overset{\|}{C}}\text{—}\overset{O}{\overset{\|}{C}}OH$

$\xleftarrow[-CO_2]{\text{heat}} CH_3CH_2CH(CO_2H)\text{—}\overset{O}{\overset{\|}{C}}\text{—}\overset{O}{\overset{\|}{C}}OH \xleftarrow[(2)\ H_3O^+]{(1)\ OH^-,\ H_2O,\ \text{heat}}$ product of (d)

(f) $C_6H_5CH(\overset{O}{\overset{\|}{C}}OC_2H_5)\text{—}CH{=}O \xleftarrow[NaOC_2H_5,\ (2)\ H^+]{(1)\ H\overset{O}{\overset{\|}{C}}OC_2H_5} C_6H_5CH_2\overset{O}{\overset{\|}{C}}OC_2H_5$

(g) 2-acetylcyclopentanone ($\overset{O}{\overset{\|}{C}}CH_3$) $\xleftarrow{H_2O}$ iminium (pyrrolidinium, $\overset{+}{N}$) with $\overset{O}{\overset{\|}{C}}CH_3$ $\xleftarrow[(R_3N)]{CH_3\overset{O}{\overset{\|}{C}}Cl}$ 1-pyrrolidinylcyclopentene $\xleftarrow[H^+,\ -H_2O]{\text{pyrrolidine (N—H)}}$ cyclopentanone

(h) 2-methyl-2-acetylcyclopentanone (CH_3, $\text{—}\overset{\|}{C}\text{—}CH_3$ with O) $\xleftarrow[(CH_3)_3COK]{CH_3I}$ product of (g)

(i) 2-ethylcyclohexanone (ring C=O, CH_2CH_3) $\xleftarrow[-CO_2]{\text{heat}}$ 2-ethyl-2-carboxycyclohexanone (ring C=O, CH_2CH_3, CO_2H) $\xleftarrow[(2)\ H_3O^+]{(1)\ OH^-,\ H_2O,\ \text{heat}}$

2-ethyl-2-(ethoxycarbonyl)cyclohexanone (ring C=O, CH_2CH_3, $CO_2C_2H_5$) $\xleftarrow[NaOC_2H_5]{CH_3CH_2Br}$ 2-(ethoxycarbonyl)cyclohexanone (ring C=O, $\overset{O}{\overset{\|}{C}}OC_2H_5$)

20.26

(a) $CH_3\overset{O}{\overset{\|}{C}}-\underset{CH_3}{\underset{|}{\overset{CH_3}{\overset{|}{C}}}}-CH_3 \xleftarrow{Zn,\ H^+} CH_3\overset{O}{\overset{\|}{C}}-\underset{CH_3}{\underset{|}{\overset{CH_3}{\overset{|}{C}}}}-CH_2Br \xleftarrow{PBr_3} CH_3\overset{O}{\overset{\|}{C}}-\underset{CH_3}{\underset{|}{\overset{CH_3}{\overset{|}{C}}}}-CH_2OH$

$\xleftarrow[(2)\ H_3O^+]{(1)\ LiAlH_4}$ CH_3C(ethylene ketal, $O\!-\!CH_2CH_2\!-\!O$)$-\underset{CH_3}{\underset{|}{\overset{CH_3}{\overset{|}{C}}}}CO_2C_2H_5 \xleftarrow[H^+]{HOCH_2CH_2OH} CH_3\overset{O}{\overset{\|}{C}}-\underset{CH_3}{\underset{|}{\overset{CH_3}{\overset{|}{C}}}}-CO_2C_2H_5 \xleftarrow[NaOC(CH_3)_3]{CH_3I}$

$CH_3\overset{O}{\overset{\|}{C}}-\underset{CH_3}{\underset{|}{CH}}-CO_2C_2H_5 \xleftarrow[NaOC_2H_5]{CH_3I} CH_3\overset{O}{\overset{\|}{C}}CH_2\overset{O}{\overset{\|}{C}}OC_2H_5$

(b) $CH_3\overset{O}{\overset{\|}{C}}CH_2CH_2CH_2CH_3 \xleftarrow[-CO_2]{\text{heat}} CH_3\overset{O}{\overset{\|}{C}}\underset{CH_2CH_2CH_3}{\underset{|}{CH}}\overset{O}{\overset{\|}{C}}OH \xleftarrow[(2)\ H_3O^+]{(1)\ OH^-,\ H_2O,\ \text{heat}}$

$CH_3\overset{O}{\overset{\|}{C}}\underset{CH_2CH_2CH_3}{\underset{|}{CH}}\overset{O}{\overset{\|}{C}}OC_2H_5 \xleftarrow[CH_3CH_2CH_2Br]{NaOC_2H_5} CH_3\overset{O}{\overset{\|}{C}}CH_2\overset{O}{\overset{\|}{C}}OC_2H_5$

(c) $CH_3\overset{O}{\overset{\|}{C}}CH_2CH_2\overset{O}{\overset{\|}{C}}CH_3 \xleftarrow[-CO_2]{\text{heat}} CH_3\overset{O}{\overset{\|}{C}}\underset{CH_2-\underset{CH_3}{\underset{|}{C}}=O}{\underset{|}{CH}}\overset{O}{\overset{\|}{C}}OH \xleftarrow[(2)\ H_3O^+]{(1)\ OH^-,\ H_2O,\ \text{heat}}$

$$CH_3\overset{O}{\overset{\|}{C}}\underset{\underset{\underset{CH_3}{|}}{\overset{|}{C}=O}}{\underset{|}{\underset{CH_2}{\overset{|}{}}}}CH\overset{O}{\overset{\|}{C}}OC_2H_5 \xleftarrow[CH_3COCH_2Br]{NaOC_2H_5} CH_3\overset{O}{\overset{\|}{C}}CH_2\overset{O}{\overset{\|}{C}}OC_2H_5$$

(d) $$CH_3\overset{OH}{\overset{|}{C}}HCH_2CH_2CO_2H \xleftarrow{NaBH_4} CH_3\overset{O}{\overset{\|}{C}}CH_2CH_2\overset{O}{\overset{\|}{C}}OH \xleftarrow[-CO_2]{heat}$$

$$CH_3\overset{O}{\overset{\|}{C}}\underset{\underset{CO_2H}{|}}{\underset{CH_2}{|}}{CH}\overset{O}{\overset{\|}{C}}OH \xleftarrow[(2)\ H_3O^+]{(1)\ OH^-,\ H_2O,\ heat} CH_3\overset{O}{\overset{\|}{C}}\underset{\underset{\underset{\underset{O}{\|}}{COC_2H_5}}{|}}{\underset{CH_2}{|}}{CH}\overset{O}{\overset{\|}{C}}OC_2H_5$$

$$\xleftarrow[BrCH_2CO_2C_2H_5]{NaOC_2H_5} CH_3\overset{O}{\overset{\|}{C}}CH_2\overset{O}{\overset{\|}{C}}OC_2H_5$$

(e) $$CH_3\overset{OH}{\overset{|}{C}}H\underset{C_2H_5}{\underset{|}{C}}HCH_2OH \xleftarrow[(2)\ H^+]{(1)\ LiAlH_4} CH_3\overset{O}{\overset{\|}{C}}\underset{C_2H_5}{\underset{|}{C}}H\overset{O}{\overset{\|}{C}}OC_2H_5 \xleftarrow[C_2H_5Br]{NaOC_2H_5} CH_3\overset{O}{\overset{\|}{C}}CH_2\overset{O}{\overset{\|}{C}}OC_2H_5$$

(f) $$CH_3\overset{OH}{\overset{|}{C}}HCH_2\overset{OH}{\overset{|}{C}}HC_6H_5 \xleftarrow{NaBH_4} CH_3\overset{O}{\overset{\|}{C}}CH_2\overset{O}{\overset{\|}{C}}C_6H_5 \longleftarrow \text{compare Problem 20.13}$$

20.27

(a) $$CH_3CH_2\underset{CH_3}{\underset{|}{C}}HCO_2H \xleftarrow[-CO_2]{} (CH_3CH_2)(CH_3)C(CO_2H)_2 \xleftarrow[(2)\ H_3O^+]{(1)\ OH^-,\ H_2O,\ heat}$$

$$(CH_3CH_2)(CH_3)C(CO_2C_2H_5)_2 \xleftarrow[NaOC_2H_5]{CH_3I} CH_3CH_2CH(CO_2C_2H_5)_2$$

$$\xleftarrow[NaOC_2H_5]{CH_3CH_2Br} \underset{\underset{CO_2C_2H_5}{|}}{\overset{\overset{CO_2C_2H_5}{|}}{CH_2}}$$

(b) $$CH_3\underset{CH_3}{\underset{|}{C}}HCH_2CH_2CH_2OH \xleftarrow[(2)\ H^+]{(1)\ LiAlH_4} CH_3\underset{CH_3}{\underset{|}{C}}HCH_2CH_2CO_2H$$

[from Problem 20.16(c)]

(c) $CH_3CH_2CH(CH_2OH)CH_2OH \xleftarrow[(2)\ H^+]{(1)\ LiAlH_4} CH_3CH_2CH(CO_2C_2H_5)_2 \longleftarrow$ compare Section 20.4

(d) $HOCH_2CH_2CH_2CH_2OH \xleftarrow[(2)\ H^+]{(1)\ LiAlH_4} HO_2CCH_2CH_2CO_2H \xleftarrow[-CO_2]{heat}$

$(HO_2C)_2CHCH_2CO_2H \xleftarrow{HCl,\ heat} (C_2H_5O_2C)_2CHCH_2CO_2C_2H_5$

$\longleftarrow CH_2(CO_2C_2H_5)_2 + NaOC_2H_5 + BrCH_2CO_2C_2H_5$

20.28 The following reaction took place,

$$CH_3\overset{O}{\overset{\|}{C}}CH_2\overset{O}{\overset{\|}{C}}OC_2H_5 + BrCH_2CH_2CH_2Br \xrightarrow{NaOC_2H_5} BrCH_2CH_2CH_2CH(COCH_3)(\overset{O}{\overset{\|}{C}}OC_2H_5)$$

$$\xrightarrow[-H^+]{NaOC_2H_5} C_2H_5O_2C{-}C(=C(CH_3){-}O^-)(CH_2CH_2CH_2{-}Br) \longrightarrow C_2H_5O\overset{O}{\overset{\|}{C}}{-}C{=}C(CH_3){-}O{-}CH_2{-}CH_2{-}CH_2 \text{ (ring)}$$

Perkin's ester

$$\xrightarrow[(2)\ H_3O^+]{(1)\ OH^-,\ H_2O,\ heat} HO_2C{-}C{=}C(CH_3){-}O{-}CH_2{-}CH_2{-}CH_2 \text{ (ring)}$$

Perkin's acid

20.29

(a) $BrCH_2CH_2Br + CH_2(CO_2C_2H_5)_2 + NaOC_2H_5 \longrightarrow$

$$\left[BrCH_2CH_2{-}CH(CO_2C_2H_5)_2\right] \xrightarrow[-H^+]{NaOC(CH_3)_3} \left[BrCH_2CH_2{-}\ddot{C}^-(CO_2C_2H_5)_2\right]$$

$\longrightarrow$ (cyclopropane-1,1-diester: CH_2–CH_2 ring with C bearing two $CO_2C_2H_5$) $\xrightarrow[\text{(3) heat, } -CO_2]{\text{(1) } OH^-, H_2O, \text{ heat; (2) } H_3O^+}$ cyclopropane–CO_2H

(b) $2NaCH(CO_2C_2H_5)_2 + BrCH_2CH_2CH_2Br \longrightarrow$

$(C_2H_5O_2C)_2C(H)CH_2CH_2CH_2C(H)(CO_2C_2H_5)_2$

A

$\xrightarrow[Br_2]{NaOC_2H_5}$ $\left[(C_2H_5O_2C)_2C(H)CH_2CH_2CH_2C(Br)(CO_2C_2H_5)_2\right]$ $\xrightarrow{NaOC_2H_5}$

(cyclopentane-1,1,2,2-tetraester: $CO_2C_2H_5$, $C_2H_5O_2C$, $CO_2C_2H_5$, $CO_2C_2H_5$) **B** $\xrightarrow[\text{(2) } H_3O^+]{\text{(1) } OH^-, H_2O}$ (cyclopentane-1,1,2,2-tetracarboxylic acid: CO_2H, CO_2H, CO_2H, CO_2H) **C** $\xrightarrow[-2CO_2]{\text{heat}}$

(trans-cyclopentane-1,2-dicarboxylic acid: CO_2H, CO_2H) + (enantiomer: CO_2H, HO_2C) — **D** Racemic form

(cis-cyclopentane-1,2-dicarboxylic acid: HO_2C, CO_2H) — **E** Meso compound

(c) $BrCH_2CH_2CH_2CH_2Br \xrightarrow{NaCH(CO_2C_2H_5)_2} BrCH_2CH_2CH_2CH_2CH(CO_2C_2H_5)_2$

$\xrightarrow{NaOC(CH_3)_3}$ (cyclopentane-1,1-diester: $CO_2C_2H_5$, $CO_2C_2H_5$) $\xrightarrow[\text{(3) heat}]{\text{(1) } OH^-, H_2O; \text{ (2) } H_3O^+}$ cyclopentane–CO_2H

20.30 (a) $CH_2(CO_2C_2H_5)_2 + {}^-OC_2H_5 \rightleftharpoons {}^-{:}CH(CO_2C_2H_5)_2 + C_2H_5OH$

$$C_6H_5CH{=}CH{-}\overset{O}{\overset{\|}{C}}OC_2H_5 + {}^-{:}CH(CO_2C_2O_5)_2 \rightleftharpoons C_6H_5\underset{CH(CO_2C_2H_5)_2}{\underset{|}{C}}HCH{=}\overset{O^-}{\overset{|}{C}}OC_2H_5$$

$$\xrightleftharpoons{+H^+} C_6H_5\underset{CH(CO_2C_2H_5)_2}{\underset{|}{C}}HCH_2\overset{O}{\overset{\|}{C}}OC_2H_5$$

(b)

$$CH_3\ddot{N}H_2 + CH_2{=}CH{-}\overset{O}{\overset{\|}{C}}OCH_3 \rightleftharpoons CH_3{-}\overset{H}{\overset{|}{\underset{H}{\underset{|}{N^+}}}}{-}CH_2{-}CH{=}\overset{O^-}{\overset{|}{C}}OCH_3 \rightleftharpoons$$

$$CH_3\underset{H}{\underset{|}{N}}{-}CH_2{-}CH_2{-}\overset{O}{\overset{\|}{C}}OCH_3 \xrightleftharpoons{CH_2=CH-\overset{O}{\overset{\|}{C}}OCH_3} CH_3N(CH_2CH_2CO_2CH_3)_2$$

$$\xrightleftharpoons{\text{base}} CH_3{-}N\begin{cases} CH_2{-}\ddot{C}H^-{-}CO_2CH_3 \\ CH_2{-}CH_2{-}C(=O)OCH_3 \end{cases} \xrightarrow[\text{(several steps)}]{\text{Dieckmann condensation}} CH_3{-}N\text{(piperidine ring with } CO_2CH_3 \text{ and } {=}O\text{)}$$

(c)

$$CH_3{-}\overset{CH_3}{\overset{|}{\underset{CH(CO_2C_2H_5)_2}{\underset{|}{C}}}}{-}CH_2{-}\overset{O}{\overset{\|}{C}}OC_2H_5 + C_2H_5O^- \rightleftharpoons CH_3{-}\overset{CH_3}{\overset{|}{\underset{CH(CO_2C_2H_5)_2}{\underset{|}{C}}}}{-}CH{=}\overset{O^-}{\overset{|}{C}}OC_2H_5 + C_2H_5OH$$

$$CH_3{-}\overset{CH_3}{\overset{|}{\underset{CH(CO_2C_2H_5)_2}{\underset{|}{C}}}}{-}CH{=}\overset{O^-}{\overset{|}{C}}OC_2H_5 \rightleftharpoons CH_3{-}\overset{CH_3}{\overset{|}{C}}{=}CH{-}\overset{O}{\overset{\|}{C}}OC_2H_5 + {}^-{:}CH(CO_2C_2H_5)_2$$

The Michael reaction is reversible and the reaction just given is an example of a reverse Michael reaction.

20.31 Two reactions take place. The first is a normal Knoevenagel condensation,

$$R{-}\underset{R'}{\underset{|}{C}}{=}O + CH_2(COCH_3)_2 \xrightarrow[-H_2O]{\text{base}} R{-}\underset{R'}{\underset{|}{C}}{=}C\begin{cases}\overset{O}{\overset{\|}{C}}CH_3 \\ \underset{O}{\underset{\|}{C}}CH_3\end{cases}$$

Then the α, β-unsaturated diketone reacts with a second mole of the active methylene compound in a Michael addition.

$$\text{R–C(R')=C(COCH}_3)_2 + CH_2(COCH_3)_2 \xrightarrow{\text{base}} \text{R–CR'[CH(COCH}_3)_2]–CH(COCH_3)_2$$

20.32

$$CH_3CH_2\overset{O}{\overset{\|}{C}}OC_2H_5 \xrightarrow[C_2H_5ONa]{H\overset{O}{\overset{\|}{C}}OC_2H_5} CH_3CH(CHO)\overset{O}{\overset{\|}{C}}OC_2H_5$$

$$\xrightarrow[C_2H_5ONa]{H_2N\overset{O}{\overset{\|}{C}}NH_2} \left[\text{cyclic intermediate: } C_2H_5O–C(O^-) \text{ in ring with HN, C(=O), N–H, CH, C–CH}_3\right] \longrightarrow \text{thymine}$$

20.33

$$\text{cycloheptanone} \xrightarrow[(2)\ H_3O^+]{(1)\ CH_3MgI} \underset{\textbf{A}}{\text{1-methylcycloheptanol (CH}_3\text{, OH)}} \xrightarrow{H^+,\ \text{heat}} \underset{\textbf{B}}{\text{1-methylcycloheptene}} \xrightarrow[(2)\ Zn,\ H_2O]{(1)\ O_3}$$

$$\underset{\textbf{C}}{CH_3\overset{O}{\overset{\|}{C}}(CH_2)_5CHO} \xrightarrow[\text{pyridine}]{CH_2(CO_2H)_2} \underset{\text{Queen substance}}{CH_3\overset{O}{\overset{\|}{C}}(CH_2)_5CH{=}CHCO_2H}$$

$$\xrightarrow[Pd]{H_2,} \underset{\textbf{D}}{CH_3\overset{O}{\overset{\|}{C}}(CH_2)_7CO_2H} \xrightarrow[(2)\ H_3O^+]{(1)\ I_2/NaOH} \underset{\textbf{E}}{HO_2C(CH_2)_7CO_2H}$$

20.34

$$CH_2{=}C(CH_3){-}CH{=}CH_2 + HBr \longrightarrow \underset{\textbf{F}}{CH_3C(CH_3){=}CHCH_2Br} \xrightarrow{Na^{+}\ {}^{-}:CH(\overset{O}{\overset{\|}{C}}CH_3)(CO_2C_2H_5)}$$

$$\underset{\textbf{G}}{CH_3\underset{\underset{CH_3}{|}}{C}{=}CHCH_2\underset{\underset{CO_2C_2H_5}{|}}{CH}\overset{\overset{O}{\|}}{C}CH_3} \xrightarrow[\text{(2) } H_3O^+\text{, (3) heat}]{\text{(1) dil. NaOH}} \underset{\textbf{H}}{CH_3\underset{\underset{CH_3}{|}}{C}{=}CHCH_2CH_2\overset{\overset{O}{\|}}{C}CH_3} \xrightarrow[\text{(2) } H_3O^+]{\text{(1) LiC}\equiv\text{CH}}$$

$$\underset{\textbf{I}}{CH_3\underset{\underset{CH_3}{|}}{C}{=}CHCH_2CH_2\overset{\overset{OH}{|}}{\underset{\underset{CH_3}{|}}{C}}C{\equiv}CH} \xrightarrow[\text{Lindlar's catalyst}]{H_2} \text{linalool}$$

20.35

$$(C_2H_5O_2C)_2HC{:}^-\,Na^+ + BrCH_2CH_2CH_2Br \longrightarrow \left[(C_2H_5O_2C)_2CH{-}CH_2CH_2CH_2Br\right]\ (C_{10}H_{17}BrO_4)$$

$$\xrightarrow{NaOC_2H_5} \text{1,1-bis(ethoxycarbonyl)cyclobutane}\ (C_{10}H_{16}O_4) \xrightarrow[\text{(2) } H_2O]{\text{(1) LiAlH}_4} \text{1,1-bis(hydroxymethyl)cyclobutane, } (CH_2OH)_2\ (C_6H_{12}O_2) \xrightarrow{HBr}$$

$$\text{1,1-bis(bromomethyl)cyclobutane, } (CH_2Br)_2\ (C_6H_{10}Br_2) \xrightarrow[2NaOC_2H_5]{CH_2(CO_2C_2H_5)_2} \text{spiro[3.3]heptane-2,2-}(CO_2C_2H_5)_2\ (C_{13}H_{20}O_4)$$

$$\xrightarrow[\text{(2) } H^+]{\text{(1) } OH^-,\ H_2O} \text{spiro[3.3]heptane-2,2-}(CO_2H)_2\ (C_9H_{12}O_4) \xrightarrow{\text{heat}} \underset{\textbf{J}}{\text{spiro[3.3]heptane-2-}CO_2H}\ (C_8H_{12}O_2) + CO_2$$

20.36 (a) $ClCH_2CO_2C_2H_5 + C_2H_5O^- \rightleftharpoons Cl{-}\ddot{C}^-HCO_2C_2H_5 + C_2H_5OH$

$$R{-}\overset{\overset{R'}{|}}{\underset{\underset{O}{\|}}{C}} + {:}^-\overset{\overset{Cl}{|}}{C}HCO_2C_2H_5 \rightleftharpoons \left[R{-}\overset{\overset{R'}{|}}{\underset{\underset{O^-}{|}}{C}}{-}\overset{\overset{Cl}{|}}{C}HCO_2C_2H_5\right] \longrightarrow R{-}\overset{\overset{R'}{|}}{C}\overset{O}{\frown}CHCO_2C_2H_5$$

(b) Decarboxylation of the epoxy acid gives an enol anion which, on protonation, gives an aldehyde.

$$\text{R}-\overset{\text{R}'}{\overset{|}{\text{C}}}\underset{\text{O}}{\diagdown\!\diagup}\text{CH}-\underset{\text{O}}{\underset{\|}{\text{C}}}-\text{O}-\text{H} + \text{C}_5\text{H}_5\text{N}: \xrightarrow{-\text{CO}_2} \text{R}-\overset{\text{R}'}{\overset{|}{\text{C}}}=\underset{\text{O}^-}{\underset{|}{\text{CH}}} \longrightarrow \text{R}-\overset{\text{R}'}{\overset{|}{\text{CH}}}-\underset{\text{O}}{\underset{\|}{\text{CH}}}$$

$C_5H_5\overset{+}{N}H$

(c) β-Ionone $\xrightarrow[C_2H_5ONa]{ClCH_2CO_2C_2H_5}$ (epoxy ester, $CO_2C_2H_5$)

$\xrightarrow[(2)\ H_3O^+]{(1)\ OH^-,\ H_2O}$ (epoxy acid, CO_2H) $\xrightarrow[\text{heat},\ -CO_2]{C_5H_5N}$

[(enolate $\overset{|}{\text{CH}}$–O^-) + $C_5H_5\overset{+}{N}H$] $\longrightarrow$ (aldehyde, $\underset{\text{O}}{\underset{\|}{\text{CH}}}$)

20.37

(a) $C_6H_5-\overset{H}{\overset{|}{C}}=O + (CH_3CH_2CO)_2O \xrightarrow{CH_3CH_2CO_2K} C_6H_5-CH=\overset{CH_3}{\overset{|}{C}}-CO_2H + CH_3CH_2CO_2H$

(b) $Cl-C_6H_4-\overset{H}{\overset{|}{C}}=O + (CH_3CO)_2O \xrightarrow{CH_3CO_2K} Cl-C_6H_4-CH=CHCO_2H + CH_3CO_2H$

20.38

(a) $CH_2=\overset{CH_3}{\overset{|}{C}}-CO_2CH_3$

(b) $KMnO_4$, OH^-, then H_3O^+

(c) CH_3OH, H^+

(d) CH_3ONa, then H^+

(e) and (f) 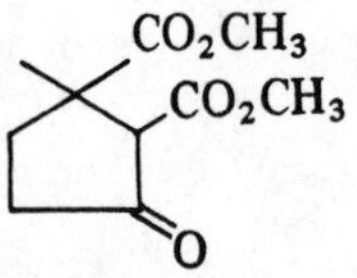 [1,1-bis(CO_2CH_3)-2-(CO_2CH_3)cyclopentan-3-one structure] and [CO_2CH_3 / CH_3O_2C cyclopentanone structure]

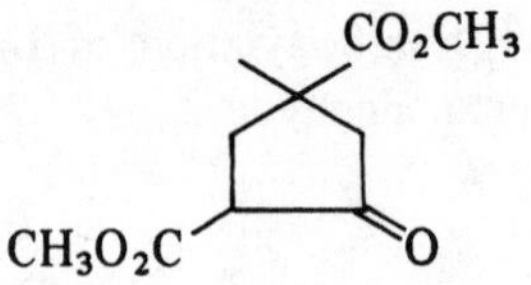

(g) OH^-, H_2O, then H_3O^+

(h) heat ($-CO_2$)

(i) CH_3OH, H^+

(j) $BrCH_2CO_2CH_3$, Zn, then H_3O^+

(k) 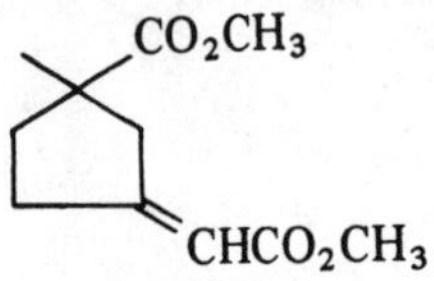 [cyclopentane structure with CO_2CH_3 and $=CHCO_2CH_3$]

(l) H_2, Pt

(m) CH_3ONa, then H^+

(n) 2 $NaNH_2$ + 2 CH_3I

SECTION REFERENCES FOR THE ADDITIONAL PROBLEMS

20.25 17.10, 20.2, 20.3, 20.5, 20.10

20.26 16.7, 17.10, 20.2, 20.3

20.27 17.10, 20.4

20.28 20.3

20.29 20.4

20.30 17.9, 20.2, 20.8, 20.10

20.31 17.9, 20.7, 20.8

20.32 20.2, 20.11, 24.2

20.33 6.4, 6.10, 7.13, 14.12, 17.4, 20.7

20.34 16.11, 20.3

20.35 20.4

20.36 17.2, 17.5, 20.2

20.37 18.6, 17.5, 20.2

20.38 16.11, 17.1, 18.7, 18.10, 20.2

SELF-TEST

20.1 Supply the structural formulas of the missing reactants and major organic products. If no reaction occurs, write N.R.

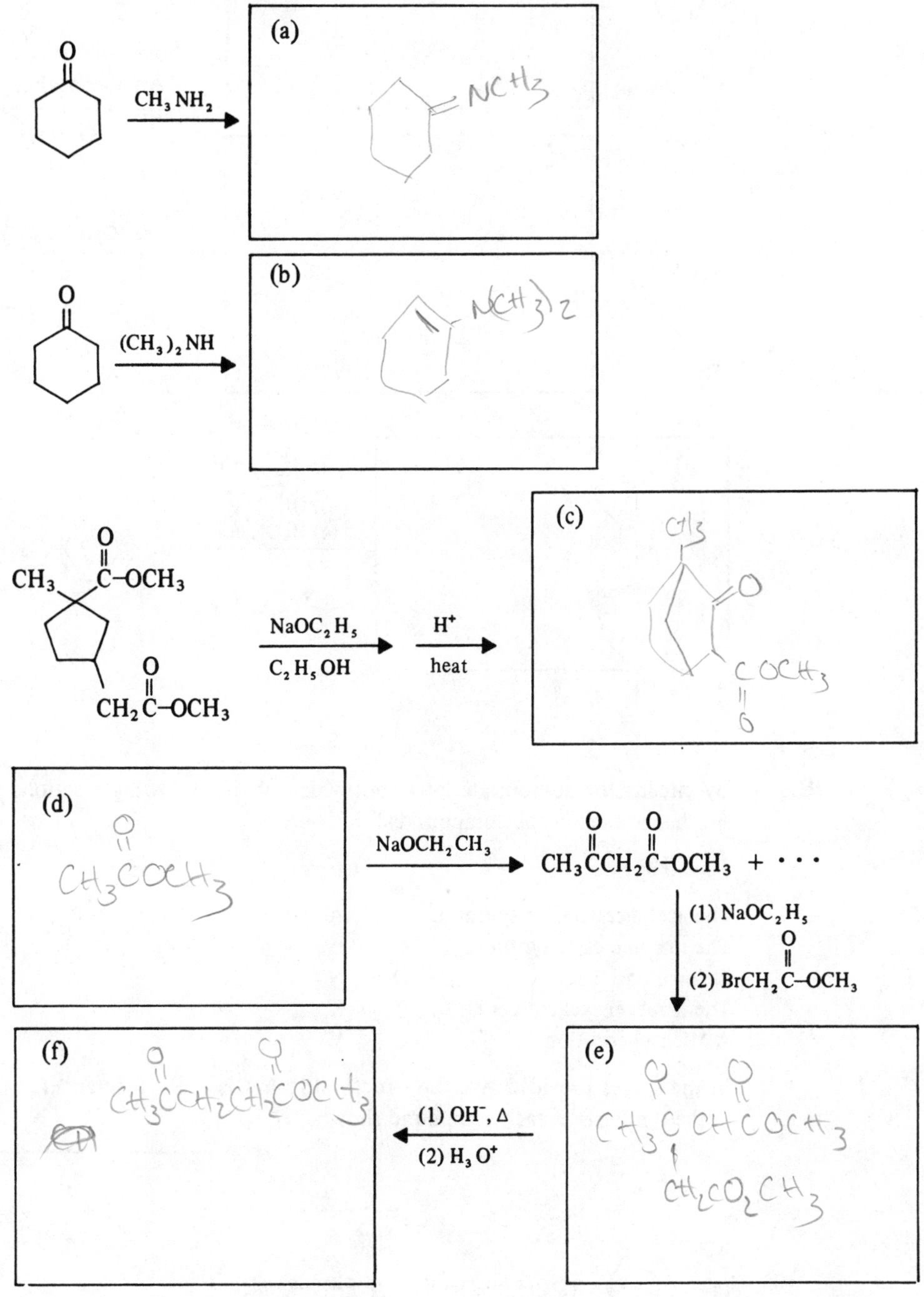

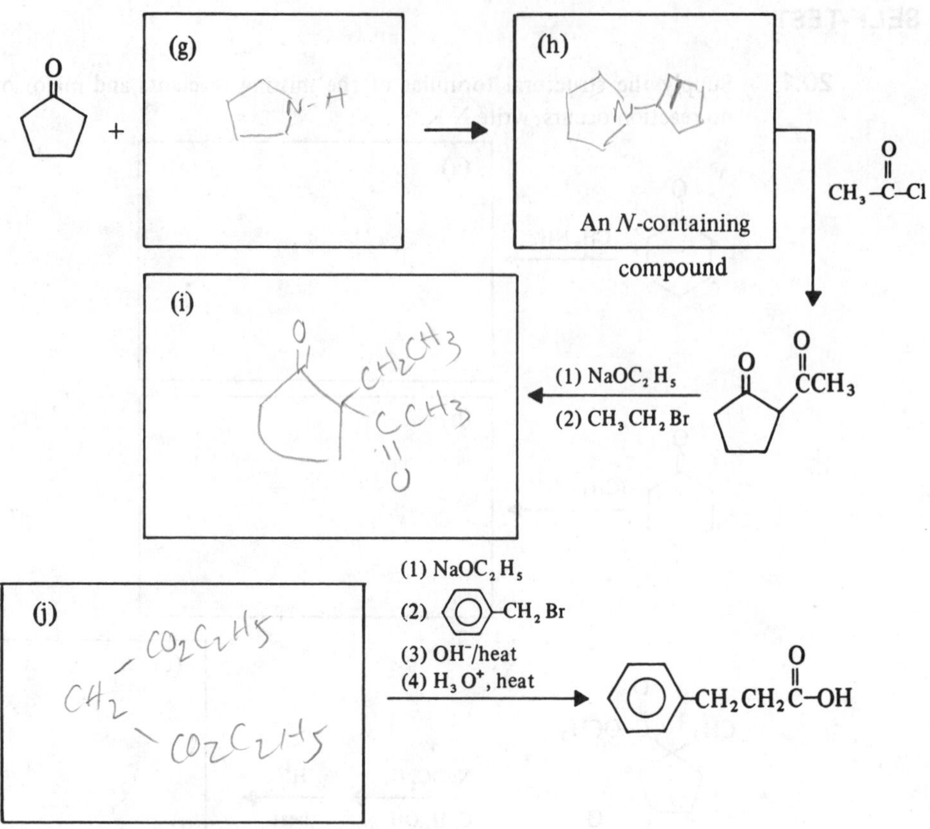

20.2 By circling the appropriate letter tell which of the following reactions you would use to synthesize each of the compounds listed here.

If your answer is	*Circle:*
The acetoacetic ester synthesis	A
The malonic ester synthesis	E
An enamine	N
The Knoevenagel condensation	K
A Michael addition	Mi

In the spaces provided give the structural formulas of the reactants (not reagents) needed in the synthesis of each compound shown.

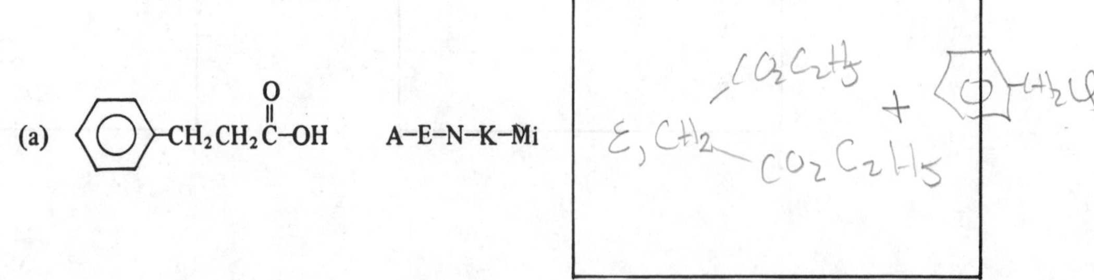

(b) $C_6H_5CH_2CH_2\overset{O}{\overset{\|}{C}}CH_3$ A–E–N–K–Mi

(c) 2-($CH_2\overset{O}{\overset{\|}{C}}CH_3$)cyclopentanone A–E–N–K–Mi

20.3 Which hydrogen atoms in the following ester are most acidic?

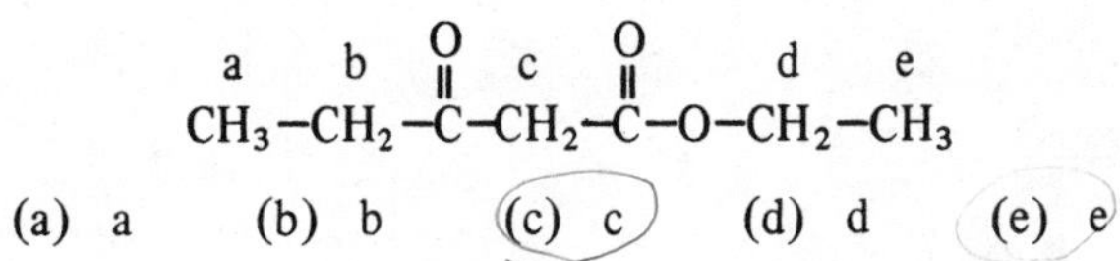

$$\overset{a}{CH_3}-\overset{b}{CH_2}-\overset{O}{\overset{\|}{C}}-\overset{c}{CH_2}-\overset{O}{\overset{\|}{C}}-O-\overset{d}{CH_2}-\overset{e}{CH_3}$$

(a) a (b) b (c) c (d) d (e) e

20.4 What would be the product of the following reaction?

$$CH_3CH_2\overset{O}{\overset{\|}{C}}OEt \xrightarrow[(2)\ H^+]{(1)\ NaOEt} ?$$

(a) $CH_3CH_2\overset{O}{\overset{\|}{C}}CH_2CH_2\overset{O}{\overset{\|}{C}}OEt$

(b) $CH_3CH_2\overset{O}{\overset{\|}{C}}CH_2\overset{O}{\overset{\|}{C}}CH_3$

(c) $CH_3CH_2\overset{O}{\overset{\|}{C}}CH_2\overset{O}{\overset{\|}{C}}OEt$

(d) $CH_3\overset{O}{\overset{\|}{C}}CH_2CH_2\overset{O}{\overset{\|}{C}}OEt$

(e) $CH_3CH_2\overset{O}{\overset{\|}{C}}\underset{CH_3}{\underset{|}{C}H}\overset{O}{\overset{\|}{C}}OEt$

20.5 Which starting materials could be used in a crossed Claisen condensation to prepare following compound?

$$EtO-\overset{O}{\overset{\|}{C}}-\overset{O}{\overset{\|}{C}}-\underset{CH_3}{\underset{|}{CH}}-\overset{O}{\overset{\|}{C}}OEt$$

(a) CH_3CO_2Et and $EtO-\overset{O}{\overset{\|}{C}}-\overset{O}{\overset{\|}{C}}CH_2CH_3$

(b) $CH_3CH_2CO_2Et$ and EtO_2C-CO_2Et

(c) $CH_3CH_2CO_2Et$ and HCO_2Et

(d) $EtO_2C\underset{CH_3}{\underset{|}{C}}HCO_2Et$ and HCO_2Et

(e) More than one of the above

ANSWERS TO SECOND REVIEW PROBLEM SET

The problems review concepts from Chapters 1-20.

1. Increasing acidity →

(a) CH_3COCH_3 < CH_3CH_2OH < $CH_3OCOCH_2COOCH_3$ < CH_3COOH

(b) 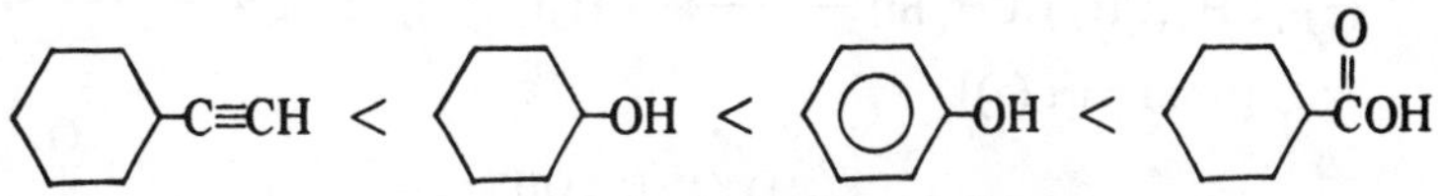 cyclohexyl–C≡CH < cyclohexanol < phenol (C_6H_5OH) < cyclohexanecarboxylic acid ($C_6H_{11}COOH$)

(c) $(CH_3)_3C–C_6H_4–COOH$ < C_6H_5COOH < $(CH_3)_3\overset{+}{N}–C_6H_4–COOH$

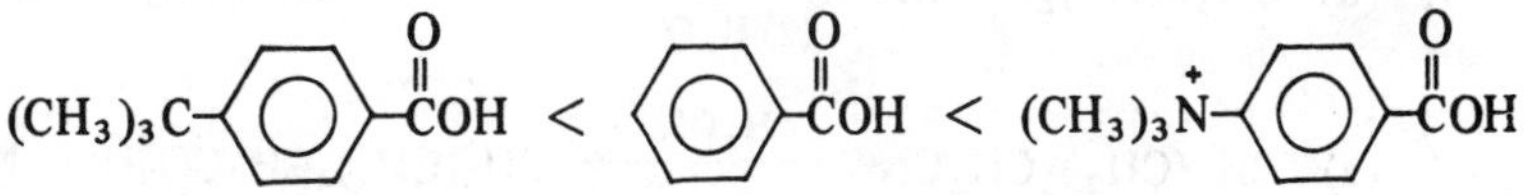

(d) CH_3CH_2COOH < $CH_3CHClCOOH$ < CH_3CCl_2COOH

(e) 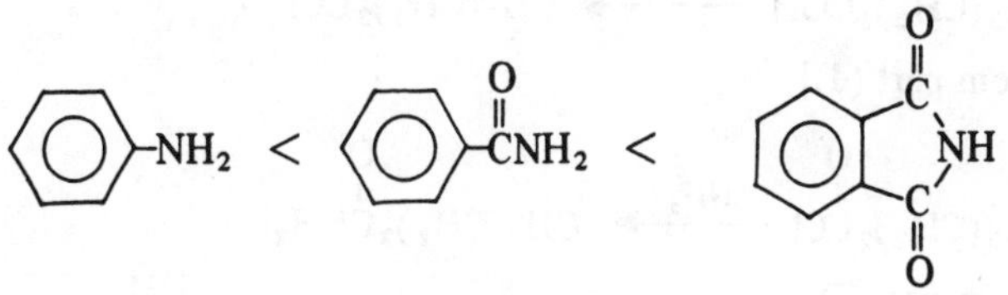 $C_6H_5NH_2$ < $C_6H_5CONH_2$ < phthalimide

2. Increasing basicity →

(a) CH_3CONH_2 < NH_3 < $CH_3CH_2NH_2$

(b) 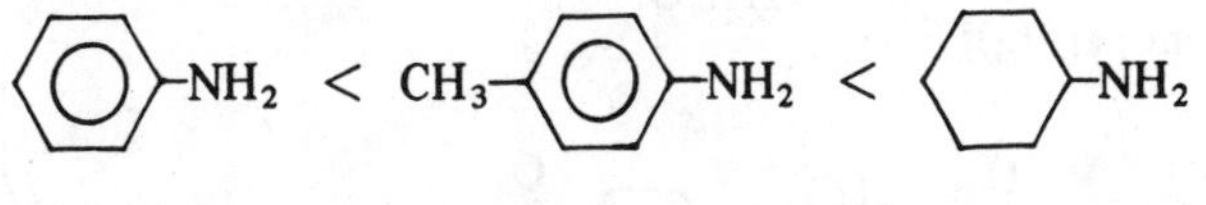 $C_6H_5NH_2$ < $CH_3–C_6H_4–NH_2$ < cyclohexyl–NH_2

(c) $O_2N–C_6H_4–NH_2$ < $C_6H_5NH_2$ < $CH_3–C_6H_4–NH_2$

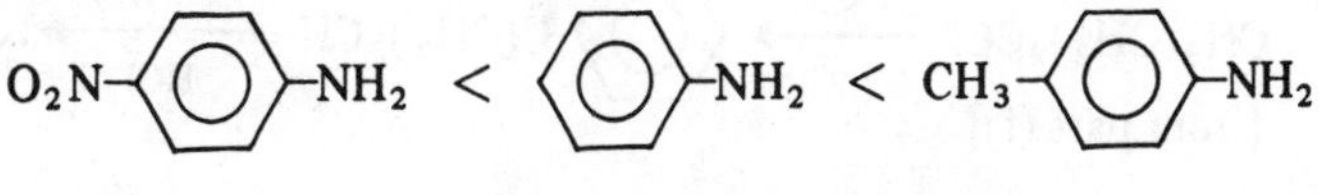

(d) $CH_3CH_2CH_3$ < CH_3OCH_3 < CH_3NHCH_3

3.

(a) $CH_3(CH_2)_2CH_2OH \xrightarrow[\text{(or HBr)}]{PBr_3} CH_3(CH_2)_2CH_2Br$

(b) $CH_3(CH_2)_2CH_2Br$ [from part (a)] $\xrightarrow{\text{potassium phthalimide}} CH_3(CH_2)_2CH_2\text{–N(phthalimide)}$

$\xrightarrow[\text{heat}]{H_2NNH_2} CH_3(CH_2)_2CH_2NH_2 + \text{phthalhydrazide}$

(c) $CH_3(CH_2)_2CH_2Br$ [from part (a)] $\xrightarrow{NaCN} CH_3(CH_2)_2CH_2CN \xrightarrow{LiAlH_4} CH_3(CH_2)_3CH_2NH_2$

(d) $CH_3(CH_2)_2CH_2OH \xrightarrow[(2)\ H_3O^+]{(1)\ KMnO_4,\ OH^-,\ heat} CH_3(CH_2)_2\overset{O}{\overset{\|}{C}}OH$

(e) $CH_3(CH_2)_2CH_2CN$ [from part (c)] $\xrightarrow[H_2O,\ heat]{H_3O^+} CH_3(CH_2)_2CH_2CO_2H + NH_4^+$

(f) $CH_3(CH_2)_2\overset{O}{\overset{\|}{C}}OH$ [from part (d)] $\xrightarrow{SOCl_2} CH_3(CH_2)_2\overset{O}{\overset{\|}{C}}Cl$

(g) $CH_3(CH_2)_2\overset{O}{\overset{\|}{C}}Cl$ [from part (f)] $\xrightarrow{NH_3} CH_3(CH_2)_2\overset{O}{\overset{\|}{C}}NH_2$

(h) $CH_3(CH_2)_2\overset{O}{\overset{\|}{C}}Cl \xrightarrow[\text{base}]{CH_3(CH_2)_2CH_2OH} CH_3(CH_2)_2\overset{O}{\overset{\|}{C}}OCH_2(CH_2)_2CH_3$

(i) $CH_3CH_2CH_2\overset{O}{\overset{\|}{C}}NH_2$ [from part (g)] $\xrightarrow[(2)\ H_3O^+]{(1)\ Br_2,\ OH^-} CH_3CH_2CH_2NH_2 + CO_3^{2-}$

(j) $CH_3(CH_2)_2\overset{O}{\overset{\|}{C}}Cl$ [from part (f)] $\xrightarrow[AlCl_3]{C_6H_6} C_6H_5\overset{O}{\overset{\|}{C}}(CH_2)_2CH_3 \xrightarrow[HCl]{Zn(Hg)} C_6H_5CH_2(CH_2)_2CH_3$

(k) $CH_3(CH_2)_2\overset{O}{\overset{\|}{C}}Cl \xrightarrow{CH_3(CH_2)_2\overset{O}{\overset{\|}{C}}ONa} [CH_3(CH_2)_2\overset{O}{\overset{\|}{C}}]_2O$

[from part (f)]

(l) $CH_3(CH_2)_2CH_2Br \xrightarrow{\overset{+-}{Na}:CH(CO_2Et)_2} CH_3(CH_2)_2CH_2CH(CO_2Et)_2 \xrightarrow[(2)\ H_3O^+]{(1)\ OH^-,\ H_2O,\ heat}$

[from part (a)]

$CH_3(CH_2)_2CH_2CH(CO_2H)_2 \xrightarrow[-CO_2]{heat} CH_3(CH_2)_2CH_2CH_2CO_2H$

4.

(a) $CH_3-C_6H_5 \xrightarrow{Br_2,\ FeBr_3} CH_3-C_6H_4-Br \xrightarrow{Mg,\ ether} CH_3-C_6H_4-MgBr$

(separate from ortho isomer)

$\xrightarrow[(2)\ H^+]{(1)\ H_2C\overset{O}{—}CH_2} CH_3-C_6H_4-CH_2CH_2OH \xrightarrow[CH_2Cl_2]{PCC} CH_3-C_6H_4-CH_2\overset{O}{\overset{\|}{C}}H$

$\xrightarrow[\text{(Aldol condensation)}]{OH^-} CH_3-C_6H_4-CH_2CH=\underset{C_6H_4-CH_3}{C}-\overset{O}{\overset{\|}{C}}H$

(b) $C_6H_6 \xrightarrow[HF]{CH_2=CHCH_3} C_6H_5-CH(CH_3)_2 \xrightarrow[h\nu]{NBS,\ CCl_4} C_6H_5-C(CH_3)_2-Br$

$\xrightarrow{base} C_6H_5-\underset{CH_3}{C}=CH_2 \xrightarrow[(2)\ H_2O_2,\ OH^-]{(1)\ THF:BH_3} C_6H_5-\underset{CH_3}{C}HCH_2OH \xrightarrow[(2)\ CH_3CH_2Br]{(1)\ NaH}$

$C_6H_5-\underset{CH_3}{C}HCH_2OCH_2CH_3$

(c) $C_6H_5NH_2 \xrightarrow{(CH_3\overset{O}{\overset{\|}{C}})_2O} C_6H_5NH\overset{O}{\overset{\|}{C}}CH_3 \xrightarrow[\text{(separate from ortho isomer)}]{Cl_2,\ FeCl_3} Cl-C_6H_4-NH\overset{O}{\overset{\|}{C}}CH_3$

$$\xrightarrow[\text{H}_2\text{O, heat}]{\text{OH}^-} \text{Cl-C}_6\text{H}_4\text{-NH}_2 \xrightarrow[\text{H}_2\text{O}]{\text{Br}_2,} \text{Cl-C}_6\text{H}_2\text{Br}_2\text{-NH}_2 \xrightarrow[(2)\ \text{H}_3\text{PO}_2]{(1)\ \text{HONO}} \text{Cl-C}_6\text{H}_3\text{Br}_2$$

(d) $$\text{CH}_3\text{-C}_6\text{H}_5 \xrightarrow[\text{(separate from ortho isomer)}]{\text{HNO}_3,\ \text{H}_2\text{SO}_4} \text{CH}_3\text{-C}_6\text{H}_4\text{-NO}_2 \xrightarrow[(2)\ \text{H}_3\text{O}^+]{(1)\ \text{KMnO}_4,\ \text{OH}^-,\ \text{heat}}$$

$$\text{O}_2\text{N-C}_6\text{H}_4\text{-COH (C=O)} \xrightarrow{\text{SOCl}_2} \text{O}_2\text{N-C}_6\text{H}_4\text{-CCl (C=O)} \xrightarrow[\text{ether}]{\text{LiAlH[OC(CH}_3)_3]_3}$$

$$\text{O}_2\text{N-C}_6\text{H}_4\text{-CH (C=O)} \xrightarrow[\text{OH}^-]{\text{CH}_3\text{CC}_6\text{H}_5\ (\text{C=O})} \text{O}_2\text{N-C}_6\text{H}_4\text{-CH=CHC(=O)-C}_6\text{H}_5$$

(e) $$\text{C}_6\text{H}_5\text{-CH}_3 \xrightarrow[h\nu]{\text{NBS, CCl}_4} \text{C}_6\text{H}_5\text{-CH}_2\text{Br} \xrightarrow{\text{NaC}\equiv\text{CH}} \text{C}_6\text{H}_5\text{-CH}_2\text{C}\equiv\text{CH}$$

$$\xrightarrow[\text{H}_2\text{O}]{\text{Hg}^{2+},\ \text{H}_3\text{O}^+} \text{C}_6\text{H}_5\text{-CH}_2\text{CCH}_3\ (\text{C=O}) \xrightarrow{\text{HCN}} \text{C}_6\text{H}_5\text{-CH}_2\text{C(OH)(CN)CH}_3 \xrightarrow[\text{heat}]{\text{H}_3\text{O}^+,}$$

$$\left[\text{C}_6\text{H}_5\text{-CH}_2\text{C(OH)(CO}_2\text{H)CH}_3\right] \xrightarrow{-\text{H}_2\text{O}} \text{C}_6\text{H}_5\text{-CH=C(CO}_2\text{H)CH}_3$$

5.

$$\text{2-Methyl-1,3-butadiene} + \text{Diethyl fumarate (EtOC(=O)CH=CHC(=O)OEt)} \xrightarrow[\text{reaction}]{\text{Diels-Alder}} \text{A}$$

A: trans-4-methyl-4-cyclohexene-1,2-dicarboxylic acid diethyl ester (CH$_3$, COEt, COEt) + enantiomer

$$\xrightarrow[(2)\ \text{H}_2\text{O}]{(1)\ \text{LiAlH}_4} \text{B (CH}_3\text{, CH}_2\text{OH, CH}_2\text{OH)} + \text{enantiomer} \xrightarrow{\text{PBr}_3} \text{C (CH}_3\text{, CH}_2\text{Br, CH}_2\text{Br)} + \text{enantiomer} \xrightarrow[\text{H}^+]{\text{Zn}}$$

2-Methyl-1,3-butadiene | Diethyl fumarate | + enantiomer **A** | + enantiomer **B** | + enantiomer **C**

CH_3 / CH_3 / CH_3 (trimethylcyclohexene structure)

+

enantiomer

D

6.

(a) **A** is $CH_2{=}CH\overset{CH_3}{\underset{OH}{C}}C{\equiv}CH$ **C** is $BrMgOCH_2CH{=}\overset{CH_3}{C}C{\equiv}CMgBr$

(b) **A** is an allylic alcohol and thus forms a carbocation readily. **B** is a conjugated enyne and is therefore more stable than **A**.

$$\underset{\mathbf{A}}{CH_2{=}CH{-}\overset{CH_3}{\underset{OH}{C}}{-}C{\equiv}CH} \xrightarrow[-H_2O]{H^+} CH_2{=}CH{-}\overset{CH_3}{\underset{+}{C}}{-}C{\equiv}CH \longleftrightarrow$$

$$\underset{+}{CH_2}{-}CH{=}\overset{CH_3}{C}{-}C{\equiv}CH \xrightarrow[-H^+]{H_2O} \underset{\mathbf{B}}{HOCH_2{-}CH{=}\overset{CH_3}{C}{-}C{\equiv}CH}$$

7.

$$R{-}\overset{O}{\overset{\|}{C}}H + BrMgOCH_2CH{=}\overset{CH_3}{C}C{\equiv}CMgBr \xrightarrow[(2)\ H^+]{}$$

$$\underset{\mathbf{D}}{R{-}\underset{OH}{CH}{-}C{\equiv}C{-}\overset{CH_3}{C}{=}CHCH_2OH} \xrightarrow{H_2,\ Ni_2B\ (P\text{-}2)}$$

$$\underset{\mathbf{E}}{R{-}\underset{OH}{CH}{-}CH{=}CH{-}\overset{CH_3}{C}{=}CHCH_2OH} \xrightarrow{(CH_3\overset{O}{\overset{\|}{C}})_2O}$$

$$\underset{\mathbf{F}}{R{-}\underset{OH}{CH}{-}CH{=}CH{-}\overset{CH_3}{C}{=}CHCH_2O\overset{O}{\overset{\|}{C}}CH_3} \xrightarrow{-H_2O}$$

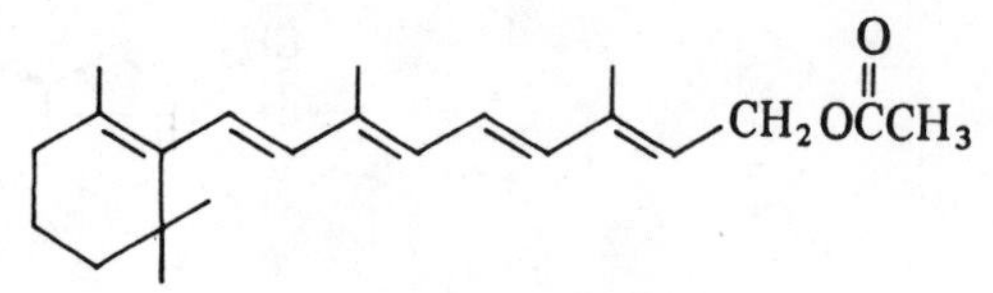

CH_2OCCH_3 (with C=O)

Vitamin A acetate

8.

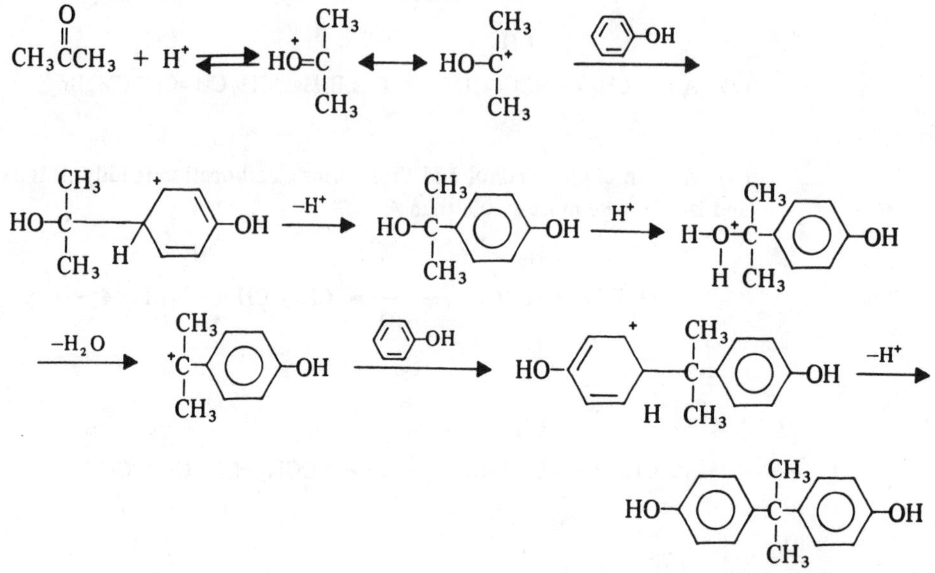

$$CH_3\overset{O}{\overset{\|}{C}}CH_3 + H^+ \rightleftharpoons HO\overset{+}{=}C(CH_3)_2 \longleftrightarrow HO{-}C^+(CH_3)_2 \xrightarrow{C_6H_5OH}$$

$$HO{-}C(CH_3)_2{-}C_6H_5^+(H){-}OH \xrightarrow{-H^+} HO{-}C(CH_3)_2{-}C_6H_4{-}OH \xrightarrow{H^+} H{-}\overset{+}{O}(H){-}C(CH_3)_2{-}C_6H_4{-}OH$$

$$\xrightarrow{-H_2O} {}^+C(CH_3)_2{-}C_6H_4{-}OH \xrightarrow{C_6H_5OH} HO{-}C_6H_5^+(H){-}C(CH_3)_2{-}C_6H_4{-}OH \xrightarrow{-H^+}$$

$$HO{-}C_6H_4{-}C(CH_3)_2{-}C_6H_4{-}OH$$

Bisphenol A

9.

$$O_2N{-}C_6H_4{-}CH_3 \xrightarrow[(2)\ H_3O^+]{(1)\ KMnO_4,\ OH^-,\ heat} O_2N{-}C_6H_4{-}\overset{O}{\overset{\|}{C}}OH \xrightarrow{SOCl_2}$$

A

$$O_2N{-}C_6H_4{-}\overset{O}{\overset{\|}{C}}Cl \xrightarrow{HOCH_2CH_2N(C_2H_5)_2} O_2N{-}C_6H_4{-}\overset{O}{\overset{\|}{C}}OCH_2CH_2N(C_2H_5)_2$$

B C

$$\xrightarrow{H_2,\ cat.} H_2N{-}C_6H_4{-}\overset{O}{\overset{\|}{C}}OCH_2CH_2N(C_2H_5)_2$$

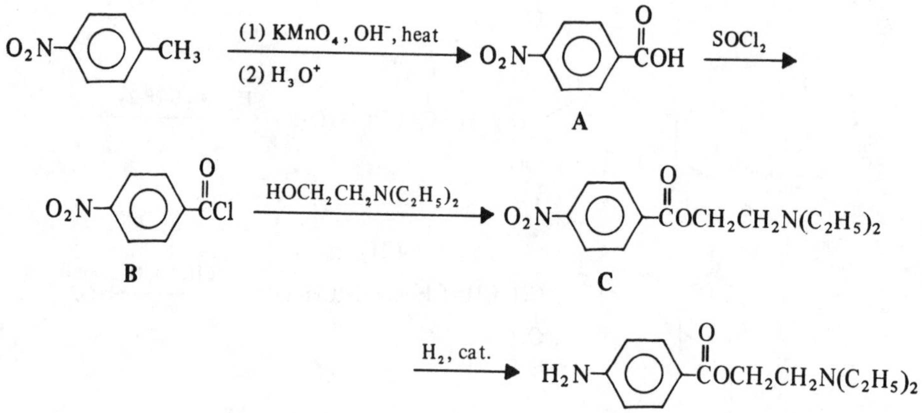

Procaine

10.

$$\text{Cyclohexanone} \xrightarrow[\text{(2) } H^+]{\text{(1) } HC{\equiv}CNa} \underset{\mathbf{A}}{\text{1-(HC}{\equiv}\text{C)-1-OH-cyclohexane}} \xrightarrow{Cl-\overset{O}{\overset{\|}{C}}-Cl}$$

$$\underset{\mathbf{B}}{\text{1-(HC}{\equiv}\text{C)-1-(O-}\overset{O}{\overset{\|}{C}}\text{-Cl)-cyclohexane}} \xrightarrow{NH_3} \underset{\text{Ethinamate}}{\text{1-(HC}{\equiv}\text{C)-1-(O-}\overset{O}{\overset{\|}{C}}\text{-}NH_2\text{)-cyclohexane}}$$

11.

$$C_6H_5\overset{O}{\overset{\|}{C}}H \xrightarrow[\text{(2) } H_3O^+]{\text{(1) } C_6H_5MgBr} \underset{\mathbf{A}}{C_6H_5\underset{C_6H_5}{\underset{|}{C}}HOH} \xrightarrow{PBr_3} \underset{\mathbf{B}}{C_6H_5\underset{C_6H_5}{\underset{|}{C}}HBr} \xrightarrow[-HBr]{(CH_3)_2NCH_2CH_2OH}$$

$$C_6H_5\underset{C_6H_5}{\underset{|}{C}}HOCH_2CH_2N(CH_3)_2$$

Diphenhydramine

The last step probably takes place by an S_N1 mechanism. Diphenylmethyl bromide, **B**, ionizes readily because it forms the resonance-stabilized benzylic carbocation,

$$C_6H_5\underset{C_6H_5}{\underset{|}{C}}H^+$$

12.

(a) For this synthesis we need to prepare the benzylic halide, $Br-C_6H_4-\underset{C_6H_5}{\underset{|}{C}}HBr$, and then allow it to react with $(CH_3)_2NCH_2CH_2OH$ as in Problem 11.

This benzylic halide can be made as follows

$$Br-C_6H_4-\overset{O}{\overset{\|}{C}}H \xrightarrow[\text{(2) } H_3O^+]{\text{(1) } C_6H_5MgBr} Br-C_6H_4-\underset{C_6H_5}{\underset{|}{C}}HOH \xrightarrow{PBr_3} Br-C_6H_4-\underset{C_6H_5}{\underset{|}{C}}HBr$$

(b) For this synthesis we can prepare the requisite benzylic halide in two ways:

$$o\text{-}CH_3C_6H_4-\overset{O}{\overset{\|}{C}}H \xrightarrow[\text{(2) } H_3O^+]{\text{(1) } C_6H_5MgBr} o\text{-}CH_3C_6H_4-\underset{C_6H_5}{\underset{|}{C}}HOH \xrightarrow{PBr_3} o\text{-}CH_3C_6H_4-\underset{C_6H_5}{\underset{|}{C}}HBr$$

or

$$C_6H_5\overset{O}{\overset{\|}{C}}H \xrightarrow[(2)\ H_3O^+]{(1)\ m\text{-}CH_3C_6H_4MgBr} m\text{-}CH_3C_6H_4\underset{C_6H_5}{CH}OH \xrightarrow{PBr_3} m\text{-}CH_3C_6H_4\underset{C_6H_5}{CH}Br$$

We then allow the benzylic halide to react with $(CH_3)_2NCH_2CH_2OH$ as in Problem 11.

13.

$$CH_3\underset{Br}{CH}CO_2C_2H_5 \xrightarrow{CH_2(CO_2C_2H_5)_2,\ EtO^-} \underset{\textbf{A}}{(C_2H_5O_2C)(CH_3)CH{-}CH(CO_2C_2H_5)_2}$$

$$\xrightarrow[\text{(Michael addition)}]{CH_2{=}CHCN,\ EtO^-} \underset{\textbf{B}}{(C_2H_5O_2C)(CH_3)CH{-}C(CO_2C_2H_5)_2{-}CH_2CH_2C{\equiv}N} \xrightarrow[\text{(converts }{-}C{\equiv}N\text{ to }{-}CO_2C_2H_5)]{C_2H_5OH,\ H^+}$$

$$\underset{\textbf{C}}{(C_2H_5O_2C)(CH_3)CH{-}C(CO_2C_2H_5)_2{-}CH_2CH_2CO_2C_2H_5} \xrightarrow[\text{(Dieckmann condensation)}]{EtO^-} \underset{\textbf{D}}{\text{cyclic }\beta\text{-keto ester: }C_2H_5O_2C{-}CH{-}\overset{O}{\overset{\|}{C}}{-}CH(CH_3){-}C(CO_2C_2H_5)_2{-}CH_2\text{ (ring)}}$$

$$\xrightarrow[(2)\ H_3O^+]{(1)\ OH^-,\ H_2O,\ heat} \left[HO_2C{-}CH{-}\overset{O}{\overset{\|}{C}}{-}CH(CH_3){-}C(CO_2H)_2{-}CH_2\text{ (ring)}\right] \xrightarrow[-2CO_2]{(3)\ heat} \text{2-methyl-3-carboxycyclopentanone (}CH_3\text{, }CO_2H\text{)}$$

14.

$$CH_3\overset{O}{\overset{\|}{C}}CH_3 \xrightarrow[\text{(acid-catalyzed aldol condensation)}]{HCl} \underset{\textbf{A}}{CH_3\overset{CH_3}{\overset{|}{C}}{=}CH\overset{O}{\overset{\|}{C}}CH_3} \xrightarrow[\text{(Michael addition)}]{CH_3\overset{O}{\overset{\|}{C}}CH_2\overset{O}{\overset{\|}{C}}OEt,\ base}$$

$$\left[C_2H_5O_2C-CH(COCH_3)-C(CH_3)_2-CH_2-CO-CH_3 \right] \text{ (B)} \xrightarrow[\text{(intramolecular aldol condensation)}]{\text{base}} \text{C}$$

C: ethyl 2,2,4-trimethyl-6-oxocyclohex-4-ene-1-carboxylate ($C_2H_5O_2C$, CH_3, CH_3, CH_3, O)

$$\text{C} \xrightarrow[\text{(hydrolysis and decarboxylation of } \beta\text{-keto ester)}]{H^+, H_2O, \text{ heat}} \text{D} + CO_2$$

D: 3,5,5-trimethylcyclohex-2-en-1-one (CH_3, CH_3, CH_3, O)

15.

$$C_6H_5CHO + CH_2(NO_2)CH_2OH \xrightarrow[\text{(aldol-type addition)}]{EtO^-} C_6H_5CH(OH)CH(NO_2)CH_2OH \text{ (A)} \xrightarrow{H_2\text{, catalyst}}$$

$$C_6H_5CH(OH)CH(NH_2)CH_2OH \text{ (B)} \xrightarrow{Cl_2CHCOCl} C_6H_5CH(OH)CH(NHCOCHCl_2)CH_2OH \text{ (C)} \xrightarrow{\text{excess } (CH_3CO)_2O}$$

$$C_6H_5CH(OCOCH_3)CH(NHCOCHCl_2)CH_2OCOCH_3 \text{ (D)} \xrightarrow[H_2SO_4]{HNO_3} O_2N\text{-}C_6H_4\text{-}CH(OCOCH_3)CH(NHCOCHCl_2)CH_2OCOCH_3 \text{ (E)}$$

$$\xrightarrow[\text{(ester groups hydrolyze more rapidly than amide groups)}]{OH^-, H_2O} O_2N\text{-}C_6H_4\text{-}CH(OH)CH(NHCOCHCl_2)CH_2OH$$

Chloramphenicol

16.

$$CH_3CH_2CH_2CH(CH_3)CHO \xrightarrow[\text{(aldol addition)}]{HCHO, OH^-} CH_3CH_2CH_2C(CH_3)(CH_2OH)CHO \text{ (A)} \xrightarrow[\text{(Cannizzaro reaction)}]{HCHO, OH^-}$$

$$\underset{\mathbf{B}}{HOCH_2C(CH_3)(CH_2CH_2CH_3)CH_2OH} \xrightarrow{ClCOCl} \underset{\mathbf{C}}{ClCOOCH_2C(CH_3)(CH_2CH_2CH_3)CH_2OCOCl} \xrightarrow{NH_3} \underset{\text{Meprobamate}}{H_2NCOOCH_2CH(CH_3)(CH_2CH_2CH_3)CHCH_2OCONH_2}$$

17. (succinic anhydride) $\xrightarrow{CH_3CH_2CH_2OH}$ $\underset{\mathbf{A}}{CH_3CH_2CH_2OCOCH_2CH_2COOH}$ $\xrightarrow{SOCl_2}$

$$\underset{\mathbf{B}}{CH_3CH_2CH_2OCOCH_2CH_2COCl} \xrightarrow{(CH_3CH_2)_2NH} \underset{\mathbf{C}}{CH_3CH_2CH_2OCOCH_2CH_2CON(CH_2CH_3)_2}$$

18. cyclopentadiene + $C_6H_5CH{=}CHNO_2$ $\xrightarrow[\text{reaction)}]{\text{(Diels-Alder}}$ **A** (bicyclic, C_6H_5, NO_2) $\xrightarrow{H_2,\ Pt}$

B (bicyclic, C_6H_5, NH_2) $\xrightarrow{CH_3CHO}$ [**C** (bicyclic, C_6H_5, $NH{=}CHCH_3$)] $\xrightarrow{H_2,\ Ni}$ Fencamfamine (bicyclic, C_6H_5, $NHCH_2CH_3$)

19. 1-methylcyclohexene (CH_3) $\xrightarrow[(2)\ NaHSO_3]{(1)\ OsO_4}$ **A** (1-methylcyclohexane-1,2-diol: CH_3, OH, OH) $\xrightarrow[CH_3CO_2H]{CrO_3}$

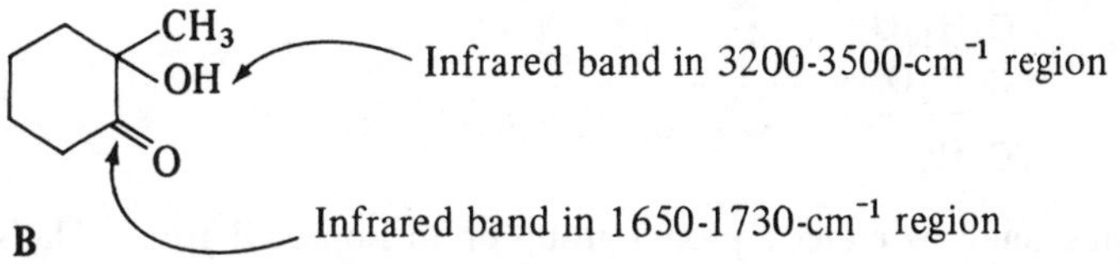

CH_3 OH — Infrared band in 3200-3500-cm^{-1} region

O — Infrared band in 1650-1730-cm^{-1} region

B

Notice that the second step involves the oxidation of a secondary alcohol in the presence of a tertiary alcohol. This selectivity is possible because tertiary alcohols do not undergo oxidation readily (Section 15.3).

20. Working backward, we notice that methyl *trans*-4-isopropylcyclohexanecarboxylate has both large groups equatorial and is, therefore, more stable than the corresponding cis isomer. This stability of the trans isomer means that if we were to synthesize the cis isomer or a mixture of both the cis and trans isomers we could obtain the desired trans isomer by a base-catalyzed isomerization:

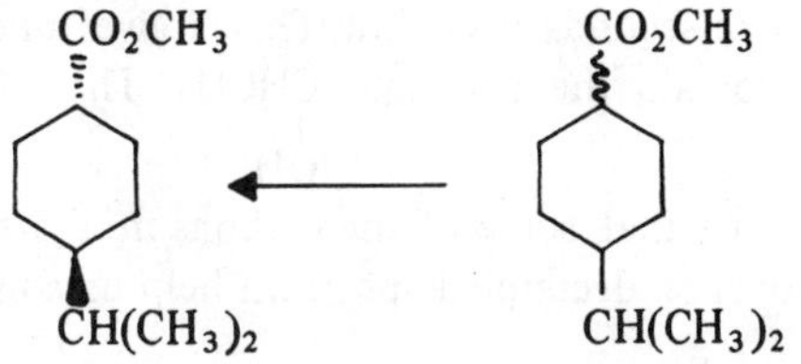

CO_2CH_3 / $CH(CH_3)_2$ (more stable trans isomer) ← CO_2CH_3 / $CH(CH_3)_2$ (cis isomer or mixture of cis and trans isomers)

We could synthesize a mixture of the desired isomers from phenol in the following way:

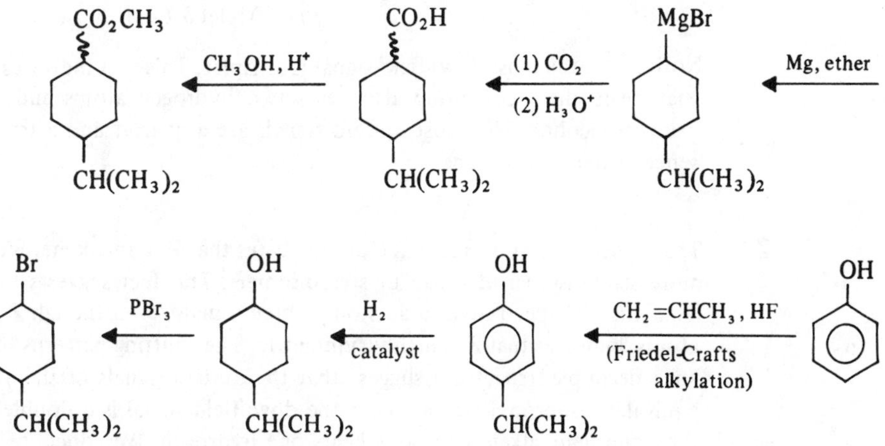

CO_2CH_3 / $CH(CH_3)_2$ ← (CH_3OH, H^+) — CO_2H / $CH(CH_3)_2$ ← ((1) CO_2 (2) H_3O^+) — MgBr / $CH(CH_3)_2$ ← (Mg, ether)

Br / $CH(CH_3)_2$ ← (PBr_3) — OH / $CH(CH_3)_2$ ← (H_2, catalyst) — OH / $CH(CH_3)_2$ ← ($CH_2{=}CHCH_3$, HF (Friedel-Crafts alkylation)) — OH

21. The positive iodoform test and the strong infrared absorption of **X** indicate that it contains a $-\underset{\underset{O}{\|}}{C}CH_3$ group. Subtracting this from the molecular formula, $C_5H_{10}O$, leaves only C_3H_7.

$$\begin{array}{l} C_5H_{10}O \\ -C_2H_3O \\ \hline C_3H_7 \end{array}$$

This could be either a propyl group or an isopropyl group. The splitting patterns of the proton off-resonance decoupled spectrum are consistent only with an isopropyl group, hence **X** is isopropyl methyl ketone. The assignments are the following:

$(CH_3)_2CHC(=O)CH_3$ — *(a)* $(CH_3)_2$, *(c)* CH, *(d)* C=O, *(b)* CH_3

(a) Quartet δ 18.1

(b) Quartet δ 27.3

(c) Doublet δ 41.5

(d) Singlet δ 211.8

22. That **Y** gives a green opaque solution when treated with CrO_3 in aqueous H_2SO_4 indicates that **Y** is a primary or secondary alcohol. That **Y** gives a negative iodoform test indicates that **Y** does not contain the grouping $-\underset{OH}{\underset{|}{C}}HCH_3$. The ^{13}C spectrum of **Y** contains only four signals indicating that some of the carbons in **Y** are equivalent. The splitting patterns of the off-resonance decoupled spectrum help us conclude that **Y** is 2-ethyl-1-butanol.

$(CH_3CH_2)_2CHCH_2OH$ — *(a)* CH_3, *(b)* CH_2, *(c)* CH, *(d)* CH_2OH

(a) Quartet δ 11.1

(b) Triplet δ 23.0

(c) Doublet δ 43.6

(d) Triplet δ 64.6

Notice that the most downfield signal is a triplet. This fact indicates that the carbon atom that bears the $-OH$ group also bears two hydrogen atoms and, therefore that **Y** is a primary alcohol. The most upfield signals are a quartet and a triplet indicating the presence of the ethyl groups.

23. That **Z** decolorizes bromine in CCl_4 indicates that **Z** is an alkene. We are told that **Z** is the more stable isomer of a pair of stereoisomers. This fact suggests that **Z** is a trans alkene. That the ^{13}C spectrum contains only three signals, even though **Z** contains eight carbon atoms, indicates that **Z** is highly symmetric. The splitting patterns of the proton off-resonance decoupled spectrum suggest that the upfield signals of the alkyl groups arise from equivalent isopropyl groups. That the downfield signal is a doublet suggests that each of the equivalent alkenyl carbons bears one hydrogen. We conclude, therefore, and that **Z** is *trans*-2,5-dimethyl-3-hexene.

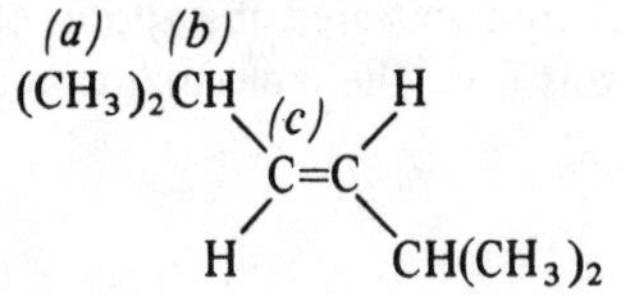

$(CH_3)_2CH-CH=CH-CH(CH_3)_2$ (trans) — *(a)* $(CH_3)_2$, *(b)* CH, *(c)* C=C

(a) Quartet δ 22.8

(b) Doublet δ 31.0

(c) Doublet δ 134.5

L SPECIAL TOPIC Alkaloids

SOLUTIONS TO PROBLEMS

L.1 (a) The first step is similar to a crossed Claisen condensation (see Section 20.2A):

$$\text{3-pyridyl}-\overset{O}{\overset{\|}{C}}OC_2H_5 + \text{N-methyl-2-pyrrolidinone} \xrightarrow{C_2H_5ONa} \text{3-pyridyl}-\overset{O}{\overset{\|}{C}}-\text{(3-position of N-methyl-2-pyrrolidinone)}$$

(b) This step involves hydrolysis of an amide (lactam) and can be carried out with either acid or base. Here we use acid.

$$\xrightarrow[H_2O]{H_3O^+} \text{3-pyridyl}-\overset{O}{\overset{\|}{C}}-\underset{CO_2H}{\underset{|}{C}}HCH_2CH_2NHCH_3 \xrightarrow{\text{heat}} \text{3-pyridyl}-\overset{O}{\overset{\|}{C}}-CH_2CH_2CH_2NHCH_3$$

(c) This step is the decarboxylation of a substituted malonic acid; it requires only the application of heat and takes place during the acid hydrolysis of step (b).

(d) This is the reduction of a ketone to a secondary alcohol. A variety of reducing agents can be used, sodium borohydride, for example.

$$\xrightarrow{NaBH_4} \text{3-pyridyl}-\overset{OH}{\overset{|}{C}}HCH_2CH_2CH_2NHCH_3$$

(e) Here we convert the secondary alcohol to an alkyl bromide with hydrogen bromide; this reagent also gives a hydrobromide salt of the aliphatic amine.

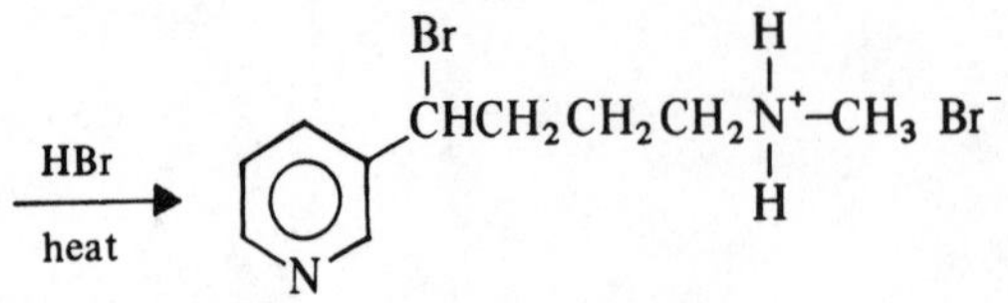

$\xrightarrow[\text{heat}]{\text{HBr}}$ 3-pyridyl–$\overset{\text{Br}}{\text{CH}}CH_2CH_2CH_2\overset{\text{H}}{\underset{\text{H}}{N^+}}$–$CH_3$ Br^-

(f) Treating the salt with base produces the secondary amine; it then acts as a nucleophile and attacks the carbon atom bearing the bromine. This reaction leads to the formation of a five-membered ring and (±) nicotine.

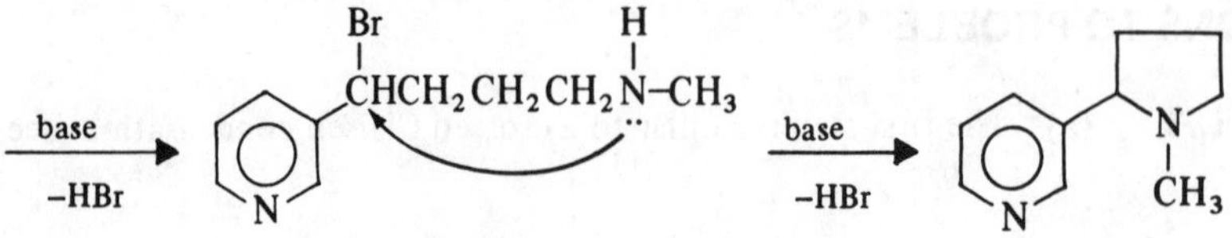

$\xrightarrow[\text{-HBr}]{\text{base}}$ 3-pyridyl–$\overset{\text{Br}}{\text{CH}}CH_2CH_2CH_2\overset{\text{H}}{\ddot{N}}$–$CH_3$ $\xrightarrow[\text{-HBr}]{\text{base}}$ nicotine (N–CH_3)

L.2 (a) The stereocenter adjacent to the ester carbonyl group is racemized by base (probably through the formation of an anion that can undergo inversion of configuration, cf. Section 17.3).

(b) [Structure: N–CH_3 bicyclic tropane ester; O–C(=O)–C(H)(C_6H_5)(CH_2OH)]

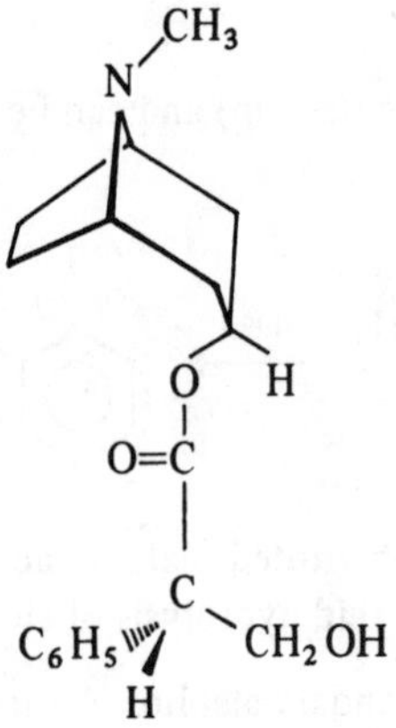

L.3

(a) [Structure: Tropine — N–CH_3 bicyclic with HO, H]

$C_6H_5CHCO_2H$ with CH_2OH on the CH

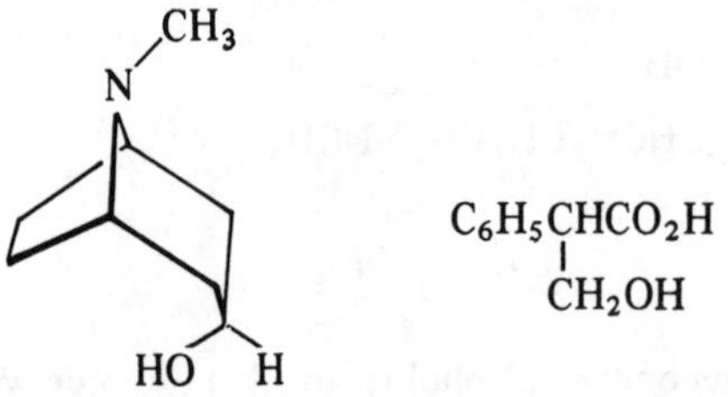

Tropine (±) Tropic acid

(b) Tropine is a meso compound; it has a plane of symmetry that passes through the >CHOH group, the >NCH_3 group, and between the two –CH_2– groups of the five-membered ring.

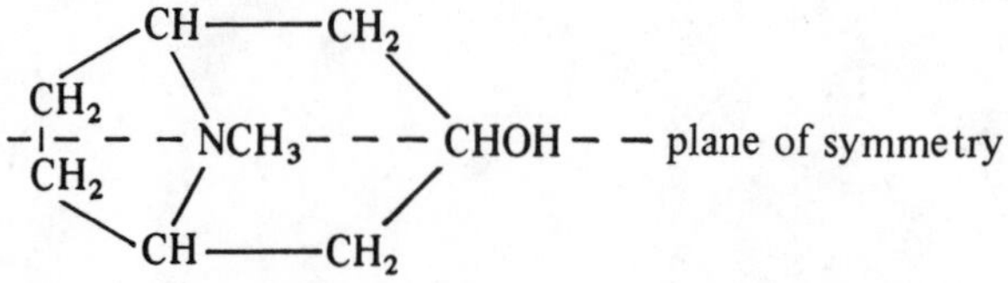

CH—CH_2 / CH_2 -- NCH_3 -- CHOH -- plane of symmetry / CH_2 / CH—CH_2

(c)

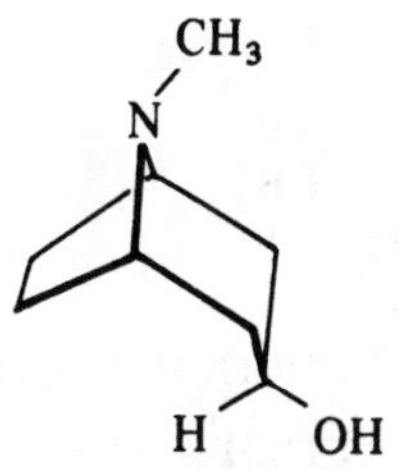

ψ-Tropine

L.4

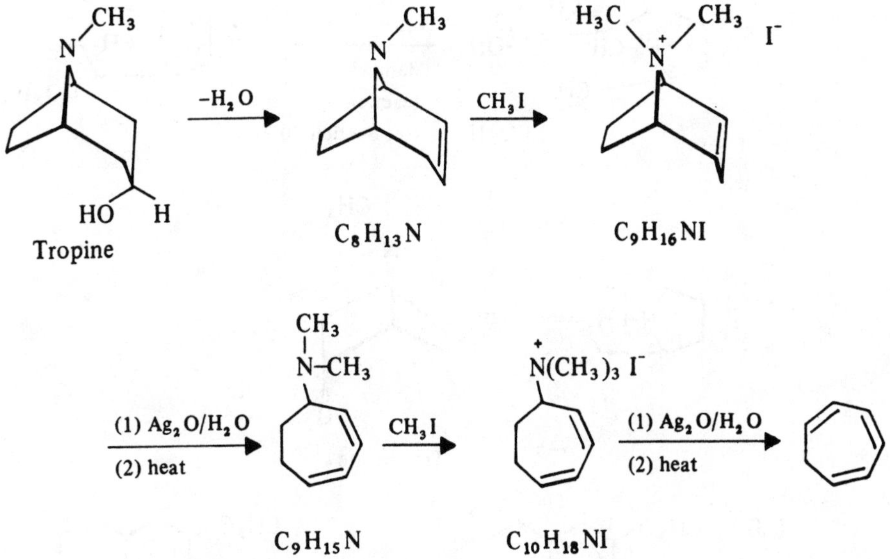

Tropine $\xrightarrow{-H_2O}$ $C_8H_{13}N$ $\xrightarrow{CH_3I}$ $C_9H_{16}NI$ (I^-)

$\xrightarrow[(2)\ \text{heat}]{(1)\ Ag_2O/H_2O}$ $C_9H_{15}N$ $\xrightarrow{CH_3I}$ $C_{10}H_{18}NI$ ($\overset{+}{N}(CH_3)_3\ I^-$) $\xrightarrow[(2)\ \text{heat}]{(1)\ Ag_2O/H_2O}$ cycloheptatriene

L.5 One possible sequence of steps is the following:

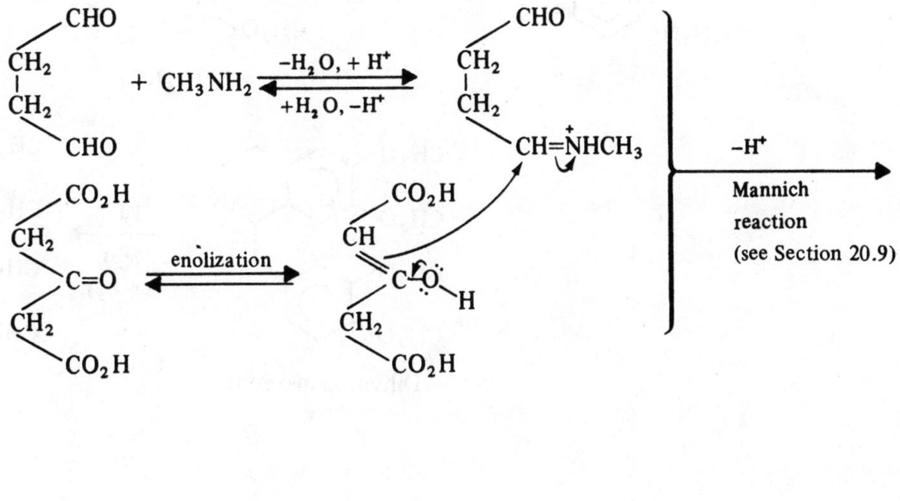

$$OHC{-}CH_2{-}CH_2{-}CHO + CH_3NH_2 \underset{+H_2O,\ -H^+}{\overset{-H_2O,\ +H^+}{\rightleftharpoons}} OHC{-}CH_2{-}CH_2{-}CH{=}\overset{+}{N}HCH_3$$

$$HO_2C{-}CH_2{-}C(=O){-}CH_2{-}CO_2H \overset{\text{enolization}}{\rightleftharpoons} HO_2C{-}CH{=}C(OH){-}CH_2{-}CO_2H$$

$\xrightarrow{-H^+}$ Mannich reaction (see Section 20.9)

CHO
CH_2
CH_2 NHCH$_3$
CH O
CH–C–CH_2CO_2H
CO_2H

$+H^+$, $-H_2O$ / $+H_2O$, $-H^+$

CH_3
N
CHCOCH$_2$CO$_2$H
CO_2H

enolization

CO_2H
CH
N$^+$–CH_3 C–OH
CH
CO_2H

$-H^+$
Mannich reaction
(see Section 20.9)

CO_2H
N–CH_3 =O
CO_2H

$-2CO_2$

N–CH_3 =O ≡

CH_3
N
O

Tropinone

L.6 CH_3O, CH_3O, NH_2 + COCl, CH_3O, CH_3O $\xrightarrow{OH^-}$

CH_3O, CH_3O, O, N, H, CH_3O, CH_3O

$C_{20}H_{25}NO_5$

P_2O_5
heat
$(-H_2O)$

CH_3O, CH_3O, N, CH_3O, CH_3O

Dihydropapaverine

Pd
heat
$(-H_2)$

CH_3O, CH_3O, N, CH_3O, CH_3O

Papaverine

L.7 A Diels-Alder reaction was carried out using 1,3-butadiene as the diene component.

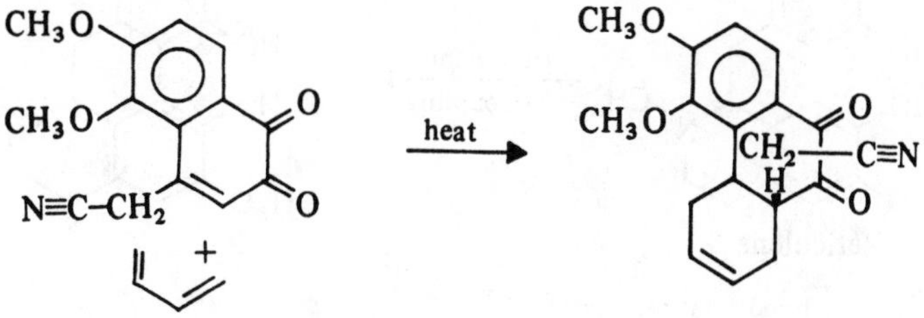

CH_3O, CH_3O, $N{\equiv}C{-}CH_2$, O, O, + (butadiene) —heat→ CH_3O, CH_3O, CH_2, $C{\equiv}N$, H, O, O

L.8 Acetic anhydride acetylates both –OH groups.

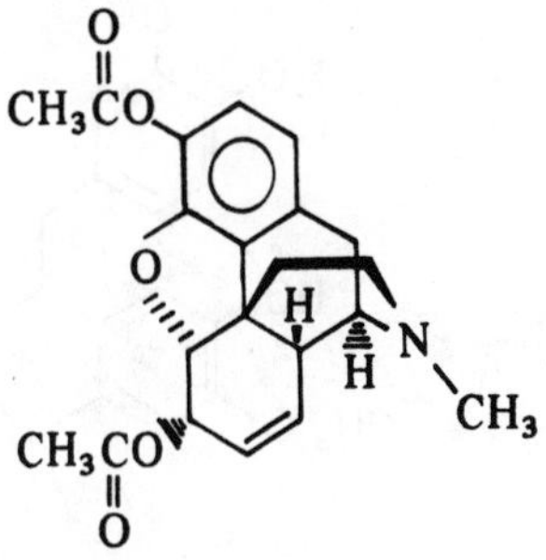

O, CH_3CO, O, H, H, N, CH_3, CH_3CO, O

Heroin

L.9 (a) A Mannich reaction (see Section 20.9).

(b) $CH_2O + HN(CH_3)_2 \underset{}{\overset{+H^+ \atop -H_2O}{\rightleftharpoons}} CH_2{=}\overset{+}{N}(CH_3)_2$

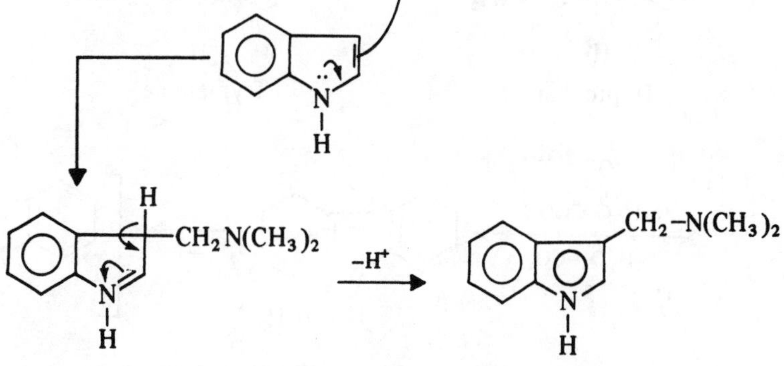

N–H (indole)

H, $-CH_2N(CH_3)_2$, N, H $\xrightarrow{-H^+}$ $CH_2{-}N(CH_3)_2$, N, H

Gramine

L.10

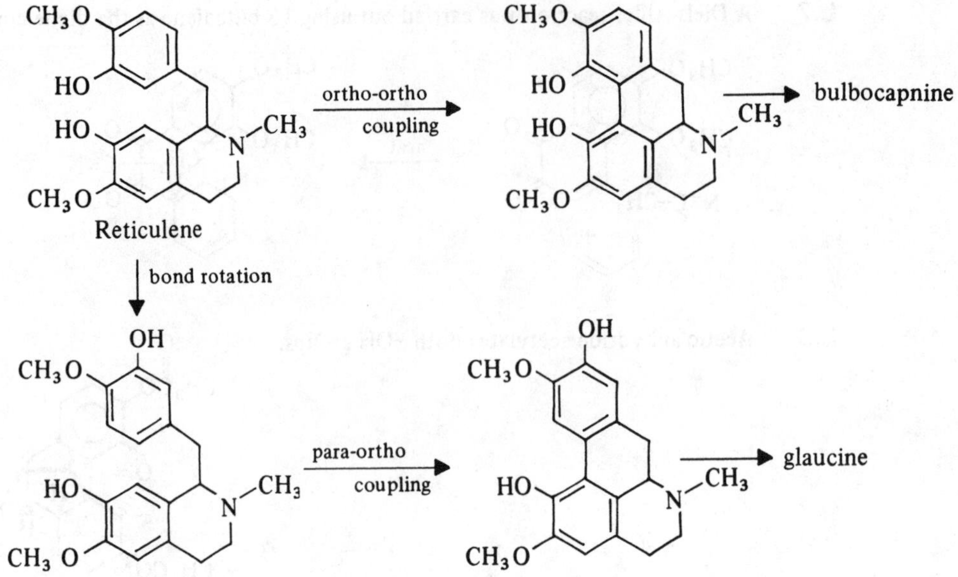

CH_3O, HO, HO, CH_3O — Reticulene — ortho-ortho coupling → CH_3O, HO, HO, CH_3O, N–CH_3 → bulbocapnine

Reticulene ↓ bond rotation

OH, CH_3O, HO, CH_3O, N–CH_3 — para-ortho coupling → OH, CH_3O, HO, CH_3O, N–CH_3 → glaucine

L.11 * and o are ^{14}C labels
▪ is an ^{15}N label

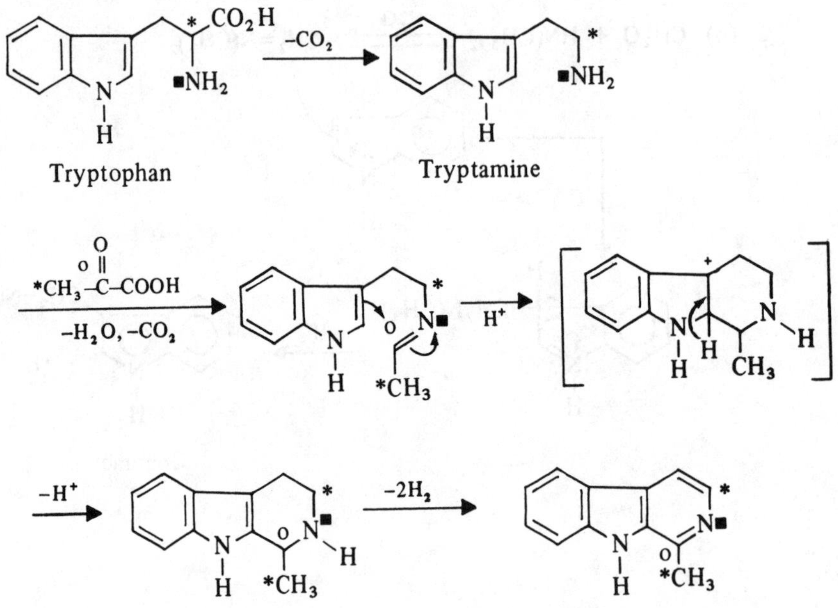

Tryptophan (*CO_2H, ▪NH_2, N–H) —$-CO_2$→ Tryptamine (*, ▪NH_2, N–H)

$\xrightarrow[-H_2O,\ -CO_2]{{}^{*}CH_3-\overset{o}{C}(=O)-COOH}$ (N–H, *, N▪, o, *CH_3) $\xrightarrow{H^+}$ [+, N–H, H, N–H, CH_3]

$\xrightarrow{-H^+}$ (N–H, *, N▪–H, o, *CH_3) $\xrightarrow{-2H_2}$ (N–H, *, N▪, o, *CH_3)

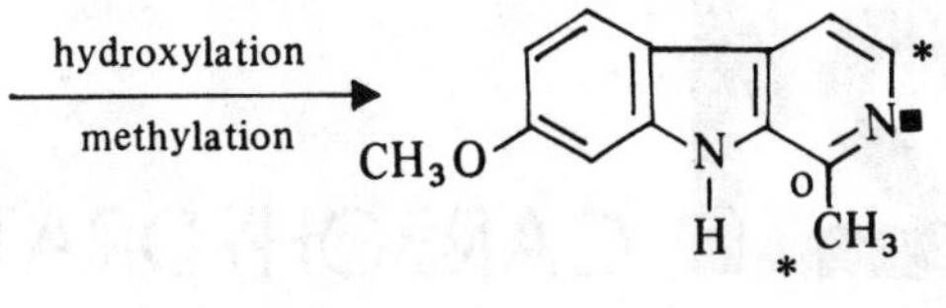

hydroxylation / methylation →

CH_3O ... N ... $N^{\blacksquare}$ ... H ... CH_3 (Harmine structure)

Harmine

(Cf. T. A. Geissman and D. H. G. Crout, *Organic Chemistry of Secondary Plant Metabolism,* Freeman, Cooper and Co., San Francisco, 1969, pp. 473-474.)

21 CARBOHYDRATES

SUMMARY OF SOME REACTIONS OF MONOSACCHARIDES

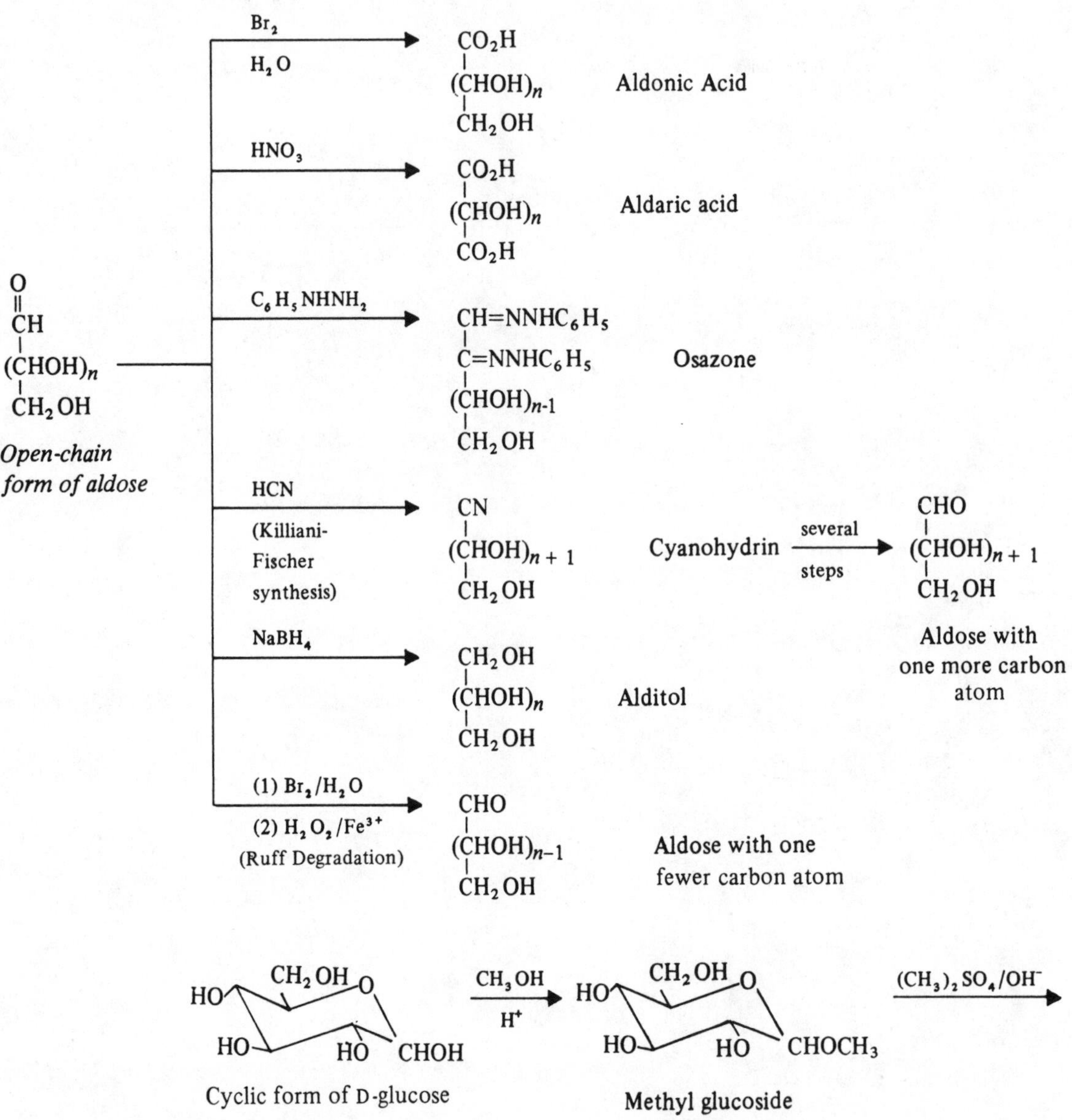

HO, CH_2OH, O, HO, HO, $CHOH$ — Cyclic form of D-glucose $\xrightarrow[H^+]{CH_3OH}$ HO, CH_2OH, O, HO, HO, $CHOCH_3$ — Methyl glucoside $\xrightarrow{(CH_3)_2SO_4/OH^-}$

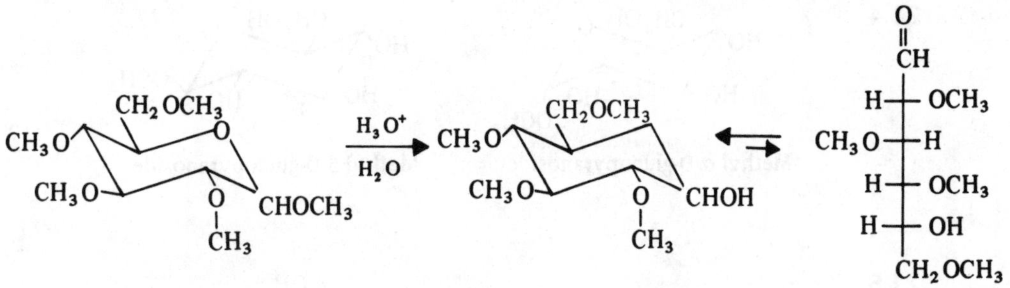

SOLUTIONS TO PROBLEMS

21.1 (a) Two,

CHO
*CHOH
*CHOH
CH_2OH

(b) Two,

CH_2OH
C=O
*CHOH
*CHOH
CH_2OH

(c) There would be four stereoisomers (two sets of enantiomers) with each general structure: $2^2 = 4$.

21.2

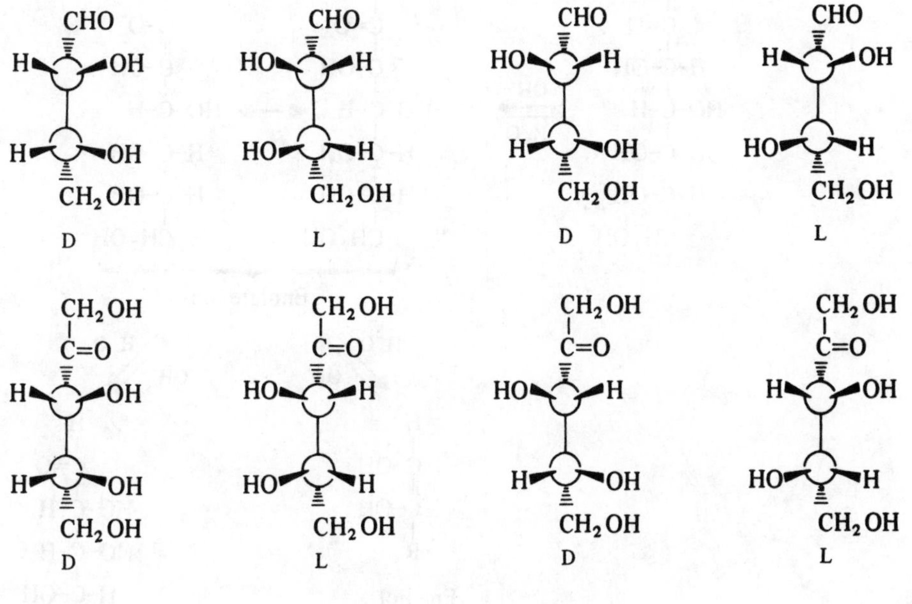

21.3 Since glycosides are acetals they undergo hydrolysis in aqueous acid to form cyclic hemiacetals that then undergo mutarotation.

21.4

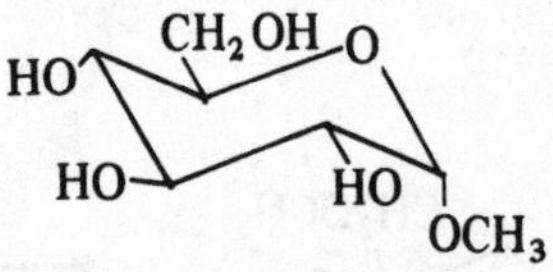

Methyl α-D-glucopyranoside

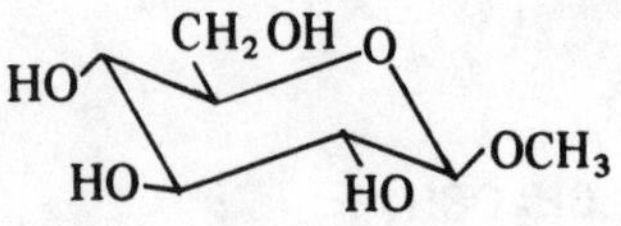

Methyl β-D-glucopyranoside

21.5

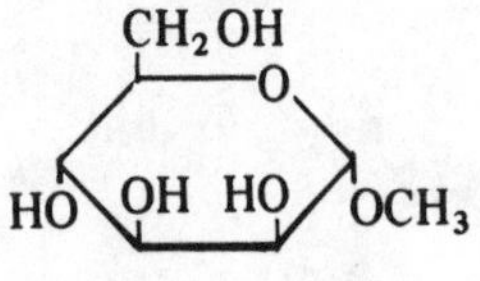

Haworth formula

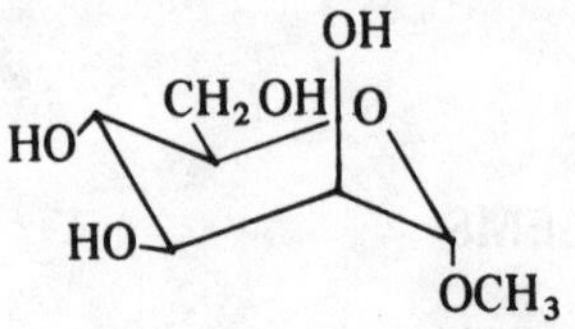

Conformational formula

Methyl-α-D-mannopyranoside

21.6 α-D-Glucopyranose will give a positive test with Benedict's or Tollens' solution because it is a cyclic hemiacetal. Methyl α-D-glucopyranoside, because it is a cyclic acetal, will not.

21.7

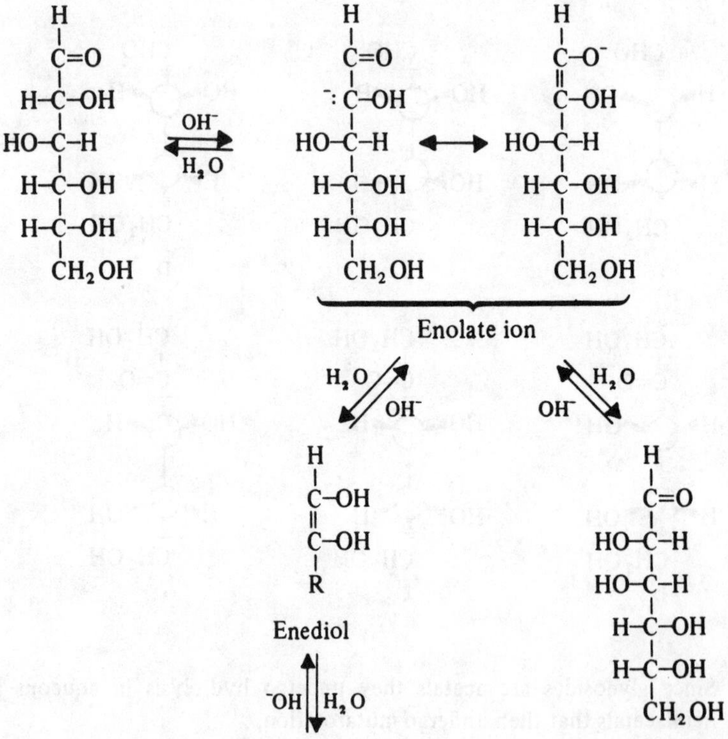

Enolate ion

Enediol

D-Mannose

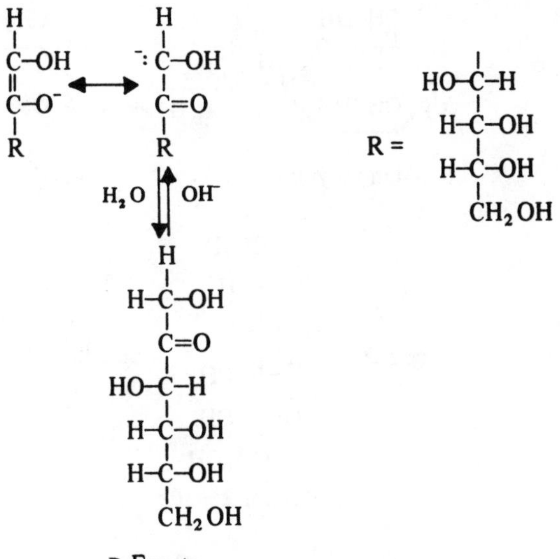

H
C–OH
C–O^-
R

↔

H
$^-$: C–OH
C=O
R

R = HO–C–H, H–C–OH, H–C–OH, CH_2OH

H_2O ⇅ OH^-

H
H–C–OH
C=O
HO–C–H
H–C–OH
H–C–OH
CH_2OH

D-Fructose

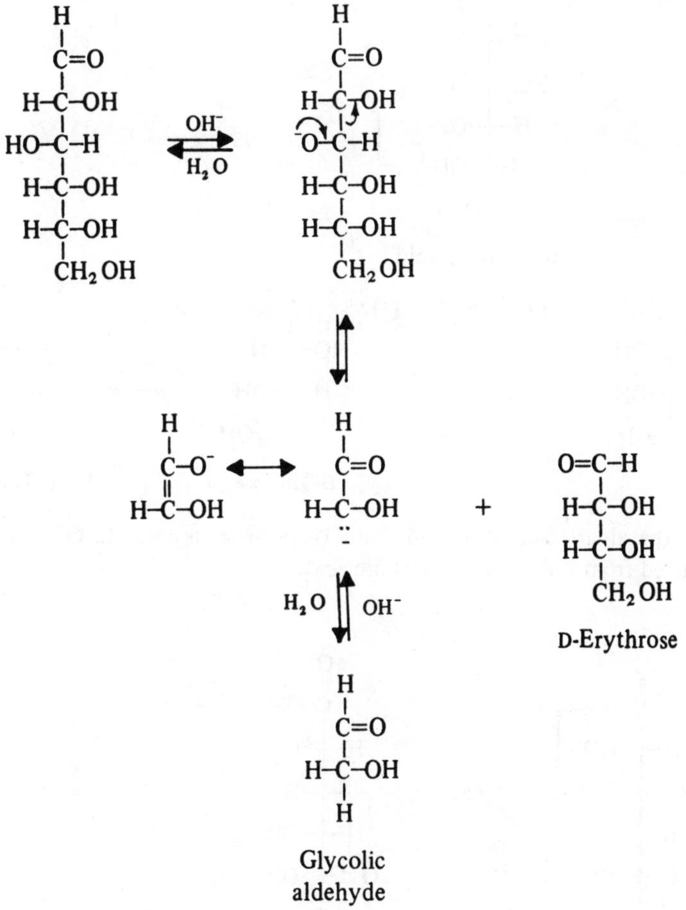

H
C=O
H–C–OH
HO–C–H
H–C–OH
H–C–OH
CH_2OH

OH^- ⇄ H_2O

H
C=O
H–C–OH
$^-$O–C–H
H–C–OH
H–C–OH
CH_2OH

⇅

H
C–O^-
H–C–OH

↔

H
C=O
H–C–OH
:

\+

O=C–H
H–C–OH
H–C–OH
CH_2OH

D-Erythrose

H_2O ⇅ OH^-

H
C=O
H–C–OH
H

Glycolic
aldehyde

21.8

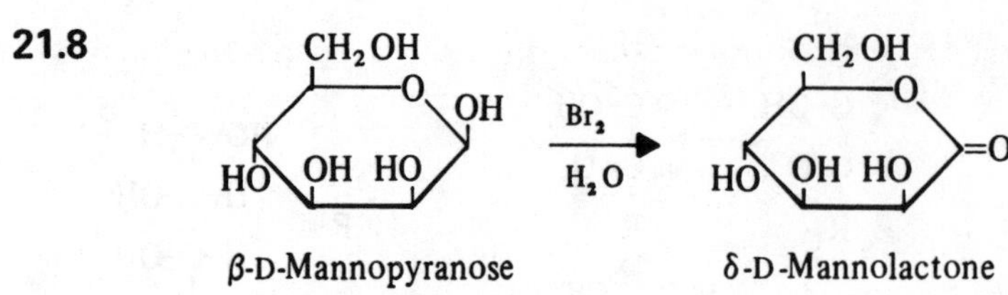

β-D-Mannopyranose δ-D-Mannolactone

⇄

CO_2H
HO—H
HO—H
H—OH
H—OH
CH_2OH

D-Mannonic acid

⇄

CH_2OH
HO—H
O
=O
OH HO

γ-D-Mannolactone

21.9 (a) Yes (b)

CO_2H
HO—H
HO—H
H—OH
H—OH
CO_2H

D-Mannaric acid

(c) Yes

(d)

CO_2H
H—OH
H—OH
CO_2H

(e) No (f)

CHO
HO—H
H—OH
CH_2OH

D-Threose

$\xrightarrow{HNO_3}$

CO_2H
HO—H
H—OH
CO_2H

D-Tartaric acid

(g) The aldaric acid obtained from D-erythrose is *meso*-tartaric acid; the aldaric acid obtained from D-threose is D-tartaric acid.

21.10

O
‖
C
H—OH
HO—H O
H—
H—OH
C—OH
‖
O

and

O
‖
C—OH
H—OH
—H
H—OH
O H—OH
C
‖
O

21.11 One way of predicting the products from a periodate oxidation is to place an —OH group on each carbon atom at the point where C—C bond cleavage has occurred:

$$\begin{array}{l} -\overset{|}{C}-OH \\ -\underset{|}{C}-OH \end{array} \xrightarrow{IO_4^-} \begin{array}{c} -\overset{|}{C}-OH \\ OH \\ + \\ OH \\ -\underset{|}{C}-OH \end{array}$$

Then if we recall (Section 8.16) that *gem*-diols are usually unstable and lose water to produce carbonyl compounds, we get the following results:

$$-\overset{|}{\underset{OH}{C}}-O-H \longrightarrow -\overset{|}{C}=O + H_2O$$

$$-\overset{OH}{\underset{|}{C}}-O-H \longrightarrow -\underset{|}{C}=O + H_2O$$

Let us apply this procedure to several examples here while we remember that for every C—C bond that is broken 1 mole of HIO_4 is consumed.

(a)

$$\begin{array}{c} CH_3 \\ H-C-OH \\ ---|---- \\ H-C-OH \\ CH_3 \end{array} + HIO_4 \longrightarrow \begin{array}{c} CH_3 \\ H-C-O-H \\ OH \\ + \\ OH \\ H-C-O-H \\ CH_3 \end{array} \xrightarrow[-2H_2O]{} 2CH_3\overset{O}{\overset{\|}{C}}-H$$

(b)

$$\begin{array}{c} H \\ H-C-OH \\ ---|---- \\ H-C-OH \\ ---|---- \\ H-C-OH \\ CH_3 \end{array} + 2HIO_4 \longrightarrow \begin{array}{c} H \\ H-C-O-H \\ OH \\ + \\ O-H \\ H-C-OH \\ OH \\ + \\ OH \\ H-C-O-H \\ CH_3 \end{array} \xrightarrow[-3H_2O]{} \begin{array}{c} H \\ H-C=O \\ + \\ O \\ \| \\ H-C-OH \\ + \\ H-C=O \\ CH_3 \end{array}$$

(c)

$$\begin{array}{c} H \\ | \\ H\text{-}C\text{-}OH \\ | \\ H\text{-}C\text{-}OH \\ | \\ H\text{-}C\text{-}OCH_3 \\ | \\ OCH_3 \end{array} + HIO_4 \longrightarrow \begin{array}{c} H \\ | \\ H\text{-}C\text{-}OH \\ | \\ OH \\ + \\ OH \\ | \\ H\text{-}C\text{-}OH \\ | \\ H\text{-}C\text{-}OCH_3 \\ | \\ OCH_3 \end{array} \xrightarrow[-2H_2O]{} \begin{array}{c} H \\ | \\ H\text{-}C{=}O \\ + \\ O \\ \| \\ H\text{-}C \\ | \\ H\text{-}C\text{-}OCH_3 \\ | \\ OCH_3 \end{array}$$

(d)

$$\begin{array}{c} H \\ | \\ H\text{-}C\text{-}OH \\ | \\ H\text{-}C\text{-}OH \\ | \\ C{=}O \\ | \\ CH_3 \end{array} + 2HIO_4 \longrightarrow \begin{array}{c} H \\ | \\ H\text{-}C\text{-}OH \\ | \\ OH \\ + \\ OH \\ | \\ H\text{-}C\text{-}OH \\ | \\ OH \\ + \\ OH \\ | \\ C{=}O \\ | \\ CH_3 \end{array} \xrightarrow[-2H_2O]{} \begin{array}{c} H \\ | \\ H\text{-}C{=}O \\ + \\ O \\ \| \\ H\text{-}C\text{-}OH \\ + \\ O \\ \| \\ CH_3COH \end{array}$$

(e)

$$\begin{array}{c} CH_3 \\ | \\ C{=}O \\ | \\ H\text{-}C\text{-}OH \\ | \\ C{=}O \\ | \\ CH_3 \end{array} + 2HIO_4 \longrightarrow \begin{array}{c} CH_3 \\ | \\ C{=}O \\ | \\ OH \\ + \\ OH \\ | \\ H\text{-}C\text{-}OH \\ | \\ OH \\ + \\ OH \\ | \\ C{=}O \\ | \\ CH_3 \end{array} \xrightarrow[-2H_2O]{} 2CH_3\overset{O}{\overset{\|}{C}}OH + H\overset{O}{\overset{\|}{C}}OH$$

(f)

$$H_2C\begin{array}{l} \diagup CH_2\diagdown \overset{H}{\overset{|}{C}}\text{-}OH \\ \diagdown CH_2\diagup \underset{H}{\underset{|}{C}}\text{-}OH \end{array} + HIO_4 \longrightarrow H_2C\begin{array}{l} \diagup CH_2\text{-}\overset{H}{\overset{|}{C}}\text{-}OH \\ \qquad\quad | \\ \qquad\quad OH \\ \qquad\quad OH \\ \qquad\quad | \\ \diagdown CH_2\text{-}\underset{H}{\underset{|}{C}}\text{-}OH \end{array} \xrightarrow[-2H_2O]{}$$

$$H\overset{O}{\overset{\|}{C}}CH_2CH_2CH_2\overset{O}{\overset{\|}{C}}H$$

(g)

$$\begin{array}{c} H \\ | \\ H-C-OH \\ \text{-----}|\text{-----} \\ CH_3-C-OH \\ | \\ CH_3 \end{array} + HIO_4 \longrightarrow \begin{array}{c} H \\ | \\ H-C-OH \\ | \\ OH \\ + \\ OH \\ | \\ CH_3-C-OH \\ | \\ CH_3 \end{array} \xrightarrow{-2H_2O} \begin{array}{c} H \\ | \\ H-C=O \\ + \\ CH_3-C=O \\ | \\ CH_3 \end{array}$$

(h)

$$\begin{array}{c} O \\ \| \\ H-C \\ \text{-----}|\text{-----} \\ H-C-OH \\ \text{-----}|\text{-----} \\ H-C-OH \\ \text{-----}|\text{-----} \\ H-C-OH \\ | \\ H \end{array} + 3HIO_4 \longrightarrow \begin{array}{c} O \\ \| \\ H-C-OH \\ + \\ OH \\ | \\ H-C-OH \\ | \\ OH \\ + \\ OH \\ | \\ H-C-OH \\ | \\ OH \\ + \\ OH \\ | \\ H-C-OH \\ | \\ H \end{array} \xrightarrow{-3H_2O} 3H\overset{O}{\overset{\|}{C}}OH + H\overset{O}{\overset{\|}{C}}H$$

D-Erythrose

21.12 Oxidation of an aldohexose and a ketohexose would each require 5 moles of HIO_4 but would give different results.

$$\begin{array}{c} CHO \\ \text{-----}|\text{-----} \\ CHOH \\ \text{-----}|\text{-----} \\ CHOH \\ \text{-----}|\text{-----} \\ CHOH \\ \text{-----}|\text{-----} \\ CHOH \\ \text{-----}|\text{-----} \\ CH_2OH \end{array} + 5HIO_4 \longrightarrow \begin{array}{c} HCO_2H \\ + \\ HCO_2H \\ + \\ HCO_2H \\ + \\ HCO_2H \\ + \\ HCO_2H \\ + \\ HCHO \end{array} \quad (5\ HCO_2H + HCHO)$$

Aldohexose

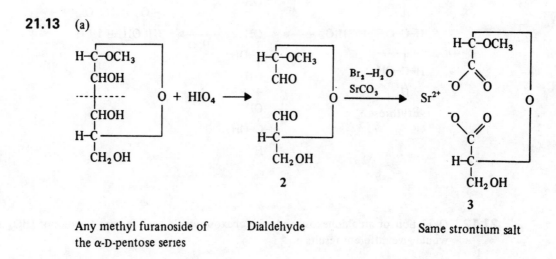

$$
\begin{array}{l}
CH_2OH \\
\text{- - - - - -} \\
C{=}O \\
\text{- - - - - -} \\
CHOH \\
\text{- - - - - -} \\
CHOH \\
\text{- - - - - -} \\
CHOH \\
\text{- - - - - -} \\
CH_2OH
\end{array}
+ 5HIO_4 \longrightarrow
\begin{array}{c}
HCHO \\ + \\ CO_2 \\ + \\ HCO_2H \\ + \\ HCO_2H \\ + \\ HCO_2H \\ + \\ HCHO
\end{array}
\;(3HCO_2H, 2HCHO + CO_2)
$$

Ketohexose

21.13 (a)

$$
\begin{array}{l}
H\text{-}C\text{-}OCH_3 \\
CHOH \\
\text{- - - - - -} \\
CHOH \\
H\text{-}C \\
CH_2OH
\end{array}
O + HIO_4 \longrightarrow
\begin{array}{l}
H\text{-}C\text{-}OCH_3 \\
CHO \\
\\
CHO \\
H\text{-}C \\
CH_2OH
\end{array}
O
\xrightarrow[SrCO_3]{Br_2\text{-}H_2O}
Sr^{2+}
\begin{array}{l}
H\text{-}C\text{-}OCH_3 \\
C(O^-)(=O) \\
\\
C(O^-)(=O) \\
H\text{-}C \\
CH_2OH
\end{array}
O
$$

2 3

Any methyl furanoside of the α-D-pentose series — Dialdehyde — Same strontium salt

(b) Although both compounds yield the same dialdehyde **2** (and Strontium salt **3**), periodate oxidation of a methyl α-D-pentofuranoside consumes only 1 mole of HIO_4 and produces no formic acid.

21.14 (a)

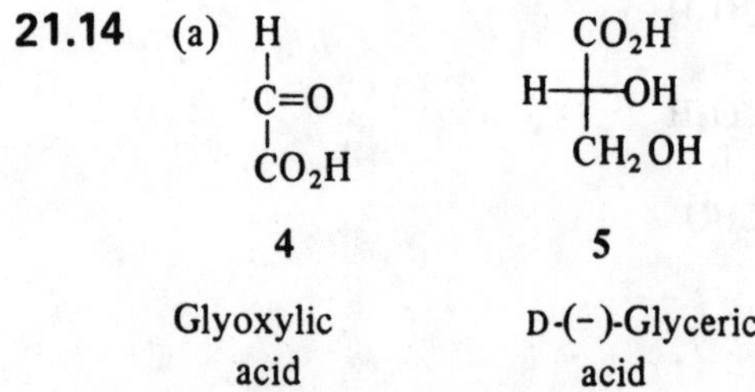

Glyoxylic acid — D-(−)-Glyceric acid

(b) This relates the configuration of the highest numbered carbon atom of the aldose to that of D-(+)-glyceraldehyde, and thus allows us to place the aldose in the D-family.

21.15 (a) Yes, D-glucitol would be optically active; only those alditols whose molecules possess a plane of symmetry would be optically inactive.

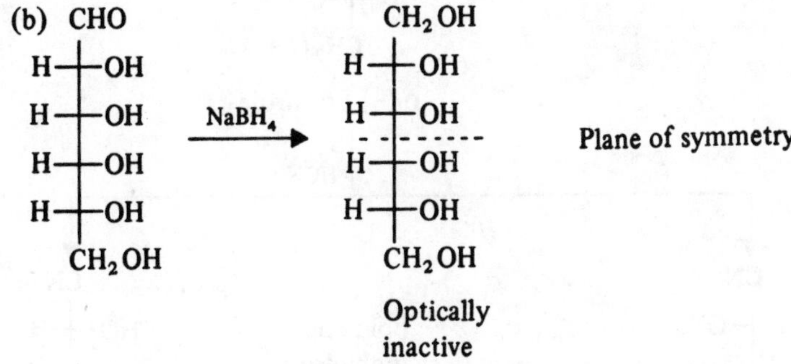

CHO
H—OH
HO—H
HO—H
H—OH
CH_2OH

$\xrightarrow{NaBH_4}$

CH_2OH
H—OH
HO—H
- - - - - - - - - - Plane of symmetry
HO—H
H—OH
CH_2OH
Optically inactive

21.16 (a)

CH_2OH
C=O
HO—H
H—OH
H—OH
CH_2OH

$\xrightarrow{C_6H_5NHNH_2}$

$CH{=}NNHC_6H_5$
$C{=}NNHC_6H_5$
HO—H
H—OH
H—OH
CH_2OH

(b) This experiment shows that D-glucose and D-fructose have the same configurations at C-3, C-4, and C-5.

21.17 (a)

CHO
HO—H
HO—H
CH_2OH
L-Erythrose

CHO
H—OH
HO—H
CH_2OH
L-Threose

(b) L-Glyceraldehyde

CHO
HO—H
CH_2OH

21.18 (a)

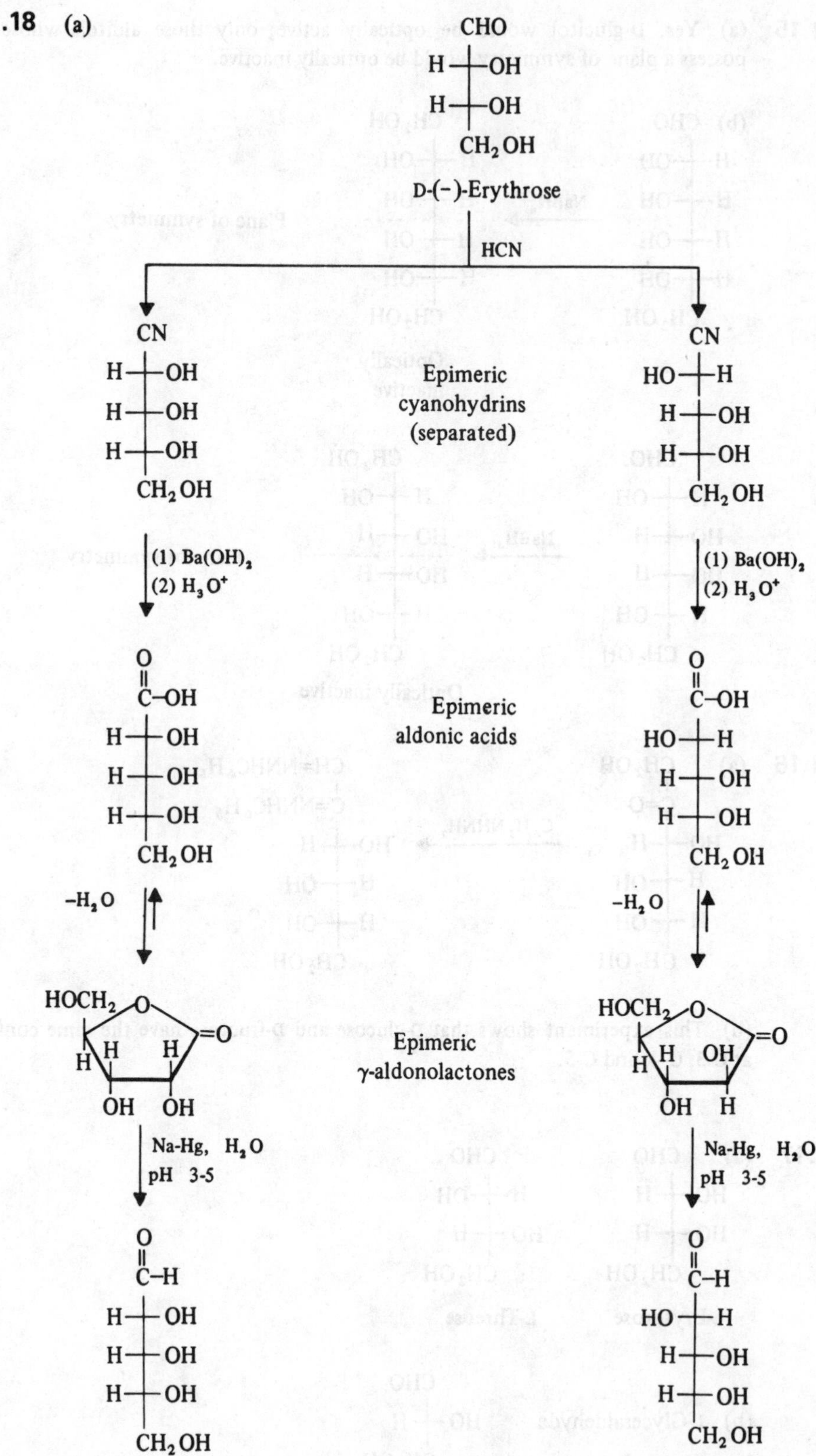
CHO
H—OH
H—OH
CH₂OH
D-(−)-Erythrose
HCN
CN
H—OH
H—OH
H—OH
CH₂OH
Epimeric
cyanohydrins
(separated)
CN
HO—H
H—OH
H—OH
CH₂OH
(1) Ba(OH)₂
(2) H₃O⁺
(1) Ba(OH)₂
(2) H₃O⁺
O
C—OH
H—OH
H—OH
H—OH
CH₂OH
Epimeric
aldonic acids
O
C—OH
HO—H
H—OH
H—OH
CH₂OH
−H₂O
−H₂O
HOCH₂
O
=O
H
H
H
OH
OH
Epimeric
γ-aldonolactones
HOCH₂
O
=O
H
H
OH
OH
H
Na-Hg, H₂O
pH 3-5
Na-Hg, H₂O
pH 3-5
O
C—H
H—OH
H—OH
H—OH
CH₂OH
O
C—H
HO—H
H—OH
H—OH
CH₂OH

(b)

```
      O                              O
      ‖                              ‖
      C–H                            C–OH
   H–┼–OH                         H–┼–OH
   H–┼–OH        HNO3  →          H–┼–OH
   H–┼–OH                         H–┼–OH
      CH2OH                          C–OH
                                     ‖
                                     O
```

D-(−)-Ribose   Optically inactive

```
      O                              O
      ‖                              ‖
      C–H                            C–OH
  HO–┼–H                         HO–┼–H
   H–┼–OH        HNO3  →          H–┼–OH
   H–┼–OH                         H–┼–OH
      CH2OH                          C–OH
                                     ‖
                                     O
```

D-(−)-Arabinose   Optically active

21.19 A Kiliani-Fischer synthesis starting with D-(−)-threose would yield **I** and **II**.

```
      CHO            CHO
   H–┼–OH        HO–┼–H
  HO–┼–H         HO–┼–H
   H–┼–OH         H–┼–OH
      CH2OH          CH2OH
      I              II
```

D-(+)-Xylose   D-(−)-Lyxose

I must be D-(+)-xylose because when oxidized by nitric acid, it yields an optically inactive aldaric acid:

```
                      CO2H
                   H–┼–OH
  I   HNO3  →     HO–┼–H
                   H–┼–OH
                      CO2H
```

Optically inactive

II must be D-(−)-lyxose because when oxidized by nitric acid it yields an optically active aldaric acid:

$$\textbf{II} \xrightarrow{HNO_3} \begin{array}{c} CO_2H \\ HO-\!\!\!-H \\ HO-\!\!\!-H \\ H-\!\!\!-OH \\ CO_2H \end{array}$$

Optically active

21.20

$$\begin{array}{cccc} CHO & CHO & CHO & CHO \\ HO-\!\!\!-H & H-\!\!\!-OH & HO-\!\!\!-H & H-\!\!\!-OH \\ HO-\!\!\!-H & HO-\!\!\!-H & H-\!\!\!-OH & H-\!\!\!-OH \\ HO-\!\!\!-H & HO-\!\!\!-H & HO-\!\!\!-H & HO-\!\!\!-H \\ CH_2OH & CH_2OH & CH_2OH & CH_2OH \\ \text{L-(+)-Ribose} & \text{L-(+)-Arabinose} & \text{L-(−)-Xylose} & \text{L-(+)-Lyxose} \end{array}$$

21.21 Since D-(+)-galactose yields an optically inactive aldaric acid it must have either structure **III** or structure **IV**.

$$\begin{array}{c} CHO \\ H-\!\!\!-OH \\ H-\!\!\!-OH \\ H-\!\!\!-OH \\ H-\!\!\!-OH \\ CH_2OH \\ \textbf{III} \end{array} \xrightarrow{HNO_3} \begin{array}{c} CO_2H \\ H-\!\!\!-OH \\ H-\!\!\!-OH \\ \hdashline H-\!\!\!-OH \\ H-\!\!\!-OH \\ CO_2H \\ \text{Optically inactive} \end{array} \qquad \begin{array}{c} CHO \\ H-\!\!\!-OH \\ HO-\!\!\!-H \\ HO-\!\!\!-H \\ H-\!\!\!-OH \\ CH_2OH \\ \textbf{IV} \end{array} \xrightarrow{HNO_3} \begin{array}{c} CO_2H \\ H-\!\!\!-OH \\ HO-\!\!\!-H \\ \hdashline HO-\!\!\!-H \\ H-\!\!\!-OH \\ CO_2H \\ \text{Optically inactive} \end{array}$$

A Ruff degradation beginning with **III** would yield D-(−)-ribose

$$\textbf{III} \xrightarrow[H_2O]{Br_2} \xrightarrow[Fe_2(SO_4)_3]{H_2O_2} \begin{array}{c} CHO \\ H-\!\!\!-OH \\ H-\!\!\!-OH \\ H-\!\!\!-OH \\ CH_2OH \end{array}$$

D-(−)-Ribose

A Ruff degradation beginning with **IV** would yield D-(−)-lyxose: thus D-(+)-galactose must have structure **IV**.

IV $\xrightarrow[H_2O]{Br_2}$ $\xrightarrow[Fe_2(SO_4)_3]{H_2O_2}$

CHO
HO—H
HO—H
H—OH
CH_2OH

D-(−)-Lyxose

21.22 D-(+)-glucose, as shown here.

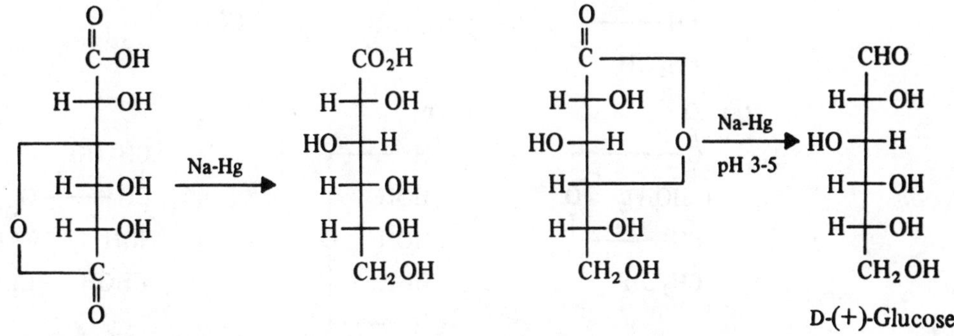

The other γ-lactone of D-glucaric acid

D-(+)-Glucose

21.23 If the methyl glucoside had been a furanoside, hydrolysis of the methylation product would have given:

CHO
H—OCH_3
CH_3O—H
H—OH
H—OCH_3
CH_2OCH_3

And, oxidation would have given:

^{1}CHO
H^2—OCH_3
CH_3O^3—H
H^4—OH
H^5—OCH_3
$^6CH_2OCH_3$

$\xrightarrow{HNO_3}$

$^1CO_2H$
H^2—OCH_3
CH_3O^3—H
$^4CO_2H$

A dimethoxysuccinic acid

$^4CO_2H$
H^5—OCH_3
+ $^6CH_2OCH_3$

A dimethoxypropanoic acid

+ $^1CO_2H$
H^2—OCH_3
$^3CO_2H$

Methoxymalonic acid

+ $^5CO_2H$
$^6CH_2OCH_3$

Methoxymalonic acid $\xrightarrow{-CO_2}$

CO_2H
$^2CH_2OCH_3$

Methoxyacetic acid

21.24 (a) CHO–CHOH–CHOH–CHOH–CH_2OH

(b) CH_2OH–C=O–CHOH–CHOH–CHOH–CH_2OH

(c) CHO–$(CHOH)_n$–C(HO)(H)–CH_2OH or CH_2OH–C=O–$(CHOH)_n$–C(HO)(H)–CH_2OH

(d) CHOR–$(CHOH)_n$–CH–CH_2OH (ring closed through O between CHOR and CH)

(e) CO_2H–$(CHOH)_n$–CH_2OH

(f) CO_2H–$(CHOH)_n$–CO_2H

(g) O=C–$(CHOH)_n$–CH–CH_2OH (ring closed through O between C and CH)

(h) OH–CH–CHOH–CHOH–CHOH–CH–CH_2OH (ring closed through O between the two CH) or CH_2OH–CH–O–CHOH–CHOH–CHOH–CHOH (six-membered ring)

(i) OH–CH–CHOH–CHOH–CH–CHOH–CH_2OH (ring closed through O between the two CH) or CH_2OH–CHOH–CH–O–CHOH–CHOH–CHOH (five-membered ring)

(j) Any sugar that has a free aldehyde or ketone group or one that exists as a cyclic hemiacetal or hemiketal. The following are examples:

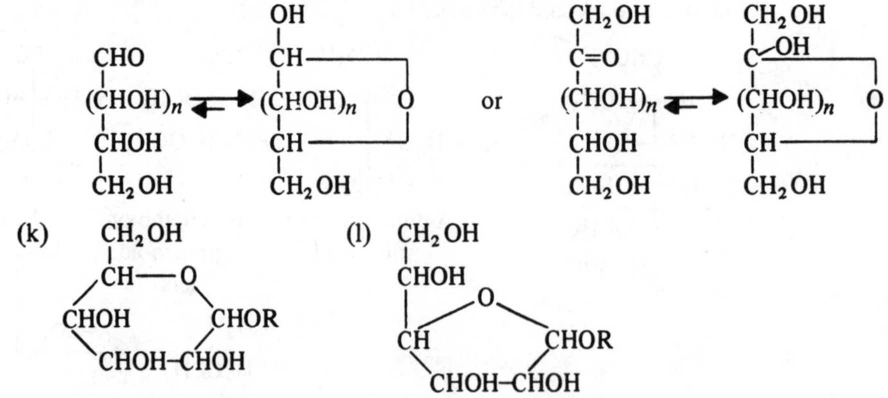

(m) Any two aldoses that differ only in configuration at C-2. (See also Section 21.6 for a broader definition.) D-Erythrose and D-threose are examples.

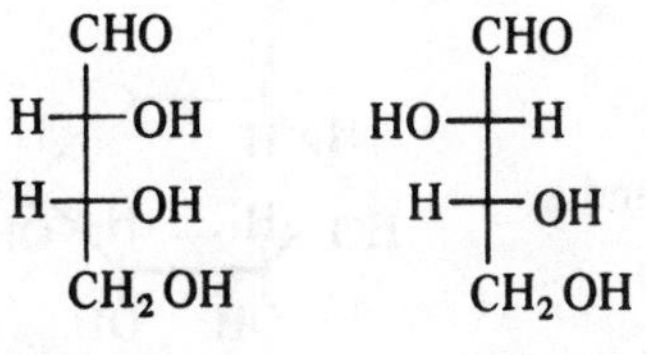

D-Erythrose D-Threose

(n) Cyclic sugars that differ only in the configuration of C-1. The following are examples:

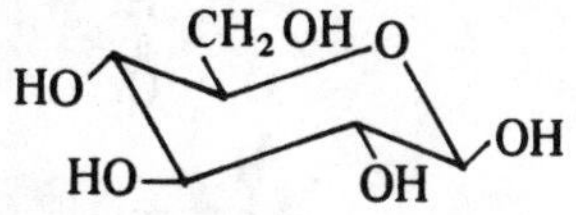

and

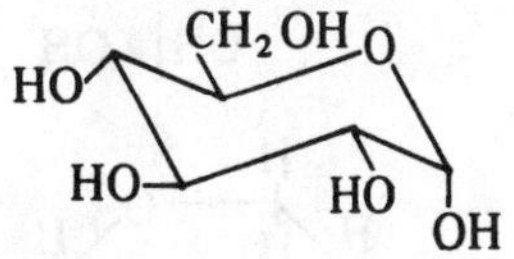

(o) $CH{=}NNHC_6H_5$
$C{=}NNHC_6H_5$
$(CHOH)_n$
CH_2OH

(p) Maltose is an example:

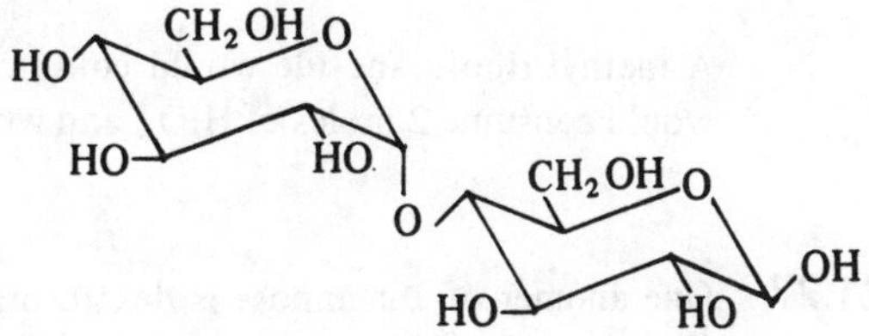

(q) Amylose is an example:

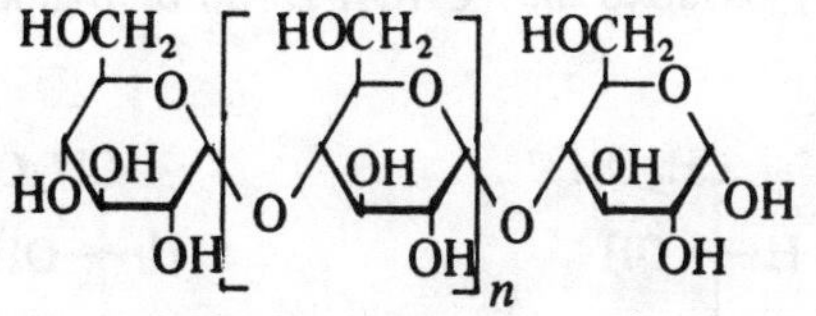

(r) Any sugar in which all potential carbonyl groups are present as acetals or ketals (i.e., as glycosides). Sucrose (Section 21.11A) is an example of a nonreducing disaccharide; the methyl D-glucopyranosides (Section 21.3) are example of nonreducing monosaccharides.

21.25

(a)

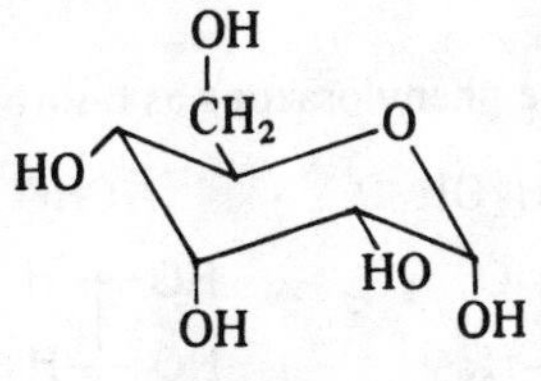

(b)

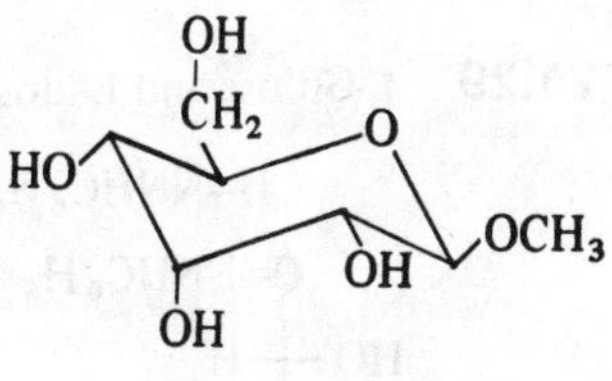

(c)

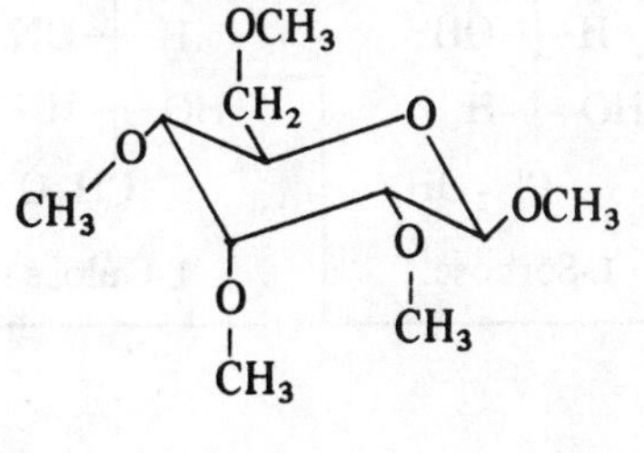

21.26

(a) HOCH₂, O, H; H H H OH; OH OH (methyl furanose ring) and H; H H O H; HO H H OH; OH OH (pyranose ring)

(b) HOCH₂ O H; H H H OCH₃; OH OH $+ HIO_4 \longrightarrow$ HOCH₂ O H; H OCH₃; CH HC; ‖ ‖; O O

H; H H O H; HO H H OCH₃; OH OH $+ 2HIO_4 \longrightarrow$ H; H O H; HC ‖ O; OCH₃; HC ‖ O $+ HCOH$ (with C=O)

A methyl ribofuranoside would consume only 1 mole of HIO_4; a methyl ribopyranoside would consume 2 moles of HIO_4 and would also produce 1 mole of formic acid.

21.27 One anomer of D-mannose is dextrorotatory ($[\alpha]_D^{25} = +29.3°$), the other is levorotatory ($[\alpha]_D^{25} = -17.0°$).

21.28 The microorganism selectively oxidizes the —CHOH group of D-glucitol that corresponds to C-5 of D-glucose.

D-Glucose		D-Glucitol		L-Sorbose		
¹CHO		CH₂OH		CH₂OH		CH₂OH
H—²—OH		H—OH		H—OH		C=O
HO—³—H	$\xrightarrow{H_2,\ Ni}$	HO—H	$\xrightarrow[\textit{Acetobacter suboxydans}]{O_2}$	HO—H	≡	HO—H
H—⁴—OH		H—OH		H—OH		H—OH
H—⁵—OH		H—OH		C=O		HO—H
⁶CH₂OH		CH₂OH		CH₂OH		CH₂OH

21.29 L-Gulose and L-idose would yield the same phenylosazone as L-sorbose.

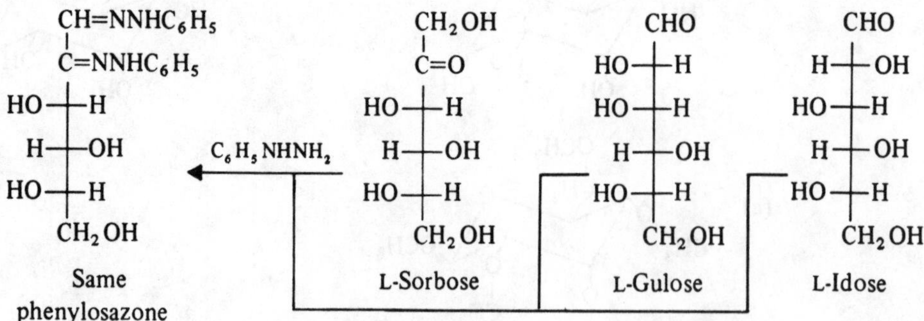

21.30

CH_2OH
C=O
H—OH
H—OH
H—OH
CH_2OH
D-Psicose

$\xrightarrow{C_6H_5NHNH_2}$

$CH{=}NNHC_6H_5$
$C{=}NNHC_6H_5$
H—OH
H—OH
H—OH
CH_2OH

$\xleftarrow{C_6H_5NHNH_2}$

CHO
H—OH
H—OH
H—OH
H—OH
CH_2OH
D-Allose

CH_2OH
C=O
HO—H
HO—H
H—OH
CH_2OH
D-Tagatose

$\xrightarrow{C_6H_5NHNH_2}$

$CH{=}NNHC_6H_5$
$C{=}NNHC_6H_5$
HO—H
HO—H
H—OH
CH_2OH

$\xleftarrow{C_6H_5NHNH_2}$

CHO
H—OH
HO—H
HO—H
H—OH
CH_2OH
D-Galactose

21.31 **A** is D-altrose, **B** is D-talose, and **C** is D-galactose:

CHO
HO—H
H—OH
H—OH
H—OH
CH_2OH
D-Altrose
A

$\xrightarrow[Ni]{H_2,}$

CH_2OH
HO—H
H—OH
H—OH
H—OH
CH_2OH

≡

CH_2OH
HO—H
HO—H
HO—H
H—OH
CH_2OH

Same alditol

$\xleftarrow[Ni]{H_2,}$

CHO
HO—H
HO—H
HO—H
H—OH
CH_2OH
D-Talose
B

A $\xrightarrow{C_6H_5NHNH_2}$

$CH{=}NNHC_6H_5$
$C{=}NNHC_6H_5$
H—OH
H—OH
H—OH
CH_2OH

B $\xrightarrow{C_6H_5NHNH_2}$

$CH{=}NNHC_6H_5$
$C{=}NNHC_6H_5$
HO—H
HO—H
H—OH
CH_2OH

⟺ Different phenylosazones

CHO		$CH{=}NNHC_6H_5$		CHO
H—OH		$C{=}NNHC_6H_5$		HO—H
HO—H	$C_6H_5NHNH_2$ →	HO—H	← $C_6H_5NHNH_2$	HO—H
HO—H		HO—H		HO—H
H—OH		H—OH		H—OH
CH_2OH		CH_2OH		CH_2OH
D-Galactose C		Same phenylosazone		D-Talose B
↓ H_2, Ni				↓ H_2, Ni
CH_2OH				CH_2OH
H—OH				HO—H
HO—H		Different alditols		HO—H
HO—H				HO—H
H—OH				H—OH
CH_2OH				CH_2OH

(*Note:* If we had designated D-talose as **A**, and D-altrose as **B**, then **C** is D-allose)

21.32

CHO		CO_2H		CO_2H
H—OH		H—OH		HO—H
HO—H	Br_2, H_2O →	HO—H	pyridine (epimerization) →	HO—H
H—OH		H—OH		H—OH
H—OH		H—OH		H—OH
CH_2OH		CH_2OH		CH_2OH
D-Glucose				

	O=C— (lactone)		CHO
	HO—H		HO—H
$-H_2O$ →	HO—H O	Na-Hg, pH 3-5 →	HO—H
	H— (ring O)		H—OH
	H—OH		H—OH
	CH_2OH		CH_2OH
			D-Mannose

21.33 The conformation of D-idopyranose with four equatorial —OH groups and an axial —CH_2OH group is more stable than the one with four axial —OH groups and an equatorial —CH_2OH group.

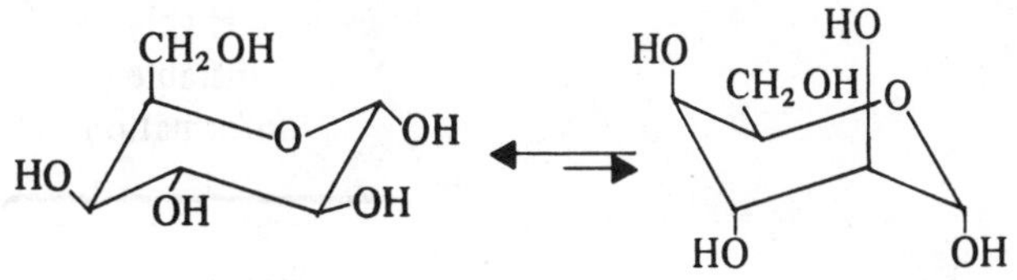

More stable

4 Equatorial —OH groups
1 Axial —CH_2OH

Less stable

4 Axial —OH groups
1 Equatorial —CH_2OH

21.34 (a) The anhydro sugar is formed when the axial —CH_2OH group reacts with C-1 to form a cyclic acetal.

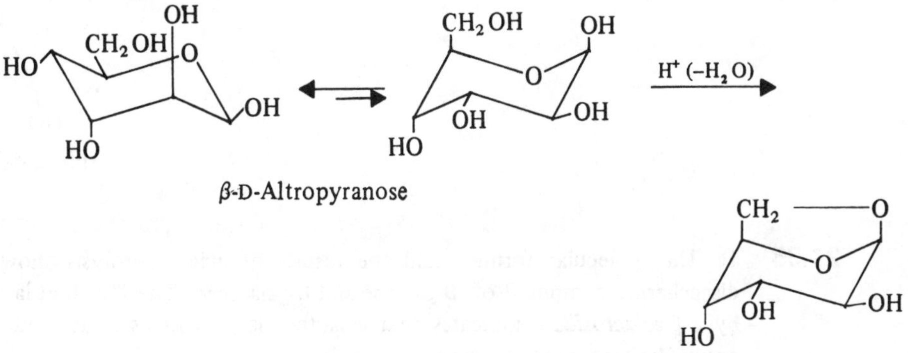

$\xrightarrow{H^+ (-H_2O)}$

β-D-Altropyranose

Anhydro sugar

Because the anhydro sugar is an acetal (i.e., an internal glycoside), it is a nonreducing sugar.

Methylation followed by acid hydrolysis converts the anhydro sugar to 2,3,4-tri-*O*-methyl-D-altrose:

$\xrightarrow[OH^-]{(CH_3)_2SO_4}$ $\xrightarrow{H^+, H_2O}$

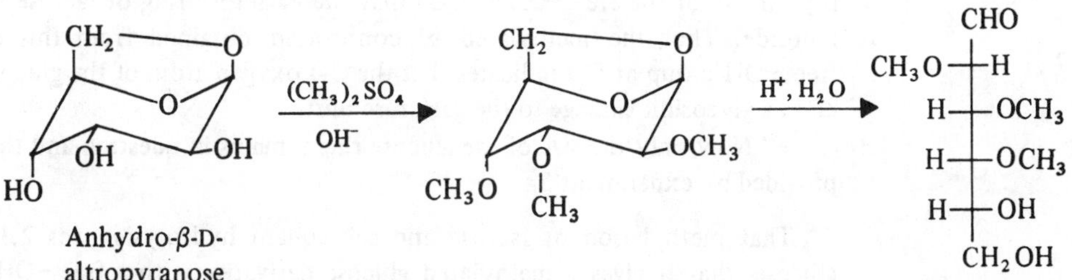

Anhydro-β-D-altropyranose

2,3,4-Tri-*O*-methyl-D-altrose

(b) Formation of an anhydro sugar requires that the monosaccharide adopt a chair conformation with the —CH_2OH group axial. With β-D-altropyranose this requires that two —OH groups be axial as well. With β-D-glucopyranose, however, it requires that all four —OH groups become axial, and thus that the molecule adopt a very unstable conformation:

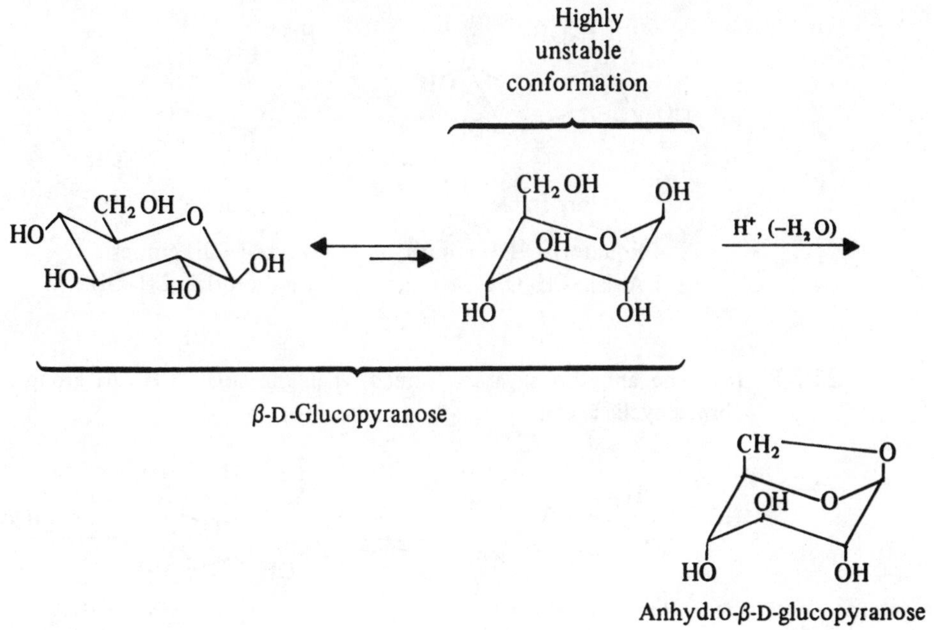

Highly
unstable
conformation

$\xrightarrow{H^+, (-H_2O)}$

β-D-Glucopyranose

Anhydro-β-D-glucopyranose

21.35 1. The molecular formula and the results of acid hydrolysis show that lactose is a disaccharide composed of D-glucose and D-galactose. The fact that lactose is hydrolyzed by a *β-galactosidase* indicates that galactose is present as a glycoside and that the glycosidic linkage is beta to the galactose ring.

2. That lactose is a reducing sugar, forms a phenylosazone, and undergoes mutarotation indicates that one ring (presumably that of D-glucose) is present as a hemiacetal and thus is capable of existing to a limited extent as an aldehyde.

3. This experiment confirms that the D-glucose unit is present as a cyclic hemiacetal and that the D-galactose unit is present as a cyclic glycoside.

4. That 2,3,4,6-tetra-*O*-methyl-D-galactose is obtained in this experiment indicates (by virture of the free —OH at C-5) that the galactose ring of lactose is present as a pyranoside. That the methylated gluconic acid obtained from this experiment has a free —OH group at C-4 indicates that the C-4 oxygen atom of the glucose unit is connected in a glycosidic linkage to the galactose unit.

Now only the size of the glucose ring remains in question and the answer to this is provided by experiment 5.

5. That methylation of lactose and subsequent hydrolysis gives 2,3,6-tri-*O*-methyl-D-glucose—that it gives a methylated glucose derivative with a free —OH at C-4 and C-5—demonstrates that the glucose ring is present as a pyranose. (We know already that the oxygen at C-4 is connected in a glycosidic linkage to the galactose unit; thus a free —OH at C-5 indicates that the C-5 oxygen atom is a part of the hemiacetal group of the glucose unit and that the ring is six membered.)

21.36 Melibiose has the following structure:

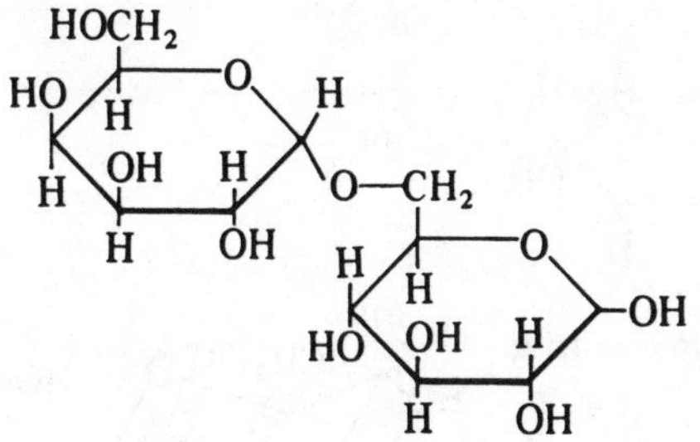

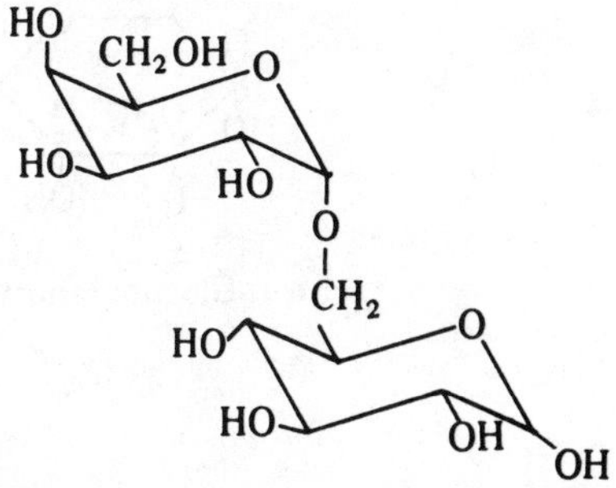

6-*O*-(α-D-galactopyranosyl)-D-glucopyranose

We arrive at this conclusion from the data given:

1. That melibiose is a reducing sugar, that it undergoes mutarotation and forms a phenylosazone indicates that one monosaccharide is present as a cyclic hemiacetal.

2. That acid hydrolysis gives D-galactose and D-glucose indicates that melibiose is a disaccharide composed of one D-galactose unit and one D-glucose unit. That melibiose is hydrolyzed by an α-galactosidase suggests that melibiose is an α-D-galactosyl-D-glucose.

3. Oxidation of melibiose to melibionic acid and subsequent hydrolysis to give D-galactose and D-gluconic acid confirms that the glucose unit is present as a cyclic hemiacetal and that the galactose unit is present as a glycoside. (Had the reverse been true, this experiment would have yielded D-glucose and D-galactonic acid.)

Methylation and hydrolysis of melibionic acid produces 2,3,4,6-tetra-*O*-methyl-D-galactose and 2,3,4,5-tetra-*O*-methyl-D-gluconic acid. Formation of the first product—a galactose derivative with a free —OH at C-5—demonstrates that the galactose ring is six membered; formation of the second product—a gluconic acid derivative with a free —OH at C-6—demonstrates that the oxygen at C-6 of the glucose unit is joined in a glycosidic linkage to the galactose unit.

4. That methylation and hydrolysis of melibiose gives a glucose derivative (2,3,4-tri-*O*-methyl-D-glucose) with free —OH groups at C-5 and C-6 shows that the glucose ring is also six membered. Melibiose is, therefore, 6-*O*-(α-D-galactopyranosyl)-D-glucopyranose.

21.37 Trehalose has the following structure:

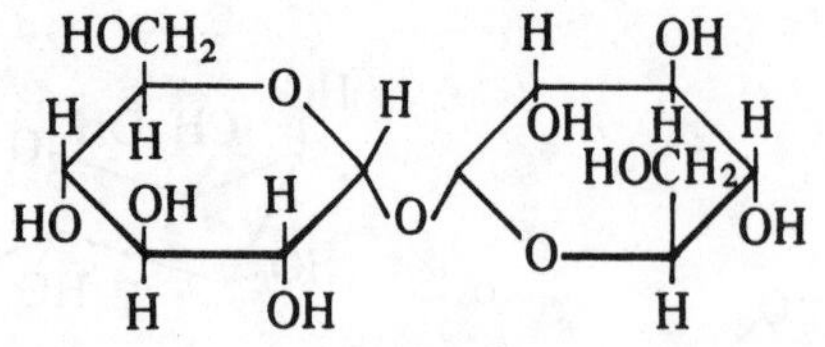

HOCH$_2$ ... H OH ... or

α-D-Glucopyranosyl-α-D-glucopyranoside

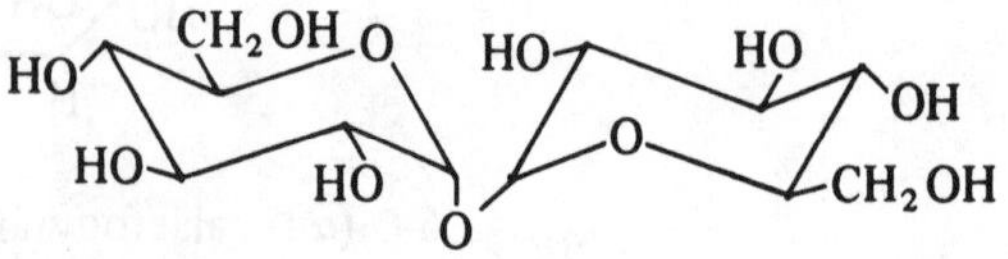

We arrive at this structure in the following way:

1. Acid hydrolysis shows that trehalose is a disaccharide consisting only of D-glucose units.

2. Hydrolysis by α-glucosidases and not by β-glucosidases shows that the glycosidic linkages are alpha.

3. That trehalose is a nonreducing sugar, that it does not form phenylosazone, and that it does not react with bromine water indicate that no hemiacetal groups are present. This means that C-1 of one glucose unit and C-1 of the other must be joined in a glycosidic linkage. Fact 2 (just cited) indicates that this linkage is alpha to each ring.

4. That methylation of trehalose followed by hydrolysis yields only 2,3,4,6-tetra-*O*-methyl-D-glucose demonstrates that both rings are six membered.

21.38 (a) Tollens' reagent or Benedict's reagent will give a positive test with D-glucose but will give no reaction with D-glucitol.

(b) D-Glucaric acid will give an acidic aqueous solution that can be detected with blue litmus paper. D-Glucitol will give a neutral aqueous solution.

(c) D-Glucose will be oxidized by bromine water and the red brown color of bromine will disappear. D-Fructose will not be oxidized by bromine water since it does not contain an aldehyde group.

(d) Nitric acid oxidation will produce an *optically active* aldaric acid from D-glucose but an *optically inactive* aldaric acid will result from D-galactose.

(e) Maltose is a reducing sugar and will give a positive test with Tollens' or Benedict's solution. Sucrose is a nonreducing sugar and will not react.

(f) Maltose will give a positive Tollens' or Benedict's test; maltonic acid will not.

(g) 2,3,4,6-Tetra-*O*-methyl-β-D-glucopyranose will give a positive test with Tollens' or Benedict's solution; methyl β-D-glucopyranoside will not.

(h) Periodic acid will react with methyl α-D-ribofuranoside because it has hydroxyl groups on adjacent carbons. Methyl 2-deoxy-α-D-ribofuranoside will not react.

21.39 That the Schardinger dextrins are nonreducing shows that they have no free aldehyde or hemiacetal groups. This lack of reaction strongly suggests the presence of a *cyclic* structure. That methylation and subsequent hydrolysis yields only 2,3,6-tri-*O*-methyl-D-glucose indicates that the glycosidic linkages all involve C-1 of one glucose unit and C-4 of the next. That α-glucosidases cause hydrolysis of the glycosidic linkages indicates that they are α-glycosidic linkages. Thus we are led to the following general structure.

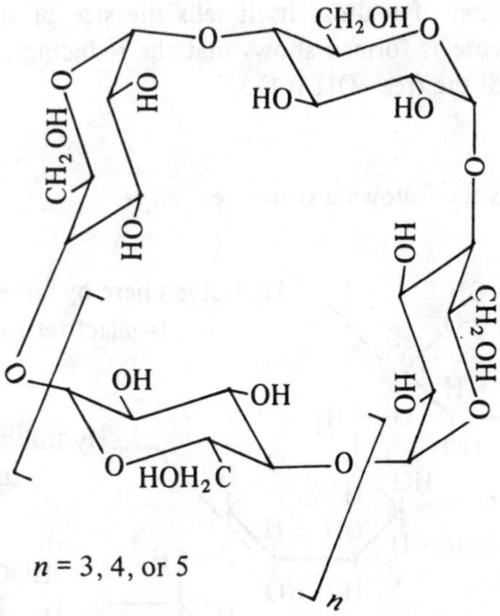

n = 3, 4, or 5

Note: Schardinger dextrins are extremely interesting compounds. They are able to form complexes with a wide variety of compounds by incorporating these compounds in the cavity in the middle of the cyclic dextrin structure. Complex formation takes place, however, only when the cyclic dextrin and the guest molecule are the right size. Anthracene molecules, for example, will fit into the cavity of a cyclic dextrin with eight glucose units but will not fit into one with seven. For more information about these fascinating compounds, see R. J. Bergeron, "Cycloamyloses," *J. Chem. Educ.*, **54**, 204 (1977).

21.40 Isomaltose has the following structure:

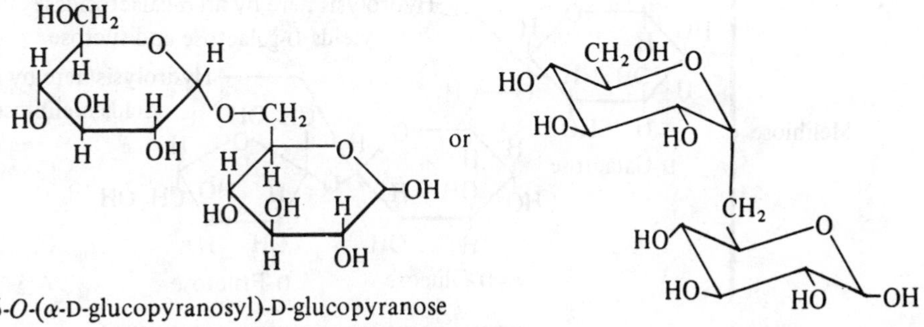

6-*O*-(α-D-glucopyranosyl)-D-glucopyranose

(1) The acid and enzymic hydrolysis experiments tell us that isomaltose has two glucose units linked by an α linkage.

(2) That isomaltose is a reducing sugar indicates that one glucose unit is present as a cyclic hemiacetal.

(3) Methylation of isomaltonic acid followed by hydrolysis gives us information about the size of the nonreducing pyranoside ring and about its point of attachment to the reducing ring. The formation of the first product (2,3,4,6-tetra-*O*-methyl-D-glucose)—a compound with an —OH at C-5—tells us that the nonreducing ring is present as a pyranoside. The formation of 2,3,4,5-tetra-*O*-methyl-D-gluconic acid—a compound with an —OH at C-6—shows that the nonreducing ring is linked to C-6 of the reducing ring.

(4) Methylation of maltose itself tells the size of the reducing ring. That 2,3,4-tri-*O*-methyl-D-glucose is formed shows that the reducing ring is also six membered; we know this because of the free —OH at C-5.

21.41 Stachyose has the following structure:

CH_2OH HO H H D-Galactose H OH H O—CH_2 H HO
Hydrolysis here by an α-galactosidase yields D-galactose and raffinose

HO H H D-Galactose H OH H O—CH_2 H HO
Hydrolysis here by an α-galactosidase yields sucrose

Raffinose
H H O H $HOCH_2$ O H
D-Glucose HO OH H O H HO CH_2OH
H HO OH H D-Fructose

Sucrose

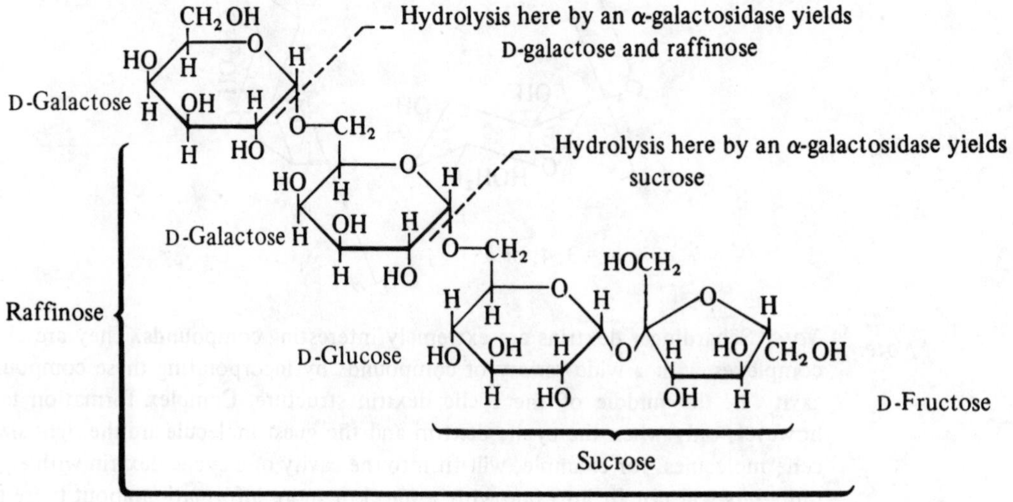

Raffinose has the following structure:

CH_2OH HO H H H OH H O—CH_2 H HO D-Galactose
Hydrolysis here by an α-galactosidase yields D-galactose and sucrose

Melibiose
Hydrolysis here by an invertase yields melibiose and fructose

H H O H CH_2OH O H
HO OH H O H HO CH_2OH
H OH OH H
D-Glucose D-Fructose

Sucrose

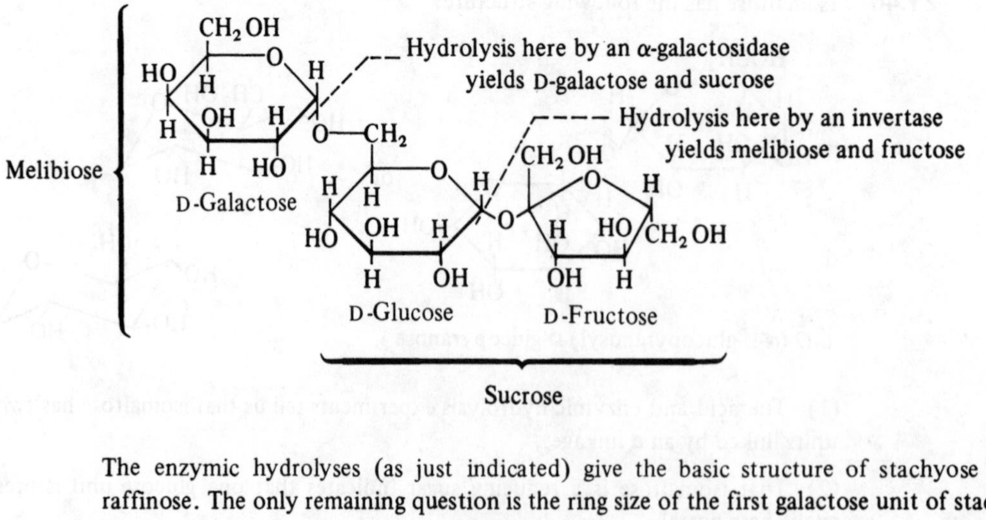

The enzymic hydrolyses (as just indicated) give the basic structure of stachyose and raffinose. The only remaining question is the ring size of the first galactose unit of stachyose. That methylation of stachyose and subsequent hydrolysis yields 2,3,4,6-tetra-*O*-methyl-D-galactose establishes that this ring is a pyranoside.

21.42 Arbutin has the following structure.

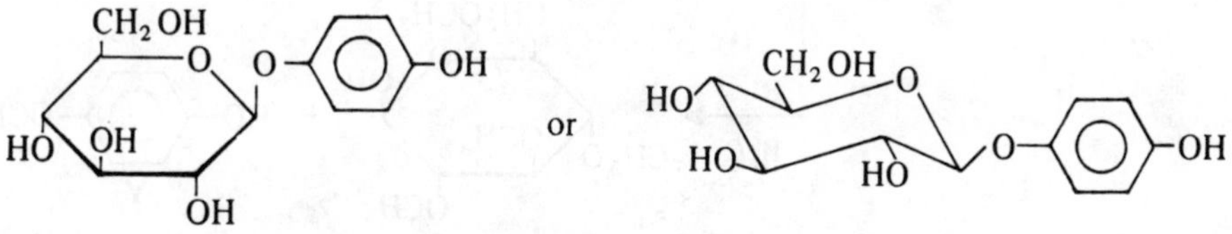

CH_2OH — O — O — C₆H₄ — OH (ring: HO, OH, OH) or (chair: HO, CH_2OH, O, HO, HO, O — C₆H₄ — OH)

p-Hydroxyphenyl-β-D-glucopyranoside

Compounds **X**, **Y**, and **Z** are hydroquinone, *p*-methoxyphenol, and *p*-dimethoxybenzene, respectively.

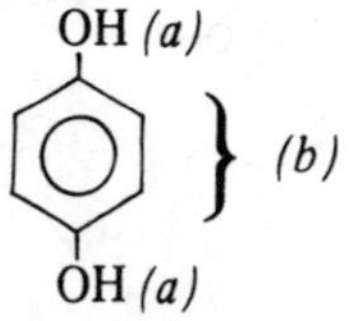

OH *(a)*, ring } *(b)*, OH *(a)*

(a) Singlet δ 7.9 (2H)
(b) Singlet δ 6.8 (4H)

X
Hydroquinone

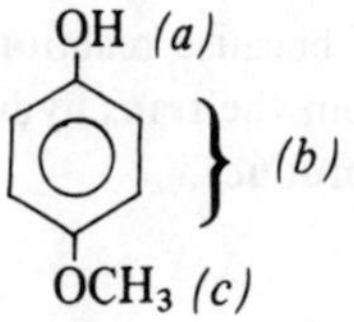

OH *(a)*, ring } *(b)*, OCH_3 *(c)*

(a) Singlet δ 4.8 (1H)
(b) Multiplet δ6.8 (4H)
(c) Singlet δ 3.9 (3H)

Y
p-Methoxyphenol

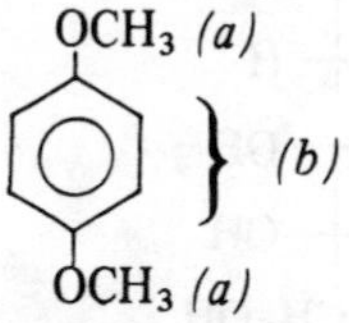

OCH_3 *(a)*, ring } *(b)*, OCH_3 *(a)*

(a) Singlet δ 3.75 (6H)
(b) Singlet δ 6.8 (4H)

Z
p-Dimethoxybenzene

The reactions that take place are the following:

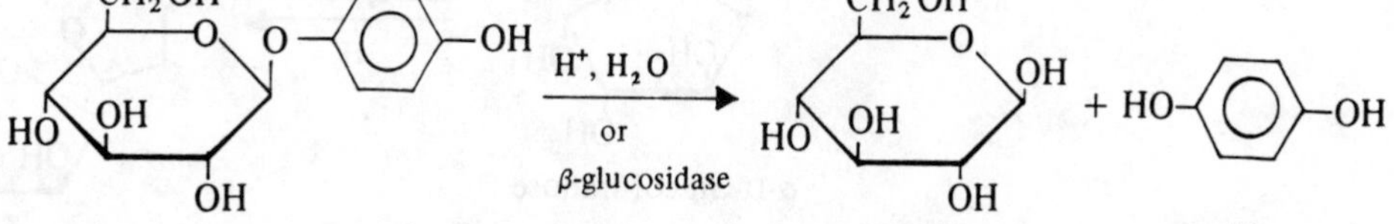

Arbutin (CH_2OH, HO, OH, OH; O — C₆H₄ — OH) $\xrightarrow[\text{or } \beta\text{-glucosidase}]{H^+, H_2O}$ (CH_2OH, HO, OH, OH, OH) + HO — C₆H₄ — OH

D-Glucose — **X** Hydroquinone

Arbutin $\xrightarrow[\text{OH}^-]{(CH_3)_2SO_4 \text{ (excess)}}$ [methylated arbutin: CH_2OCH_3, OCH_3, CH_3O, OCH_3 pyranoside with O–C$_6$H$_4$–OCH_3]

$\xrightarrow[H_2O]{H^+}$ 2,3,4,6-Tetra-*O*-methyl-D-glucose + $HO-C_6H_4-OCH_3$ (Y, *p*-Methoxyphenol)

p-Methoxyphenol $\xrightarrow[\text{OH}^-]{(CH_3)_2SO_4}$ $CH_3O-C_6H_4-OCH_3$ (Z, *p*-Dimethoxybenzene)

21.43 (a) and (b) Two molecules of acetone react with four hydroxyl groups of D-glucose co yield a compound (see following compound) containing two cyclic ketal linkages. Reaction with cis hydroxyl groups is preferred in reactions like this. Thus D-glucose reacts with acetone preferentially in the furanose form because reaction in the pyranose form would require the formation of a cyclic ketal from the trans hydroxyl groups at C-3 and C-4. This would introduce greater strain into the product.

α-D-Glucopyranose ⇌ CHO, H–C–OH, HO–C–H, H–C–OH, H–C–OH, CH_2OH

α-D-Glucofuranose $\xrightarrow[H^+]{2CH_3\overset{O}{\overset{\|}{C}}CH_3}$ Diacetone glucose

D-Galactose reacts similarly, but it can react in the pyranose form because the hydroxyl groups at C-3 and C-4 are cis (as are those at C-1 and C-2).

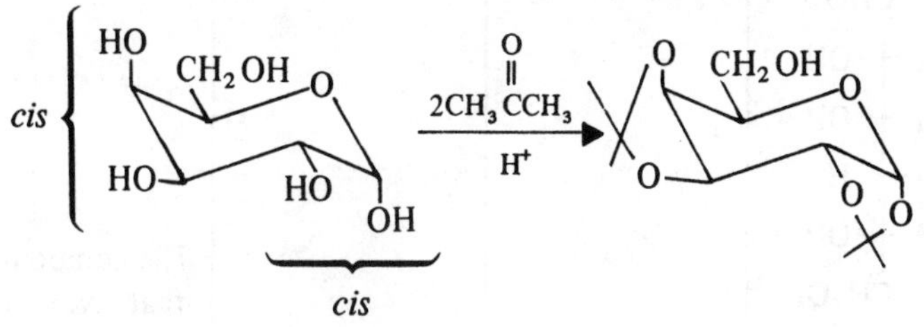

cis HO CH$_2$OH O HO HO OH cis $\xrightarrow[H^+]{2CH_3\overset{O}{\overset{\|}{C}}CH_3}$ O CH$_2$OH O O O O

α-D-Galactopyranose

SECTION REFERENCES FOR ADDITIONAL PROBLEMS

	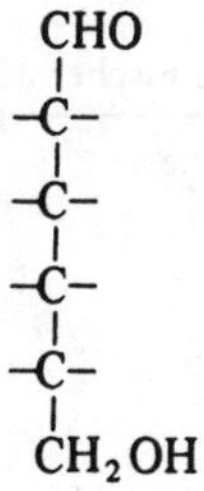	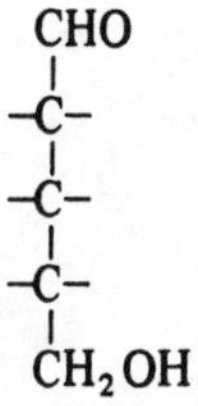	
21.22	21.1-21.4, 21.6, 21.11, 21.12	**21.31**	21.3, 21.8
21.23	21.1, 21.3, 21.8, 21.10	**21.32**	16.7, 21.3, 21.8
21.24	21.4, 21.8	**21.33**	21.10, 21.11
21.25	21.3, 21.8	**21.34**	21.10, 21.11
21.26	15.6, 15.7	**21.35**	21.10, 21.11
21.27	21.6, 21.8, 21.9	**21.36**	17.13, 21.4
21.28	21.6	**21.37**	21.10, 21.11
21.29	21.5, 21.6, 21.8	**21.38**	21.10, 21.11
21.30	19.1, 19.3, 21.4, 21.7, 21.9	**21.39**	21.10, 21.11

SELF-TEST

21.1 Supply the appropriate structural formula or complete the partial formula for each of the following:

(a)	(b)	(c)	(d)
	CHO —C— —C— —C— —C— CH$_2$OH	CHO —C— —C— —C— CH$_2$OH	
A ketotetrose	A D-sugar	An L-sugar	An aldose

	(e)	(f)	(g)	(h)
CHO H—OH H—OH HO—H H—OH CH_2OH D-Gulose	α-D-Gulopyranose	β-D-Gulopyranose	The compound that gives the same osazone as D-gulose	The compound that gives the same aldaric acid as D-gulose

21.2 Which of the following monosaccharides yields an optically inactive alditol on $NaBH_4$ reduction?

Answer: ______

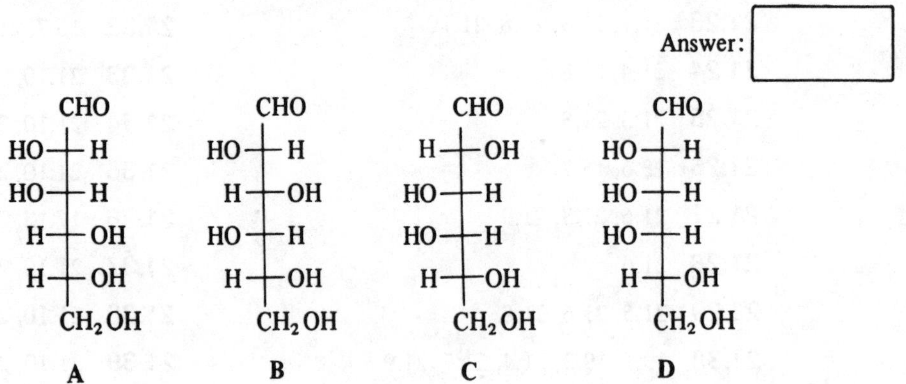

21.3 Give the structural formula of the monosaccharide that you could use as starting material in the Kiliani-Fischer synthesis of the following compound:

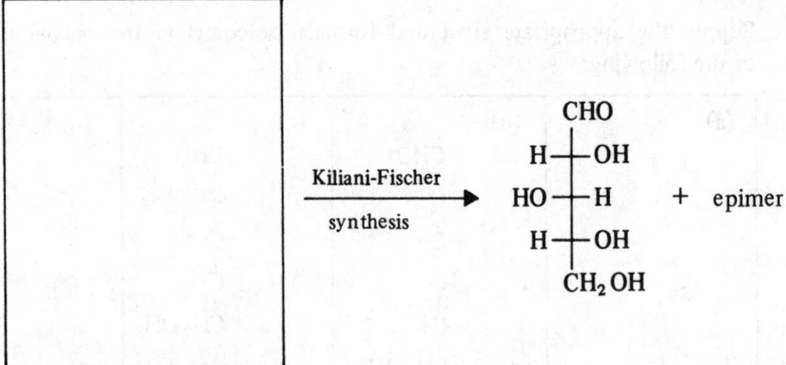

21.4 The D-aldopentose, (a), is oxidized to an aldaric acid, (b), which is optically active. Compound (a) undergoes a Ruff degradation to form an aldotetrose, (c), which undergoes oxidation to an optically inactive aldaric acid, (d). Supply the reagents for these transformations and the structural formulas of (a), (b), (c), and (d).

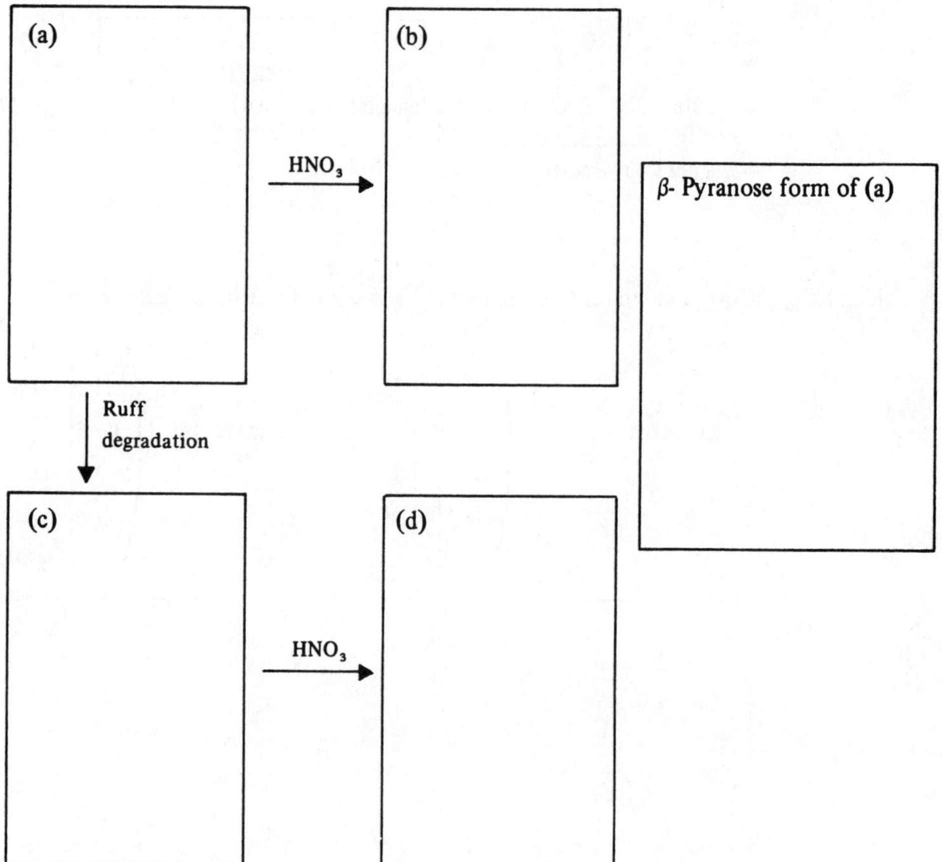

21.5 Give the structural formula of the β-pyranose form of (a) in the space just given.

21.6 Complete the following skeletal formulas and statements by filling in the blanks and circling the words that make the statements true.

The Haworth and conformational formulas of the β-cyclic hemiacetal

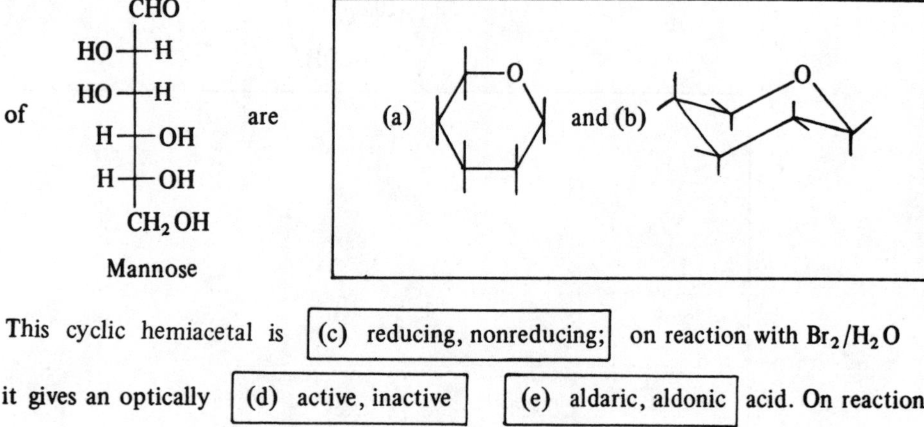

This cyclic hemiacetal is (c) reducing, nonreducing; on reaction with Br_2/H_2O it gives an optically (d) active, inactive (e) aldaric, aldonic acid. On reaction

with dilute HNO_3 it gives an optically (f) active, inactive (g) aldaric, aldonic acid. Reaction of the cyclic hemiacetal with (h) ______ converts it into an optically (i) active, inactive alditol.

21.7 Outline chemical tests that would allow you to distinguish between:

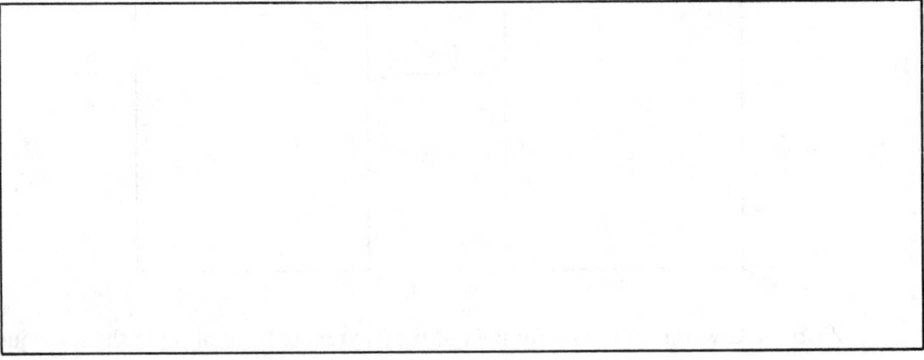

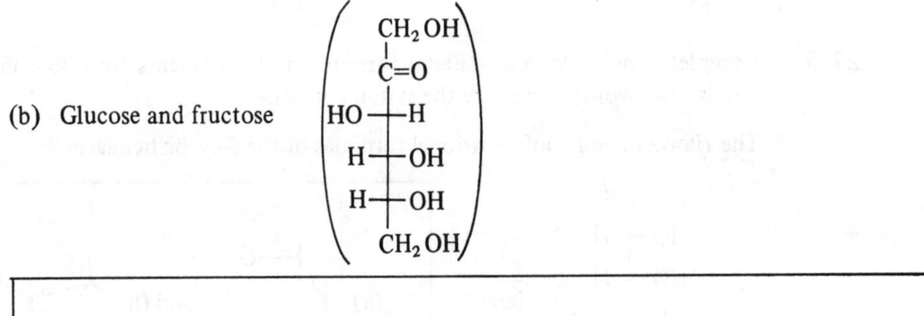

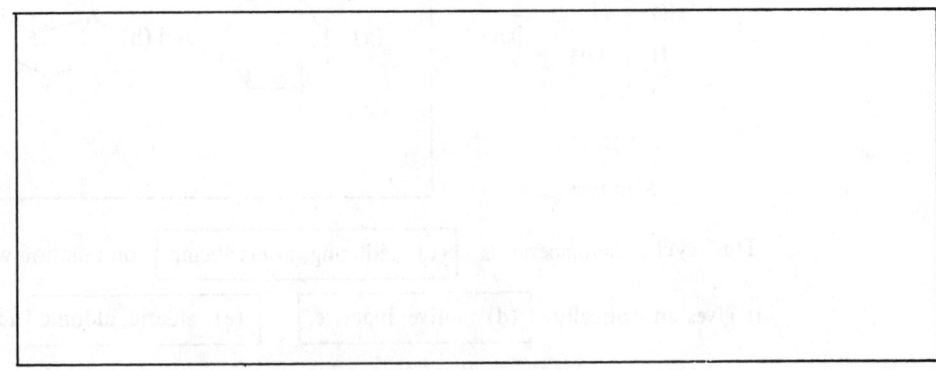

21.8 Hydrolysis of (+)-sucrose (ordinary table sugar) yields:

(a) D-glucose

(b) D-mannose

(c) D-fructose

(d) D-galactose

(e) More than one of the above

21.9 Select the reagent needed to perform the following transformation:

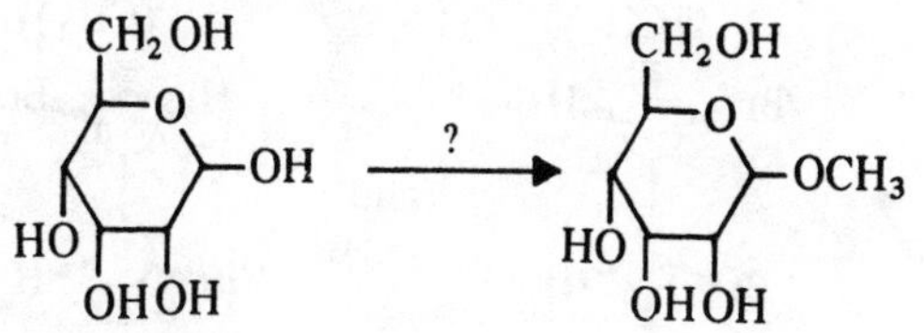

[Pyranose ring with CH_2OH, HO, OH, OH and anomeric –OH] $\xrightarrow{?}$ [Same ring with anomeric –OCH_3]

(a) CH_3OH, KOH

(b) $(CH_3C(=O))_2O$

(c) $(CH_3)_2SO_4$, OH^-

(d) CH_3OH, HCl

(e) CH_3OCH_3, HCl

22 LIPIDS

SOLUTIONS TO PROBLEMS

22.1 (a) There are two sets of enantiomers, giving a total of four stereoisomers

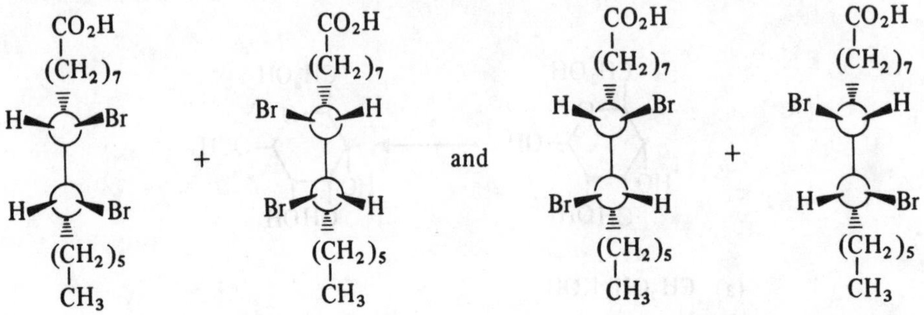

CO_2H / $(CH_2)_7$ / H, Br / H, Br / $(CH_2)_5$ / CH_3 + CO_2H / $(CH_2)_7$ / Br, H / Br, H / $(CH_2)_5$ / CH_3

erythro

and

CO_2H / $(CH_2)_7$ / H, Br / Br, H / $(CH_2)_5$ / CH_3 + CO_2H / $(CH_2)_7$ / Br, H / H, Br / $(CH_2)_5$ / CH_3

threo

(b) $CH_3(CH_2)_5CH{=}CH(CH_2)_7CO_2H$ (H, H) + Br_2 ⟶ bromonium ion [$CH_3(CH_2)_5$, H; $(CH_2)_7CO_2H$, H; Br^+] + Br^-, attack by paths (a) and (b)

(a): Br, $CH_3(CH_2)_5$, H; $(CH_2)_7CO_2H$, H, Br ≡ CO_2H / $(CH_2)_7$ / H, Br / Br, H / $(CH_2)_5$ / CH_3

(b): $CH_3(CH_2)_5$, H, Br; Br, $(CH_2)_7CO_2H$, H ≡ CO_2H / $(CH_2)_7$ / Br, H / H, Br / $(CH_2)_5$ / CH_3

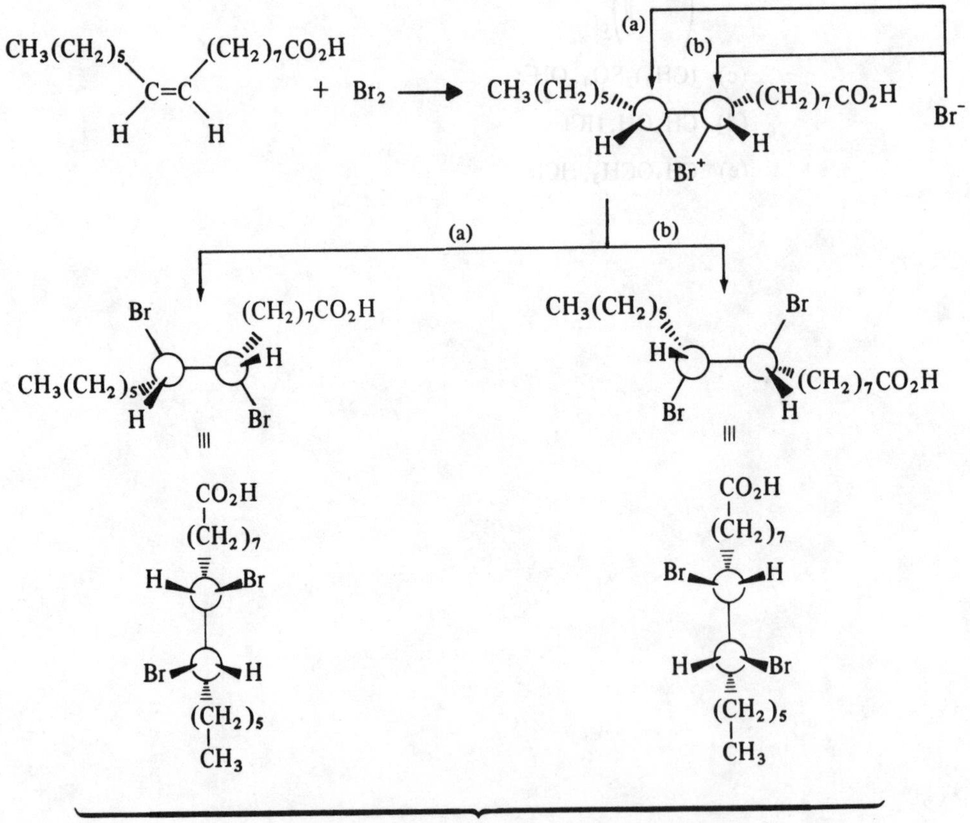

(±)-*threo*-9, 10-Dibromohexadecanoic acids

Formation of a bromonium ion at the other face of palmitoleic acid gives a result such that the *threo* enantiomers are the only products formed (obtained as a racemic modification).

The designations *erythro* and *threo* come from the names of the sugars called *erythrose* and *threose* (Section 21.7A).

22.2

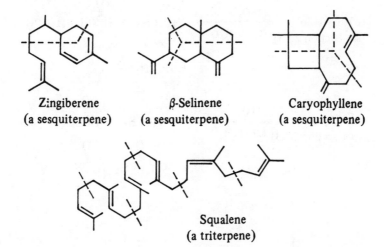

Zingiberene (a sesquiterpene)

β-Selinene (a sesquiterpene)

Caryophyllene (a sesquiterpene)

Squalene (a triterpene)

22.3

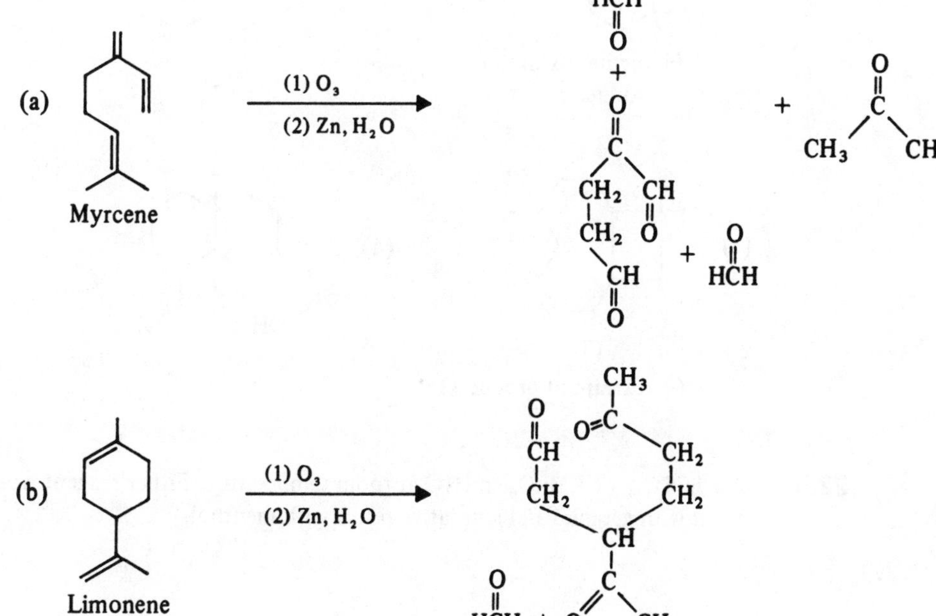

(c) α-Farnesene (see Section 22.3) $\xrightarrow[\text{(2) Zn, H}_2\text{O}]{\text{(1) O}_3}$ $CH_3\overset{O}{\overset{\|}{C}}CH_3 + H\overset{O}{\overset{\|}{C}}CH_2CH_2\overset{O}{\overset{\|}{C}}CH_3$

$+ H\overset{O}{\overset{\|}{C}}CH_2\overset{O}{\overset{\|}{C}}H + H\overset{O}{\overset{\|}{C}}{-}\overset{O}{\overset{\|}{C}}CH_3 + H\overset{O}{\overset{\|}{C}}H$

(d) Geraniol (see Section 22.3) $\xrightarrow[\text{(2) Zn, H}_2\text{O}]{\text{(1) O}_3}$ $CH_3\overset{O}{\overset{\|}{C}}CH_3 + H\overset{O}{\overset{\|}{C}}CH_2CH_2\overset{O}{\overset{\|}{C}}CH_3$

$+ H\overset{O}{\overset{\|}{C}}CH_2OH$

(e) Squalene (see Section 22.3) $\xrightarrow[\text{(2) Zn, H}_2\text{O}]{\text{(1) O}_3}$ $2CH_3\overset{O}{\overset{\|}{C}}CH_3 + H\overset{O}{\overset{\|}{C}}CH_2CH_2\overset{O}{\overset{\|}{C}}H$

$+ 4CH_3\overset{O}{\overset{\|}{C}}CH_2CH_2\overset{O}{\overset{\|}{C}}H$

22.4

(a) 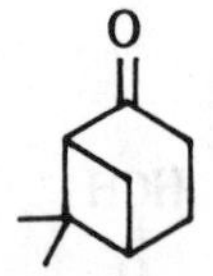 [structure] + CO_2

(+ further oxidation products)

(b) 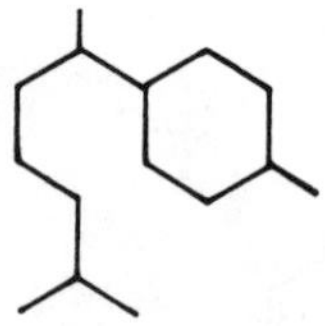 [structure]

(c) 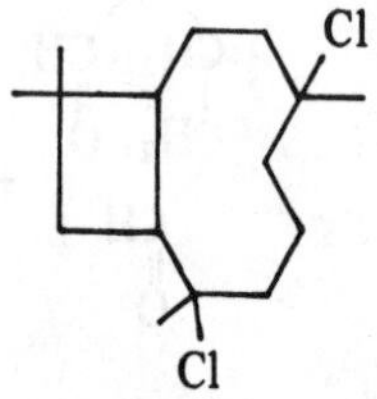 [structure with Cl, Cl]

(+ rearranged products)

(d) 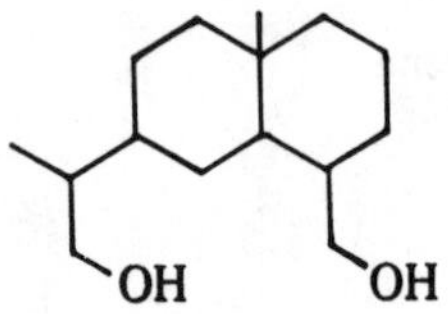 [structure with OH, OH]

22.5 Br_2 in CCl_4 or $KMnO_4$ in H_2O at room temperature. Either reagent would give a positive result with geraniol and a negative result with menthol.

22.6

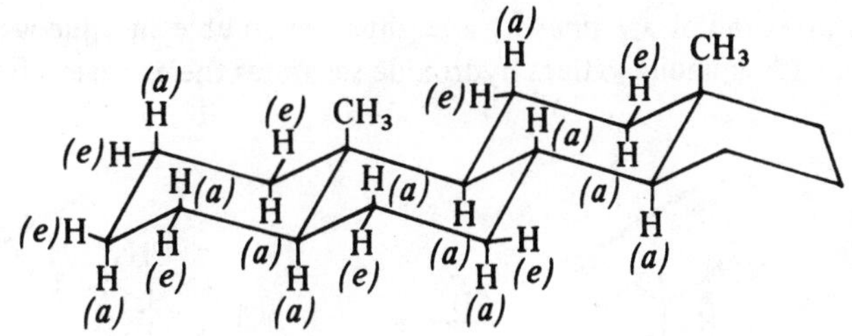

5α-Series

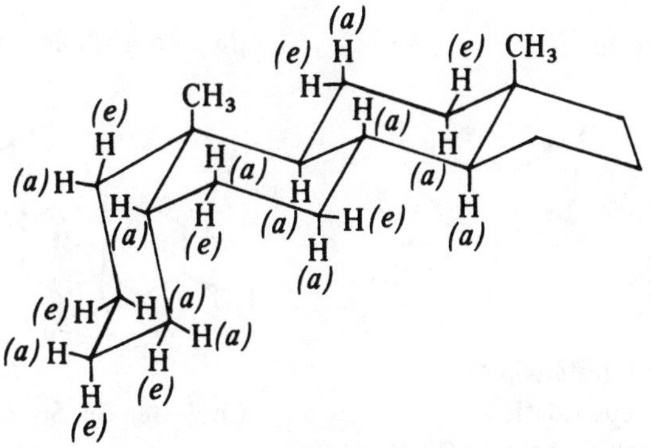

5β-Series

22.7

(a)

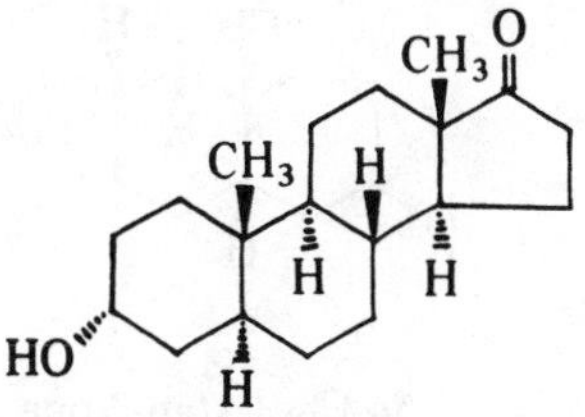

3α-Hydroxy-5α-androstan-17-one
(androsterone)

(b)

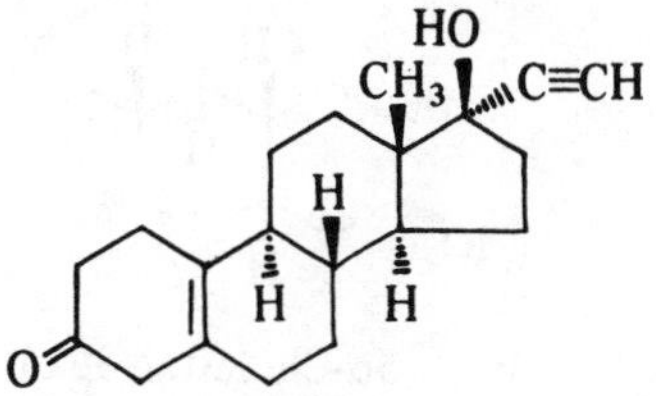

17α-Ethynyl-17β-hydroxy-5(10)-estren-3-one
(norethynodrel)

22.8

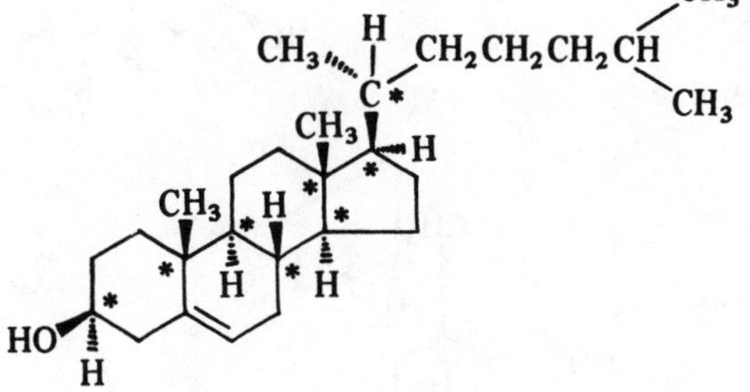

Absolute configuration of cholesterol
(5-cholesten-3β-ol)

22.9 Estrone and estradiol are *phenols* and thus are soluble in aqueous sodium hydroxide. Extraction with aqueous sodium hydroxide separates the estrogens from the androgens.

22.10

(a) Cholesterol $\xrightarrow[CCl_4]{Br_2}$ 5α, 6β-Dibromocholestan-3β-ol

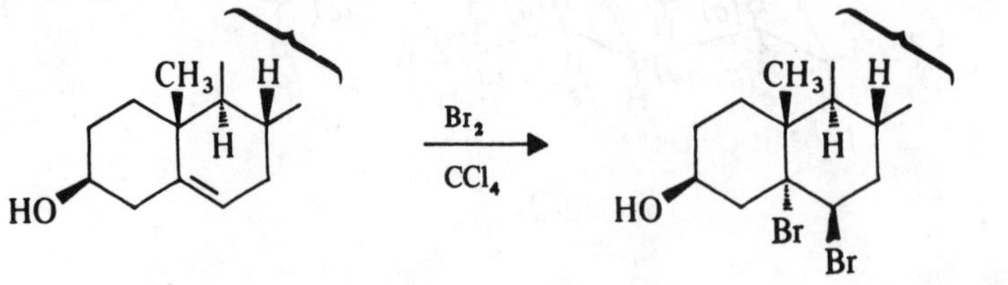

(b) 5α, 6α-Epoxycholestan-3β-ol (prepared by epoxidation of cholesterol, cf. Section 22.4G) $\xrightarrow[H_2O]{H^+,}$ Cholestan-3β, 5α, 6β-triol

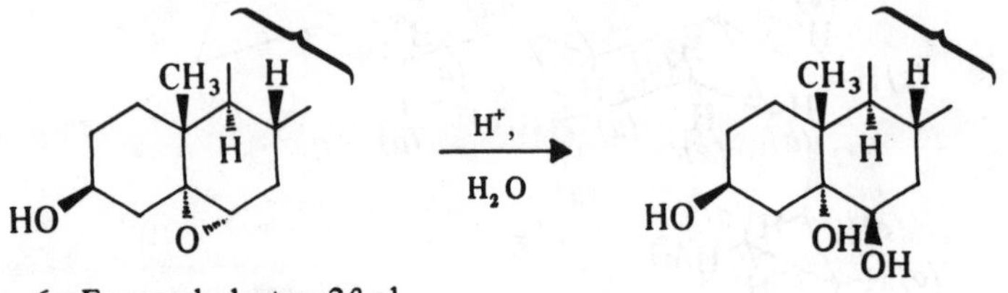

(c) 5α-Cholestan-3β-ol (prepared by hydrogenation of cholesteral, cf. Section 22.4G) $\xrightarrow[H_2SO_4]{CrO_3}$ 5α-Cholestan-3-one

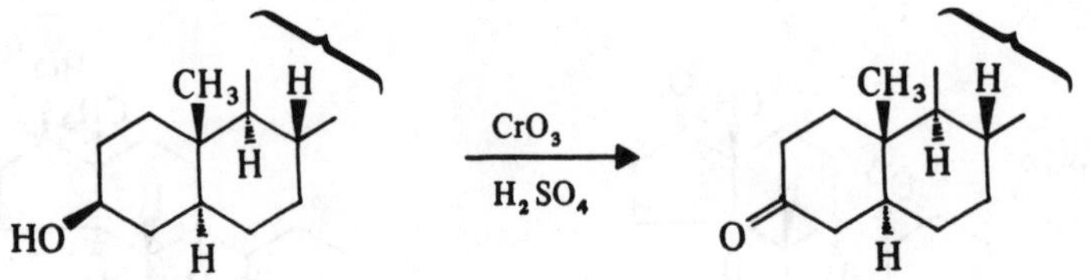

(d) Cholesterol $\xrightarrow[\text{(cf. p. 1062)}]{THF:BH_3}$ $\xrightarrow{CH_3CO_2D}$ 6α-Deuterio-5α-cholestan-3β-ol

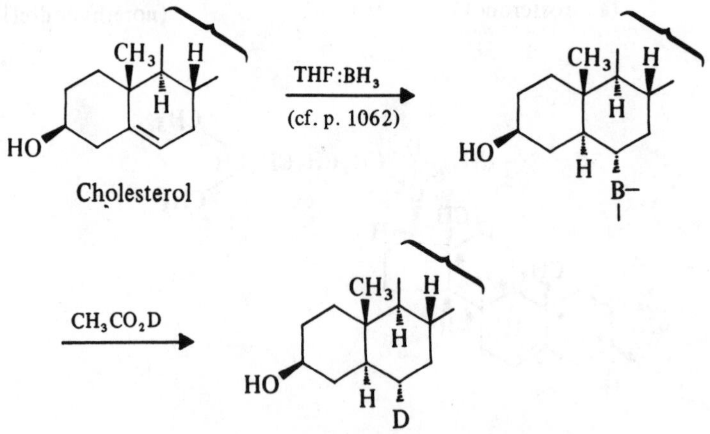

(e) 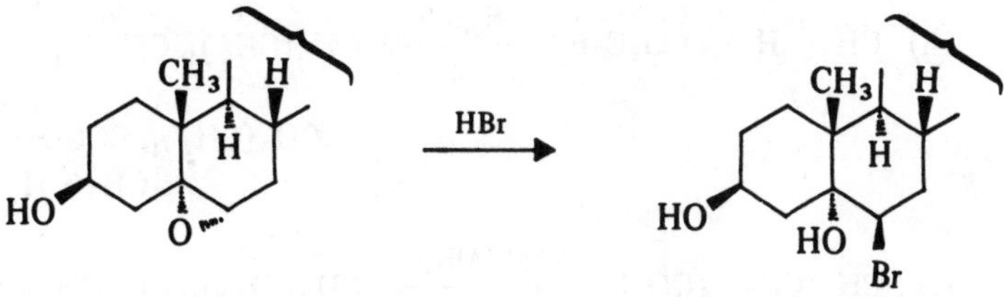 $\xrightarrow{HBr}$

5α, 6α-Epoxycholestan-3β-ol → 6β-Bromocholestan-3β, 5α-diol

22.11

(a) $\underset{\displaystyle CH_2OH}{\overset{\displaystyle CH_2OH}{CHOH}} + \overset{O}{\overset{\|}{RCOH}} + \overset{O}{\overset{\|}{R'COH}} + H_3PO_4 + HOCH_2CH_2\overset{+}{N}(CH_3)_3\ \ X^-$

(b) $\underset{\displaystyle CH_2OH}{\overset{\displaystyle CH_2OH}{CHOH}} + \overset{O}{\overset{\|}{RCOH}} + \overset{O}{\overset{\|}{R'COH}} + H_3PO_4 + HOCH_2CH_2NH_2$

(c) $\underset{\displaystyle CH_2OH}{\overset{\displaystyle CH_2OH}{CHOH}} + CH_3(CH_2)_nCH_2\overset{O}{\overset{\|}{C}}H + \overset{O}{\overset{\|}{R'COH}} + H_3PO_4$

$+ HOCH_2CH_2\overset{+}{N}(CH_3)_3\ \ X^-$

22.12

(a) $CH_3(CH_2)_{16}CO_2H + C_2H_5OH \overset{H^+}{\rightleftharpoons} CH_3(CH_2)_{16}CO_2C_2H_5 + H_2O$

$CH_3(CH_2)_{16}CO_2H \xrightarrow{SOCl_2} CH_3(CH_2)_{16}COCl \xrightarrow{C_2H_5OH} CH_3(CH_2)_{16}CO_2C_2H_5$

(b) $CH_3(CH_2)_{16}COCl \xrightarrow{(CH_3)_3COH} CH_3(CH_2)_{16}CO_2C(CH_3)_3$

(c) $CH_3(CH_2)_{16}COCl \xrightarrow{NH_3} CH_3(CH_2)_{16}CONH_2$

(d) $CH_3(CH_2)_{16}COCl \xrightarrow{(CH_3)_2NH} CH_3(CH_2)_{16}CON(CH_3)_2$

(e) $CH_3(CH_2)_{16}CONH_2 \xrightarrow{LiAlH_4} CH_3(CH_2)_{16}CH_2NH_2$

(f) $CH_3(CH_2)_{16}CONH_2 \xrightarrow{Br_2,\ OH^-} CH_3(CH_2)_{15}CH_2NH_2$

(g) $CH_3(CH_2)_{16}COCl \xrightarrow{LiAlH[OC(CH_3)_3]_3} CH_3(CH_2)_{16}CHO$

(h) $CH_3(CH_2)_{16}CO_2C_2H_5 \xrightarrow{H_2, Ni} CH_3(CH_2)_{16}CH_2OH$

$CH_3(CH_2)_{16}CH_2OH + CH_3(CH_2)_{16}COCl \longrightarrow CH_3(CH_2)_{16}CO_2CH_2(CH_2)_{16}CH_3$

(i) $CH_3(CH_2)_{16}CO_2H \xrightarrow[(2)\ H_2O]{(1)\ LiAlH_4} CH_3(CH_2)_{16}CH_2OH$

$CH_3(CH_2)_{16}CO_2C_2H_5 \xrightarrow{H_2, Ni} CH_3(CH_2)_{16}CH_2OH$

(j) $CH_3(CH_2)_{16}COCl + (CH_3)_2CuLi \longrightarrow CH_3(CH_2)_{16}COCH_3$

(k) $CH_3(CH_2)_{16}CH_2OH \xrightarrow{PBr_3} CH_3(CH_2)_{16}CH_2Br$

(l) $CH_3(CH_2)_{16}CH_2Br \xrightarrow[(2)\ H^+,\ H_2O,\ heat]{(1)\ NaCN} CH_3(CH_2)_{16}CH_2CO_2H$

22.13

(a) $CH_3(CH_2)_{11}CH_2CO_2H \xrightarrow{Br_2, P} CH_3(CH_2)_{11}CH(Br)CO_2H$

(b) $CH_3(CH_2)_{11}CH(Br)CO_2H \xrightarrow[(2)\ H^+]{(1)\ OH^-,\ heat} CH_3(CH_2)_{11}CH(OH)CO_2H$

(c) $CH_3(CH_2)_{11}CH(Br)CO_2H \xrightarrow[(2)\ H^+]{(1)\ NaCN} CH_3(CH_2)_{11}CH(CN)CO_2H$

(d) $CH_3(CH_2)_{11}CH(Br)CO_2H \xrightarrow[(2)\ H^+]{(1)\ NH_3\ (excess)} CH_3(CH_2)_{11}CH(NH_2)CO_2H$

or $CH_3(CH_2)_{11}CH(NH_3^+)CO_2^-$

22.14

(a) $CH_3(CH_2)_5CH{=}CH(CH_2)_7CO_2H \xrightarrow{I_2} CH_3(CH_2)_5CHICHI(CH_2)_7CO_2H$

(b) $CH_3(CH_2)_5CH{=}CH(CH_2)_7CO_2H \xrightarrow{H_2, Ni} CH_3(CH_2)_{14}CO_2H$

(c) $CH_3(CH_2)_5CH{=}CH(CH_2)_7CO_2H \xrightarrow{KMnO_4} CH_3(CH_2)_5CHOHCHOH(CH_2)_7CO_2H$

(d) $CH_3(CH_2)_5CH{=}CH(CH_2)_7CO_2H \xrightarrow{HCl} CH_3(CH_2)_5CH_2CHCl(CH_2)_7CO_2H$

+

$CH_3(CH_2)_5CHClCH_2(CH_2)_7CO_2H$

22.15 Elaidic acid is *trans*-9-octadecenoic acid:

$$\begin{array}{ccc} CH_3(CH_2)_7 & & H \\ & C=C & \\ H & & (CH_2)_7CO_2H \end{array}$$

It is formed by the isomerization of oleic acid.

22.16 (a)

$$\begin{array}{ccc} CH_3(CH_2)_9 & & (CH_2)_7CO_2H \\ & C=C & \\ H & & H \end{array} \quad \text{and} \quad \begin{array}{ccc} CH_3(CH_2)_9 & & H \\ & C=C & \\ H & & (CH_2)_7CO_2H \end{array}$$

(b) Infrared spectroscopy

(c) A peak in the 675-730 cm^{-1} region would indicate that the double bond is cis; a peak in the 960-975 cm^{-1} region would indicate that it is trans.

22.17 A reverse Diels-Alder reaction takes place.

heat

22.18

CH_3 CH_2

α-Phellandrene β-Phellandrene

Note: On permanganate oxidation, the $=CH_2$ group of β-phellandrene is converted to CO_2 and thus is not detected in the reaction.

22.19

$$CH_3(CH_2)_5C{\equiv}CH + NaNH_2 \xrightarrow[NH_3]{} \underset{\mathbf{A}}{CH_3(CH_2)_5C{\equiv}CNa}$$

$$\xrightarrow{ICH_2(CH_2)_7CH_2Cl} \underset{\mathbf{B}}{CH_3(CH_2)_5C{\equiv}CCH_2(CH_2)_7CH_2Cl} \xrightarrow{NaCN}$$

$$CH_3(CH_2)_5C{\equiv}CCH_2(CH_2)_7CH_2CN \xrightarrow{KOH,\ H_2O} CH_3(CH_2)_5C{\equiv}CCH_2(CH_2)_7CH_2CO_2K$$

C → D

$$\xrightarrow{H_3O^+} CH_3(CH_2)_5C{\equiv}CCH_2(CH_2)_7CH_2CO_2H \xrightarrow{H_2,\ Pd\text{-}BaSO_4}$$

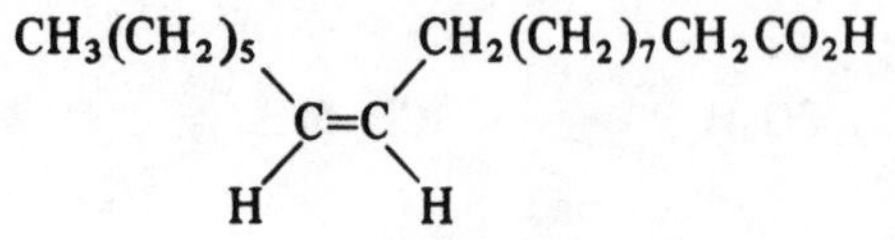

E

$$CH_3(CH_2)_5CH{=}CHCH_2(CH_2)_7CH_2CO_2H \text{ (cis)}$$

Vaccenic acid

22.20 $FCH_2(CH_2)_6CH_2Br + HC{\equiv}CNa \longrightarrow FCH_2(CH_2)_6CH_2C{\equiv}CH$

F

$$\xrightarrow[(2)\ I(CH_2)_7Cl]{(1)\ NaNH_2} FCH_2(CH_2)_6CH_2C{\equiv}C(CH_2)_7Cl \xrightarrow{NaCN}$$

G

$$FCH_2(CH_2)_6CH_2C{\equiv}C(CH_2)_7CN \xrightarrow[(2)\ H^+]{(1)\ KOH} FCH_2(CH_2)_6CH_2C{\equiv}C(CH_2)_7CO_2H$$

H → I

$$\xrightarrow[Ni_2B\ (P\text{-}2)]{H_2} FCH_2(CH_2)_6CH_2CH{=}CH(CH_2)_7CO_2H \text{ (cis)}$$

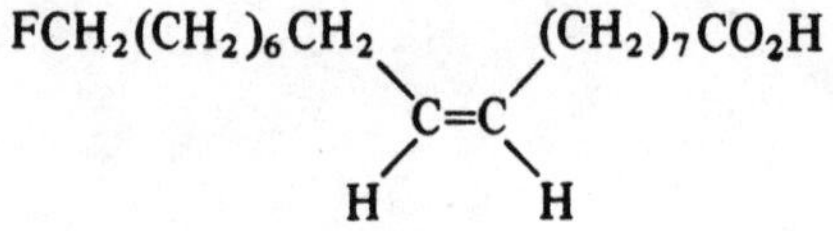

22.21

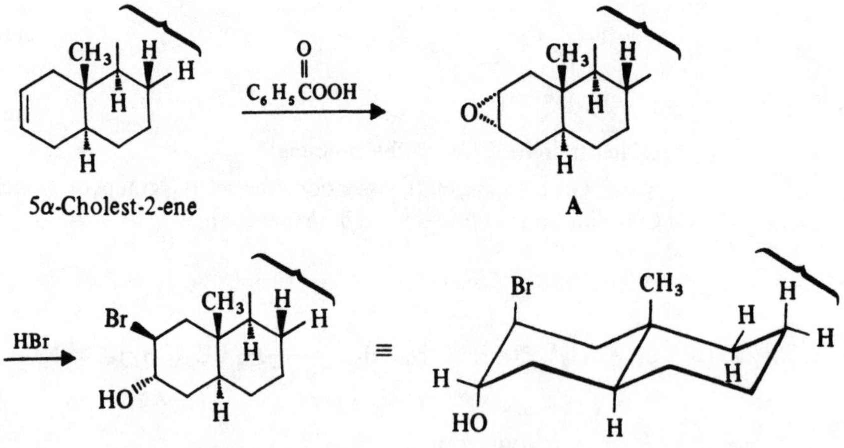

5α-Cholest-2-ene $\xrightarrow{C_6H_5CO_3H}$ A (2α,3α-epoxide) $\xrightarrow{HBr}$ B (2β-bromo-3α-hydroxy, diaxial)

Here we find that epoxidation takes place at the less hindered α face (cf. Section 22.4G). Ring opening by HBr takes place in an anti fashion to give a product with diaxial substituents.

22.22 (a) $CH_2{=}CH{-}CH{=}CH_2$

(b) OH^- (Removal of the α hydrogen atom allows isomerization to the more stable compound with a trans ring junction.)

(c) $LiAlH_4$

(d) H_3O^+ and heat. (Hydrolysis of the enol ether is followed by dehydration of one alcohol group.)

(e) $HCO_2C_2H_5$, C_2H_5ONa

(f) OsO_4, then $NaHSO_3$

(g) $CH_3\overset{\overset{\displaystyle O}{\|}}{C}CH_3$, H^+

(h) H_2, Pd catalyst

(i) H_3O^+, H_2O

(j) HIO_4

(k) Base and heat. (This reaction is an aldol condensation.)

(l) and (m) Na_2CrO_4, CH_3CO_2H to oxidize the aldehyde to an acid, followed by esterification.

(n) H_2 and Pt. (Hydrogen addition takes place from the less hindered α face of the molecule.)

(o), (p), (q) $NaBH_4$ to reduce the keto group; OH^-, H_2O to hydrolyze the ester; and acetic anhydride to esterify the OH^- at the 3-position.

(r) and (s) $SOCl_2$ to make the acid chloride, followed by treatment with $(CH_3)_2Cd$.

(t) $CH_3\overset{\overset{\displaystyle CH_3}{|}}{C}HCH_2CH_2CH_2MgBr$, followed by H_3O^+.

(u), (v), (w) Acetic acid and heat to dehydrate the tertiary alcohol; followed by acetic anhydride to acetylate the secondary alcohol; followed by H_2, Pt to hydrogenate the double bond.

22.23 (a) $CH_3(CH_2)_4\overset{\overset{\displaystyle O}{\|}}{C}H$

(b) C_4H_9Li

(c) 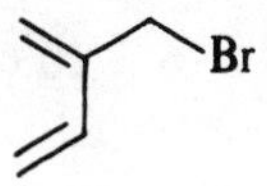 Br

(d) 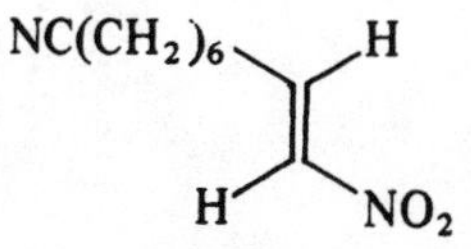 $NC(CH_2)_6$ H H NO_2

(e) Michael addition using a basic catalyst.

22.24 First an elimination takes place,

$$R_3\overset{+}{N}CH_2CH_2\overset{\overset{O}{\|}}{C}CH_2CH_3 + NH_2^- \longrightarrow CH_2{=}CH\overset{\overset{O}{\|}}{C}CH_2CH_3 + R_3N + NH_3$$

then a conjugate addition occurs, followed by an aldol addition:

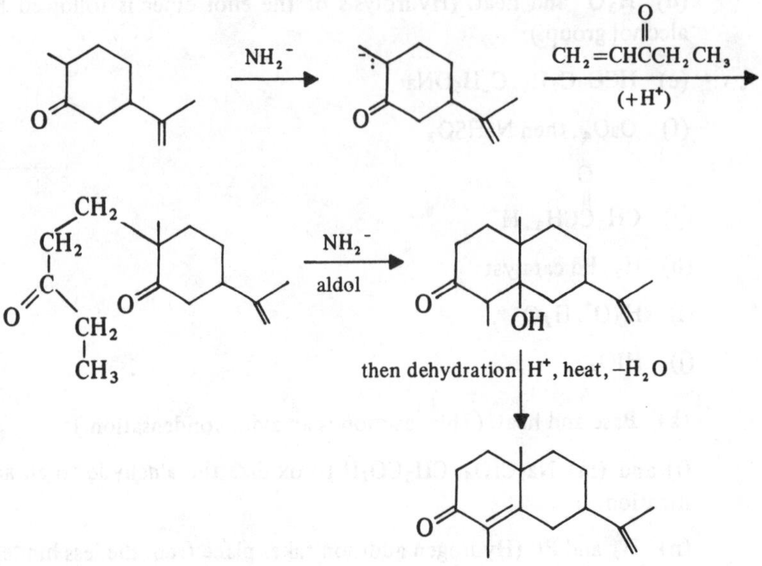

NH$_2^-$; $CH_2{=}CH\overset{\overset{O}{\|}}{C}CH_2CH_3$ (+H$^+$)

NH$_2^-$, aldol; OH

then dehydration H$^+$, heat, $-H_2O$

SECTION REFERENCES FOR ADDITIONAL PROBLEMS

22.12 16.4, 16.5, 18.3, 18.5, 18.7, 18.8, 19.5

22.13 18.3, 18.9, 19.5, 22.2

22.14 22.2

22.15 6.9

22.16 13.11, 21.4D

22.17 10.12, 22.3

22.18 13.2, 22.3

22.19 6.19, 18.3

22.20 6.19, 18.3

22.21 7.9, 22.4G

22.22 (a) 10.12, (b) 17.1, (c) 15.7, (d) 17.2, 6.13, (e) 20.2, (f) 7.10, (g) 16.7, (h) 6.5, (i) 16.7, (j) 21.4D, (k) 17.5, (l) 18.3, (m) 18.7, (n) 22.4G, (o) 15.2, (p) 18.7, (q) 18.7, (r) 18.4, (s) 16.5, (t) 15.7, (u) 6.13, (v) 18.7, (w) 6.4, 6.5, 22.4G

22.23 16.7, 20.6, 20.8

SELF-TEST

22.1 Write an appropriate formula in each box

(a)

A naturally occurring fatty acid

(b)

A soap

(c)

A solid fat

(d)

An oil

(e)

A synthetic detergent

(f)

5α-Estran-17-one

22.2 Give a reagent that would distinguish between each of the following:

(a) Pregnane and 20-pregnanone

(b) Stearic acid and oleic acid

(c) 17α-Ethynyl-1,3,5(10)-estratriene-3,17β-diol (ethynylestradiol) and 1,3,5(10)-estratriene-3,17β-diol (estradiol)

22.3 What product would be obtained by catalytic hydrogenation of 4-androstene.

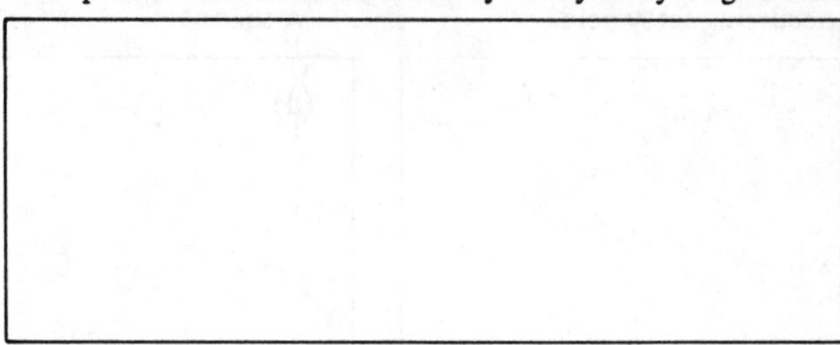

22.4 Supply the missing compounds

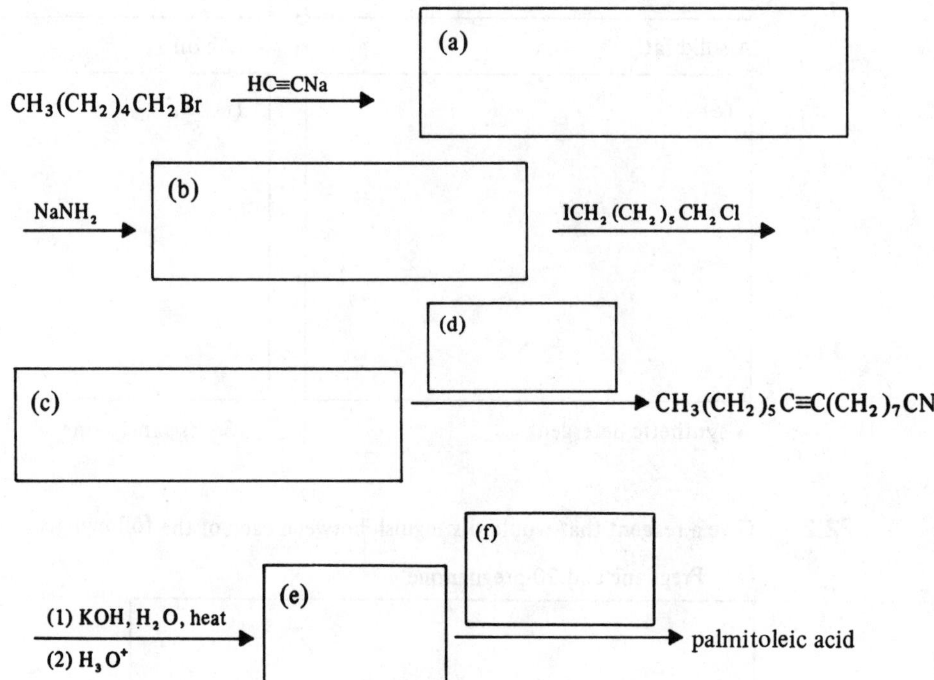

22.5 Circle the correct answer.

This compound is a
- Monoterpene
- Sesquiterpene
- Diterpene
- Triterpene

22.6 Mark off the isoprene units in the previous compound.

22.7 Which is a systematic name for the steroid shown here?

(a) 5α-Androstan-3α-ol

(b) 5β-Androstan-3β-ol

(c) 5α-Pregnan-3α-ol

(d) 5β-Pregnan-3β-ol

(e) 5α-Estran-3α-ol

M

SPECIAL TOPIC
Thiol Esters and Lipid Biosynthesis

SOLUTIONS TO PROBLEM

M.1

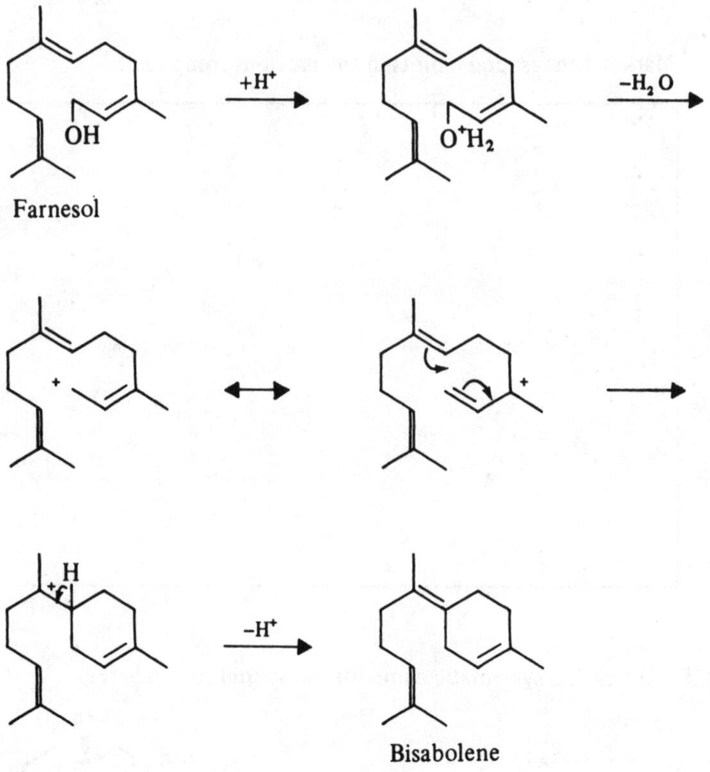

OH

Farnesol

$+H^+$

O^+H_2

$-H_2O$

$-H^+$

Bisabolene

23 AMINO ACIDS AND PROTEINS

SOLUTIONS TO PROBLEMS

Note: In the solutions given here all amino acid structures are drawn as the zwitterion regardless of the acidity or basicity of their solutions.

23.1 (a) $HO_2CCH_2CH_2CH(NH_3^+)CO_2H$

(b) $^-O_2CCH_2CH_2CH(NH_2)CO_2^-$

(c) $HO_2CCH_2CH_2CH(NH_3^+)CO_2^-$ predominates at the isoelectric point rather than $^-OOCCH_2$-$CH_2CH(NH_3^+)CO_2H$ because of the acid-strengthening inductive effect of the α-ammonio group.

(d) Since glutamic acid is a dicarboxylic acid, acid must be added (i.e., the pH must be made lower) to suppress the ionization of the second carboxyl group and thus achieve the isoelectric point. Glutamine, with only one carboxyl group, is similar to glycine or phenylalanine and has its isoelectric point at a higher pH.

23.2 The conjugate acid is highly stabilized by resonance.

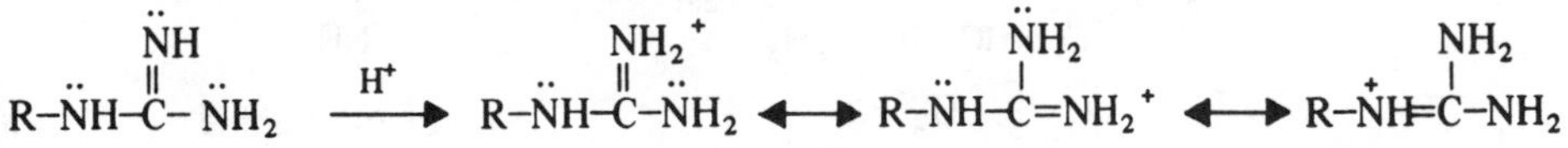

$$R\text{-}\ddot{N}H\text{-}C(=\ddot{N}H)\text{-}\ddot{N}H_2 \xrightarrow{H^+} R\text{-}\ddot{N}H\text{-}C(=NH_2^+)\text{-}\ddot{N}H_2 \longleftrightarrow R\text{-}\ddot{N}H\text{-}C(\ddot{N}H_2)\text{=}NH_2^+ \longleftrightarrow R\text{-}\overset{+}{N}H\text{=}C(NH_2)\text{-}NH_2$$

23.3

(a) Potassium phthalimide (NK) $+ BrCH(CO_2C_2H_5)_2 \longrightarrow$

N-phthalimido-$CH(CO_2C_2H_5)_2$ $\xrightarrow[(CH_3)_2CHCH_2Br]{NaOCH_2CH_3}$

N-phthalimido-$C(CO_2C_2H_5)_2CH_2CH(CH_3)_2$ $\xrightarrow[heat]{NaOH,}$ $C_6H_4(CO_2^-)$-$C(=O)$-$NHC(CO_2^-)_2CH_2CH(CH_3)_2$

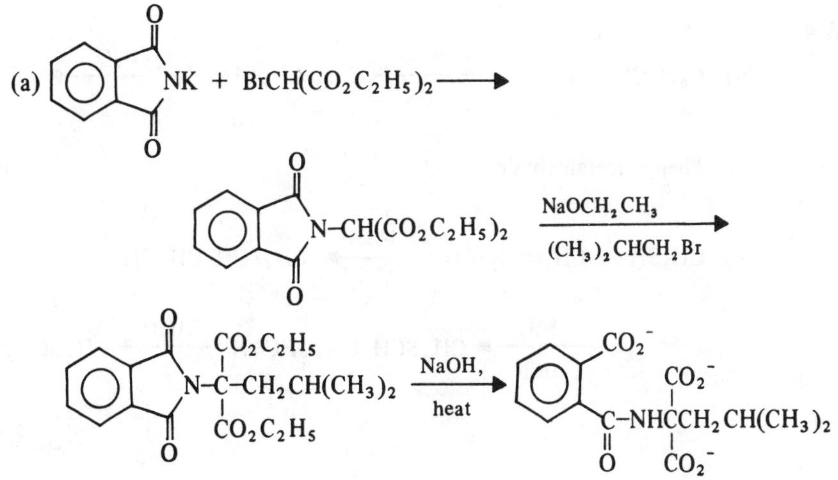

$$\xrightarrow[\text{heat}]{\text{HCl}} (CH_3)_2CHCH_2\underset{\underset{NH_3^+}{|}}{C}HCO_2^- + CO_2 + C_6H_4(CO_2H)_2$$

DL-Leucine

(b)

$$C_6H_4(CO)_2N\text{—}CH(CO_2C_2H_5)_2 \xrightarrow[CH_3I]{NaOCH_2CH_3} C_6H_4(CO)_2N\text{—}C(CH_3)(CO_2C_2H_5)_2 \xrightarrow[\text{heat}]{NaOH} C_6H_4(CO_2^-)\text{—}C(=O)\text{—}NHC(CH_3)(CO_2^-)_2$$

$$\xrightarrow[\text{heat}]{\text{HCl}} CH_3\underset{\underset{NH_3^+}{|}}{C}HCO_2^- + CO_2 + C_6H_4(CO_2H)_2$$

DL-Alanine

(c)

$$C_6H_4(CO)_2NCH(CO_2C_2H_5)_2 \xrightarrow[C_6H_5CH_2Br]{NaOCH_2CH_3} C_6H_4(CO)_2N\text{—}C(CH_2C_6H_5)(CO_2C_2H_5)_2 \xrightarrow[\text{heat}]{NaOH}$$

$$C_6H_4(CO_2^-)\text{—}C(=O)\text{—}NHC(CH_2C_6H_5)(CO_2^-)_2 \xrightarrow[\text{heat}]{\text{HCl}} C_6H_5CH_2\underset{\underset{NH_3^+}{|}}{C}HCO_2^- + CO_2 + C_6H_4(CO_2H)_2$$

DL-Phenylalanine

23.4

(a)

$$C_6H_5CH_2\overset{\overset{O}{\|}}{C}H \xrightarrow[HCN]{NH_3} C_6H_5CH_2\underset{\underset{NH_2}{|}}{C}HC{\equiv}N \xrightarrow{H_3O^+} C_6H_5CH_2\underset{\underset{NH_3^+}{|}}{C}HCO_2^-$$

Phenyl acetaldehyde — DL-Phenylalanine

(b)

$$CH_3SH + CH_2{=}CH\text{—}\overset{\overset{O}{\|}}{C}H \xrightarrow{\text{base}} CH_3SCH_2CH_2\overset{\overset{O}{\|}}{C}H$$

$$\xrightarrow[HCN]{NH_3} CH_3SCH_2CH_2\underset{\underset{NH_2}{|}}{C}HC{\equiv}N \xrightarrow{H_3O^+} CH_3SCH_2CH_2\underset{\underset{NH_3^+}{|}}{C}HCO_2^-$$

DL-Methionine

23.5 Because of the presence of an electron-withdrawing 2,4-dinitrophenyl group, the labeled amino acid is relatively nonbasic and is, therefore, insoluble in dilute aqueous acid. The other amino acids (those that are not labeled) dissolve in dilute aqueous acid.

23.6

(a) $H_3\overset{+}{N}CH(CH(CH_3)CH_3)CONHCH(CH_3)CONHCH_2CO_2^-$ $\xrightarrow[HCO_3^-]{O_2N\text{-}C_6H_3(NO_2)\text{-}F}$

Val•Ala•Gly

$O_2N\text{-}C_6H_3(NO_2)\text{-}NHCH(CH(CH_3)CH_3)CONHCH(CH_3)CONHCH_2CO_2^-$ $\xrightarrow[heat]{H_3O^+}$

$O_2N\text{-}C_6H_3(NO_2)\text{-}NHCH(CH(CH_3)CH_3)CO_2H$ + $H_3\overset{+}{N}CH(CH_3)CO_2^-$ + $H_3\overset{+}{N}CH_2CO_2^-$

Labeled valine (separate and identify) — Alanine — Glycine

(b) $O_2N\text{-}C_6H_3(NO_2)\text{-}NHCH(CH(CH_3)CH_3)CO_2H$ + $O_2N\text{-}C_6H_3(NO_2)\text{-}NHCH_2CH_2CH_2CH_2CH(NH_3^+)CO_2^-$

α-Labeled valine — ε-Labeled lysine

+ $H_3\overset{+}{N}CH_2CO_2^-$

Glycine

23.7

$C_6H_5\text{-}N{=}C{=}S$ + $H_2\ddot{N}\text{-}CH(CH_2CH_2SCH_3)CO\text{-}NHCH(CH(CH_3)CH_2CH_3)CO\text{-}NHCH(CH_2CH_2CH_2NHC(=NH)NH_2)CO_2^-$ $\xrightarrow{OH^-}$

Phenylisothiocyanate — Met•Ile•Arg

$$C_6H_5\text{-}NH\text{-}\overset{S}{\overset{\|}{C}}\text{-}NH\text{-}\underset{CH_2CH_2SCH_3}{CH}CO\text{-}NH\underset{CHCH_3CH_2CH_3}{CH}CO\text{-}NH\underset{CH_2CH_2CH_2NHC(=NH)NH_2}{CH}CO_2^- \xrightarrow{H^+}$$

$$\text{(phenylthiohydantoin: } C_6H_5\text{-}N\text{, }C{=}S\text{, NH, C-CH, }O\text{, } CH_2CH_2SCH_3) + H_2N\underset{CHCH_3CH_2CH_3}{CH}CO\text{-}NH\underset{CH_2CH_2CH_2NHC(=NH)NH_3^+}{CH}CO_2^-$$

Phenylthiohydantoin derived from methionine

$$\xrightarrow[(2)\ H^+]{(1)\ C_6H_5\text{-}N{=}C{=}S,\ OH^-} \text{(phenylthiohydantoin: } C_6H_5\text{-}N\text{, }C{=}O\text{, NH, C-CH, }O\text{, } CHCH_2CH_3(CH_3)) + H_2N\text{-}\underset{CH_2CH_2CH_2NHC(=NH)NH_3^+}{CH}\text{-}CO_2^-$$

Phenylthiohydantoin derived from isoleucine

23.8 (a) Two structures are possible with the sequence Glu·Cys·Gly. Glutamic acid may be linked to cysteine through its α-carboxyl group,

$$HO_2CCH_2CH_2\underset{NH_3^+}{CH}CO\text{-}NH\underset{CH_2SH}{CH}CO\text{-}NHCH_2CO_2^-$$

or through its γ-carboxyl group,

$$H_3\overset{+}{N}\underset{CO_2^-}{CH}CH_2CH_2CO\text{-}NH\underset{CH_2SH}{CH}CO\text{-}NHCH_2CO_2^-$$

(b) This result shows that the second structure is correct, that in glutathione the γ-carboxyl group is linked to cysteine.

23.9 We look for points of overlap to determine the amino acid sequence in each case.

(a)

Ser · Thr
Thr · Hyp
Pro · Ser

Pro · Ser · Thr · Hyp

(b)

Ala · Cys
Cys · Arg
Arg · Val
Leu · Ala

Leu · Ala · Cys · Arg · Val

23.10 Sodium in liquid ammonia brings about reductive cleavage of the disulfide linkage of oxytocin to two thiol groups, then air oxidizes the two thiol groups back to a disulfide linkage:

$$R{-}CH_2{-}S{-}S{-}CH_2{-}R \xrightarrow[NH_3]{Na} R{-}CH_2{-}SH \;\; HS{-}CH_2{-}R \xrightarrow{O_2} R{-}CH_2{-}S{-}S{-}CH_2{-}R$$

See also Special Topic F.

23.11

$$\underset{\text{Glycine}}{H_3\overset{+}{N}CH_2CO_2^-} + \underset{\textit{tert}\text{-Butoxycarbonyl azide}}{(CH_3)_3CO\overset{O}{\overset{\|}{C}}N_3} \xrightarrow[25°C]{OH^-} \xrightarrow{H_3O^+}$$

$$\underset{\text{Boc-Gly}}{(CH_3)_3C{-}O\overset{O}{\overset{\|}{C}}NHCH_2CO_2H} \xrightarrow[(2)\ ClCO_2C_2H_5]{(1)\ (C_2H_5)_3N}$$

$$(CH_3)_3C\overset{O}{\overset{\|}{C}}OCNHCH_2\overset{O}{\overset{\|}{C}}O\overset{O}{\overset{\|}{C}}OC_2H_5 \xrightarrow[-CO_2,\ -C_2H_5OH]{H_3\overset{+}{N}CHCO_2^-(CH(CH_3)CH_3)\ \text{Valine}}$$

Mixed anhydride

$$(CH_3)_3CO\overset{O}{\overset{\|}{C}}NHCH_2\overset{O}{\overset{\|}{C}}NH\underset{\underset{CH_3}{|}}{\underset{CHCH_3}{\underset{|}{C}}}HCO_2H \xrightarrow[(2)\ ClCO_2C_2H_5]{(1)\ (C_2H_5)_3N}$$

Boc-Gly·Val

$$(CH_3)_3CO\overset{O}{\overset{\|}{C}}NHCH_2\overset{O}{\overset{\|}{C}}NH\underset{\underset{CH_3}{|}}{\underset{CHCH_3}{\underset{|}{C}}}H\overset{O}{\overset{\|}{C}}O\overset{O}{\overset{\|}{C}}OC_2H_5 \xrightarrow{H_3\overset{+}{N}CHCO_2^-(CH_3)\ \text{alanine}}$$

Mixed anhydride

$$(CH_3)_3CO\overset{O}{\overset{\|}{C}}NHCH_2\overset{O}{\overset{\|}{C}}NH\underset{\underset{CH_3}{|}}{\underset{CHCH_3}{\underset{|}{C}}}H\overset{O}{\overset{\|}{C}}NH\underset{CH_3}{\underset{|}{C}}HCO_2H \xrightarrow[CH_3CO_2H,\ 25°C]{CF_3CO_2H}$$

Boc-Gly·Val·Ala

$$(CH_3)_2C{=}CH_2 + CO_2 + H_3\overset{+}{N}CH_2\overset{O}{\overset{\|}{C}}NH\underset{\underset{CH_3}{|}}{\underset{CHCH_3}{\underset{|}{C}}}H\overset{O}{\overset{\|}{C}}NH\underset{CH_3}{\underset{|}{C}}HCO_2^-$$

Gly·Val·Ala

23.12

(a)
$$2C_6H_5CH_2O\overset{O}{\overset{\|}{C}}Cl + H_2NCH_2CH_2CH_2CH_2\underset{NH_2}{\underset{|}{C}}HCO_2^- \xrightarrow[25°C]{OH^-}$$

Benzyl chlorocarbonate; Lysine

$$C_6H_5CH_2O\overset{O}{\overset{\|}{C}}NHCH_2CH_2CH_2CH_2\underset{\underset{C_6H_5CH_2OC=O}{|}}{\underset{NH}{\underset{|}{C}}}HCO_2H \xrightarrow[(2)\ ClCO_2C_2H_5]{(1)\ (C_2H_5)_3N}$$

$$C_6H_5CH_2OC(=O)NHCH_2CH_2CH_2CH_2CH(NHC(=O)OCH_2C_6H_5)C(=O)OC(=O)OC_2H_5 \xrightarrow[-CO_2,\ -C_2H_5OH]{CH_3CH_2CH(CH_3)CH(NH_3^+)CO_2^-}$$

$$C_6H_5CH_2OC(=O)NHCH_2CH_2CH_2CH_2CH(NHC(=O)OCH_2C_6H_5)C(=O)NHCH(CO_2^-)CH(CH_3)CH_2CH_3 \xrightarrow[CH_3CO_2H,\ cold]{HBr}$$

$$2C_6H_5CH_2Br + 2CO_2 + H_3\overset{+}{N}CH_2CH_2CH_2CH_2CH(NH_2)C(=O)NHCH(CO_2^-)CH(CH_3)CH_2CH_3$$

Lys•Ile

(b)

$$3C_6H_5CH_2OC(=O)Cl + H_2NC(=NH)NHCH_2CH_2CH_2CH(NH_2)CO_2^- \xrightarrow[25°C]{OH^-}$$

$$C_6H_5CH_2OC(=O)NHC(=NH)N(C(=O)OCH_2C_6H_5)CH_2CH_2CH_2CH(NHC(=O)OCH_2C_6H_5)CO_2H \xrightarrow[(2)\ ClCO_2C_2H_5]{(1)\ (C_2H_5)_3N}$$

$$C_6H_5CH_2OC(=O)NHC(=NH)N(C(=O)OCH_2C_6H_5)CH_2CH_2CH_2CH(NHC(=O)OCH_2C_6H_5)C(=O)OC(=O)OC_2H_5 \xrightarrow[-CO_2,\ -C_2H_5OH]{CH_3CH(NH_3^+)CO_2^-}$$

$$C_6H_5CH_2OC(=O)NHC(=NH)N(C(=O)OCH_2C_6H_5)CH_2CH_2CH_2CH(NHC(=O)OCH_2C_6H_5)C(=O)NHCH(CH_3)CO_2H \xrightarrow[CH_3CO_2H,\ cold]{HBr}$$

$$3C_6H_5CH_2Br + 3CO_2 + {}^{+}H_3N\overset{NH}{\overset{\|}{C}}NHCH_2CH_2CH_2\underset{NH_2}{\underset{|}{C}H}CONH\underset{CH_3}{\underset{|}{C}H}CO_2^-$$

Arg·Ala

23.13 The weakness of the benzyl-oxygen bond allows these groups to be removed by catalytic hydrogenolysis.

23.14 (a) An electrophilic aromatic substitution reaction:

$$\left(\underset{C_6H_5}{CH_2CH}\right)_n + CH_3OCH_2Cl \xrightarrow{BF_3} \left(\underset{C_6H_4CH_2Cl}{CH_2CH}\right)_n + CH_3OH$$

(b) The linkage between the resin and the polypeptide is a benzylic ester. It is cleaved by HBr in CF_3CO_2H at room temperature because the carbocation that is formed initially is the relatively stable , benzylic cation.

23.15

$$\text{Ⓟ}{-}CH_2Cl + HO\overset{O}{\overset{\|}{C}}\underset{CH_3}{\underset{|}{C}H}NH\overset{O}{\overset{\|}{C}}OC(CH_3)_3$$

↓ base — 1 Add Boc·Ala

$$\text{Ⓟ}{-}CH_2O\overset{O}{\overset{\|}{C}}\underset{CH_3}{\underset{|}{C}H}NH\overset{O}{\overset{\|}{C}}OC(CH_3)_3$$

2 Purify by washing

3 Remove protecting group

↓ CF_3CO_2H, CH_2Cl_2

$$\text{Ⓟ}{-}CH_2O\overset{O}{\overset{\|}{C}}\underset{CH_3}{\underset{|}{C}H}NH_2$$

4 Purify by washing

↓ $HO\overset{O}{\overset{\|}{C}}\underset{CH_2C_6H_5}{\underset{|}{C}H}NH\overset{O}{\overset{\|}{C}}OC(CH_3)_3$ and dicyclohexylcarbodiimide — 5 Add Boc·Phe

$$\text{(resin)}\!-\!CH_2O\overset{O}{\overset{\|}{C}}\underset{CH_3}{\underset{|}{C}}HNH\overset{O}{\overset{\|}{C}}\underset{\underset{C_6H_5}{|}}{\underset{CH_2}{\underset{|}{C}}}HNH\overset{O}{\overset{\|}{C}}OC(CH_3)_3$$

6 Purify by washing

↓ CF_3CO_2H, CH_2Cl_2

7 Remove protecting group

$$\text{(resin)}\!-\!CH_2O\overset{O}{\overset{\|}{C}}\underset{CH_3}{\underset{|}{C}}HNH\overset{O}{\overset{\|}{C}}\underset{\underset{C_6H_5}{|}}{\underset{CH_2}{\underset{|}{C}}}HNH_2$$

8 Purify by washing

↓ $HO\overset{O}{\overset{\|}{C}}\underset{\underset{O=COC(CH_3)_3}{|}}{\underset{NH}{\underset{|}{C}}}HCH_2CH_2CH_2CH_2NH\overset{O}{\overset{\|}{C}}OC(CH_3)_3$ and dicyclohexylcarbodiimide

9 Add protected Lys

$$\text{(resin)}\!-\!CH_2O\overset{O}{\overset{\|}{C}}\underset{CH_3}{\underset{|}{C}}HNH\overset{O}{\overset{\|}{C}}\underset{\underset{C_6H_5}{|}}{\underset{CH_2}{\underset{|}{C}}}HNH\overset{O}{\overset{\|}{C}}\underset{(CH_2)_4NH\overset{\|}{\underset{O}{C}}OC(CH_3)_3}{\underset{|}{C}}HNH\overset{O}{\overset{\|}{C}}OC(CH_3)_3$$

10 Purify by washing

↓ CF_3CO_2H, CH_2Cl_2

11 Remove protecting groups

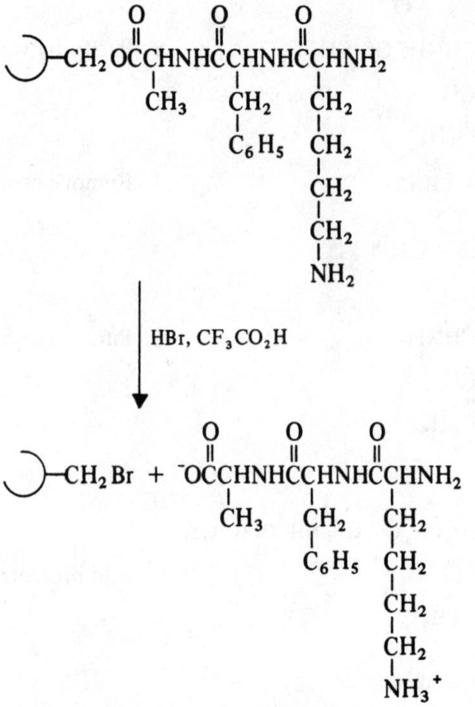

$$\text{(resin)}-CH_2O\overset{O}{\overset{\|}{C}}\underset{CH_3}{\underset{|}{C}H}NH\overset{O}{\overset{\|}{C}}\underset{\underset{C_6H_5}{|}}{\underset{CH_2}{\underset{|}{C}H}}NH\overset{O}{\overset{\|}{C}}\underset{CH_2CH_2CH_2CH_2NH_2}{\underset{|}{C}H}NH_2$$

12 Purify by washing

$\downarrow$ HBr, CF_3CO_2H

13 Detach tripeptide

$$\text{(resin)}-CH_2Br + {}^{-}O\overset{O}{\overset{\|}{C}}\underset{CH_3}{\underset{|}{C}H}NH\overset{O}{\overset{\|}{C}}\underset{\underset{C_6H_5}{|}}{\underset{CH_2}{\underset{|}{C}H}}NH\overset{O}{\overset{\|}{C}}\underset{CH_2CH_2CH_2CH_2NH_3^+}{\underset{|}{C}H}NH_2$$

14 Isolate product

Lys•Phe•Ala

23.16 (a) Isoleucine, threonine, hydroxyproline, and cystine.

(b)

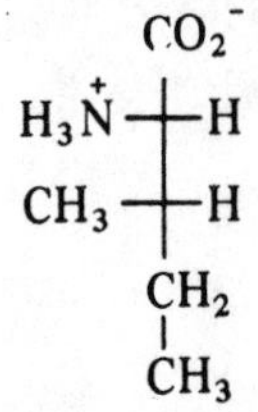

Fischer projections (top to bottom):

CO_2^-; $H_3\overset{+}{N}$—C—H; CH_3—C—H; CH_2; CH_3 and CO_2^-; $H_3\overset{+}{N}$—C—H; H—C—CH_3; CH_2; CH_3

CO_2^-; $H_3\overset{+}{N}$—C—H; H—C—OH; CH_3 and CO_2^-; $H_3\overset{+}{N}$—C—H; HO—C—H; CH_3

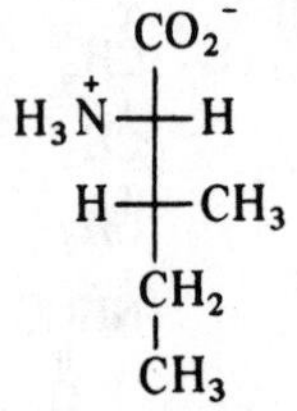

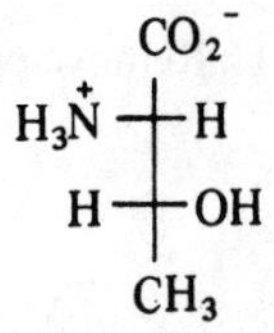

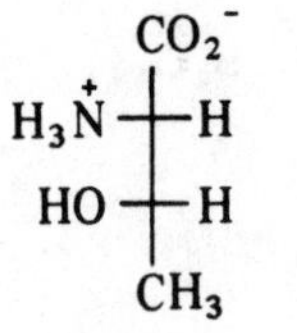

$$\begin{array}{c} CO_2^- \\ H_2\overset{+}{N}\!-\!\!\!+\!\!\!-H \\ CH_2 \quad CH_2 \\ H \diagdown / \\ OH \end{array} \quad \text{and} \quad \begin{array}{c} CO_2^- \\ H_2\overset{+}{N}\!-\!\!\!+\!\!\!-H \\ CH_2 \quad CH_2 \\ OH \diagdown / \\ H \end{array}$$

(With cystine, both stereocenters are α carbon atoms, thus according to the problem, both must have the L-configuration, and no isomers of this type can be written.)

(c) Diastereomers

23.17 (a) Alanine

$$CH_3\underset{\underset{NH_3^+}{|}}{C}HCO_2^- + HONO \longrightarrow CH_3\underset{\underset{OH}{|}}{C}HCO_2H + N_2$$

(b) Proline and hydroxyproline. All of the other amino acids have at least one primary amino group.

(c) HO–(3,5-dibromophenyl)–$CH_2\underset{\underset{NH_3^+}{|}}{C}HCO_2^-$ (Br at both positions ortho to OH)

(d) C_6H_5–$CH_2\underset{\underset{NH_3^+}{|}}{C}HCO_2C_2H_5$

(e) $CH_3\underset{|}{C}HCO_2^-$ with NH–C(=O)–C_6H_5 on the central carbon:

$$\begin{array}{l} CH_3CHCO_2^- \\ \quad\;\; | \\ \quad\; NH \\ \quad\;\; | \\ \quad\; C{=}O \\ \quad\;\; | \\ \quad\; C_6H_5 \end{array}$$

23.18 (a)

$$\begin{array}{c} CO_2^- \\ {}^+H_3N\!-\!\!\!+\!\!\!-H \\ CH_2OH \\ (-)\text{-Serine} \end{array} \xrightarrow[CH_3OH]{HCl} \begin{array}{c} CO_2CH_3 \\ {}^+H_3N\!-\!\!\!+\!\!\!-H \quad Cl^- \\ CH_2OH \\ \mathbf{A} \\ (C_4H_{10}ClNO_3) \end{array}$$

$$\xrightarrow{PCl_5} \begin{array}{c} CO_2CH_3 \\ {}^+H_3N\!-\!\!\!+\!\!\!-H \quad Cl^- \\ CH_2Cl \\ \mathbf{B} \\ (C_4H_9Cl_2NO_2) \end{array} \xrightarrow[(2)\ OH^-]{(1)\ H_3O^+,\ H_2O,\ heat}$$

$$\underset{\substack{\textbf{C}\\(C_3H_6ClNO_2)}}{{}^+H_3N\text{—}\underset{CH_2Cl}{\overset{CO_2^-}{C}}\text{—}H} \xrightarrow[\text{dil. } H_3O^+]{\text{Na-Hg}} \underset{\text{L-(+)-Alanine}}{{}^+H_3N\text{—}\underset{CH_3}{\overset{CO_2^-}{C}}\text{—}H}$$

(b)

$$\mathbf{B} \xrightarrow{OH^-} \underset{\substack{\textbf{D}\\(C_4H_8ClNO_2)}}{H_2N\text{—}\underset{CH_2Cl}{\overset{CO_2CH_3}{C}}\text{—}H} \xrightarrow{\text{NaSH}} \underset{\substack{\textbf{E}\\(C_4H_9NO_2S)}}{H_2N\text{—}\underset{CH_2SH}{\overset{CO_2CH_3}{C}}\text{—}H} \xrightarrow[(2)\ OH^-]{(1)\ H_3O^+,\ H_2O,\ \text{heat}} \underset{\text{L-(+)-Cysteine}}{{}^+H_3N\text{—}\underset{CH_2SH}{\overset{CO_2^-}{C}}\text{—}H}$$

(c)

$$\underset{\text{L-(−)-Asparagine}}{H_3\overset{+}{N}\text{—}\underset{CH_2CNH_2\ (C{=}O)}{\overset{CO_2^-}{C}}\text{—}H} \xrightarrow[\substack{\text{Hofmann}\\\text{rearrangement}}]{\text{NaOBr, } OH^-} \underset{\substack{\textbf{F}\\(C_3H_7N_2O_2)}}{H_2N\text{—}\underset{CH_2NH_2}{\overset{CO_2^-}{C}}\text{—}H}$$

$$\underset{\substack{\textbf{C}\\\text{[from part (a)]}}}{{}^+H_3N\text{—}\underset{CH_2Cl}{\overset{CO_2^-}{C}}\text{—}H} \xrightarrow{NH_3} \textbf{F}$$

23.19

(a)

$$CH_3\overset{O}{\overset{\|}{C}}NHCH(CO_2C_2H_5)_2 + CH_2{=}CH\text{—}C{\equiv}N \xrightarrow[C_2H_5OH]{NaOC_2H_5}$$

$$\underset{\textbf{G}}{CH_3\overset{O}{\overset{\|}{C}}NH\text{—}\underset{CO_2C_2H_5}{\overset{CO_2C_2H_5}{C}}\text{—}CH_2CH_2C{\equiv}N} \xrightarrow[\substack{\text{reflux 6h}\\(66\%\ \text{yield})}]{\text{conc. HCl}}$$

$$\underset{\text{DL-Glutamic acid}}{HO_2CCH_2CH_2\underset{NH_3^+}{C}HCO_2^-} + CH_3CO_2H + 2C_2H_5OH + NH_4^+ + CO_2$$

(b)

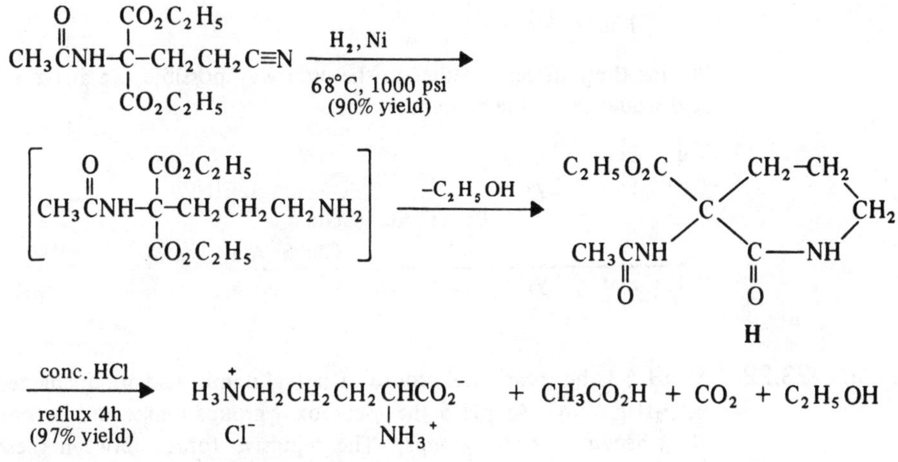

$$\mathrm{CH_3\overset{O}{\overset{\|}{C}}NH{-}\underset{CO_2C_2H_5}{\underset{|}{\overset{CO_2C_2H_5}{\overset{|}{C}}}}{-}CH_2CH_2C{\equiv}N \xrightarrow[\text{68°C, 1000 psi (90\% yield)}]{H_2,\,Ni}}$$

$$\left[\mathrm{CH_3\overset{O}{\overset{\|}{C}}NH{-}\underset{CO_2C_2H_5}{\underset{|}{\overset{CO_2C_2H_5}{\overset{|}{C}}}}{-}CH_2CH_2CH_2NH_2}\right] \xrightarrow{-C_2H_5OH}$$

(cyclic lactam: $C_2H_5O_2C$ and $CH_3\overset{O}{\overset{\|}{C}}NH$ on C, ring C–CH_2–CH_2–CH_2–NH–$\overset{O}{\overset{\|}{C}}$–C; NH bears H)

$$\xrightarrow[\text{reflux 4h (97\% yield)}]{\text{conc. HCl}} \mathrm{\overset{+}{H_3N}CH_2CH_2CH_2\underset{NH_3^+}{\underset{|}{C}}HCO_2^-\ \ Cl^-} + CH_3CO_2H + CO_2 + C_2H_5OH$$

DL-Ornithine hydrochloride

23.20 We look for points of overlap:

				Phe ·	Ser			
		Pro ·	Gly ·	Phe				
	Pro ·	Pro			Ser ·	Pro ·	Phe	
Arg ·	Pro						Phe ·	Arg
Arg ·	Pro ·	Pro ·	Gly ·	Phe ·	Ser ·	Pro ·	Phe ·	Arg

Bradykinin

23.21 1. This experiment shows that valine is the *N*-terminal amino acid and that valine is attached to leucine. (Lysine labeled at the ε-amino group is to be expected if lysine is not the *N*-terminal amino acid and if it is linked in the polypeptide through its α-amino group.)

2. This experiment shows that alanine is the *C*-terminal amino acid and that it is linked to glutamic acid.

At this point, then, we have the following information about the structure of the heptapeptide.

Val · Leu (Ala, Lys, Phe) Glu · Ala

The sequence here is unknown

3. (a) This experiment shows that the dipeptide, **A**, is

Leu · Lys

(b) The carboxypeptidase reaction shows that the *C*-terminal amino acid of the tripeptide, **B**, is glutamic acid; the DNP labeling experiment shows that the *N*-terminal amino acid is phenylalanine. Thus the tripeptide **B** is

Phe · Ala · Glu

Putting these pieces together in the only way possible, we arrive at the following amino acid sequence for the heptapeptide.

Val ·	Leu					
	Leu ·	Lys				
			Phe ·	Ala ·	Glu	
					Glu ·	Ala
Val ·	Leu ·	Lys ·	Phe ·	Ala ·	Glu ·	Ala

23.22 At pH 2-3 the γ-carboxyl groups of polyglutamic acid are uncharged (they are present as $-CO_2H$ groups). At pH 5 the γ-carboxyl groups ionize and become negatively charged (they become γ-CO_2^- groups). The repulsive forces between these negatively charged groups cause an unwinding of the α-helix and the formation of a random coil.

23.23 The observation that the proton nmr spectrum taken at room temperature shows two different signals for the methyl groups suggests that they are in different environments. This would be true if rotation about the carbon-nitrogen bond was not taking place.

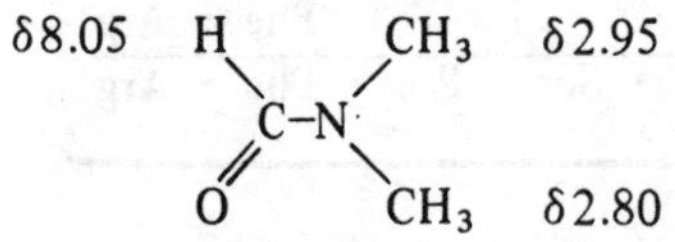

δ8.05 H(–C(=O)–N) ; N–CH_3 δ2.95 ; N–CH_3 δ2.80

We assign the δ2.80 signal to the methyl group that is on the same side as the electronegative oxygen atom.

The fact that the methyl signals appear as doublets (and that the formyl signal is a multiplet) indicates that long-range coupling is taking place between the methyl protons and the formyl proton.

That the two doublets are not simply the result of spin-spin coupling is indicated by the observation that the distance that separates one doublet from the other changes when the applied magnetic field strength is lowered. [*Remember!* The magnitude of a chemical shift is proportional to the strength of the applied magnetic field while the magnitude of a coupling constant is not.]

That raising the temperature (to 111°C) causes the doublets to coalesce into a single signal indicates that at higher temperatures the molecules have enough energy to surmount the energy barrier of the carbon-nitrogen bond. Above 111°C, rotation is taking place so rapidly that the spectrometer is unable to discriminate between the two methyl groups.

SECTION REFERENCES FOR ADDITIONAL PROBLEMS

23.17	8.3, 8.9, 23.2	**23.21**	23.6, 23.7
23.18	23.2, 19.7, 15.15, 18.7, 18.8	**23.22**	23.6, 23.7
23.19	3.16, 5.6, 8.12, 14.7, 18.7, 19.5	**23.23**	23.2, 23.9
23.20	18.7, 18.8, 19.3, 20.8, 23.2		

SELF-TEST

23.1 Write the structural formula of the principal ionic species present in aqueous solutions at pH 2, 7, and 12 of isoleucine (2-amino-3-methylpentanoic acid)

At pH = 2	At pH = 7	At pH = 12
(a)	(b)	(c)

23.2 A hexapeptide gave the following products:

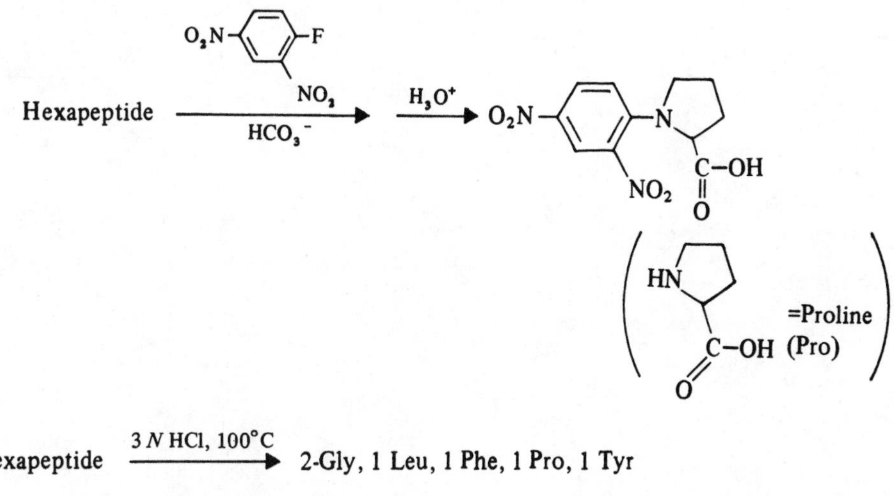

Hexapeptide $\xrightarrow[HCO_3^-]{O_2N\text{-}C_6H_3(NO_2)\text{-}F}$ $\xrightarrow{H_3O^+}$ O_2N–(2,4-dinitrophenyl)–N(pyrrolidine)–C(=O)–OH (NO_2)

(HN(pyrrolidine)–C(=O)–OH =Proline (Pro))

Hexapeptide $\xrightarrow{3\ N\ HCl,\ 100°C}$ 2-Gly, 1 Leu, 1 Phe, 1 Pro, 1 Tyr

Hexapeptide $\xrightarrow{1\ N\ HCl,\ 80°C}$ Phe•Gly•Tyr + Gly•Phe•Gly + Pro•Leu•Gly + Leu•Gly•Phe

The structure of the hexapeptide (using abbreviations such as Gly•Leu•etc) is

24 NUCLEIC ACIDS AND PROTEIN SYNTHESIS

SOLUTIONS TO PROBLEMS

24.1 Adenine:

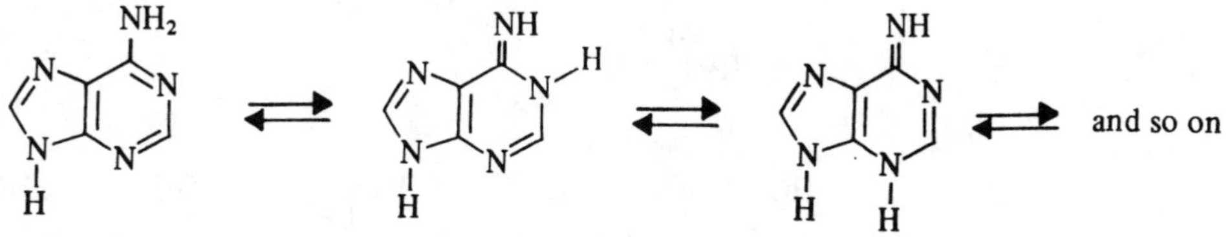

and so on

Guanine:

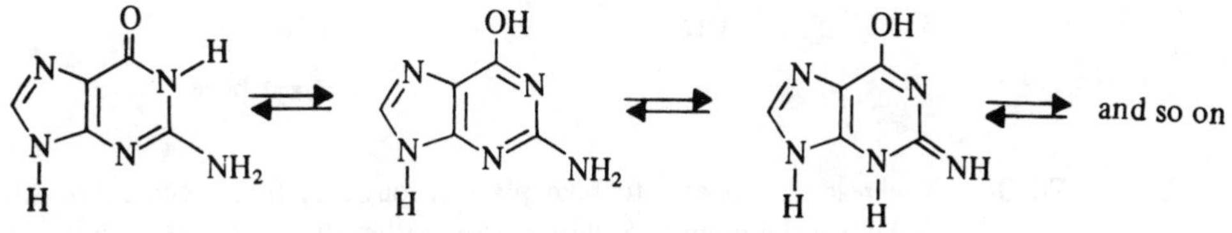

and so on

Cytosine:

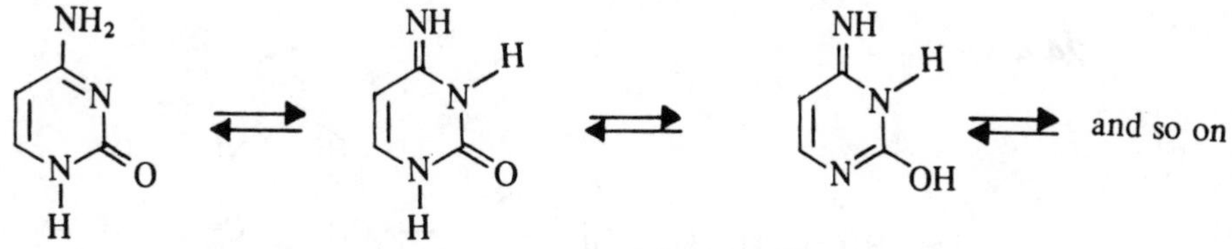

and so on

Thymine ($R = CH_3$) or uracil ($R = H$):

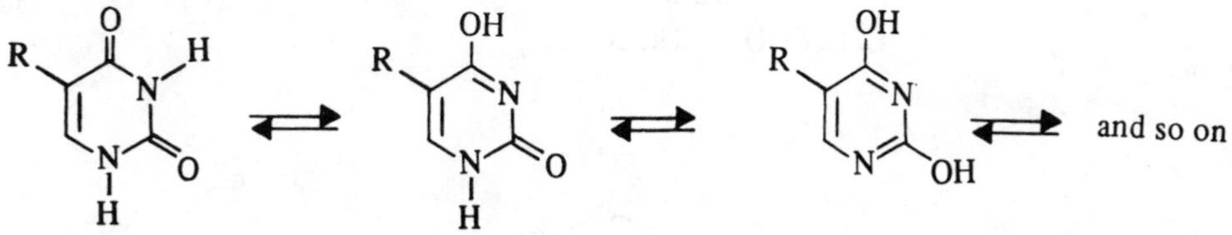

and so on

24.2 (a) The nucleosides have an *N*-glycosidic linkage that (like an *O*-glycosidic linkage) is rapidly hydrolyzed by aqueous acid but is one that is stable in aqueous base.

(b)

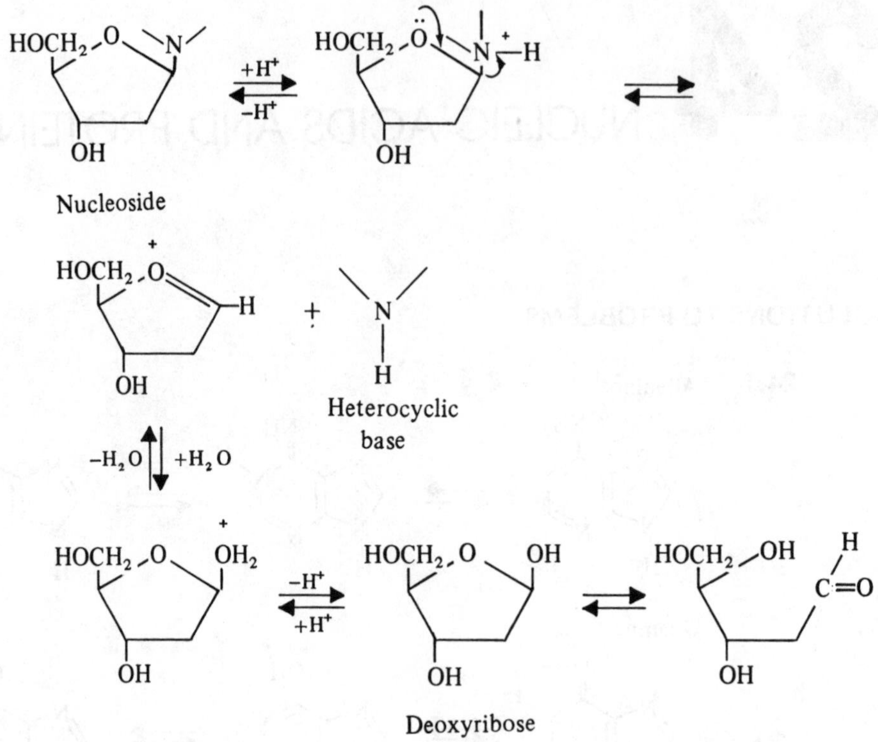

Nucleoside

Heterocyclic base

$-H_2O$ $+H_2O$

Deoxyribose

24.3 The reaction appears to take place through an S_N2 mechanism. Attack occurs preferentially at the primary 5′-carbon atom rather than at the secondary 3′-carbon atom.

24.4

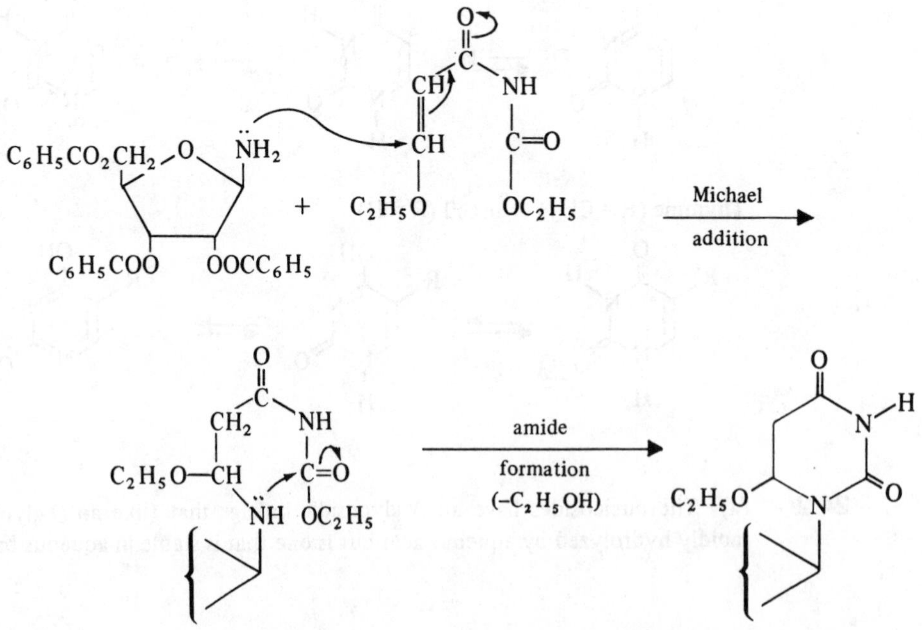

Michael addition

amide formation ($-C_2H_5OH$)

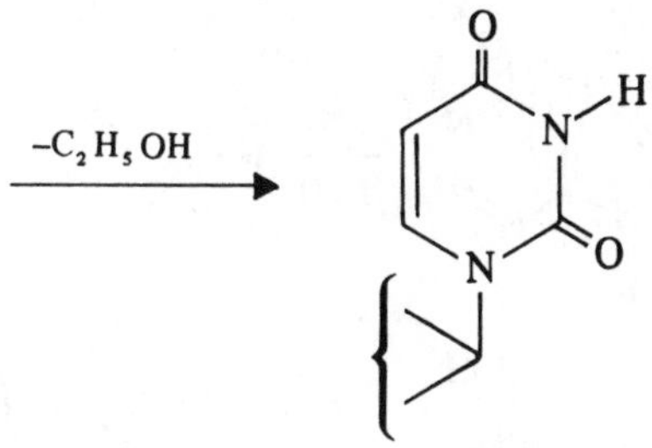

$\xrightarrow{-C_2H_5OH}$

24.5 (a) The isopropylidene group is part of a cyclic ketal and is thus susceptible to hydrolysis by mild acid.

(b) It can be installed by treating the nucleoside with acetone and a trace of acid and by simultaneously removing the water that is produced.

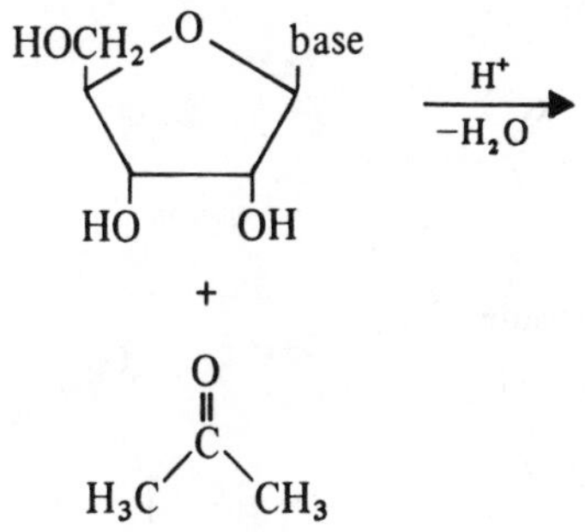

$\xrightarrow[-H_2O]{H^+}$

24.6 (a) $6 \times 10^9 \text{ base pairs} \times \dfrac{34\text{ Å}}{10 \text{ base pairs}} \times \dfrac{10^{-10}\text{ m}}{\text{Å}} \cong 2\text{ m}$

(b) $6 \times 10^{-12} \dfrac{\text{g}}{\text{ovum}} \times 3 \times 10^9 \text{ ova} = 1.8 \times 10^{-2}\text{ g}$

24.7 (a)

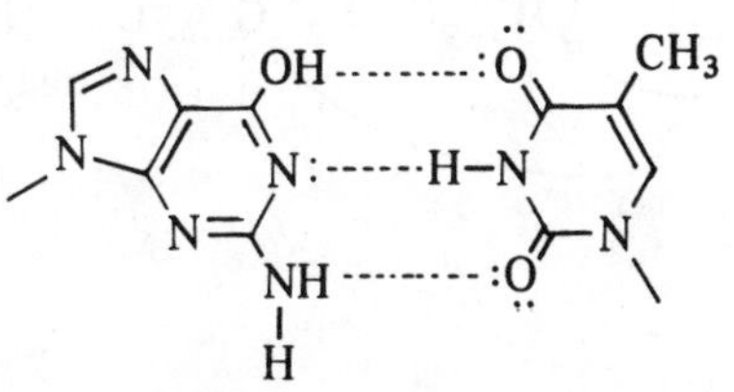

Lactim form of guanine   Thymine

(b) Thymine would pair with adenine and thus adenine would be introduced into the complementary strand where guanine should occur.

24.8 (a) A diazonium salt and a heterocyclic analog of a phenol.

$\xrightarrow{HONO}$ $\xrightarrow[H_2O]{-N_2}$ $\rightleftarrows$

Hypoxanthine nucleotide

(b)

Hypoxanthine Cytosine

(c) Original double strand

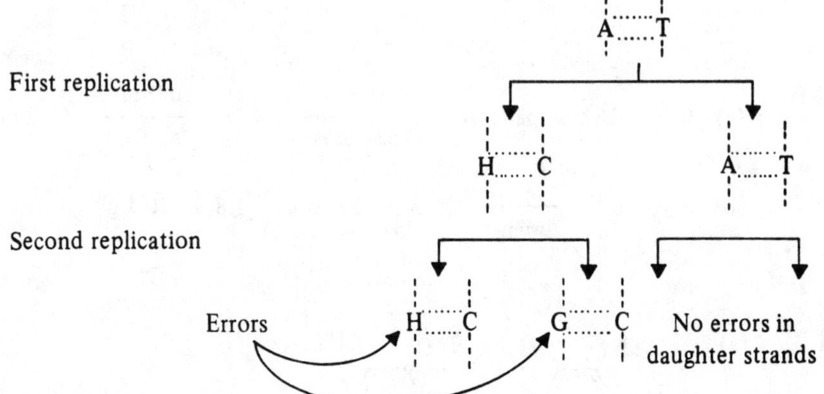

24.9

Uracil (in mRNA) Adenine (in DNA)

24.10

(a)	UGG	GGG	UUU	UAC	AGC	*m*RNA
(b)	Tyr	Gly	Phe	Tyr	Ser	Amino acids
(c)	ACC	CCC	AAA	AUG	UCG	Anticodons

24.11

	Arg	Ile	Cys	Tyr	Val	Amino acids
(a)	AGA	AUA	UGC	UGG	GUA	*m*RNA
(b)	TCT	TAT	ACG	ACC	CAT	DNA
(c)	UCU	UAU	ACG	ACC	CAU	Anticodons

24.12 A change from C–T–T to C–A–T or a change from C–T–C to C–A–C.

O SPECIAL TOPIC
Reactions Controlled by Orbital Symmetry

SOLUTIONS TO PROBLEMS

O.1 Conrotatory motion of the type shown would lead to increasingly unfavorable interaction of the methyl groups as the transition state is approached. Thus this path is not followed to any appreciable extent.

O.2 According to the Woodward-Hoffmann rule for electrocyclic reactions of $4n$ π-electron systems (Section 0.2A), the photochemical cyclization of *cis,trans*-2,4-hexadiene should proceed with *disrotatory motion*. Thus it should yield *trans*-3,4-dimethylcyclobutene:

H, CH_3, CH_3, H; $h\nu$ disrotatory; H, CH_3, CH_3, H; + enantiomer

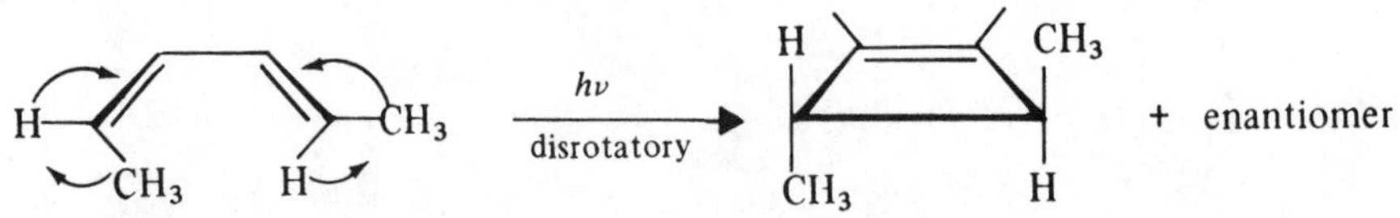

cis,trans-2,4-Hexadiene *trans*-3,4-Dimethylcyclobutene

O.3

(a) CH_3, CH_3, H_3C, CH_3, (+), (+), H, H; 200°C; CH_3 (+) (−) CH_3, (+), (−), H_3C, (−), (+), H, (−), CH_3 (+)

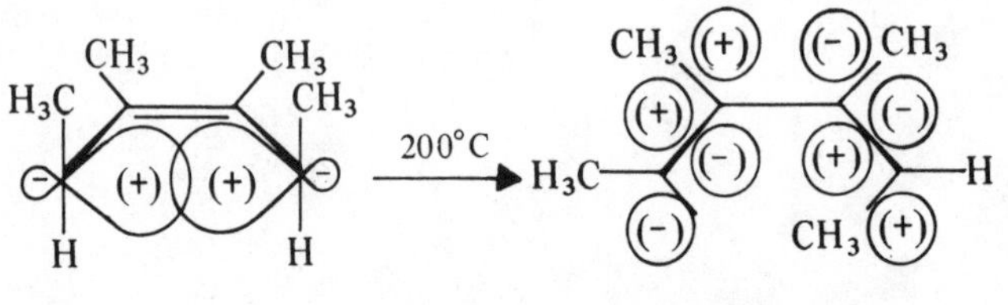

ψ_2 of a hexadiene (Section O.2A)

(b) This is a thermal electrocyclic reaction of a $4n$ π-electron system; it should, *and does*, proceed with conrotatory motion.

O.4

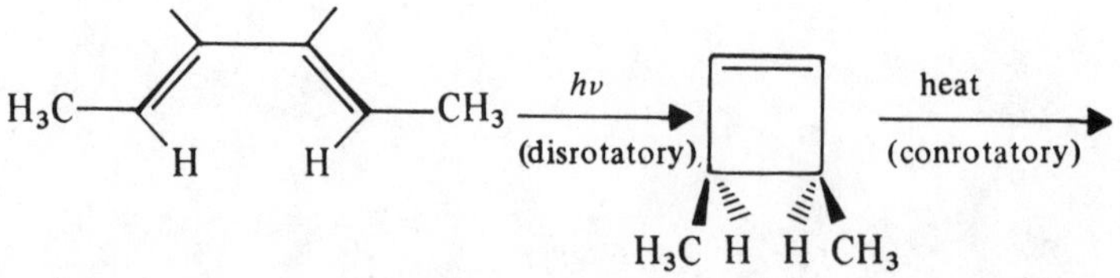

H_3C, CH_3, H, H; $h\nu$ (disrotatory); heat (conrotatory); H_3C H H CH_3

trans,trans,-2,4-hexadiene *cis*-3,4-Dimethylcyclobutene

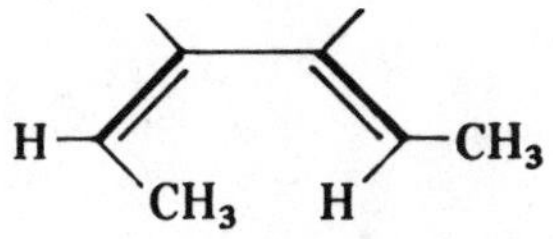

H CH_3
CH_3 H

cis, trans-2, 4-Hexadiene

Here we find that two consecutive electrocyclic reactions (the first photochemical, the second thermal), provide a stereospecific synthesis of *cis,trans*-2,4-hexadiene from *trans, trans*-2,4-hexadiene.

O.5 (a) This is a photochemical electrocyclic reaction of an eight π-electron system—a $4n$ π system where $n = 2$. It should, therefore, proceed with disrotatory motion.

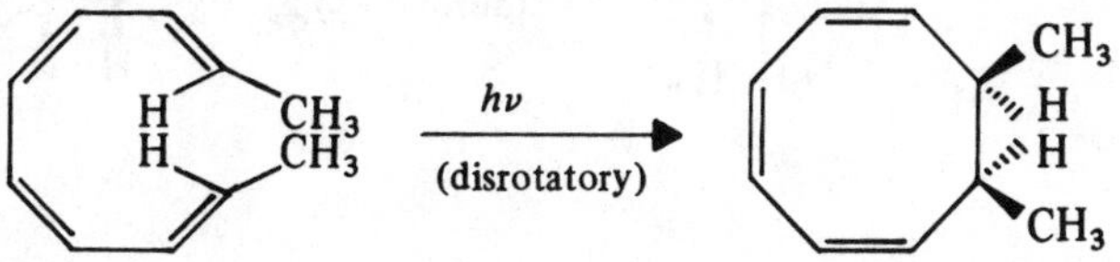

H CH_3 H CH_3 —$h\nu$ (disrotatory)→ CH_3 H H CH_3

cis-7, 8-Dimethyl-1, 3, 5-cyclooctatriene

(b) This is a thermal electrocyclic reaction of the eight π-electron system. It should proceed with conrotatory motion.

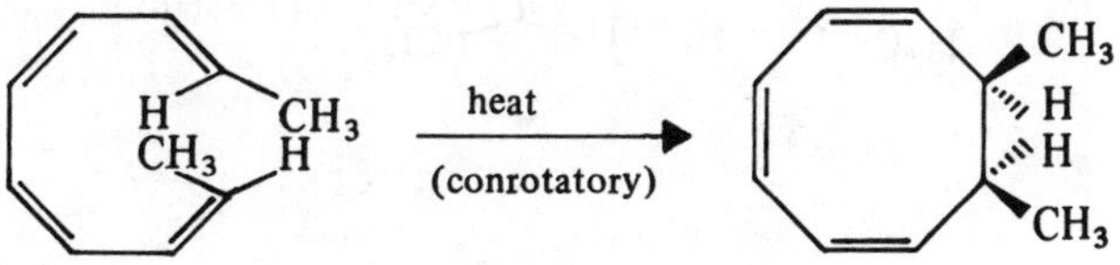

H CH_3 CH_3 H —heat (conrotatory)→ CH_3 H H CH_3

cis-7, 8-Dimethyl-1, 3, 5-cyclooctatriene

O.6 (a) This is conrotatory motion and since this is a $4n$ π-electron system (where $n = 1$) it should occur under the influence of heat.

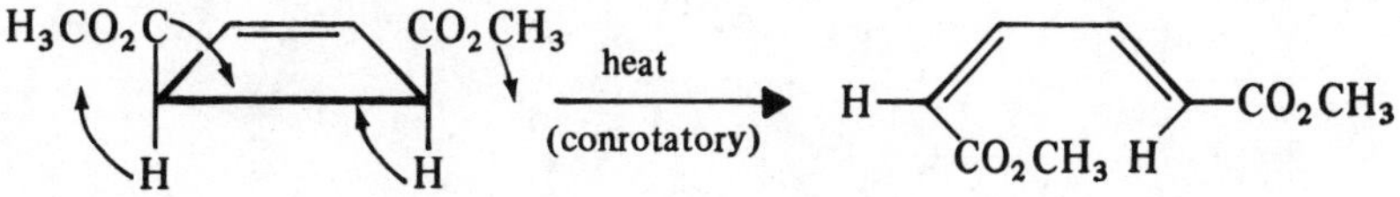

H_3CO_2C CO_2CH_3 H H —heat (conrotatory)→ H CO_2CH_3 CO_2CH_3 H

(b) This is conrotatory motion and since this is also a $4n$ π-electron system (where $n = 2$) it should occur under the influence of heat.

H₃C H H CH₃ —heat (conrotatory)→ H H₃C CH₃ H

+ enantiomer

(c) This is disrotatory motion. This, too is a $4n$ π-electron system (where $n = 1$), thus it should occur under the influence of light.

H H —$h\nu$ (disrotatory)→ H H

O.7 (a) This is a $(4n + 2)$ π-electron system (where $n = 1$); a thermal reaction should take place with disrotatory motion:

H₃C H H CH₃ ⇄ heat (disrotatory) CH₃ H₃C H H

(b) This is also a $(4n + 2)$ π-electron system; a photochemical reaction should take place with conrotatory motion.

CH₃ H₃C H H ⇄ $h\nu$ (conrotatory) H CH₃ H CH₃

O.8 Here we need a conrotatory ring opening of *trans*-5,6-dimethyl-1,3-cyclohexadiene (to produce *trans,cis,trans*-2,4,6-octatriene), then we need a disrotatory cyclization to produce *cis*-5,6-dimethyl-1,3-cyclohexadiene.

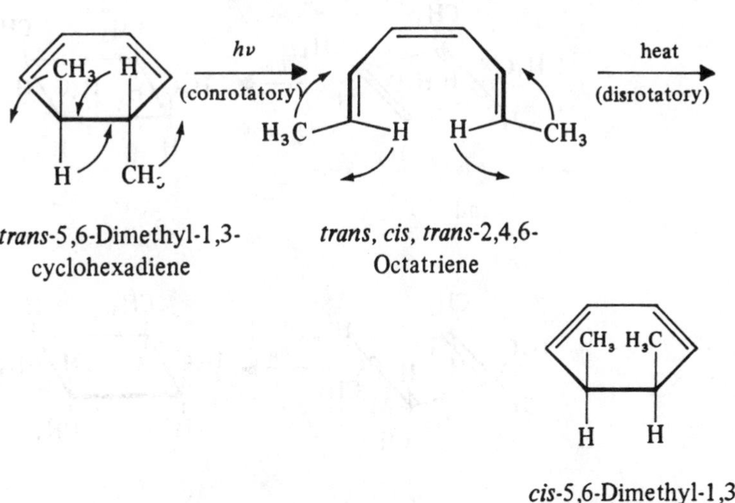

Since both reactions involve $(4n + 2)$ π electron systems we apply light to accomplish the first step and heat to accomplish the second. It would also be possible to use heat to produce *trans,cis,cis*-2,4,6-octatriene then use light to produce the desired product.

O.9 The first electrocyclic reaction is a thermal, conrotatory ring opening of a $4n$ π-electron system. The second electrocyclic reaction is a thermal, disrotatory ring closure of a $(4n + 2)$ π-electron system.

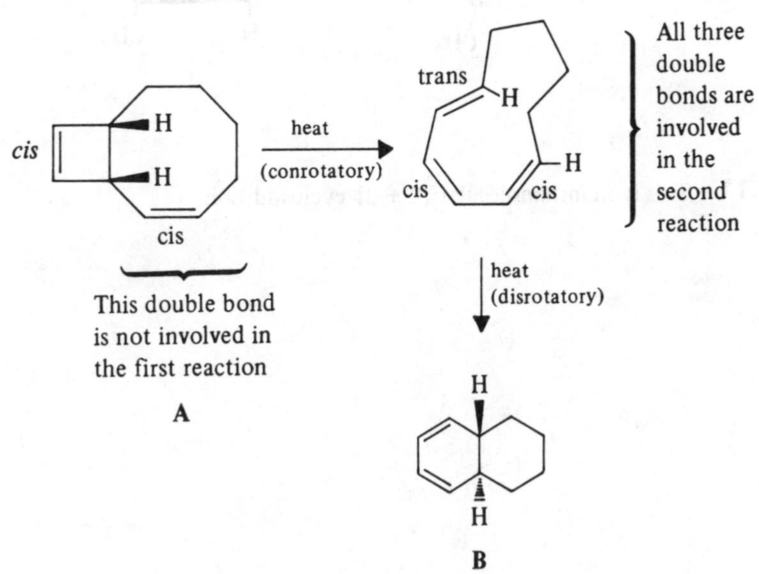

O.10 (a) There are two possible products that can result from a concerted cycloaddition. They are formed when *cis*-2-butene molecules come together in the following ways:

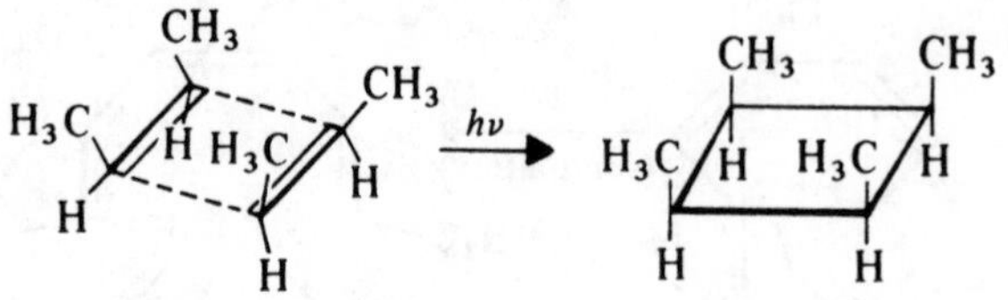

and

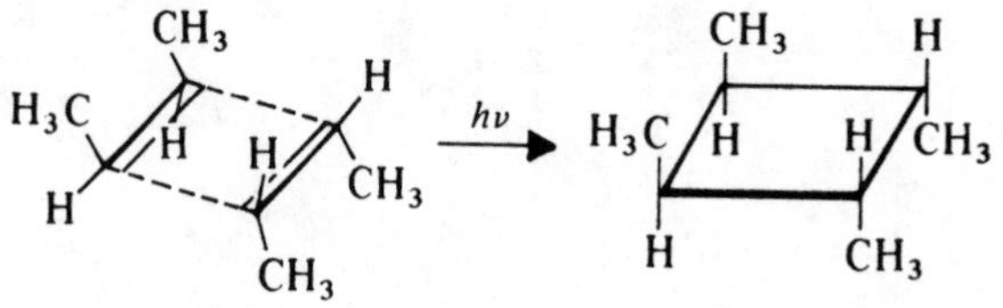

(b) There are two possible products that can be obtained from *trans*-2-butene as well.

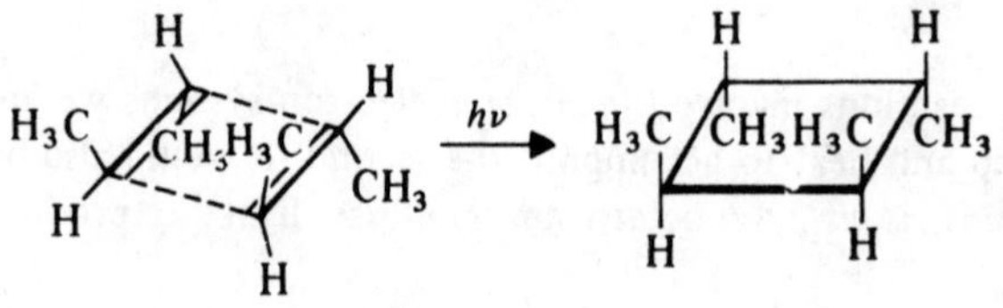

and

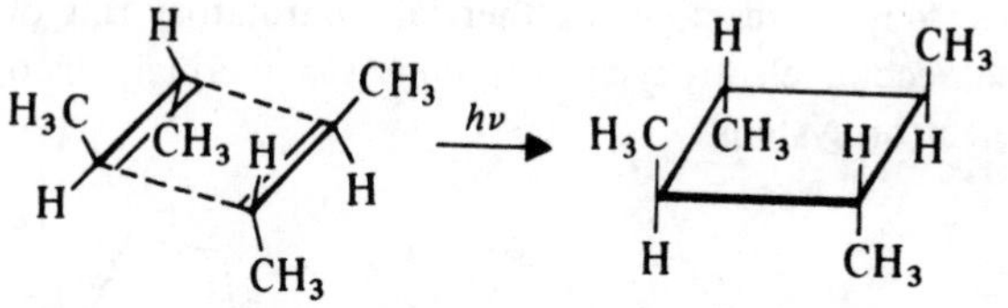

O.11 This is an intramolecular [2 + 2] cycloaddition.

O.12

(a) CN CN H H CN NC H_3C CH_3

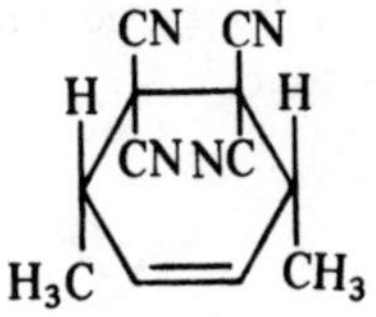

(b) CN CN H_3C H CN NC H CH_3

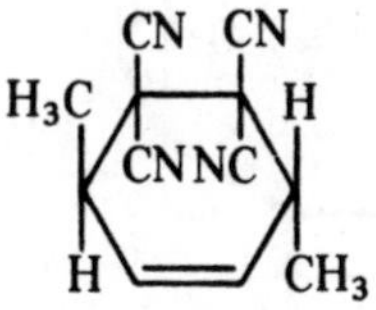

Enantiomers

O.13

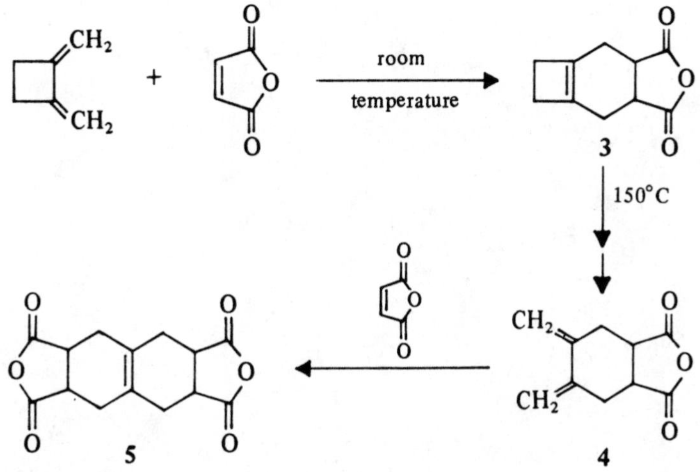

CH_2 CH_2 + O O O — room temperature → 3

150°C

4 CH_2 CH_2 O O O

5

O.14

A is

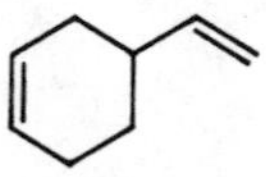

B and **C** are H H and H H

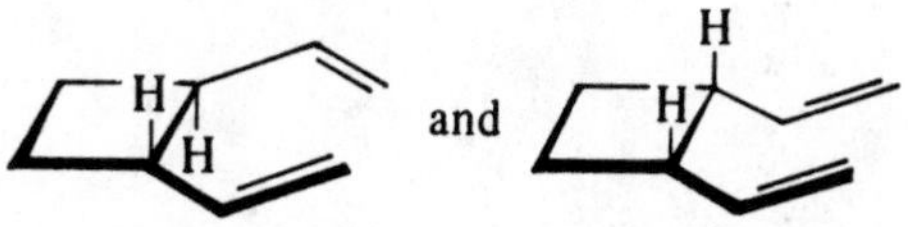

O.15 **A** is the product of a disrotatory thermal electrocyclic reaction involving a 6π electron segment of cyclooctatetraene. **B** is the Diels-Alder adduct.

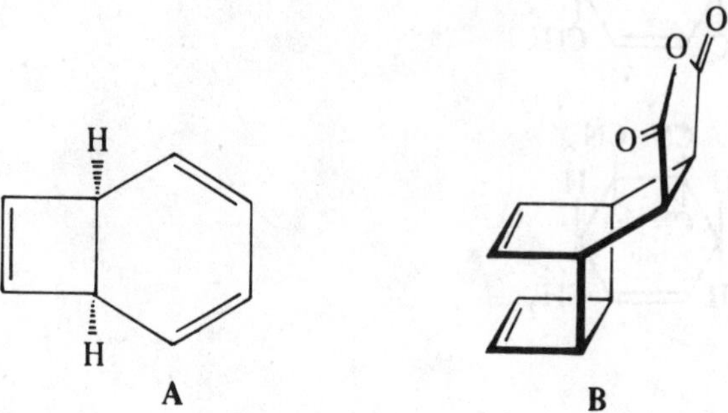

P SPECIAL TOPIC Nucleophilic Substitution Reactions—Another Look

SOLUTIONS TO PROBLEMS

P.1 (a) and (b)

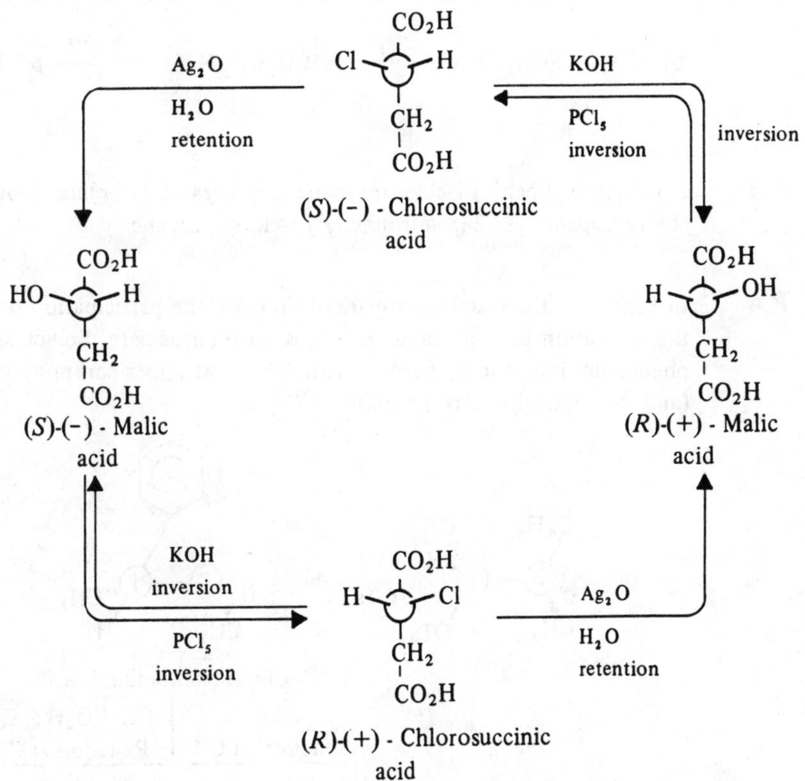

(c) The reaction takes place with retention of configuration.

(d)

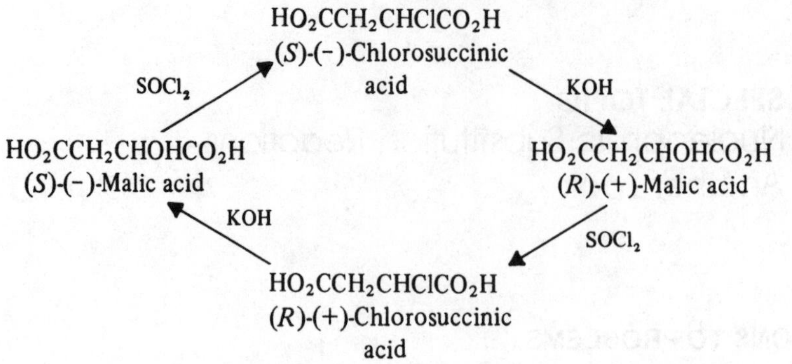
HO₂CCH₂CHClCO₂H
(S)-(−)-Chlorosuccinic acid
SOCl₂
KOH
HO₂CCH₂CHOHCO₂H
(S)-(−)-Malic acid
HO₂CCH₂CHOHCO₂H
(R)-(+)-Malic acid
KOH
SOCl₂
HO₂CCH₂CHClCO₂H
(R)-(+)-Chlorosuccinic acid

P.2 (a) Reaction of an alkene with halogen in water solution.

(b)

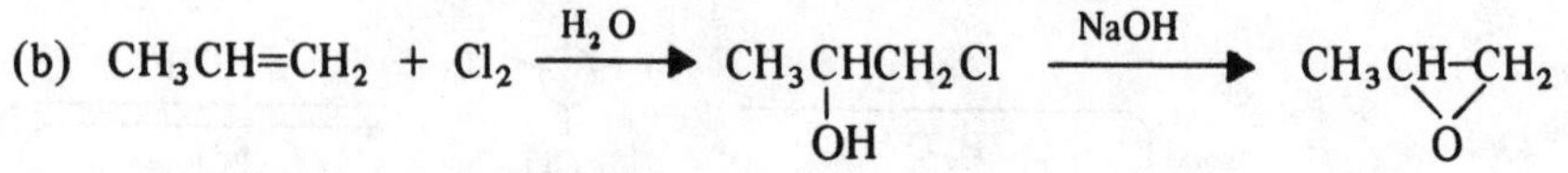
CH₃CH=CH₂ + Cl₂
H₂O
CH₃CHCH₂Cl
OH
NaOH
CH₃CH—CH₂
O

P.3 2-(*p*-Hydroxyphenyl)-1-chloropropane > 2-(*p*-tolyl)-1-chloropropane > 2-phenyl-1-chloropropane > 2-(*p*-nitrophenyl)-1-chloropropane

P.4 In each case, the reactions apparently involve the participation of the phenyl group and the formation of a phenonium ion as an intermediate. Solvolysis of **A** yields a chiral phenonium ion—one that reacts with solvent at either carbon to produce the same chiral (and thus optically active) acetate.

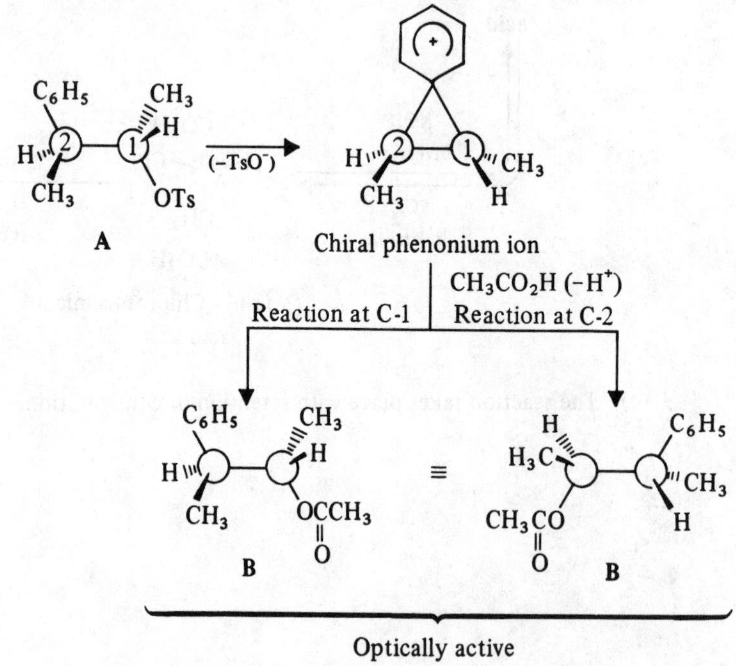
C₆H₅
CH₃
H
H
CH₃
OTs
A
(−TsO⁻)
Chiral phenonium ion
CH₃CO₂H (−H⁺)
Reaction at C-1
Reaction at C-2
OCCH₃
O
B
≡
H₃C
CH₃CO
B
Optically active

The phenonium ion produced from **C** gives enantiomers when it reacts with acetate.

A APPENDIX
Empirical and Molecular Formulas

In Section 1.2B, we discussed briefly the pioneering work of Berzelius, Dumas, Liebig, and Cannizzaro in devising methods for determining the formulas of organic compounds. Although the experimental procedures for these analyses have been refined, the basic methods for determining the elemental composition of an organic compound today are not substantially different from those used in the nineteenth century. A carefully weighed quantity of the compound to be analyzed is oxidized completely to carbon dioxide and water. The weights of carbon dioxide and water are carefully measured and used to find the percentages of carbon and hydrogen in the compound. The percentage of nitrogen is usually determined by measuring the volume of nitrogen (N_2) produced in a separate procedure.

Special techniques for determining the percentage composition of other elements typically found in organic compounds have also been developed, but the direct determination of the percentage of oxygen is difficult. However, if the percentage composition of all the other elements is known, then the percentage of oxygen can be determined by difference. The following examples will illustrate how these calculations can be carried out.

EXAMPLE A

A new organic compound is found to have the following elemental analysis.

Carbon	67.95%
Hydrogen	5.69
Nitrogen	26.20
Total:	99.84%

Since the total of these percentages is very close to 100% (within experimental error) we can assume that no other element is present. For the purpose of our calculation it is convenient to assume that we have a 100-g sample. If we did, it would contain the following:

67.95 g of carbon
5.69 g of hydrogen
26.20 g of nitrogen

In other words, we use percentages *by weight* to give us the ratios *by weight* of the elements in the substance. To write a formula for the substance, however, we need *ratios by moles.*

We now divide each of these weight-ratio numbers by the atomic weight of the particular element and obtain the number of moles of each element, respectively, in 100 g of the compound. This operation gives us the ratios *by moles* of the elements in the substance:

$$\text{C} \quad \frac{67.95 \text{ g}}{12.01 \text{ g/mole}} = 5.66 \text{ moles}$$

$$\text{H} \quad \frac{5.69 \text{ g}}{1.008 \text{ g/mole}} = 5.64 \text{ moles}$$

$$\text{N} \quad \frac{26.20 \text{ g}}{14.01 \text{ g/mole}} = 1.87 \text{ moles}$$

One possible formula for the compound, therefore, is $C_{5.66}H_{5.64}N_{1.87}$.

By convention, however, we use *whole* numbers in formulas. Therefore, we convert these fractional numbers of moles to whole numbers by dividing each by 1.87, the smallest number.

$$\text{C} \quad \frac{5.66}{1.87} = 3.03 \text{ which is} \sim 3$$

$$\text{H} \quad \frac{5.64}{1.87} = 3.02 \text{ which is} \sim 3$$

$$\text{N} \quad \frac{1.87}{1.87} = 1.00$$

Thus within experimental error, the ratios by moles are 3C to 3H to 1N, and C_3H_3N is the *empirical formula.* By empirical formula we mean the formula in which the subscripts are the smallest integers that give the ratio of atoms in the compound. In contrast, a *molecular* formula discloses the complete composition of one molecule. The molecular formula of this particular compound could be C_3H_3N or some whole number multiple of C_3H_3N; that is, $C_6H_6N_2$, $C_9H_9N_3$, $C_{12}H_{12}N_4$, and so on. If, in a separate determination, we find that the molecular weight of the compound is 108 ± 3, we can be certain that the *molecular formula* of the compound is $C_6H_6N_2$.

FORMULA	MOLECULAR WEIGHT
C_3H_3N	53.06
$C_6H_6N_2$	106.13 (which is within the range 108 ± 3)
$C_9H_9N_3$	159.19
$C_{12}H_{12}N_4$	212.26

The most accurate method for determining molecular weights is by mass spectroscopy; this method (which can also be used to determine molecular formulas and structures) is described in Special Topic E. A variety of other methods based on freezing point depression, boiling point elevation, osmotic pressure, and vapor density can also be used to determine molecular weights.

EXAMPLE B

Histidine, an amino acid isolated from protein, has the following elemental analysis:

Carbon	46.38%
Hydrogen	5.90
Nitrogen	27.01
Total:	79.29
Difference	20.71 (assumed to be oxygen)
	100.00%

Since no elements, other than carbon, hydrogen, and nitrogen, are found to be present in histidine the difference is assumed to be oxygen. Again, we assume a 100-g sample and divide the weight of each element by its gram-atomic weight. This gives us the ratio of moles (A).

$$\begin{array}{llll} & \text{(A)} & \text{(B)} & \text{(C)} \\ \text{C} & \dfrac{46.38}{12.01} = 3.86 & \dfrac{3.86}{1.29} & = 2.99 \times 2 = 5.98 \sim 6 \text{ carbon atoms} \\ \text{H} & \dfrac{5.90}{1.008} = 5.85 & \dfrac{5.85}{1.29} & = 4.53 \times 2 = 9.06 \sim 9 \text{ hydrogen atoms} \\ \text{N} & \dfrac{27.01}{14.01} = 1.94 & \dfrac{1.94}{1.29} & = 1.50 \times 2 = 3.00 = 3 \text{ nitrogen atoms} \\ \text{O} & \dfrac{20.71}{16.00} = 1.29 & \dfrac{1.29}{1.29} & = 1.00 \times 2 = 2.00 = 2 \text{ oxygen atoms} \end{array}$$

Dividing each of the moles (A) by the smallest of them does not give a set of numbers (B) that is close to a set of whole numbers. Multiplying each of the numbers in column (B) by 2 does, however, as seen in column (C). The empirical formula of histidine is, therefore, $C_6H_9N_3O_2$.

In a separate determination the molecular weight of histidine was found to be 158 ± 5. The empirical formula weight of $C_6H_9N_3O_2$ (155.15) is within this range; thus the molecular formula for histidine is the same as the empirical formula.

PROBLEMS

A.1 What is the empirical formula of each of the following compounds?

(a) Hydrazine, N_2H_4
(b) Benzene, C_6H_6
(c) Dioxane, $C_4H_8O_2$
(d) Nicotine, $C_{10}H_{14}N_2$
(e) Cyclodecane, $C_{10}H_{20}$
(f) Acetylene, C_2H_2

A.2 The empirical formulas and molecular weights of several compounds are given next. In each case calculate the molecular formula for the compound.

EMPIRICAL FORMULA	MOLECULAR WEIGHT
(a) CH_2O	179 ± 5
(b) CHN	80 ± 5
(c) CCl_2	410 ± 10

A.3 The widely used antibiotic, penicillin G, gave the following elemental analysis: C, 57.45%; H, 5.40%; N, 8.45%; S, 9.61%. The molecular weight of penicillin G is 330 ± 10. Assume that no other elements except oxygen are present and calculate the empirical and molecular formulas for penicillin G.

ADDITIONAL PROBLEMS

A.4 Calculate the percentage composition of each of the following compounds.
(a) $C_6H_{12}O_6$
(b) $CH_3CH_2NO_2$
(c) $CH_3CH_2CBr_3$

A.5 An organometallic compound called *ferrocene* contains 30.02% iron. What is the minimum molecular weight of ferrocene?

A.6 A gaseous compound gave the following analysis: C, 40.04%; H, 6.69%. At standard temperature and pressure, 1.00 g of the gas occupied a volume of 746 mL. What is the molecular formula of the compound?

A.7 A gaseous hydrocarbon has a density of 1.251 g/L at standard temperature and pressure. When subjected to complete combustion, a 1.000-L sample of the hydrocarbon gave 3.926 g of carbon dioxide and 1.608 g of water. What is the molecular formula for the hydrocarbon?

A.8 Nicotinamide, a vitamin that prevents the occurrence of pellagra, gave the following analysis: C, 59.10%; H, 4.92%; N, 22.91%. The molecular weight of nicotinamide was shown in a separate determination to be 120 ± 5. What is the molecular formula for nicotinamide?

A.9 The antibiotic chloramphenicol gave the following analysis: C, 40.88%; H, 3.74%; Cl, 21.95%; N, 8.67%. The molecular weight was found to be 300 ± 30. What is the molecular formula for chloramphenicol?

SOLUTIONS TO PROBLEMS OF APPENDIX A

A.1 (a) NH_2 (b) CH (c) C_2H_4O (d) C_5H_7N (e) CH_2 (f) CH

A.2

	EMPIRICAL FORMULA	EMPIRICAL FORMULA WEIGHT	$\left(\frac{\text{MOLECULAR WEIGHT}}{\text{EMP. FORM. WT.}}\right)$	MOLECULAR FORMULA
(a)	CH_2O	30	$\frac{179}{30} \cong 6$	$C_6H_{12}O_6$
(b)	CHN	27	$\frac{80}{27} \cong 3$	$C_3H_3N_3$
(c)	CCl_2	83	$\frac{410}{83} \cong 5$	C_5Cl_{10}

A.3 If we assume that we have a 100-g sample, the amounts of the elements are

	WEIGHT	Moles (A)	B
C	57.45	$\frac{57.45}{12.01} = 4.78$	$\frac{4.78}{0.300} = 15.9 = 16$
H	5.40	$\frac{5.40}{1.008} = 5.36$	$\frac{5.36}{0.300} = 17.9 = 18$
N	8.45	$\frac{8.45}{14.01} = 0.603$	$\frac{0.603}{0.300} = 2.01 = 2$
S	9.61	$\frac{9.61}{32.06} = 0.300$	$\frac{0.300}{0.300} = 1.00 = 1$
O*	19.09	$\frac{19.09}{16.00} = 1.19$	$\frac{1.19}{0.300} = 3.97 = 4$
	100.00		

(* by difference from 100)

The empirical formula is thus $C_{16}H_{18}N_2SO_4$. The empirical formula weight (334.4) is within the range given for the molecular weight (330 ± 10), thus the molecular formula for penicillin G is the same as the empirical formula.

A.4 (a) To calculate the percentage composition from the molecular formula, first determine the weight of each element in 1 mole of the compound. For $C_6H_{12}O_6$,

$$C_6 = 6 \times 12.01 = 72.06 \qquad \frac{72.06}{180.2} = 0.400 = 40.0\%$$

$$H_{12} = 12 \times 1.008 = 12.10 \qquad \frac{12.10}{180.2} = 0.0671 = 6.7\%$$

$$O_6 = 6 \times 16.00 = 96.00 \qquad \frac{96.00}{180.2} = 0.533 = 53.3\%$$

$$\text{MW} \qquad 180.16$$

(MW = molecular weight)

Then determine the percentage of each element using the formula.

$$\text{Percentage of A} = \frac{\text{Weight of A}}{\text{Molecular Weight}} \times 100$$

(b)

$$C_2 = 2 \times 12.01 = 24.02 \qquad \frac{24.02}{75.07} = 0.320 = 32.0\%$$

$$H_5 = 5 \times 1.008 = 5.04 \qquad \frac{5.04}{75.07} = 0.067 = 6.7\%$$

$$N = 1 \times 14.01 = 14.01 \qquad \frac{14.01}{75.07} = 0.187 = 18.7\%$$

$$O_2 = 2 \times 16.00 = 32.00 \qquad \frac{32.00}{75.07} = 0.426 = 42.6\%$$

$$\text{Total} = 75.07$$

(c)

$$C_3 = 3 \times 12.01 = 36.03 \qquad \frac{36.03}{280.77} = 0.128 = 12.8\%$$

$$H_5 = 5 \times 1.008 = 5.04 \qquad \frac{5.04}{280.77} = 0.018 = 1.8\%$$

$$Br_3 = 3 \times 79.90 = 239.70 \qquad \frac{239.70}{280.77} = 0.854 = 85.4\%$$

$$\text{Total} = 280.77$$

A.5 If the compound contains iron, each molecule must contain at least one atom of iron, and 1 mole of the compound must contain at least 55.85 g of iron. Therefore,

$$\text{MW of ferrocene} = 55.85\ \frac{\text{g of Fe}}{\text{mole}} \times \frac{1.000\ \text{g}}{0.3002\ \text{g of Fe}}$$

$$= 186.0\ \frac{\text{g}}{\text{mole}}$$

A.6 First we must determine the empirical formula. Assuming that the difference between the percentages given in 100% is due to oxygen, we calculate:

$$C \quad 40.04 \qquad \frac{40.04}{12.01} = 3.33 \qquad \frac{3.33}{3.33} = 1$$

$$H \quad 6.69 \qquad \frac{6.69}{1.008} = 6.64 \qquad \frac{6.64}{3.33} \cong 2$$

$$O \quad 53.27 \qquad \frac{53.27}{16.00} = 3.33 \qquad \frac{3.33}{3.33} = 1$$

$$100.00$$

The empirical formula is thus CH_2O.

To determine the molecular formula we must first determine the molecular weight. At standard temperature and pressure, the volume of 2 moles of an ideal gas is 22.4 L. Assuming ideal behavior,

$$\frac{1.00 \text{ g}}{0.746 \text{ L}} = \frac{\text{MW}}{22.4 \text{ L}} \quad \text{Where MW = Molecular weight}$$

$$\text{MW} = \frac{(1.00)(22.4)}{0.746} = 30.0 \text{ g}$$

The empirical formula weight (30.0) equals the molecular weight, thus the molecular formula is the same as the empirical formula.

A.7 As in Problem A.6, the molecular weight is found by the equation

$$\frac{1.251 \text{ g}}{1.00 \text{ L}} = \frac{\text{MW}}{22.4 \text{ L}}$$

$$\text{MW} = (1.251)(22.4)$$
$$\text{MW} = 28.02$$

To determine the empirical formula, we must determine the amount of carbon in 3.926 g of carbon dioxide, and the amount of hydrogen in 1.608 g of water.

$$\text{C} \qquad \left(3.926 \cancel{\text{g } CO_2}\right)\left(\frac{12.01 \text{ g C}}{44.01 \cancel{\text{g } CO_2}}\right) = 1.071 \text{ g carbon}$$

$$\text{H} \qquad \left(1.608 \cancel{\text{g } H_2O}\right)\left(\frac{2.016 \text{ g H}}{18.016 \cancel{\text{g } H_2O}}\right) = \frac{0.179 \text{ g hydrogen}}{1.250 \text{ g sample}}$$

The weight of C and H in a 1.250-g sample is 1.250 g. Therefore there are no other elements present.

To determine the empirical formula we proceed as in Problem A.6 except that the sample size is 1.250 instead of 100 g.

$$\text{C} \qquad \frac{1.071}{12.01} = 0.0892 \qquad \frac{0.0892}{0.0892} = 1$$

$$\text{H} \qquad \frac{0.179}{1.008} = 0.178 \qquad \frac{0.178}{0.0892} = 2$$

The empirical formula is thus CH_2. The empirical formula weight (14) is one half the molecular weight. Thus the molecular formula is C_2H_4.

A.8 Use the procedure of Problem A.3.

$$\text{C} \qquad 59.10 \qquad \frac{59.10}{12.01} = 4.92 \qquad \frac{4.92}{0.817} = 6.02 \cong 6$$

$$\text{H} \qquad 4.92 \qquad \frac{4.92}{1.008} = 4.88 \qquad \frac{4.88}{0.817} = 5.97 \cong 6$$

$$\text{N} \quad 22.91 \quad \frac{22.91}{14.01} = 1.64 \quad \frac{1.64}{0.817} = 2$$

$$\text{O} \quad \underline{13.07} \quad \frac{13.07}{16.00} = 0.817 \quad \frac{0.817}{0.817} = 1$$

$$100.00$$

The empirical formula is thus $C_6H_6N_2O$. The empirical formula weight is 123.13, which is equal to the molecular weight within experimental error. The molecular formula is thus the same as the empirical formula.

A.9

$$\text{C} \quad 40.88 \quad \frac{40.88}{12.01} = 3.40 \quad \frac{3.40}{0.619} = 5.5 \quad 5.5 \times 2 = 11$$

$$\text{H} \quad 3.74 \quad \frac{3.74}{1.008} = 3.71 \quad \frac{3.71}{0.619} = 6 \quad 6 \times 2 = 12$$

$$\text{Cl} \quad 21.95 \quad \frac{21.95}{35.45} = 0.619 \quad \frac{0.619}{0.619} = 1 \quad 1 \times 2 = 2$$

$$\text{N} \quad 8.67 \quad \frac{8.67}{14.01} = 0.619 \quad \frac{0.619}{0.619} = 1 \quad 1 \times 2 = 2$$

$$\text{O} \quad \underline{24.76} \quad \frac{24.76}{16.00} = 1.55 \quad \frac{1.55}{0.619} = 2.5 \quad 2.5 \times 2 = 5$$

$$100.00$$

The empirical formula is thus $C_{11}H_{12}Cl_2N_2O_5$. The empirical formula weight (323) is equal to the molecular weight, therefore the molecular formula is the same as the empirical formula.

APPENDIX
Molecular Model Set Exercises

The exercises in this appendix are designed to help you gain an understanding of the three-dimensional nature of molecules. You are encouraged to perform these exercises with a model set as described.

These exercises should be performed as part of the study of the chapters shown below.

Chapter in Text	Accompanying Exercises
3	1,3,4,5,6,8,10,11,12,14,15,16,17,18,20,21
4	2,7,9,13,24,25,26,27
6	9,19,22,28
10	31
11	23
21	29
23	30
Special Topic O	31,32

The following molecular model set exercises were developed by Ronald Starkey for use with the Theta Molecular Model Set (J. Wiley & Sons, Inc.).

Refer to the instruction booklet that accompanies the model set for details of molecular model assembly.

EXERCISE 1 (Chapter 3)

Assemble a molecular model of methane, CH_4. Note that the hydrogen atoms describe the apexes of a regular tetrahedron with the carbon atom at the center of the tetrahedron. Demonstrate by attempted superposition that two models of methane are identical.

Replace any one hydrogen atom on each of the two methane models with a halogen (a green atom-center in the Theta Molecular Model Set) to form two molecules of CH_3X. Are the two structures identical? Does it make a difference which of the four hydrogen atoms on a methane molecule you replace? How many different configurations of CH_3X are possible?

Repeat the same considerations for two disubstituted methanes with two identical substituents (CH_2X_2), and then with two different substituents (CH_2XY). Two shades of green atom-centers could be used for the two different substituents.

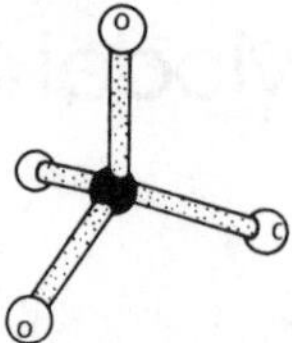

Methane, CH_4

EXERCISE 2 (Chapter 4)

Construct a model of a trisubstituted methane molecule (CHXYZ). Four different colored atom-centers (red, blue, yellow, and white) are attached to a central tetrahedral black carbon atom center. Note that the carbon now has four different substituents. Compare this model with a second model of CHXYZ. Are the two structures identical (superposable)?

Interchange any two substituents on one of the carbon atoms. Are the two CHXYZ molecules identical now? Does the fact that interchange of any two substituents on the carbon interconverts the stereoisomers indicate that there are only two possible configurations of a tetrahedral carbon atom?

Compare the two models that were not identical. What is the relationship between them? Do they have a mirror-image relationship? That is, are they related as an object and its mirror reflection?

EXERCISE 3 (Chapter 3)

Make a model of ethane, CH_3CH_3. Does each of the carbon atoms retain a tetrahedral configuration? Can the carbon atoms be rotated with respect to each other without breaking the carbon-carbon bond?

Rotate about the carbon-carbon bond until the carbon-hydrogen bonds of one carbon atom are aligned with those of the other carbon atom. This is the eclipsed conformation. When the C—H bond of one carbon atom bisects the H—C—H angle of the other carbon atom the conformation is called staggered. Remember conformations are arrangements of atoms in a molecule that can be interconverted by bond rotations.

In which of the two conformations of ethane you made are the hydrogen atoms of one carbon closer to those of the other carbon?

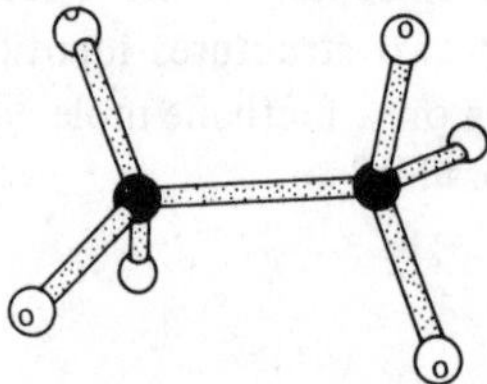

Ethane, CH_3CH_3

EXERCISE 4 (Chapter 3)

Prepare a second model of ethane. Replace one hydrogen, any one, on each ethane model with a substituent such as a halogen (a green atom-center), to form two models of CH_3CH_2X. Are the structures identical? If not, can they be made identical by rotation about the C—C bond? With one of the models demonstrate that there are three equivalent staggered conformations (see Exercise 3) of CH_3CH_2X. How many equivalent eclipsed conformations are possible?

EXERCISE 5 (Chapter 3)

Assemble a model of a 1,2-disubstituted ethane molecule, CH_2XCH_2X. Note how the orientation of and the distance between the X groups changes with rotation of the carbon-carbon bond. The arrangement in which the X substituents are at maximum separation is the *anti*-staggered conformation. The other staggered conformations are called *gauche*. How many *gauche* conformations are possible? Are they energetically equivalent? Are they identical?

EXERCISE 6 (Chapter 3)

Construct two models of butane, $CH_3CH_2CH_2CH_3$. Note that the structures can be viewed as dimethyl substituted ethanes. Show that rotations of the C-2, C-3 bond of butane produce eclipsed, *anti*-staggered, and *gauche*-staggered conformations. Measure the distance between C-1 and C-4 in the conformations just mentioned. The scale of the Theta Molecular Model Set is 3 cm in a model corresponds to approximately 1.0 Å (0.1 nm) on a molecular scale. In which eclipsed conformation are the C-1 and C-4 carbon atoms closest to each other? How many eclipsed conformations are possible?

EXERCISE 7 (Chapter 4)

Using two models of butane verify that the two hydrogen atoms on C-2 are not stereochemically equivalent. Replacement of one hydrogen leads to a product that is not identical to that obtained by replacement of the other C-2 hydrogen atom. Both replacement products have the same molecular formula $CH_3CHXCH_2CH_3$. What is the relationship of the two products?

EXERCISE 8 (Chapter 3)

Make a model of hexane, $CH_3CH_2CH_2CH_2CH_2CH_3$. Extend the six carbon chain as far as it will go. This puts C-1 and C-6 at maximum separation. Notice that this *straight-chain* structure maintains the tetrahedral bond angles at each carbon atom and therefore the carbon chain adopts a zigzag arrangement. Does this extended chain adopt staggered or eclipsed conformations of the hydrogen atoms? How could you describe the relationship of C-1 and C-4?

EXERCISE 9 (Chapter 6)

Prepare models of the four isomeric butenes, C_4H_8. Note that the restricted rotation about the double bond is responsible for the cis-trans stereoisomerism. Verify this by observing that breaking the π bond of *cis*-2-butene allows rotation and thus conversion to *trans*-2-butene. Are any of the four isomeric butenes chiral (nonsuperposable with its mirror image)? Indicate pairs of butene isomers that are stuctural (constitutional) isomers. Indicate pairs that are diastereoisomers. How does the distance between the C-1 and C-4 carbon atoms in *trans*-2-butene compare with that of the *anti* conformation of butane? Compare the C-1 and C-4 distance in *cis*-2-butene and the conformation of butane in which the methyls are eclipsed.

1-Butene	*cis*-2-Butene	*trans*-2-Butene	2-Methylpropene
CH_3CH_2 H	CH_3 CH_3	CH_3 H	CH_3 H
C=C	C=C	C=C	C=C
H H	H H	H CH_3	CH_3 H

EXERCISE 10 (Chapter 3)

Make a model of cyclopropane. The Theta Molecular Model Set requires the use of 1.5-cm flexible tubing for the carbon-carbon bonds of the cyclopropane ring. The flexible tubes illustrate quite well the "bend-bond" nature of the ring bonds. It should be apparent that the ring carbon atoms must be coplaner. What is the relationship of the hydrogen atoms on adjacent carbon atoms? Are they staggered, eclipsed, or skewed?

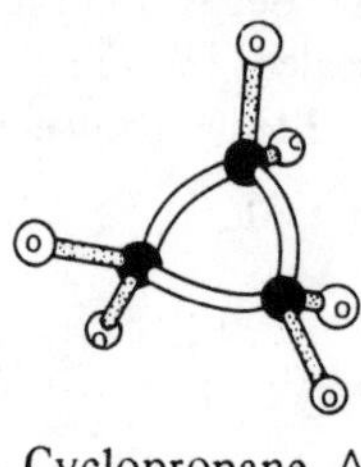

Cyclopropane, Δ

EXERCISE 11 (Chapter 3)

A model of cyclobutane can be assembled in a conformation that has the four carbon atoms coplaner. For this exercise the rigid 2.0-cm tubes of the Theta Molecular Model Set should be used for the carbon-carbon bonds of the ring. How many eclipsed hydrogen atoms are there in the conformation? Torsional strain (strain due to deviations from an eclipsed conformation) can be relieved at the expense of increased angle strain by a slight folding of the ring. The deviation of one ring carbon from the plane of the other three carbon atoms is about 25°. This folding compresses the C—C—C bond angle to about 88°. Rotate the ring carbon bonds of the planar conformation to obtain the folded conformation. Are the hydrogen atoms on adjacent carbon atoms eclipsed or skewed? Considering both structural and stereoisomer forms how many dimethylcyclobutane structures are possible? Do deviations of the ring from planarity have to be considered when determining the number of possible dimethyl structures?

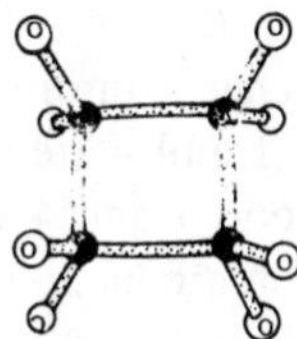

Cyclobutane, □

EXERCISE 12 (Chapter 3)

Cyclpentane is a more flexible ring system than cyclobutane or cyclopropane. A model of cyclopentane in a conformation with all the ring carbon atoms coplaner exhibits minimal deviation of the C—C—C bond angles from the normal tetrahedral bond angle. How many eclipsed hydrogen interactions are there in this planer conformation? If one of the ring carbon atoms is pushed slightly above (or below) the plane of the other carbon atoms a model of the envelope conformation is obtained. Does the envelope conformation relieve some of the torsional strain? How many eclipsed hydrogen interactions are there in the envelope conformation?

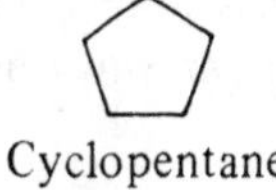

Cyclopentane

EXERCISE 13 (Chapter 4)

Make a model of 1,2-dimethylcyclopentane. How many stereoisomers are possible for this compound? Identify each of the possible structures as either cis or trans. Is it apparent that cis-trans isomerism is possible in this compound because of restricted rotation? Are any of the stereoisomers chiral? What are the relationships of the 1,2-dimethylcyclopentane stereoisomers?

EXERCISE 14 (Chapter 3)

Assemble the six-membered ring compound cyclohexane. Is the ring flat or puckered? Place the ring in a chair conformation and then in a boat conformation. Demonstrate that the chair and boat are indeed conformations of cyclohexane—that is, they may be interconverted by rotations about the carbon-carbon bonds of the ring.

Chair form

Boat form

Note that in the chair conformation carbon atoms 2, 3, 5, and 6 are in the same plane and carbon atoms 1 and 4 are above and below the plane, respectively. In the boat conformation carbon atoms 1 and 4 are both above (they could also both be below) the plane described by carbon atoms 2, 3, 5, and 6. Is it apparent why the boat is sometimes associated with the flexible form? Are the hydrogen atoms in the chair conformation staggered or eclipsed? Are any hydrogen atoms eclipsed in the boat conformation? Do carbon atoms 1 and 4 have an *anti* or a *gauche* relationship in the chair conformation? (*Hint:* Look down the C-2, C-3 bond).

A twist conformation of cyclohexane may be obtained by slightly twisting carbon atoms 2 and 5 of the boat conformation as shown:

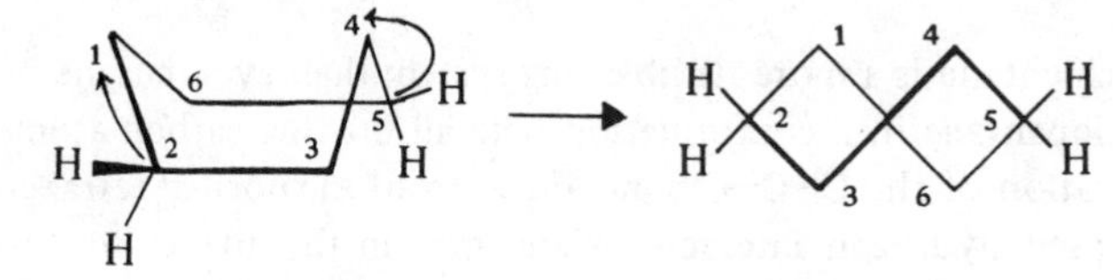

Boat form → Twist form

Note that the C-2, C-3 and the C-5, C-6 sigma bonds no longer retain their parallel orientation in the twist conformation. If the ring system is twisted too far, another boat conformation results. Compare the nonbonded (van der Waals repulsion) interactions and the torsional strain present in the boat, twist, and chair conformations of cyclohexane. Is it apparent why the relative order of thermodynamic stabilities is chair > twist > boat?

EXERCISE 15 (Chapter 3)

Construct a model of methylcyclohexane. How many chair conformations are possible? How does the orientation of the methyl group change in each chair conformation?

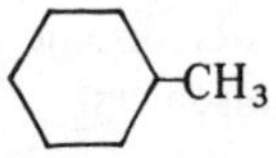

Identify carbon atoms in the chair conformation of methylcyclohexane that have intramolecular interactions corresponding to those found in the *gauche* and *anti* conformations of butane. Which of the chair conformations has the greatest number of *gauche* interactions? How many more? If we assume, as is the case for butane, that the *anti* interaction is 0.8 kcal/mole more favorable than *gauche*, then what is the relative stability of the two chair conformations of methylcyclohexane? *Hint:* Identify the relative number of *gauche* interactions in the two conformations.

EXERCISE 16 (Chapter 3)

Compare models of the chair conformations of monosubstituted cyclohexanes in which the substituent alkyl groups are methyl, ethyl, isopropyl, and *t*-butyl.

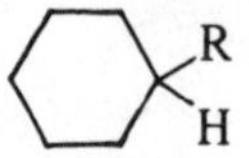

Rationalize the relative stability of axial and equatorial conformations of the alkyl group given in the table for each compound. The chair conformation with the alkyl group equatorial is more stable by the amount shown.

ALKYL GROUP	ΔH *(kcal/mole)* EQUATORIAL $\rightleftharpoons$ AXIAL
CH_3	1.6
CH_2CH_3	1.7
$CH(CH_3)_2$	2.1
$C(CH_3)_3$	5.0 (approximate)

EXERCISE 17 (Chapter 3)

Make a model of 1,2-dimethylcyclohexane. Answer the questions posed in Exercise 13 with regard to 1,2-dimethylcyclohexane.

EXERCISE 18 (Chapter 3)

Compare models of the neutral and charged molecules shown next. Identify the structures that are isoelectronic, that is, those that have the same electronic structure. How do those structures that are isoelectronic compare in their molecular geometry?

CH_3CH_3	CH_3NH_2	CH_3OH
$CH_3CH_2^-$	$CH_3NH_3^+$	$CH_3OH_2^+$

EXERCISE 19 (Chapter 6)

Prepare a model of cyclohexene. Note that chair and boat conformations are no longer possible, as carbon atoms 1, 2, 3, and 6 lie in a plane. Are cis and trans stereoisomers possible for the double bond? Attempt to assemble a model of *trans*-cyclohexene. Can it be done? Are cis and trans stereoisomers possible for 2,3-dimethylcyclohexene? For 3,4-dimethylcyclohexene?

6 5 1 4 2 3

Cyclohexene

Assemble a model of *trans*-cyclooctene. Observe the twisting of the π-bond system. Would you expect the cis stereoisomer to be more stable than *trans*-cyclooctene? Is *cis*-cyclooctene chiral? Is *trans*-cyclooctene chiral?

EXERCISE 20 (Chapter 3)

Construct models of *cis*-decalin (*cis*-bicyclo[4.4.0]decane) and *trans*-decalin. Observe how it is possible to interconvert one conformation of *cis*-decalin in which both rings are in chair conformations to another all chair conformation. This interconversion is not possible in the case of the *trans*-decalin isomer. Suggest a reason for the difference in behavior of the cis trans isomers. *Hint:* What would happen to carbon atoms 7 and 10 of *trans*-decalin if the other ring (indicated by carbon atoms numbered 1 to 6) is converted to the alternative chair conformation. Is the situation the same for *cis*-decalin?

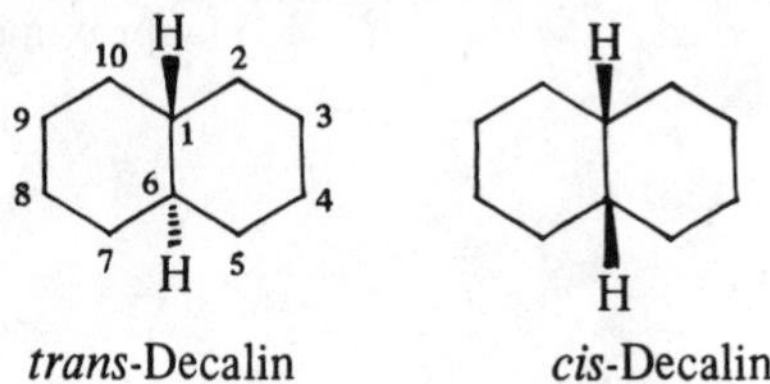

trans-Decalin *cis*-Decalin

EXERCISE 21 (Chapter 3)

Assemble a model of norbornane (bicyclo[2.2.1]heptane). Observe the two cyclopentane ring systems in the molecule. The structure may also be viewed as a methylene (CH_2) bridge between carbon atoms 1 and 4 of cyclohexane. Describe the conformation of the cyclohexane ring system in norbornane. How many eclipsing interactions are present?

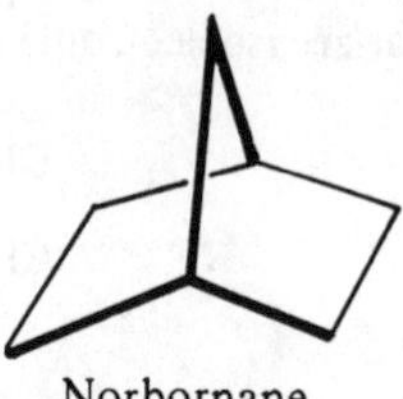

Norbornane

Using a model of twistane identify the cyclohexane ring systems held in twist conformations. In adamantane find the chair conformation cyclohexane systems. How many are present? Evaluate the torsional and angle strain in adamantane. Which of the three compounds in this exercise are chiral?

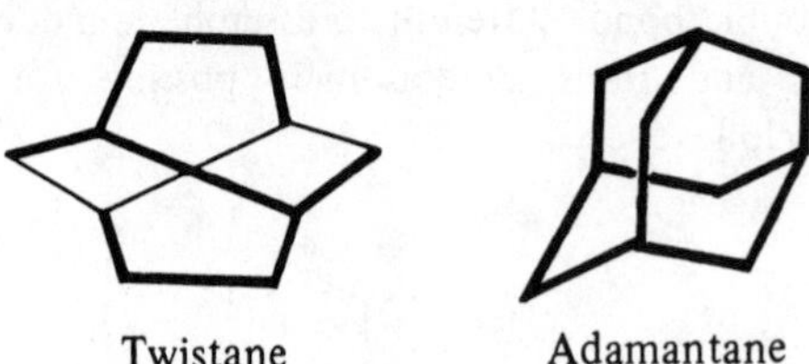

Twistane Adamantane

EXERCISE 22 (Chapter 6)

An hypothesis known as "Bret's Rule" states that a double bond to a bridgehead of a small-ring bridged bicyclic compound is not possible. The basis of this rule can be seen if you attempt to make a model of bicyclo[2.2.1]hept-1-ene, **A**. One approach . . . to the assembly of this model is to try to bridge the number 1 and number 4 carbon atoms of cyclohexene with a methylene (CH_2) unit. Compare this bridging with the ease of installling a CH_2 bridge between the 1 and 4 carbon atoms of cyclohexane to form a model of norbornane (see Exercise 21). Explain the differences in ease of assembly of these two models.

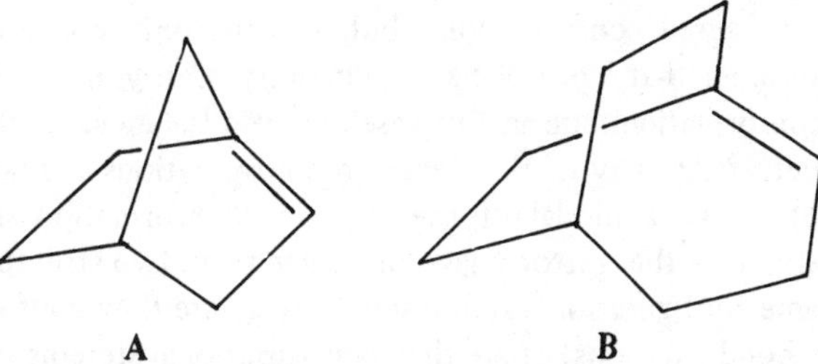

Bridgehead double bonds can be accommodated in larger ring bridged bicyclic compounds such as bicyclo[3.2.2]non-1-ene, **B**. Although this compound has been prepared in the laboratory it is an extremely reactive alkene. The Theta Molecular Model Set model of bicyclo[3.2.2]non-1-ene clearly shows the strained (twisted) double bond system.

EXERCISE 23 (Chapter 11)

Not all cyclic structures with alternating double and single bonds are aromatic. Cyclooctatetraene shows none of the aromatic characteristics of benzene. From examination of molecular models of cyclooctatetraene and benzene explain why there is π-electron delocalization in benzene but not in cyclooctatetraene. *Hint:* Can the carbon atoms of the eight-membered ring readily adopt a planar arrangement?

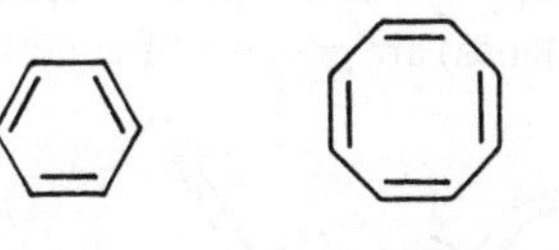

Benzene Cyclooctatetraene

Note that benzene can be represented several different ways with the Theta Molecular Model Set. In this exercise the Kekulé representation with alternating double and single bonds is appropriate. Alternative representations of benzene are shown in the model set instruction booklet.

EXERCISE 24 (Chapter 4)

Consider the $CH_3CHXCHYCH_3$ system. A butane that has at C-2 and C-3 different shades of green atom-centers is representative. Assemble all possible stereoisomers of this structure. How many are there? Indicate the relationship among them. Are they all chiral?

Repeat the analysis with the $CH_3CHXCHXCH_3$ system. The green atom-centers are suitable for representation of the X substituent.

EXERCISE 25 (Chapter 4)

The $CH_3CHXCHXCH_3$ molecule can exist as the stereoisomers shown here. In the eclipsed conformation (meso) shown on the left the molecule has a plane of symmetry that bisects the C-2, C-3 bond. This is a more energetic conformation than any of the three staggered conformations, but it is the only conformation of this configurational stereoisomer that has a plane of symmetry. Can you consider a molecule achiral if only one conformation, and in this case not even the most stable conformation, has a plane of symmetry? Are any of the staggered conformations achiral (superposable on their mirror image)? Make a model of the staggered conformation shown here and make another model that is the mirror image of it. Are these two structures different conformations of the same configurational stereoisomers (e.g., are they conformers that can be interconverted by bond rotations) or are they configurational stereoisomers? Based on your answer to the last question suggest an explanation for the fact that the molecule is not optically active?

E S

EXERCISE 26 (Chapter 4)

Not all molecular chirality is a result of a center of chirality, such as CHXYZ. Cumulated dienes (1,2-dienes or allenes) are capable of generating molecular chirality.

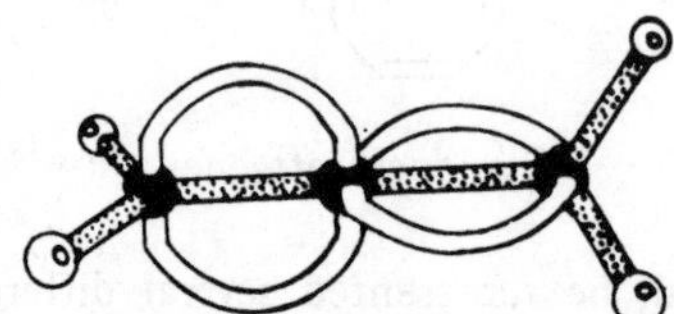

1,2-Propadiene (allene), $H_2C{=}C{=}CH_2$

Identify, using models, which of the following cumulated dienes are chiral.

A B C

Are the following compounds chiral? How are they structurally related to cumulated dienes?

D E

Is the cumulated triene **F** chiral? Explain the presence or absence of molecular chirality. More than one stereoisomer is possible for triene **F**. What are the structures, and what is the relationship between those structures?

$$\begin{matrix} H & & & & & & CH_3 \\ & \diagdown & & & & \diagup & \\ & & C{=}C{=}C{=}C & & & & \\ & \diagup & & & & \diagdown & \\ CH_3 & & & & & & H \end{matrix}$$

F

EXERCISE 27 (Chapters 4 and 11)

Substituted biphenyl systems can produce molecular chirality if the rotation about the bond connecting the two rings is restricted. Which of the three biphenyl compounds indicated here are chiral and would be expected to be optically active?

a f

b e

J. a = f = CH_3	K. a = b = CH_3	L. a = f = CH_3
b = e = $N(CH_3)_3^+$	e = f = $N(CH_3)_3^+$	b = e = H

EXERCISE 28 (Chapter 6)

Assemble a model of ethyne (acetylene). The linear geometry of the molecule should be readily apparent. Note that the Theta Molecular Model Set depicts the σ and both the π bonds of the triple bond system. Based on attempts to assemble cycloalkynes, predict the smallest ring cycloalkyne that is stable.

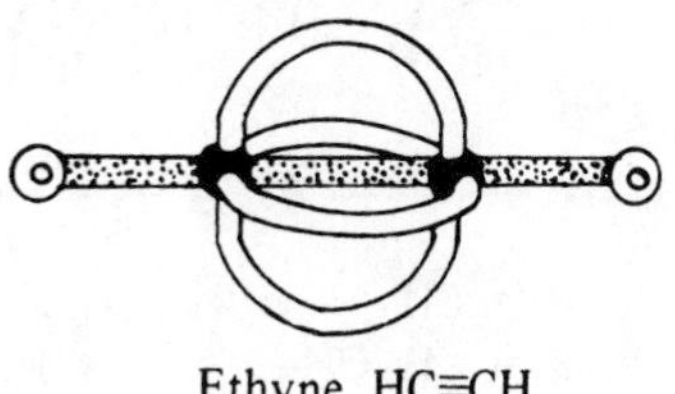

Ethyne, $HC{\equiv}CH$

EXERCISE 29 (Chapter 1)

Construct a model of β-D-glucopyranose. Note that in one of the chair conformations all the hydroxyl groups and the CH_2OH group are in an equatorial orientation. Convert the structure of β-D-glucopyranose to α-D-glucopyranose, to β-D-mannopyranose, and to β-D-galactopyranose. Indicate the number of large ring substituents (OH or CH_2OH) that are axial in the more favorable chair conformation of each of these sugars. Is it reasonable that the β-anomer is more stable than the α-anomer of D-glucopyranose?

Make a model of β-L-glucopyranose. What is the relationship between the D and L configurations? Which is more stable?

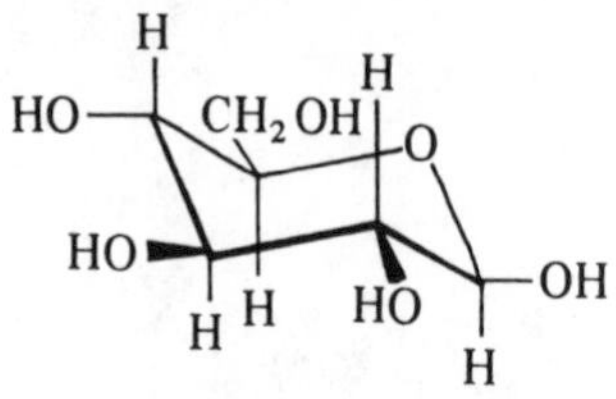

β-D-Glucopyranose

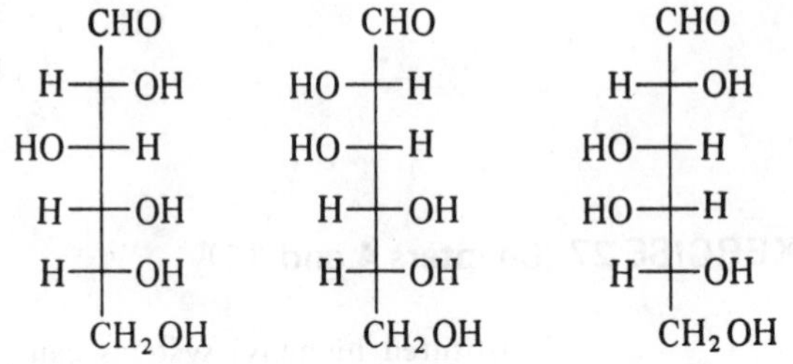

D-(+)-Glucose D-(+)-Mannose D-(+)-Galactose

EXERCISE 30 (Chapter 23)

Assemble a model of tripeptide **A** shown here. If the model is made according to the representation of a peptide shown in the Theta Molecular Model Set, you will be able to observe the restricted rotation of the C–N bond in the amide linkage. Note the planarity of the six atoms associated with the amide portions of the molecule. Which bonds along the peptide chain are free to rotate? The amide linkage can either be cisoid or transoid. How does the length (from the *N*-terminal nitrogen atom to the *C*-terminal carbon atom) of the tripeptide chain that is transoid compare with one that is cisoid? Which is more "linear"? Convert a model of tripeptide **A** in the transoid arrangement to a model of tripeptide **B**. Which tripeptide has a longer chain?

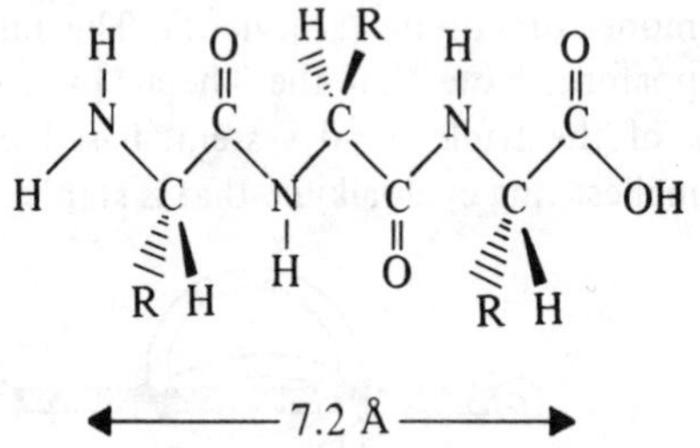

← 7.2 Å →

Tripeptide **A**	R = CH_3	(L-alanine)
Tripeptide **B**	R = CH_2OH	(L-serine)

EXERCISE 31 (Chapter 10 and Special Topic O)

Make models of the π molecular orbitals for the following compounds. Use the phase representation of each contributing atomic orbital shown in the Theta Molecular Model Set instruction booklet. Compare each model with π molecular orbital diagrams shown in the textbook.

(a) π_1 and π_2 of ethene ($CH_2{=}CH_2$)

(b) π_1 thru π_4 of 1,3-butadiene ($CH_2{=}CH{-}CH{=}CH_2$)

(c) π_1, π_2 and π_3 of the allyl (propenyl) radical ($CH_2{=}CH{-}CH_2$)

EXERCISE 32 (Special Topic O)

Explain the observed stereochemistry of the pericyclic reactions shown here. The course of the reactions are controlled by orbital symmetry.

(a) An electrocyclic reaction.

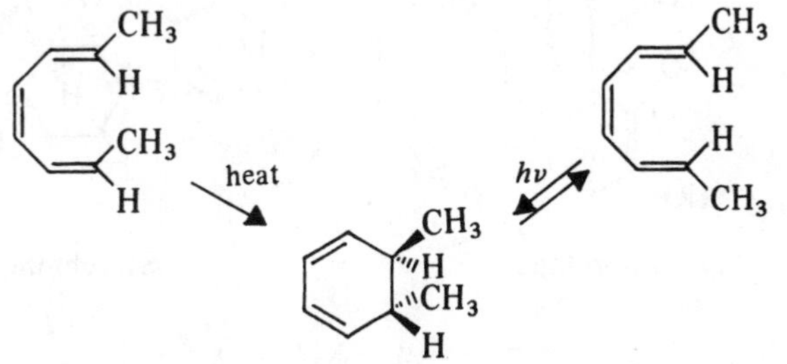

(b) (4+2) Cycloaddition reactions.

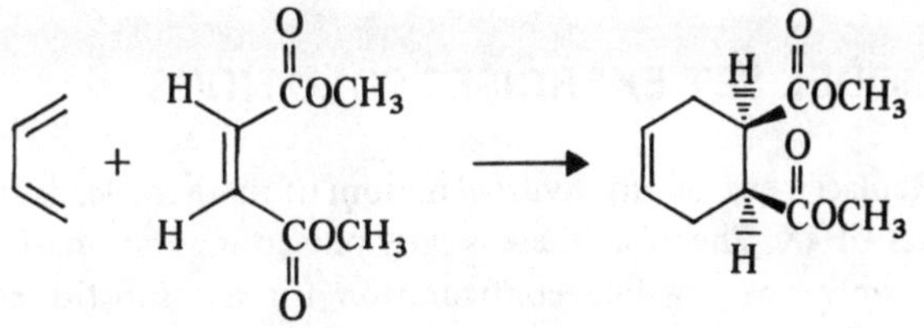

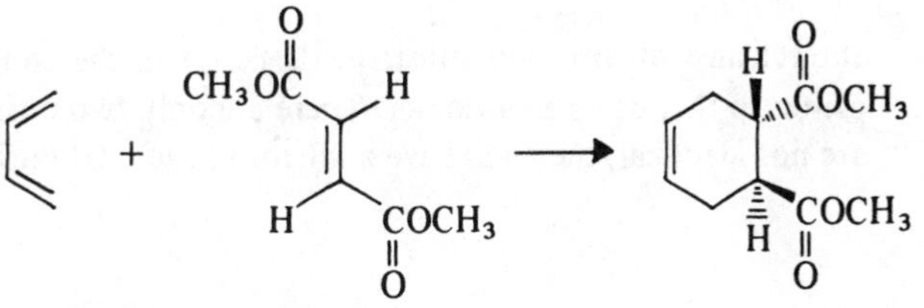

EXERCISE 33

The Theta Molecular Model Set is well suited for the assembly of many fairly complex natural products. Several interesting representative natural product structures, suitable for your model making pleasure, are shown here.

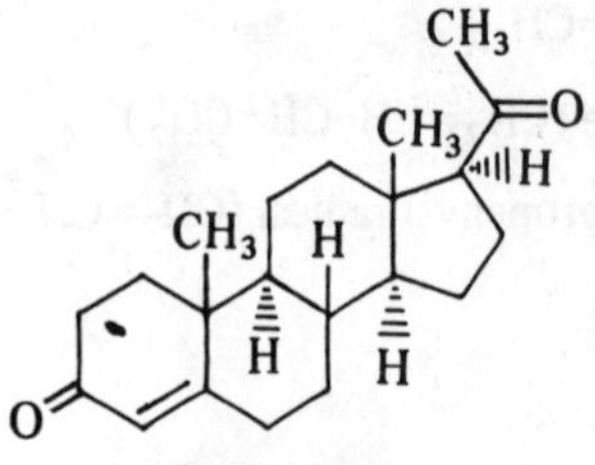

Progesterone

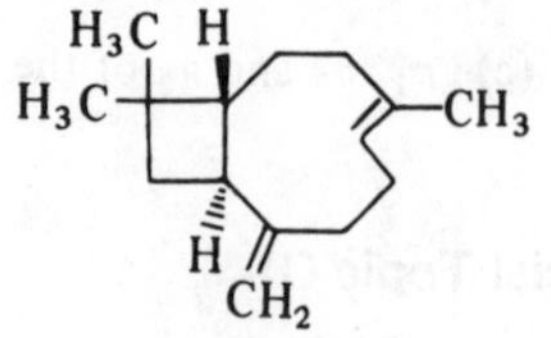

Caryophyllene

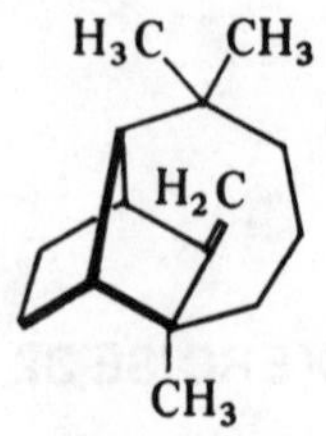

Longifolene

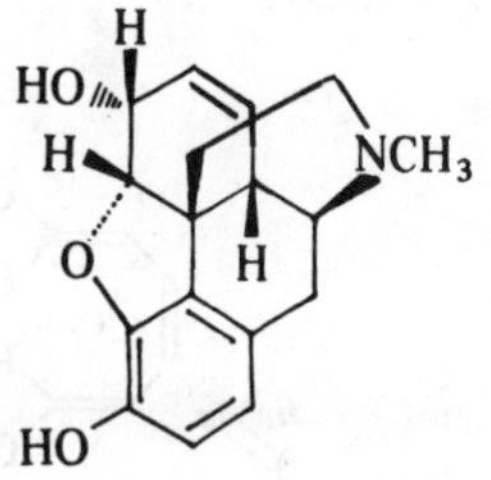

Morphine

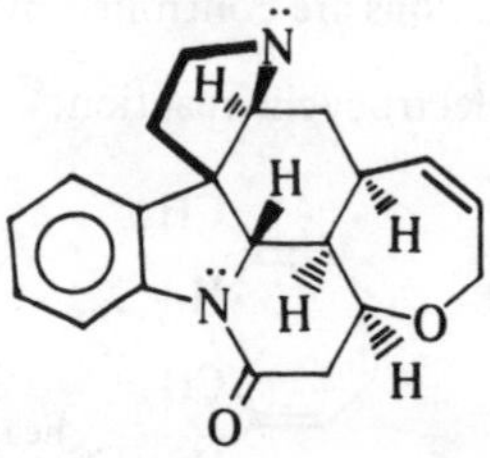

Strychnine

MOLECULAR MODEL SET EXERCISES SOLUTIONS

Solution 1 Replacement of any hydrogen atom of methane leads to the same monosubstituted product CH_3X. Therefore there is only one configuration of a monosubstituted methane. There is only one possible configuration for a disubstituted methane of either the CH_2X_2 or CH_2XY type.

Solution 2 Interchange of any two substituents converts the configuration of a tetrahedral stereocenter to that of its enantiomer. There are only two possible configurations. If the models are not identical, they will have a mirror-image relationship.

Solution 3 The tetrahedral carbon atoms may be rotated without breaking the carbon-carbon bond. There is no change in the carbon-carbon bond orbital overlap during rotation. The eclipsed conformation places the hydrogen atoms closer together than they are in the staggered conformation.

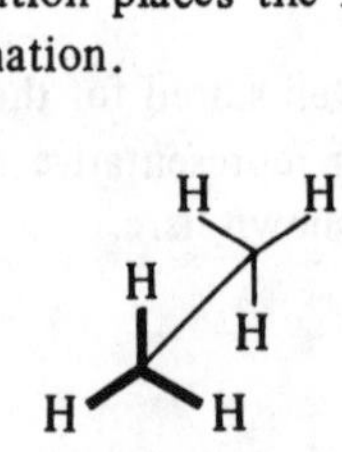

Staggered conformation

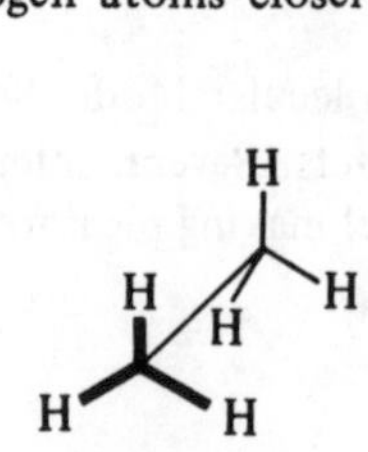

Eclipsed conformation

Solution 4 All mono substituted ethanes (CH_3CH_2X) may be made into identical structures by rotations about the C—C bond. The following structures are three energetically equivalent staggered conformations.

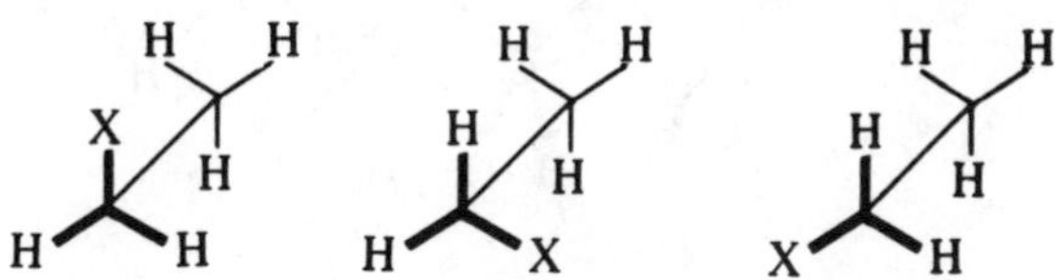

The three equivalent eclipsed conformations are

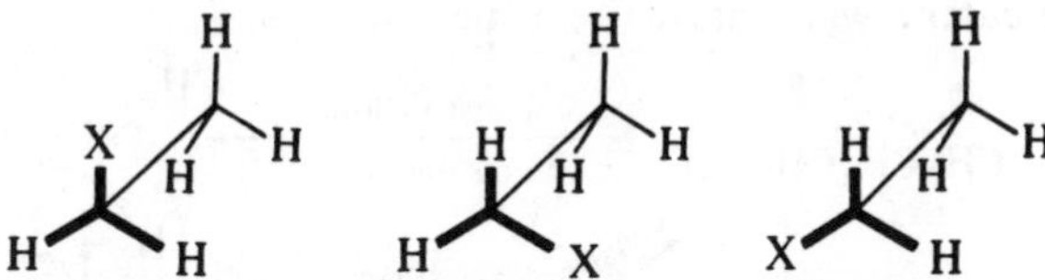

Solution 5 The two *gauche* conformations are energetically equivalent, but not identical (superposable) since they are conformational enantiomers. They bear a mirror-image relationship and are interconvertable by rotation about the carbon-carbon bond.

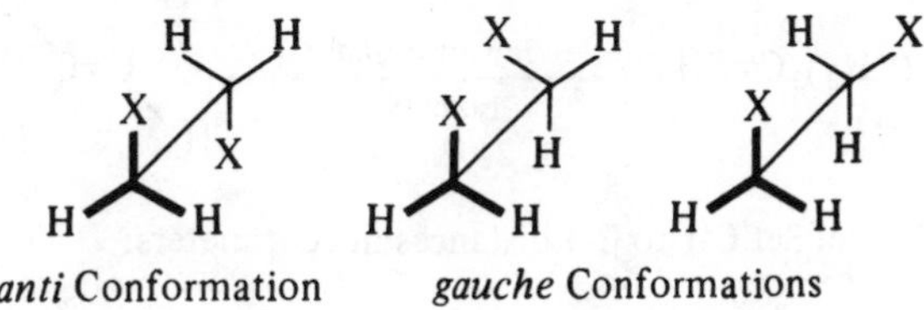

anti Conformation   *gauche* Conformations

Solution 6 There are three eclipsed conformations. The methyl groups (C-1 and C-4) are closest together in the methyl-methyl eclipsed conformation. The carbon-carbon internuclear distances between C-1 and C-4 are shown in the following table. The number of conformations of each type, the model distances, and the corresponding molecular distance in angstroms (Å) are shown.

CONFORMATION	NUMBER	cm	Å
Eclipsed (CH_3, CH_3)	1	7.4	2.5
Gauche	2	8.5	2.8
Eclipsed (H, CH_3)	2	10.0	3.3
Anti	1	11.0	3.7

Solution 7 The enantiomers formed from replacement of the C-2 hydrogen atoms of butane are

CH_3 / H►C◄X / CH_2CH_3 CH_3 / X►C◄H / CH_2CH_3

Solution 8 The extended chain assumes a staggered arrangement. The relationship of C-1 and C-4 is *anti*.

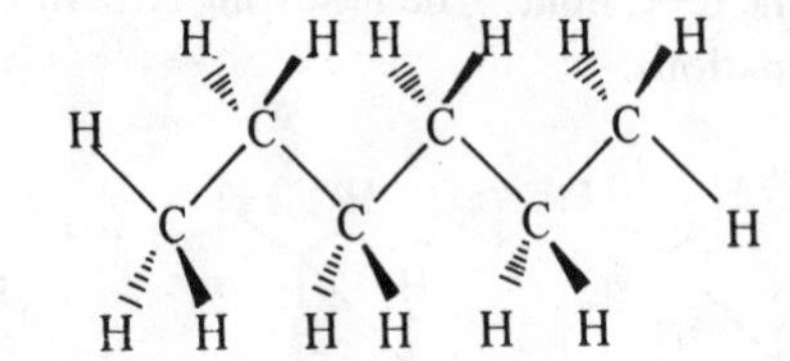

Solution 9 None of the isomeric butenes is chiral. They all have a plane of symmetry. All the isomeric butenes are related as constitutional (or structural isomer) except *cis*-2-butene and *trans*-2-butene, which are diastereomers.

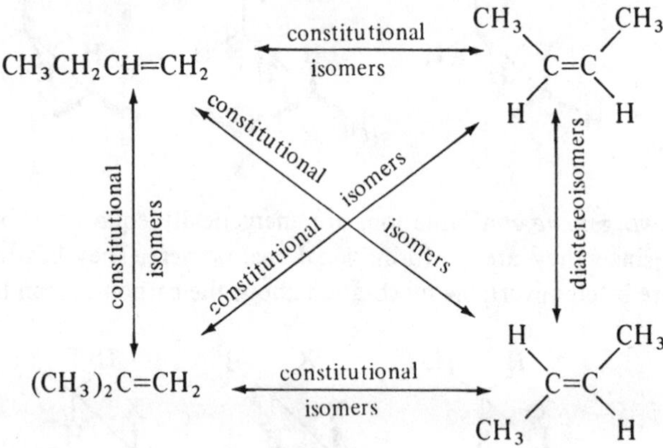

Molecular Model Set C-1 to C-4 distances in centimeters:

COMPOUND	DISTANCES (cm)
cis-1-Butene	6.0
trans-2-Butene	11.0
Butane (*gauche*)	8.5
Butane (*anti*)	11.0

Solution 10 The hydrogen atoms are all eclipsed in cyclopropane.

Solution 11 All the hydrogen atoms are eclipsed in the planar conformation of cyclobutane. The folded ring system has skew hydrogen interactions. There are six possible isomers of dimethylcyclobutane. Since the ring is not held in one particular folded conformation deviations of the ring planarity need not be considered in determining the number of possible dimethyl structures.

Solution 12 In the planar conformation of cyclopentane all five methylene pairs of hydrogen atoms are eclipsed. That produces 10 eclipsed hydrogen interactions. Some torsional strain is relieved in the envelope conformation since there are only 6 eclipsed bydrogen interactions.

Solution 13 The three configurational stereoisomers of 1,2-dimethylcyclopentane are shown here. Both trans stereoisomers are chiral, while the cis configuration is an achiral meso compound.

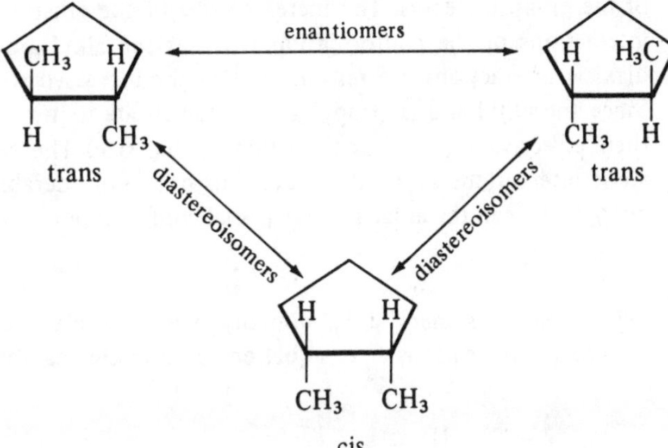

Solution 14 The puckered ring of the chair and the boat conformation may be interconverted by rotation about the carbon-carbon bonds. The chair is more rigid than the boat conformation. All hydrogen atoms in the chair conformation have a staggered arrangement. In the boat conformation there are eclipsed relationships between the hydrogen atoms on C-2 and C-3, and also between those on C-5 and C-6. Carbon atoms that are 1,4 to each other in the chair conformation have a *gauche* relationship. An evaluation of the three conformations confirms the relative stability: chair > twist > boat. The boat conformation has considerable eclipsing strain and nonbonded (van der Waals repulsion) interactions, the twist conformation has slight eclipsing strain, and the chair conformation has a minimum of eclipsing and nonbonded interactions.

Solution 15 Interconversion of the two chair conformations of methylcyclohexane changes the methyl group from an axial to a less crowded equatorial orientation, or the methyl that is equatorial to the more crowded axial position.

CH_3 5 6 1 H 4 3 2 ⇄ CH_3 H

Axial methyl Equatorial methyl

The conformation with the axial methyl group has two *gauche* (1,3 diaxial) interactions that are not present in the equatorial methyl conformation. These *gauche* interactions are axial methyl to C-3 and axial methyl to C-5. The methyl to C-3 and methyl to C-5 relationships with methyl groups in an equatorial orientation are anti.

Solution 16 The $\Delta H°$ value reflects the relative energies of the two chair conformations for each structure. The crowding of the alkyl group in an axial orientation becomes greater as the bulk of the group increases. The increased size of the substituent has little effect on the steric interactions of the conformation that has the alkyl group equatorial. The *gauche* (1,3-diaxial) interactions are responsible for the increased strain for the axial conformation. Since the ethyl and isopropyl groups can rotate to minimize the nonbonded interactions their effective size is less than their actual size. The *t*-butyl group cannot relieve the steric interactions by rotation and thus has a considerably greater difference in potential energy between the axial and equatorial conformation.

Solution 17 All four stereoisomers of 1,2-dimethylcyclohexane are chiral. The *cis*-1,2-dimethylcyclohexane conformations have equal energy and are readily interconverted, as shown here.

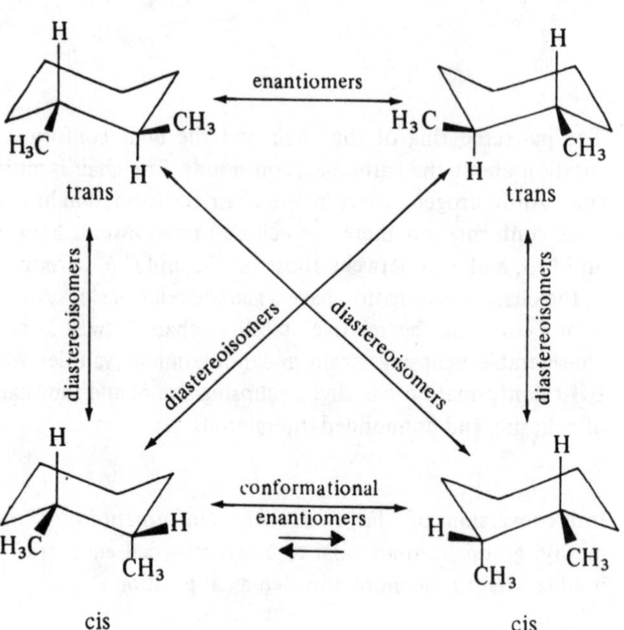

Solution 18 The structures that are isoelectronic have the same geometry. Isoelectronic structures are

CH_3CH_3 and $CH_3NH_3^+$

CH_3NH_2 $CH_3CH_2^-$ and $CH_3OH_2^+$

Structure CH_3NH^- would be isoelectronic to CH_3OH.

Solution 19 Cis-trans stereoisomers are possible only for 3,4-dimethylcyclohexene. The ring size and geometry of the double bond prohibit a trans configuration of the double bond. Two configurational isomers (they are enantiomers) are possible for 2,3-dimethylhexene.

cis-Cyclooctene is more stable because it has less strain than the *trans*-cyclooctene structure. The relative stability of cycloalkene stereoisomers in rings larger than cyclodecane generally favors trans. The *trans*-cyclooctene structure is chiral.

trans-Cyclooctene
(one enantiomer)

Solution 20 The ring fusion in *trans*-decalin is equatorial, equatorial. That is, one ring is attached to the other as 1,2-diequatorial substituents would be. Interconversion of the chair conformations of one ring (carbon atoms 1 thru 6) in *trans*-decalin would require the other ring to adopt a 1,2-diaxial orientation. Carbon atoms 7 and 10 would both become axial substituents to the other ring. The four carbon atoms of the *substituent* ring (carbon atoms 7 thru 10) cannot bridge the diaxial distance. In *cis*-decalin both conformations have an axial, equatorial ring fusion. Four carbon atoms can easily bridge the axial, equatorial distance.

Solution 21 The cyclohexane ring in norbornane is held in a boat conformation, and therefore has four hydrogen eclipsing interactions. All the six-membered ring systems in twistane are in twist conformations. All four of the six-membered ring systems in adamantane are chair conformations.

Solution 22 Bridging the 1 and 4 carbon atoms of cyclohexane is relatively easy since in the boat conformation the flagpole hydrogen atoms (on C-1 and C-4) are fairly close and their C–H bonds are directed toward one another. With cyclohexene the geometry of the double bond and its inability to rotate freely, make it impossible to bridge the C-1, C-4 distance with a single methylene group. Note however, that a cyclohexene ring can accommodate a methylene bridge between C-3 and C-6. This bridged bicyclic system (bicyclo[2.2.1] hept-2-ene) does not have a bridgehead double bond.

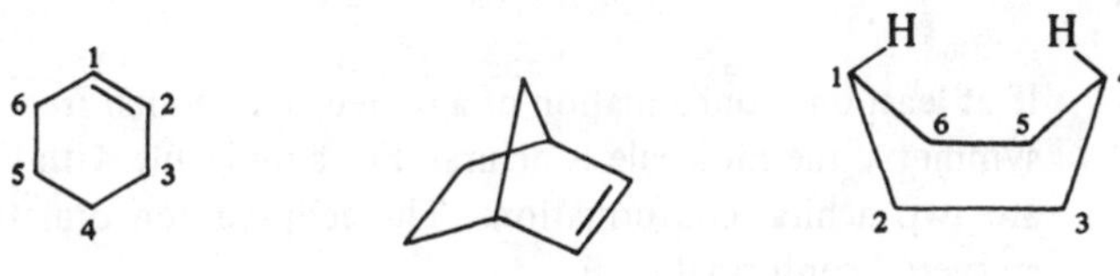

Bicyclo[2.2.1]hept-2-ene

Solution 23 The 120° geometry of the double bond is ideal for incorporation into a planar six-membered ring, as the internal angle of a regular hexagon is 120°. Cyclooctatetraene cannot adopt a planar ring system without considerable angle strain. The eight-membered ring adopts a "tub" conformation that minimizes angle strain and does not allow significant p-orbital overlap other than that of the four double bonds in the system. The cyclooctatetraene thus has four isolated double bonds and is not a delocalized π-electron system.

Cyclooctatetraene (tub conformation)

Solution 24 In the $CH_3CHXCHYCH_3$ system there are four stereoisomers, all of which are chiral.

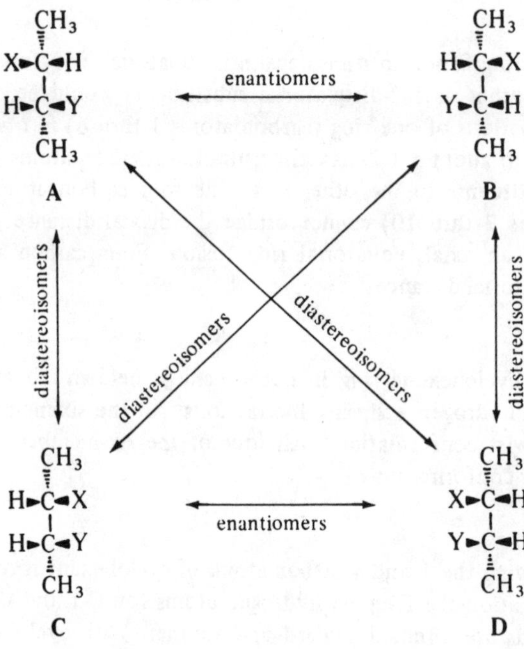

In the $CH_3CHXCHXCH_3$ system there are three stereoisomers, two of which are chiral. The third stereoisomer (**G**) (shown on page 637) is an achiral meso structure.

Solution 25 If at least one conformation of a molecule in which free rotation is possible has a plane of symmetry the molecule is achiral. For a molecule with the configurations specified, there are two achiral conformations. The eclipsed conformation **E** shown in the exercise and staggered conformation **F**.

F T

A model of **F** is identical with its mirror image. It is achiral, although it does not have a plane of symmetry, due to the presence of a *center* of symmetry that is located between C-2 and C-3. A center of symmetry, like a plane of symmetry, is a reflection symmetry element. A center of symmetry involves reflection thru a point, a plane of symmetry requires reflection about a plane. A model of the mirror image of **S** (structure **T**) is not identical to **S**, but is a conformational enantiomer of **S**. They can be made identical by rotation about the C-2, C-3 bond. Since **S** and **T** are conformational enantiomers each will be present in equal amounts in a solution of this configurational stereoisomer. Both conformation **S** and conformation **T** are chiral and therefore should rotate the plane of plane polarized light. Since they are enantiomeric the rotations of light will be equal in magnitude but *opposite* in direction. The net result is a racemic form of conformational enantiomers, and thus optically inactive. A similar argument can be made for any other chiral comformation of this configuration of $CH_3CHXCHXCH_3$.

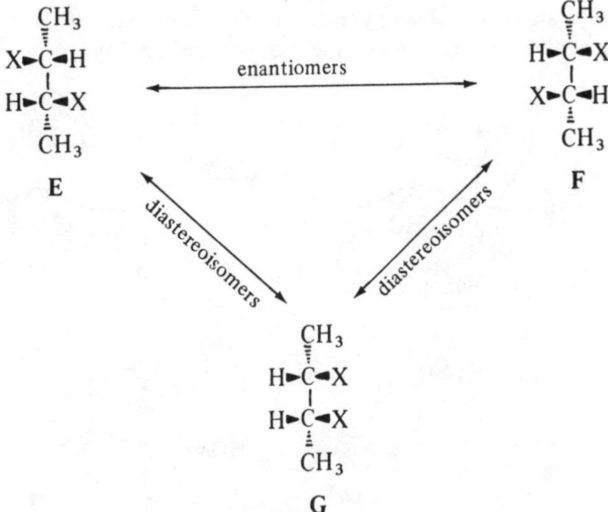

Solution 26 Structures B and C are chiral. Structure A has a plane of symmetry and is therefore achiral. Compounds D and E are both chiral. The relative orientation of the terminal groups in D and E is perpendicular as is the case in the cumulated dienes.

Cumulated triene F is achiral. It has a plane of symmetry passing thru all six carbon atoms. Structure F has a trans configuration. The cis diastereomer is the only other possible stereoisomer.

Solution 27 Structure J can be isolated as a chiral stereoisomer because of the large steric barrier to rotation about the bond connecting the rings. Biphenyl K has a plane of symmetry and is therefore achiral. The symmetry plane of **K** is shown here. Compound **I** has a low energy barrier to rotation and thus would not be optically active. Any chiral conformation of L can easily be converted to its enantiomer by rotation. It is only when a ≠ b and f ≠ e and rotation is restricted by bulky groups that chiral (optically active) stereoisomers can be isolated.

A plane of symmetry

Solution 28 The smallest ring stable cycloalkyne is the nine-membered ring cyclononyne. A model of this alkyne can easily be assembled with the Theta Molecular Model Set.

Solution 29 As shown here, the alternative chair conformation of β-D-glucopyranose has all large substituents in an axial orientation. The structures α-D-glucopyranose, β-D-mannopyranose, β-D-galactopyranose all have one large axial substituent in the most favorable conformation. β-L-Glucopyranose is the enantiomer (mirror image) of β-D-glucopyranose. Enantiomers are of equal thermodynamic stability.

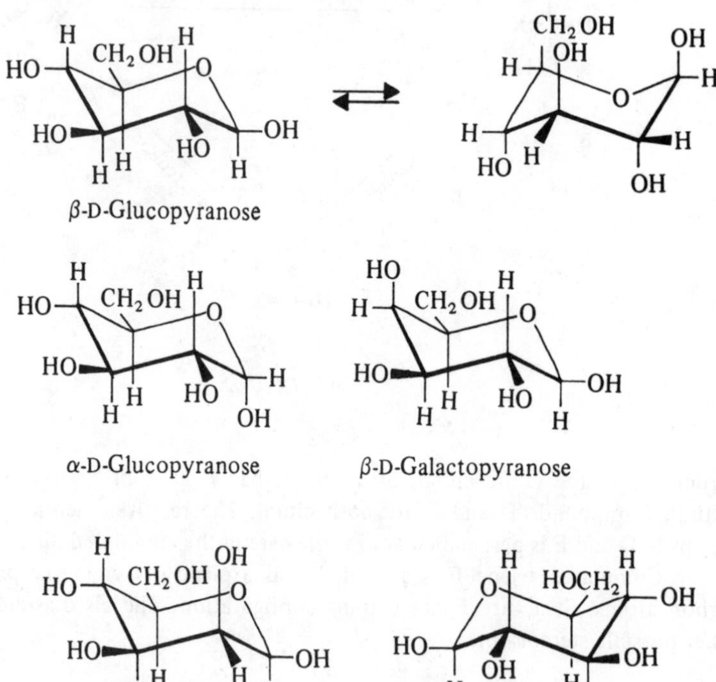

β-D-Glucopyranose

α-D-Glucopyranose β-D-Galactopyranose

β-D-Mannopyranose β-L-Glucopyranose

Solution 30 The peptide chain bonds not free to rotate are indicated by the bold lines in the structures shown here. The transoid arrangement produces a more linear tripeptide chain. The length of the tripeptide chain does not change if you change the substituent R groups.

Solution 31 The models of the π molecular orbitals for ethene are shown here. A representation of these orbitals can be found in the text on pages 57 and 58.

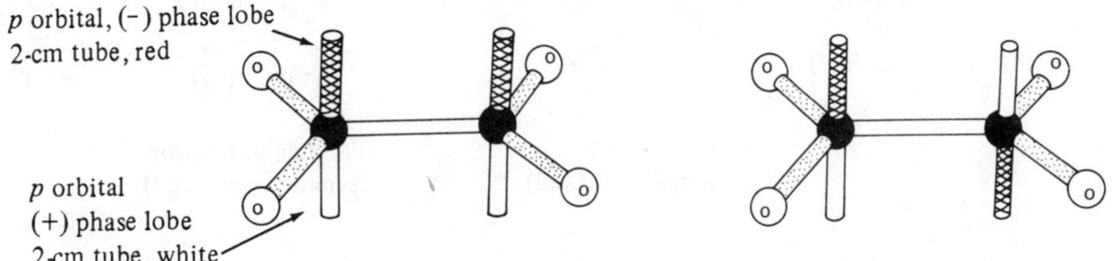

Ethene, π bonding molecular orbital

Ethene, π^* antibonding molecular orbital

The π molecular orbitals for 1,3-butadiene are shown in the text on page 464. A model of one of the π molecular orbitals of 1,3-butadiene is shown in the model set instruction booklet. The phases of the contributing atomic orbitals to the molecular orbitals of the allyl radical can be found in the text on page 449. The π molecular orbital of the allyl radical has a node at C-2. This can be illustrated with the Theta Molecular Model Set by not placing red or white p-orbital tubes on the C-2 atom center prongs. The absence of tubes indicates an orbital phase of zero.

Solution 32 The complete solution to this exercise is given in the text pages 1161-1165 and 1174-1175. The orbitals involved are shown below.

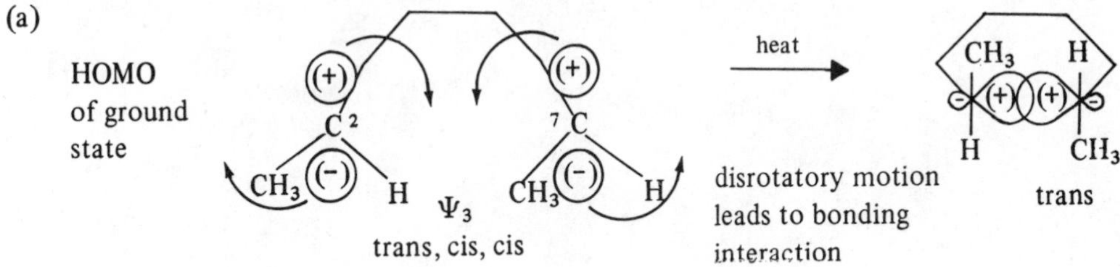

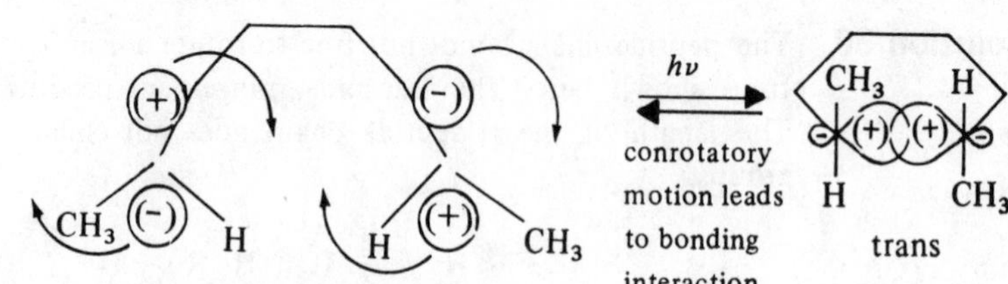

Ψ_4 of *trans, cis, trans*-2,4,6-Octatriene

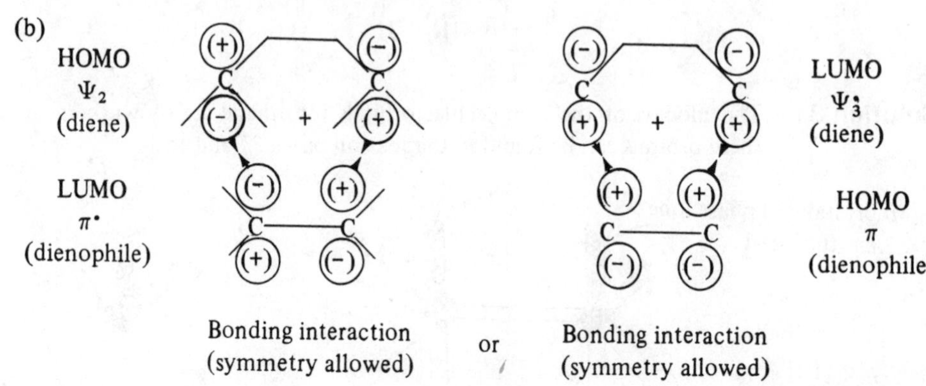

Bonding interaction (symmetry allowed) or Bonding interaction (symmetry allowed)

APPENDIX
Glossary of Important Terms

Acidity Constant (K_a). The acidity constant is a measure of the strength of an acid in water. For the acid HA, the equilibrium is

$$HA + H_2O \rightleftharpoons H_3O^+ + A^-$$

and $K_a = \dfrac{[H_3O^+][A^-]}{[HA]}$

Therefore, the larger K_a is, the stronger the acid is. (Section 2.15B).

Addition Reaction. A reaction in which the product molecule contains all the atoms that were present in the reactant molecules (Section 7.1).

Example:

```
 H H                     H  H
 | |                     |  |
H-C=C-H + Br-Br ───▶ H-C──C-H
                         |  |
                         Br Br
```

Alkanes. Hydrocarbons with the general formula C_nH_{2n+2} are called alkanes. Molecules of alkanes have no rings (i.e., they are acyclic) and they have only single bonds between carbon atoms. Their carbon atoms are sp^3 hybridized.

Aliphatic Hydrocarbon. A hydrocarbon such as an alkane, alkene, alkyne, cycloalkane, cycloalkene; that is, all hydrocarbons that are not aromatic (Section 11.1).

Alkyl Group. The molecular fragment that remains when a hydrogen atom is removed from a hydrocarbon (Section 2.8A).

Example:

```
  H H          H H
  | |          | |
H-C-C-H      H-C-C-
  | |          | |
  H H          H H
Hydrocarbon  Alkyl group
```

Allylic Hydrogen. A hydrogen on a carbon atom that is adjacent to a C=C double bond (Section 10.2).

Allylic Substitution. Substitution of an allylic hydrogen or group by another atom or group (Section 10.2).

Annulene. A monocyclic compound that can be represented by a structure having alternating single and double bonds. For example, cyclobutadiene is [4] annulene and benzene is [6] annulene (Sections 11.6A, 11.12)

Anti Addition. See **Syn and Anti Addition.**

Antibonding Molecular Orbital. The molecular orbital formed when atomic orbitals of opposite phase sign overlap. The antibonding orbital has higher energy than a bonding orbital. The electron probability density of the region between the nuclei is small and it contains a **node**-a region where $\psi = 0$. Thus, having electrons in an antibonding orbital does not help hold the nuclei together. The internuclear repulsions tend to make them fly apart.

Aprotic Solvents. Solvents that lack an —H bonded to a strongly electronegative element. Common aprotic solvents are acetone, $CH_3\overset{O}{\overset{\|}{C}}CH_3$; acetonitrile $CH_3C{\equiv}N$, sulfur dioxide, SO_2; dimethylsulfoxide, CH_3SOCH_3; and trimethylamine, $N(CH_3)_3$ (Section 5.13C).

Aromatic Compound. Certain cyclic conjugated compounds. Aromatic compounds have a stability significantly greater than that of a hypothetical resonance structure (e.g., a Kekulé structure). Many aromatic compounds react with electrophilic reagents (Br_2, HNO_3, H_2SO_4) by substitution rather than addition even though they are unsaturated. A modern definition of an aromatic compound is one that has planar monocyclic rings with 2, 6, 10, 14, . . ., and so on, delocalized π electrons (Sections 11.6, 11.12)

Atomic Orbital (AO). A region of space about the nucleus of a single atom where there is a high probability of finding an electron. Atomic orbitals called *s* orbitals are spherical; those called *p* orbitals are like two almost-touching spheres. Orbitals can hold a maximum of two electrons when their spins are paired. Orbitals are described by a wave funcation, Ψ, and each orbital has a characteristic energy. The phase sign associated with an orbital may be (+) or (−) (Sections 1.11, 1.17).

Aufbau Principle. Orbitals are filled so that those of lowest energy are filled first (Section 1.11).

Axial Group. A group that is attached to a carbon of cyclohexane and that is oriented in a direction that is generally perpendicular to the average plane of the ring (Section 3.12).

Benzenoid. An aromatic compound whose molecules contain benzene rings or fused benzene rings. Examples are benzene, naphthalene, anthracene, and phenanthrene (Sections 11.7A, 11.12).

Boat Conformation. The conformation of cyclohexane (see following structure) in which there is torsional strain but no angle strain:

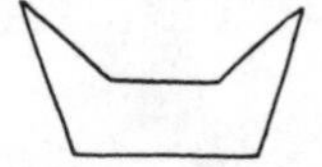

(Section 3.10)

Bonding Molecular Orbital. The molecular orbital formed when atomic orbitals with the same phase sign interact. The electron probability density of a bonding molecular orbital is large in the region of space between the two nuclei (Sections 1.12, 1.17).

Bond-Line Formula. Formula that shows only the carbon skeleton. The number of hydrogen atoms necessary to fulfill the carbon atoms' valences are assumed to be present, but we do not write them in. Other atoms are written in (Section 1.20D).

Example: $CH_3CHClCH_2CH_3$ is written

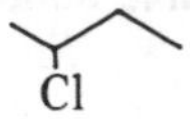

Cl

Branched Alkane. Alkane in which at least one carbon atom is bonded to three or four other carbon atoms (Section 3.2).

Brønsted-Lowry Acid-Base Theory. An acid is a substance that can donate a proton. A base is a substance that can accept a proton (Section 2.15A).

Chain Reactions. Reactions whose mechanisms involve a series of steps with each step producing a reactive intermediate that causes the next step to occur. The halogenation of an alkane is a chain reaction (Sections 9.3, 9.12).

Chair Conformation. The most stable conformation of cyclohexane:

(Section 3.10)

Chirality. Equivalent to "handedness." A chiral molecule is one that is not superposable on its mirror reflection. An *achiral* molecule is one that can be superposed on its mirror reflection. Any tetrahedral atom that has four different groups attached to it is a **stereocenter**. A pair of enantiomers will be possible for all molecules that contain *a single* stereocenter. For molecules with more than one stereocenter, the number of stereoisomers will not exceed 2^n where n is the number of stereocenters (Section 4.3, 4.17).

Circle-and-Line Formula. Three-dimensional formula in which carbon atoms are shown by circles (Section 1.20E).

Example:

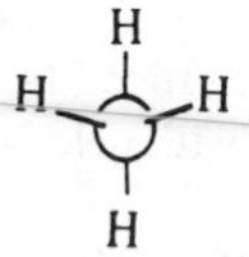

H H H H

cis-trans Isomers. Isomers that differ only in the orientation in space of groups attached to doubly bonded atoms (Section 2.4B) or to rings (Sections 3.13, 3.19).

Example:

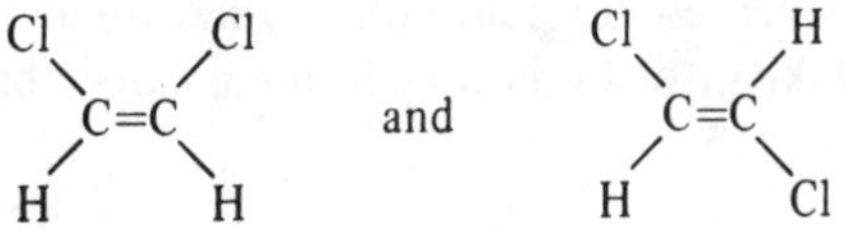

Cl Cl C=C H H and Cl H C=C H Cl

cis-trans Isomerism results because rotation about a carbon-carbon double bond or in a ring is restricted.

Classification of Alkadienes. Cumulated double bonds are double bonds that share a carbon atom: C=C=C. **Conjugated double bonds** are double bonds that are separated by a single bond: C=C–C=C.

Isolated double bonds are double bonds that are separated by at least one saturated carbon atom C=C–$\overset{|}{\underset{|}{\text{C}}}$–C=C (Sections 10.1, 10.8).

Collision Theory. For a chemical reaction to take place, the reacting particles must collide. The rate of a reaction is given by the equation

$$\text{Reaction rate} = Z \times P \times f$$

where Z is the frequency of collision between reacting molecules, P is the probability that the colliding molecules are oriented in a way that allows reaction to take place, and f is the fraction of collisions in which the collision energy is greater than the energy of activation.

In terms of the concentrations of reactants A and B,

$$\text{Reaction rate} = k\,[\text{A}]\,[\text{B}]$$

where k is a proportionality constant called the rate constant (Section 5.4A).

Condensed Structural Formula. A formula in which the atoms that are attached to a particular carbon atom are written immediately after that atom (Section 1.20B).

Example: $CH_3CHClCH_2CH_3$

Configuration. The particular arrangement of atoms (or groups) in space that is characteristic of a given stereoisomer. The configuration at each stereocenter can be designated as (*R*) or (*S*) using the rules given in Sections 4.3, 4.5.

Conformational Analysis. Analysis of the energy changes that a molecule undergoes as groups rotate about single bonds (Sections 3.6, 3.19).

Conjugated Unsaturated System. A system that has a *p* orbital on an atom adjacent to a double bond—a molecule with delocalized π bonds. (Section 10.1).

Constitutional Isomers (formerly called **structural isomers**). Isomers that have their atoms joined in a different order (Sections 1.4 and 4.2, 4.17).

Covalent Bond. A bond that results when atoms share electrons (Sections 1.6 and 1.6B).

Cracking. A process for converting hydrocarbons into other hydrocarbons by heating (thermal cracking) or by heating in the presence of a catalyst (catalytic cracking) (Section 3.15B).

Cycloalkane. Hydrocarbons with the general formula C_nH_{2n} whose molecules have their carbon atoms arranged into a ring are called cycloalkanes. They have only single bonds between carbon atoms, and their carbon aotms are sp^3 hybridized (Sections 3.1, 3.19).

Dash Formula. The structural formula in which bonding electron pairs are represented by dashes (Section 1.20A).

Example:

```
  H
  |
H-C-H
  |
  H
```

Dash-Line-Wedge Formula. Atoms that project out of the plane of the paper are connected by a wedge (◀), those that lie behind the plane are connected with a dash (⋯), and those atoms in the plane of the paper are connected by a line (Section 1.20E).

Example:

```
     H
     |
H ⁄ C ''''H
      ◥
       H
```

Debromination. Elimination of Br_2 from a *vic*-dihalide (Section 6.16).

Dehydration. Elimination of H_2O from an alcohol (Section 6.13).

Dehydrogenation. Elimination of H_2 from a molecule (Sections 6.17 and 15.3).

Dehydrohalogenation. Elimination of HX (X = Cl, Br, I) from an alkyl halide (Sections 6.12, 6.17).

Diastereomers are stereoisomers that are not enantiomers, that is, they are stereoisomers that are *not* related as an object and its mirror reflection (Sections 4.2, 4.17).

Dielectric Constant. A measure of the polarity of a solvent. Also described as a measure of the ability of a solvent to insulate charges from each other (Section 5.13D).

Dimerization. The combination of two identical molecules (Section 7.11).

Dipolar Ion. When a molecule contains both a basic group ($-NH_2$) and an acidic group ($-CO_2H$), both groups exist primarily in the ionic form; that is, $-NH_3^+$ and $-CO_2^-$. Such an ion is called a dipolar ion or **zwitterion** (Section 23.2C)

Dipole-Dipole Forces. Weakly attractive forces between molecules that possess permanent dipole moments (Section 2.17B).

Dipole Moment. The product of the magnitude of the charge in electrostatic units and the distance that separates them in centimeter units (Section 1.9).

E1 Reaction. A unimolecular elimination. The first step of an E1 reaction, formation of a carbocation, is the same as that of an S_N1 reaction, consequently E1 and S_N1 reactions compete with each other. E1 reactions are important when tertiary halides are subjected to solvolysis in polar solvents especially at higher temperatures. The steps in the E1 reaction of *tert*-butyl chloride (Sections 5.17, 5.20) are

$$\textbf{Step 1}\quad CH_3-\underset{CH_3}{\overset{CH_3}{\underset{|}{\overset{|}{C}}}}-Cl \xrightarrow{\text{slow}} CH_3-\underset{CH_3}{\overset{CH_3}{\underset{|}{\overset{|}{C^+}}}} + :Cl^-$$

$$\textbf{Step 2}\quad Sol-\ddot{O}H + H-CH_2-\underset{CH_3}{\overset{CH_3}{\underset{|}{\overset{|}{C^+}}}} \longrightarrow CH_2=C(CH_3)_2 + Sol-\overset{+}{\ddot{O}}H_2$$

E2 Reaction. A bimolecular elimination that often competes with S_N2 reactions. E2 reactions are favored by the use of a high concentration of a strong, bulky, and slightly polarizable base. The order of reactivity of alkyl halides toward E2 reactions is $3^\circ >> 2^\circ > 1^\circ$. The mechanism of the E2 reaction (Sections 5.16, 5.20) involves a single step:

$$B:^- + -\overset{H}{\overset{|}{C}}-\underset{X}{\underset{|}{C}}- \longrightarrow B-H + >C=C< + :X^-$$

Electrical Effect (or **Electronic Effect**). An effect on relative reaction rates when some molecular feature stabilizes the electrical charge on a transition state or an intermediate. The electron-releasing ability of methyl groups stabilizes the transition state and the carbocation formed by ionization of a 3° alkyl halide in the slow step of an S_N1 reaction, for example (Section 5.13).

Electronegativity. The ability of an atom to attract electrons that it is sharing in a covalent bond (Section 1.9).

1,2-Elimination Reaction or **a β Elimination.** A reaction in which the pieces of some molecule are eliminated from adjacent atoms of the reactant leading to the introduction of a multiple bond. Dehydrohalogenation is an elimination reaction in which HX is eliminated from an alkyl halide, leading to the formation of an alkane (Section 5.15A).

$$-\overset{H}{\overset{|}{C}}-\underset{X}{\underset{|}{C}}- + :B^- \longrightarrow >C=C< + H:B + :X^-$$

Enantiomers. Enantiomers are stereoisomers that are related like an object and its mirror reflection. Enantiomers only occur with compounds whose molecules are chiral, that is, with molecules that are *not* superposable on their mirror reflections. Separate enantiomers rotate the plane of polarized light and are said to be *optically active*. They have equal but opposite specific rotations (Sections 4.2, 4.3, 4.5, 4.6, 4.17).

Energy of Activation. The minimum amount of energy (on a molar basis) that must be provided for a reaction to take place. It is the potential energy difference between the reactants and the transition state (Section 5.4D).

Epimers. Any pair of diastereomers that differ only in the configuration at a single atom (Section 21.6).

Equatorial Group. A group that is attached to a carbon of cyclohexane and that is oriented in a direction that is generally in the average plane of the ring (Section 3.12).

Formal Charge. Calculated by taking the group number of that atom (from the Periodic Table) and subtracting the number of electrons associated with it using the formula (Section 1.7A),

$$\text{Formal charge} = \text{group number} - [\tfrac{1}{2}\,(\text{number of shared electrons}) + (\text{number of unshared electrons})]$$

Free Radicals (also called **Carbon Radicals**). Free radicals are formed by homolysis of a bond to a carbon atom. Carbon radicals have an unpaired electron and show the following order of stabilities (Sections 5.1A, 9.12).

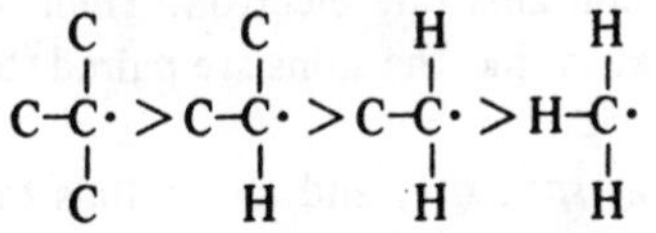

$$\begin{array}{ccccccc} \text{C} & & \text{C} & & \text{H} & & \text{H} \\ | & & | & & | & & | \\ \text{C–C}\cdot & > & \text{C–C}\cdot & > & \text{C–C}\cdot & > & \text{H–C}\cdot \\ | & & | & & | & & | \\ \text{C} & & \text{H} & & \text{H} & & \text{H} \\ 3^\circ & & 2^\circ & & 1^\circ & & \text{Methyl} \end{array}$$

Functional Group. A grouping of atoms that effectively determines the properties of the compound (Section 2.8).

Halogenations of Alkanes. Substitution reactions in which a halogen replaces one (or more) of the alkane's hydrogen atoms (Section 3.17A).

$$RH + X_2 \longrightarrow RX + HX$$

The reactions occur by a free radical chain mechanism (Section 9.3).

Step 1 $X_2 \longrightarrow 2X\cdot$
Step 2 $RH + X\cdot \longrightarrow R\cdot + HX$
Step 3 $R\cdot + X_2 \longrightarrow RX + X\cdot$

Hammond-Lefler Postulate. A postulate which holds that the structure of the transition state of an endothermic step of a reaction resembles the products of that step more than it does the reactants. Conversely, the structure of the transition state of an exothermic step is more like the reactants than the products (Section 5.13A).

Heat of Combustion. The heat evolved on complete combustion of 1 mole of a substance at 25°C and 1 atm pressure. This heat, called ΔH°, is negative for exothermic reactions and positive for endothermic reactions (Section 6.9B).

Heat of Hydrogenation. The heat of reaction (ΔH°) for the addition of H_2 to 1 mole of a compound (Section 6.9A).

Heat of Reaction. The enthalpy change (ΔH°) for a chemical reaction equal to $H^\circ_{\text{products}} - H^\circ_{\text{reactants}}$. For an exothermic reaction ΔH° is negative; for an endothermic reaction ΔH° is positive (Section 9.2).

Heterocyclic Compound. A compound whose molecules have a ring containing an element other than carbon (Sections 11.10, 11.12).

Heterolysis. Cleavage of a covalent bond that leads to ions, that is, $A{:}B \longrightarrow A^+ + {:}B^-$ (Section 5.1A).

Homolysis. Cleavage of a covalent bond that leads to radicals, that is, $A{:}B \longrightarrow A\cdot + B\cdot$ (Section 5.1A).

Hückel's Rule. A rule that states that planar monocyclic conjugated rings with $(4n + 2)$-π electrons (i.e., with 2, 6, 10, 14, 18, or 22 π electrons) should be aromatic. Hückel's rule has an upper limit. Systems with more than 22 π electrons are not aromatic. (Sections 11.6, 11.12).

Hund's Rule. When we fill orbitals of equal energy (degenerate orbitals) such as the three *2p* orbitals, we add one electron to each orbital with their spins unpaired until each of the degenerate orbitals contains one electron. Then we begin adding a second electron to each degenerate orbital so that the spins are paired (Section 1.11).

Hybrid Orbitals. Orbitals such as sp^3, sp^2, and *sp* orbitals that are formed by mixing (hybridizing) the wave functions for orbitals of a different type (i.e., *s* orbitals and *p* orbitals) but from the same atom. (Sections 1.14 and 1.17).

Hydrocarbon. A compound whose molecules contain only carbon and hydrogen (Section 2.2).

Hydrogenation. Chemical addition of H_2 to an unsaturated compound (Section 3.18).

Hydrogen Bond. The relatively strong dipole-dipole attraction that occurs between a hydrogen that is bonded to a strongly electronegative atom and the nonbonding electron pairs on another electronegative atom (Section 2.17C).

X–H····· : Y (X and Y are strongly electronegative usually, O, N, or halogen)

Index of Hydrogen Deficiency. The number of pairs of hydrogen atoms that must be subtracted from the molecular formula of the corresponding alkane to give the molecular formula of the compound under consideration (Section 6.8).

Example: the index of hydrogen deficiency of C_5H_8 is *two* because it contains two pairs of hydrogen atoms less than the corresponding alkane, C_5H_{12}.

Ionic (or Electrovalent) Bond. A force of attraction between oppositely charged ions formed by the transfer of one or more electrons from one atom to another (Sections 1.6 and 1.6A).

Isomers are different compounds that have the same molecular formula. All isomers fall into either of two groups: *constitutional* isomers or *stereoisomers* (Sections 4.2, 4.17).

SUBDIVISION OF ISOMERS

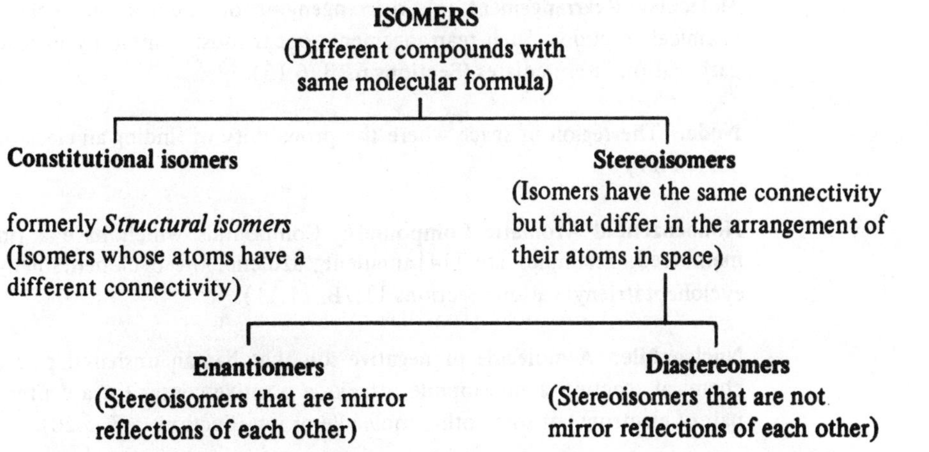

ISOMERS
(Different compounds with same molecular formula)

Constitutional isomers
formerly *Structural isomers*
(Isomers whose atoms have a different connectivity)

Stereoisomers
(Isomers have the same connectivity but that differ in the arrangement of their atoms in space)

Enantiomers
(Stereoisomers that are mirror reflections of each other)

Diastereomers
(Stereoisomers that are not mirror reflections of each other)

Leaving Group. The group that is displaced by a nucleophile in a substitution reaction (Section 5.3A).

Lewis Acid-Base Theory. An acid is an electron-pair acceptor. A base is an electron-pair donor (Section 2.16).

Markovnikov's Rule. In the addition of HX to an alkene, the hydrogen adds to the carbon atom of the double bond with the greater number of hydrogen atoms (Section 7.2).

Example: $CH_2{=}CHCH_3 + H{-}Cl \longrightarrow CH_3CHClCH_3$ (i.e., $CH_3\underset{|}{C}HCH_3$ with Cl on the central carbon)

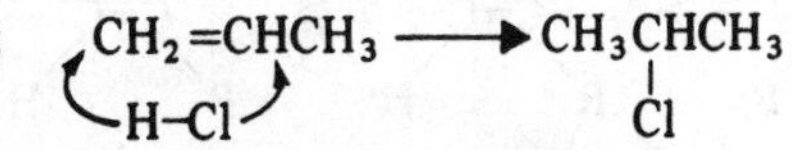

Markovnikov's rule may be stated in mechanistic terms: In the ionic addition of an unsymmetrical reagent to a double bond, the positive portion of the adding reagent attaches itself to a carbon atom of the double bond so as to yield the more stable carbocation (Section 7.2B).

The H^+ ion adds as it does because the carbocation $CH_3\overset{+}{C}HCH_3$ is secondary and is therefore more stable than $\overset{+}{C}H_2CH_2CH_2$ (1°), which would be formed if H^+ added to the central carbon atom.

Meso Compound. An optically inactive compound whose molecules are achiral even though they contain stereocenters (Sections 4.9A, 4.17).

Meta. A prefix used to designate 1,3-disubstituted benzenes (Section 11.8).

Molecular Formula. The formula that shows only the number of each kind of atom in the molecule (Section 1.2B).

Example: $C_2H_4O_2$.

Molecular Orbitals (MO). When atomic orbitals overlap, they combine to form molecular orbitals. Molecular orbitals correspond to regions of space encompassing two (or more) nuclei where electrons are to be found. Like atomic orbitals, molecular orbitals can hold up to two electrons if their spins are paired (Sections 1.12, 1.17).

Molecular Rearrangement. A rearrangement of the carbon skeleton during certain chemical reactions. Such rearrangements occur most commonly in reactions that involve carbocation intermediates (Sections 6.13, 6.15).

Node. The region in space where the probability of finding an electron is zero (Sections 1.11, 1.17)

Nonbenzenoid Aromatic Compounds. Compounds which have a ring that is not six membered. Examples are [14]annulene, azulene, the cyclopentadienyl anion, and the cycloheptatrienyl cation (Sections 11.7B, 11.11).

Nucleophile. A molecule or negative ion that has an unshared pair of electrons. In a chemical reaction a nucleophile attacks a positive center (or a center that can accept a pair of electrons) of some other molecule or ion (Sections 5.3, 5.20).

Nucleophilic Substitution Reaction (abbreviated as S_N reaction). A substitution reaction brought about when a nucleophile reacts with a *substrate* that bears a *leaving group* (Sections 5.3, 5.20).

Order of Alkene Stability. The stability of an alkene depends on the number of alkyl groups bonded to the $>C{=}C<$ group (Section 6.9).
Relative stabilities:

$$R_2C{=}CR_2 > R_2C{=}CHR > R_2C{=}CH_2 > (R)HC{=}CH(R)\ (trans) >$$

$$(R)HC{=}CH(R)\ (cis) > RHC{=}CH_2 > H_2C{=}CH_2$$

Ortho. A prefix used to designate 1,2-disubstituted benzenes (Section 11.8).

Para. A prefix used to designate 1,4-disubstituted benzenes (Section 11.8).

Pauli Exclusion Principle. A maximum of two electrons may be placed in each orbital but only when the spins of the electrons are paired (Section 1.11).

Peptide Linkage. The amide linkages ($-\overset{O}{\overset{\|}{C}}-NH-$) that join α amino acid residues in proteins (Section 23.5).

Plane of Symmetry. An imaginary plane that bisects a molecule in such a way that the two halves of the molecule are mirror reflections of each other. Any molecule that has a plane of symmetry will be achiral (Sections 4.4, 4.17).

Polar Covalent Bond. A covalent bond between two atoms of unequal electronegativity in which the atom with greater electronegativity draws the electron pair closer to it (Section 1.9).

Polarizability. The ability of electrons in an atom to respond to a changing electric field (Section 2.17D).

Protic Solvent. Solvents that have an —H bonded to an oxygen or nitrogen atom (or to another strongly electronegative atom). Common protic solvents are formic acid, $\overset{\overset{\displaystyle O}{\|}}{HC}OH$; formamide $\overset{\overset{\displaystyle O}{\|}}{HC}NH_2$; water, H_2O; alcohols, ROH; ammonia, NH_3, and ethylene glycol, $HOCH_2CH_2OH$ (Section 5.13C).

Racemic Modification or Racemic Form. An equimolar mixture of enantiomers (Section 4.7A).

Reaction Mechanism. A description of how a chemical reaction takes place. If the mechanism is multistep, it includes the steps involved and the *intermediates* that form (Section 5.1).

Reaction Rate. The rate at which reactants are converted to products in a chemical reaction. The rate of a reaction can be determined experimentally by measuring the rate at which reactants disappear from the mixture or the rate products form in the mixture (Section 5.7).

Regioselective Reaction. A reaction that yields only one (or a predominance of one) constitutional isomer when more than one structural isomer can potentially be produced (Section 7.2D).

Resolution. The separation of the enantiomers of a racemic form (Sections 4.14, 4.17).

Resonance Energy of an aromatic compound (sometimes called the **stabilization energy** or **delocalization** energy) is the difference in energy between the actual aromatic compound and that calculated for one of the hypothetical resonance structures, for example, a Kekulé structure) (Sections 11.4, 11.12).

Resonance Theory. Whenever a molecule or ion can be represented by two or more Lewis structures that differ only in the positions of the electrons, (a) none of these structures (called resonance structures) is satisfactory, and (b) the actual molecule or ion will be best represented by a hybrid of these structures (Section 1.8).

Ring Strain. A strain that gives certain cycloalkanes greater potential energy than others. The principal sources of ring strain are angle strain and torsional strain (Sections 3.8, 3.9, 3.19). (In larger rings ring strain may also arise from van der Waals repulsions across rings.)

Saturated Compound. A compound whose molecules contain only single bonds (Section 2.2).

S_N1 Reaction. A nucleophilic substitution reaction for which the rate-determining step is *unimolecular*. The hydrolysis of *tert*-butyl chloride is an S_N1 reaction that takes place in three steps as follows. The rate-determining step is step 1.

Step 1 $(CH_3)_3CCl \xrightarrow{\text{slow}} (CH_3)_3C^+ + Cl^-$

Step 2 $(CH_3)_3C^+ + H_2\ddot{O}: \xrightarrow{\text{fast}} (CH_3)_3COH_2^+$

Step 3 $(CH_3)_3COH_2^+ + H_2O \xrightarrow{\text{fast}} (CH_3)_3COH + H_3O^+$

S_N1 reactions are important with tertiary halides and with other substrates (benzylic and allylic) that can form relatively stable carbocations (Sections 5.10, 5.20).

S_N2 Reaction. A nucleophilic substitution reaction for which the rate-determining step is *bimolecular* (i.e., the transition state involves two species). The reaction of methyl chloride with hydroxide ion is an S_N2 reaction. According to the Ingold mechanism it takes place in a *single step* as follows (Sections 5.5, 5.6, 5.8, 5.20):

$$HO:^- + CH_3{-}Cl \longrightarrow \overset{\delta-}{HO}\cdots\underset{H\ \ H}{\overset{H}{C}}\cdots\overset{\delta-}{Cl} \longrightarrow HO{-}CH_3 + :Cl^-$$

Transition state

The order of reactivity of alkyl halides in S_N2 reactions is

$CH_3{-}X > RCH_2X > R_2CHX$

Methyl 1° 2°

Solvolysis. A nucleophilic substitution reaction in which the nucleophile is a molecule of the solvent (Sections 5.12B, 5.20).

Stereochemistry. Chemical studies that take into account the spatial aspects of molecules (Sections 4.2, 4.17).

Stereoisomers. Stereoisomers have their atoms joined in the same order but differ in the way their atoms are arranged in space. Stereoisomers can be subdivided into two categories: *enantiomers* and *diastereomers* (Sections 4.2, 4.17).

Stereoselective Reaction. One that yields exclusively (or predominantly) one of a set of stereoisomers (Sections 4.8A, 4.17).

Steric Effect. An effect on relative reaction rates caused by the space-filling properties of those parts of a molecule attached at or near the reacting site. *Steric hindrance* is an important steric effect in S_N2 reactions. It explains why methyl halides are most reactive and tertiary halides are least reactive (Sections 5.13A, 5.20).

Substitution Reaction. A reaction in which one group replaces another (Section 5.2).

Syn and Anti Addition. Addition of both parts of the adding reagent to the same face of the molecule (syn addition), or one part to each of opposite faces (anti addition) (Sections 7.7, 7.10A).

Torsional Strain. This refers to a small barrier to free rotation about the carbon-carbon single bond. For ethane this barrier is 2.8 kcal/mole (11.7 kJ/mole). (Sections 3.6 and 3.7A, 3.19).

Unbranched Alkane. Alkane in which each carbon atom is bonded to no more than two other carbon atoms (Section 3.2).

Unsaturated Compound. A compound whose molecules contain multiple bonds (Sections 2.2 and 6.5).

Van der Waals Forces. Weakly attractive forces between nonpolar molecules or between parts of the same molecule. These forces are caused by temporary dipoles in one molecule being induced by similar temporary dipoles in surrounding molecules (Sections 2.17D, and 3.17). If the molecules or parts of the molecule are too close together then the van der Waals forces become repulsive.

VSEPR (Valence Shell Electron Pair Repulsion) Model. Within the confines of the molecule, electron pairs of the valence shell tend to stay as far apart as possible (Section 1.18).

D APPENDIX
Answers to Self-Tests

CHAPTER 1

1.1 (a) $:\overset{..}{\underset{-}{N}}::\underset{+}{N}::\overset{..}{O}:$ (b) $:N:::\overset{+}{N}:\overset{..}{\underset{..}{O}}:^{-}$

1.2 (a) sp^2 (planar) (b) sp^3 (tetrahedral) (c) sp^3 (tetrahedral)

1.3 (a) 2 (b) Decrease

1.4 $H:C:::N:$ $H:\overset{+}{C}::\overset{..}{N}:^{-}$

1.5 $H:\overset{..}{N}::C::\overset{..}{O}:$

1.6 (a) sp^2 (b) sp^3 (c) sp^3

1.7

```
    H H H        H H H         H H H         H Cl H
    | | |        | | |         | | |         | |  |
  H-C-C-C-Cl   H-C-C-C-Cl   Cl-C-C-C-Cl    H-C-C--C-H
    | | |        | | |         | | |         | |  |
    H H Cl       H Cl H        H H H         H Cl H
```

1.8 (a) $\underline{1s}$ ⥮ $\underline{2s}$ ⥮ $\underline{2p_x}$ ↓ $\underline{2p_y}$ $2p_z$ (b) BF_3 (c) $\overset{\delta+\ \delta-}{B\text{–}F}$ (d) sp^2

1.9 (d) **1.10** (d) **1.11** (e) **1.12** (d) **1.13** (c)

CHAPTER 2

2.1 (a) $(H)_2C{=}C(Br)_2$ (b) $CH_3\overset{O}{\overset{\|}{C}}\text{–}OH$ (c) [skeletal structure of 2-butanol, OH] (d) –N– [skeletal structure]

(e) $CH_3\overset{O}{\overset{\|}{C}}\text{–}OCH_3$

2.2 (a) 3° (b) 2° (c) 1° (d) 2° (e) 3° (f) 1°

2.3 Alcohol hydroxyl, amide, alkene, ether

2.4 HNO_3

2.5 $H_3O^+ + CH_3\overset{O}{\overset{\|}{C}}\text{–}O^- \rightleftharpoons CH_3\overset{O}{\overset{\|}{C}}\text{–}OH + H_2O$

2.6 $HA + H_2O \rightleftharpoons H_3O^+ + A^-$

2.7 (a) 2.8 (e) 2.9 (a) 2.10 (b) 2.11 (c)

CHAPTER 3

3.1 2,4,5-Trimethylheptane

3.2

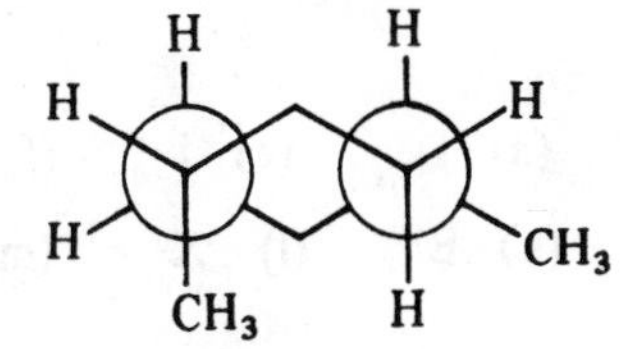

3.3 (a) 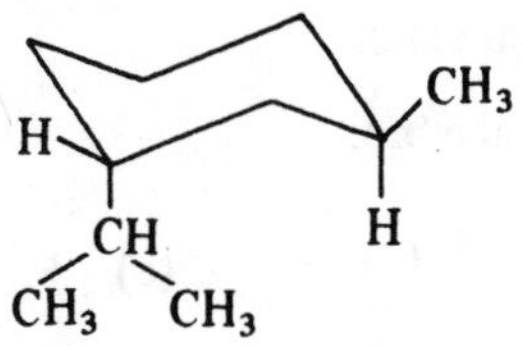 (b) I

3.4 (a) (b) 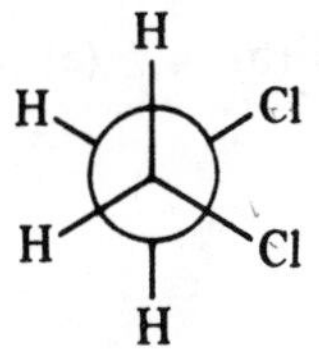 (c)

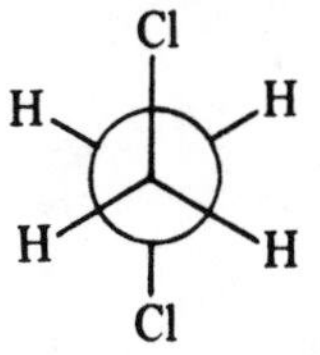

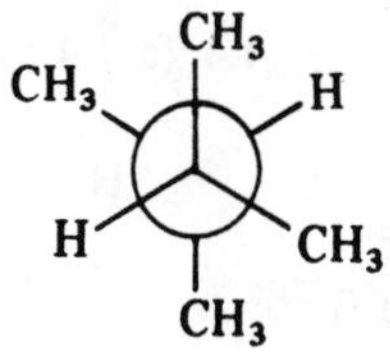

3.5 (a) 3-Methylhexane (b) 3,3-Dimethylheptane

3.6

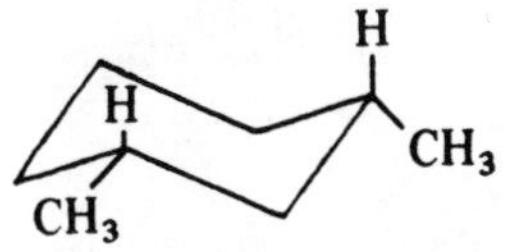

3.7 (a) cis (b) 2 (c)

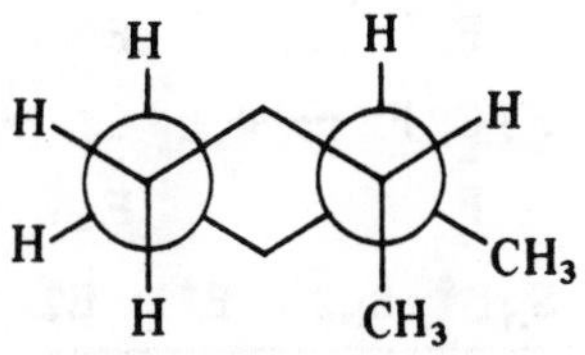

3.8 6-Cyclobutyl-2,3,5-trimethyloctane

3.9 (a) $CH_3CHCl_2 + ClCH_2CH_2Cl$

(b) H_2/Ni

(c) $CH_3CH_2CH_2CH_3$

(d)

3.10 (c) **3.11** (c) **3.12** (b) **3.13** (a) **3.14** (b) **3.15** (a) **3.16** (a)

CHAPTER 4

4.1 (a) X (b) I (c) E (d) D (e) I (f) S (g) I (h) E (i) I (j) S (k) E (l) X (m) D

4.2 (a) *cis*-Dimethylcyclopentane; no.
(b) *cis*-1,2-Dihydroxycyclohexane; no.
(c) *trans*-1,2-Dibromocyclobutane; no.

4.3 (a) - (b) - (c) - (d) - (e) + (f) + (g) - (h) + (i) + (j) + (k) -

4.4 (a) **4.5** (b) **4.6** (b) **4.7** (e) **4.8** (b)

CHAPTER 5

5.1 (a) $F_3C{:}\ddot{F}{:} \longrightarrow F_3C\cdot + \cdot\ddot{F}{:}$

(b) $F_3C{:}\ddot{F}{:} \longrightarrow F_3C^+ + {:}\ddot{F}{:}^-$

(c) $F_3C{:}\ddot{F}{:} \longrightarrow F_3C{:}^- + \ddot{F}{:}^+$

5.2

	S_N1	S_N2	E1	E2
(a)	+	-	+	-
(b)	-	+	-	+
(c)	-	+	-	+
(d)	+	-	+	-
(e)	-	+	-	+
(f)	+	+	+	+

5.3 (a) 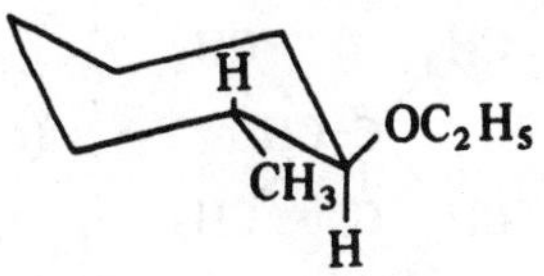 [chair cyclohexane structure with H, CH_3, OC_2H_5, H]

(b)

EXPERIMENTAL NUMBER	INITIAL RATE
2	0.04
3	0.02

(c) 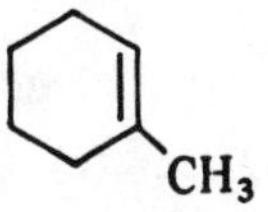 [1-methylcyclohexene structure with CH_3]

(d) Elimination

5.4 (a) + (b) - (c) - (d) + 5.5 (b) 5.6 (b) 5.7 (a)

CHAPTER 6

6.1 (a) *trans*-2-Hexene (b) 3-Methylcyclohexene

6.2 (a) Zn (b) 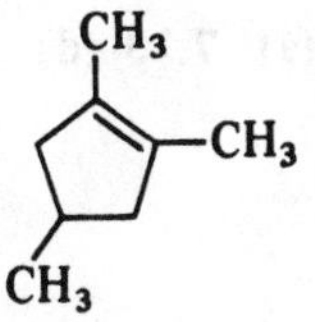 [cyclopentene structure with three CH_3 groups] (c) $CH_3\overset{CH_3}{\overset{|}{C}}HCH_2CH_2Br$

(d) $CH_3\overset{CH_3}{\overset{|}{C}}{=}CHCH_2CH_3$

6.3 (b) (e) (f) (g)

6.4 1, 3, 2

6.5 [cyclopentene structure] 6.6 (c) 6.7 (d) 6.8 (a)

CHAPTER 7

7.1 (a) $CH_3\overset{CH_3}{\overset{|}{C}}(OH){-}\overset{CH_3}{\overset{|}{C}}HCH_3$ (b) (1) HBr, peroxides (2) OH^-/H_2O

(c) 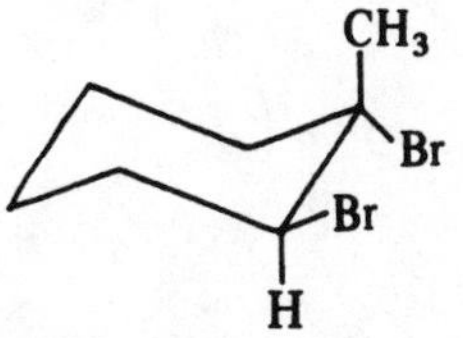 [chair cyclohexane structure with CH_3, Br, Br, H]

(d) Cl_2/H_2O (e) (1) $H\overset{O}{\overset{\|}{C}}{-}O{-}OH$ (2) H_3O^+ (f) $KMnO_4/25°C$

7.2 (a) 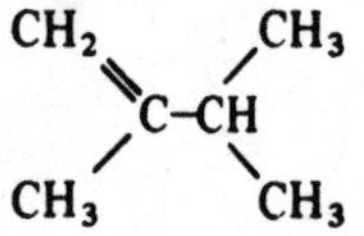 $CH_2{=}C(CH_3){-}CH(CH_3)_2$ (b) $(CH_3)_2C(Cl){-}CH(CH_3)_2$ (c) $(CH_3)_2C{=}C(CH_3)_2$

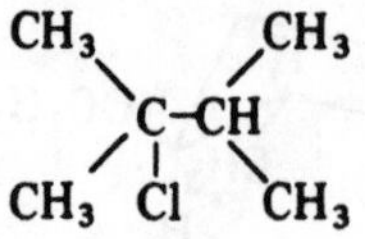

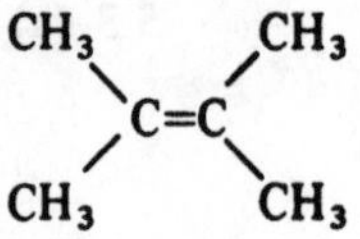

(d) $CH_3\overset{O}{\overset{\|}{C}}CH_3$ (e) $H\overset{O}{\overset{\|}{C}}H$ (f) $CH_3\overset{O}{\overset{\|}{C}}CH(CH_3)_2$

7.3 (a) 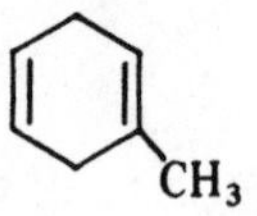 1-methylcyclohexene ring, CH_3 (b) 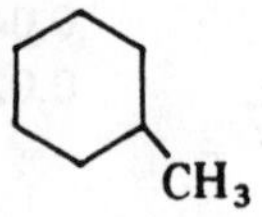 cyclohexane ring, CH_3

7.4 (a) $CH_3CH_2\underset{I}{\underset{|}{C}}{=}CH_2$ (b) H_2/Ni_2B (P-2) (c) 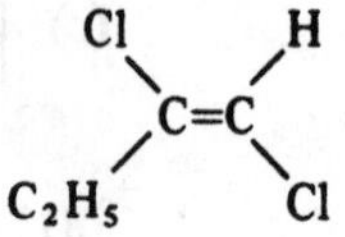 $(Cl)(C_2H_5)C{=}C(H)(Cl)$

(d) $Li/C_2H_5NH_2$ (e) $CH_3CCl_2CH_3$ (f) $CH_3\overset{O}{\overset{\|}{C}}CH_2CH_3$

7.5 (a) 0 (b) +

7.6 (e) 7.7 (c) 7.8 (e) 7.9 (a) 7.10 (d) 7.11 (c) 7.12 (d) 7.13 (b)

CHAPTER 8

8.1 (a) 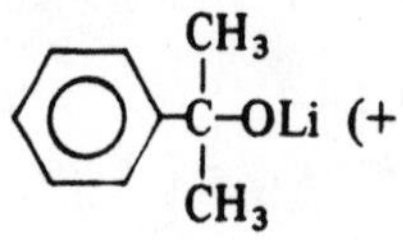 $C_6H_5{-}C(CH_3)_2{-}OLi$ $(+\,H_2)$ (b) 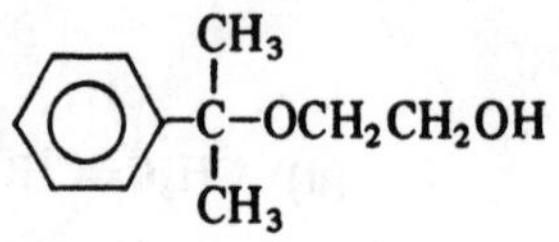 $C_6H_5{-}C(CH_3)_2{-}OCH_2CH_2OH$

(c) $C_6H_5{-}C(CH_3)_2{-}OH$ (d) $2\ CH_3\overset{CH_3}{\overset{|}{C}}HBr$

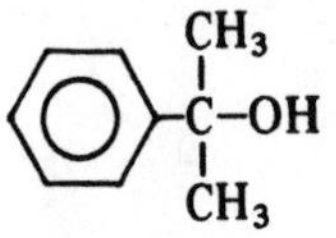

(e) N.R.; that is, alcohol is insoluble (f) 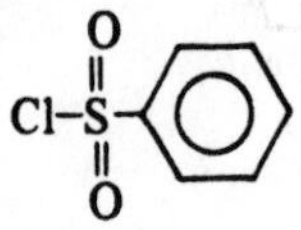 $Cl{-}\underset{O}{\underset{\|}{\overset{O}{\overset{\|}{S}}}}{-}C_6H_5$

(g) $CH_4 + CH_3CH_2CH_2O^-$

8.2 (a) B (b) A

8.3 $CH_3CH_2\overset{+}{O}H_2 + Br^-$

8.4 (d) 8.5 (a)

CHAPTER 9

9.1. (a) $CH_3C(CH_3)(Br)CH_3$ (b) CH_3CHCl_2 + $ClCH_2CH_2Cl$ (c) 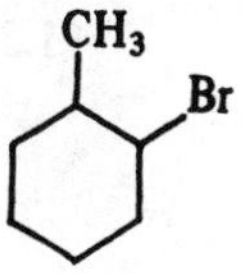 1-bromo-2-methylcyclohexane (structure: cyclohexane ring with CH_3 and Br on adjacent carbons)

9.2 (a) -31.5 (b) +11.5 (c) -69

9.3 $CH_3CH_2CH_2CH_2Br$ and $CH_3CH_2CH(Br)CH_3$

9.4 (a) $CH_3\dot{C}H_2$ and $\dot{C}H_3$ (b) +85 (c) $CH_3CH(Br)CH_3$ (d) -15

9.5 (a) + 10 kcal/mole butane (b) No

9.6 (d) 9.7 (b) 9.8 (c) 9.9 (b)

CHAPTER 10

10.1 (a) 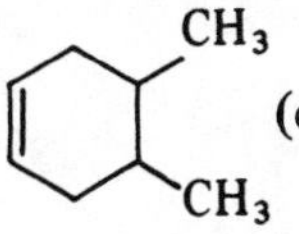 cyclohexene ring with CH_3 and CH_3 on adjacent carbons (cis) (b) $CH_3CH_2CH{=}CHCH(Cl)CH_3$

(c) $CH_3CH(Cl)CH(Cl)\dot{C}H{=}CHCH_3$ + $CH_3CH(Cl)CH{=}CHCH(Cl)CH_3$

(d) 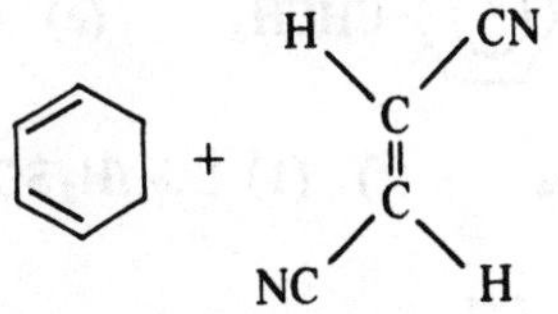 1,3-cyclohexadiene + $(H)(NC)C{=}C(CN)(H)$ (e) $CH_3CH{=}CHCH_2Br$ (f)  3-bromocyclopentene (cyclopentene ring with Br)

10.2 (a) 0 (b) + (c) 0 (d) +

10.3 B 10.4 (a) 10.5 (d) 10.6 (c) 10.7 (c) 10.8 (c) 10.9 (b)

CHAPTER 11

11.1 (a) D, F (b) B, C, D, E, F (c)  cyclopropenyl cation $^+$ ⟷ $^+$ cyclopropenyl cation (resonance structures)

11.2 (a) *o*-Xylene (b) Ethylbenzene (c) 4-Bromo-1-isopropylbenzene

(d) 2-Chloro-1,3-diphenylpropane. (e) 3,4,5-Trinitrotoluene

11.3 (a) [cycloheptatrienyl cation structure] (b) [cyclohexadiene structure] (and others)

(c) [cyclopentadienyl anion structure] (d) [cyclopentadienyl cation structure]

11.4 (a) 3 (b) [1,2,4-trimethylbenzene structure: CH_3, CH_3, CH_3] (c) 1,2,4-Trimethylbenzene

11.5 (e) **11.6** (a) **11.7** (b) **11.8** (b) **11.9** (a)

CHAPTER 12

12.1 (a) [ring with $NHCOCH_3$, NO_2 ortho, CH_3 para] (+ some ortho) (b) $CH_3\overset{O}{\overset{\|}{C}}$–Cl/$AlCl_3$ (c) HSO_3–C_6H_4–NH–$\overset{O}{\overset{\|}{C}}$–$C_6H_5$

(d) $C_6H_5\overset{Br}{\overset{|}{C}}HCH_3$ (e) (1) $KMnO_4/OH^-$/heat (2) H_3O^+ (3) HNO_3/H_2SO_4 (f) (1) SO_3/H_2SO_4 (2) HNO_3/H_2SO_4/heat (3) H_3O^+/heat (g) [m-bromobenzonitrile: ring with CN and Br meta]

(h) [three resonance structures of the arenium ion with CN, H, Br] ⟷ ⟷

(i) CH_3O–C_6H_4–SO_3H (+ ortho) (j) CH_3–C_6H_4–$\overset{O}{\overset{\|}{C}}CH_3$ (+ ortho)

(k) 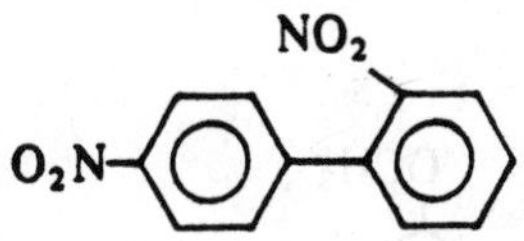 $O_2N-C_6H_4-C_6H_4-NO_2$ (2-nitro) (+ ortho) (l) 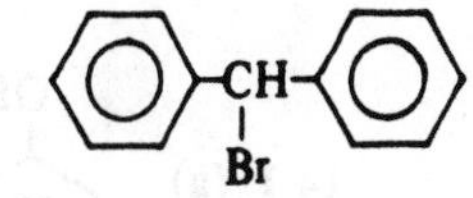 $C_6H_5-\underset{Br}{CH}-C_6H_5$

12.2 (cyclohexadienyl cation: CH_3, H, Br, +)

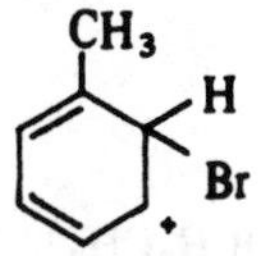

12.3 (cyclohexadienyl cation, H, H, H, +) ⟷ (H, H, H, +)

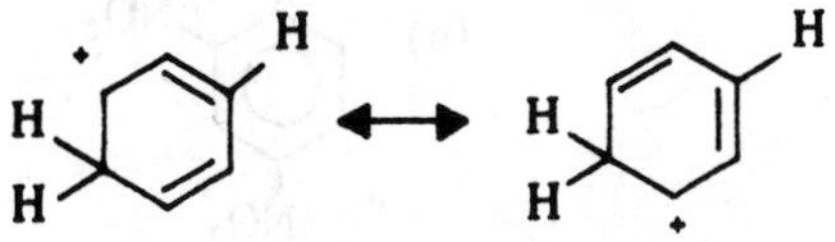

12.4 A $C_6H_5-CH_2CH{=}CH_2$ B $C_6H_5-CH_2\overset{Cl}{C}HCH_3$

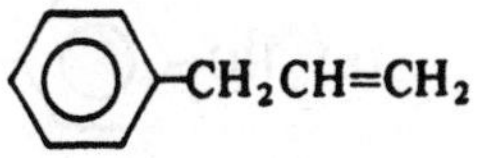

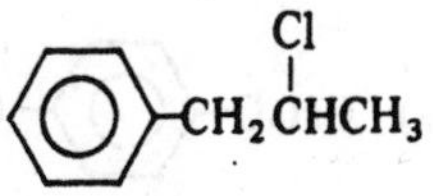

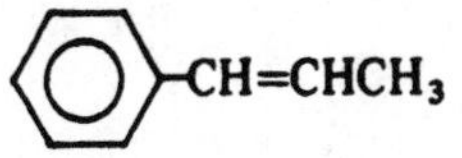

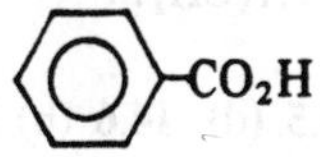

C $C_6H_5-CH{=}CHCH_3$ D $C_6H_5-CO_2H$

12.5 $C_6H_6 \xrightarrow[AlCl_3]{CH_3CH_2\overset{O}{\overset{\|}{C}}-Cl} C_6H_5\overset{O}{\overset{\|}{C}}CH_2CH_3 \xrightarrow[HCl]{Zn(Hg)} C_6H_5CH_2CH_2CH_3$

$C_6H_5CH_2CH_2CH_3 \xrightarrow{SO_3,\ H_2SO_4}$ $HSO_3-C_6H_4-CH_2CH_2CH_3$ + ortho (ortho $\xrightarrow{H_3O^+,\ heat}$ back to $C_6H_5CH_2CH_2CH_3$)

$HSO_3-C_6H_4-CH_2CH_2CH_3 \xrightarrow[H_2SO_4]{HNO_3}$ $HSO_3-C_6H_3(NO_2)-CH_2CH_2CH_3 \xrightarrow[heat]{H_3O^+}$ $C_6H_4(NO_2)CH_2CH_2CH_3$ (ortho)

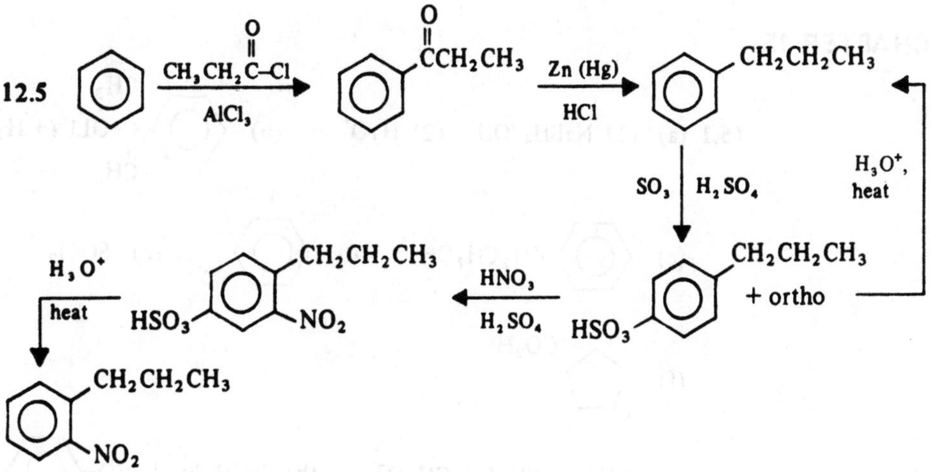

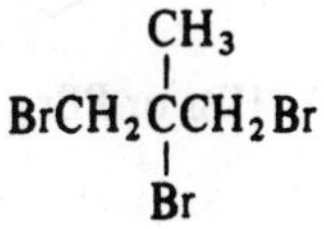

12.6 (a) 12.7 (a) 12.8 (b)

CHAPTER 13

13.1 (a) 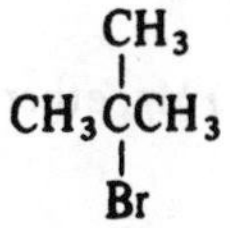 $CH_3\overset{CH_3}{\underset{Br}{C}}CH_3$ (b) $BrCH_2\overset{CH_3}{\underset{Br}{C}}CH_2Br$ (c) $CH_2{=}\underset{CH_3}{C}CH_2\overset{CH_3}{\underset{CH_3}{C}}CH_3$

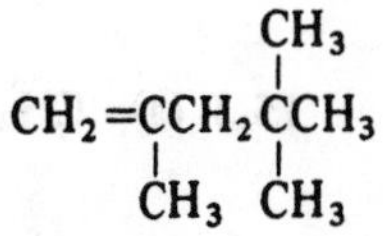

(d) 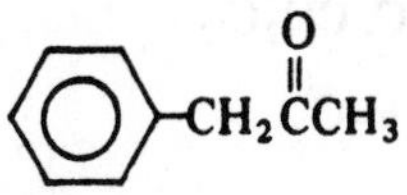 $C_6H_5-CH_2\overset{O}{\overset{\|}{C}}CH_3$ (e) $CH_3CH_2C{\equiv}CCH_2NO_2$

13.2 (c) 13.3 (a)

CHAPTER 14

14.1 (a) 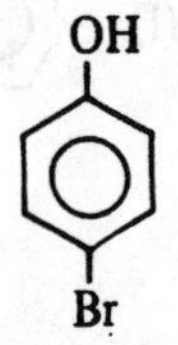 OH, Br (b) 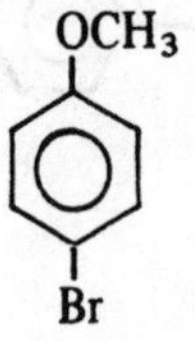 OCH_3, Br (c) 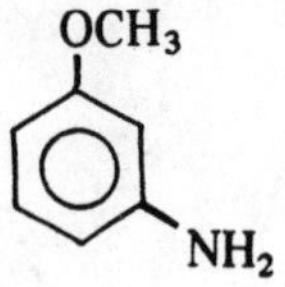 OCH_3, NH_2

14.2 (a) 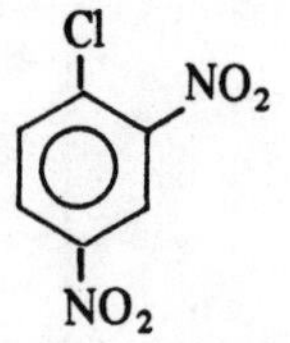 Cl, NO_2, NO_2 (b) 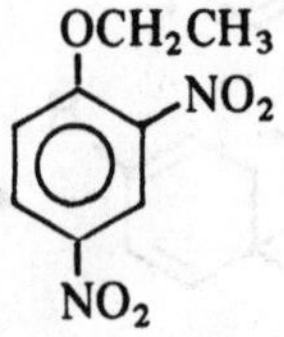 OCH_2CH_3, NO_2, NO_2

14.3 (a) 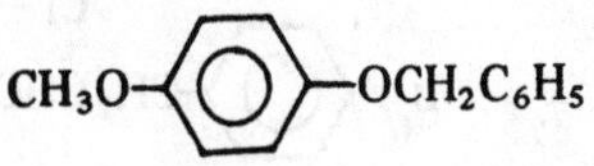 CH_3O–⟨benzene⟩–$OCH_2C_6H_5$ (b) 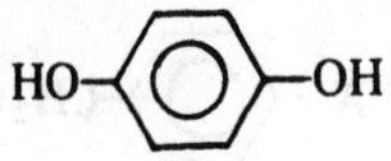 HO–⟨benzene⟩–OH (c) CH_3Br

(d) $C_6H_5CH_2Br$

14.4 (a) 14.5 (d) 14.6 (b) 14.7 (e)

CHAPTER 15

15.1 (a) (1) $NaBH_4/OH^-$ (2) H_3O^+ (b) 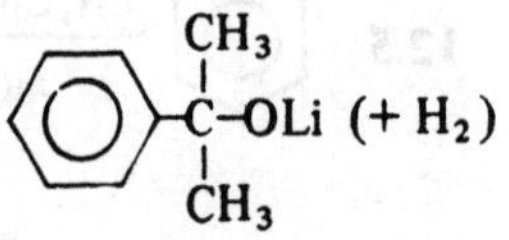 C_6H_5–C(CH_3)(CH_3)–OLi (+ H_2)

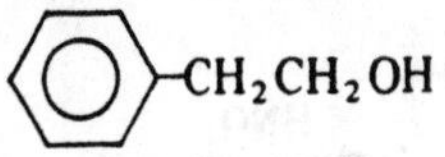

(c) C_6H_5–CH_2CH_2OH (d)  ⟨benzene⟩ (e) $SOCl_2$

(f) 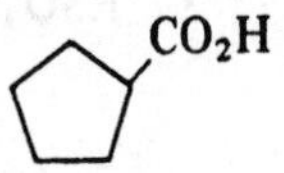 ⟨cyclopentane⟩–CO_2H

(g) CH_4 + $CH_3CH_2CH_2O^-$ (h) 2 CH_3Br + HO–⟨benzene⟩–OH

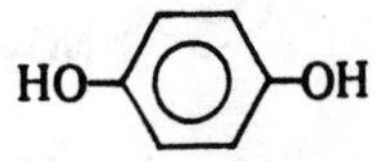

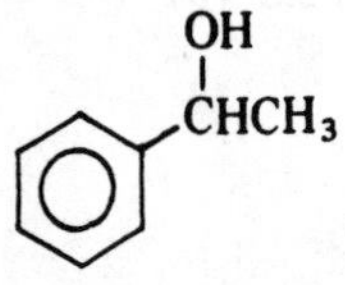

15.2 (a) 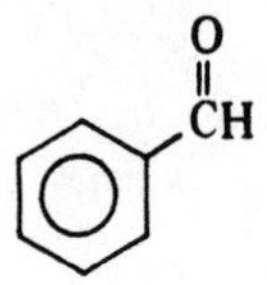 C_6H_5–CH(=O) (b) C_6H_5–CH(OH)CH_3 (c) C_6H_5–C(=O)–CH_3

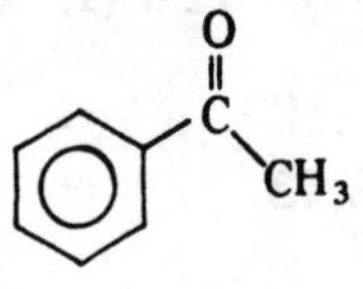

15.3 (a) CH_3MgI (b) HBr or PBr_3 (c) (1) Mg, ether

(d) (2) HCH(=O) (3) H_3O^+ (e) PCC, CH_2Cl_2

(f) (1) HC≡CNa, ether (2) H_3O^+, H_2O

15.4 (b) 15.5 (a)

CHAPTER 16

16.1 5-Hydroxy-3-methylhextanal

16.2 (d)

16.3 (a) $Ag(NH_3)_2OH$ (b) $Ag(NH_3)_2OH$

16.4 (a) $CH_2{=}P(C_6H_5)_3$ (b) HCN (c) $LiAlH[OC(CH_3)_3]_3$

(d) $CH_3CH(OCH_3)_2$ (e) $C_6H_5C(CH_3)_2OH$

(f) $CH_3CH_2COC_6H_5$ (C=O) (g) (1) $BrCH_2CO_2CH_3$, Zn (2) H_3O^+

16.5 (a) $HC(=O)C_6H_4C(=O)OH$ —(1) $CH_3OH/HCl(g)$ (2) neutralize→ $(CH_3O)_2CH{-}C_6H_4{-}C(=O)OH$ —(1) $LiAlH_4$ (2) H_3O^+→ $HC(=O){-}C_6H_4{-}CH_2OH$

(b) cyclopentene —HBr→ cyclopentyl-Br —(1) $P(C_6H_5)_3$ (2) base→ cyclopentylidene$=P(C_6H_5)_3$ —cyclopentanone→ cyclopentylidenecyclopentane

cyclopentene —H_3O^+→ cyclopentyl-OH —H_2CrO_4→ cyclopentanone

16.6 (d) **16.7** (b) **16.8** (e)

CHAPTER 17

17.1 (a) $CH_3CH_2CH(OH)CH(CH_3)CHO$ (b) $CH_3CH_2CH(OH)CH(CH_3)CH_2OH$ (c) $CH_3CH_2CH{=}C(CH_3)CHO$

(d) $LiAlH_4$ (e) H_2, Ni (f) $CH_3OH_{(excess)}$, H^+

(g) $CH_3CH_2CH_2CH(CH_3)CH(OCH_3)_2$ (h) $CH_3CH_2CH_2CH(CH_3)CHO$

(i) (1) $CH_3CHBrCO_2CH_2CH_3$, Zn (2) H_3O^+

17.2 (a) $CH_3\overset{O}{\overset{\|}{C}}C_6H_5$ (b) $C_6H_5CH{=}CH\overset{O}{\overset{\|}{C}}C_6H_5$ (c) CN^-, CH_3CO_2H, CH_3CH_2OH

17.3 (a) 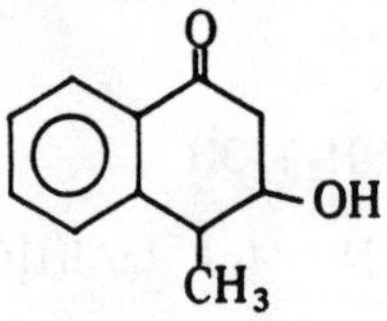 O, OH, CH_3 (b) 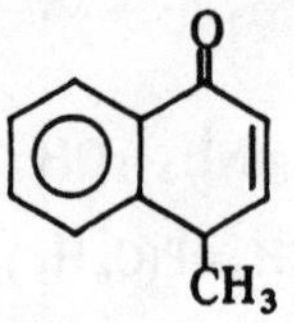 O, CH_3

(c) $(CH_3)_2CuLi$ (d) 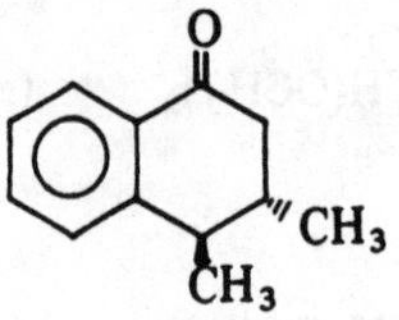 O, CH_3, CH_3

(e) Zn(Hg)/HCl

17.4 (e) 17.5 (a)

CHAPTER 18

18.1 (a) 4-Nitrobenzoic acid, (b) 3-Chlorobenzoic acid (c) 3-Chlorobutanoic acid

18.2 (a) D (b) B (c) D

18.3 (a) 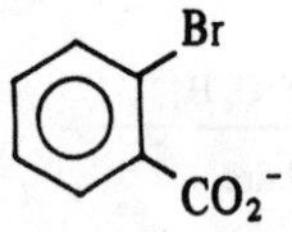 Br, CO_2^- (b) $C_2H_5{-}\underset{CH_3}{\overset{CH_3}{C}}{-}Br$ (c) $CH_3CH_2CH_2\overset{O}{\overset{\|}{C}}{-}Cl$

(d) $SOCl_2$ (e) $LiAlH_4$ (f) 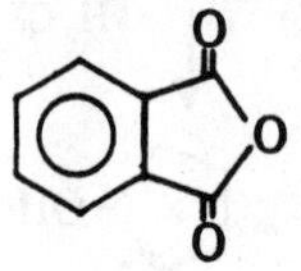 O, O, O (g) $CH_3CH_2CH_2\overset{O}{\overset{\|}{C}}{-}NH_2$

$+\ CH_3CH_2OH$ (h) 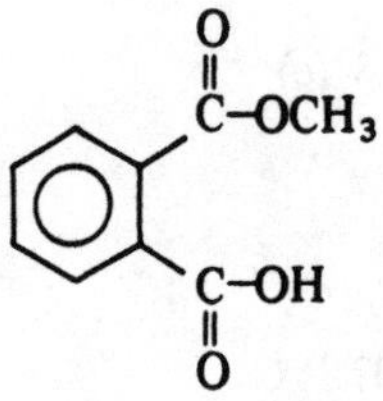 $\overset{O}{\overset{\|}{C}}{-}OCH_3$, $\underset{O}{\underset{\|}{C}}{-}OH$ (i) 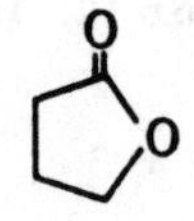 O, O

(j) (1) Mg/ether (2) CO_2 (3) H_3O^+ (k) 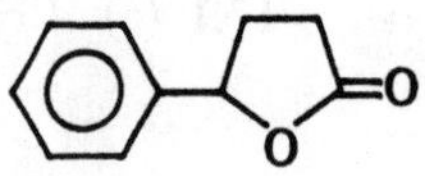 O, O

(l) 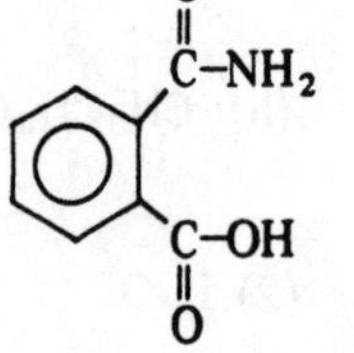 $\overset{O}{\overset{\|}{C}}{-}NH_2$, $\underset{O}{\underset{\|}{C}}{-}OH$ (m) 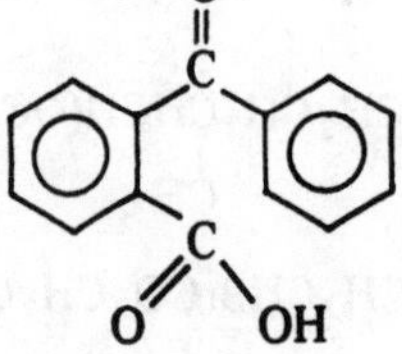 O, C, C, O, OH (n) 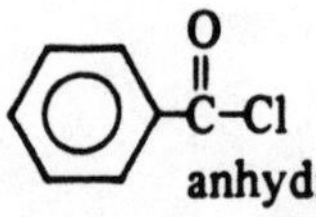 $\overset{O}{\overset{\|}{C}}{-}Cl$ (or anhydride)

18.4 (a) Aqueous $NaHCO_3$ (b) Aqueous $NaHCO_3$

18.5 (b) **18.6** (b) **18.7** (d)

CHAPTER 19

19.1 (a) 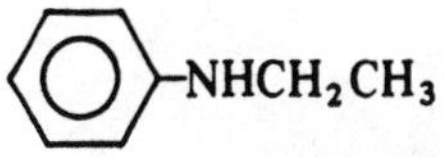 C_6H_5–$NHCH_2CH_3$ (b) C_6H_5–CH_2NH_2

(c) CH_3O–C_6H_4–NH_2 (d)  cyclohexane–NH_2

19.2 (a) 1 (b) 4 (c) 3 (d) 2

19.3 (a) H_2/Ni (b) $NaNO_2/HCl$ (c) 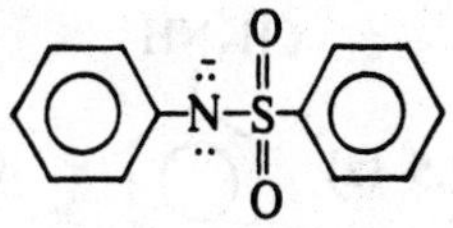 C_6H_5–$\ddot{N}$–SO_2–C_6H_5 (d) CuCN

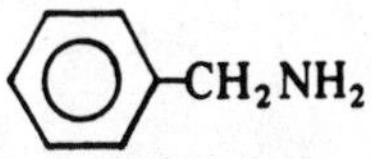

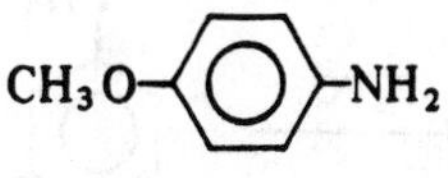

(e) 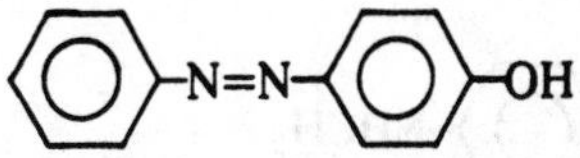 C_6H_5–N=N–C_6H_4–OH (f) 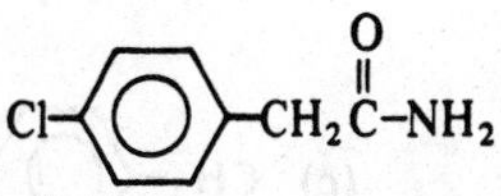 Cl–C_6H_4–$CH_2\overset{O}{\overset{\|}{C}}$–$NH_2$

(g) H_3PO_2 (h) H_2O/heat (i) $Br_2/NaOH$ (j) $LiAlH_4$

(k) 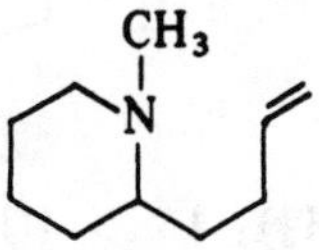 N-methyl-2-(but-3-enyl)piperidine (CH_3 on N) (l) 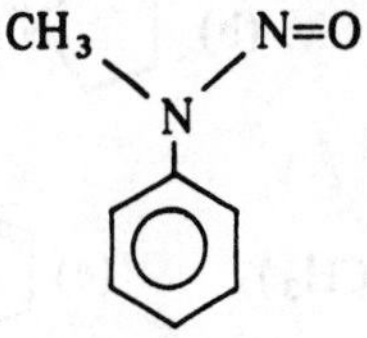 $C_6H_5N(CH_3)$–N=O (m) 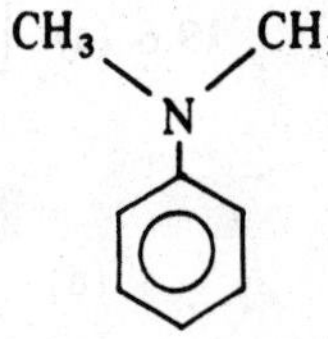 $C_6H_5N(CH_3)CH_3$

(n) N.R. [same as (m)] (o) 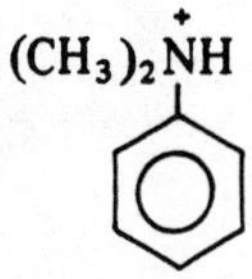 $(CH_3)_2\overset{+}{N}H$–C_6H_5

19.4 (a) 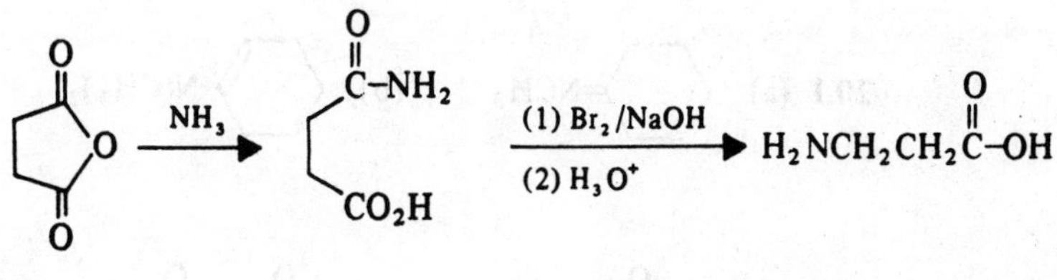 succinic anhydride $\xrightarrow{NH_3}$ $H_2N\overset{O}{\overset{\|}{C}}CH_2CH_2CO_2H$ $\xrightarrow[(2)\ H_3O^+]{(1)\ Br_2/NaOH}$ $H_2NCH_2CH_2\overset{O}{\overset{\|}{C}}$–OH

(b) 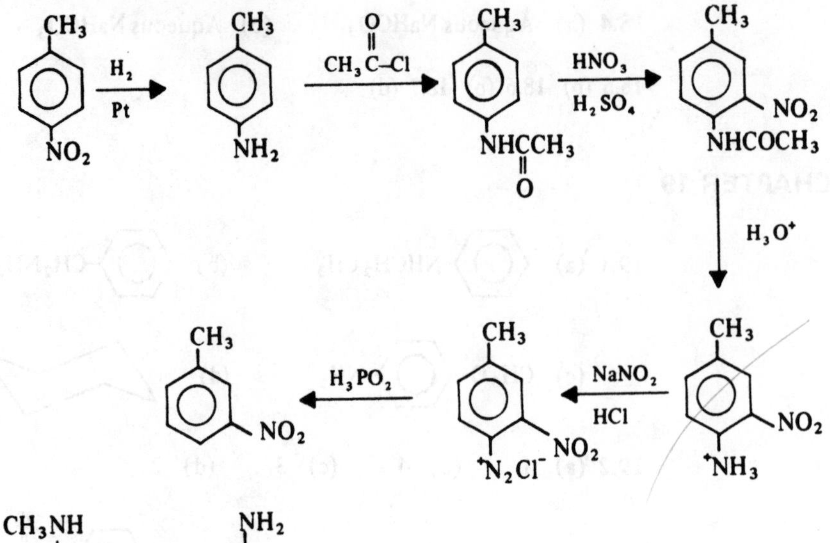 p-nitrotoluene (CH_3, NO_2) $\xrightarrow[Pt]{H_2}$ p-toluidine (CH_3, NH_2) $\xrightarrow{CH_3\overset{O}{\overset{\|}{C}}-Cl}$ (CH_3, $NH\overset{O}{\overset{\|}{C}}CH_3$) $\xrightarrow[H_2SO_4]{HNO_3}$ (CH_3, NO_2, $NHCOCH_3$) $\xrightarrow{H_3O^+}$ (CH_3, NO_2, $^+NH_3$) $\xrightarrow[HCl]{NaNO_2}$ (CH_3, NO_2, $^+N_2Cl^-$) $\xrightarrow{H_3PO_2}$ m-nitrotoluene (CH_3, NO_2)

19.5 (a)  $C_6H_5NHCH_3$ (b)  p-$CH_3C_6H_4NH_2$ (or *o* or *m*),

(c) 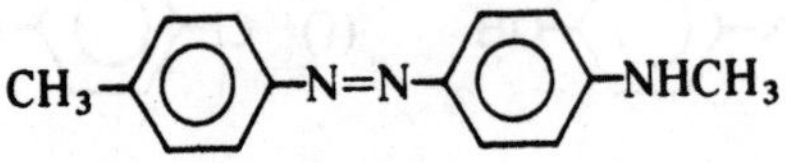 $CH_3-C_6H_4-N{=}N-C_6H_4-NHCH_3$

19.6 (a) 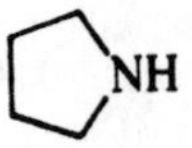 pyrrolidine (ring NH) (b) 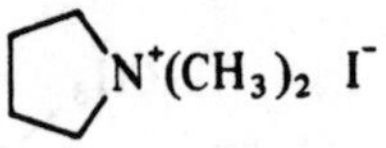 ring $N^+(CH_3)_2$ I^- (c) 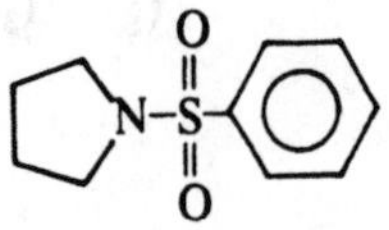 ring $N-\overset{O}{\overset{\|}{\underset{O}{\underset{\|}{S}}}}-C_6H_5$

(d) 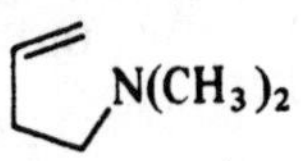 $CH_2{=}CHCH_2CH_2N(CH_3)_2$ (e) 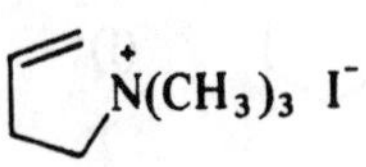 $CH_2{=}CHCH_2CH_2\overset{+}{N}(CH_3)_3$ I^- (f) $CH_2{=}CH-CH{=}CH_2$ + $(CH_3)_3N$

19.7 (d) 19.8 (e)

CHAPTER 20

20.1 (a) 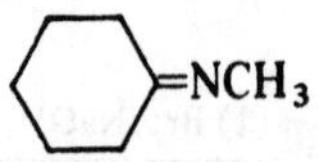 cyclohexylidene $=NCH_3$ (b) 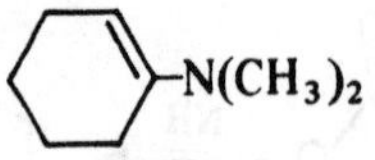 cyclohexenyl $-N(CH_3)_2$ (c) 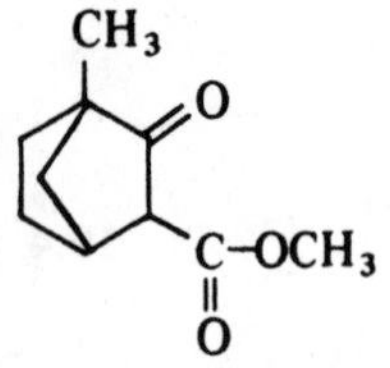 bicyclic ketone with CH_3, $=O$, and $\overset{}{C}(=O)-OCH_3$

(d) 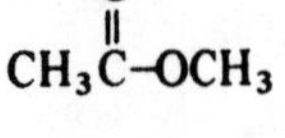 $CH_3\overset{O}{\overset{\|}{C}}-OCH_3$ (e) 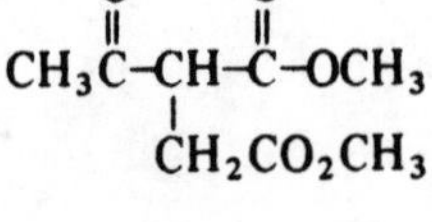 $CH_3\overset{O}{\overset{\|}{C}}-\underset{CH_2CO_2CH_3}{\underset{|}{CH}}-\overset{O}{\overset{\|}{C}}-OCH_3$

(f) $CH_3\overset{O}{\overset{\|}{C}}CH_2CH_2\overset{O}{\overset{\|}{C}}-OCH_3$ (g) 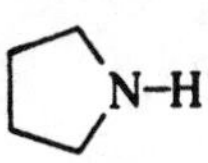 pyrrolidine (N–H) (h) 1-(1-cyclopentenyl)pyrrolidine

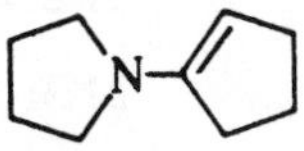

(i) 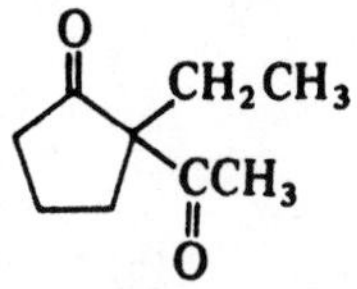 2-ethyl-2-acetylcyclopentanone (CH_2CH_3, $\overset{\|}{C}CH_3$ with O) (j) 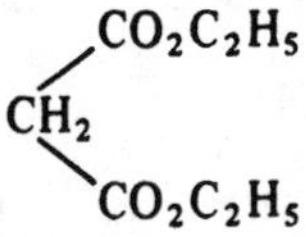 $CH_2(CO_2C_2H_5)_2$

20.2 (a) E, $CH_2(CO_2C_2H_5)_2$ + $C_6H_5CH_2Cl$

(b) A, $CH_3\overset{O}{\overset{\|}{C}}CH_2CO_2C_2H_5$ + $C_6H_5CH_2Cl$

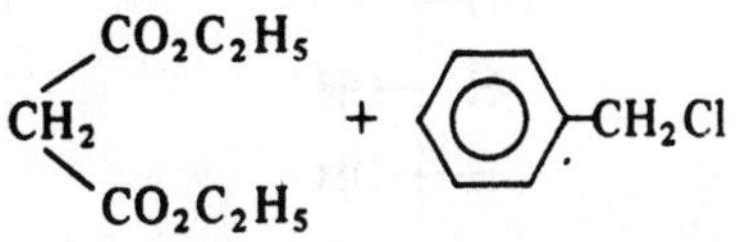

(c) N, 1-(1-cyclopentenyl)pyrrolidine + $BrCH_2\overset{O}{\overset{\|}{C}}CH_3$

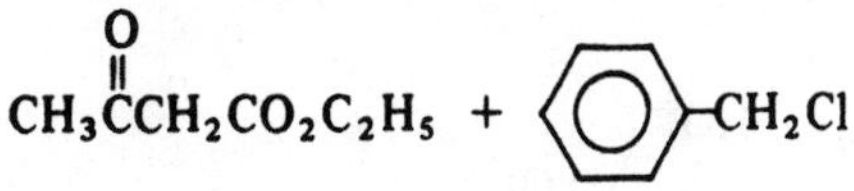

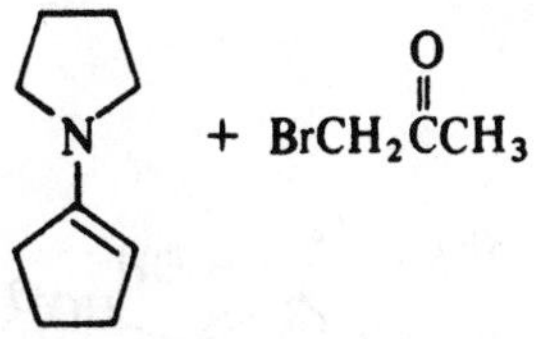

20.3 (c) 20.4 (e) 20.5 (b)

CHAPTER 21

21.1 (a) 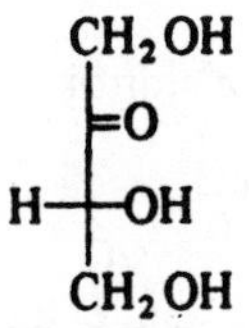 CH_2OH / $=O$ / $H-C-OH$ / CH_2OH

(b) 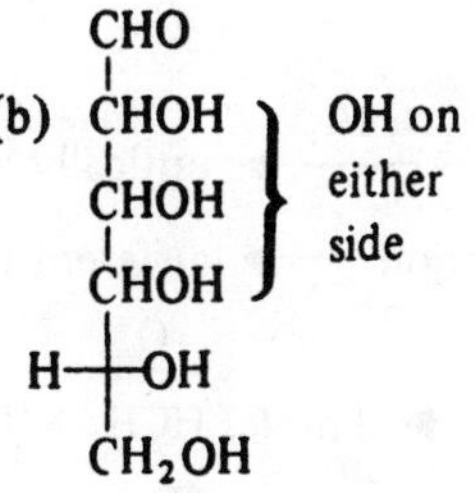 CHO / CHOH / CHOH / CHOH / $H-C-OH$ / CH_2OH (OH on either side, for the three CHOH)

(c) CHO / CHOH / CHOH / HO–C–H / CH_2OH (OH on either side, for the two CHOH)

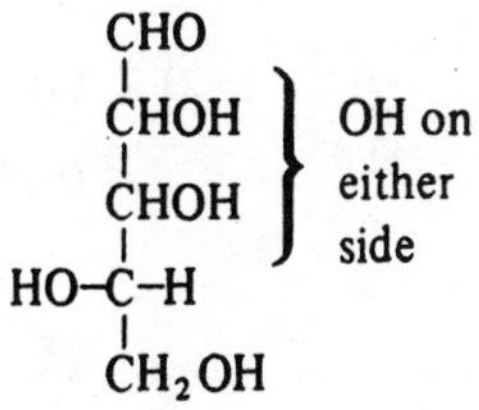

(d) CHO / $(CHOH)_n$ / CH_2OH

$n = 1,2,3,...$

(e) 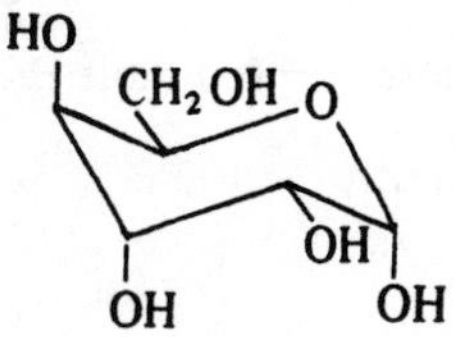 HO, CH_2OH, O, OH, OH, OH (pyranose ring, chair form)

(f) 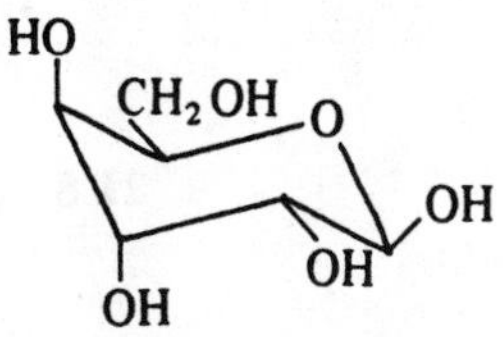 HO, CH_2OH, O, OH, OH, OH (pyranose ring, chair form)

(g) 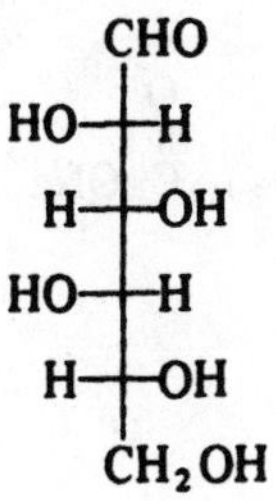 CHO / HO–C–H / H–C–OH / HO–C–H / H–C–OH / CH_2OH

(h) CHO / HO–C–H / H–C–OH / HO–C–H / HO–C–H / CH_2OH

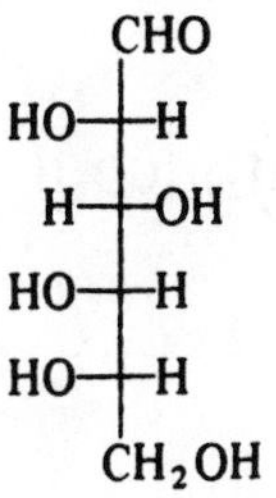

21.2 C

21.3
$$\begin{array}{c} CHO \\ HO\!-\!\!\!+\!\!\!-\!H \\ H\!-\!\!\!+\!\!\!-\!OH \\ CH_2OH \end{array}$$

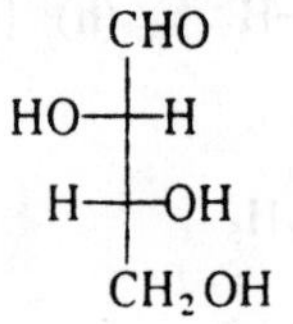

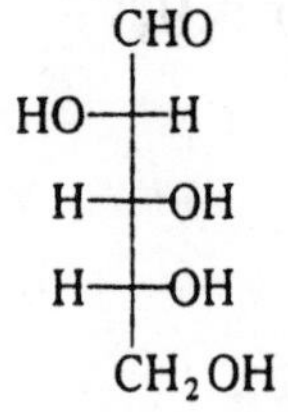

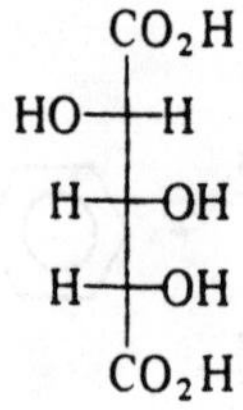

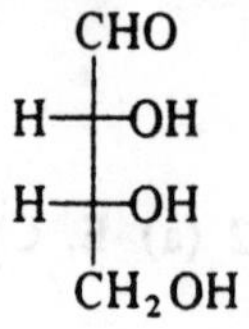

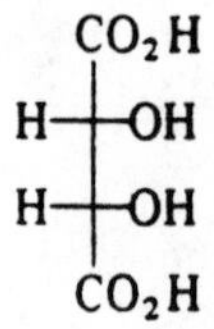

21.4 (a)
$$\begin{array}{c} CHO \\ HO\!-\!\!\!+\!\!\!-\!H \\ H\!-\!\!\!+\!\!\!-\!OH \\ H\!-\!\!\!+\!\!\!-\!OH \\ CH_2OH \end{array}$$
(b)
$$\begin{array}{c} CO_2H \\ HO\!-\!\!\!+\!\!\!-\!H \\ H\!-\!\!\!+\!\!\!-\!OH \\ H\!-\!\!\!+\!\!\!-\!OH \\ CO_2H \end{array}$$
(c)
$$\begin{array}{c} CHO \\ H\!-\!\!\!+\!\!\!-\!OH \\ H\!-\!\!\!+\!\!\!-\!OH \\ CH_2OH \end{array}$$
(d)
$$\begin{array}{c} CO_2H \\ H\!-\!\!\!+\!\!\!-\!OH \\ H\!-\!\!\!+\!\!\!-\!OH \\ CO_2H \end{array}$$

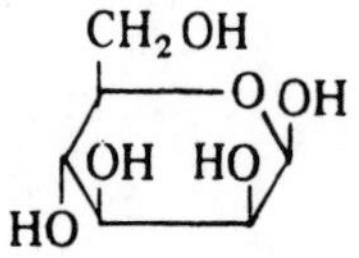

21.5 [Chair structure: HO, HO, O, OH, OH]

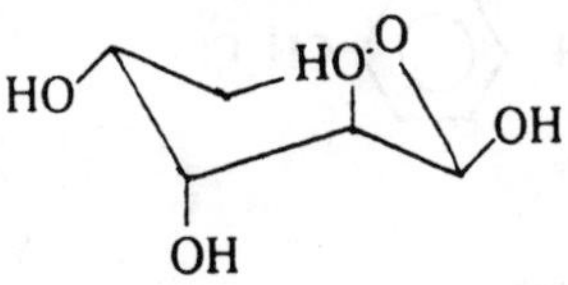

21.6 (a) [Haworth structure: CH_2OH, O, OH, OH, HO, HO] (b) [Chair structure: CH_2OH, HO, O, HO, HO, OH] (c) Reducing

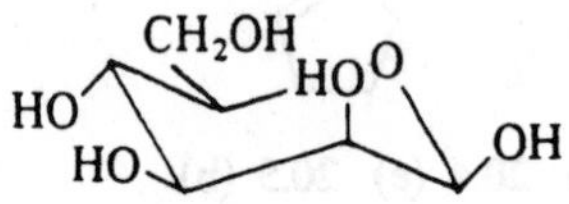

(d) Active (e) Aldonic (f) Active (g) Aldaric (h) $NaBH_4$

(i) Active

21.7 (a) Galactose $\xrightarrow{NaBH_4}$ optically *inactive* alditol

(b) HIO_4 oxidation ⟶ different products:

$$\text{Fructose} \longrightarrow 2\ \text{moles}\ \overset{O}{\overset{\|}{H C H}} + CO_2 + 3\ \overset{O}{\overset{\|}{H C}}{-}OH$$

$$\text{Glucose} \longrightarrow 1\ \text{mole}\ \overset{O}{\overset{\|}{H C H}} + 5\ \overset{O}{\overset{\|}{H C}}{-}OH$$

21.8 (e) 21.9 (d)

CHAPTER 22

22.1 (a) $CH_3(CH_2)_{12}CO_2H$ (b) $CH_3(CH_2)_{12}\overset{O}{\overset{\|}{C}}{-}ONa$

(c) $CH_2OC(=O)(CH_2)_{12}CH_3$
$CHOC(=O)(CH_2)_{12}CH_3$
$CH_2OC(=O)(CH_2)_{12}CH_3$

(d) $CH_2OC(=O)(CH_2)_7CH{=}CH(CH_2)_5CH_3$
$CHOC(=O)(CH_2)_7CH{=}CH(CH_2)_5CH_3$
$CH_2OC(=O)(CH_2)_7CH{=}CH(CH_2)_5CH_3$

(e) $CH_3(CH_2)_{13}SO_3Na$

(f) 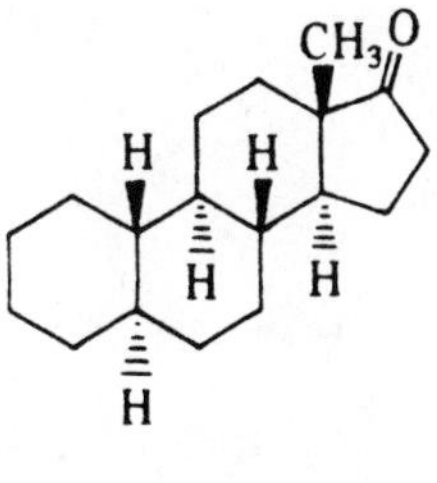 CH_3 O H H H H H

22.2 (a) I_2/OH^- (iodoform test) (b) Br_2/CCl_4 (c) $Ag(NH_3)_2OH$

22.3 5α-Androstane

22.4 (a) $CH_3(CH_2)_4CH_2C{\equiv}CH$ (b) $CH_3(CH_2)_5C{\equiv}CNa$

(c) $CH_3(CH_2)_5C{\equiv}C(CH_2)_6CH_2Cl$ (d) KCN

(e) $CH_3(CH_2)_5C{\equiv}C(CH_2)_7CO_2H$ (f) H_2/Pd

22.5 Sesquiterpene

22.6

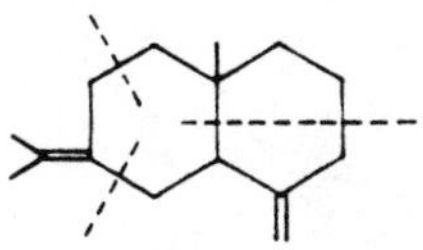

22.7 (e)

CHAPTER 23

23.1 (a) $CH_3CH_2CH(CH_3)CH(NH_3^+)CO_2H$ (b) $CH_3CH_2CH(CH_3)CH(NH_3^+)CO_2^-$

(c) $CH_3CH_2CH(CH_3)CH(NH_2)CO_2^-$

23.2 Pro·Leu·Gly·Phe·Gly·Tyr